Main groups

Transition metals

Main groups

1 1A	2 2A	3 3B	4 4B	5 5B	6 6B	7 7B	8	9 8B	10	11 1B	12 2B	13 3A	14 4A	15 5A	16 6A	17 7A	18 8A
1 H 1.00794																	2 He 4.00260
3 Li 6.941	4 Be 9.01218											5 B 10.81	6 C 12.011	7 N 14.0067	8 O 15.9994	9 F 18.998403	10 Ne 20.1797
11 Na 22.98977	12 Mg 24.305											13 Al 26.98154	14 Si 28.0855	15 P 30.97376	16 S 32.066	17 Cl 35.453	18 Ar 39.948
19 K 39.0983	20 Ca 40.078	21 Sc 44.9559	22 Ti 47.88	23 V 50.9415	24 Cr 51.996	25 Mn 54.9380	26 Fe 55.847	27 Co 58.9332	28 Ni 58.69	29 Cu 63.546	30 Zn 65.39	31 Ga 69.72	32 Ge 72.61	33 As 74.9216	34 Se 78.96	35 Br 79.904	36 Kr 83.80
37 Rb 85.4678	38 Sr 87.62	39 Y 88.9059	40 Zr 91.224	41 Nb 92.9064	42 Mo 95.94	43 Tc (98)	44 Ru 101.07	45 Rh 102.9055	46 Pd 106.42	47 Ag 107.8682	48 Cd 112.41	49 In 114.82	50 Sn 118.710	51 Sb 121.757	52 Te 127.60	53 I 126.9045	54 Xe 131.29
55 Cs 132.9054	56 Ba 137.33	57 *La 138.9055	72 Hf 178.49	73 Ta 180.9479	74 W 183.85	75 Re 186.207	76 Os 190.2	77 Ir 192.22	78 Pt 195.08	79 Au 196.9665	80 Hg 200.59	81 Tl 204.383	82 Pb 207.2	83 Bi 208.9804	84 Po (209)	85 At (210)	86 Rn (222)
87 Fr (223)	88 Ra 226.0254	89 †Ac 227.0278	104 Rf (261)	105 Db (262)	106 Sg (266)	107 Bh (264)	108 Hs (269)	109 Mt (268)	110 (271)	111 (272)	112 (277)	114 (289)	116 (289)	118 (293)			

*Lanthanide series

58 Ce 140.12	59 Pr 140.9077	60 Nd 144.24	61 Pm (145)	62 Sm 150.36	63 Eu 151.96	64 Gd 157.25	65 Tb 158.9254	66 Dy 162.50	67 Ho 164.9304	68 Er 167.26	69 Tm 168.9342	70 Yb 173.04	71 Lu 174.967

†Actinide series

90 Th 232.0381	91 Pa 231.0359	92 U 238.0289	93 Np 237.048	94 Pu (244)	95 Am (243)	96 Cm (247)	97 Bk (247)	98 Cf (251)	99 Es (252)	100 Fm (257)	101 Md (258)	102 No (259)	103 Lr (262)

To my organic chemistry students,
who made this book possible

Microscale Operational Organic Chemistry

A Problem-Solving Approach to the Laboratory Course

John W. Lehman

Professor Emeritus
Lake Superior State University

Prentice Hall, Upper Saddle River, New Jersey 07458

Library of Congress Cataloging-in-Publication Data

Lehman, John W.
    Microscale operational organic chemistry : a problem-solving approach
to the laboratory course / John W. Lehman.
      p. cm.
    "Based on the third edition of Operational organic chemistry"—Pref.
    Includes bibliographical references and index.
    ISBN 0-13-033518-5 (alk. paper)
    1. Chemistry, Organic—Laboratory manuals. I. Lehman, John W.
    Operational organic chemistry. 3rd ed. II. Title.
    QD261.L384 2003
    547'.0078—dc21
        2003048646

Senior Editor: *Nicole Folchetti*
Editor in Chief: *John Challice*
Editorial Assistant: *Chris Cella*
Production Editor: *Erin Connaughton*
Vice President and Director of Production and Manufacturing, ESM: *David W. Riccardi*
Director of Creative Services: *Paul Belfanti*
Creative Director: *Carole Anson*
Art Director: *Jayne Conte*
Cover Designer: *Bruce Kenselaar*
Art Editor: *Connie Long*
Manufacturing Manager: *Trudy Pisciotti*
Manufacturing Buyer: *Lynda Castillo*
Senior Marketing Manager: *Steve Sartori*

© 2004 Pearson Education, Inc.
Upper Saddle River, NJ 07458

The spectra in this book were reproduced from *The Aldrich Library of FT-IR Spectra*, 2nd ed., and
The Aldrich Library of ^{13}C and ^{1}H FT-NMR Spectra with the permission of the Aldrich Chemical Company.

Referenced trademarks:
Advil, Alconox, Anacin, Bakelite, Bausch & Lomb, BenGay, Carbowax, Chromosorb, Crisco, Deet, Drierite,
Drummond Microcaps, Enovid Excedrin, Fluorolube, Freon, Kimax, Kool-Aid, Lucite, Luer-Lok, Mel-Temp, Mini-Press,
Neosporin, Nochromix, No-Doz, Norit, Nutra-Sweet, Oblivon, Parafilm, Perclene, Perkin-Elmer, Plexiglas, Pyrex, Scoopula,
Spectronic, Styrofoam, Tabasco Sauce, Teflon, Thermowell, Tylenol, Whatman.

Printed in the United States of America

10 9 8 7 6 5 4 3 2 1

ISBN: 0-13-033518-5

Pearson Education LTD., *London*
Pearson Education Australia PTY, Limited, *Sydney*
Pearson Education Singapore, Pte. Ltd
Pearson Education North Asia Ltd., *Hong Kong*
Pearson Education Canada, Ltd., *Toronto*
Pearson Educación de Mexico, S.A. de C.V.
Pearson Education—Japan, *Tokyo*
Pearson Education Malaysia, Pte. Ltd.

Microscale Operational Organic Chemistry

A Problem-Solving Approach to the Laboratory Course

Contents

Part III Minilabs   459

Part IV Qualitative Organic Analysis 513

Part V The Operations 559

Appendixes and Bibliography   787

Preface

To the Instructor

This book is a microscale laboratory textbook based on the third edition of *Operational Organic Chemistry: A Problem-Solving Approach to the Laboratory Course*. Every experiment and minilab in the book can be performed by students using the microscale glassware available in a Mayo/Pike-style microscale lab kit with 14/10 standard-taper joints and threaded connectors. Most of the experiments could also be performed successfully with alternative microscale glassware, such as that provided in a Williamson lab kit, but some might need to be scaled down further and the instructor would have to provide additional instructions regarding the use of the glassware.

Some organic chemists regard a "microscale" experiment as one involving approximately 0.1 g of the limiting reactant. Dealing with such small quantities can be discouraging to students with standard scale fingers, who end up with a drop or a few grains of product, if any. Therefore, I have adopted a working definition of a microscale experiment as one that can be performed using the glassware available in a typical microscale lab kit, along with appropriate locker supplies as recommended in the *Instructor's Manual*.

In writing this book, I have been guided by my convictions that students (1) perform better in the organic laboratory course if they master the major lab operations early and apply them throughout the course, (2) learn organic chemistry better if they keep their minds engaged by approaching each experiment as a problem-solving exercise, and (3) perform any task better if they are sufficiently motivated.

Part I is devoted to experiments designed to teach the basic laboratory operations, where an *operation*, as used here, is a process that utilizes one or more basic lab techniques, such as heating, cooling, and vacuum filtration, to accomplish some end, such as the purification of a solid. Once students have mastered the major operations by completing the appropriate experiments in Part I, they should be ready to apply those operations in Part II, which contains a large selection of experiments that are correlated with topics found in most organic chemistry lecture textbooks. This operational approach helps students understand that an organic synthesis, for example, is not a unique phenomenon that can be experienced only by mechanically following a detailed "recipe." Rather, it is the outcome of a logical sequence of interrelated operations adapted to the requirements of the synthesis.

In addition to teaching lab skills, the experiments in this book are designed to help students develop the observational and critical thinking skills that are essential prerequisites for a successful career in science and in virtually every other professional field. Each major experiment requires the student to solve a specific scientific problem through the application of sound scientific methodology. Before an experiment, the student must first define the problem based on information provided in a hypothetical Scenario. After a preliminary reading of the experiment, the student should be able to develop a working hypothesis regarding its outcome. During the experiment, the student gathers and evaluates evidence bearing on the problem and, as necessary, reevaluates and revises the hypothesis based on experimental observations and data. Finally, the student tests the hypothesis by obtaining a melting point, a spectrum, a gas chromatogram, or by some other means, and arrives at a conclusion. Because of the level at which most undergraduate organic chemistry courses are taught, the problems must, of necessity, be kept relatively simple and (with a few exceptions) should not be compared to "real" research problems tackled by professional chemists. It is not the intent of this book to make every student a research chemist; most students who take an organic chemistry course have no intention of going into the field. But the critical thinking and methodological skills required to solve the problems are comparable to those applied by research scientists, and applying those skills should give the student a better understanding of the nature and practice of science.

As a motivational device and to provide a frame of reference for the problems, students are asked to regard themselves as "consulting chemists" working for an institute operated by their college or university. Various individuals and organizations come to the institute with their scientific problems and the problems are relayed to the "project group" comprising each lab section, to be solved individually or (sometimes) through collaboration. Although some of the Scenarios are a bit contrived, most of them describe tasks similar to those a practicing chemist might be called upon to perform.

To implement this problem-solving approach in the organic chemistry lab, each experiment includes a section, "Applying Scientific Methodology," intended to help the student understand the problem, formulate a meaningful hypothesis, and solve the problem. The Introduction and Experiment 1 describe in some detail how the student can apply scientific methodology to the solution of a problem, so students should read at least the section "Problem Solving in the Organic Chemistry Lab" in the Introduction and should be asked to read the Scenario and "Applying Scientific Methodology" sections of Experiment 1 even if you choose not to assign it. Because each experiment is designed as a problem for the student to solve, the outcome is not explicitly stated in the experiment itself. In a few experiments, such as Experiments 31 and 43, the identity of an organic reactant is unknown as well. Therefore, it is imperative that the instructor or laboratory coordinator obtain a copy of the *Instructor's Manual*, which is provided free of charge by Prentice Hall to adopters of this book.

Part III contains a number of minilabs to provide additional flexibility. These short experiments can be used to fill in those gaps in a lab course when, for example, the students finish a two-period experiment during the first hour or so of the second period. They can also be used to teach techniques (such as paper chromatography in Minilab 9) that are seldom used in the major experiments or to introduce additional theoretical topics, such as photochemistry (Minilab 24). Part IV is a self-contained introduction to qualitative organic analysis, which is also incorporated into some of the Part II experiments, particularly Experiments 38 and 49. Part V contains detailed descriptions of the laboratory operations, which are flagged in the experiments by operation numbers in brackets, such as [OP-30].

Acknowledgments

Although I have personally lab-tested all of the experiments in this book, many students and faculty members have been involved in their development and testing as well. The procedures have also been class-tested by undergraduate organic chemistry students at Lake Superior State University and elsewhere. I am especially grateful to the students who occasionally suffered through the earlier versions of some experiments. I learned from their misadventures and improved the procedures to smooth the path of future students using this book. For this version of *Operational Organic Chemistry*, I would like to acknowledge the contributions of Patricia Walker, who lab-tested many of the procedures, and Teresa Hill, whose students at Rochester Community College class-tested five experiments. I owe a debt of gratitude to the late Miles Pickering, whose articles in the *Journal of Chemical Education* stimulated me to develop the problem-solving approach used in this book. I also wish to express my appreciation to James G. Vogel and Roger Kugel, from whom I got the idea for the Consulting Chemists Institute, and to Anthony Winston, John Penn, and the other organizers of West Virginia University's microscale chemistry workshop, who helped me accomplish the conversion of my own organic chemistry lab to a microscale lab. I am indebted to my editor at Prentice

Hall, Nicole Folchetti, who provided encouragement and assistance throughout the project, and to the staff at Prentice Hall, nSight, Inc., and Laserwords, Inc. for their contributions to the production of this book. Most of the spectra in this book and in the *Instructor's Manual* are reproduced from the spectral libraries of the Aldrich Chemical Company, whose generosity is gratefully acknowledged. I would also like to thank the following people, who reviewed the manuscript and provided helpful suggestions and comments: Daniel Blanchard (Kutztown University), Dana Chatellier (University of Delaware), Sergio Cortes (University of Texas, Dallas), Mark Hemric (Oklahoma Baptist University), Joseph Hornback (University of Denver), Joe LeFevre (State University of New York-Oswego), Richard Pagni (University of Tennessee-Knoxville), Albert Payton (Broward Community College), Michael Pelter (Purdue University-Calumet), Maria Vogt (Bloomfield College), and Catherine Woytowicz (George Washington University).

All textbooks can benefit from comments and criticism by their readers, so I welcome correspondence to point out errors or to suggest improvements in the book.

John W. Lehman
jlehman@lssu.edu

Microscale Operational Organic Chemistry

A Problem-Solving Approach to the Laboratory Course

Introduction

Problem Solving in the Organic Chemistry Lab

Organic chemistry is not most people's idea of a "fun" course, but that is no reason not to enjoy your organic chemistry lab experience. Many experiences can be enjoyable if they give you the opportunity to use your imagination and to test your mental and manual skills. During this lab course, you will play the role of a consultant in a Consulting Chemists Institute operated by your college or university. When D. K. Little wants to know what happens to his company's food preservative in stomach acid, when Rusty Tappet accidentally pours diesel fuel into a barrel of racing fuel, when Gilda Lilly wants to know the color of a synthetic dye, or when the Olfactory Factory needs a way to convert an oversupply of anisole to a perfume ingredient, you and the other members of your "project group" (lab section) will be called upon to solve their problems.

To solve such a problem, you must *think* before you act. In other words, you will need to read the experiment, understand the problem, and try to predict a likely outcome of the experiment before you actually carry out the experiment in the laboratory. Your prediction, stated clearly in writing, becomes your *working hypothesis*. In many cases the most likely outcome will become apparent after you read the experiment, especially if you apply the concepts you have learned in the organic chemistry lecture. In other cases there may be several reasonable outcomes, and you will have to make an "educated guess" as to the most likely outcome. During the experiment you will need to gather evidence that may support your hypothesis—or prove it wrong. That means making careful observations and gathering data that relate to the problem. As you evaluate the evidence, you may decide that your original hypothesis was wrong, or at least incomplete, and needs to be revised or replaced by a new one. By the time you finish the experiment, you will have tested your hypothesis and arrived at a conclusion. In this way, each experiment will help you develop your observational and critical thinking skills, as well as your lab skills, as you apply them to the solution of the problem posed in that experiment. The next section will tell you, in more detail, how to approach and solve a scientific problem.

Scientific Methodology

If you are taking an organic chemistry course, you are probably planning a career in some field of science or technology or a field that is based on scientific knowledge and principles, such as medicine. To succeed in such a field, you must learn to think and work like a scientist. Scientists follow certain basic principles that are often lumped together under the expression "the scientific method." In fact, there is no universal scientific method that all scientists follow rigorously. But most of them go through at least some of the following steps when dealing with a scientific problem:

- Define the problem
- Plan a course of action
- Gather evidence
- Evaluate the evidence
- Develop a hypothesis

- Test the hypothesis
- Reach a conclusion
- Report the results

Defining the Problem.   Most people think of a "problem" in a negative sense, as in "We've got a problem here," or "What's your problem?" To a scientist, a problem is not a perceived difficulty but an *opportunity* for exploring and learning more about some aspect of the physical world. A problem may be inherent in an assigned task, or it may arise from anything that the scientist is curious about, such as an unexplained phenomenon or an unexpected observation. A problem is often defined in the form of a question: What is the identity of the liquid my instructor gave me? What is the mercury concentration in a Lake Michigan salmon? How do fireflies generate light? In this laboratory course, the problem associated with each experiment will be described in the *Scenario* that leads off the experiment.

Planning a Course of Action.   A scientist must plan his or her own course of action for solving a scientific problem. This often requires that the scientist carry out a literature search to glean information and data relating to the problem, decide which experimental methods and instruments to use, and develop a detailed procedure to be followed. In most lab courses, the procedure is "in the book," and the student simply follows the procedure as if it were a recipe for baking a cake. That is true of a few basic experiments in this textbook, but for most of them, you will have to perform some calculations and develop your own experimental plan based on the information and directions given in the experiment. In a few cases, you may be required to develop a procedure of your own.

Gathering Evidence.   Scientists will gather as much evidence as they feel is needed to solve a problem and convince other scientists that their solution is correct. Evidence is gathered by making careful *observations* and *measurements*.

   To make valid observations you must be *objective*, reporting only what you actually saw and not what you expected to see. A wildlife biologist who expects wild chimpanzees to behave just like chimpanzees at the zoo is not likely to make any important discoveries about chimpanzees! Keep the following points in mind when you make observations.

1 Don't confuse an *observation* with an *inference*. An observation is whatever you perceive with your senses (sight, smell, touch, taste, or hearing) during an event. An inference is a guess about the *cause* of the event. Writing "The solution turned brown upon addition of 0.1 M KMnO$_4$" records an observation. Writing "The solution must have contained an alkene because it turned brown upon addition of 0.1 M KMnO$_4$" is an inference.

2 Be prepared to be surprised. If you observe something you didn't expect, don't simply disregard the observation or report what you thought you should have seen. Consider an unexpected observation as an opportunity to learn something you didn't know before—something that might lead to a new discovery.

3 Write down your observations as you make them or shortly afterward. If you wait too long, you are likely to leave out important details.

4 Record your observations clearly, completely, and systematically. If you are doing repetitive measurements or carrying out the same test on a series of samples, for example, record your observations in a table.

Making accurate measurements, such as determining the mass or melting point of the product of a chemical synthesis, requires a certain amount of skill and know-how. You can obtain such skills by, for example, watching your instructor demonstrate the operation of an instrument, and then practicing on the instrument until you obtain consistent and accurate results *before* you use it to make a measurement you intend to report. If you are not sure how to use an instrument properly, ask the instructor to show you.

Evaluating the Evidence. Evaluating the evidence involves assessing the reliability of your experimental results and looking for clues among your results that may point to a solution to the problem. For example, searching the infrared spectrum of an unknown liquid for evidence of a specific functional group and comparing its boiling point with the boiling points of known compounds may help you solve the problem "What is the identity of the liquid my instructor gave me?" Solving such problems often requires clear and logical thinking; in other cases a more creative, intuitive approach can be valuable. In either case, all of your reasoning and intuition may be fruitless if your experimental results are unreliable. The validation of experimental results requires first asking yourself whether the results make sense physically. If you obtain a melting point that is much lower than the literature value, a product mass that is higher than the theoretical yield for a synthesis, or any other result that seems suspect, then you need to find out whether the result is, in fact, erroneous. You should review everything you did that led to the result, using notes from your lab notebook to jog your memory as necessary. Perhaps you only need to repeat a melting point or dry a product longer, but in any case, you should find out what you did wrong and correct it. In some kinds of experiments (but none in this book), validation of results may also require a statistical analysis of experimental data. If you perform such an experiment, your instructor will tell you how to do that.

Developing and Testing Hypotheses. A hypothesis can be regarded as an "educated guess" about the cause of some phenomenon or the outcome of an experiment. Hypotheses can help us see the significance of an object or event that would otherwise mean little. For example, the movements of the planets seemed erratic and mysterious before Copernicus developed his hypothesis that the earth revolves around the sun. A hypothesis must be *testable* to have validity; that is, it must be formulated in such a way that experiments can be devised whose outcome might prove the hypothesis *wrong*. John Dalton's hypothesis that atoms are indivisible was proved wrong after it was shown that bombarding uranium atoms with neutrons caused them to split into smaller atoms. But Dalton's more fundamental hypothesis that all matter is made of atoms has been tested repeatedly over the years and has never been proven wrong, so it is generally accepted as true.

Most of the experiments in this book require you to formulate and test a working hypothesis based on a problem outlined in the Scenario. A working hypothesis can be a prediction about the outcome of an experiment based on information available to the experimenter, which, for your lab course, will usually be found in the write-up for the experiment. Such a hypothesis can be proposed and revised at any time during the course of an experiment. An appropriate working hypothesis and the method of testing it may become apparent upon reading the experiment. For example, after reading Experiment 9, you should be able to develop a working hypothesis such as "The red pigment in Brand X tomato paste was chemically altered when the tomatoes were processed," and then test your hypothesis by recording a spectrum of the isolated pigment in solution.

One drawback of a working hypothesis is that the scientist may become so attached to it that he or she will overlook or ignore evidence that contradicts it. For this reason, some scientists prefer to explore a problem without any preconceived ideas about the outcome—which is not always easy to do. For most of the experiments in this book, you can formulate a working hypothesis after reading the experiment, but a few experiments require that you gather some experimental evidence first. In either case, if the experimental evidence does not support your initial hypothesis, you should be ready to revise or abandon it without regret.

Reaching Conclusions. If a hypothesis passes all of the tests you carry out, then you are ready to state a conclusion, such as "The liquid my instructor gave me is benzaldehyde." This does not necessarily mean that your conclusion is correct; you may have not carried out enough tests, or some of your test results may be faulty. But if you have performed an experiment carefully and reasoned logically, you will probably arrive at a valid conclusion.

Reporting Results. You should report all results of an experiment as clearly, completely, and unambiguously as possible. Write up your results in correct English using complete sentences and correct spelling. Be as specific as you can, avoiding such generalities as "My yield was lower than expected due to human error." Label any tables clearly, giving names or standard abbreviations for all physical properties and the units in which they are measured.

If an experiment requires that you graph your data, use accurately ruled graph paper, never notebook paper or paper you have ruled yourself. Plot the dependent variable on the *y*-axis and the independent variable on the *x*-axis. Label both axes with the physical quantities being graphed and their units, if any. Select appropriate uniform scale intervals so that your data points extend most of the way up and across the graph paper. If the relationship you are graphing is linear, draw the straight line that best fits the data points.

Applying Scientific Methodology. How can you apply scientific methodology in your organic chemistry lab course? Consider Experiment 5 in this book, which involves the reaction of isopentyl alcohol with acetic acid. By reading the Scenario, you learn that you are expected to prepare isopentyl acetate (also known as "banana oil") by this reaction and then analyze your product by gas chromatography to see whether it meets the specified outcome of containing less than 10% isopentyl alcohol and 2% acetic acid. From this information, you can state the problem as, for example: "When isopentyl alcohol is heated with acetic acid in the presence of a catalyst, will the reaction yield isopentyl acetate containing less than 10% isopentyl alcohol and 2% acetic acid?" You can then formulate one or more working hypotheses, sometimes by simply restating the problem in a form that predicts the outcome of the experiment. Such a hypothesis might be, for example: "The reaction of isopentyl alcohol with acetic acid *will* yield isopentyl acetate containing less than 10% isopentyl alcohol and 2% acetic acid." This statement actually contains two predictions; that there will be a chemical reaction that yields some isopentyl acetate, and that the product's purity will meet the specifications stated. Thus you will have to gather evidence to help you prove or disprove each prediction.

As you read the directions for the experiment, consider the kinds of evidence you can gather by measurement or observation. During the experiment itself, make careful observations of everything that might provide evidence relating to the problem. What changes occur when the chemicals are combined and heated together? Do the changes indicate that a chemical reaction is taking place? Do they give any clues about the possible identity of the product? What physical properties of the product can you observe (such as odor) or measure (such as boiling point) during the experiment, and what bearing do they have on the identity and purity of the product? What do your results from the analysis of the product by gas chromatography tell you about the purity of the product?

After you have completed the experiment, you can evaluate the evidence, decide whether the results confirm or disprove your hypothesis, arrive at a conclusion that answers the question proposed in your problem statement, and write a report describing your findings. Appendixes II and III suggest ways of recording and reporting your experimental results. Your instructor will let you know what kind of report he or she prefers.

Organization of This Book

Microscale Operational Organic Chemistry is divided into five parts. Part I contains 12 experiments whose main purpose is to help you learn basic laboratory operations by applying them to the solution of a scientific problem. In the dictionary, an *operation* is defined as a process or series of acts performed to effect a certain purpose or result. For example, the *recrystallization* operation, which is used to purify a solid substance, involves at least four separate steps: (1) *heating* the solid in a boiling solvent until it dissolves, (2) *cooling* the resulting solution until the solid crystallizes, (3) *filtering* the crystals to separate them from the solvent, and (4) *washing* the crystals to remove residual impurities. Such operations are often called *techniques*, but the latter term is also used to refer to a single step, such as filtration, in a complex operation. In Part I, you will learn most of the operations in this book while carrying out experiments designed to accomplish other outcomes. For example, you will measure a melting point, not just to learn how to measure melting points but also to establish the identity of a substance you have isolated or synthesized.

Once you have mastered the operations in Part I, you will apply them in experiments from Part II, which are correlated with topics in your lecture textbook. Performing these experiments will not only increase your proficiency in the laboratory but also give you the opportunity to apply—in a hands-on environment—what you have learned in the lecture course to the solution of specific scientific problems. In the process, you should find that learning by doing can be much more enjoyable and effective than learning by memorization and other purely mental processes.

Each experiment in Parts I and II contains information under most or all of the following headings.

Operations. This heading is followed by a list of the operations to be used in the experiment, each preceded by an operation number, such as OP-25 for recrystallization. The description for each operation can be located quickly by noting the large operation numbers in the right-hand corners of the odd-numbered pages in Part V as you flip through the pages. Operations being used for the first time are emphasized by boldface type; you should read their descriptions thoroughly before you come to the laboratory. After you have used an operation once or twice, you should not have to reread the entire description the next time you use it, but you should at least read the Summary and review the General Directions section (if provided) to refresh your memory. Eventually, you shouldn't need to refer to the operation description at all, unless you encounter experimental difficulties or are applying the operation in a new situation.

Before You Begin. Under this heading, you will find a *prelab assignment:* a list of things to do before you come to the laboratory. The prelab assignment always includes reading the experiment and reading or reviewing the operations. Starting with Experiment 5 you will also be expected to write an experimental plan for each experiment, as described in Appendix V, and you will usually need to carry out some calculations. Your instructor may require that you have your experimental plan and calculations approved before you begin an experiment.

Scenario. The Scenario presents a hypothetical situation involving the scientific problem you are to solve. A typical Scenario will describe a chemistry-related problem posed by a company or individual and tell you what role you will play in its solution.

Applying Scientific Methodology. This section is intended to help you formulate and solve the problem posed in the Scenario by following the approach described in the previous sections, "Problem Solving in the Organic Chemistry Lab" and "Scientific Methodology."

Background Essay. Each background essay appears under a different descriptive heading, such as "Crime and Chemistry." The essay will often show the relation between the lab work and related concepts from the lecture course. It may also relate historical sidelights or interesting facts that show the "real-world" relevance of the experiment.

Understanding the Experiment. This section describes the main purpose of the experiment, explains the theoretical basis of the experiment when appropriate, and helps you understand the experimental methodology. It may also provide information that will help you interpret your results or cope with unexpected complications as they arise.

Reactions and Properties. For most experiments, this section gives balanced equations for synthetic reactions and tabulates the relevant physical properties of reactants, products, and other chemicals. The Properties table usually contains any data needed for the prelab calculations.

Directions. This section describes the course of action you will follow to carry out the experiment. You should not follow it mechanically, as you would a recipe, but try to understand the purpose of each operation you are performing. The section entitled "Understanding the Experiment" will help you do that. The Directions for most Part I experiments are more detailed than those for Part II. By the time you get to Part II, you will be expected to know how to perform most lab operations proficiently without the aid of frequent reminders.

Safety Notes.   Characteristics of some hazardous chemicals and precautions for their use are described in the Directions under this heading. See the following "Laboratory Safety" section for general information about laboratory hazards.

Exercises.   Your instructor will assign exercises to be completed and turned in with your laboratory report.

Other Things You Can Do.   The other things may include additional experiments or minilabs that you can perform, or library research projects that you can complete. You must have your instructor's permission to start any project marked by an asterisk. Read the relevant sections of Appendix VII, "The Chemical Literature," and scan the Bibliography for possible sources before you begin a library research project.

Part III of the text contains 46 short experiments called minilabs. Minilabs take only part of a lab period and do not require extensive reports. They are designed to add flexibility to the lab course and "fill in the gaps" when, for example, a two-week experiment can be finished in less than two full lab periods.

Part IV is a comprehensive, self-contained introduction to qualitative organic analysis that will help you learn how to identify organic compounds by using chemical and spectral methods.

Part V contains descriptions of all of the operations, which are referred to in each experiment's directions by number, in the form [OP-25].

Appendixes I through V contain illustrations of laboratory equipment and information about keeping lab notebooks, writing lab reports, developing experimental plans, and performing stoichiometric calculations. Appendix VI contains tables of properties for qualitative organic analysis. Appendix VII is a guide to the chemical literature that describes some of the most important works in organic chemistry and, in some cases, tells you how to use them.

The Bibliography lists a large number of useful works in organic chemistry, ranging from enormous multivolume sets such as *Chemical Abstracts* to short papers from the *Journal of Chemical Education*. References to entries listed in the Bibliography are made throughout the text in the form [Bibliography, F20], where the letter refers to a category and the number to a location within that category; for example, F20 is the 20th book listed under category F, Spectrometry.

A Guide to Success in the Organic Chemistry Lab

What to Expect in Your Organic Chemistry Lab Course

Students are sometimes apprehensive about having to work in an organic chemistry lab. They may have heard that organic chemistry is difficult or that the lab is a dangerous place because of the hazardous nature of organic chemicals. While some students do find the subject matter of organic chemistry difficult, most of them have little trouble completing the laboratory experiments successfully. In fact, many students enjoy the lab far more than the lecture course, and they generally receive higher grades in lab than in lecture.

It is true that some organic chemicals are quite hazardous and can cause serious injury, or even death, if not handled properly. But if the chemistry lab were a very dangerous place to work, you would expect professional chemists to have short life spans. In fact, chemists tend to live longer than professionals in most other fields; in one listing of occupational risks, chemists rank near the bottom (lowest risk), right between school administrators and ticket agents. In more than 30 years of teaching organic chemistry, I've observed only a few lab accidents that caused significant injury, most of them involving broken glass. Only a handful of students were injured by contact with chemicals, and none of them suffered permanent injury. That is not to say that you can afford to be careless in the lab—the potential for a serious accident is always there. But if you follow the safety rules in the following "Laboratory Safety" section and take reasonable precautions when handling chemicals, you should have little reason for concern about lab accidents.

Getting Started

By the end of the first week of your organic chemistry course, you should have read the "Laboratory Safety" section of this book and any other safety rules or data provided by your instructor. Before you begin working in the laboratory, your instructor should review the safety rules and tell you what safety supplies, such as safety goggles and protective gloves and aprons, you will need to work in the lab. During the first laboratory period, the instructor will show you where safety equipment is located and tell you how to use it. As you locate each item, check it off the following list and make a note of its location. (Your instructor may suggest additions or changes to the list.)

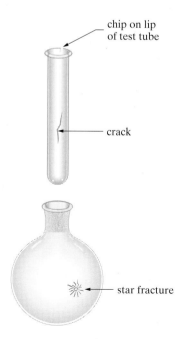

chip on lip
of test tube

crack

star fracture

Glassware defects

- Fire extinguishers
- Fire blanket
- Safety shower
- _____

- Eyewash fountain
- First aid supplies
- Spill clean-up supplies
- _____

You should also learn the locations of chemicals, consumable supplies (such as filter paper and boiling chips), waste containers, and various items of equipment such as balances and drying ovens.

Your instructor will assign you a locker and provide a list of locker supplies. You will then need to check into the laboratory. This usually involves getting a locker key or a combination lock and checking your locker for missing or damaged items. You will find illustrations of typical locker supplies in Appendix I at the back of this book. If you find any glassware items with chips, cracks, or star fractures, you should have them replaced; they may cause cuts, break on heating, or shatter under stress. If necessary, clean up any dirty glassware in your locker (see OP-1) and organize it neatly at this time.

Working Efficiently

Because of wide variations in individual working rates, it is usually not possible to schedule experiments so that everyone can be finished in the allotted time. If all labs were geared to the slowest student, the objectives of the course could not be accomplished in the limited time available. If you fall behind in the lab, you may need to put in extra hours outside your scheduled laboratory period in order to complete the course. The following suggestions should help you work more efficiently and finish each experiment on time.

1 *Be prepared to start the experiment the moment you reach your work area.* Don't waste the precious minutes at the start of a laboratory period doing calculations, reading the experiment, washing glassware, or carrying out other activities that should have been done at the end of the previous period or during the intervening time. The first half hour of any lab period is the most important—if you use it to collect the necessary materials, set up the apparatus, and get the initial operation (reflux, distillation, etc.) under way, you should have no trouble completing the experiment on time.

2 *Organize your time efficiently.* Schedule a time each week to read the experiment and operation descriptions and complete the prelab assignment—an hour before the lab period begins is too late! Plan ahead so that you know approximately what you will be doing at each stage of the experiment. A written experimental plan, prepared as described in Appendix V, is invaluable for this purpose.

3 *Organize your work area.* Before performing any operation, arrange all of the equipment and supplies you will need during the operation neatly on your bench top, in the approximate order in which they will be used. Place small objects and any items that might be contaminated by contact with the bench top on a paper towel, laboratory tissue, or mat. After you use each item, move it to an out-of-the-way location where it can be cleaned and returned to its proper location when time permits; for example, put dirty glassware in a washing trough in the sink. Keep your locker well organized, placing each item in the same location after use so that you can immediately find the equipment you need. This will also help you notice whether any items are missing so that you can hunt for them before you leave the lab; otherwise the items will probably disappear before the next lab period, and you may be charged for them.

Getting Along in the Laboratory

You will get along much better in the laboratory if you can maintain peace and harmony with your coworkers—or at least keep from aggravating them—and stay on good terms with your instructor. Following these common-sense rules will help you do that.

1 *Leave all chemicals where you found them.* You will understand the reason for this rule once you experience the frustration of hunting high and low for a reagent, only to find it at another student's station in a far corner of the lab. Containers should be taken to the reagent bottles to be filled; reagent bottles should never be taken to your lab station.

2 *Take only what you need.* Liquids and solutions should ordinarily be obtained using bottle-top dispensers, automatic pipets, or other measuring devices so that you will take no more than you expect to use for a given operation. Solids can be weighed out directly from stock bottles.

3 *Prevent contamination of chemicals.* Don't use your own pipet to remove liquids from stock bottles, and don't return unused chemicals to stock bottles. Be sure to close all bottles tightly after use—particularly those containing anhydrous chemicals and drying agents.

4 *If you must use a burner, inform your neighbors*—unless they are already using burners. This will allow them to cover any containers of flammable solvents and take other necessary precautions. In some circumstances, you may have to use a different heat source, move your operation to a safe location (for instance, under a fume hood), or find something else to do while flammable solvents are in use.

5 *Return all community equipment to the designated locations.* This may include ring stands, lab kits, clamps, condenser tubing, and other items that are not in your own locker. Since such items will be needed by students in other lab sections, they should always be returned to the proper storage area at the end of the period.

6 *Clean up for the next person.* Few experiences are more annoying than finding that the lab kit you just checked out is full of dirty glassware or that your lab station is cluttered with paper towels, broken glass, and spilled chemicals. The last 15 minutes or so of every laboratory period should be set aside for cleaning up your lab station and the glassware used during the experiment. Put things away so that your work station is uncluttered. Clean off the bench top with a towel or wet sponge; remove condenser tubing, other supplies, and debris from the sink; and thoroughly wash any dirty glassware that is to be returned to the stockroom, as well as that from your locker. Clean up any spills and broken glassware immediately. If you spill a corrosive or toxic chemical, such as sulfuric acid or aniline, inform the instructor before you attempt to clean it up.

7 *Heed Gumperson's Second Law,* which advises you to maximize the labor and minimize the oratory while in the **lab***oratory*. This does not mean that all conversation must come to a halt. Quiet conversation during a lull in the experimental activity is okay, but a constant stream of chatter directed at a student performing a delicate operation is distracting and can lead to an accident. For the same reason, radios, tape or CD players, and other audio devices should not be brought into the laboratory.

Laboratory Safety

Safety Standards

In the United States, a Federal Laboratory Safety Standard designated as OSHA 29 CFR 1910.1450 requires that any laboratory in which hazardous chemicals are handled should have a written safety plan called a *chemical hygiene plan*. Your institution's chemical hygiene plan should describe safe operating procedures to be followed in carrying out various laboratory operations, and emergency response procedures to be followed in case of accident. According to the OSHA standard, laboratory instructors are required to see that students know and follow established safety rules, have access to and know how to use appropriate emergency equipment, and are aware of hazards associated with specific experiments. The lab instructor alone cannot prevent laboratory accidents, however. You also have a responsibility to follow safe laboratory practices while performing experiments and to be ready to respond in case of accident.

Preventing Laboratory Accidents

Most organic lab courses are completed without incident, aside from minor cuts or burns, and serious accidents are rare. Nevertheless, the potential for a serious accident always exists. To reduce the likelihood of an accident, *you must learn the following safety rules and observe them at all times.* Anyone failing to do so may be expelled from the laboratory.

1 *Wear approved eye protection in the laboratory at all times.* Even when you are not working with hazardous materials, another student's actions could endanger your eyes, so never remove your safety goggles or safety glasses until you leave the lab. Don't wear contact lenses in the laboratory, because chemicals splashed into an eye may get underneath a contact lens and cause damage before the lens can be removed. Learn the location of the eyewash fountain nearest to you at the first laboratory session, and learn how to use it.

2 *Never smoke in the laboratory or use open flames in operations involving low-boiling flammable solvents.* Anyone smoking in an organic chemistry laboratory is subject to immediate expulsion. Before you light a burner or even strike a match, inform your neighbors of your intention to use a flame. If anyone nearby is using flammable solvents, either wait until they are finished or move to a safer location, such as a fume hood. Diethyl ether and petroleum ether are extremely flammable, but other common solvents, such as acetone and ethanol, can be dangerous as well. When ventilation is inadequate, the vapors of diethyl ether and other highly volatile liquids can travel a long distance across a bench top, so igniting a flame at one end of a lab bench can actually start an ether fire at its opposite end! Learn the location and operation of the fire extinguishers, fire blankets, and safety showers at the first laboratory session.

3 *Consider all chemicals to be hazardous and minimize your exposure to them.* Never taste chemicals, do not inhale the vapors of volatile chemicals or the dust of finely divided solids, and prevent contact between chemicals and your skin, eyes, and clothing. Many chemicals can cause poisoning by ingestion, inhalation, or absorption through the skin. Strong acids and bases, bromine, thionyl chloride, and other corrosive

materials can produce severe burns and require special precautions, such as wearing gloves and lab aprons. Some chemicals cause severe allergic reactions, and others may be carcinogenic (tending to cause cancer) or teratogenic (tending to cause birth defects) by inhalation, ingestion (swallowing), or skin absorption. To prevent accidental ingestion of toxic chemicals, do not bring food or drink into the laboratory or use mouth suction for pipetting, and wash your hands thoroughly after handling any chemical. Clean up chemical spills immediately, using a neutralizing agent and plenty of water for acids and bases, and an absorbent for solvents. In case of a major spill, or if the chemical spilled is very corrosive or toxic, notify your instructor before you try to clean it up. Read the "Safety Notes" section in each experiment and use protective gloves or a fume hood when directed.

4 *Exercise great care when working with glass and when inserting or removing glass tubing.* Among the most common accidents in a chemistry lab are cuts from broken glass and burns from touching hot glass. Protect your hands with gloves or a towel when inserting a glass tube into a stopper or removing it; grasp the glass close to the stopper and gently twist it in or out. Remember that hot glass remains hot for some time, so after you have fire-polished a glass rod or completed another glass-working operation, give the glass plenty of time to cool before you touch it. Refer to OP-3 for more information about safe glass-working procedures.

5 *Wear appropriate clothing in the laboratory.* Wear clothing that is substantial enough to offer some protection against accidental chemical spills, such as long-sleeved shirts or blouses and long pants or dresses. Wear shoes that can protect you from spilled chemicals and broken glass, not open sandals or cloth-topped athletic shoes. Human hair is very flammable, so wear a hair net while using a burner if you have long hair. To protect your clothing (and yourself) from chemical spills, it is a good idea to wear a lab jacket or apron in the lab.

6 *Dispose of chemicals properly.* For reasons of safety and environmental protection, most organic chemicals should not be washed down the drain. Except when your instructor or an experiment's directions indicate otherwise, place used organic chemicals and solutions in designated waste containers. Some aqueous solutions can be safely poured down the drain, but consult your instructor if there is any question as to the best method for disposing of a particular chemical or solution. See "Disposal of Hazardous Wastes" at the end of this section for additional information.

7 *Never work alone in the laboratory or perform unauthorized experiments.* If you wish to work in the laboratory when no formal lab period is scheduled, you must obtain written permission from the instructor and be certain that others will be present while you are working.

Reacting to Accidents: First Aid

If you have or witness a serious accident involving poisoning or injury, report it to the instructor as soon as possible. Serious accidents should be treated by a competent physician, but applying some basic first aid procedures before a physician arrives can help minimize the damage.

If you have an accident that requires quick action to prevent permanent injury, take the appropriate action as described here, if you can, and see that the instructor is informed of the accident. If you *witness* an accident, call the instructor immediately and leave the first aid to him or her, unless: (1) no instructor or assistant is in the laboratory area; (2) the victim requires immediate attention because of stopped breathing, heavy bleeding, etc.; (3) you have had appropriate emergency response training.

If an accident victim stops breathing or goes into shock as a result of any kind of accident, standard procedures for artificial respiration and treating shock should be applied. Descriptions of these procedures can be found in Chapter 2 of *The CRC Handbook of Laboratory Safety*, 4th ed. [Bibliography, C1].

Eye Injuries

If any chemical enters your eyes, flush them *immediately* with water from an eyewash fountain while holding your eyelids open. If you are wearing contact lenses, remove them first. Continue irrigation for at least 15 minutes (or until a nurse or physician arrives); then have your eyes examined by a physician. If foreign bodies such as glass particles are propelled into your eye, seek immediate medical attention. Removal of such particles is a job for a specialist.

Chemical Burns

If a corrosive chemical is spilled on your skin or clothing, remove any contaminated clothing and flush the affected area *immediately* with a large amount of water until the chemical is completely removed, using a safety shower if the area of injury is extensive or if it is inaccessible to washing from a tap. Speed and thoroughness in washing are the most important factors in reducing the extent of injury. Dry the area gently with a clean, soft towel. If you or another victim experiences pain, or if the skin is red or swollen, immerse the injured area in cold water or apply cold, wet dressings. Do not use neutralizing solutions, ointments, or greases on chemical burns, unless they are specifically called for in a first aid procedure. Unless the skin is only reddened over a small area, a chemical burn should be examined by a nurse or physician. First aid procedures for burns caused by specific chemicals, such as bromine, are given in the *Sigma-Aldrich Library of Chemical Safety Data* [Bibliography, C4] and in the *First Aid Manual for Chemical Accidents* [Bibliography, C3].

If a chemical burn is very extensive or severe, the victim should lie down with the head and chest a little lower than the rest of the body. If the victim is conscious and able to swallow, he or she should be provided with plenty of nonalcoholic liquid (water, tea, coffee, etc.) to drink until a physician or an ambulance arrives.

Thermal Burns

If you are burned by hot glass, another hot object, or a flame, try to determine the type and extent of the burn and then take the appropriate action as described next.

- *First-degree burn*—the skin is reddened but there are no blisters or broken skin. Immerse the affected area in clean, cold water or apply ice to reduce the pain and facilitate healing.

- *Second-degree burn*—blisters are raised. Immerse the burned area in clean, cold water or apply ice; then cover the area with sterile gauze or another clean dressing. If legs or arms are burned, keep them elevated above the trunk of the body. Never puncture blisters raised by a second-degree burn.
- *Third-degree burn*—the skin is broken and underlying tissue is damaged. Place a thick sterile dressing or clean cloths over the affected area and have the burn examined by a nurse or physician as soon as possible. Do not remove burned clothing, immerse the burned area in cold water, or cover the burn with greasy ointment. If legs or arms are burned, keep them elevated above the trunk of the body.

In case of an extensive thermal burn, the burned area should be covered with the cleanest available cloth material and the victim should lie down, with the head and chest lower than the rest of the body, until a physician or an ambulance arrives. If the injured person is conscious and able to swallow, he or she should be provided with plenty of nonalcoholic liquid (water, tea, coffee, etc.) to drink.

Bleeding, Cuts, and Abrasions

In case of a cut or abrasion that does not involve heavy bleeding, cleanse the wound and surrounding skin with soap and lukewarm water, applying it by wiping away from the wound. Try to remove imbedded glass shards, if there are any, by using tweezers. Hold a sterile gauze pad over the wound until bleeding stops. If the damage is confined to the skin, apply an antibacterial cream such as Neosporin to prevent infection. Place a fresh gauze pad over the wound and secure it loosely with a triangular or rolled bandage. Replace the pad and bandage as necessary with clean, dry ones. Avoid contact between the wound and the mouth, fingers, handkerchiefs, or other non-sterile objects. If the wound is deep or extensive, it should be treated further by a nurse or physician.

In case of a major wound that involves heavy bleeding, *immediately* apply pressure directly over the wound with a cloth pad (such as a clean handkerchief or other clean cloth) or sterile dressing, pressing firmly with one or both hands to reduce the bleeding as much as possible. Then call for assistance. A compression bandage should be applied on top of the original dressing if profuse bleeding continues. (Removing the original dressing may delay clotting.) If no compression bandage is available, you can roll a clean cork in sterile gauze and bind it to apply pressure directly over the wound. The victim should lie down with the bleeding part higher than the heart, and the dressing should be held in place with heavy gauze or other cloth strips. A physician or ambulance should be called as soon as possible, and the victim should be kept warm with a blanket or coat. If the injured person is conscious and able to swallow, he or she should be provided with plenty of nonalcoholic liquid (water, tea, coffee, etc.) to drink until a physician arrives.

Poisoning

If, after contact with, inhalation of, or accidental ingestion of a chemical, you experience a burning sensation in the throat, discoloration of the lips or mouth, stomach cramps, nausea and vomiting, or confusion, *immediately* seek treatment for chemical poisoning.

If a poison has been *ingested* (swallowed), an ambulance and a poison control center should be called immediately. If the victim is conscious, loosen tight clothing around the neck and waist, have the victim rinse his or her mouth several times with cold water and spit it out, then give the victim 1–2 cups of water or milk to drink. Unless otherwise advised by the poison control center, induce vomiting by tickling the back of the throat or by giving 2 tablespoons (~30 mL) of ipecac syrup followed by a cup of water. Vomiting should not be induced if the victim is unconscious, in convulsions, or has severe pain and burning sensations in the mouth or throat, or if the poison is a petroleum product or a strong acid or alkali. When vomiting begins, lower the victim's head to prevent vomit from reentering the mouth, and collect a sample of the vomit for possible analysis. If convulsions occur, remove any objects that might cause injury, or move the victim away from such objects. Watch out for and remove any obstructions in the victim's mouth (including dentures). If necessary, insert a soft pad between the victim's teeth to protect the tongue from being bitten. If the victim has stopped breathing, clear the airway and administer artificial respiration. When there is time, the poison should be identified (if possible) and an appropriate antidote given (see *The Merck Index*, 9th ed., pp. MISC-22ff., or call a poison control center). If the poison cannot be identified and a medical professional is not present, a heaping tablespoon (15 g) of a universal antidote can be given in a half glass of warm water. The universal antidote can be prepared by combining 2 parts activated charcoal, 1 part magnesium oxide, and 1 part tannic acid. If it is known, a sample of the poison should be saved for the physician.

If a poison has been *inhaled*, the victim must be taken to fresh air and a physician called immediately. Loosen any tight clothing around the neck and waist, and use a tongue depressor or other device, if necessary, to keep the victim's airway open. Administer manual (not mouth-to-mouth) artificial respiration if the victim has stopped breathing. Keep the victim warm and as quiet as possible until a physician arrives. If the poison is a highly toxic gas, such as hydrogen cyanide, hydrogen sulfide, or phosgene, the persons attempting to rescue the victim should wear self-contained respirators while they are in contact with the vapors.

In case of *skin contact* with a toxic substance, follow the same general procedure as that for chemical burns. The *Sigma-Aldrich Library of Chemical Safety Data* [Bibliography, C4] or another appropriate source should be consulted for specific procedures to be used for injury by certain substances.

Reacting to Accidents: Fire

Fires are unlikely in most modern organic chemical laboratories, which use flameless heat sources for nearly all operations. But burners may be used for special applications, such as bending glass, and some chemicals may ignite on a hot surface, such as the top of a hot plate.

In case of fire, your first response should ordinarily be to *get away* as quickly as possible and let the instructor deal with the fire. However, if a fire is small and is confined to a container such as a flask or beaker, you may be able to extinguish it by placing a watch glass over the mouth of the container. If no instructor is in the laboratory, obtain a fire extinguisher of the appropriate type and attempt to put out the fire by aiming the extinguisher at the base of the fire, while maintaining a safe distance. Most labs should be equipped with class BC or ABC dry chemical extinguishers, which are effective against solvent and electrical fires. A class D fire, which involves burning metals or metal hydrides such as sodium metal and lithium aluminum hydride, can be extinguished with an appropriate class D fire extinguisher or by smothering the fire in dry sand, sodium chloride, or sodium carbonate. If a fire is too large to be put out by a fire extinguisher, sound the nearest fire alarm and evacuate the area.

If your hair or clothing catches on fire, *don't panic. Walk* (don't run) directly to the nearest fire blanket or safety shower and attempt to extinguish the fire. Don't wrap yourself in a fire blanket while you are standing, as that may direct the flames around your face; instead *drop* to the floor and *roll* as you wrap the blanket around your body. To use a safety shower, get directly under the shower head and pull the chain. If another person's hair or clothing has caught on fire, try to prevent panic as you lead him or her to a safety shower or fire blanket.

Chemical Hazards

For your own health and safety, it is essential that you *exercise caution while handling chemicals* and *minimize your exposure to them.* Most academic chemistry departments have policies and procedures for dealing with hazardous chemicals, which should be incorporated into the institution's chemical hygiene plan. Your instructor will inform you of any departmental or institutional rules relating to the safe handling and disposal of hazardous chemicals. The experiments in this book describe specific chemical hazards and handling precautions under the heading "Safety Notes." The hazard descriptions are meant to inform you of the potential danger posed by certain chemicals when they are not handled properly. If you take reasonable precautions and follow the directions in the Safety Notes, there is little likelihood that you will suffer ill effects from working with any chemical.

There are several different kinds of chemical hazards. A chemical may be toxic and therefore capable of causing illness or death when ingested, inhaled, or allowed to contact the skin. Certain chemicals can burn when exposed to a spark, an open flame, or a high temperature. A few chemicals can explode when heated, subjected to shock, or mixed with certain organic materials. Some chemicals may react spontaneously or when combined with other chemicals, generating heat that could cause a fire.

The severity of any hazards associated with a chemical can be indicated by a labeling system, such as the one established by the National Fire Protection Association (NFPA), which rates the health, flammability, and reactivity hazards of chemicals. This book uses a similar system in which the general health hazard number is usually replaced by two numbers indicating the hazards posed by contact with and inhalation of a chemical. Hazard information on selected chemicals (usually the most hazardous ones) used in the experiments is provided in the form of octagonal signs containing hazard index numbers. Each hazard index is a number from 0 to 4, where 0 indicates no known hazard and 4 indicates a very great health or safety hazard. Letters such as "W" are sometimes included to warn of other hazards, such as reactivity with water. The significance of the numbers and letters is explained in Table 1.

Table 1 Meanings of numbers and letters on hazard symbols

Symbol	Health (Contact or Vapor)	Fire	Reactivity
0	No known hazard	Will not burn	Stable
1	May cause irritation if not treated	Ignites after strong preheating	Unstable only at high temperature and pressure
2	May cause injury; requires treatment	Ignites after moderate heating	Unstable, but won't detonate
3	May cause serious injury despite treatment	Ignites at normal temperatures	Detonates or explodes with difficulty
4	May cause death or major injury despite treatment	Very flammable	Readily detonates or explodes
CA	Carcinogen (cancer-causing agent)		
OX			Strong oxidant; may react violently with combustible material
P			Polymerizes readily
W			Reacts violently with water

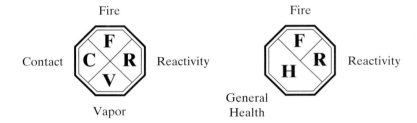

You should learn the location of each category on the hazard signs. Going clockwise from the top, the order is Fire → Reactivity → Vapor → Contact. The *fire* index number in the upper quadrant rates the fire danger posed by the chemical. The *reactivity* index number in the right quadrant assesses the danger of a violent reaction or explosion. The *vapor* index number in the lower quadrant rates the potential health effect of inhaling the vapors of the chemical. The *contact* index number in the left quadrant rates the adverse health effects that may result from skin contact (the eye contact hazard, which is not rated, may be even greater in some cases). For some chemicals, the vapor and contact index numbers are replaced by a single *health* index number that represents the overall health hazard of the chemical. If any quadrant is left blank, it means that no hazard index number was reported in the sources consulted; it does *not* mean that no hazard exists.

The meaning of the hazard signs is illustrated by the examples in the margin. Carbon tetrachloride is nonflammable and thermally stable, so its Fire and Reactivity hazard index numbers are both zero. Inhaling its vapors may be extremely harmful, and there is a moderate risk of injury from skin contact. Carbon tetrachloride has also been identified as a carcinogen in tests with laboratory animals and is suspected of causing cancer in humans. For these reasons, carbon tetrachloride is not used in the experiments in this book. The second hazard sign shows that hexane is highly flammable but very stable and its health risk is comparatively low.

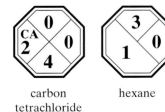

carbon
tetrachloride

hexane

Chemicals with a high *fire* hazard number should obviously be kept away from ignition sources, including flames, sparks, and hot surfaces. For example, diethyl ether will ignite if it is spilled on a hot plate at a temperature of 160°C or higher.

Chemicals with a high *reactivity* hazard number should be handled with great care, following any precautions given in the Safety Notes. Such chemicals may ignite or explode if subjected to shock or brought into contact with metal spatulas or materials that may catalyze their decomposition.

The value of a *vapor* or *contact* hazard number may suggest the appropriate response to inhalation of or contact with a chemical. If you inhale significant amounts of a chemical's vapor or the dust of a solid chemical, you should go to a window or other area where you can breathe fresh air, unless the chemical's vapor hazard number is zero. If the chemical has a high vapor hazard number, or if inhalation was prolonged, see your instructor to determine whether treatment is required. In case of skin contact with a chemical having a contact hazard number other than zero, you should wash the area of contact with soap and water. If the chemical has a high contact hazard number, or if the exposure is extensive, see your instructor to determine whether treatment is required. If you get a chemical in your eyes, follow the procedures described under the heading "Eye Injuries" on page 16. The higher the chemical's contact hazard number, the more likely it is that eye damage will result without prompt medical treatment.

Chemicals designated as strong oxidants (OX) must not be allowed to contact other chemicals, except for the ones specified in the experimental directions. In fact, you should never mix *any* chemicals together unless directed to, since mixing incompatible chemicals may result in the generation of toxic gases, fire, or an explosion.

Chemicals that polymerize readily (P) may sometimes undergo spontaneous, rapid polymerization, which generates heat. If this reaction occurs in a capped reagent bottle it could cause the bottle to shatter violently. Most such chemicals are stabilized by the addition of a small amount of antioxidant or other stabilizer, so they are unlikely to react unless the stabilizer has been removed.

Water-sensitive chemicals (W) react violently and exothermically upon contact with water, often generating toxic fumes. Some water-sensitive chemicals generate toxic fumes even when exposed to moist air. Such chemicals should be kept in tightly closed containers that are opened only for transfers and closed immediately afterward. They should be used under fume hoods at some distance from any source of water.

Carcinogens

A few chemicals used in the experiments are suspected *carcinogens*— agents suspected of causing cancer (CA). Although this label certainly should not be disregarded, it is important to recognize that such chemicals present little risk of cancer to students if used as directed. The carcinogenic activity of a compound is generally established by animal tests in which high doses of the chemical are administered by various routes for prolonged periods. For example, phenacetin taken orally has been found to cause cancer in laboratory animals, but it is highly unlikely that anyone will swallow enough phenacetin during a laboratory experiment to cause cancer. Chromium(VI) compounds have been shown to cause cancers of the lungs, nasal cavity, and sinuses in humans, but the persons at risk are industrial workers who have been continuously exposed to the dust of chromium compounds in the workplace. In this course, you will use chromic acid solutions [prepared from chromium(VI) oxide] a drop or so at a time, so there will be no possibility of inhaling chromium dust.

It is prudent to take appropriate precautions when handling any potential carcinogen, such as wearing protective gloves and clothing and working under a hood. Some halogenated hydrocarbons are suspected of causing cancer if inhaled, so avoid breathing the vapors of chloroform, dichloromethane and any other chlorinated solvents. Compounds such as benzene and carbon tetrachloride, which might present a significant risk of cancer under conditions likely to be encountered in an undergraduate laboratory, are not used in this book.

Teratogens

If you are pregnant or think you may be pregnant, inform your instructor before you enter the organic chemistry lab for the first time. Some organic chemicals are known or suspected *teratogens*, meaning that they may harm a developing fetus. As is the case for carcinogens, handling a chemical designated as a teratogen does not necessarily represent a danger to the fetus. Ethanol (ethyl alcohol) is a teratogen when ingested, but people are far more likely to ingest ethanol in a bar, a restaurant, or at home than in a chemistry laboratory. Nevertheless, some chemicals that are routinely used in organic chemistry laboratories may represent a significant danger to a developing fetus. Therefore, it is important to contact your instructor, the laboratory coordinator, or a designated chemical hygiene officer to discuss your options. The best option may be to take the laboratory course after your child is delivered. If you prefer to remain in the lab course, you should obtain the consent of your physician and then make arrangements with your instructor to minimize your exposure to potential teratogens. Such arrangements might include substituting less hazardous chemicals for teratogenic ones; performing alternative experiments that do not require the use of teratogens; or using gloves, protective clothing, and a hood when handling any potentially teratogenic chemical.

Sources of Information about Chemical Hazards

Because of space limitations, the chemical hazard information provided in this book is, of necessity, incomplete. The most complete source of information about the hazards associated with any chemical is its *Material Safety Data Sheet (MSDS)*, which is provided by the manufacturer or vendor of the chemical. Chemical manufacturers, for reasons related to legal liability, tend to list every potential mishap—no matter how unlikely—that might result from the use of their chemicals. Because MSDSs are intended primarily for chemical and industrial workers, whose exposure to chemicals may be intense and prolonged, they often recommend protective apparatus and measures that may not be necessary when chemicals are used in small quantities for short periods of time. This makes it difficult to determine which hazards associated with the chemicals used in a given organic chemistry experiment are truly significant. Nevertheless, the MSDS for a given chemical can be helpful if the information it provides is used appropriately. The MSDSs for chemicals used in your organic chemistry laboratory should be available from the chemistry department office, the chemical hygiene officer, or some other designated source at your institution. Other useful sources of chemical hazard information include *The Sigma-Aldrich Library of Chemical Safety Data* [Bibliography, C4], *Sax's Dangerous Properties of Industrial Materials* [C5], *Hazards in the Chemical Laboratory* [C6], and *Bretherick's Handbook of Reactive Chemical Hazards* [C8].

Disposal of Hazardous Wastes

Some of the chemicals and solutions generated in organic chemistry labs are regarded as hazardous wastes, which are regulated by various state and federal agencies. Institutions classified as hazardous waste generators are required to collect and dispose of hazardous wastes in ways that pose minimum potential harm to human health and to the environment. Some hazardous wastes can be purified and reused, or converted to nonhazardous materials. Other hazardous wastes are placed in landfills, incinerated, or otherwise disposed of by commercial waste disposal firms.

Except when precluded by local regulations, moderate quantities of many common chemicals can be safely and acceptably disposed of down a laboratory drain, if certain procedures are followed. Small quantities (less than 100 g) of most water-soluble organic compounds, including water-soluble alcohols, aldehydes, amides, amines, carboxylic acids, esters, ethers, and ketones, can be disposed of in this way. The organic compound should be mixed or flushed with at least 100 volumes of excess water. Dilute aqueous solutions of inorganic compounds can be disposed of down the drain if they contain cations and anions from the following list and no other cations or anions.

Cations of these metals: Al, Ca, Cu, Fe, Li, Mg, Mo, Pd, K, Na, Sn, Zn
Anions: HSO_3^-, BO_3^{3-}, $B_4O_7^{2-}$, Br^-, CO_3^{2-}, Cl^-, OCN^-, OH^-, I^-, O^{2-}, PO_4^{3-}, SO_4^{2-}, SO_3^{2-}

Some less common cations and anions, such as Cs^+ and SCN^-, are also permissible.

Substances that should *not* be poured down the drain include the following.

- Water-soluble organic compounds that are highly flammable or boil below 50°C
- Hydrocarbons, halogenated hydrocarbons, and other water-insoluble organic compounds
- Flammable or explosive solids, liquids, or gases
- Phenols and other taste- or odor-producing substances
- Wastes containing poisons in toxic concentrations
- Corrosive wastes capable of damaging the sewer system

These and any other chemical wastes designated by your instructor should be placed in specially labeled waste containers, which should be provided in the laboratory.

Mastering the Operations

The experiments in Part I will help you learn the basic laboratory operations of organic chemistry and become proficient in their use. The first time an operation is used in Part I, its number is highlighted in boldfaced type in the **Operations** list at the beginning of the experiment. The section titled **Before You Begin** lists some prelab assignments that you must complete before you arrive at the laboratory to start the experiment.

Learning Basic Operations
The Effect of pH
on a Food Preservative

EXPERIMENT 1

Laboratory Orientation. Acid–Base Reactions. Reaction Stoichiometry.

Operations

OP-1   Cleaning and Drying Glassware
OP-4   Weighing
OP-5   Measuring Volume
OP-6   Making Transfers
OP-13 Vacuum Filtration
OP-23 Drying Solids   *p55c*

Before You Begin

1 Read the Introduction and Laboratory Safety sections of this book.
2 Read the experiment carefully, particularly the section titled "Understanding the Experiment" and the Directions.
3 Read the sections in Part V describing operations OP-l, OP-4, OP-5, OP-6, OP-13, and OP-23. Most of the descriptions are quite brief.
4 If you are required to write up your experiments in a formal laboratory notebook, prepare the notebook as directed by your instructor after reading Appendix II.

Scenario

For this and the other experiments in this book, you will play the role of a consultant in a Consulting Chemists Institute operated by your college or university, with your instructor as the laboratory supervisor and other students in your lab section as members of your project group. Most of the Scenarios in this book describe purely hypothetical situations. Except where otherwise indicated, any similarity between a person or organization named herein and an actual person or organization is coincidental.

Fresh Foods Incorporated (FFI) is a small, family-owned chemical company that manufactures sodium benzoate and other food preservatives for sale to food processors. One of its competitors has launched an ad campaign claiming that FFI's sodium benzoate changes to a different chemical in the stomach, implying that the chemical may be harmful or ineffective. D. K. Little, FFI's marketing director, has asked your Institute to find out whether the competitor's claim is valid. Your assignment is to place sodium benzoate into a medium that simulates stomach acid, which is an aqueous hydrochloric acid solution having a pH range of 1–3, to see whether or not a new substance forms in that environment. Since this is your first day on the job, your supervisor will show you around the laboratory and also help you develop some basic lab skills involving measurements of mass and volume.

Applying Scientific Methodology

Read "Scientific Methodology" in the Introduction before you read this section.

As in the other experiments in this book, the scientific *problem* to be solved is described in the Scenario. In this experiment the fundamental problem, stated as a question, is "Does a new substance form when sodium benzoate is placed into a simulated stomach acid?" After reading the experiment thoroughly, you should be ready to formulate a *working hypothesis* such as "A new substance will (or will not) form when sodium benzoate is placed into a simulated stomach acid." Your *course of action*, described in detail in the Directions, is then fairly obvious; you will have to place some sodium benzoate in an aqueous solution of hydrochloric acid having a pH in the range 1–3. You can do this by dissolving the sodium benzoate in water and adding enough dilute hydrochloric acid (aqueous HCl) to lower the pH to about 2. During this process you will have to *gather evidence* about whether or not a chemical reaction has occurred. Evidence for a chemical reaction may include a color change, formation of a gas, formation of a precipitate, evolution of heat, or any combination of these. As you *evaluate the evidence* (i.e., think about what you've observed), you may decide to stay with your working hypothesis (if the evidence supports it), modify it, or discard it and formulate a new one. You then should be able to *arrive at a conclusion* based on the experimental evidence and any clues you came across while reading the experiment. You will *report your results* after referring to the "Report" section that follows the Directions.

Sodium Benzoate as a Food Preservative

When resin from the Sumatran tree *Styrax benzoin* is heated to 100°C, white vapors rise and condense to form needlelike crystals of benzoic acid, which was named for the tree and its resin, gum benzoin. Benzoic acid can easily be converted to its salt, sodium benzoate, which is used widely as a food preservative, particularly in acidic foods such as fruits and fruit juices. Benzoic acid is so widely distributed in plants that the urine of all plant-eating animals (including humans) contains hippuric acid, a compound synthesized in the kidneys by the combination of benzoic acid with the amino acid glycine. Significant amounts of benzoic acid can be isolated from such diverse natural sources as anise seed, cranberries, prunes, cherry bark, cloves, and the scent glands of the beaver. In cranberries and other plant products, benzoic acid is a natural preservative that inhibits the growth of bacteria, yeasts, and molds, and thus retards spoilage.

If you read the labels on cans and bottles in a grocery store, you will find sodium benzoate (sometimes called benzoate of soda) listed on many of them. Sodium benzoate is used as a preservative in such food products as jams and jellies, soft drinks, fruit juices, pickles, condiments, margarine, and canned and frozen seafood, and even in such nonfood items as toothpaste and tobacco. Benzoic acid is also an effective food preservative, but sodium benzoate is a more popular food additive because it is much more soluble in water than is benzoic acid, making it easier to blend into water-containing food products.

A hexagon with a circle in it represents a benzene ring, whose molecular formula in these compounds is C_6H_5.

benzoic acid   sodium benzoate

hippuric acid   glycine

Understanding the Experiment

This experiment will help you become familiar with the laboratory environment and with some fundamental laboratory operations, such as measuring mass and volume, separating solids from liquids by vacuum filtration, and drying solids.

In part **A**, you will add hydrochloric acid to a solution of sodium benzoate and, if a different substance forms, recover it and measure its mass. To understand what is happening, you need to know something about the properties of the substances involved and remember what you have previously learned about acid-base chemistry and stoichiometry. When sodium benzoate dissolves in water, it dissociates into benzoate ions (which are weakly basic) and sodium ions, according to the reaction

$$C_6H_5COONa \longrightarrow C_6H_5COO^- + Na^+$$
$$\text{benzoate ion}$$

Hydrochloric acid is a solution of hydrogen chloride (HCl) in water. The strongest acid present in this solution is the hydronium ion, formed by the transfer of a proton from an HCl molecule to a water molecule.

$$HCl + H_2O \longrightarrow Cl^- + H_3O^+$$
$$\text{hydronium ion}$$

Key Concept: In a Lowry-Brønsted acid-base reaction under standard conditions, the stronger acid transfers protons to a stronger base to yield a weaker acid and a weaker base.

If the hydronium ion concentration is high enough, protons will be transferred from the strong acid H_3O^+ to the basic benzoate ions. This will yield benzoic acid, which, being quite insoluble in water (sodium benzoate is about 200 times more soluble), should precipitate from solution. The reaction is

$$C_6H_5COO^- + H_3O^+ \longrightarrow C_6H_5COOH + H_2O \qquad \textbf{(1)}$$
$$\text{benzoic acid}$$

The net reaction is the sum of these three reaction steps.

$$C_6H_5COONa + HCl \longrightarrow C_6H_5COOH + NaCl \qquad \textbf{(2)}$$

Stop and Think: What is the relationship between hydronium ion concentration and pH?

The hydronium ion concentration depends on the pH of the solution; the lower the pH, the higher $[H_3O^+]$ will be. By carrying out the experiment, you will discover whether or not the pH of stomach acid is low enough to cause the conversion of sodium benzoate to benzoic acid.

In any experiment involving the conversion of one substance to another, it is important to understand the stoichiometry of the reaction. (Read Appendix IV if you need a review of stoichiometric calculations.) Equation **2** shows that if the reaction were complete at the pH used, a mole of benzoic acid would be formed for each mole of sodium benzoate in the reaction mixture. You will start with about 0.400 g of sodium benzoate, which is 2.78 mmol, so the theoretical yield of benzoic acid (if any forms) should be 2.78 mmol as well, corresponding to a mass of 0.339 g. Keep in mind that you will probably not measure out *exactly* the mass of sodium benzoate specified, in which case you will need to calculate the theoretical yield (to the appropriate number of significant figures) based on your measured mass of sodium benzoate.

Whether or not your yield approaches the theoretical value will depend on a number of factors, including the following:

- You may make errors in measuring mass or volume.
- The proposed reaction may not occur, or if it does, it may not be complete.
- Some of the product may remain dissolved in the reaction mixture and may not be recovered.
- You may lose some product while transferring it from one vessel to another.

Always try to minimize your losses by taking great pains to perform accurate measurements and to make nearly *quantitative transfers*; that is, to scrape or rinse the last traces of solid from the glassware. This includes rinsing stirring rods, spatulas, and other items that come in contact with the reaction mixture. Making efficient transfers is particularly important in the microscale lab. Losing 40 milligrams of product by transfer during a standard scale preparation having a theoretical yield of 4.0 g is not a serious problem because it represents a loss of only 1.0% of the product, but losing the same amount from a typical microscale preparation having a theoretical yield of 0.20 g amounts to a 20% loss. Although you can't possibly scrape the last molecule of product from one vessel to another, you should try to transfer virtually all of the product visible to the naked eye. Get into the habit of doing the kind of careful, meticulous work that will increase your yields and save you time and effort in the long run.

The product of a chemical preparation should always be dried to *constant mass*, meaning that its mass after drying should not change significantly between two successive weighings. For most preparations in this book, you can assume that a product is sufficiently dry if—following the initial drying period—its mass does not decrease by more than 0.5% after an additional 5 minutes of oven drying. If the product is being dried at room temperature in a desiccator, the interval between weighings should be longer.

In part **B** you will determine the density of an unknown liquid by using a graphical method. The validity of your results will depend on the care you take in measuring the mass and volume of the liquid and in graphing your results.

Laboratory Safety and Orientation

Before coming to lab, you should have read the Introduction, which includes a section entitled "A Guide to Success in the Organic Chemistry Lab," and the "Laboratory Safety" section of this book. Before you begin to work in the laboratory, your instructor will review the safety rules and tell you what safety supplies you must have, such as safety goggles and protective gloves and aprons. During the first laboratory period, the instructor will show you where safety equipment is located and tell you how to use it. You should also learn the locations of chemicals, consumable supplies (such as filter paper and boiling chips), waste containers, and various items of equipment such as balances and drying ovens. Obtain a locker and a supply list and check into the laboratory as directed by your instructor. You will find illustrations of typical locker supplies in Appendix I at the back of this book. If necessary, clean up (see OP-1) any dirty glassware and replace any damaged glassware in your locker at this time.

Reactions and Properties

Table 1.1 Physical properties

	M.W.	mp	Water solubility
benzoic acid	122.1	122	0.34
sodium benzoate	144.1	—	61.2

Note: M.W. = molecular weight; melting points (mp) are in °C; solubilities are in grams of solute per 100 mL of water at room temperature (usually 25°C).

Directions

Your instructor will demonstrate the operation of balances, automatic pipets, and any other special equipment you will use during this experiment.

A. *The Effect of pH on Sodium Benzoate*

Safety Notes

> Sodium benzoate and benzoic acid are mild irritants. Do not get them in your eyes or on your skin or clothing.

When "water" is specified in a procedure, use distilled or deionized water unless otherwise indicated.

top laading balance

Observe and Note: Can you detect any evidence for a chemical reaction? If so, describe it in your lab notebook.

Stop and Think: What is the purpose of cooling the solution?

Stop and Think: What is the substance?

Reaction.   Obtain a short ($5\frac{3}{4}$-inch) Pasteur pipet, attach a latex rubber bulb to it, and calibrate the pipet as described in OP-5. Weigh [OP-4] approximately 0.400 g of sodium benzoate to the maximum accuracy of your balance and transfer [OP-6] it quantitatively to a 5-mL conical reaction vial. Measure [OP-5] 3 mL of water into the vial, and stir with a thin glass stirring rod until the sodium benzoate dissolves. Measure [OP-5] 1.0 mL of 3 *M* hydrochloric acid with your calibrated Pasteur pipet and transfer [OP-6] it slowly, while stirring, to the vial until you have added about 0.8 mL. Then, while stirring, add more 3 *M* HCl drop by drop, testing the solution frequently with pH paper, until its pH is 2. To test the pH, use your stirring rod or the closed end of a capillary melting-point tube to transfer a small drop of the supernatant liquid (liquid at the surface) to a strip of pH paper. Adding a little excess HCl will do no harm. Cool the solution for five minutes or more by setting the reaction vial in a small beaker containing cracked ice and a small amount of tap water. To keep the vial from tipping over, attach an air condenser to it and clamp the condenser to a ringstand.

Separation.   If a new substance has formed, separate it from the reaction mixture by vacuum filtration [OP-13] with a Hirsch funnel. Use about 1 mL of ice-cold water to transfer any solid that adheres to the walls of the vial and wash the solid on the filter as directed in OP-13. Let the solid air dry on the filter for a few minutes with the vacuum turned on, then dry [OP-23] it to constant mass. If an oven is used for drying, its temperature should be 90°C or lower because the solid may sublime above that temperature. Weigh [OP-4] the dry product in a tared (preweighed) 1-dram screw-cap vial. Label the vial with the experiment number, the name of the product, its mass, your name, the current date, and any other information required by your instructor. Unless your instructor indicates otherwise, *always* turn in your product when you finish an experiment. If you do not, you may not get credit for the experiment.

B. *Measuring the Density of an Unknown Liquid*

> **The unknown liquid may be harmful if inhaled or absorbed through the skin. Avoid contact with the liquid and do not breathe its vapors.**

At your instructor's request, first measure the density of room-temperature water by the procedure described here. If your measured density is not within 2% of the expected density of water (~0.997 g/mL at room temperature), repeat the measurement.

Have ready two clean, dry, 1-dram screw-cap vials. Use a measuring pipet or a bottle-top dispenser to measure [OP-5] about 1.1 mL of the unknown liquid into one of the vials, and take both vials to a balance. Weigh [OP-4] the empty vial and its cap to the nearest milligram (0.001 g) and record the mass. (Alternatively, zero the balance with the empty capped vial and record the mass of the liquid directly.) If one is available, use an automatic pipet set at 200 μL (0.200 mL) to measure the volume [OP-5] of the liquid; otherwise, use a 1-mL measuring pipet. (Do not try to adjust the automatic pipet! If it does not read 200 μL, see your instructor.) Use the pipet to accurately measure 0.200 mL of the liquid into the empty vial, cap it immediately, and record the mass. Without delay, measure an additional 0.200-mL portion of the liquid into the vial, cap it, and again record the mass. Repeat this process until you have added a total of five 0.200-mL portions. Use the same balance for all readings.

Prepare a graph of your data, plotting cumulative volume on the *x*-axis and cumulative mass (of the liquid) on the *y*-axis. Draw the best straight line through the data points. From the slope of the graph calculate the density of the liquid to three decimal places. Show your graph and your calculated density to your instructor; if either is unacceptable, you may be asked to repeat your measurements or redraw your graph.

Cleanup Routine.   After completing this and all subsequent experiments you should

- Clean [OP-1] the glassware you used during the experiment.
- Clear off your work area, wipe the bench top with a sponge or wet towel, and remove any refuse or equipment from the sink.
- Turn in any items you may have checked out of the stockroom and return any community supplies to their proper locations.
- Turn in your labeled product to the instructor, when applicable.
- See that all the items on the locker list are safely inside your locker (you may have to pay for missing supplies), then lock it.

Waste Disposal: The liquid in the filter flask may be poured down the drain.

Safety Notes

Take Care!   Avoid contact with the liquid and do not breathe its vapors.

Waste Disposal: Put the liquid in a solvent recovery container, as directed by your instructor.

The cumulative volume (or mass) is the total volume (or mass) added up to a particular time; for example, after three additions the cumulative volume is 3 × 0.200 mL, or 0.600 mL.

Stop and Think: What is the relationship between the slope of your graph and the density? Does your calculated density make sense physically?

Report.   Read Appendix III before your write your report. (Your instructor may ask you to write up your results as if you were submitting them to D. K. Little of Fresh Foods Incorporated.) For this and all subsequent experiments you should

- State the problem(s) in your own words, and write down your working hypothesis (or hypotheses).
- Report and interpret the relevant evidence, which may include observations, lists or tables of numerical data, etc.
- State your conclusion, and tell how the evidence supports it.

For a synthesis, as in part **A** of this experiment, calculate the theoretical yield of the product based on your measured mass and the percent yield of your preparation. For part **B** of this experiment, include the density of your unknown liquid, a table showing your data, and your graph. Follow your instructor's directions regarding other items from Appendix III to include in your report.

Exercises

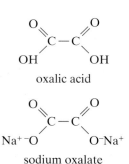

oxalic acid

sodium oxalate

1 You can confirm that a substance has been converted to a different substance by showing that the substances have different properties. Your observations should have revealed a difference in at least one property of sodium benzoate and benzoic acid. What property was that, and what observation revealed the difference?

2 Label the stronger acid, stronger base, weaker acid, and weaker base in Equation **1** for the proton transfer reaction in part **A** of the experiment.

3 When you lower the pH of a solution containing aqueous sodium benzoate, a precipitate eventually forms. (a) What will happen if you then raise the pH by adding aqueous NaOH? Why will this happen? (b) Write a balanced equation for the proton-transfer reaction involved and label the stronger acid, stronger base, weaker acid, and weaker base.

4 The water solubilities of oxalic acid and sodium oxalate at room temperature are 10 g/100 mL and 3.7 g/100 mL, respectively. Could you prepare oxalic acid by adding HCl to a solution of sodium oxalate, cooling it to room temperature, and filtering the resulting mixture? Explain why or why not.

5 The equation for a straight line can be written in the form $y = mx + b$, where x and y are variables, m is the slope, and b is the intercept. Write an equation for the straight line you obtained in your graph for part **B**, and show how it yields the equation for density $d = m/v$. (Remember that m in this equation stands for mass, not the slope.)

6 (a) Calculate the ratio of dissolved benzoic acid to benzoate ion that will exist in solution at equilibrium at a pH of 2.00. The acid equilibrium constant (K_a) for benzoic acid is 6.46×10^{-5}. Note that this calculation does not account for the benzoic acid that has precipitated from solution, but you can assume that the higher the ratio, the more benzoic acid will precipitate. (b) Carry out the same calculation for a pH of 4.00 and explain why it was important to reduce the pH to below 4 in this experiment.

Other Things You Can Do

(Starred items require your instructor's permission.)

*1 Make a flat-bottomed stirring rod and other useful laboratory items as described in Minilab 1 in Part III of this book.

*2 Try converting some other salts of organic acids, such as sodium salicylate and sodium oxalate, to the corresponding acids. You may need to look up the solubilities of the substances involved to explain your results.

3 To test the preservative effect of sodium benzoate, prepare a solution containing 0.15 g of sodium benzoate in 1 mL of water. Cut an apple into halves and use a brush to apply the solution to *one* of the cut surfaces. Leave the apple halves at room temperature and observe them over a period of several days.

4 Write a short research paper about food additives after consulting such sources as the *Kirk-Othmer Encyclopedia of Chemical Technology* [Bibliography, A8] and appropriate sources listed in section L of the Bibliography.

Extraction and Evaporation
Separating the Components of "Panacetin"

EXPERIMENT **2**

Separation Methods.

Operations

OP-7   Heating
OP-14 Centrifugation
OP-15 Extraction
OP-16 Evaporation
OP-4   Weighing
OP-5   Measuring Volume
OP-6   Making Transfers
OP-13   Vacuum Filtration
OP-23   Drying Solids

Before You Begin

1   Read the experiment carefully.
2   Read operations OP-14, OP-15, OP-16, and the section titled "Heat Sources" in OP-7. Review the other listed operations as necessary.

Scenario

See Experiment 1 if you don't understand the purpose of the Scenarios.

Your supervisor has been e-mailed the following message from a drug watchdog agency, the Association for Safe Pharmaceuticals (ASP).

Greetings:

Our roving agent in Southern California, Sam Surf, recently purchased some Panacetin—an analgesic drug preparation—at a drugstore in San Diego. According to the label on the bottle, the Panacetin tablets were manufactured in the United States by a legitimate pharmaceutical company, but Sam detected some discrepancies on the label and flaws in the tablets themselves that made him suspect they might be counterfeit. Such illegal knockoffs of a domestic drug can be manufactured cheaply elsewhere and smuggled into the United States, where they are sold at a big profit margin.

The label on the bottle lists the ingredients per tablet as aspirin (200 mg), acetaminophen (250 mg), and sucrose (50 mg). The sucrose is an inactive ingredient used to make the tablets more palatable to children. We have reason to believe that Panacetin does contain aspirin, sucrose, and another active component, but we're not sure what that component is or whether the amounts listed on the label are accurate. The unknown component is probably a chemical relative of acetaminophen, either acetanilide or phenacetin. Both of these kill pain as effectively as acetaminophen, so the consumer wouldn't notice their presence in an analgesic drug. But acetanilide and phenacetin are banned in the United States because of their toxicity, and we would like to keep them off the market.

We want your Consulting Chemists Institute to analyze this drug preparation to find out what percentages of aspirin, sucrose, and the unknown component it contains, and whether the unknown is acetanilide or phenacetin. You have two weeks to complete your investigation.

> purpose

Les Payne, Director of Operations, ASP

Applying Scientific Methodology

The Scenario presents two problems for you to solve in this experiment and the next: (1) Is the composition of Panacetin as stated on the label accurate? (2) What is the identity of the unknown component in Panacetin? You will concentrate on the first problem in this experiment, following the course of action described in the Directions. Because of possible material losses, you must allow some margin for error in deciding whether the percentage composition derived from the label (10% sucrose, 40% aspirin, 50% unknown component) is accurate. For the purposes of the experiment, a range of 6–14% sucrose, 30–50% aspirin, and 40–60% of the unknown component are close enough to indicate that the label is reasonably accurate. As in the previous experiment, you should start with a working hypothesis, gather and interpret evidence, change your hypothesis if the evidence does not support it, arrive at a conclusion, and report your results.

Painkilling Drugs, from Antifebrin to Tylenol

Analgesic drugs reduce pain; *antipyretic* drugs reduce fever. Some drugs, including aspirin and acetaminophen, do both. Many of the common over-the-counter analgesic-antipyretic drug preparations contain aspirin, acetaminophen, or combinations of these substances with other ingredients. For example, acetaminophen is the active ingredient of Tylenol, and Extra Strength Excedrin contains aspirin, acetaminophen, and caffeine. From their molecular structures, you can see that acetaminophen is chemically related to both acetanilide and phenacetin, whose painkilling effects were discovered late in the nineteenth century.

In 1886, a pair of clinical assistants, Arnold Cahn and Paul Hepp, were looking for something that would rid their patients of a particularly unpleasant intestinal worm. The trick was to find a drug that would kill the worm but not the patient, and their method—not a very scientific one—was to test the chemicals in their stockroom until they found one that worked. When someone came across an ancient bottle labeled NAPHTHALENE, they tried it out on a patient who had every malady in the book, including worms. It didn't faze the worms, but it reduced the patient's fever dramatically. Before Cahn and Hepp went out on a limb and endorsed naphthalene as a cure-all for fevers, someone noticed that the white substance in the bottle was nearly odorless. Since naphthalene has a strong mothball-like odor, Hepp suspected that the

Some compounds with
analgesic-antipyretic properties

naphthalene

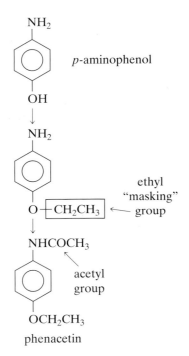

NH₂

p-aminophenol

OH

↓

NH₂

O─[CH₂CH₃] ← ethyl "masking" group

↓

NHCOCH₃

acetyl group

OCH₂CH₃

phenacetin

Synthesis of phenacetin

bottle was mislabeled and sent it to his cousin, a chemist at a nearby dye factory, for analysis. The tests showed that the new drug was not naphthalene at all, but acetanilide.

Acetanilide proved to have painkilling as well as fever-reducing properties and was soon being marketed under the proprietary name Antifebrin. Unfortunately, some patients using Antifebrin developed a serious form of anemia called methemoglobinemia, in which hemoglobin molecules are altered in a way that reduces their ability to transport oxygen through the bloodstream. Even though Antifebrin is now considered too toxic for medicinal use, its discovery did much to stimulate the development of safer and more effective analgesic-antipyretic drugs.

About six months after the discovery of Antifebrin, a similar drug was developed as the result of a storage problem. Carl Duisberg, director of research for the Friedrich Bayer Company, had to get rid of 50 tons of *para*-aminophenol—a seemingly useless yellow powder that was a by-product of dye manufacturing. Rather than pay a teamster to haul the stuff away, Duisberg decided to change it into something Bayer could sell. After reading about Antifebrin, he reasoned that a compound with a similar molecular structure might have similar therapeutic uses. Duisberg knew that a hydroxyl (OH) group attached to a benzene ring is characteristic of many toxic substances (for example, phenol), so he decided to "mask" the hydroxyl group in *para*-aminophenol with an ethyl (CH_3CH_2-) group, as shown in the margin. Incorporation of an acetyl (CH_3CO-) group then yielded phenacetin, which proved to be a remarkably effective and inexpensive analgesic-antipyretic drug. Until recently phenacetin was used in APC tablets (which contained aspirin, phenacetin, and caffeine) and in other analgesic/antipyretic drug preparations. It is no longer approved for medicinal use in the United States because it may cause kidney damage, hemolytic anemia, or even cancer in some patients.

Ironically, the substance Duisberg would have obtained had he not masked the hydroxy group is acetaminophen, which has proven to be a safer drug than either acetanilide or phenacetin. In the body, acetanilide and phenacetin are both converted to acetaminophen, which is believed to be the active form of all three drugs.

Understanding the Experiment

Most natural products and many commercial preparations are mixtures containing a number of different substances. To obtain a pure compound from such a mixture, the desired compounds are separated from the other components of the mixture by taking advantage of differences in their physical and chemical properties. For example, substances having very different solubilities in a given solvent may be separated by extraction or filtration, and liquids with different boiling points can be separated by distillation. Acidic or basic substances are often converted to water-soluble salts, which can then be separated from the water-insoluble components of a mixture.

In this experiment, you will separate the components of a simulated pharmaceutical preparation, Panacetin, making use of their solubilities and acid-base properties. Panacetin contains aspirin, sucrose, and an unknown component that may be either acetanilide or phenacetin. These substances have the following characteristics:

- Sucrose is insoluble in the organic solvent dichloromethane (CH_2Cl_2, also called methylene chloride).
- Aspirin, acetanilide, and phenacetin are soluble in dichloromethane, but relatively insoluble in water.
- Aspirin reacts with sodium bicarbonate to form a salt, sodium acetylsalicylate, which is insoluble in dichloromethane and soluble in water.
- Acetanilide and phenacetin are not converted to salts by sodium bicarbonate.

Mixing the Panacetin with dichloromethane should therefore dissolve the aspirin and the unknown component, leaving the sucrose behind as an insoluble solid that can be removed by filtration or (as in this experiment) by centrifugation. Aspirin can be removed from the dichloromethane solution by extraction with an aqueous solution of sodium bicarbonate. The base converts aspirin to its sodium salt, as shown in the margin. This salt will migrate from the dichloromethane layer, in which it is insoluble, to the aqueous layer, in which it is soluble. The unknown component will stay behind in the dichloromethane layer. After separation of the layers, aspirin can be recovered by the same method you used in Experiment 1 to prepare benzoic acid from its salt—precipitation from the aqueous layer with hydrochloric acid followed by vacuum filtration. The unknown component can then be isolated by evaporating the solvent from the dichloromethane solution.

This separation process is summarized by the following flow diagram:

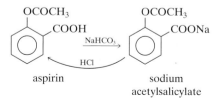

aspirin   sodium
                           acetylsalicylate

Minilab 2 in Part III will help you understand how the extraction operation works; you can carry out this minilab at your instructor's request.

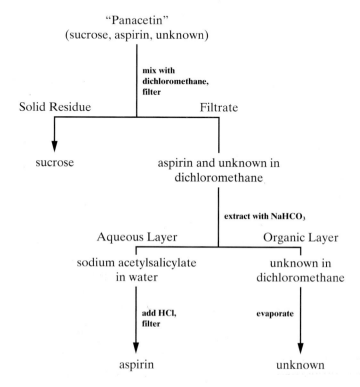

You can estimate the percentage composition of Panacetin from the masses of the dried components. Careful work is required to obtain accurate results in this experiment; errors can arise from incomplete mixing with dichloromethane, incomplete extraction or precipitation of aspirin, incomplete drying of the recovered components, and losses in transferring substances from one container to another.

Directions

Safety Notes

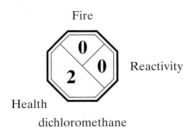

Fire

Reactivity

Health

dichloromethane

Take Care! Avoid contact with dichloromethane, do not breathe its vapors.

Stop and Think: Does all of the solid dissolve? If not, why not?

Stop and Think: What components of Panacetin are in the decanted liquid?

Stop and Think: Which component of Panacetin is in the aqueous layer? In the dichloromethane layer? If you don't shake the centrifuge tube long enough, how might that affect your results?

Observe and Note: Can you detect any evidence for a chemical reaction? If so, describe it in your lab notebook.

See "Chemical Hazards" in the Laboratory Safety section for an explanation of the hazard symbols.

Separation of Sucrose. Accurately weigh [OP-4] about 0.400 g of Panacetin and transfer it to a clean, *dry* 15-mL screw-cap centrifuge tube (tube **A**) that has been tared (preweighed). Measure [OP-5] 8.0 mL of dichloromethane into the centrifuge tube. Dissolve as much solid as possible in this solvent by shaking vigorously for a minute or so. If necessary, use a wooden applicator stick to break up any solid lodged in the tip. Balancing the centrifuge tube against one of equal mass, centrifuge [OP-14] the mixture for at least three minutes, and then transfer [OP-6] the liquid to a second screw-cap centrifuge tube (tube **B**), leaving all of the solid behind. Leave centrifuge tube **A** (uncapped) under a fume hood until you have finished the rest of the experiment to allow any remaining dichloromethane to evaporate. Then weigh it and calculate the mass of the sucrose it contains. Record the mass of the sucrose in your laboratory notebook. If requested, submit the sucrose to your instructor in a tared and labeled vial.

Separation of Aspirin. Extract [OP-15] the dichloromethane in centrifuge tube **B** with two *separate* 4-mL portions of 5% sodium bicarbonate. For each extraction, use a stirring rod to stir the liquid layers until any fizzing subsides before you stopper and shake the centrifuge tube. After the first extraction, transfer [OP-6] the bottom (dichloromethane) layer to a labeled 20-mL beaker, transfer the top (aqueous) layer to a second small beaker, and return the bottom layer to the centrifuge tube without delay. After the second extraction, transfer the dichloromethane layer to a tared 4-dram screw-cap vial, and combine the second aqueous layer with the first one. Save the dichloromethane solution in the vial for the following step, "Isolation of the Unknown Component."

Slowly add 1.0 mL of 6 *M* hydrochloric acid to the combined aqueous extracts while stirring with a glass rod. Test the pH of the solution with pH paper as described in Experiment 1, and add more acid, if necessary, to bring the pH to 2 or lower. Cool the mixture in an ice/water bath for at least five minutes, collect the aspirin by vacuum filtration [OP-13] with a Hirsch funnel, and wash it on the filter with cold water. Let the aspirin dry on the filter for a few minutes with the aspirator running, then dry [OP-23] it to constant mass. Weigh the aspirin and record its mass in your lab notebook. Submit the aspirin to your instructor in a vial labeled as described in Experiment 1.

Waste Disposal: The filtrate may be poured down the drain.

Isolation of the Unknown Component. *Under the hood*, evaporate [OP-16] the solvent from the dichloromethane solution with a *gentle* stream of dry air while heating [OP-7] it in a warm water bath. (Alternatively, you can evaporate the solvent under vacuum and recover it as described in OP-16.) Discontinue evaporation when only a solid residue remains in the vial or when no more solvent evaporates. Let the unknown component dry [OP-23] to constant mass. Weigh it before you begin Experiment 3.

Calculate your percent recovery, dividing the sum of the masses of all components by the mass of Panacetin that you started with. Calculate the approximate percentage composition of Panacetin, based on the total mass of components recovered. (These percentages should add up to 100%.) In your report, be sure to include the information specified in the "Report" section of Experiment 1 and any other information requested by your instructor.

If it is quite impure, the unknown may remain liquid after all of the solvent is removed. It should solidify upon cooling.

Exercises

1 (a) Describe any evidence that a chemical reaction occurred when you added 6 *M* HCl to the solution of sodium acetylsalicylate. (b) Explain why the changes that you observed took place.

2 Describe and explain the possible effect on your results of the following experimental errors or variations. In each case specify the component(s) whose percentage(s) would be too high or too low. (a) After adding dichloromethane to Panacetin, you didn't stir or shake the mixture long enough. (b) During the $NaHCO_3$ extraction, you failed to mix the aqueous and organic layers thoroughly. (c) You mistakenly extracted the dichloromethane solution with 5% HCl rather than 5% $NaHCO_3$. (d) Instead of using pH paper, you neutralized the $NaHCO_3$ solution to pH 7 using litmus paper.

3 Although acetanilide and phenacetin are not appreciably acidic, acetaminophen (like aspirin) is a stronger acid than water. What problem would you encounter if the unknown component were acetaminophen rather than acetanilide or phenacetin and you extracted the aspirin with 5% NaOH? Explain, giving equations for any relevant reactions.

4 Acetaminophen is a weaker acid than carbonic acid (H_2CO_3), but aspirin is a stronger acid than carbonic acid. Prepare a flow diagram like the one in this experiment, showing a procedure for separating a mixture of sucrose, aspirin, and acetaminophen.

5 Write balanced reaction equations for the reactions involved (a) when aspirin dissolves in aqueous $NaHCO_3$ and (b) when aspirin is precipitated from a sodium acetylsalicylate solution by HCl. Assuming that both reactions are spontaneous under standard conditions, label the stronger acid, stronger base, weaker acid, and weaker base in each equation.

Other Things You Can Do

(Starred items require your instructor's permission.)

*1 Carry out Minilab 2 in Part III to help you visualize what happens during an extraction.
 2 Write a short research paper about the chemistry and physiological effects of analgesic-antipyretic drugs using such sources as the *Kirk-Othmer Encyclopedia of Chemical Technology* [Bibliography, A8] and titles from section L of the Bibliography.

Recrystallization and Melting-Point Measurement
Identifying a Component of "Panacetin"

Purification Methods.

Operations

OP-25 Recrystallization
OP-30 Melting Point
OP-4   Weighing
OP-7   Heating
OP-23   Drying Solids

Before You Begin

1 Read the experiment and operations OP-25 and OP-30. Review the other operations as necessary.
2 Read "Using the Bibliography" in Appendix VII, and familiarize yourself with the layout of the Bibliography.
3 Calculate the minimum volume of boiling water that will be needed to dissolve all of your unknown compound if it is acetanilide *and* if it is phenacetin (see "Understanding the Experiment").

Scenario

The Scenario for this experiment is given in Experiment 2.

Applying Scientific Methodology

The problem you will be trying to solve in this experiment is "What is the identity of the unknown component of Panacetin?" Your course of action is described in the Directions. During the experiment, you should be gathering and evaluating evidence that can help you solve the problem. Any working hypothesis can be no more than a guess, so you might want to wait until you carry out the *Purification* step before you formulate your initial hypothesis. You will then test your hypotheses in the *Analysis* step, which will enable you to reach a conclusion. Keep a careful record of your observations and your interpretation of the evidence so that you will be able to describe them in your report.

Using Chemical Reference Books

Knowing the physical properties of a compound—such as its melting or boiling point, refractive index, and spectral absorption bands—can help you identify the compound and estimate its degree of purity. Whenever a new

This and other general reference books are described under Category A in Appendix VII.

compound is discovered or synthesized, its physical properties are measured and reported in one or more scientific journals to enable future investigators to identify that compound when they encounter it. Physical properties and other information about the more familiar compounds are reported in chemical handbooks and other reference books. For instance, from its entry in *The Merck Index*, we learn that acetaminophen is known by at least 8 chemical names and 45 proprietary drug names, including Alpiny, Bickiemol, Cetadol, Dial-a-gesic, Enelfa, Finimal, Gelocatil, Homoolan, and so on, down the alphabet. The same source lists the compound's important physical properties and uses, gives references describing its preparation from various starting materials, and tells where to find more information about it. For example, *The Merck Index* refers to an evaluation of acetaminophen's effect on the kidneys located on page 1238 of volume 320 (published in 1989) of the *New England Journal of Medicine*, as indicated by the following entry:

Evaluation of renal effects: D. P. Sandler *et al.*, *N. Engl. J. Med.* **320**, 1238 (1989).

In such citations, the abbreviated journal name is followed by the volume number, the page number on which the article begins, and the year of publication. Currently the preferred convention is to list the year first (in boldface type), followed by the volume number (in italics) and page number, as illustrated for the following citation to an article in the *Journal of Organic Chemistry*.

J. Org. Chem.	**1994**,	*59*,	2546.
journal abbrev.	year	volume	page

The full names of scientific periodicals can be found in the *Chemical Abstracts Service Source Index (CASSI)*.

Figure 3.1 reproduces the entry for acetanilide from the 15th edition of *Lange's Handbook of Chemistry*. This handbook reports that acetanilide has a melting point of 114°C and that it is only slightly soluble in water (aq) at 25°C, dissolving to the extent of 0.56 g per 100 mL of water at that temperature. *The Merck Index* reports that a gram of acetanilide dissolves in about 185 mL of cold (near room temperature) water or 20 mL of boiling water, and that a gram of phenacetin dissolves in 1310 mL of cold water or 82 mL of boiling water. Such entries often tell you what you need to know to purify a compound and identify or characterize the pure substance. These and other physical properties of acetanilide and phenacetin are summarized in Table 3.1 (see "Structures and Properties") where solubilities from *The Merck Index* are expressed in grams of solute per 100 mL of water.

No.	Name	Formula	Formula weight	Beilstein reference	Density, g/mL	Refractive index	Melting point, °C	Boiling point, °C	Flash point, °C	Solubility in 100 parts solvent
a18	Acetanilide	$CH_3CONHC_6H_5$	135.17	12, 237	1.219_4^{15}		114	304–305	173	0.56 aq^{25}; 25 acet; 29 alc; 2 bz; 27 chl; 5 eth

Figure 3.1 *Lange's Handbook of Chemistry* entry for acetanilide (Reprinted with permission from *Lange's Handbook of Chemistry*, 15th ed., by N. A. Lange, edited by J. A. Dean. Copyright McGraw-Hill, Inc., New York, 1999.)

Understanding the Experiment

In this experiment, you will purify and identify the unknown component of Panacetin from Experiment 2.

No separation is perfect; traces of impurities will always remain in a substance that has been separated from a mixture. Therefore, some kind of purification process is needed to remove them. Solids can be purified by such operations as recrystallization, chromatography, and sublimation; liquids are usually purified by distillation or chromatography. The solubility information from Table 3.1 indicates that both acetaminophen and phenacetin are relatively soluble in boiling water, but insoluble in cold water. This suggests that the unknown component can be purified by recrystallization, in which an impure solid dissolves in a hot (usually boiling) solvent and then crystallizes from the cooled solution in a purer form.

Key Concept: The solubilities of most substances decrease as the temperature is lowered.

Based on the mass of unknown that you recovered from Experiment 2, you can estimate the volume of boiling water needed to dissolve that mass of either acetanilide or phenacetin. For example, suppose your unknown is acetanilide and you recovered 0.18 g of the crude solid. Table 3.1 shows that the solubility of acetanilide in boiling water is 5.0 g per 100 mL water, so the volume of boiling water you will need to dissolve it is approximately 3.6 mL.

$$\text{✳ } 0.18 \text{ g acetanilide} \times \frac{100 \text{ mL water}}{5.0 \text{ g acetanilide}} = 3.6 \text{ mL water } \text{✳}$$

Phenacetin, which is less soluble in boiling water, will require more water to dissolve. You should begin the recrystallization operation using the smaller volume of water (don't add it all at once; see OP-25) and add more only if your compound does not dissolve in that amount of water at its boiling point.

After a compound has been purified, it should be *analyzed* to establish its identity and degree of purity. Although sophisticated instruments such as NMR spectrometers and mass spectrometers are now used to determine the structures of most newly discovered organic compounds, an operation as simple as a melting-point measurement can help identify a compound whose properties have already been reported in the chemical literature. By itself, the melting point of a compound is not sufficient proof of its identity because thousands of compounds may share the same melting point. But when an unknown compound is thought to be one of a small number of possible compounds, its identity can often be determined by mixing the unknown with an authentic sample of each known compound and measuring the melting points of the mixtures. The use of a *mixture melting point* for identification is based on the fact that the melting point of a pure compound is lowered and its melting point range broadened when it is combined with a different compound. For example, if your unknown is phenacetin, it should melt sharply near 135°C, and a mixture of the unknown with an authentic sample of phenacetin should have essentially the same melting point. But a 1:1 mixture of phenacetin with acetanilide should melt at a considerably lower temperature over a much broader range.

Key Concept: Impurities lower the melting point of a pure substance.

The melting point of a compound can also give a rough indication of its purity. If your compound melts over a narrow range (1–2°C or less) at a temperature close to the literature value, it is probably quite pure. If its melting point range is broad and substantially lower than the literature value, it is probably contaminated by water or other impurities.

Structures and Properties

acetanilide

phenacetin

Table 3.1 Physical properties

	M.W.	mp	Solubility, c.w.	Solubility, b.w.
acetanilide	135.2	114	0.54	5.0
phenacetin	179.2	135	0.076	1.22

Note: Melting points are in °C; solubilities are in grams of solute per 100 mL of cold water (c.w.) or boiling water (b.w.).

Directions

Safety Notes

Acetanilide and phenacetin can irritate the skin and eyes, so minimize contact with your unknown compound.

Observe and Note: How much water was needed to dissolve it? What happens as the solution cools?

Purification. Recrystallize [OP-25] the unknown drug component from Experiment 2 by boiling [OP-7] it with just enough water to dissolve it completely, then letting it cool slowly to room temperature. Use a container (such as a test tube or a small Erlenmeyer flask) that will easily accommodate the largest volume of recrystallization solvent that you calculated. If necessary, induce crystallization by scratching the sides of the container with a glass stirring rod. Then cool the container further in a beaker containing ice and tap water to increase the yield of product. Collect the solid by vacuum filtration, and wash it with a small amount of ice-cold water. Dry [OP-23] the product to constant mass and weigh [OP-4] it in a tared vial.

Waste Disposal: The filtrate may be poured down the drain.

Analysis. Grind a small amount of the dry unknown component to a fine powder on a watch glass using a spatula or a flat-bottomed stirring rod (see Minilab 1). Divide the solid into four nearly equal portions. Combine portions 1 and 2. Thoroughly mix portion 3 with an approximately equal amount of finely ground acetanilide, and thoroughly mix portion 4 with an approximately equal amount of finely ground phenacetin. Measure the melting-point ranges [OP-30] for the purified unknown (portions 1 + 2), the mixture with acetanilide, and the mixture with phenacetin. In each case, record the temperature at which you see the first trace of liquid and the temperature at which the sample is completely liquid. Unless your instructor indicates otherwise, you should carry out at least two measurements with each of these. Turn in the remaining product to your instructor in a vial labeled as shown by the example in the margin (or as directed by your instructor).

Exp. 3
phenacetin
0.183 g
mp 112–114°C
Ann A. Liszt
9/24/03

Stop and Think: Is your unknown reasonably pure? If not, what should you do?

Exercises

1 (a) What is the minimum volume of boiling water needed to dissolve 0.200 g of phenacetin (see Table 3.1)? (b) About how much phenacetin will remain dissolved when the water is cooled to room temperature?

(c) Calculate the maximum mass of solid (undissolved) phenacetin that can be recovered when the cooled solution is filtered.

2 An unknown compound **X** is one of the four compounds listed in Table 3.2. A mixture of **X** with benzoic acid melts at 89°C, a mixture of **X** with phenyl succinate melts at 120°C, and a mixture of **X** with *m*-aminophenol melts at 102°C. Give the identity of **X** and explain your reasoning.

Table 3.2   Melting points for Exercise 2

Compound	mp, °C
o-toluic acid	102
benzoic acid	121
phenyl succinate	121
m-aminophenol	122

3 Tell how each of the following experimental errors will affect your experimental results (yield, purity, or both) and explain why. (a) You failed to dry the product completely. (b) You used enough water to recrystallize phenacetin, but your unknown was acetanilide. (c) In Experiment 2, you didn't extract all of the aspirin from the dichloromethane solution.

4 Tell whether each of the experimental errors in Exercise 3 will affect the melting point of the unknown component. If it will, tell how it will affect the melting point and explain why.

5 Using one or more of the reference books listed in Appendix VII, give the following information about the analgesic drug ibuprofen: chemical names, molecular weight, molecular formula, structural formula, melting point, and a suitable recrystallization solvent.

6 Locate citations for one or more journal articles that give procedures for preparing (a) aspirin and (b) propoxyphene (Darvon). Give the full name of each journal, the volume number, the page number(s), and the year.

Other Things You Can Do

(Starred items require your instructor's permission.)

*1 Purify an unknown solid by recrystallization as described in Minilab 3.

*2 Carry out the purification and melting-point analysis of an acetanilide-salicylic acid mixture as described in *J. Chem. Educ.* **1989**, *66*, 1063.

3 Calibrate your thermometer by measuring the melting points of compounds listed in Table F1 of OP-30.

4 Look up the properties of a common organic compound in at least five different reference books and compare the kind of information provided by each.

Heating Under Reflux
Synthesis of Salicylic Acid from Wintergreen Oil

EXPERIMENT 4

Preparation and Purification of Solids.

Operations

OP-2   Using Specialized Glassware
OP-4   Weighing
OP-7   Heating
OP-13 Vacuum Filtration
OP-23 Drying Solids
OP-25 Recrystallization
OP-30 Melting Point

Before You Begin

1 Read the experiment and operation OP-2. Read "Smooth Boiling Devices" and "Heating Under Reflux" in OP-7, and review the other operations as necessary.

2 Read Appendix IV, "Calculations for Organic Synthesis." Then calculate the mass and volume of 2.00 mmol of methyl salicylate, the theoretical yield of salicylic acid from that much methyl salicylate, and the minimum volume of water needed to recrystallize that much salicylic acid.

3 Complete the experimental plan (page 51) by specifying all sizes and quantities indicated by asterisks in the Chemicals and Supplies list. Note that reaction flasks (pear-shaped and round-bottom flasks) usually come in 5- and 10-mL sizes; conical vials in 3- and 5-mL sizes; Erlenmeyer flasks in 10-, 25-, and 50-mL sizes; and beakers in 10-, 20-, 30-, 50-, 100-, and 150-mL sizes.

Be sure to distinguish millimoles (mmol) from moles (mol); most Prelab Assignments will specify amounts of chemicals in millimoles.

Scenario

The new-age pharmaceutical company Natural Nostrums manufactures drugs from "natural" starting materials. For example, the company manufactures a painkilling drug it advertises as "organic aspirin" starting with methyl salicylate, which occurs naturally in wintergreen oil. Most commercially marketed aspirin is manufactured starting with benzene, a product of petroleum refining. An intermediate in both of these syntheses is salicylic acid.

Natural Nostrums claims that its aspirin, which is supposedly more natural than aspirin made from benzene, has fewer side effects than ordinary aspirin. Critics have accused the company of false and misleading advertising,

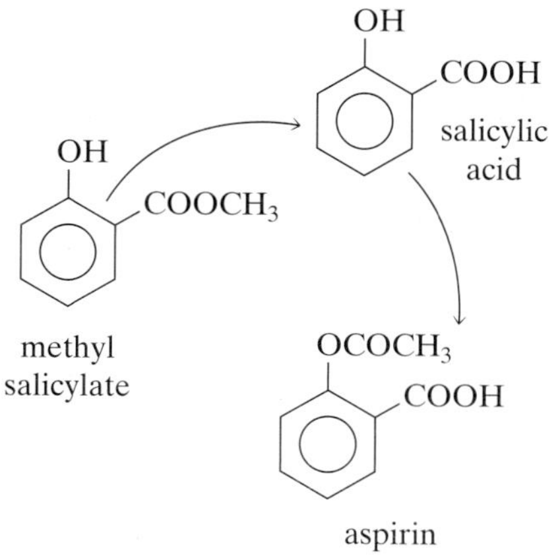

OH

COOCH$_3$

methyl
salicylate

OH

COOH

salicylic
acid

OCOCH$_3$

COOH

aspirin

synthesis of aspirin from
methyl salicylate

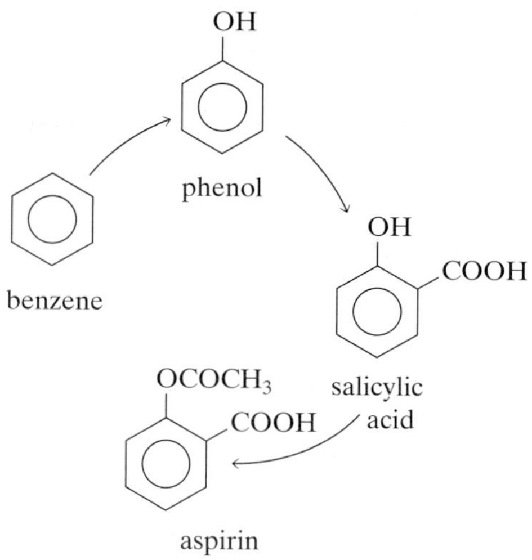

benzene

OH

phenol

OH

COOH

salicylic
acid

OCOCH$_3$

COOH

aspirin

synthesis of aspirin
from benzene

asserting that salicylic acid made from methyl salicylate is no different than salicylic acid made from benzene, and that the resulting aspirin is therefore no better than any other aspirin.

Les Payne, Director of Operations for the Association for Safe Pharmaceuticals (ASP), is investigating the company. He just shipped your supervisor a sample of salicylic acid manufactured from benzene and a bottle of methyl salicylate that one of his agents obtained from the chemical stockroom at Natural Nostrums. Your assignment is to prepare salicylic acid from this methyl salicylate and find out whether or not it differs from salicylic acid made from benzene.

Applying Scientific Methodology

As in the previous experiments, you need to state the problem as a question, formulate a working hypothesis, follow the course of action described in the Directions, gather and evaluate evidence, test your hypothesis, arrive at a conclusion, and report your findings.

Wintergreen Oil

For many years, commercial methyl salicylate was obtained from wintergreen oil, an aromatic liquid distilled from the leaves of the wintergreen plant (*Gaultheria procumbens*) or from the bark of sweet birch trees (*Betula lenta*). When cheap raw materials became available from petroleum, a more economical commercial process was developed for synthesizing methyl salicylate from salicylic acid and methanol (CH$_3$OH, methyl alcohol).

wintergreen plant

salicylic acid (from benzene) → methyl salicylate (CH₃OH)

Regardless of its source, methyl salicylate is methyl salicylate. There is no difference between its pure natural and synthetic forms because both are composed of the same kind of molecules and must therefore have the same properties. Any differences between oil of wintergreen obtained from natural sources and synthetic methyl salicylate are due mainly to impurities in the natural oil.

A natural substance is one obtained directly from plant, animal, or mineral sources. A synthetic substance is one prepared by chemically altering other substances, which may themselves be either natural or synthetic.

natural methyl salicylate   synthetic methyl salicylate

Methyl salicylate, both natural and synthetic, has been used for many years as a flavoring agent, because of its pleasant penetrating odor and flavor. According to Euell Gibbons, author of *Stalking the Healthful Herbs* and other books about edible and medicinal wild plants, you can prepare a good wintergreen tea by pouring boiling water into a jar filled with freshly picked wintergreen leaves and letting it steep overnight or longer. The distinctive flavor of wintergreen is found in a variety of commercial products, including candy, chewing gum, root beer, and toothpaste. Medicinally, wintergreen oil has some of the painkilling properties of other *salicylates*, a group of related organic compounds that includes salicylic acid and aspirin. When absorbed through the skin, it produces an astringent but soothing sensation, and it is frequently used in preparations such as BenGay to alleviate muscular aches and arthritis.

Understanding the Experiment

In this experiment, you will be carrying out an *organic synthesis*, the preparation of salicylic acid from methyl salicylate. An organic synthesis can be described as the preparation of a desired organic compound by chemically modifying the molecules of another organic compound, which is called the *starting material*. Until 1874, all commercial salicylic acid was synthesized from natural wintergreen oil. In this experiment, you will reproduce this nineteenth-century synthesis of salicylic acid starting with methyl salicylate, the major constituent of wintergreen oil.

Organic reactions are slower than most inorganic reactions. For example, when aqueous solutions of silver nitrate and sodium chloride are mixed, the resulting reaction is almost instantaneous, taking place as rapidly as silver and chloride ions can come together to form silver chloride.

$$Ag^+ + Cl^- \longrightarrow \textbf{AgCl}$$

In an organic reaction, the reacting molecules may come together as frequently as the ions in a precipitation reaction, but they may not collide with enough energy or in the right orientation to react. An ion-combination reaction has such a low activation energy that ions need only "stick together" to form a product. Covalent molecules must ordinarily undergo a process of bond breaking and bond formation to be converted to product molecules, and such processes have relatively high activation energies.

Some of the bond-breaking and bond-forming steps involved in this experiment's synthesis are shown in the following *mechanism* for the conversion of methyl salicylate to salicylic acid.

Key Concept: A mechanism is a step-by-step description of a chemical reaction, showing what bonds are broken and formed as reactant molecules are converted to product molecules.

Mechanism for the Reaction. Note that the —OH group of methyl salicylate is ionized at the high pH of the reaction mixture.

According to this mechanism, the formation of a bond between a hydroxide ion and a carbonyl (C=O) carbon atom is accompanied by the breaking of a pi bond in the C=O group. That pi bond is then re-formed while a bond to an —OCH$_3$ group is broken.

The *reaction*, then, is the crucial step in an organic synthesis—it may be over in a few minutes or require a few weeks, but enough time should be allowed to bring the reaction as close to completion as possible before proceeding to the next steps. These steps have already been described in Experiments 2 and 3; the desired product must be *separated* from the reaction mixture, *purified* to remove residual contaminants, and *analyzed* to verify its identity and purity.

Completion implies that as many reactant molecules as possible have been converted to product molecules under the conditions of the reaction.

Methyl salicylate is the limiting reactant in this synthesis; sodium hydroxide is used in excess to ensure a reasonably fast, complete reaction. It is best to measure limiting reactants that are liquids by mass, because the mass of a liquid can be measured more accurately than its volume. To avoid waste and minimize spillage at the balance, you should dispense the estimated *volume* of liquid from the stock bottle before you take it to the balance to weigh it. This is why you were asked (in "Before You Begin") to calculate both the mass and volume of methyl salicylate required. If the mass of the methyl salicylate is not within about 5% of your calculated value, you can add or remove liquid with a Pasteur pipet.

See Appendix IV for a discussion of limiting reactants.

You will carry out the reaction by boiling a mixture of methyl salicylate and sodium hydroxide in a conical reaction vial equipped with a condenser. A hot plate with an aluminum block or sand bath is a suitable heat source for the reaction. The purpose of the condenser is to convert water and methyl salicylate vapors back to liquids and return them to the reaction flask so that they don't boil away.

Stop and Think: Why? (Consider the pH of the reaction mixture.)

The product of the initial reaction will be a salt of salicylic acid—disodium salicylate—and not salicylic acid itself. Adding 3 M sulfuric acid will then precipitate salicylic acid. When you recover the product by vacuum filtration, remember that the *solvent* in the reaction mixture is water, so use cold water (*not* 3 M sulfuric acid) to wash the product on the filter. The solubility of salicylic acid in water is about 0.18 g/100 mL at 20°C and 6.7 g/100 mL at the boiling point, so it can be purified by recrystallization from water. The melting point of your product should give you a good indication of its purity; the melting range of pure salicylic acid is reported to be 158–160°C. You can then carry out a mixed melting point (see OP-30) with the benzene-derived salicylic acid provided by your instructor to see whether it is identical to or different from your own.

To help you organize your time efficiently, an experimental plan for the preparation of salicylic acid is provided. You will need to complete the plan as directed in "Before You Begin." For subsequent experiments, you should write your own experimental plans (described in Appendix V). Note that the 30-minute reaction period should be used to collect the supplies and assemble the apparatus needed for upcoming operations.

Reactions and Properties

Table 4.1 Physical properties

	M.W.	mp	bp	d
methyl salicylate	152.1	−8	223	1.174
salicylic acid	138.1	159		

Note: Melting points and boiling points are in °C; density (d) is in g/mL.

Experimental Plan

Chemicals and Supplies Needed

An asterisk (*) indicates a number (size or quantity) to be filled in.

Heating under reflux [OP-7]

* g (* mL) of methyl salicylate (avoid contact, inhalation)
* mL of 6 M NaOH (wear gloves, avoid contact)
*-mL conical vial
heat source (specify type), condenser (specify type), boiling chip,
ring stand, clamps

Addition of sulfuric acid

* mL of 3 M H_2SO_4
*-mL beaker
pH paper, stirring rod, Pasteur pipet

Vacuum filtration [OP-13]

* mL of cold water for washing
Hirsch funnel, filter flask, filter paper, flat-bladed microspatula,
watch glass

Recrystallization [OP-25]

* mL of water for recrystallization
* mL of cold water for washing
two *-mm test tubes
*-mL graduated cylinder
Hirsch funnel, filter flask, filter paper, flat-bladed microspatula,
watch glass

Drying [OP-23] and weighing [OP-4]

drying container, desiccator (or oven), tared vial

Melting points [OP-30]

mp tubes, flat-bladed microspatula, watch glass

Lab Checklist

- Collect and clean supplies for reflux.
- Obtain heat source.
- Assemble reflux apparatus.
- Measure methyl salicylate and 6 M NaOH solution.
- Add reactants and boiling chip to reaction vial.
- Heat reactants under reflux for 30 minutes.
- Collect supplies for H_2SO_4 addition (during reflux).
- Collect vacuum filtration supplies and assemble apparatus (during reflux).
- Collect supplies for recrystallization (during reflux).
- Measure 3 M H_2SO_4 solution (during reflux).
- Precipitate product with 3 M H_2SO_4.
- Filter and wash product; air dry on filter.

- Measure and boil water for recrystallization.
- Recrystallize product from boiling water.
- Filter and wash product; air dry on filter.
- Dry product.
- Weigh product.
- Measure melting point.
- Turn in product.
- Clean up.

Directions

Safety Notes

> Concentrated sodium hydroxide solutions are very corrosive and can cause severe damage to skin and eyes. Wear protective gloves and goggles and avoid contact with the 6 *M* NaOH. (Always wear your safety goggles in the organic chemistry lab!)
> Methyl salicylate can irritate the eyes and skin; avoid contact and inhalation.

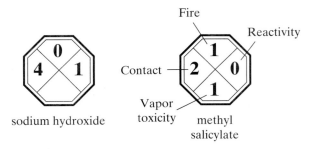

sodium hydroxide   methyl salicylate

Alternatively, you can use a stirrer and stir bar and omit the boiling chip. See OP-10.

Take Care: Wear gloves, avoid contact with the NaOH.

Observe and Note: What did you observe when you combined methyl salicylate with 6 *M* NaOH? When you heated the reaction mixture to boiling?

Take Care! Wear gloves.

Observe and Note: What happened?

Reaction: Obtain an appropriate heat source and assemble [OP-2] an apparatus for heating under reflux [OP-7] consisting of an air condenser attached to a conical reaction vial of appropriate size. Measure 2.00 mmol of methyl salicylate into the reaction vial; then add 2.5 mL of aqueous 6 *M* sodium hydroxide and a boiling chip or two. Heat the reaction mixture under gentle reflux for 30 minutes, measuring from the time the solution starts to boil. Be sure that the reflux ring of condensing vapor remains at least 2 cm below the top of the condenser. If its level goes higher than this, reduce the heating rate. If, after 30 minutes under reflux, the reaction mixture is cloudy or contains an oily upper layer, continue heating under reflux until the cloudiness (or oil) disappears and the reaction mixture no longer smells like wintergreen.

When the reaction is complete, remove the condenser as soon as it is cool enough to handle; otherwise the ground joints might freeze. Remove the boiling chip(s) and transfer the reaction mixture to a beaker large enough to contain it and the sulfuric acid to be added. Slowly add 2.7 mL of 3 *M* sulfuric acid while stirring, and test the pH of the supernatant liquid (liquid at the surface) using your stirring rod and a strip of pH paper. If the pH is above 2, add enough additional sulfuric acid drop by drop to bring it down to 2. Cool the mixture in an ice/water bath for five minutes or more.

Separation.   Collect the salicylic acid by vacuum filtration [OP-13] using a Hirsch funnel, and wash it on the filter with ice-cold water.

Purification and Analysis.   Purify the salicylic acid by recrystallization [OP-25] from boiling water. There should be no need to filter the hot solution. Collect the product by vacuum filtration, and wash it with a little ice-cold water. Dry [OP-23] the product to constant mass and measure its mass [OP-4]. Measure the melting points [OP-30] of the dry product *and* of a 1:1 mixture of the product with salicylic acid synthesized from benzene. In your report, try to account for any significant material losses (see Exercise 3).

Waste Disposal: The filtrate may be poured down the drain.

Observe and Note: How much water was required?

Waste Disposal: The filtrate may be poured down the drain.

Stop and Think: Do the results support your initial hypothesis?

Exercises

1 (a) Calculate the volume of 6 *M* NaOH required to react completely with the amount of methyl salicylate you used. How much of the 6 *M* NaOH that you used was in excess of the theoretical amount? (b) What volume of 3 *M* H_2SO_4 is needed to neutralize all of the disodium salicylate and the excess NaOH present after the initial reaction? How much sulfuric acid was in excess?

2 Refer to the reaction equations in the "Reactions and Properties" section as you answer the following questions. (a) Which two hydrogen atoms of salicylic acid are most likely to be acidic? Which hydrogen atom(s) of methyl salicylate would you expect to be acidic? (b) Based on your answer to (a), draw the structure of the white solid that forms immediately after NaOH and methyl salicylate are combined, and write an equation for its formation. (*Hint:* Acid-base reactions are much faster than most organic reactions.)

3 Based on the amount of water you used during the recrystallization, estimate the amount of salicylic acid that was lost as a result of being dissolved in the filtrate. Assume that the recrystallization solution was cooled to 10°C; the solubility of salicylic acid at that temperature is 0.14 g per 100 mL of water.

4 Describe and explain the possible effect on your results of the following experimental errors or variations. (a) You added only enough 3 *M* H_2SO_4 to bring the pH down to 4. (b) Your reaction mixture had an oily layer on top when you added the 3 *M* sulfuric acid. (c) Because of a label-reading error by a lab assistant, the bottle labeled "salicylic acid from benzene" actually contained acetylsalicylic acid (aspirin).

5 From the equations in the Reactions and Properties section, you can see that methanol and sodium sulfate are by-products of the synthesis of salicylic acid. During which step of the synthesis would these compounds have been separated from the final product? Explain your answer, based on any relevant properties of salicylic acid and the by-products. (If necessary, you can look up the properties in a reference book.)

Other Things You Can Do

(Starred items require your instructor's permission.)

*1 Develop and test a hypothesis based on observations you make while performing Minilab 4. You can also test your product from this experiment with ferric chloride as described in Minilab 4.

*2 Convert your salicylic acid to aspirin by following the procedure in Experiment 33. (You can omit the spectral analysis.)

3 Find some examples of organic compounds besides methyl salicylate that are available commercially in both natural and synthetic forms. You might start searching at a local drugstore or health food store. Look up the compounds in *The Merck Index* and give their chemical structures.

4 Write a research paper about salicylic acid and the salicylates using references from the Bibliography.

Simple Distillation, Gas Chromatography
Preparation of Synthetic Banana Oil

EXPERIMENT **5**

Preparation, Purification, and Analysis of Liquids. Gas Chromatography.

Operations

OP-10 Mixing
OP-21 Washing Liquids
OP-22 Drying Liquids
OP-27 Simple Distillation
OP-34 Gas Chromatography
OP-2   Using Specialized Glassware
OP-4   Weighing
OP-6   Making Transfers
OP-7   Heating

Before You Begin

1   Read the experiment and operations OP-10, OP-21, OP-22, OP-27, and OP-34. Review the other operations as necessary. (Reading OP-27 and OP-34 may be deferred until the second lab period for this experiment.)
2   Calculate the mass and volume of 20.0 mmol of isopentyl alcohol and the theoretical yield of isopentyl acetate from this amount of the alcohol.
3   Read Appendix V, "Planning an Experiment," and then write an experimental plan like the one you used in Experiment 4.

Scenario

Your supervisor has just received the following message from the Cavendish Distilling Company.

Greetings:

We have a problem. Cavendish Distilling Company markets the popular liqueur Banana Elixir, which is flavored with a natural banana extract from fruit grown on our Caribbean banana plantations. Last June, Hurricane Floyd blew down all of our banana trees.

Our stock of banana extract is running low, and we have no alternative source of bananas at this time. As a temporary solution, we have decided to add a synthetic banana flavoring to our remaining stock of the natural extract until our plantations start producing again.

Synthetic banana flavorings are formulated mainly from isopentyl acetate, with smaller amounts of other esters. I understand that esters can be prepared economically by a process called the Fischer esterification, which involves the combination of an acid such as acetic acid with an alcohol. According to our technical staff, one problem with this method is that Fischer esterifications do not go to completion, leaving considerable amounts of starting material in the product. We can tolerate

banana tree

*The banana plant (*Musa cavendishii *and other species) is actually a gigantic herb and not a true tree. Each year, the foliage-bearing part withers away and is replaced by new growth from an underground stem.*

up to 10% isopentyl alcohol in our isopentyl acetate, since the alcohol is a component of natural banana extract. But we cannot tolerate more than 2% acetic acid, because it would tend to degrade the flavor of the liqueur. Will you have your consulting chemists prepare some isopentyl acetate and analyze it to see if it falls within our tolerances?
        Amy Lester, CEO

Applying Scientific Methodology

By now you should be familiar with the steps involved in applying scientific methodology to the solution of a problem, so they will not all be repeated hereafter. In this experiment, the problem described in the Scenario does not involve the identity of the product, but its purity, so your course of action will include the use of an instrumental method, gas chromatography, to determine the composition of the product.

Esters and Artificial Flavorings

The word *flavor* is used to describe the overall sensory effect of a substance taken into the mouth. Flavor may involve tactile, temperature, and pain sensations, as well as smell and taste. Many fruits, flowers, and spices contain esters that contribute to their characteristic flavors, where an *ester* is an organic compound that contains the functional group (characteristic combination of atoms) shown in the margin.

Most volatile esters have strong, pleasant odors that can best be described as "fruity." Some esters with flavors characteristic of real and "fantasy" fruits are shown in Table 5.1. The ester you will prepare in this

functional group
of an ester

Table 5.1 Flavor notes of some esters used in artificial flavorings

Name	Structure	Flavor note
propyl acetate	$\overset{\overset{\displaystyle O}{\|\|}}{CH_3C}-OCH_2CH_2CH_3$	pears
octyl acetate	$\overset{\overset{\displaystyle O}{\|\|}}{CH_3C}-O(CH_2)_7CH_3$	oranges
benzyl acetate	$\overset{\overset{\displaystyle O}{\|\|}}{CH_3C}-OCH_2-\bigcirc$	peaches, strawberries
isopentenyl acetate	$\overset{\overset{\displaystyle O}{\|\|}}{CH_3C}-OCH_2CH=\overset{\overset{\displaystyle CH_3}{\|}}{C}-CH_3$	"Juicy Fruit"
isobutyl propionate	$\overset{\overset{\displaystyle O}{\|\|}}{CH_3CH_2C}-OCH_2\overset{\overset{\displaystyle CH_3}{\|}}{CH}-CH_3$	rum
ethyl butyrate	$\overset{\overset{\displaystyle O}{\|\|}}{CH_3CH_2CH_2C}-OCH_2CH_3$	pineapples

experiment, isopentyl acetate, has a strong banana odor when undiluted and an odor reminiscent of pears in dilute solution. It is used as an ingredient in artificial coffee, butterscotch, and honey flavorings, as well as in pear and banana flavorings.

Many different esters are included in the basic repertoire of the flavor chemist, who combines natural and synthetic ingredients to prepare artificial flavorings. These ingredients may include natural products, synthetic organic compounds identical to those found in nature, and synthetic compounds not found in nature but accepted as safe for use in food. Each flavor ingredient is characterized by one or more flavor *notes* that suggest the predominant impact the ingredient makes on the senses of taste and smell. Although the flavor note of a single ingredient may seem unrelated to the overall character of a natural flavor, the combination of carefully selected ingredients in the right proportions can often yield a good approximation of that flavor. A high-boiling *fixative*, such as glycerine or benzyl benzoate, is usually added to an artificial flavoring to retard vaporization of volatile components, and the flavor notes of the individual components are blended by dissolving them in a solvent called the *vehicle*. The most frequently used vehicle is ethanol (ethyl alcohol).

The formulation of artificial flavorings is perhaps as much an art as a science. The components of a strawberry flavoring, for example, may vary widely depending on the manufacturer and the specific application. Because natural flavors are usually very complex, a cheap artificial flavoring may be a poor imitation of its natural counterpart, but advances in flavor chemistry have made possible the production of superior flavorings that reproduce natural flavors very closely. Superior flavorings may contain natural oils or extracts that have been fortified with a few synthetic ingredients to enhance the overall effect and to replace flavor elements lost during the distillation or extraction process. Even the most experienced flavor chemist can't hope to do as well as a strawberry plant, which may combine several hundred different flavor components in its berries. But a superior strawberry flavoring with a few dozen ingredients may be hard to distinguish from the real thing—except by the most discriminating of strawberry aficionados.

$$\underset{\text{isopentyl acetate}}{CH_3\overset{\overset{\displaystyle O}{\|}}{C}OCH_2CH_2\underset{\underset{\displaystyle CH_3}{|}}{C}HCH_3}$$

Some food products, such as cola beverages and Juicy Fruit gum, are characterized by fantasy flavors that have no counterparts in nature, but most artificial flavorings are meant to resemble natural flavors.

Fixatives

$$\underset{\text{glycerine}}{\overset{\overset{\displaystyle OH\ OHOH}{|\ \ \ |\ |}}{CH_2CHCH_2}}$$

benzyl benzoate

Vehicle

$$\underset{\substack{\text{ethyl alcohol} \\ \text{(ethanol)}}}{CH_3CH_2OH}$$

Understanding the Experiment

In this experiment you will prepare synthetic banana oil, which is known by several chemical names, including isopentyl acetate, isoamyl acetate, and 3-methylbutyl ethanoate. Esters are often prepared by the Fischer esterification method, which involves heating a carboxylic acid with an alcohol in the presence of an acid catalyst, as shown by the following general equation:

$$\underset{\substack{\text{carboxylic} \\ \text{acid}}}{R\overset{\overset{\displaystyle O}{\|}}{C}OH} + \underset{\text{alcohol}}{HOR'} \overset{H^+}{\longrightarrow} \underset{\text{ester}}{R\overset{\overset{\displaystyle O}{\|}}{C}OR'} + H_2O$$

The acid catalyst is used to increase the rate of the reaction, which would otherwise require a much longer reaction time.

You will synthesize isopentyl acetate by combining isopentyl alcohol (3-methyl-1-butanol) with acetic acid and sulfuric acid and then heating the reaction mixture under reflux for an hour. The alcohol is the limiting reactant,

Reactions carried out with acid catalysts often yield polymeric, tarlike by-products that have high boiling points and are insoluble in water.

so it should be weighed; the acids can be measured by volume. The esterification reaction is reversible, having an equilibrium constant of approximately 4.2. If you were to start with equimolar amounts of acetic acid and isopentyl alcohol, only about two-thirds of each reactant would be converted to isopentyl acetate by the time equilibrium was reached. Your highest attainable yield in that case would be only 67% of the theoretical value. To increase the yield of isopentyl acetate, you will apply Le Châtelier's principle, using a 100% excess of acetic acid—the less expensive reactant—to shift the equilibrium toward the products. Even then, the reaction will not be complete at equilibrium, so the reaction mixture will contain some unreacted isopentyl alcohol as well as the excess acetic acid.

As you learned in Experiment 2, a pure component can be obtained from a mixture by separating it from all other components of the mixture, using procedures that take advantage of differences in solubility, boiling points, acid–base properties, and other characteristics of the components. Because isopentyl acetate is a liquid, the separation and purification operations will differ from those used previously for solid products.

At the end of the reflux period, the reaction mixture will contain (in addition to the ester) unreacted acetic acid, sulfuric acid, water, unreacted isopentyl alcohol, and some unwanted by-products (see Figure 5.1). Isopentyl acetate is quite insoluble in water, whereas both acetic acid and sulfuric acid are water soluble and acidic. This makes it easy to separate the two acids from the product by washing the reaction mixture with water and then with aqueous sodium bicarbonate. Water does not remove the acids entirely because they are somewhat soluble in the ester as well, but it removes the bulk of them and thus helps prevent a violent reaction with sodium bicarbonate in the second washing step. The aqueous sodium bicarbonate converts the acids to their salts, sodium acetate and sodium sulfate, which are insoluble in the ester but very soluble in water; these salts migrate to the aqueous layer where they can be removed. The rule of thumb given in OP-21 will help you estimate the quantity of aqueous sodium bicarbonate needed.

The water that forms during the reaction will be separated from the ester along with the wash liquids. Any traces of water that remain are then removed by a drying agent, either magnesium sulfate or sodium sulfate.

Because isopentyl alcohol has a lower boiling point than that of isopentyl acetate and the by-products have higher boiling points, it should be possible—in principle—to remove the alcohol and by-products from the ester by distillation. Isopentyl alcohol should distill first, followed by the ester, and any by-products should remain behind in the pot—the vessel in which the reaction mixture is boiled. For the reasons described in OP-27, the separation is incomplete, so you will still have some isopentyl alcohol in your isopentyl acetate after the purification step.

You will determine the composition of your distillate by injecting a very small amount into an instrument called a gas chromatograph. (Your instructor will demonstrate the operation of this instrument.) Inside the gas chromatograph, the liquid will vaporize, and the vapors of different components will travel through a packed column at different rates. As the vapors exit the column, their presence will be detected and recorded on a graph called a gas chromatogram, which should display several peaks of different sizes. The area under the peak for each component will be proportional to the amount

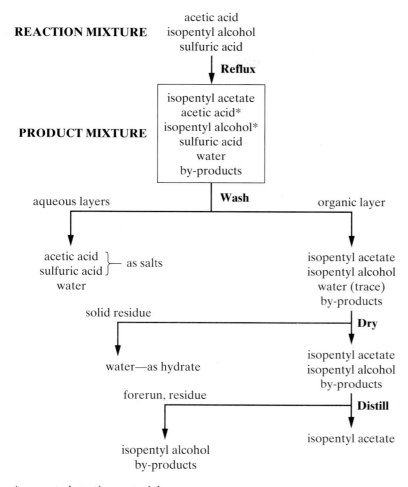

Figure 5.1   Flow diagram for the synthesis of isopentyl acetate

of that component present, so by measuring the peak areas you can estimate the percentage of each component in the distillate. This will tell you
how much (if any) isopentyl alcohol and acetic acid remain in your product.

The procedure for the synthesis of isopentyl acetate is summarized in
the flow diagram in Figure 5.1, which illustrates the transformations or separations that occur during each operation. Appendix V tells you how to interpret such a flow diagram.

Isopentyl acetate is known to be an *alarm pheromone* of the honeybee.
A pheromone is a "molecular messenger" that produces a response, such as
mating behavior or aggression toward a perceived threat, in another member of the same species. When a worker honeybee stings someone, it releas-
es a tiny amount (about 1 μg) of isopentyl acetate in its stinger, attracting
more honeybees to the scene. Although isopentyl acetate alone does not
cause the bees to sting (other pheromones in the stinger do that), it agitates
them and puts them on guard. So it might be wise to steer clear of beehives
on your way home from the lab!

Reactions and Properties

$$CH_3\overset{\overset{\displaystyle O}{\|}}{C}-OH + HOCH_2CH_2\overset{\overset{\displaystyle CH_3}{|}}{C}HCH_3 \rightleftharpoons CH_3\overset{\overset{\displaystyle O}{\|}}{C}-OCH_2CH_2\overset{\overset{\displaystyle CH_3}{|}}{C}HCH_3 + H_2O$$

acetic acid   isopentyl alcohol   isopentyl acetate

Table 5.2   Physical properties

	M.W.	bp	*d*	Solubility
acetic acid	60.1	118	1.049	miscible
isopentyl alcohol	88.1	130	0.815	2.7
isopentyl acetate	130.2	142	0.876	0.25
sulfuric acid	98.1	290	1.84	miscible

Note: Boiling points are in °C; densities in g/mL; solubilities are in g/100 mL solvent.

Directions

Safety Notes

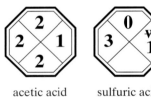

acetic acid   sulfuric acid

Acetic acid causes chemical burns that can seriously damage skin and eyes; its vapors are highly irritating to the eyes and respiratory tract. Wear gloves, dispense under a hood, avoid contact, and do not breathe its vapors.
Sulfuric acid causes chemical burns that can seriously damage skin and eyes. Wear gloves and avoid contact.
Isopentyl alcohol and isopentyl acetate can irritate the skin, eyes, and respiratory tract.

Glacial acetic acid is a pure grade of acetic acid that freezes at about 17°C.

Take Care! Wear gloves, avoid contact with acetic acid and sulfuric acid, do not breathe their vapors.

Observe and Note: What evidence do you observe that suggests that a chemical reaction is taking place?

Stop and Think: What is the density of isopentyl acetate? Which layer should initially be on top, the aqueous layer or the organic layer?

Stop and Think: What gas is evolved during the NaHCO₃ wash?

Reaction.   Accurately weigh 20.0 mmol of isopentyl alcohol into an appropriate reaction vessel and add a magnetic stir bar [OP-10]. *Under a hood*, add 2.3 mL (~40 mmol) of glacial acetic acid; then stir the mixture as you carefully add 0.15 mL (or about 6 drops) of concentrated sulfuric acid. Connect a water-cooled condenser to the reaction flask, turn on the cooling water, and heat the reaction mixture under reflux [OP-7] with magnetic stirring [OP-10] for one hour after boiling begins. Reduce the heating rate if the reflux ring rises to within 2 cm of the top of the condenser.

Separation.   When the reaction time is up, allow the reaction mixture to cool to about room temperature. Turn off the cooling water and remove the condenser, then remove the stir bar with a forceps and rinse it over the reaction flask with a few drops of water. Transfer the reaction mixture to a screw-cap centrifuge tube, and add 6 mL of water, using some of the water for the transfer. Wash [OP-21] the reaction mixture with the water and remove the aqueous layer, leaving the organic layer in the centrifuge tube. Then carefully wash the organic layer with two separate portions of 5% aqueous sodium bicarbonate, removing the aqueous layer after each washing. (**Take Care!** Pressure may build up in the capped centrifuge tube.) During the first washing, stir the layers until gas evolution subsides before you cap the centrifuge tube, and vent it frequently thereafter. Try to remove all of the aqueous layer after the last washing to minimize the amount of water in the product. Dry [OP-22] the crude isopentyl acetate by adding a small amount of anhydrous sodium sulfate (about as much as you can get on the grooved end of a Hayman-style microspatula), shaking the capped tube vigorously, and letting the mixture stand for at least ten minutes. Carefully transfer [OP-6] the liquid to a dry 5-mL conical vial using a dry filter-tip pipet. Try not to transfer any drying agent with the liquid.

Waste Disposal: Dissolve the spent drying agent in the combined wash solvents and pour the mixture down the drain.

Purification and Analysis.   Using Mayo-Pike-style microware [OP-2], assemble an apparatus for microscale simple distillation [OP-27] incorporating a Hickman still equipped with a water-cooled condenser and a thermometer. Be sure the thermometer bulb is positioned as shown in Figure E9, OP-27, and have your instructor check your setup before you start. Distill the crude product *slowly*, collecting any liquid that distills between 136°C and 143°C. Record the actual boiling range you observe; wait until the entire thermometer bulb is moist with condensing vapors, liquid has begun to appear in the well of the Hickman still, and the temperature is stable before you record the initial temperature reading. As the well fills with distillate, use a Pasteur pipet to transfer it to a tared, labeled 1-dram vial. Stop the distillation when only a drop or so of liquid remains in the conical vial, *or* when the temperature reaches 143°C. If the distillate is cloudy or contains water droplets, dry it [OP-22]. Weigh [OP-4] the distillate. Using a Carbowax column or another suitable column, obtain a gas chromatogram [OP-34] of the distillate and measure the peak areas as directed by your instructor. Unless your instructor indicates otherwise, assume that the components of the distillate appear on the gas chromatogram in the order (1) isopentyl alcohol, (2) isopentyl acetate, and (3) acetic acid. From the peak areas, calculate the percentage of each component in the distillate, then calculate your percent yield of isopentyl acetate based on its percentage in the distillate.

Stop and Think: What does the boiling range of the distillate tell you about the purity of your product?

Waste Disposal: Put any forerun and any residue left in the boiling flask into a designated waste container.

Exercises

1 (a) Calculate the amount of isopentyl acetate (isoamyl acetate) that should be present in the reaction mixture at equilibrium, based on the quantities of starting materials you used and a value of 4.2 for the equilibrium constant. (Use the quadratic equation; because volumes cancel out, moles can be used in place of molar concentrations.) (b) Estimate the mass of isopentyl acetate that was lost (1) as a result of incomplete reaction, (2) during the washings, and (3) during the distillation (see OP-27). Assume that the ester's solubility in aqueous $NaHCO_3$ is about the same as in water. Compare the sum of these estimated losses with your actual product loss and try to account for any significant differences.

2 What gas escaped during the sodium bicarbonate washing? Write balanced equations for two reactions that took place during this operation.

3 Tell how the procedure for the preparation of isopentyl acetate could be modified to increase the yield and purity of the product.

4 Describe and explain how each of the following experimental errors or variations might affect your results. (a) You failed to dry the reaction flask after washing it with water. (b) You forgot to add the sulfuric acid. (c) You used twice the amount of acetic acid specified in the procedure.

Isoamyl acetate equilibrium

$$i\text{-AmOH} + \text{HOAc} \rightleftharpoons$$
$$i\text{-AmOAc} + H_2O$$

$$K = \frac{[i\text{-AmOAc}][H_2O]}{[i\text{-AmOH}][\text{HOAc}]}$$

Ac = acetyl, CH_3CO-
i-Am = isoamyl, $CH_3CHCH_2CH_2-$
$$|$$
$$CH_3$$

(d) You left out the sodium bicarbonate washing step. (e) Your thermometer bulb was 1 cm higher than it should have been.

5   (a) In the Methodology section, it was stated that the reaction of an equimolar mixture of isopentyl alcohol and acetic acid will produce, at most, 67% of the theoretical amount of isopentyl acetate. Verify this with an equilibrium constant calculation, using $K = 4.2$. (b) Compare this with the corresponding percentage for the conditions used in this experiment (see Exercise 1). Are your results consistent with Le Châtelier's principle? Explain.

6   Based on the procedure that you used in this experiment and using the same molar quantities of reactants, develop a procedure that would be suitable for the preparation of isobutyl propionate. Specify the amounts of all materials required and a distillation range for the product. Obtain the necessary physical properties from one of the reference books listed in the Bibliography.

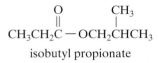

isobutyl propionate

Other Things You Can Do

(Starred items require your instructor's permission.)

*1   You and your coworkers can prepare a series of esters and compare their odors as described in Minilab 5.

*2   Prepare another ester of a primary alcohol, such as butyl acetate or isobutyl propionate, by the general method described for isopentyl acetate. Work out a procedure for the synthesis (see Exercise 6) and have it approved by your instructor. In some cases, a longer reflux time may be necessary for satisfactory results.

3   Read about the isolation of isopentyl acetate (isoamyl acetate) in the alarm pheromone of the honeybee in *Nature* **1962**, *195*, 1018.

4   Read about various types of pheromones in *Chemical Communication: The Language of Pheromones* [Bibliography, L3].

Fractional Distillation
Separation of Petroleum Hydrocarbons

EXPERIMENT **6**

Separation Methods.

Operations

OP-29   Fractional Distillation
OP-4   Weighing
OP-7   Heating
OP-34   Gas Chromatography

Before You Begin

1   Read the experiment and operation OP-29. Review the other operations as necessary.
2   Prepare a brief experimental plan following the directions in Appendix V.

Scenario

The Northern Pines Chemical Company specializes in manufacturing chemicals from wood products such as turpentine. For example, they use a major component of turpentine, α-pinene, to prepare organic compounds such as isobornyl acetate and terpineol, which impart a "piney" note to perfumes. To obtain pure α-pinene, it must be separated from the other major component of turpentine, β-pinene. They have been carrying out this separation by fractional distillation using an expensive column packing material that has to be replaced frequently. They would like to switch to a cheaper and longer-lasting packing material, but because the new packing material will be less efficient than their current type, they will need to purchase a longer column. From the difference between the boiling points of the pinenes (10°C), they estimate that such a column will need to have at least 20 theoretical plates (hypothetical column segments) to provide adequate separation. But the height of each theoretical plate, and thus the total length of the column, depends on the efficiency of the packing material.

Forrest Greenwood, the operations engineer at Northern Pines, has sent your supervisor some of the new packing material for testing. Your assignment is to determine its HETP (a measure of column efficiency) and to *purpose* estimate how long their new column must be to separate α-pinene from β-pinene efficiently. To do this, you will fractionally distill a mixture of two petroleum hydrocarbons, toluene and cyclohexane, and measure the composition of the fractions by gas chromatography.

See OP-29 for information about fractional distillation and distilling columns.

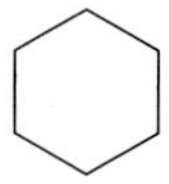

cyclohexane

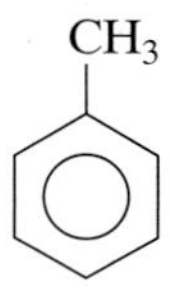

CH$_3$

toluene

Applying Scientific Methodology

This experiment differs from the previous ones in that no hypothesis is being tested; your job is to determine the value of a physical quantity, the

HETP (height equivalent to a theoretical plate) of the column, and to report the value as your conclusion. The accuracy of your results—and thus the validity of your conclusion—will depend on the care you take in performing the distillation.

Distillation in Petroleum Refining

Distillation has been used since antiquity to separate the components of mixtures—the ancient Egyptians were distilling wood to make an embalming fluid more than 3500 years ago. Today, distillation is used to manufacture perfumes, flavor ingredients, liquors, charcoal, coke, and a host of organic chemicals, but its most important application is in the refining of petroleum to produce fuels, lubricants, and petrochemicals.

The first step in petroleum refining is the separation of petroleum into different hydrocarbon fractions by distilling it through huge fractionating columns, called distillation towers, which may be up to 200 feet high. Since components having different numbers of carbon atoms usually have significantly different boiling points, this process separates the petroleum into fractions containing hydrocarbons of similar carbon content. Thus a lower boiling fraction might contain hydrocarbons with 5 or 6 carbon atoms, while a higher boiling fraction contains hydrocarbons having from 12 to 18 carbon atoms. Vapors from the lower boiling hydrocarbons rise to the top of the tower where they are condensed and collected as "top fractions." Hydrocarbons in top fractions, such as the one called straight-run gasoline, can be chemically modified and used in gasoline. The less volatile middle fractions are collected partway down the tower; these include kerosene and a gas-oil fraction used to produce diesel fuel, jet fuel, and home heating oil. The bottom fraction is usually subjected to vacuum distillation to produce vacuum gas oils, which can be used as fuel oils or converted to hydrocarbons suitable for gasoline. Such petroleum products as paraffin wax, lubricating grease, and asphalt also come from this fraction.

The octane number of a fuel is a measure of its antiknock properties. Most gasolines have octane numbers ranging from 85 to the low 90s.

With an overall octane number range of 30–50, straight-run gasoline is not a suitable motor fuel, but its octane number can be increased to nearly 100 by a process known as catalytic reforming. In this process, the straight-run gasoline is heated to about 500°C at high pressure in the presence of a suitable catalyst. Catalytic reforming converts alkanes to cycloalkanes and cycloalkanes to aromatic compounds; the result is a mixture rich in high-octane aromatics. For example, hexane may be *cyclized* to yield cyclohexane, which may then lose hydrogen atoms in a process called *dehydrogenation* to yield the aromatic compound benzene. Cyclohexane cannot easily be separated from petroleum by distillation alone, so pure cyclohexane is generally obtained by catalytic hydrogenation of benzene, the reverse of the second step shown here.

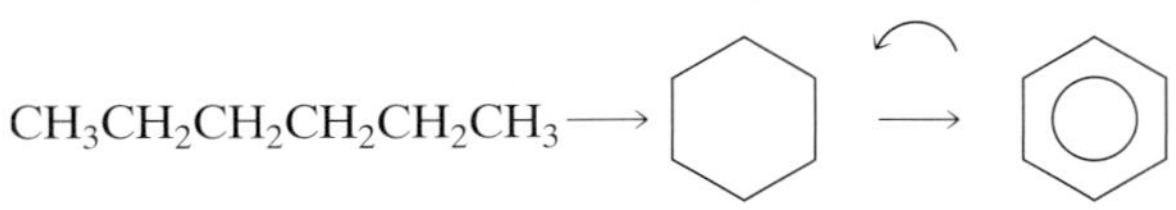

$$CH_3CH_2CH_2CH_2CH_2CH_3 \longrightarrow \bighexagon \longrightarrow \bighexagon$$

hexane   cyclohexane   benzene

Toluene is obtained by dehydrogenation of methylcyclohexane, which can be formed by cyclization of heptane and other alkanes.

$$CH_3CH_2CH_2CH_2CH_2CH_2CH_3 \longrightarrow \quad \longrightarrow$$

heptane   methylcyclohexane   toluene

Understanding the Experiment

In this experiment, you will separate the components of an equimolar mixture of cyclohexane and toluene by fractional distillation and assess the efficiency of the separation by measuring the composition of the fractions. The more completely the cyclohexane and toluene are separated from one another, the more efficient is the separation. The degree of separation depends not only on the column packing you use, but also on such factors as the stability of the heat source, the rate of distillation, and the way the column is packed. Good separation requires a low rate of distillation to maintain a high *reflux ratio*—the ratio of liquid returned to the boiling flask to liquid that distills into the receiving vessel—so patience is required if you are to attain good results. The efficiency of the column will be reduced if it is not packed uniformly, so it is important to distribute the packing material as evenly as possible.

You will measure the composition of each fraction you collect by gas chromatography. Your gas chromatograms should display two peaks, the cyclohexane peak being the first to appear. The areas of the peaks can be converted to relative masses by multiplying them by the appropriate correction factors from Table 6.1. You will then determine the HETP of the column packing from the composition of the first few drops of liquid you collect, the *HETP sample*. The number of theoretical plates provided by your distillation apparatus can be calculated by using the Fenske equation, expressed as follows for an equimolar mixture of two components, A and B:

See OP-29 for definitions of theoretical plate and HETP and for a discussion of the Fenske equation.

$$\text{total number of theoretical plates} = \frac{\log \dfrac{n_A}{n_B}}{\log \alpha} \qquad (1)$$

In Equation **1**, n_A/n_B is the ratio of the number of moles of cyclohexane to those of toluene in the HETP sample. The volatility factor, α, for the

Table 6.1 Gas chromatography correction factors for cyclohexane and toluene

	TC Detector	FI Detector
cyclohexane	1.11	0.942
toluene	1.05	1.02

Note: Benzene = 1.00. Your instructor will tell you what kind of detector your gas chromatograph has.

cyclohexane-toluene mixture is 2.33. The boiling flask furnishes one theoretical plate, so you will have to subtract 1 from the total number of theoretical plates to obtain the number of plates provided by the column itself. From this number and the length of your column packing, you can calculate the HETP of the column packing; the lower the HETP, the more efficient the packing.

Properties

Table 6.2 Physical properties of cyclohexane and toluene

	M.W.	bp	d
cyclohexane	84.2	81	0.774
toluene	92.2	111	0.867

Note: Boiling points are in °C and densities in g/mL.

Directions

Safety Notes

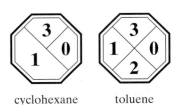

cyclohexane   toluene

Cyclohexane is very flammable and may irritate the skin, eyes, and respiratory tract.
Do not use open flames during the experiment.
Toluene is flammable, and inhalation, ingestion, or skin absorption may be harmful. Avoid contact with the liquid and do not breathe its vapors.

Take Care! Avoid contact with the liquid mixture and do not breathe its vapors.

Stop and Think: Will the thermometer record the boiling temperature of the liquid as soon as it begins to boil? Why or why not?

Observe and Note: When does the thermometer begin to record the true boiling temperature? Record that temperature and describe what you observe at that point.

Separation.   All components of the fractional distillation apparatus must be clean and dry. Pack an air condenser uniformly with stainless-steel sponge (see OP-29) or another packing material specified by your instructor, and measure the height of the packing to the nearest millimeter. Insulate the column as directed by your instructor. Obtain five clean, dry fraction collectors, such as 1-dram screw-cap vials. Label one vial "HETP" and number the other vials from 1 to 4, then weigh vials #1–4 with their caps on. Measure 5.0 mL of an equimolar mixture of cyclohexane and toluene into a 10-mL round-bottom flask (keep the flask stoppered to prevent evaporation) and add a stir bar. Clamp the flask to a ring stand over an appropriate heat source [OP-7] and assemble an apparatus for microscale fractional distillation [OP-29], taking care to place the thermometer bulb correctly.

Heat the mixture to a gentle boil with the stirrer turned on, and adjust the heating rate so that the vapors rise slowly up the column. Reduce the heating rate if the column begins to flood (fill with liquid); the packing should be moistened by condensing vapors, but should not contain any flowing liquid. Distillate should begin collecting not long after the rising vapors reach the top of the column. Adjust the heating rate as necessary so that liquid collects *slowly* in the well of the Hickman still. Use a Pasteur pipet to transfer distillate from the well to the appropriate collecting vial. Collect the first 2–4 drops of distillate in the HETP vial, cap it tightly, then begin collecting in vial #1. Record the distillation temperature (it should be ~81°C or higher) and continue to distill slowly. You may have to gradually increase the heating rate to keep the distillation rate more or less uniform. Begin collecting in vial #2 when the temperature reaches 85°C and in vial #3 when it reaches 97°C. When the temperature reaches 107°C, remove the heat source and let the liquid that remains in the column drain into the boiling flask. When the boiling flask has cooled, transfer its contents to vial #4. At this point, you should have four fractions covering the following approximate boiling ranges:

1 81–84°C
2 85–96°C
3 97–106°C
4 107–111°C

Analysis.   Weigh [OP-4] fractions #1–4 (or whichever fractions your instructor specifies) and analyze them and the HETP sample by gas chromatography [OP-34]. Measure the peak areas on each gas chromatogram and use the appropriate correction factors to convert the areas to relative masses. Calculate the percentage (by mass) and the actual mass of cyclohexane and toluene in each fraction you analyzed, the number of theoretical plates provided by your fractional distillation apparatus, the HETP of your column, and the length of the column (in cm) that will provide 20 theoretical plates. At your instructor's request, plot the component masses for each fraction (on the *y*-axis) as a function of boiling temperature, using the midpoints of the appropriate boiling ranges on the *x*-axis (use different symbols, such as × and ∘, for different components). Then draw a smooth curve connecting the data points for cyclohexane and another one (overlapping the first) connecting the data points for toluene.

Stop and Think: Why does the boiling temperature rise during the distillation?

Waste Disposal: Combine all fractions and place them in the hydrocarbon solvent recovery container.

Exercises

1 A certain fractional distillation apparatus contains a 24-cm Vigreux column stacked on top of a 30-cm glass tube half-filled with 4 × 4-mm porcelain saddles. This assembly is inserted into a boiling flask. How many theoretical plates does the entire apparatus provide if the HETP of the saddles is 5 cm, the HETP of the Vigreux column is 8 cm, and the HETP of the empty glass tube is 15 cm?

2 Describe and explain how each of the following experimental errors or variations would affect your HETP value and the efficiency of your separation. (a) You didn't collect the HETP sample until midway through the distillation. (b) All of your liquid distilled within five minutes of the time you began heating. (c) You stacked two packed air condensers over the boiling flask rather than using one.

3 Using the data in Table 6.3, construct a temperature-composition diagram like that shown in Figure E16 of OP-29. (a) From your diagram, estimate the initial composition of the distillate obtained by simple distillation of a mixture containing 20 mole percent cyclohexane and 80 mole percent toluene. (b) Estimate the initial composition if the same mixture is distilled through a three-plate column.

4 If you were to return your 97–106°C fraction to the empty boiling flask and redistill it, you might expect it all to distill between 97°C and 106°C, as it did the first time. It actually yields some distillate in all four boiling ranges. Explain.

Table 6.3 Temperature-composition data for cyclohexane-toluene

	Mol % cyclohexane	
T, °C	Liquid	Vapor
110.7	0	0
108.3	4.1	10.2
105.5	9.1	21.2
103.9	11.8	26.4
101.8	16.4	34.8
99.5	21.7	42.2
97.4	27.3	49.2
95.5	32.3	54.7
93.8	37.9	59.9
91.9	45.2	66.2
89.8	53.3	72.4
88.0	59.9	77.4
86.6	67.2	81.1
84.8	76.3	86.4
83.8	81.4	89.5
82.7	87.4	92.6
81.1	96.4	97.3
80.7	100.0	100.0

5   Derive Equation **1** (for an equimolar mixture of two components) from the Fenske equation in OP-29.

6   Suggest a chemical method that could be used to remove small amounts of toluene from cyclohexane.

7   Show how nylon-66 can be synthesized using cyclohexane as the starting material.

Other Things You Can Do

(Starred items require your instructor's permission.)

*1 Analyze some or all of your fractions with a refractometer [OP-32] rather than a gas chromatograph, using a graph that plots refractive index *vs.* mole fraction. (Assume a linear relationship between these variables.) The refractive indexes at 20°C of pure cyclohexane and toluene are 1.4260 and 1.4968, respectively.

*2 Carry out a gas chromatographic analysis of commercial xylene as described in Minilab 6.

3   Write a research paper about distillation and its uses based on information from *Kirk-Othmer* [Bibliography, A8] and other sources listed in the Bibliography.

Addition, Sublimation
Preparation of Camphor

Preparation and Purification of Solids. Oxidation.

Operations

OP-8   Cooling
OP-11   Addition of Reactants
OP-26   Sublimation
OP-4   Weighing
OP-10   Mixing
OP-13   Vacuum Filtration
OP-23   Drying Solids
OP-30   Melting Point

Before You Begin

1 Read the experiment and read OP-8, OP-11, and OP-26. Read or review the other operations as necessary.
2 Calculate the mass of 2.50 mmol of isoborneol and the theoretical yield of camphor from that amount of isoborneol.
3 Prepare an experimental plan for the experiment following the directions in Appendix V.

Scenario

The Northern Pines Chemical Company was pleased with your analysis of their column packing material (see Experiment 6), so they have requested your services for another project. One of the many chemicals they manufacture from α-pinene is camphor. The last step in the preparation of camphor from α-pinene is the oxidation of isoborneol, for which Northern Pines currently uses the powerful oxidizing agent chromic acid. But the chromium-containing by-products of this reaction are classified as hazardous wastes and disposing of them properly is very costly, so Northern Pines wants your Consulting Chemists Institute to develop a more environmentally acceptable oxidation process for them to use. In searching the chemical literature, your supervisor came across an article in the *Journal of Chemical Education* that describes the use of common laundry bleach (aqueous sodium hypochlorite) to oxidize a secondary alcohol, cyclohexanol, to a ketone, cyclohexanone. The only significant by-products of this process are water and sodium chloride.

See J. Chem. Educ. **1985**, *62*, 519.

cyclohexanol   cyclohexanone

The oxidation of isoborneol also involves the conversion of a secondary alcohol to a ketone, so your supervisor thinks the laundry-bleach method may be just the kind of environmentally friendly process Northern Pines is looking for. The camphor that the company currently manufactures is about 95% pure, the remaining 5% being unreacted isoborneol. Your assignment is to carry out the oxidation of isoborneol with laundry bleach to find out if the reaction does in fact produce any camphor, and if so, to determine whether you can obtain camphor having a purity of 95% or better.

Applying Scientific Methodology

You should be able to identify the problem(s) and develop one or more working hypotheses after reading the experiment. Make careful observations during the experiment to detect any evidence that a reaction is taking place. You will test your hypotheses by measuring the melting point of the product, which is very sensitive to the presence of impurities.

Camphor and the Camphoraceous Odor

The Chinese camphor tree, *Cinnamomum camphora*, is a tall, striking evergreen tree with dark shiny leaves. When steam is forced through the chopped-up wood of a camphor tree, camphor distills with the steam and crystallizes as a translucent white solid. Just as there are left-handed and right-handed gloves, scissors, and corkscrews, there are so-called left-handed and right-handed camphor molecules that are mirror images of one another.

camphor tree

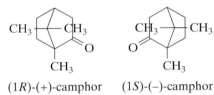

(1R)-(+)-camphor   (1S)-(–)-camphor

Isomers whose molecules differ only in their "handedness" are called *enantiomers*. (1R)-(+)-camphor, the enantiomer obtained from the camphor tree, is composed of the "right-handed" molecules shown here. The less common "left-handed" enantiomer, (1S)-(–)-camphor, has been isolated from feverfew (*Chrysanthemum parthenium*), a daisylike plant unrelated to the camphor tree. Camphor is synthesized commercially from α-pinene by the pathway outlined here.

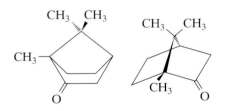

Perspective drawings showing the structure of camphor

α-pinene ⟶ pinene hydrochloride ⟶ camphene ⟶ isobornyl acetate ⟶ isoborneol ⟶ camphor

Most synthetic camphor contains an equal number of left-handed and right-handed molecules.

The history of camphor is longer and more involved than that of perhaps any other natural product. Scientific speculations about camphor have appeared in print since the time of Libavius (*Alchymia*, 1595). Although its

molecular formula ($C_{10}H_{16}O$) was determined in 1833, its complicated bicyclic structure baffled nineteenth-century scientists. Over the next 60 years,
they proposed more than 30 different structures for camphor, all of them
wrong, before Julius Bredt finally came up with the correct structure in 1893.
    The penetrating "camphoraceous" odor of camphor is shared by many
compounds of similar molecular shape and size. Compounds as diverse in
structure as the ones shown here all have roughly spherical molecules and
similar camphoraceous odors:

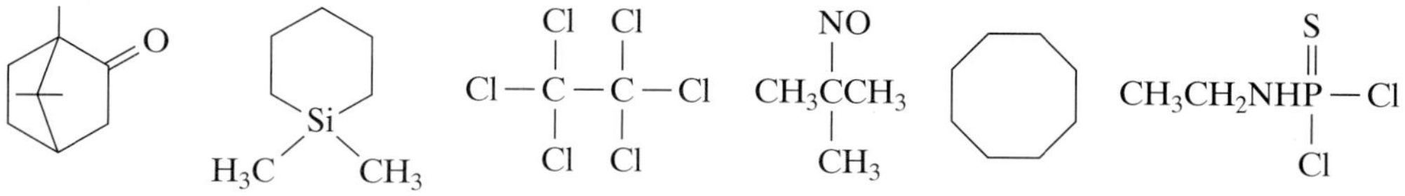

*The story of Bredt's quest is described
in J. Chem. Educ.* **1983**, *60*, 341.

Most scientists believe that the sense of smell is based on the presence in
the nasal passageways of a large number of odor receptors, each of which is
programmed to detect a specific kind of odor. This idea gained support in
1996, when scientists who had induced bacteria to grow an odor receptor
normally found in rats discovered that molecules of two compounds with
floral odors, lilial and lyral, became strongly attached to the receptors. The
similarities among molecules having camphoraceous odors suggest that the
odor of a substance may, at least in part, depend on the size and shape of its
molecules. A spherical molecule, for example, might fit nicely inside a hemi-
spherically convex odor receptor, causing it to transmit a neural message
that the brain interprets as a camphorlike odor. But scientists still do not
fully understand how humans and other animals can detect and recognize a
multitude of different odors.

Understanding the Experiment

In this experiment, you will oxidize a secondary alcohol, isoborneol, to a ketone, camphor. Secondary alcohols can be converted to ketones by powerful oxidizing agents such as chromic acid, but many chromium compounds
are highly toxic and corrosive, and some are known to cause cancer. They
also present a difficult disposal problem because they cannot legally be discharged into waterways or other places where they might harm the environment. For these reasons, you will use a safer and more environmentally
friendly oxidizing agent, the familiar laundry bleach that is sold under such
trade names as Clorox and Javex. Most chlorine bleaches contain about
5.25% sodium hypochlorite (NaOCl) in an aqueous solution. Adding a little
acetic acid facilitates oxidation by converting sodium hypochlorite to
hypochlorous acid (HOCl), which is probably the active oxidizing agent.
    In some previous experiments, you heated the reaction mixture to
speed up the reaction. In this experiment, you will need to slow it down instead, because the oxidation of isoborneol is exothermic and the heat
evolved can lead to the formation of unwanted by-products such as camphoric acid. You will control the reaction rate by adding the sodium
hypochlorite a little at a time from a Pasteur pipet rather than combining
all of the reactants at once.

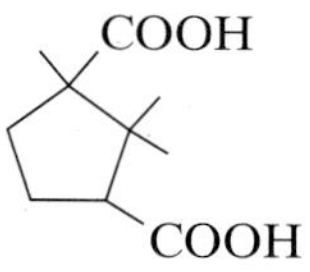

camphoric acid

When a reaction takes place under reflux, the boiling action helps mix the reactants. In this experiment, the reactants are mixed using a magnetic stirrer. To ensure a complete reaction, you must add enough sodium hypochlorite solution to keep the oxidizing agent in excess throughout the reaction. When HOCl is present in excess, a drop of the acidic reaction mixture placed on an indicator paper impregnated with starch and potassium iodide will oxidize iodide ions to iodine, which turns the starch a deep blue-black color. The color near the center of the drop may be bleached white, but some color should remain around its edges. Any excess HOCl that remains after the reaction is over can be destroyed by treatment with the reducing agent sodium bisulfite, according to the following equation:

$$HOCl + HSO_3^- \longrightarrow HCl + HSO_4^-$$

Because of its compact and symmetrical molecular structure, camphor changes directly from a solid to a vapor when heated, which allows it to be purified by sublimation. Isoborneol also sublimes at elevated temperatures, so the sublimed camphor will probably contain some unreacted isoborneol. Assuming that isoborneol is the only significant impurity in the product, its purity can be estimated with good accuracy from its melting point, because camphor has an unusually large freezing-point depression constant. The product can be regarded as a solid solution with camphor as the solvent and isoborneol as the solute, so you can use the following equation to calculate the molal concentration (m) of isoborneol in the product:

The melting point of a solid equals the freezing point of the corresponding liquid.

$$\Delta T = K_f \times m$$

ΔT = melting-point depression (reported mp − observed mp)
K_f = freezing-point depression constant for camphor = 40°C kg mol^{-1}
m = molal concentration of isoborneol (mol isoborneol/kg camphor)

Knowing m, the number of moles of isoborneol per kilogram of camphor, you can calculate the mass percent of isoborneol and camphor in the product.

Reactions and Properties

isoborneol   camphor

Table 7.1 Physical properties

	M.W.	mp	bp	d
isoborneol	154.3	212		
camphor	152.2	179	204	
sodium hypochlorite	74.4			
acetic acid	60.1	17	118	1.049

Note: mp and bp are in °C; *d* is in g/mL.

Directions

Safety Notes

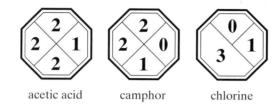

acetic acid   camphor   chlorine

Reaction.   *Under the hood* combine 2.50 mmol of isoborneol with 0.20 mL of glacial acetic acid in a small Erlenmeyer flask and drop in a stir bar. Measure 5.0 mL of 5.25% sodium hypochlorite solution (Clorox or another hypochlorite laundry bleach) into a screw-cap vial, and keep the vial capped when not in use. While stirring the reaction mixture, use a calibrated Pasteur pipet to add [OP-11] about 4.5 mL of the NaOCl solution, a few drops at a time, at a rate of about 1 mL per minute.

Take Care! Wear gloves, avoid contact with acetic acid and the NaOCl solution, and do not breathe their vapors.

When the addition is complete, seal the flask with Parafilm and stir [OP-10] the reactants at room temperature for 30 minutes. After the first 10 minutes and every 5 minutes thereafter, test the reaction mixture for excess hypochlorite by transferring a drop of the solution to a strip of starch-iodide paper. If at any time the test is negative, add enough 5.25% sodium hypochlorite (a few drops at a time) to the reaction mixture to give a *positive* test. When the reaction period is over, again test the reaction mixture with starch-iodide paper. If the test is positive, add enough saturated sodium bisulfite solution dropwise to give a *negative* test.

Observe and Note: What evidence can you detect that suggests that a reaction is taking place?

Separation.   Cool [OP-8] the reaction mixture for 5–10 minutes in an ice/water bath. Collect the product by vacuum filtration [OP-13], washing it on the filter with two portions of ice-cold water. Dry [OP-23] the crude product at room temperature, *not* in an oven. At your instructor's request, weigh the crude product and save a small amount of it for a melting point measurement.

Stop and Think: What is the purpose of the sodium bisulfite addition?

Waste Disposal: The filtrate may be poured down the drain.

Purification and Analysis.   Purify the crude product by sublimation [OP-26], taking care not to char the solid by overheating. Weigh [OP-4] the sublimate, dry [OP-23] it at room temperature if necessary, and measure its melting point [OP-30]. Taking as its melting point the temperature at which the solid was completely liquefied, calculate the mass percentages of isoborneol and camphor in your purified product.

Your instructor may request that you purify only part of the crude camphor.

Exercises

1 In this experiment, you started with a white, strong-smelling solid and ended up with a white, strong-smelling solid. What evidence leads you to conclude that these two solids are in fact different compounds and that you did not just isolate the unreacted starting material?

2 Following the format in Appendix V, construct a flow diagram for the synthesis of camphor.

3 The equation in the "Reactions and Properties" section shows only one isoborneol enantiomer and the camphor enantiomer it forms. Find out from your instructor if the isoborneol you used was the right-handed (R) enantiomer shown, the left-handed (S) enantiomer, or an equimolar mixture of both. Then rewrite the equation showing the correct stereochemistry of the reactants and products.

4 Describe how each of the following experimental errors or variations might affect your results. (a) You omitted the 30-minute reaction period after the addition step. (b) You added the sodium hypochlorite solution all at once. (c) You mistook a negative starch-iodide test for a positive one and stopped adding sodium hypochlorite solution midway through the reaction period.

5 Write a balanced net ionic equation for the reaction of the acidified sodium hypochlorite solution with iodide ion from the starch-iodide paper, assuming that HOCl is reduced to HCl.

6 Show which carbon-carbon bond of camphor must be broken to form camphoric acid. Use molecular models if necessary.

Other Things You Can Do

(Starred projects require your instructor's permission.)

*1 Isolate caffeine from No-Doz tablets and purify it by sublimation as described in Minilab 7.

*2 Oxidize cyclohexanol to cyclohexanone as described in *J. Chem. Educ.* **1985**, *62*, 519.

3 Write a research paper about camphor and its applications, using sources listed in the Bibliography.

Boiling Point, Refractive Index
Identification of a Petroleum Hydrocarbon

EXPERIMENT **8**

Physical Properties of Liquids. Alkanes and Cycloalkanes.

Operations

OP-31 Boiling Point
OP-32 Refractive Index
OP-4   Weighing
OP-5   Measuring Volume
OP-27   Simple Distillation

Before You Begin

1 Read the experiment and the descriptions for OP-31 and OP-32, and review the other operations as needed.
2 Prepare a brief experimental plan for this experiment following the directions in Appendix V.

Scenario

An investigative organization known as The Consumer's Advocate (TCA) publishes a monthly magazine, *Caveat Emptor*, which evaluates consumer products and exposes scams. TCA is currently investigating an auto-supplies manufacturer that markets Thrust, a gasoline additive claimed to improve engine performance. TCA's preliminary tests show that the additive has no measurable effect on either power or mileage, and they suspect that the additive is nothing more than a hydrocarbon that burns along with the gasoline. To support its case against the company, TCA needs to know the identity and octane number of the hydrocarbon. If its octane number is higher than that of a typical regular no-lead gasoline (about 87), then the company's claim that the additive improves engine performance might have some validity—although the amount of improvement would be negligible when the additive is used in the quantity recommended on the can.

Because their own chemists are busy with other projects, TCA's technical director, Patsy Haven, has farmed out the job to your Institute. Your assignment is to identify the hydrocarbon in Thrust and determine whether or not its octane number is greater than 87.

Applying Scientific Methodology

You should evaluate the evidence and formulate tentative hypotheses as you go along. For example, if you measure a boiling point of 79°C, your tentative hypothesis might be "The alkane in Thrust is 2,4-dimethylpentane" (see Table 8.2). If you then measure its density as 0.79 g/mL, you might have

Table 8.1 Octane numbers of some petroleum hydrocarbons

Hydrocarbon	Octane no.	Hydrocarbon	Octane no.
nonane	−45	2,4-dimethylpentane	82
octane	−17	methylcyclopentane	82
heptane	0	cyclopentane	83
2-methylheptane	24	2,3-dimethylpentane	89
hexane	26	2-methylbutane	89
3-methylheptane	35	butane	92
2-methylhexane	45	2,3-dimethylbutane	95
pentane	61	2,2-dimethylbutane	96
3-methylhexane	66	2,2,3-trimethylbutane	100
methylcyclohexane	71	2,2,4-trimethylpentane	100
2-methylpentane	73	2,2,3-trimethylpentane	102
3-methylpentane	75	toluene	104
cyclohexane	77	benzene	106

to change your hypothesis to "The alkane in Thrust is cyclohexane." Measuring a refractive index of 1.4262 would then confirm your second hypothesis and lead you to the conclusion that the alkane is indeed cyclohexane, whose octane number you can look up in Table 8.1.

Gasoline—A Chemical Soup

Gasoline is a kind of "chemical soup" containing an incredibly large number of ingredients that are carefully selected and blended to produce a fuel with the desired properties. Virtually all of the main fuel components of gasoline are derived either directly or indirectly from petroleum, which must be refined before a usable fuel is obtained. The word "refine" suggests a simple separation and purification process, but the refining of petroleum is a more complex operation that involves chemical as well as physical changes. As you learned in Experiment 6, petroleum is first fractionated in a distillation tower, which separates its components according to their boiling-point ranges. The fraction boiling between about 50°C and 150°C, straight-run gasoline, is not a good motor fuel by itself because it contains a large proportion of straight-chain hydrocarbons such as heptane and hexane. Straight-chain hydrocarbons burn very rapidly, generating a shock wave in the combustion chamber that reduces power and can damage the engine. This "knocking" does not occur with highly branched hydrocarbons, which burn more slowly and uniformly.

The octane number of a motor fuel is a measure of its antiknock qualities. The highly branched alkane 2,2,4-trimethylpentane (sometimes called "isooctane") is a very good motor fuel and has arbitrarily been assigned an octane number of 100. Heptane, with no branching, is assigned an octane number of zero. The performance of a particular motor fuel is measured relative to these two alkanes; for example, a fuel performing as well as a mixture containing 70% 2,2,4-trimethylpentane and 30% heptane is assigned an octane number of 70. Table 8.1 lists the octane numbers of selected hydrocarbons found in gasoline.

A major objective of petroleum refining is to convert the low-octane components of petroleum into higher-octane compounds. This can be accomplished by a variety of chemical processes, such as *isomerization*, which converts straight-chain alkanes to branched alkanes; *cracking*, which breaks down large molecules into smaller ones; *alkylation*, which combines

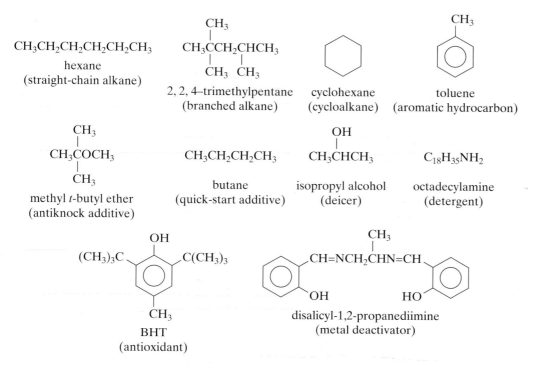

Figure 8.1 Some components of a typical gasoline

short-chain alkane and alkene molecules to form longer, branched molecules; and *catalytic reforming*, which converts alkanes to cycloalkanes and aromatic compounds. Aromatic hydrocarbons such as toluene have particularly high octane numbers and are used to increase the octane rating of no-lead fuels.

Gasoline for use in automobile engines is prepared by combining varying amounts of straight-run gasoline, cracked gasoline, alkylated gasoline, reformate, and other hydrocarbon mixtures in the right proportions to give the desired boiling-point range and octane number. The properties of the fuel are then further adjusted with a variety of additives. Antiknock additives such as methyl *t*-butyl ether (MTBE) may be included to boost the octane rating further. Quick-start additives such as butane facilitate cold weather starting. Antifreeze additives such as isopropyl alcohol reduce icing. Antioxidants such as BHT help improve fuel stability and reduce gum formation, particularly in fuels that contain appreciable amounts of alkenes. Certain metals, such as copper and iron, can catalyze gum-forming reactions, so chelating compounds such as disalicyl-1,2-propanediimine may be added to deactivate these metals. Cars with fuel injectors require detergent additives such as octadecylamine to keep their intake systems clean. Dyes are added for identification and visual appeal.

The chemistry involved in these reactions is described in many organic chemistry textbooks.

The use of MTBE has become controversial because of its environmental effects.

Understanding the Experiment

In this experiment, you will attempt to identify an unknown hydrocarbon that is one of the compounds listed in Table 8.2. Identifying an unknown organic compound is somewhat like identifying the perpetrator of a crime. The investigator compiles a list of suspects, hunts for clues that might have a bearing on the case, sifts through the evidence to eliminate most of the suspects, and then searches for additional evidence to build a case against the prime suspect. Many kinds of evidence may have a bearing on the identity of an organic compound: physical evidence, such as boiling point and density; chemical evidence, such as the appearance of a precipitate with a test reagent; and spectral evidence, such as the occurrence of an infrared band that suggests the presence of a particular functional group. Alkanes and cycloalkanes are comparatively unreactive, making it difficult to gather much chemical evidence about them, and their infrared spectra are not very revealing, so in this experiment you will identify the unknown alkane by using its physical properties alone.

Before you can measure the physical properties of a liquid accurately, the liquid must be pure. In this experiment you will purify your hydrocarbon by simple distillation. You can estimate the boiling point of the liquid as you distill it, but you will also measure its boiling point by using a capillary-tube method, described in OP-31. Because the boiling point of a liquid varies with the barometric pressure, you may have to apply a boiling-point correction, as described in OP-31.

The density of a liquid can be obtained by accurately weighing a measured volume of the liquid. The volume is measured with an appropriate pipet, and the mass should be measured to at least the nearest milligram on an accurate balance.

A good refractometer can be used to determine the refractive index of a pure liquid with great accuracy. With reasonable care, you should be able to measure the refractive index of your unknown to within 0.05% or better, so this value may be the most important clue to the identity of your hydrocarbon. The refractive index of a liquid is very sensitive to temperature, however, so you will need to correct your observed value if the temperature at the refractometer is above or below 20°C.

The physical constants of the hydrocarbons listed in Table 8.2 are different enough that an accurate determination of all three constants should allow the certain identification of an unknown as one of the twelve. Your instructor may add more hydrocarbons to your list of possibilities. If so, he or she will provide you with the appropriate physical constants or ask you to look them up.

Properties

Table 8.2 List of possible hydrocarbons

Name	bp	n_D^{20}	d^{20}
cyclopentane	49	1.4065	0.746
2,2-dimethylbutane	50	1.3688	0.649
2,3-dimethylbutane	58	1.3750	0.662
3-methylpentane	63	1.3765	0.664
hexane	69	1.3749	0.659
methylcyclopentane	72	1.4097	0.749
2,4-dimethylpentane	80	1.3815	0.673
cyclohexane	81	1.4266	0.779
2,3-dimethylpentane	90	1.3919	0.695
heptane	98	1.3877	0.684
2,2,4-trimethylpentane	99	1.3915	0.692
methylcyclohexane	101	1.4231	0.769

Note: Boiling points are in °C; n_D^{20} = refractive index at 20°C using sodium D line; d^{20} = density at 20°C.

Directions

Your instructor may suggest additional tests to carry out on your hydrocarbon.

> **Your unknown hydrocarbon is flammable; keep it away from flames and hot surfaces.**

Safety Notes

alkanes and cycloalkanes of five to eight carbons

Purification and Boiling-Point Determination.   Obtain a sample of the unknown hydrocarbon from your instructor. Record its identification number and the ambient barometric pressure in your laboratory notebook. Select a suitable heat source and assemble an apparatus for microscale simple distillation [OP-27], using a boiling chip or a stirring device. Be sure that the thermometer bulb is positioned correctly in the Hickman still. Distill the liquid slowly, setting aside a low-boiling forerun or high-boiling fraction, if any, for later disposal. Record its boiling range and the temperature when about half of it has distilled (the median boiling point). Then carry out a boiling-point measurement [OP-31] on the hydrocarbon using a capillary-tube method. The resulting boiling point should be within 1–2 degrees of the median distillation boiling point. If it is not, repeat the boiling point measurement or redistill the hydrocarbon. Apply a correction to the boiling point if the atmospheric pressure was below 750 torr.

Take Care! Keep the liquid away from flames or hot surfaces.

If the hydrocarbon boils over a broad range, it should be redistilled and a pure fraction (collected over a range of 1–2°C) should be used for analysis.

Density Measurement.   The temperature of the purified hydrocarbon should be close to 20°C. Accurately measure [OP-5] 0.200 mL of the liquid into a clean, dry, tared vial using a measuring pipet or an automatic pipet. (**Take Care!** Do not pipet by mouth.) Stopper the vial immediately and weigh [OP-4] it to the nearest milligram on an accurate balance. Calculate the density of your hydrocarbon from your results.

Stop and Think: What should happen if you put 1 or 2 drops of your hydrocarbon in a test tube containing a small amount of water and shake the test tube? Do it. Was your prediction correct?

Refractive Index Measurement.   Measure the refractive index [OP-32] of the purified hydrocarbon as directed by your instructor and record the temperature of the measurement. Apply a correction to the refractive index if the temperature of the measurement was not 20°C.

Waste Disposal: Put your hydrocarbon and any liquid saved from the distillation in a designated hydrocarbon solvent recovery container.

Exercises

1 The following properties were measured for an unknown hydrocarbon in a laboratory with an ambient temperature of 28°C and a barometric pressure of 28.9 inches of mercury (1 inch Hg = 25.4 torr):

> boiling point: 78.2°C
> refractive index: 1.3780
> mass of 0.200 mL: 0.133 g

Correct the refractive index and boiling point to 20°C and 1 atmosphere, and calculate the density of the unknown. If the unknown is one of the hydrocarbons listed in Table 8.2, what is its probable identity?

2 Give names and structural formulas for all structural (constitutional) isomers of your unknown hydrocarbon.

3 Describe and explain the possible effect on your results of the following experimental errors or variations. In each case, tell whether the resulting physical property (bp, density, or refractive index) will be too high or too low. (a) You read the temperature when the unknown liquid began to boil and recorded that temperature as its boiling point. (b) To obtain 0.200 mL of the liquid, you filled a Mohr pipet to the 0.200-mL mark and drained it completely into the weighing vial (you can ask your instructor to show you a Mohr pipet). (c) The liquid was at a temperature of 25° when you measured its mass and volume. (d) The liquid was at a temperature of 25° when you measured its refractive index, but you forgot to correct it.

4 A large oil spill that resulted from the 1989 shipwreck of the *Exxon Valdez* caused considerable damage to Alaska's wildlife. Do you think a comparable spill of a water-insoluble liquid having a density of 1.20 g/mL would have been as harmful to waterfowl and aquatic mammals, assuming the toxicity of the liquid was comparable to that of petroleum? Explain your answer.

5 (a) Write a balanced equation for the complete combustion of 2,2,4-trimethylpentane in an engine's combustion chamber. (b) Show how butane can be converted to 2,2,4-trimethylpentane using petroleum-refining processes mentioned in this experiment.

dioxane

6 (a) Dioxane has a boiling point of 101°C. Could you separate dioxane from methylcyclohexane by distillation? Explain why or why not, based on the liquid and vapor compositions during the distillation. (b) How might these liquids be separated?

Other Things You Can Do

(Starred projects require your instructor's permission.)

*1 Solve the "missing-label puzzle" described in Minilab 8.

*2 With your coworkers, obtain and compare gas chromatograms of different grades of gasoline, such as leaded, unleaded, and "gasahol." Refer to *J. Chem. Educ.* **1972**, *49*, 764 and *J. Chem. Educ.* **1976**, *53*, 51 for information and references that will help you interpret the gas chromatograms and identify some of the components.

*3 Test for lead in gasoline as follows: Saturate a piece of filter paper with gasoline and expose it to strong sunlight for several hours. Moisten the paper with 3 *M* acetic acid followed by a few drops of aqueous potassium iodide solution (16.5 g/100 mL). A yellow color after several minutes indicates the presence of lead.

4 Write a research paper about gasoline and petroleum refining using references cited in the Bibliography.

Column Chromatography, UV-VIS Spectrometry
Isolation and Isomerization of Lycopene from Tomato Paste

EXPERIMENT **9**

Isolation of Natural Products. Ultraviolet-Visible Spectrometry. Geometric Isomers.

Operations

OP-15c Liquid-Solid Extraction
OP-18 Column Chromatography
OP-38 Ultraviolet-Visible Spectrometry
OP-6   Making Transfers
OP-14 Centrifugation
OP-16 Evaporation
OP-21 Washing Liquids
OP-22 Drying Liquids

Before You Begin

1 Read the experiment and operations OP-15c, OP-18, and OP-38. Review the other operations as necessary.
2 Prepare an experimental plan following the directions in Appendix V.

Scenario

The Consumer's Advocate (see Experiment 8) is now investigating the quality of processed foods such as tomato paste. Ideally, such processed foods would contain all of the nutrients and flavor components present in the fresh fruits or vegetables, but all too often, the processing methods used tend to degrade the color, flavor, and nutritional value of a food.

The red pigment that colors ripe tomatoes is an all-*trans* form of lycopene, an antioxidant that is known to fight many kinds of cancer, including cancers of the digestive tract, cervical cancer in women, and prostate cancer in men. Heat, light, and certain chemicals may convert some of this pigment to its 13-*cis* isomer. Thus, the presence of 13-*cis*-lycopene in a canned tomato product suggests that the fresh fruit may have been subjected to excessive heat or light during processing. TCA's technical director wants your organization to assess the quality of different brands of tomato paste on the basis of their *trans*-lycopene content. Your supervisor has found a method for doing this, using ultraviolet-visible spectrometry, in the *Journal of Chemical Education*. To improve the validity of your results, you and your coworkers will work in small research teams, with each team assigned a specific brand of tomato paste.

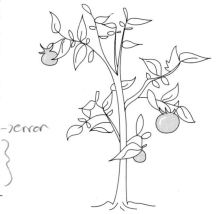

tomato plant

J. Chem. Educ. **1989**, *66*, 258.

Applying Scientific Methodology

Intro

With your instructor's permission, you or another member of your research team may bring a sample of a commercial tomato paste for testing. Your opinion about the quality of the tomato paste can then be the basis for a working hypothesis, which will be tested when you analyze your lycopene by ultraviolet–visible spectrometry. During the experiment, you should try to avoid conditions that promote isomerization or oxidation of lycopene, which would reduce the reliability of your results.

Carotenoids, Vitamin A, and Vision

Key Concept: *The wavelength of the UV-VIS radiation absorbed by a conjugated substance increases with the length of its conjugated system.*

Intro

Lycopene, with its 13 carbon-carbon double bonds, is one of the most unsaturated compounds in nature. Because most of its double bonds are conjugated, lycopene absorbs radiation at long wavelengths in the 400–500-nm region of the visible spectrum. Its resulting deep orange-red color is responsible for the redness of ripe tomatoes, rose hips, and many other fruits. An even more important plant pigment is the yellow-orange substance β-carotene, which is present not only in carrots but in all green leaves and many flowers as well. Both lycopene and β-carotene, along with most of the other natural *carotenoids*—compounds related to β-carotene—occur naturally in the all-*trans* forms shown in Figure 9.1.

Although the main function of carotenoids in plants remains somewhat of a mystery, the importance of carotenes to animals is clear—β-carotene (and, to a lesser extent, α- and γ-carotene) is converted in the intestinal wall to Vitamin A, which is then stored in the liver. Generations of children have grown up hearing the mealtime refrain "Eat your carrots—they're good for your eyes!" In fact, Vitamin A from carotenes and other sources is an essential participant in the process by which light entering your eyes causes your brain to construct a visual picture of your surroundings. One step in this process involves a configurational change analogous to the isomerization of all-*trans*-lycopene, but in reverse.

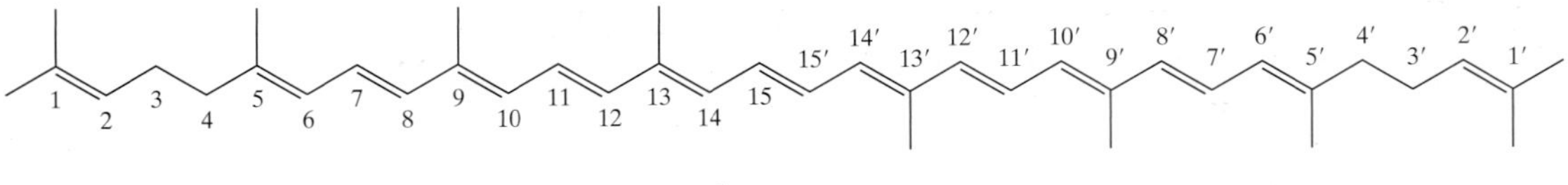

lycopene

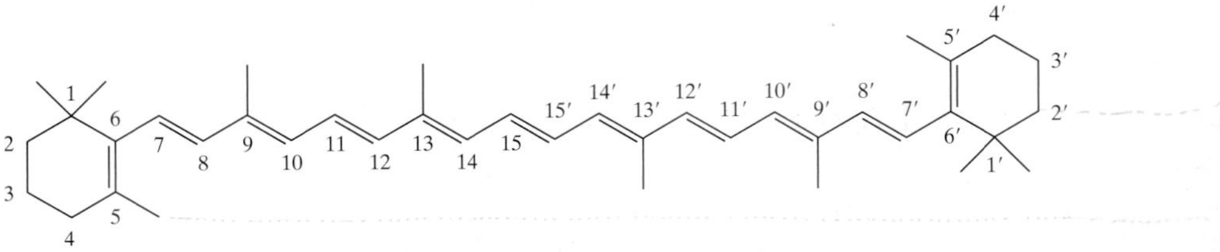

β-carotene

Figure 9.1 Structures of carotenoids
Note: A single straight line branching off from a chain or ring stands for a methyl group in these and similar formulas. A carbon atom with the requisite number of hydrogens is at each bend of the chain.

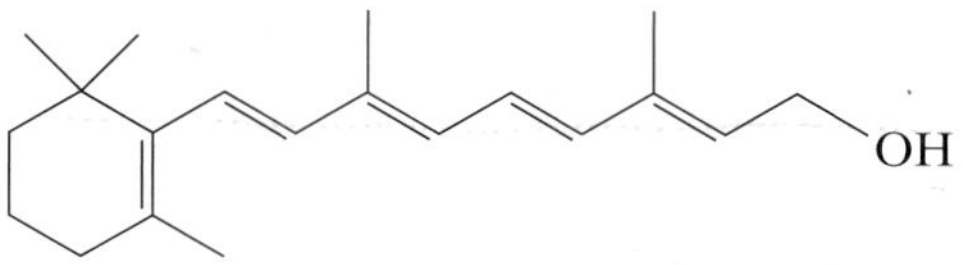

Vitamin A
(all-*trans*)

The process of vision, although extremely complex in its entirety, is based on the isomerization of an oxidized form of Vitamin A called retinal. In the rods of the retina, which are responsible for night vision, retinal occurs in combination with the complex protein opsin to form rhodopsin (visual purple). The retinal in rhodopsin assumes the shape shown in Figure 9.2A, with an 11-*cis* double bond and probably a *cisoid* conformation between the #12 and #13 carbon atoms as well. This allows a retinal molecule to fit comfortably into a cavity in an opsin molecule—much like a joey (a baby kangaroo) curled up in its mother's pouch. When a photon of light strikes a rhodopsin molecule, its 11-*cis*-retinal passenger suddenly straightens out and becomes all-*trans*-retinal. This process is incredibly fast—much faster than the blink of an eye—taking only about 0.2 trillionths of a second (200 femtoseconds). The isomerized retinal molecule no longer fits into its niche in the opsin molecule, which responds much as a mother kangaroo might when her joey creates a disturbance in her pouch—it ejects its unruly passenger, triggering the transmission of a visual message to the brain. Subsequently, the *trans*-retinal is enzymatically reduced to all-*trans*-Vitamin A, which isomerizes to 11-*cis*-Vitamin A, which is oxidized back to 11-*cis*-retinal, which promptly combines with another molecule of opsin to regenerate more rhodopsin. At this point, another photon of light can start the cycle all over again.

A similar process occurs in the cones of the retina, which are responsible for color vision.

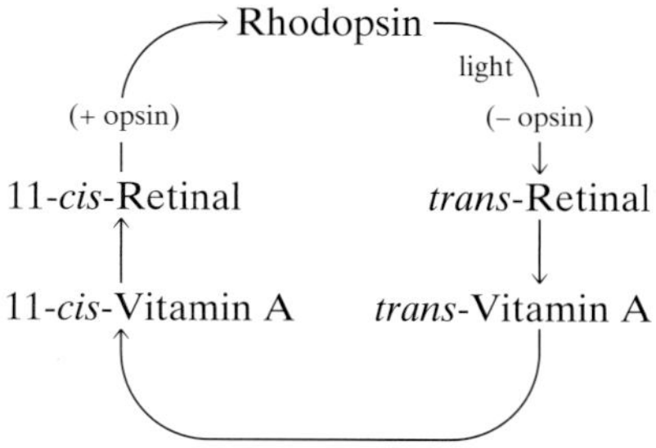

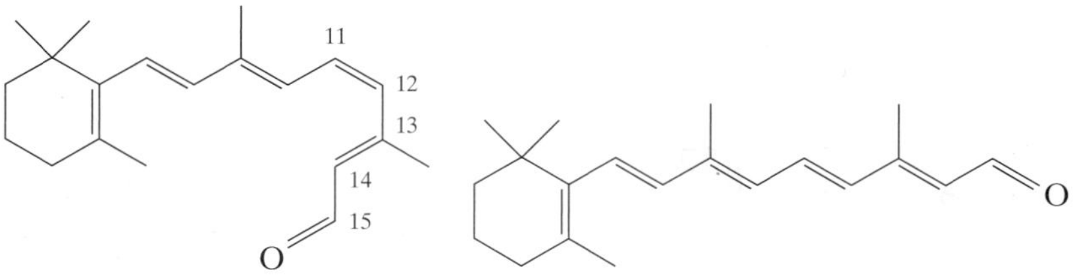

A 11-*cis*-12-*s*-*cis*-retinal   *B* all-*trans*-retinal

Figure 9.2 Retinal isomers

Understanding the Experiment

In this experiment, you will extract the carotenoid pigments (lycopene, carotenes, and xanthophylls) from canned tomato paste and separate them by column chromatography to obtain a solution containing lycopene. Then you will record the ultraviolet-visible spectrum of this solution and analyze it for evidence of isomerization. Although lycopene can be obtained directly from ripe tomatoes, it is easier to extract it from commercial tomato paste, in which the lycopene is more concentrated—one tablespoon of tomato paste yields as much lycopene as a medium ripe tomato, about 10 mg. Lycopene will isomerize if allowed to stand in solution too long, particularly in the presence of heat, light, or acids. For this reason, it is important to avoid unnecessary delays and exposure of the pigment to heat or bright light. Acids that occur naturally in tomatoes can be removed by washing the extract with aqueous potassium carbonate. Lycopene oxidizes slowly in the presence of atmospheric oxygen, so you should try to record its UV-VIS spectrum on the same day that you isolate the lycopene solution, if possible; otherwise oxidation products may alter the spectrum.

Petroleum ether is a general term used to describe volatile petroleum distillates having varying compositions and boiling ranges. Do not confuse it with diethyl ether!

You will extract lycopene and other carotenoid pigments from tomato paste with a solvent mixture that contains equal volumes of acetone and petroleum ether. Acetone is very soluble in water, so washing the extracts with water removes the acetone and leaves the pigments dissolved in the petroleum ether layer. Low-boiling petroleum ether (bp ~35–60°C) is preferred for the extraction, because it can be evaporated readily at room temperature to yield a concentrated solution of the pigments. This concentrate is then transferred to the top of a chromatography column, which should be packed with neutral alumina (Brockmann grade II–III) as described in OP-18. Be sure to read this operation carefully before you attempt to pack your column, since a poorly packed column will not provide good separation. You will elute the pigments with hexanes (a mixture of C_6H_{14} alkanes) followed by a more polar eluent containing 10% acetone with the hexanes. Ly-

High-boiling petroleum ether can be used in place of hexanes.

copene, with its 13 double bonds, is attracted to alumina somewhat more strongly than β-carotene and related carotenes, which have 11–12 double bonds. Therefore, the yellow carotene band will move down the column faster than the orange-red lycopene band. Yellow xanthophyll pigments will trail behind the lycopene band because they contain polar hydroxyl groups that are strongly attracted to alumina.

Under conditions that might be present during the processing of tomato paste, some all-*trans*-lycopene can undergo a configurational change to 13-*cis*-lycopene. You can detect such isomerization by obtaining an ultraviolet–visible spectrum of your lycopene sample, because the 13-*cis* isomer absorbs at lower wavelengths than does the all-*trans* isomer, and the relative intensities of the two largest peaks in the visible region of the spectrum change significantly (see Figure 9.3). In the ultraviolet region (if you scan it) you may see a characteristic *cis* peak that is associated with the bent geometry of the 13-*cis*-lycopene molecules. The *cis* peak occurs at about 360 nm, a wavelength at which all-*trans*-lycopene shows little absorption.

You can estimate the percentage of all-*trans*-lycopene in your sample by the following empirical method:

- Draw a straight, horizontal line tangent to the bottom of the valley between the last (highest wavelength) two peaks on the spectrum.
- Measure the height of both peaks from that line, calling the heights *a* and *b*.
- Divide the smaller height (*b*) by the larger (*a*), subtract 0.40, multiply by 100%, and divide the product by 0.40 as shown by the following equation.

This method is derived from the Journal of Chemical Education *article referred to in the Scenario.*

$$\frac{\left(\dfrac{b}{a} - 0.40\right) \times 100\%}{0.40}$$

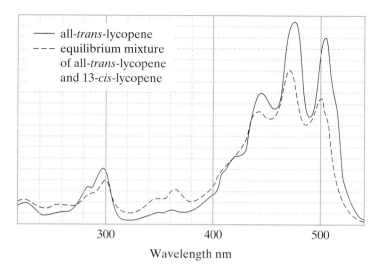

Figure 9.3 Ultraviolet–visible spectra of lycopene stereoisomers

If the percentage of all-*trans*-lycopene is significantly lower than 100%, the lycopene in your tomato paste may have isomerized while the tomatoes were being processed into tomato paste, during your isolation of the lycopene, or both. Before drawing any definite conclusion about the quality of your tomato paste, you should compare your results with those of the other members of your team. Such a comparison should help you select the result that was affected least by the experimental conditions.

Partial isomerization of all-*trans*-lycopene to 13-*cis*-lycopene is also catalyzed by iodine. To show the effect of isomerization, you can add a dilute solution of iodine to your lycopene solution and again obtain its spectrum (or your instructor may demonstrate the isomerization to the class).

Directions

Students may work together in small research teams, with each team working on a specific brand or tomato paste, agreeing upon a conclusion, and comparing its results with those of other groups working with different brands.

> **Acetone and petroleum ether are very flammable, and their vapors can irritate the eyes and upper respiratory tract. Keep petroleum ether and the petroleum ether-acetone mixture away from flames and hot surfaces, and do not breathe their vapors.**
> **The hexanes mixture is flammable, so keep it away from flames and hot surfaces.**

Safety Notes

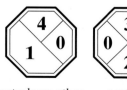

petroleum ether   acetone

Preparing the Chromatography Column.   Pack a $5\frac{3}{4}$-inch Pasteur pipet (or another suitable column) with neutral Brockmann grade II–III alumina (see OP-18). This should take about 1.5–2.0 g of alumina; see that the surface of the alumina is as nearly horizontal as possible. Obtain a 10-mL beaker and a clean, dry, 1-dram screw-cap vial to collect the eluates. Clamp the column to a ring stand over the beaker, making sure that it is as nearly vertical as possible.

Extraction of Pigments from Tomato Paste.   Protect the pigments from undue exposure to light throughout the remainder of this experiment. Weigh about 1.0 g of tomato paste into a 15-mL screw-cap centrifuge tube. Extract [OP-15c] the solid material by shaking the capped tube with 4 mL of a 50% (by volume) mixture of acetone and low-boiling petroleum ether until the solid residue looks dry and fluffy. Then use a flat-bladed microspatula to rub and crush it against the sides of the tube. Repeat the shaking and crushing steps several times. Separate the extract by centrifugation [OP-14] or with a filter-tip pipet, and transfer [OP-6] it to a second 15-mL centrifuge tube. Then repeat the extraction of the solid residue with another 4-mL portion of 50% acetone/petroleum ether, and combine the extracts in the second centrifuge tube. Wash [OP-21] the combined extracts with 5 mL of saturated sodium chloride (NaCl) solution, followed by 5 mL of 10% aqueous potassium carbonate and another 5 mL portion of saturated NaCl solution. Dry [OP-22] the lycopene-containing organic layer with anhydrous sodium sulfate or magnesium sulfate, collect it in a 5-mL conical vial, and concentrate it to a volume of 0.1–0.2 mL by evaporating [OP-16] most of the solvent under vacuum or in a stream of dry nitrogen *without* heating. If you inadvertently evaporate the solution to dryness, dissolve the residue in 0.1 mL of hexanes.

Separation and Isolation of Lycopene.   Fill the chromatography column [OP-18] with the first eluent, hexanes, and let the liquid drain until its surface *just* disappears into the alumina layer (or the upper sand layer, if there is one). Immediately transfer the lycopene extract to the top of the column with a Pasteur pipet, using a drop or two of eluent to rinse any remaining extract onto the column. When the extract surface just disappears into the alumina (or sand), fill the column nearly to the top with eluent and continue to add eluent to keep its level more or less constant. When the yellow carotene band begins to drain out of the column, fill the top of the column with the second eluent, 10% acetone in hexanes, and continue to replenish the eluent as before. When the orange-red lycopene band begins to leave the column, replace the beaker underneath it by the collection vial and collect the eluate until nearly all of this band has drained from the column (do not collect any of the subsequent yellow xanthophyll band in this vial). Then replace the vial by the beaker you used before and let any remaining eluate drain into the beaker. Use the lycopene sample for spectral analysis as soon as possible. If you cannot record its spectrum on the same day, store your sample in a tightly closed container in a refrigerator or freezer.

Spectral Analysis of Lycopene.   Transfer the lycopene eluate to an appropriate 1-cm sample cell, and add enough 10% acetone/hexanes to fill the cell about $\frac{3}{4}$ full. Cap the cell and swirl it gently to mix the contents. Use 10% acetone/hexanes in the reference cell. Record a spectrum [OP-38] of the lycopene sample over the 600–400 nm range. (With your instructor's permission you can scan the 400–250 nm UV range as well.) If necessary, dilute the lycopene solution with 10% acetone/hexanes to keep the strongest peak (at ~475 nm) on scale.

Take Care! Keep the extraction mixture away from flames and hot surfaces.

Stop and Think: What is the purpose of the K_2CO_3 wash?   use

Waste Disposal: The wash solvents can be poured down the drain.

Observe and Note: Describe what you see as the bands pass down the column.

Stop and Think: Why do the bands move down the column at different rates?

Waste Disposal: Place all eluates except the lycopene eluate in a designated solvent recovery container.

will not do

Your instructor may demonstrate the isomerization of lycopene. To do it yourself, mix a drop of a 0.025% solution of iodine in hexanes into the lycopene solution and leave the solution in the sample beam at 475 nm, monitoring its absorbance, until its absorbance remains constant (about 2 minutes). (Alternatively, leave it in bright sunlight for 15 minutes or more.) Then record another spectrum over the same wavelength range as before.

Observe and Note: Do you see any differences in the spectra? If so, describe them.

Estimate the percentage of all-*trans*-lycopene in your sample before treatment with iodine and its percentage in your (or your instructor's) sample after treatment. After consulting with the other members of your research team, decide whether the lycopene in your brand of tomato paste was significantly isomerized as a result of processing (read Exercise 1 before you do this). Compare your team's results with those of the other research teams and attempt to assess the relative quality of different brands of tomato paste.

Exercises

1 To estimate the percentage of all-*trans*-lycopene present in the original tomato paste, should you average the percentages obtained by all the members of your team? If not, what should you do, and why?

empirical formula
$C_{40}H_{56}$

2 (a) Calculate the concentration of the lycopene solution in the spectrophotometer cell before isomerization, given that the molar absorptivity of lycopene at 471 nm is 1.86×10^4. (b) Calculate the mass of lycopene in 3.0 mL of the solution having the concentration you calculated in (a).

crystals are poorly soluble lycopene

3 (a) Draw the structures of the 7-*cis*, 11-*cis*, and 13-*cis* isomers of lycopene. (b) Linus Pauling predicted that 13-*cis*-lycopene should be considerably more stable than the other two isomers. Explain.

4 Describe and explain the possible effect on your results of the following experimental errors or variations. (a) You used a can of tomato paste that had been left open in a refrigerator for several days. (b) You recorded the second spectrum immediately after adding the iodine solution. (c) You used acid-washed alumina for the chromatographic separation.

5 (a) Explain why some hydrocarbons such as lycopene and β-carotene are colored, whereas most other hydrocarbons are not. (b) The color of a lycopene solution fades and may disappear entirely if it is treated with a larger amount of iodine than you used in this experiment. Explain, giving an equation for a possible reaction.

6 (a) Write an equation for the reaction that occurred during the addition of iodine to all-*trans*-lycopene in this experiment. (b) Write a feasible mechanism for this reaction.

Other Things You Can Do

(Starred projects require your instructor's permission.)

*1 Concentrate some lycopene-containing eluate to a small volume by evaporation [OP-16] at room temperature, and cool it in ice water to obtain crystalline lycopene. Its melting point should be about 175°C.

Steam Distillation, Infrared Spectrometry
Isolation and Identification of the Major Constituent of Clove Oil

EXPERIMENT **10**

Isolation of Natural Products. Infrared Spectrometry.

Operations

OP-17 Steam Distillation
OP-36 Infrared Spectrometry
OP-4   Weighing
OP-15   Extraction
OP-16   Evaporation
OP-22   Drying Liquids

Before You Begin

1 Read the experiment and operations OP-17 and OP-36. Review the other operations as necessary.
2 Write an experimental plan following the directions in Appendix V.

Scenario

A professional aromatherapist, Rose Otto, uses the essential oil from cloves as a treatment for toothache, muscle pain, ringworm, flatulence, warts, and general exhaustion. But the latest batch of clove oil from her current supplier is darker than usual, has a harsh odor, and appears to be less effective than the oil she received previously. She suspects that the supplier has substituted some clove leaf oil for true clove oil, which is distilled from clove buds, the dried calyxes left after the flowers of the clove tree have fallen off.

Ms. Otto has asked you to provide her with an authentic sample of freshly distilled clove oil and tell her what's in it, so that she can compare the authentic clove oil with the product she has on hand. Your assignment is to isolate clove oil from ground cloves and identify its major constituent, which is known to have the molecular formula $C_{10}H_{12}O_2$. Your supervisor believes that you can identify the constituent by using infrared (IR) spectrometry.

clove bud

Applying Scientific Methodology

Reading the experiment carefully should yield a clue or two about the identity of the unknown, which is one of those illustrated in Figure 10.1. Then you should be able to develop a working hypothesis that will be tested when you obtain and interpret the infrared spectrum.

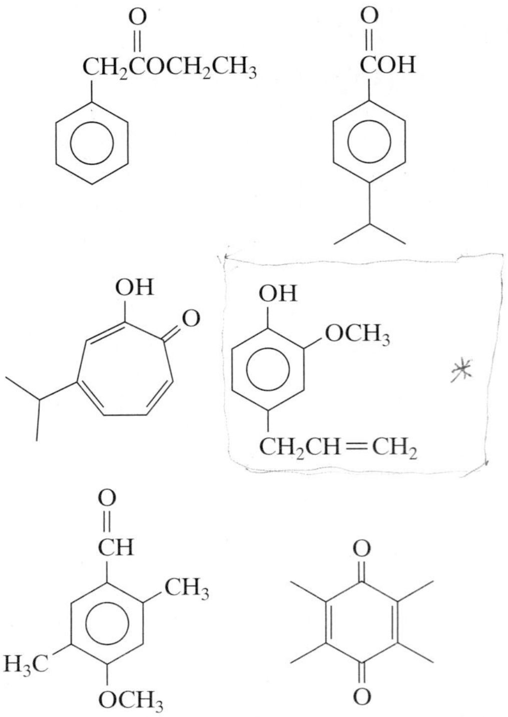

Figure 10.1 Compounds with the molecular formula $C_{10}H_{12}O_2$

Plants and Healing

As people seek alternatives to traditional medical practices, which emphasize the use of drugs and surgery to treat illness, various fields of alternative medicine are gaining adherents around the world. These include *aromatherapy*, the use of essential oils to maintain health and treat illness; *naturopathy*, a system of treating diseases by using special diets, herbs, vitamins, and other natural healing methods; and *homeopathy*, which originally relied on the use of minute doses of drugs to cure illness, but now utilizes carefully formulated mixtures of herbal medicines. Although some alternative medical practices may be associated with scientifically questionable theories, such as the idea (still accepted by some homeopathic practitioners) that the potency of a drug increases with dilution, many fields of alternative medicine utilize plant-based medicines with a long history of healing efficacy.

At a time when most physicians prescribe commercial drugs for medical conditions, we may tend to associate herbal medicine—the use of plants to treat and prevent illness—with witch doctors, shamans, or far-out medical cults. But herbal remedies have gained popularity in recent years as more and more people turn to echinacea, goldenseal, and even garlic to help them stay healthy and cope with illness. Europe is well ahead of the United States in conducting scientific research on herbal medicines. In fact, the popularity of six of the ten top-selling herbs in the United States has resulted mainly from European research. For example, a scientific team at the University of Dusseldorf, Germany, recently studied the active principals and biological effects of the purple coneflower, *Echinacea purpurea*, which is used to treat colds and flu by stimulating the immune system. One objective of their research was to improve the standardization of echinacea extracts, helping to ensure that each dose provides the same physiological activity.

After echinacea, the most widely used herbal remedy in America is garlic—one of the few remedies you are more likely to find in a grocery store than a drugstore. Although researchers disagree on the virtues of garlic, there is evidence that it lowers cholesterol and triglyceride levels in blood, helps prevent blood clots that could lead to heart attacks or strokes, and lowers blood pressure. The main active ingredient in garlic is a sulfur compound called allicin, which also gives garlic its distinctive and powerful aroma. Cooked garlic contains little if any allicin, so most of the alleged benefits of garlic are obtained only from the raw cloves or garlic capsules. Eating lots of raw garlic could limit your social life, but that may be a small price to pay for good health!

$$CH_2\!=\!CHCH_2S\overset{\displaystyle O}{\overset{\displaystyle \|}{-}}SCH_2CH\!=\!CH_2$$
allicin

Legend has it that Achilles, during the siege of Troy, used yarrow to treat the wounded Greeks. Its botanical name, *Achillea millefolium*, recognizes that tradition. The bruised leaves of yarrow help to stop bleeding, heal cuts, and relieve the pain of a wound, so the plant has been used in medical emergencies by backpackers and other outdoor adventurers. St. John's-wort (*Hypericum perforatum*) was supposedly used by the ancients to drive away evil spirits; today it is touted as a natural alternative to Prozac for treating mild to moderate depression. The indigenous North American weed boneset (*Eupatorium perfoliatum*) provides a bitter tea that was a favorite Native American remedy for fevers and other ailments. The closely related joe-pye weed (*Eupatorium purpureum*) was named after a Native American who gained fame by using it to cure typhus. It is an effective diuretic for the treatment of kidney and bladder ailments. Other popular herbal medicines include goldenseal root for treating peptic ulcers, infected gums, sore throats, and skin infections; saw palmetto berries for treating nonmalignant prostate disease; gingko leaf to improve blood flow in capillaries and arteries; aloe vera gel to heal burns, cuts, and wounds; ephedra stems to treat asthma and hay fever; and ginseng root to enhance one's general well-being and revitalize those weakened by old age or illness.

It should not surprise us that natural medicines can be effective. There are, after all, far more molecules in the world's natural life forms than have been synthesized in all the world's pharmaceutical laboratories. Many natural molecules are already known to have medicinal properties, and there must be at least as many more whose properties are yet to be discovered. Most of the drugs now prescribed by physicians were either derived from natural sources or developed by modifying the molecular structures of natural substances. For example, the heart stimulant digitalis is obtained from the foxglove plant, and the molecular structure of aspirin (acetylsalicylic acid) is similar to that of natural salicylates such as salicin from willow bark, which has been used for centuries by Native Americans to treat fevers.

The six-carbon substituent in salicin is a glucose unit.

salicin   aspirin

Although herbal medicines are generally milder and have fewer side effects than traditional prescription drugs, they are not all harmless. According to the U.S. Food and Drug Administration, herbal preparations containing *Ephedra sinica* (also known as Ma huang) can cause heart attacks, strokes, seizures, and even death if used improperly. Unlike most prescription drugs, different preparations containing the same herb may vary widely in potency and physiological effect. But when used responsibly by well-informed individuals, herbal medicine may provide a viable alternative to the use of traditional drugs for treating some medical conditions and maintaining good health.

Understanding the Experiment

The *essential oil* of a plant is a volatile mixture of water-insoluble components that exhibits the odor and other characteristics of the plant. In this experiment, you will isolate an essential oil from cloves, which are obtained from a small evergreen tree (*Syzygium aromaticum*) that grows in places such as Indonesia, Madagascar, and Zanzibar. The essential oil of cloves is a pale yellow liquid with a sweet, spicy aroma. Clove oil is unusual among essential oils in having only one major component, which makes up about 85% of the oil.

Because ground cloves lose their volatile components over time, it is best to grind fresh whole cloves just before using. Essential oils are usually isolated by steam distillation, in which steam forced through the plant material vaporizes the essential oil, which is then condensed into a receiver along with water from the condensed steam. For microscale work, it is not practical to use externally generated steam, so steam is generated internally by boiling water that contains the plant material. Steam distillation is preferable to ordinary distillation because the volatile components distill at temperatures below their normal boiling points, reducing or preventing decomposition due to overheating. During your steam distillation, the distillate should be cloudy or contain oily droplets at first and become clearer when most of the clove oil has distilled. Clove oil is separated from the distillate by extraction with dichloromethane. It is possible to separate the major component of clove oil from its minor components by extraction with aqueous sodium hydroxide, in which only the major component dissolves. This process is not very practical with microscale quantities because of material losses, but the nature of the process provides a clue that may help you solve the problem posed in the Scenario.

The major component of clove oil is a strong-smelling liquid that has the molecular formula $C_{10}H_{12}O_2$. The structures of some natural compounds that have this formula are shown in Figure 10.1. Because these compounds have different sets of functional groups, it is possible to distinguish them by using IR spectrometry. By detecting the presence or absence of IR absorption bands corresponding to specific functional groups, you should be able to arrive at the correct structure for the major component. The section "Interpretation of Infrared Spectra" in OP-36 describes the characteristic bands of organic compounds having various functional groups. Since you will not be carrying out the sodium hydroxide extraction described previously, your IR spectrum will contain a weak carbonyl ($C{=}O$) band belonging to one of the

Key Concept: Different covalent bonds vibrate at different frequencies, producing IR absorption bands at those frequencies. Thus a functional group containing a particular set of bonds produces a characteristic set of IR bands, from which it can often be identified.

minor components. This band, which provides a clue to the structure of the ~~minor component~~ (see Exercise 3), should be disregarded when you try to deduce the structure of the major component.

Directions

> **Dichloromethane may be harmful if ingested, inhaled, or absorbed through the skin. There is a possibility that prolonged inhalation of dichloromethane may cause cancer. Minimize contact with the liquid and do not breathe its vapors.**

Safety Notes

dichloromethane

If a suitable one is available, use a 20- or 25-mL round-bottom flask. Fill the flask half full with water at the start (after adding the cloves) and add no more water during the distillation.

Steam Distillation of Cloves.   If whole cloves are provided, use a spice mill to grind enough cloves to provide about 1 gram of ground cloves, grinding them into *coarse* particles to reduce frothing. Weigh [OP-4] 1.0 g of coarsely ground cloves into a 10-mL round-bottom flask. Assemble an apparatus for internal steam distillation [OP-17] using the round-bottom flask, a Hickman still, and a condenser. Do not add a stir bar or boiling chips. Calibrate a 15-mL screw-cap centrifuge tube at the 6-mL and 8-mL levels by adding the appropriate volumes of water and marking the position of each meniscus (discard the water). Add enough cold water to fill the boiling flask *no more* than half full and mark the water level with a marking pen. Begin heating the mixture to codistill the clove oil with water. During the distillation, use a 9-inch Pasteur pipet to add water through the condenser, keeping the water level relatively constant throughout the distillation; the flask should never be more than half full. Heat rapidly enough to maintain a good distillation rate, collecting about 1 mL every 5–10 minutes. If the liquid mixture threatens to froth up into the neck of the Hickman still, immediately raise the apparatus above the heat source and adjust the heating rate so that any foam stays within the flask. If liquid containing solid residue boils up into the well of the Hickman still, remove it with a long Pasteur pipet and return it to the boiling flask through the condenser. As the well fills with distillate, transfer the distillate to the calibrated centrifuge tube. Keep distilling until you have collected 6–8 mL of distillate.

Waste Disposal: Filter the clove residue through glass wool and place it in a solid-wastes container.

Isolation and Analysis of Clove Oil.   Use about 1 mL of dichloromethane to rinse out the well of your Hickman still and transfer it to the centrifuge tube. Add another 2 mL of dichloromethane, and shake the centrifuge tube to extract [OP-15] clove oil from the distillate, saving both layers. Extract the aqueous layer in the centrifuge tube with two separate 3-mL portions of dichloromethane and combine all of the dichloromethane extracts. Dry [OP-22] the dichloromethane layer with anhydrous sodium sulfate or magnesium sulfate. *Under the hood,* evaporate [OP-16] the dichloromethane under a stream of dry air or nitrogen while heating it in a warm water bath. Weigh the liquid residue, and then reweigh it after another minute or so of evaporation. If its mass decreases significantly between weighings, evaporation should be continued until the mass is essentially constant. Measure the final mass [OP-4] and obtain an infrared spectrum [OP-36] of your clove oil. Calculate the percent recovery of clove oil based on the mass of cloves you started with. Identify as many bands in the IR spectrum as you can, and

Take Care: Avoid contact with dichloromethane and do not breathe its vapors.

Waste Disposal: The aqueous layer can be poured down the drain.

If the solution is evaporated without heating, the clove oil may solidify.

deduce the identity of the major component of clove oil. Turn in your IR spectrum along with the product.

Exercises

1 Derive a systematic name for the active component of clove oil, and use this to find its common name in *The Merck Index* [Bibliography, A4] or another reference book.

2 (a) The active component of clove oil can be separated from the minor components by extraction with aqueous sodium hydroxide, followed by acidification of the aqueous extract with hydrochloric acid. What property of the active component makes this separation possible? Is this consistent with the structure you chose for it? Explain. (b) Write equations for the chemical reactions involved in the extraction and subsequent acidification of the extract.

3 (a) Clove oil contains about 10% of a minor component that can be hydrolyzed to yield the major component and acetic acid. Deduce the structure of the minor component, which has the molecular formula $C_{12}H_{14}O_3$. (b) The percentage of the major component of clove oil actually increases as the cloves are steam distilled. Explain why and give an equation for the reaction involved.

4 Describe and explain the possible effect on your results of the following experimental errors or variations. (a) You stopped the steam distillation after collecting 3 mL of distillate. (b) You forgot to wash the boiling flask, which contained a residue of sodium hydroxide left over from Experiment 4. (c) You didn't evaporate the dichloromethane long enough.

5 (a) Clove oil also contains a small amount of a substance whose systematic name is (*E*)-4,11,11-trimethyl-8-methylenebicyclo[7.2.0]undec-4-ene. Write the structure of this compound and find its common name in *The Merck Index* or another reference book.

6 Following the format in Appendix V, construct a flow diagram for this experiment.

Other Things You Can Do

(Starred projects require your instructor's permission.)

*1 Analyze your clove oil by gas chromatography, and estimate the percentage of the major component. If your instructor provides a sample of the major component, compare its IR spectrum with that of your clove oil.

*2 Obtain and analyze an essential oil from orange peel as described in Minilab 10.

*3 Steam distill the essential oils from anise seed, caraway seed, or cumin seed following the procedure in this experiment. Each of these essential oils contains a single major component that can be characterized by IR spectrometry.

4 Write a research paper about herbal medicine after referring to sources cited in the Bibliography.

Thin-Layer Chromatography, NMR Spectrometry
Identification of Unknown Ketones

Qualitative Analysis. Thin-Layer Chromatography. NMR Spectrometry. Ketones.

Operations

OP-19 Thin-Layer Chromatography
OP-37 Nuclear Magnetic Resonance Spectrometry
OP-4   Weighing
OP-5   Measuring Volume
OP-13   Vacuum Filtration
OP-23   Drying Solids
OP-25   Recrystallization
OP-30   Melting Point

Before You Begin

1 Read the experiment and operations OP-19, OP-25b, and OP-37. Review the other operations as necessary.
2 Prepare a brief experimental plan following the directions in Appendix V.

Scenario

A machine shop in your city was recently destroyed by fire under conditions that strongly suggest arson. A residue recovered at the fire's point of origin was found to contain traces of methyl ethyl ketone (MEK), one of several commercial degreasing solvents that shop employees use to clean the machinery. The primary suspect is a disgruntled ex-employee who was recently fired for sleeping on the job. While searching the suspect's garage, police discovered two unlabeled cans containing flammable liquids, which the suspect claims are charcoal starters for his grill. However, most charcoal-starting fluids are mixtures of petroleum hydrocarbons, and preliminary tests indicate that the liquids found in the suspect's possession are both ketones.

A local police detective, Spike Burns, has asked your Institute to help him solve the crime. Detective Burns provided your supervisor with samples of the two ketones. Matching one of them with the solvent recovered at the fire's point of origin will help the police discredit the suspect's claim and may result in his conviction. Your assignment is to identify the ketones and determine whether either one of them matches the solvent found at the scene of the crime. Your supervisor has requested that you use two different methods to identify your ketones.

Applying Scientific Methodology

You will, in effect, be performing two separate experiments. Your working hypothesis for the first experiment might be, for example, "The first unknown solvent is (or is not) MEK." You will then gather evidence to prove or disprove the hypothesis. In part **A**, the evidence used to test the hypothesis will be obtained by thin-layer chromatography (TLC) analysis and a melting-point determination. In part **B**, the only evidence will be provided by the nuclear magnetic resonance (NMR) spectrum of the second solvent, which should lead you to its structure. There is, of course, no guarantee that either of the two solvents will be MEK.

Crime and Chemistry

Forensic chemistry is chemistry applied to the solution of crimes. It deals with the analysis of materials that were used in committing a crime or that were inadvertently left at the crime scene. Materials used in committing a crime might include the ink on a forged document, an explosive used in a terrorist bombing, a toxic substance used in a fatal poisoning, or a flammable liquid used to start a fire. Such materials can be identified and sometimes traced to a particular source. Materials found at the scene of a crime might include paint chips, pieces of fiber from clothing, and particles of dust or soil, as well as any materials that were used in committing the crime. Chips of paint or glass found at the scene of a hit-and-run accident can be analyzed both chemically and under a microscope to determine the make and model of the car involved. Clothing can be traced by the dyes contained in fibers, and dust or soil particles may link a criminal to a particular occupation or location.

Bringing a suspect to trial requires that evidence be presented to establish, first, that a crime has actually been committed, and second, that the suspect is connected with the crime. In an arson case, for example, this requires proof that the fire was deliberately set as well as evidence implicating the suspect. One way of establishing that a fire was deliberately set is to prove that an *accelerant* (a flammable substance causing a fire to intensify and spread rapidly) was used to start and spread the fire. Because fires burn upward from the point of origin, some accelerant may soak downward into flooring, rags, paper, or other porous materials. When an investigator traces a fire to its point of origin, he or she can often collect samples of materials containing the accelerant, which are placed in airtight containers and sent to a forensic laboratory for analysis.

In the laboratory, a forensic chemist can separate an accelerant from debris collected at the crime scene by steam distillation or extraction. Once the accelerant has been isolated, it is usually classified according to chemical type (gasoline, turpentine, etc.) by an instrumental method such as gas chromatography or spectrometry. The forensic chemist may then try to match the accelerant sample with a control material, such as a liquid in the suspect's possession or a commercial material. Some flammable liquids, such as the industrial solvent MEK (whose IUPAC name is 2-butanone), contain only one major component and can be compared to the control material by using chemical or spectrometric methods. If the original accelerant was a mixture of different components, such as gasoline, its more volatile components will evaporate and burn more rapidly in a fire. For this reason, a sample of accelerant taken from the crime scene will probably not have the same composition as the original accelerant. Therefore, various control materials related to the accelerant are evaporated slowly and analyzed repeatedly by gas chromatography to determine whether the composition of a control material at any stage of evaporation matches that of the recovered material. With this procedure, it is often possible to determine the brand and grade of gasoline or other accelerant used.

Thin-layer chromatography is another important technique used by forensic chemists. TLC provides a rapid, sensitive means of analyzing many of the materials associated with various crimes. The dyes used to color gasoline can be characterized by the pattern of spots they produce on a TLC plate, making it possible in some cases to trace an arson accelerant to its source. The U.S. Treasury Department maintains a library of pen inks catalogued according to their TLC dye patterns, allowing an investigator to match the ink on a document with one on file. In a few cases, TLC analysis has proven that the ink used to fraudulently backdate a document did not even exist on the date in question! Substances suspected of being illicit drugs are frequently screened by TLC; most drugs that are mixtures of several substances, such as marijuana, produce telltale patterns that are easily recognized. A thin-layer chromatogram alone may not be sufficient to establish the identity of a suspect material, but it can narrow down the list of possibilities and thus lead to positive identification of the material by other means.

Understanding the Experiment

In this experiment, you will attempt to identify one of the unknown ketones by using TLC and the other by using NMR spectrometry. TLC can be used in the identification of pure compounds, as well as complex mixtures such as drugs and dyes. When a TLC plate (see OP-19) spotted with structurally similar organic compounds is developed with an appropriate solvent, the R_f values obtained vary more or less regularly with chain length. This correlation is illustrated in Table 11.1 for a homologous series of carboxylic acids. When an unknown compound is known to be one of a limited number of compounds, a TLC plate can be spotted with samples of the unknown and the most likely known compounds. When the plate is developed, it may be possible to match the R_f value of the unknown compound's spot with that of a known compound.

Your first unknown will be a member of a homologous series of methyl ketones represented by the formula $CH_3CO(CH_2)_nCH_3$, where n is zero or a whole number. A ketone is often converted to its 2,4-dinitrophenylhydrazone

Key Concept: The R_f value of a substance depends on its relative affinities for the adsorbent on the TLC plate and the developing solvent. A substance with a higher affinity for the adsorbent spends most of its time stuck to the adsorbent and has a low R_f value. A substance with a higher affinity for the solvent spends most of its time dissolved in the solvent and has a high R_f value.

Table 11.1 TLC R_f values for carboxylic acids

Carboxylic acid	# Carbon atoms	R_f
methanoic (formic) acid	1	0.07
ethanoic (acetic) acid	2	0.13
propanoic acid	3	0.30
butanoic acid	4	0.40
pentanoic acid	5	0.50
hexanoic acid	6	0.57
heptanoic acid	7	0.60
octanoic acid	8	0.66

Note: Developed on silica gel, with a 19:1 mixture of methyl acetate and 2.5% ammonia.

or another colored derivative for TLC analysis because the colored spots are easily located after the plate is developed, making a visualizing reagent unnecessary. The 2,4-dinitrophenylhydrazone derivative of a ketone is prepared by combining the ketone with DNPH reagent, which contains 2,4-dinitrophenylhydrazine and sulfuric acid in aqueous ethanol. Traces of sulfuric acid may remain on the derivative and catalyze isomerization reactions that will lower its melting point, so the acid is removed by washing the derivative with aqueous sodium bicarbonate after vacuum filtration. The derivative is then purified by recrystallization. Some 2,4-dinitrophenylhydrazones dissolve too readily in ethanol and too sparingly in water for either of these liquids to be a good recrystallization solvent by itself. Thus you may have to use a mixture of the two, as described in OP-25b, "Recrystallization from Mixed Solvents." To obtain a solvent mixture of the right composition, you should first dissolve the derivative in the better solvent (the one in which it is most soluble), ethanol, and then add just enough of the poorer solvent, water, to saturate the hot solution. After measuring the melting point of your purified derivative and comparing its R_f value with those of known derivatives, you should be able to identify your unknown as one of the ketones in Table 11.2.

Modern analytical instruments, such as infrared (IR) and NMR spectrometers, can greatly reduce the time and effort required for the positive identification of an unknown. In part **B** of this experiment, you will attempt to identify an unknown saturated ketone that has four to six carbon atoms from its proton NMR (^{1}H NMR) spectrum. If an NMR spectrometer (or simulator) is available, you can record the spectrum yourself, with help from your instructor. Otherwise, your instructor will provide an NMR spectrum of your unknown. The section "Interpretation of ^{1}H NMR Spectra" in OP-37 contains an introduction to NMR spectral analysis that will help you deduce the structure of the second unknown.

Reactions and Properties

ketone   2,4-dinitrophenylhydrazine   2,4-dinitrophenylhydrazone

Table 11.2 Physical properties and derivative melting points for homologous methyl ketones

Ketone	M.W.	bp	d	Derivative mp
2-propanone (acetone)	58.1	56	0.791	126
2-butanone	72.1	80	0.805	117
2-pentanone	86.1	102	0.809	143
2-hexanone	100.2	128	0.811	106
2-heptanone	114.2	151	0.811	89
2-octanone	128.2	173	0.819	58

Note: Boiling points are in °C, densities in g/mL.

Directions

With the instructor's permission, teams of two to three students can work together on some parts of this experiment, such as the preparation of derivatives of the known ketones.

> **The ketones are flammable and may be harmful if inhaled or absorbed through the skin. Avoid contact, do not breathe their vapors, and keep them away from flames.**
> **2,4-Dinitrophenylhydrazine is harmful if absorbed through the skin and will dye your hands yellow. Wear gloves and avoid contact with the DNPH reagent.**
> **Ethyl acetate and the developing solvent are flammable and may be harmful if inhaled or absorbed through the skin. Avoid contact, do not breathe their vapors, and keep them away from flames.**
> **Deuterochloroform is toxic and may be carcinogenic; avoid contact and inhalation.**

Safety Notes

ketones

ethyl acetate   TLC solvent

A. *Identification of an Unknown Methyl Ketone Using TLC*

Preliminary Work

Obtain an unknown methyl ketone from your instructor and record its identification number in your laboratory notebook. Preequilibrate the TLC developing solvent, a 3:1 mixture of toluene and petroleum ether, by filling an appropriate developing chamber containing a paper wick to a depth of about 5 mm with the solvent, covering it with a lid or plastic wrap, and sloshing the solvent up the sides of the chamber to moisten the wick (see OP-19). Then set it aside under a hood for later use.

Take Care! Avoid contact with the developing solvent, do not breathe its vapors, and keep it away from flames or hot surfaces.

Preparation of 2,4-Dinitrophenylhydrazones.
Measure [OP-5] 0.10 mL of your unknown into a 13×100 mL test tube. Dissolve it in 1.5 mL of 95% ethanol and stir in 3.5 mL of the DNPH reagent, then set the test tube aside for 15 minutes. Collect the derivative by vacuum filtration [OP-13]. After you have removed the acidic filtrate, wash the derivative on the filter with 2 mL of cold 5% sodium bicarbonate, then with cold water. Let it air dry on the filter for at least 5 minutes. Recrystallize [OP-25] the derivative from 95% ethanol or an ethanol/water mixed solvent. Collect it by vacuum filtration, and wash it on the filter with a cold 3:1 mixture of ethanol and water. Dry [OP-23] the derivative, weigh it [OP-4], and measure its melting point [OP-30]. Save enough of the derivative for the TLC separation and turn in the rest in a labeled vial.

Take Care! Do not pipet by mouth.

Take Care! Wear gloves and avoid contact with the reagent.

Waste Disposal: Place the initial filtrate in a designated DNPH waste container, or dispose of it as directed by your instructor. Pour all other filtrates down the drain.

Clean and label as many small test tubes as there are known methyl ketones available (see Table 11.2). Measure 1 mL of the DNPH reagent into each test tube and add 1 drop of the appropriate methyl ketone. Set the test tubes aside until crystallization is complete, then collect the crystalline derivatives by vacuum filtration [OP-13].

Waste Disposal: Place the filtrate in a designated DNPH waste container, or dispose of it as directed by your instructor.

TLC Separation of 2,4-Dinitrophenylhydrazones.
Read OP-19 carefully before you begin. Dissolve approximately 10 mg (0.01 g) of the unknown ketone's derivative in 0.5 mL of ethyl acetate, using a spot plate or a small,

Waste Disposal: Put the developing solvent in a hydrocarbon solvent recovery container.

labeled test tube. Do the same for each known derivative. Use the solutions to spot a silica gel TLC plate [OP-19], taking care not to touch the surface of the adsorbent layer, and label the spots with a pencil. To avoid cross-contamination, use a different micropipet (or other spotting device) for each solution. Develop the plate *under the hood* in the developing chamber you prepared. Do not move or otherwise disturb the developing chamber until the plate is developed. Mark the solvent front with a pencil before the plate dries. Measure the distances of the spots and solvent front from the starting line and calculate all R_f values. Deduce the name and structure of the unknown ketone from your results. Turn in your TLC plate, as well as the derivative of your unknown.

B. *Identification of an Unknown Ketone by NMR Spectrometry*

Take Care! Avoid contact with $CDCl_3$ and do not breathe its vapors.

Waste Disposal: Put the deuterochloroform solution in a designated solvent recovery container.

Obtain a second unknown ketone (or its ^{1}H NMR spectrum), and record its identification number in your laboratory notebook. If an NMR spectrometer is available for your use, prepare a solution of the unknown in deuterochloroform ($CDCl_3$) using a tetramethylsilane (TMS) standard, and record and integrate its proton NMR spectrum [OP-37]. Otherwise, use an NMR simulation program to generate a simulated spectrum or obtain an NMR spectrum from your instructor. Deduce the structure of the second unknown ketone from its spectrum and name it. Turn in your NMR spectrum.

Exercises

1 (a) Write a balanced equation for the reaction of your unknown methyl ketone with 2,4-dinitrophenylhydrazine. (b) The DNPH reagent contains 2.9 g of 2,4-dinitrophenylhydrazine in 100 mL of solution. What was the limiting reactant for the preparation of your 2,4-dinitrophenylhydrazone? Calculate the theoretical yield and percent yield of the reaction.

2 Describe and explain any relationship between chain length and R_f value that you observed from your TLC separation.

3 (a) The purpose of the sodium bicarbonate washing was to prevent isomerization of the derivative during the melting-point determination. Write structures for two stereoisomers of the 2,4-dinitrophenylhydrazone derivative of 2-pentanone. (b) Which one would you expect to be more stable, and why?

4 Describe and explain the possible effect on your results of the following experimental errors or variations. (a) You didn't wash the derivative of your unknown with aqueous sodium bicarbonate. (b) You let the TLC plate develop too long and you can't find the solvent front. (c) You left the TMS out of your ^{1}H NMR sample.

5 Draw structures for all ketones that have the molecular formula $C_5H_{10}O$, and sketch the ^{1}H NMR spectrum you would expect to obtain from each one. Your sketches should show the relative area, multiplicity, and approximate chemical shift of each signal.

6 Account for the fact that the 2,4-dinitrophenylhydrazone derivative of 2-octanone melts at a lower temperature than the corresponding derivative of acetone, even though the molecular weight of the 2-octanone derivative is much higher.

Other Things You Can Do

(Starred items require your instructor's permission.)

*1 Use TLC to analyze some felt-tip-pen inks as described in Minilab 11.

*2 Use NMR spectrometry to identify an alkyl chloride that has the formula $C_4H_{10}Cl$, or another compound whose molecular formula will be provided by your instructor.

3 Write a research paper about forensic chemistry, starting with sources cited in the Bibliography.

Vacuum Distillation, Optical Rotation
Optical Activity of α-Pinene

Separation of Liquids. Optical Activity. Stereoisomerism. Terpenes.

Operations

OP-28 Vacuum Distillation
OP-33 Optical Rotation
OP-4   Weighing
OP-10 Mixing
OP-31 Boiling Point

Before You Begin

1 Read the experiment and operations OP-28 and OP-33.
2 Prepare a brief experimental plan following the directions in Appendix V.

Scenario

Dick Hawkshaw, private investigator, has called on you to help him solve a mystery. A wealthy American entrepreneur, Aldo Hyde, was recently found murdered in his chalet in Cannes, France. Although the house was set on fire in an apparent attempt to cover up the murder, firefighters were able to extinguish it before the evidence was destroyed. In the house a half-filled can of paint thinner was found; it had no label, but the word TURPENTINE was written on one side with a felt-tip pen. The housekeeper informed Hawkshaw that there was no turpentine in the house before the night of the fire, so the murderer must have brought it there. Gas-chromatographic analysis of a flammable residue found at the fire's point of origin revealed that liquid from this can was used to start the blaze.

The only suspects in the case are Aldo Hyde's widow, Dr. Jacqueline Hyde, a talented but unpredictable talk-show psychologist; and Guy Framboise, a hot-tempered French businessman. Aldo and Jacqueline Hyde had lived separately for more than a year, she in New York and he in Cannes. He was planning to change his will, leaving her out of it, when his untimely death intervened, so Jacqueline stands to inherit most of his fortune. Dr. Hyde arrived in Cannes from New York on the night of the murder and checked into a hotel at 10 p.m., approximately two hours before the murder took place, but there are no witnesses who can place her at the scene of the crime. Guy Framboise made threats against Aldo's life after a joint business operation failed, and a reliable witness saw the Frenchman's Peugeot parked near Hyde's chalet on the night of the murder.

Dick Hawkshaw just shipped your supervisor a sample of the "turpentine" used in the crime. Now it is up to you to discover the crucial evidence that will identify the murderer of Aldo Hyde.

Applying Scientific Methodology

The problem, of course, is "Who murdered Aldo Hyde?" As you read the experiment, you should find some clues that will help you solve the mystery—after you have gathered the experimental evidence.

Turpentine and the Terpenes

Terpenes are among the most widely distributed natural products, occurring in nearly all plants. Terpenes are compounds that can, in principle, be broken down into two or more isopentane (2-methylbutane) units. For example, the carbon skeleton of geraniol can be separated in the middle to yield two isopentane units connected head to tail; that is, the "head" end of one unit—the end nearest the side chain—is connected to the "tail" end of the next.

These isopentane units are also called isoprene units, after a diene having the same carbon skeleton.

Terpenes having oxygen-containing functional groups are sometimes called terpenoids.

tail head

$$CH_3CH=CHCH_2-CH_2CH=CHCH_2OH$$
$$\underset{CH_3}{|}\qquad\qquad\underset{CH_3}{|}$$

geraniol

$$\underset{\underset{C}{|}}{CCCC}-\underset{\underset{C}{|}}{CCCC}\Longrightarrow \underset{\underset{C}{|}}{CCCC}+\underset{\underset{C}{|}}{CCCC}$$

carbon skeleton   isopentane units
of geraniol

Geraniol, with its rose-blossom aroma, is an important constituent of the essential oil known as rose otto. It also plays an important role in the biosynthesis of terpenes. This process involves the enzymatic isomerization of isopentenyl pyrophosphate and a subsequent reaction with its isomer to yield geranyl pyrophosphate, from which the other terpenes are produced.

geranyl pyrophosphate

isopentenyl pyrophosphate

$$-OPP = -\underset{\underset{HO}{|}}{\overset{\overset{O}{||}}{O}}P\underset{\underset{OH}{|}}{\overset{\overset{O}{||}}{O}}POH$$

Many terpenes, especially the ones that contain oxygen, have pleasant odors and flavors and are therefore important flavoring and perfume ingredients.

An important natural source of terpenes is *turpentine*, the sticky oleoresin ("pitch") obtained from conifer trees such as the Southern longleaf pine, *Pinus palustris*. The "turpentine" sold as a paint thinner, more accurately called oil of turpentine, is distilled from this oleoresin. It is the world's most abundant essential oil, being obtained as a by-product of paper production as well as from pine pitch. The major components of American oil of turpentine are (+)-α-pinene and (−)-β-pinene, with the former predominating.

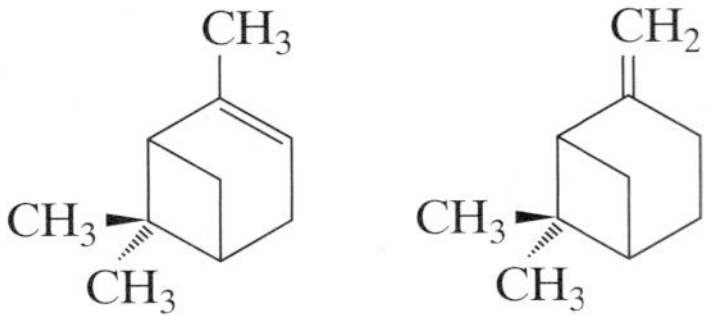

(+)-α-pinene   (−)-β-pinene

Constituents of American oil of turpentine

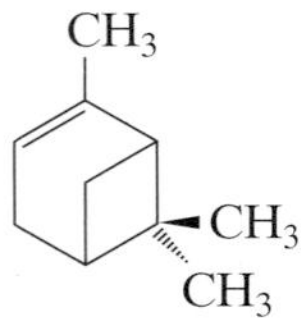

(−)-α-pinene

European oil of turpentine contains mostly the enantiomeric (−)-α-pinene and very little β-pinene. Both α- and β-pinene are used to synthesize other terpenes and their derivatives, which are used in perfumes, flavorings, and other commercial products. For example, α-pinene is a starting material in the synthesis of isobornyl acetate, which contributes a "pine needle" note to perfumes; linalool, which has a sweet woody-floral odor; and camphor, which is used in the manufacture of cosmetics, plastics, and pharmaceuticals.

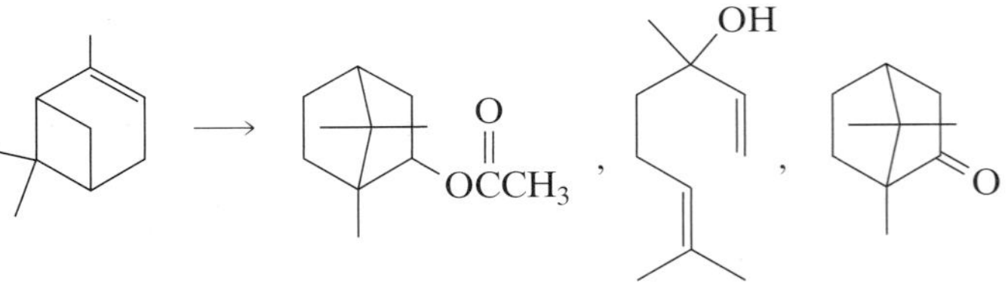

α-pinene   isobornyl acetate   linalool   camphor

Large quantities of β-pinene are converted to β-myrcene, which is then used to make perfume ingredients that have floral notes, including linalool and the aldehyde shown here.

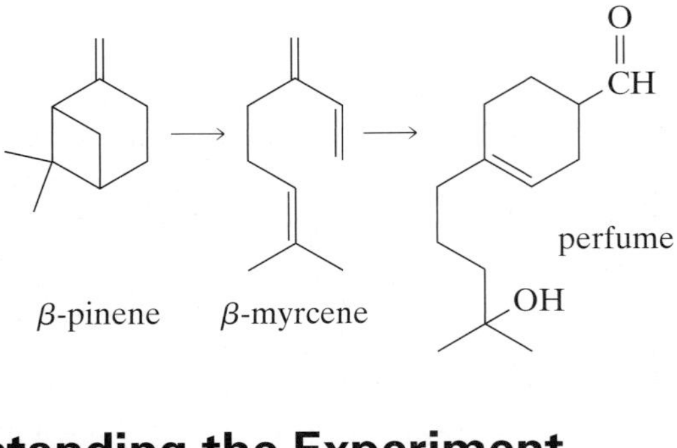

perfume aldehyde

β-pinene   β-myrcene

Understanding the Experiment

Doyle's Dr. Watson claimed that Sherlock Holmes had a profound knowledge of chemistry.

The application of chemistry to crime solving is not only the province of the forensic chemist. Mystery novels by writers from Arthur Conan Doyle to Dorothy L. Sayers contain many references to forensic chemistry. In Sayers' novel *The Documents in the Case*, a mushroom collector dies after eating a stew containing mushrooms that he has picked himself. The death is believed to be accidental, caused by the poisonous mushroom *Amanita muscaria*, until a chemist discovers that the stew contains optically inactive (and therefore synthetic) muscarine, rather than the optically active muscarine that occurs in the poisonous mushroom. Sayers was actually a step ahead of the chemists of her day (the novel was published in 1930), who—based on an incorrect achiral structure for muscarine—assumed that its natural form was optically inactive. In 1957 an X-ray crystallographer proved that Sayers had guessed right—muscarine is indeed chiral, so the premise of her novel was sound.

In this experiment, you—like the chemist in Sayers' novel—will be measuring the optical rotation of a substance to help solve a mystery. First you must obtain α-pinene in a reasonably pure form. You will purify the "turpentine" sample you are issued by distilling it, performing the distillation under vacuum to reduce the likelihood of decomposition. α-Pinene distills at about 156°C at normal atmospheric pressure, but its boiling point is reduced at lower pressure. For example, it boils at 52°C at a pressure of 20 torr. Any operation carried out under vacuum carries with it some risk of an implosion, which can cause injury from flying glass fragments, so it is important to inspect all parts of your apparatus carefully to be certain that they are free from cracks or other defects, and to have it checked by your instructor.

According to *The Merck Index*, (+)-α-pinene and (−)-α-pinene have specific rotations of +51° and −51°, respectively, at 20°C. Because your distilled α-pinene may still contain some impurities, its specific rotation may be somewhat lower (see Exercise 1).

Key Concept: As a rule, chiral compounds from natural sources are optically active. Nearly all synthetic chiral compounds are racemic mixtures and therefore optically inactive.

$$CH_3\text{-----}O\text{-----}CH_2\overset{+}{N}(CH_3)_3$$
HO
(+)-muscarine

Properties

Table 12.1 Physical properties

	M.W.	bp^{760}	bp^{20}	$[\alpha]_D$
(+)-α-pinene	136.2	156	52	+51°
(−)-α-pinene	136.2	156	52	−51°
(−)-β-pinene	136.2	163		−22°

Note: Boiling points are in °C, at 760 torr and 20 torr, respectively.

Directions

α-Pinene irritates the skin and eyes and its vapors are harmful. Avoid contact and do not breathe its vapors.
A vacuum distillation apparatus may implode if any of its components are cracked or otherwise damaged. Inspect the parts for damage and have your instructor check your apparatus. Work behind a hood sash or safety shield while the apparatus is under vacuum.

Safety Notes

α-pinene

Purification of α-Pinene.   If you are using an aspirator and do not have a manometer, measure the temperature of the aspirator water after the aspirator has run for a while and estimate the minimum pressure the aspirator can attain. Otherwise, determine the pressure as directed by your instructor. Use the nomograph in Figure E12 of OP-28 to estimate the boiling

Stop and Think: Why should the water temperature affect the pressure an aspirator can attain?

If the pressure changes during the distillation, the boiling temperature may vary.

Stop and Think: Was the initial boiling point different than you had estimated? If so, why?

Stop and Think: Which α-pinene enantiomer do you have? How do you know?

point of α-pinene at that pressure (the actual boiling temperature may be somewhat higher). After inspecting the glassware carefully, assemble an apparatus for vacuum distillation [OP-28] using a 10-mL round-bottom flask, and have it approved by your instructor. Add 2.5 mL of "turpentine" and a few microporous boiling chips or a magnetic stir bar [OP-10]. Make sure that a safety shield or hood sash is between you and the apparatus. Start the stirrer, turn on the vacuum, and adjust the heat until distillation begins, collecting the liquid that distills near or somewhat above the estimated boiling point. Whenever the well fills with distillate, release the vacuum and transfer the liquid to a tared vial, then turn on the vacuum and continue distilling. After about three-quarters of the pinene has distilled, monitor the temperature closely. When it rises or drops by 5°C or more, stop the distillation, release the vacuum, and transfer the rest of the distillate to the tared vial.

Analysis.   Measure the boiling point [OP-31] of your product by a capillary-tube method. Using a 10-mL volumetric flask, prepare a solution containing about 1.0 g of your α-pinene, weighed [OP-4] to the maximum accuracy of your balance, in absolute ethanol. (If you don't have a gram of purified α-pinene, use as much as you can, but save a small amount to turn in to your instructor.) Transfer the solution to a 1-dm polarimeter cell and measure its optical rotation [OP-33]. Then measure the optical rotation of a blank consisting of pure absolute ethanol. Calculate the specific rotation of your α-pinene and give its complete name and structure. Then decide who murdered Aldo Hyde.

Exercises

1 From its specific rotation, estimate the purity of your α-pinene, as the mass percent of the predominant α-pinene enantiomer: (a) if the impurity is (−)-β-pinene; (b) if the impurity is optically inactive.

2 The Aldrich Chemical Company manufactures a technical (low-purity) grade of pinene containing 85% (+)-α-pinene and having a specific rotation of +43°. Is the 15% (by mass) impurity in this product more likely to be (−)-α-pinene, (−)-β-pinene, or some optically inactive substance? Justify your answer with calculations.

3 Describe and explain the possible effect on your results of the following experimental errors or variations: (a) Your vacuum distillation apparatus had a leaky joint. (b) The "turpentine" contained synthetic α-pinene. (c) You prepared enough α-pinene solution to fill a 2-dm polarimeter tube, which you used for the optical rotation measurement.

4 (a) Show how α-pinene can be divided into isopentane units. (b) Do the same for linalool and camphor.

5 (a) Determine the configuration, (R) or (S), at each stereocenter of (+)-muscarine. (b) Determine the configuration at each stereocenter of (−)-β-pinene.

6 The pyrophosphate group is an excellent leaving group that can easily be lost to yield a carbocation. (a) Propose a mechanism for the synthesis of geraniol from isopentenyl pyrophosphate and its isomer (see "Turpentine

and the Terpenes") in an acidic aqueous environment. (b) Propose a mechanism for the synthesis of α-pinene from geranyl pyrophosphate under similar conditions.

7   Propose a synthesis from β-myrcene of the perfume aldehyde whose structure is shown in the section "Turpentine and the Terpenes."

Other Things You Can Do

(Starred projects require your instructor's permission.)

*1   Determine the percentage composition of turpentine by measuring its optical rotation, as described in Minilab 12.

*2   Purify technical grade (85–90%) α-terpineol, whose normal boiling point is 220°C, by vacuum distillation. Cool the purified product in ice, if necessary, to see if it will solidify (pure α-terpineol is said to solidify around 30°C).

3   Write a research paper about terpenes, starting with sources cited in the Bibliography.

Correlated Laboratory Experiments

The directions in Part I were quite detailed, to help you carry out the operations and complete the experiments successfully. Now that you have learned how to use the major laboratory operations of organic chemistry, you should be able to apply them proficiently to the experiments in Part II, which provide somewhat less detailed directions. It will be increasingly important to *plan ahead* and decide exactly how you will carry out each operation in advance. Continue to refer to the operation descriptions as necessary, particularly when an operation is to be carried out by a different method or in a different context than before. Note that frequently used elementary operations, such as weighing and measuring volume, will no longer be listed at the beginning of an experiment or flagged in the directions.

The following experiments are correlated with topics discussed in most introductory textbooks of organic chemistry. Correlations are indicated by the list of topics beginning each experiment.

PART II

Investigation of a Chemical Bond by Infrared Spectrometry

EXPERIMENT 13

Resonance. Infrared Spectrometry. Molecular Mechanics. Carbonyl Compounds. Acyl Compounds.

Operation

OP-36   Infrared Spectrometry

Before You Begin

1   Read the experiment, read or review the operations as necessary, and write an experimental plan.
2   Be prepared to predict the *relative* $C=O$ vibrational frequencies (not their actual numerical values) of compounds your instructor will assign in the lab.

Scenario

The Olfactory Factory manufactures perfumes by mixing essential oils, resins, and other ingredients as specified by secret fragrance formulas that have been kept in the Pomander family for generations. The company's quality-control manager, Dr. Hyacinth Pomander, is responsible for measuring the relative amounts of certain key aroma chemicals in the ingredients to ensure that they remain within the tolerances specified by the formulas. Most of these chemicals are aldehydes, ketones, esters, and other compounds that contain a carbonyl $(C=O)$ group. The Olfactory Factory recently purchased a number of infrared (IR) analyzers to quantify the key chemicals by measuring the intensity of the carbonyl band in each ingredient's infrared spectrum. Unfortunately, this method has not been working well because of interference by other chemicals whose molecules also possess carbonyl groups. Dr. Pomander believes that by tuning the IR analyzers to the vibrational frequency of the carbonyl group in a key substance of each perfume ingredient, she should be able to minimize such interferences, but first she needs to know how to predict their vibrational frequencies. She has sent samples of several representative carbonyl and acyl compounds to your supervisor. Your assignment is to find a way to estimate in advance their $C=O$ vibrational frequencies and then to check your predictions by measuring the vibrational frequency of each compound's carbonyl band as accurately as possible.

Applying Scientific Methodology

The basic problem in this experiment is to see if there is a way to estimate the vibrational frequencies of $C=O$ bonds in molecules that contain such bonds. After reading the experiment, you should be able to develop a hypothesis, apply it to the representative compounds assigned by your instructor, and then determine whether or not your predictions are verified by the experimental results.

The Art and Science of Perfumery

Throughout history, both men and women have used substances with pleasant odors to attract attention, to cover up unpleasant odors, and simply for the pleasure they provide. Elizabeth I used large quantities of lavender oil, in part to mask the stench of Elizabethan England, where streets doubled as sewers and bathing was considered unhealthful. The ancient Egyptians had no such aversion to personal hygiene; they used perfumes to scent their baths and sweeten their breath. Egyptian tomb paintings depict feasts at which revelers piled lumps of scented animal fat on their heads; as the evening progressed, the fat melted and coated their bodies with fragrant perfumes. A fat-based perfume found in an alabaster vase in the tomb of King Tutankhamen was still fragrant 3,000 years later, and similar vases were found in the tombs of First Dynasty kings who reigned from about 5100 B.C. The aromatic resins frankincense and myrrh were among the gifts of the Magi in the biblical account of the birth of Jesus. The Queen of Sheba reputedly used myrrh to beguile King Solomon, and frankincense is still burned during rites of the Roman Catholic and Greek Orthodox churches.

Traditional perfume ingredients include natural resins, essential oils, concretes, and absolutes, which are obtained from plant or animal sources by such processes as distillation, solvent extraction, and expression (pressing). *Resins*, such as frankincense, are solid or semisolid materials exuded by plants. *Essential oils*, such as rose otto (also called attar of roses), are volatile liquids that are usually obtained by mixing finely divided plant materials with water and then distilling the mixture with steam. *Concretes* are waxy residues obtained by extracting plant components with a volatile solvent and then evaporating the solvent. *Absolutes* are concretes that have been processed to remove insoluble materials. Jasmine, one of the most highly prized perfume ingredients, is made by extracting jasmine flowers with hydrocarbons, then treating the resulting concrete with alcohol to produce the absolute.

Because natural materials such as jasmine are often costly and may be subject to considerable variation in composition and availability, most modern perfumes contain synthetic aroma chemicals as well as natural ingredients. Aroma chemicals include synthetic versions of natural compounds and even some compounds that have never existed in nature. Most aroma chemicals have an oxygen-containing functional group; aldehydes, ketones, alcohols, esters, and ethers are well represented in perfumes. A few hydrocarbons are also used, and smelly sulfur and nitrogen compounds may be added in very small amounts. For example, the nitrogen-containing compound indole has a repulsive fecal odor when pure, but in minute amounts it imparts a fine jasmine scent to perfumes. Some representative aroma chemicals and their odors are illustrated in Figure 13.1.

A perfumer creates a new perfume in much the same way that a composer creates a symphony. Inspired by perhaps a single aroma "note," such as a recently discovered natural product or a new synthetic aroma chemical,

isocyclocitral (carnation)   muscone (musk)   phenylethyl alcohol (rose)

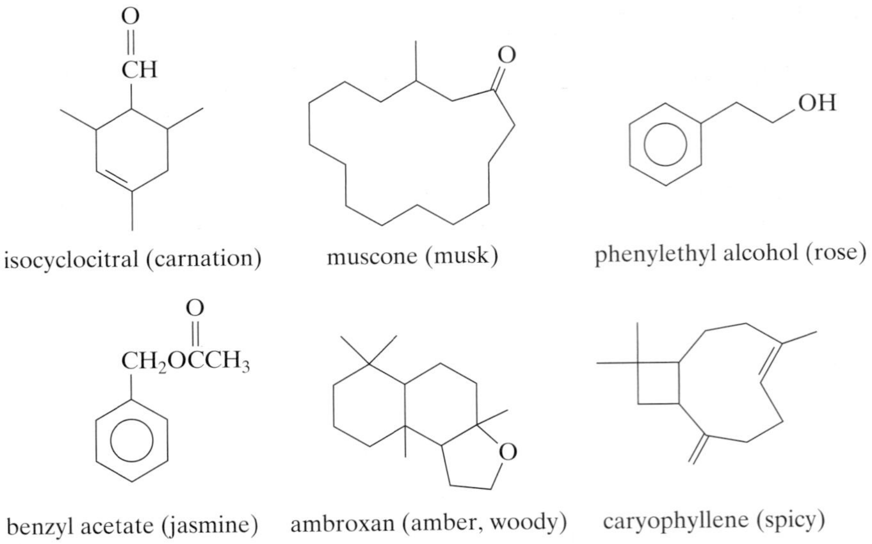

benzyl acetate (jasmine)   ambroxan (amber, woody)   caryophyllene (spicy)

Figure 13.1 Some aroma chemicals and their aroma notes

the perfumer creates a "theme," called an *accord*, which may consist of several related aroma chemicals. For example, *Chanel No. 5* is based on an accord that consists of a blend of aliphatic aldehydes. Additional ingredients with different volatilities and aroma intensities are combined with the accord to produce a "composition" consisting of top, middle, and end notes. The top note contains the most volatile and odoriferous components, whose odors predominate immediately after a perfume is applied to the skin, but fade as they evaporate. A fruity top note, for example, might consist of low-molecular-weight esters and some lactones (cyclic esters). The middle note provides the basic character of a fragrance. Mixtures of floral essential oils such as neroli oil with flowery aroma chemicals are often used as middle notes. The end note contains less volatile materials that persist longer on the skin, such as resins and certain oils. Musky aroma chemicals and woody materials such as sandalwood are among the ingredients of end notes. All perfume ingredients must be carefully selected and combined in just the right proportions to give the perfume a smooth odor profile, so that there are no abrupt changes in odor as the ingredients evaporate. For a typical perfume or cologne, the ingredients are then dissolved in ethyl alcohol or some other alcohol-containing solvent.

There is some scientific evidence that perfumes do more than just make people smell better. Pleasant odors may improve your mood, bring to mind happy memories, enhance your creativity, or make you feel more cooperative and less confrontational toward the people you work with. On the other hand, a perfume that smells good to one person may be objectionable to someone else, and not everyone wants to be subjected to olfactory assaults from such flagrantly fragrant consumer products as scented toilet paper and fabric softener. But an appreciation of pleasant and stimulating fragrances is rooted deeply in the human psyche, and any drawbacks of commercial scents may be outweighed by the benefits they provide.

Understanding the Experiment

In this experiment, your instructor will provide you (or your research team) with three or more compounds that contain a carbonyl group. Based on the discussion to follow, you will try to arrange the compounds in order of their C=O vibrational frequencies. At your instructor's option, you can also use a molecular mechanics computer program to estimate specific values for the frequencies. Then you will record the IR spectra of your compounds to determine their actual vibrational frequencies and see how accurate your predictions were.

A chemical bond is similar in some ways to a coiled spring. Just as it takes energy to stretch a spring, energy is required to stretch a chemical bond. The stronger the bond, the more energy is required. A bond stretches and contracts as it vibrates, and the frequency at which the bond vibrates (ν) is proportional to the energy associated with the vibration, as given by the equation $\Delta E = h\nu$. Thus the stronger the bond, the higher is its vibrational frequency. The C—O single bond in ethanol (CH_3CH_2OH) vibrates about 31 trillion (3.1×10^{13}) times a second, while the stronger C=O double bond in ethanal ($CH_3CH=O$) vibrates about 52 trillion times a second. A bond vibrating at a certain frequency can absorb a photon of IR radiation having exactly the same frequency. For example, a C=O bond vibrating 5.2×10^{13} times a second can absorb a photon whose frequency is 5.2×10^{13} Hz, where $1 \text{ Hz} = 1 \text{ s}^{-1}$.

It is convenient to express the vibrational frequency of infrared radiation in wave numbers ($\bar{\nu}$), where the wave number of a vibration expressed in cm^{-1} is the number of waves of electromagnetic radiation per centimeter. The wave number corresponding to the vibrational frequency of $5.2 \times 10^{13} \text{ s}^{-1}$ is about 1740 cm^{-1}, so the infrared spectrum of ethanal contains an absorption band at 1740 cm^{-1} that is generated when its carbonyl bonds absorb infrared radiation.

You can convert wave number to frequency by multiplying by the speed of light, expressed as 3.00×10^{10} cm/s.

To determine the relative vibrational frequencies for C=O bonds in different compounds, you can estimate their relative strengths using resonance theory. For example, suppose the carbonyl carbon in a compound is attached to some atom or group Z, giving the compound the general formula RCOZ. Such a compound will have at least two resonance structures. We write resonance structure **B** by moving a pair of pi electrons onto the oxygen atom.

Only those lone pairs of electrons that are involved in resonance are shown in these and subsequent resonance structures.

$$
\begin{array}{ccc}
\underset{A}{\text{R—C—Z}} & \underset{B}{\text{R—C—Z}} & \text{composite structure}
\end{array}
$$

The actual structure of the molecule can then be depicted by the composite structure shown, which represents the superposition of **A** and **B** in a single structure. The strength of the carbon-oxygen bond depends on the relative importance of the two resonance structures. If **A** and **B** were of equal importance, the carbon-oxygen bond would be halfway between a double bond and a single bond. This "one-and-a-half bond" would be considerably weaker than a full C=O double bond and would therefore vibrate at a lower

Key Concept: *Dispersion of charge stabilizes a molecule; concentration of charge destabilizes it.*

frequency. As a rule, the more electronegative Z is, the less stable (and therefore less important) resonance structure **B** is, because an electronegative substituent withdraws electrons from the positively charged carbon, increasing the concentration of charge at that point. The less important that structure **B** is, the smaller contribution it will make to the structure of the compound, and the nearer the carbon-oxygen bond will be to a full double bond.

If Z is unsaturated, additional resonance structures may be drawn. For example, if Z is a vinyl (CH_2=CH—) group there will be three significant resonance structures, two having carbon-oxygen single bonds.

$$R-\overset{\overset{O}{\|}}{C}-CH=CH_2 \longleftrightarrow R-\overset{\overset{\bar{O}:}{|}}{\underset{+}{C}}-CH=CH_2 \longleftrightarrow R-\overset{\overset{\bar{O}:}{|}}{C}=CH-\overset{+}{C}H_2$$

$$\qquad A \qquad\qquad\qquad B \qquad\qquad\qquad C$$

This will increase the contribution of resonance structures with C—O single bonds, giving the carbon-oxygen bond less double-bond character.

Additional resonance structures are possible if Z contains unpaired electrons on an atom next to the carbonyl carbon, as illustrated for an amide, in which Z is an amino (NH_2) group.

$$R-\overset{\overset{O}{\|}}{C}-\ddot{N}H_2 \longleftrightarrow R-\overset{\overset{\bar{O}:}{|}}{\underset{+}{C}}-\ddot{N}H_2 \longleftrightarrow R-\overset{\overset{\bar{O}:}{|}}{C}=\overset{+}{N}H_2$$

$$\qquad A \qquad\qquad\qquad B \qquad\qquad\qquad C$$

Stop and Think: Why?

This does not mean that resonance involving such oxygen atoms is negligible, just that the method described in this experiment works better if it is disregarded.

If resonance structure **C** is important, it will decrease the double-bond character of the carbon-oxygen bond. As a rule, structures like **C** are less important when the atom next to the carbonyl carbon is more electronegative than nitrogen. For the purposes of this experiment, you can disregard the effect of such resonance structures when the Z group has an oxygen atom at its point of attachment to the carbonyl carbon.

The method described will only provide relative frequencies; that is, it should help you arrange your assigned compounds in order of their C=O vibrational frequencies, but it will not allow you to predict the actual frequencies. With a molecular mechanics program, you can use a computer to calculate vibrational frequencies from electron distributions and other features of molecules. To do so, you will first build the structure of each molecule on the computer monitor. Then you will have the program calculate the electron distribution about the C=O bond (it may provide a pictorial representation of the electron distribution) from which it can estimate the bond's vibrational frequency, expressed in Hz or cm^{-1}.

When you record the IR spectra of your compounds, you will need to identify the carbonyl band in each spectrum and determine its wave number as accurately as possible. Generally the C=O band will be the only strong band within a region extending from about 1650 cm^{-1} to 1800 cm^{-1}. If you are using a Fourier transform infrared (FT–IR) spectrometer, the wave numbers of the important bands may be printed directly on the spectrum, or you may have to select the carbonyl band on a screen display of the spectrum and read its wave number from the monitor. With a dispersive IR spectrometer, it is essential that the chart paper be properly aligned because you will have to determine the wave number of the carbonyl band from its position on the chart paper.

Directions

This procedure is adapted from an article published in the *Journal of Chemical Education* **1996**, *73*, 188. Your instructor will provide several organic compounds for you to analyze and may assign you to a research team with other students.

The compounds you are assigned may be hazardous if inhaled or absorbed through the skin. Avoid contact and inhalation.

Safety Notes

Estimating C=O Frequencies. If the structures of your compounds are not given, write their structures from their systematic names. Write as many significant resonance structures as you can for each compound. Based on your assessment of the relative importance of the resonance structures, predict the relative strength of the carbon-oxygen bond in each compound. Then list the compounds in order of their carbonyl vibrational frequencies, from higher to lower frequency.

At your instructor's request, and following his or her instructions, use a molecular modeling program such as *MacSpartan* [Bibliography, M9] to estimate the vibrational frequency of the C=O bond in each compound.

Obtaining C=O Vibrational Frequencies from the IR Spectrum. Record the IR spectra [OP-36] of the assigned compounds. If you are working in a group, each member should record the spectrum of one compound. Accurately determine the wave number of each compound's carbonyl band from its spectrum. Calculate the vibrational frequency of each compound's carbonyl group from the wave number of its carbonyl band.

Waste Disposal: Place any unused liquids in a designated waste container.

Stop and Think: Were the results as you predicted? If not, why not?

Exercises

1 Predict the relative C=O vibrational frequencies of the aldehyde, ketone, and ester shown in Figure 13.1, listing them in order from lower to higher frequency.

2 Describe and explain the possible effect on your results of the following experimental or conceptual errors. (a) The infrared chart paper on a dispersive IR spectrometer was not aligned correctly when you recorded your spectrum. (b) One of your compounds was 3,7-dimethyl-2,6-octadienal. In writing its structure, you inadvertently put the #2 double bond between the #3 and #4 carbon atoms. (c) In comparing an aldehyde and a ketone, you assumed that an alkyl group is electron withdrawing relative to hydrogen, because in general chemistry you learned that carbon is more electronegative than hydrogen.

3 Calculate the wavelength in μm of the IR radiation absorbed by the carbonyl group of each compound whose wave number you recorded.

4 Write resonance structures for the following compounds, and predict their relative C=O vibrational frequencies, listing them in order from lower to higher frequency.

5 (a) Outline a synthesis of phenylethyl alcohol (2-phenylethanol) starting with benzene and ethylene oxide. (b) Outline a synthesis of benzyl acetate starting with toluene and ethanol.

Other Things You Can Do

(Starred items require your instructor's permission.)

1 Construct molecular models of some representative organic molecules, as described in Minilab 13.
2 Interpret one or more of your IR spectra by indicating what kind of bond is responsible for each significant IR band.
*3 Use a molecular mechanics program to determine the preferred geometry for one or more of your compounds or for another compound suggested by your instructor.
4 Write a research paper about perfumes, starting with sources listed in the Bibliography.

Properties of Common Functional Groups

Functional Group Chemistry. Qualitative Analysis. Infrared Spectrometry.

Operations

OP-5   Measuring Volume
OP-27   Simple Distillation
OP-31   Boiling Point
OP-36   Infrared Spectrometry

Before You Begin

1 Read the experiment, read or review the operations, and write a brief experimental plan.
2 Read or review the section "Interpretation of Infrared Spectra" in OP-36.

Scenario

Marvelous Molecules Incorporated (MMI) manufactures fine chemicals for use by research chemists in educational institutions and industry. One of their sales representatives, Albert Keene, just flew in from Milwaukee to meet an important client. He brought along samples of some representative chemicals to demonstrate the quality of MMI's products. Federal regulations prevented him from transporting the chemicals on the airliner so he had them sent ahead by freight carrier. On arrival, he discovered—to his dismay—that the bottles of chemicals had been exposed to high heat and humidity during transit, causing all of their labels to fall off. The labels were salvaged, but he has no idea which label belongs to which bottle, so he placed a desperate phone call to your supervisor asking for help. Your assignment is to match each chemical with the correct label. Fortunately, Al Keene selected chemicals from different families of organic compounds, so you will only have to identify the functional group present in each compound to find out what it is.

Applying Scientific Methodology

At your instructor's option, you can work in teams, with each member of the team responsible for the identification of one compound. After you observe some of your compound's physical and chemical properties, you should be able to formulate a tentative hypothesis about the identity of its functional group. You will then record the infrared (IR) spectrum of your compound to test your hypotheses and arrive at a conclusion.

Chemical Taxonomy

A botanist attempting to identify an unknown flowering plant may first examine the flowering parts to detect features that suggest the family of plants to which it belongs, and then study the whole plant systematically to determine its genus and species. For example, a botanist coming across a plant with four symmetrical flower petals and six stamens (four long and two short) might tentatively classify it into the mustard family (*Cruciferae*). Further observation of white flower petals, lyre-shaped leaves, and a red, globular root might then lead the botanist to the conclusion that the plant is a specimen of *Raphanus sativus*. On the other hand, an experienced gardener might immediately recognize the same specimen as a radish plant, based only on its general appearance.

Similarly, people experienced in handling chemicals may learn how to recognize a familiar organic compound from its odor and general characteristics, but a systematic approach is needed for positive identification of a range of organic compounds. Qualitative organic analysis—the identification of organic compounds based on their physical and chemical properties—is analogous in some ways to the identification of plants according to their *taxonomy*—their structural features and presumed natural relationships. Classifying an organic compound into a given family requires first detecting a specific functional group (characteristic set of atoms) in the molecules of the compound. Table 14.1 lists some functional groups found frequently in organic compounds.

Because functional groups influence the physical, chemical, and spectral properties of an organic compound, a chemist can usually identify a compound's functional group by a process that may involve measuring certain physical properties, observing its chemical behavior with different classification

Table 14.1   Common functional groups and their families

Functional group	Functional group name	Family
—C=C—	carbon-carbon double bond	alkene
—Cl	chlorine atom	alkyl chloride
—OH	hydroxyl group	alcohol (also phenol)
—C—H (—CHO), C=O	carbonyl group, with H on carbonyl carbon	aldehyde
—C— (—CO—), C=O	carbonyl group, no H on carbonyl carbon	ketone
—C—OH (—COOH), C=O	carboxyl group (*carb*onyl + hydr*oxyl*)	carboxylic acid
—NH$_2$	amino group	amine (primary)

Note: Condensed representations of some of the functional groups are given in parentheses.

reagents, and studying its infrared spectrum. The chemical name and structure of the compound can then be determined by methods described in Part IV of this book.

Understanding the Experiment

In this experiment, you will investigate some physical, chemical, and spectral properties of a compound that belongs to one of the families listed in Table 14.1. Then you will share your results with your coworkers, who will be assigned the remaining compounds. All of the compounds have molecules of about the same size and mass, so differences in their properties will depend primarily on their functional groups.

Physical properties that are affected by functional groups include boiling point, density, and water solubility.

Molecular Structure and Boiling Point

Boiling of a liquid occurs when the kinetic energy of its component molecules becomes high enough to overcome the forces between them, allowing them to leave the surface of the liquid and enter the gaseous state (see Figure 14.1). Since the kinetic energy of molecules increases with temperature, and the energy required to separate molecules depends on the strength of the forces between them, we can expect liquids that have strong intermolecular forces to have high boiling points as well.

The important kinds of intermolecular forces occurring between organic molecules are, in order of increasing strength, (1) dispersion forces (sometimes called van der Waals forces), (2) dipole-dipole interactions, and (3) hydrogen bonding. Dispersion forces are caused by alternating transient charge separations on the surfaces of molecules. They occur among all kinds of molecules, polar or nonpolar, causing them to "stick together" on contact—somewhat like the Styrofoam peanuts used as packing material, which cling to one another by static electricity. Dipole-dipole interactions result when molecules that have permanent bond dipoles line up so that the negative end of one molecule's dipole is opposite the positive end of another's, and vice versa. Hydrogen bonding is a special kind of dipole-dipole interaction involving the attraction of a highly polarized hydrogen atom for an electron-donating atom (such as oxygen or nitrogen) on another molecule. Among compounds of similar molecular shape and molecular weight, those capable of forming hydrogen bonds tend to have the highest boiling points, followed by compounds with polar groups capable of dipole-dipole interactions. Hydrogen bonding makes an important contribution to the boiling point only for organic compounds containing O—H and N—H bonds, with OH groups forming the strongest hydrogen bonds. Compounds with no polar groups, whose molecules are held together only by dispersion forces, tend to have the lowest boiling points.

You will distill your liquid to remove impurities that might affect your results. You can estimate its boiling point during the distillation if it is reasonably pure; otherwise you may need to redistill it or carry out an alternative boiling-point determination.

Molecular Structure and Density

The density of an organic compound depends, to some extent, on the mass/volume ratio of its constituent atoms. Atoms that have a high nuclear

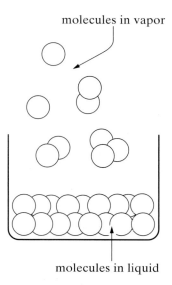

Figure 14.1   Boiling of a liquid

dipole-dipole interaction (formaldehyde)   hydrogen bonding (water)

mass confined within a small atomic volume, such as those in the top right-hand corner of the periodic table, have a high atomic density (atomic mass/atomic volume), so compounds whose molecules contain such atoms tend to have comparatively high densities. Among the atoms encountered in this experiment (other than hydrogen), oxygen has the highest atomic density, followed by chlorine, nitrogen, and carbon.

Atomic density of some "heavy" atoms, relative to carbon = 1.00
O: 1.50   Cl: 1.35   N: 1.26

The density of an organic liquid will thus depend, in part, on the number and kind of these "heavy" atoms it contains and on the fraction of its molecular weight they contribute. For example, chlorobenzene (C_6H_5Cl) has a higher density than phenol (C_6H_5OH) even though chlorine has a lower atomic density than oxygen, because the chlorine atom makes a greater contribution to the molecular weight of chlorobenzene than the oxygen atom of phenol makes to its molecular weight.

You will determine the density of your liquid by weighing a precisely measured volume of the liquid.

Molecular Structure and Solubility

Key Concept: *"Like dissolves like." Polar compounds tend to dissolve in polar solvents and nonpolar compounds in nonpolar solvents.*

$$CH_3CH_2CH_2CH_2CH_2CH_3$$

hexane

benzene

A compound will generally be soluble in a given solvent if the forces holding its own molecules together are similar to the forces holding the molecules of the solvent together *or* if the compound can form hydrogen bonds with the solvent. Thus hexane dissolves readily in benzene because both compounds are hydrocarbons whose molecules are held together by dispersion forces. Ethanol dissolves in water because both compounds contain OH groups capable of forming hydrogen bonds. Formaldehyde, although it cannot form hydrogen bonds among its own molecules, can hydrogen bond to solvents that contain OH groups, so it is also soluble in water.

Hydrogen bonding interactions

ethanol   formaldehyde
and water   and water

Keep in mind that "solubility" is a relative term; there are varying degrees of solubility. Terms commonly used to indicate the extent to which one compound dissolves in another are, in order of decreasing solubility, miscible (∞), very soluble (v), soluble (s), sparingly soluble (δ) and insoluble (i).

You will measure the solubility of your liquid by shaking it with water in a small test tube. For the purposes of this experiment, a liquid will be classified as *miscible* if it dissolves in an equal volume of water and as *soluble* if 0.10 mL of the liquid dissolves in about 3 mL of water. If the liquid dissolves, the resulting solution should be clear, like water itself. If it does not dissolve, you should observe a second liquid layer, cloudiness, or liquid droplets (not air bubbles) in the water. Since low-boiling liquids may evaporate rapidly, you should keep the test tube stoppered while you make your observations.

Classification Tests for Functional Groups

Chemists have developed a number of simple chemical tests that are positive only for compounds having certain kinds of functional groups. A litmus test, for example, can be used to detect acidic and basic functional groups. When dissolved in water, carboxylic acids turn blue litmus paper red and amines turn red litmus paper blue. Compounds that are easily oxidized react with a solution of chromium(III) oxide in sulfuric acid, commonly referred to as "chromic acid." Primary and secondary alcohols react within 2–3 seconds to form an opaque blue-green suspension. Aldehydes give the same result, but usually take 10 seconds or more to react. Aldehydes and ketones both react with 2,4-dinitrophenylhydrazine (DNPH) reagent to yield yellow or orange precipitates. Alkenes react rapidly with dilute aqueous potassium permanganate ($KMnO_4$) to form a brown precipitate as the purple color of the permanganate disappears. Alkyl halides give a green flame in the Beilstein test, which involves heating a copper wire moistened with the unknown in a burner flame.

The test results and their interpretations are summarized here.

- Blue litmus paper turns red $\Rightarrow$ carboxylic acid
- Red litmus paper turns blue $\Rightarrow$ amine
- Chromic acid forms blue-green suspension in 2–3 seconds $\Rightarrow$ 1° or 2° alcohol
- Chromic acid forms blue-green suspension in 10 seconds or more $\Rightarrow$ aldehyde
- DNPH yields orange or yellow precipitate $\Rightarrow$ aldehyde or ketone
- $KMnO_4$ yields brown precipitate $\Rightarrow$ alkene
- Burner flame turns green $\Rightarrow$ alkyl halide

Conflicting or ambiguous results may be produced with some tests because of impurities in the unknown or because the tests themselves may be open to misinterpretation. For example, an aldehyde often contains traces of the corresponding carboxylic acid as an impurity, leading to a false-positive litmus test. In such cases, it may be necessary to repeat a test or redistill the unknown. For additional information and general equations for the reactions involved, see the "Classification Tests" section of Part IV.

Infrared Spectra

Each kind of functional group is associated with one or more characteristic infrared (IR) bands that can reveal its presence. A carboxyl group (COOH), for example, contains C=O, C—O, and O—H bonds, all of which give rise to strong IR absorption bands. Infrared bands that may help you classify your unknown include the two-pronged N—H band characteristic of a primary amine, which is around $3400-3300$ cm^{-1}; the broad O—H band centered near 3300 cm^{-1} for an alcohol and 3000 cm^{-1} for a carboxylic acid; the strong C=O band near 1700 cm^{-1}; and the C—Cl band between 600 and 800 cm^{-1}. Alkenes are often characterized by vinylic C—H bending vibrations in the $650-1000$ cm^{-1} region and by a C=C band (sometimes weak or absent) near 1650 cm^{-1}. See the section "Interpretation of Infrared Spectra" in OP-36 for more detailed information about these and other IR bands associated with common functional groups.

Directions

Your instructor will give you the names that were on the labels described in the Scenario. The experiment can be performed in teams of as many as seven students, each team working with a full set of unknown liquids. Alternatively, students can work individually and report their data to the instructor for posting.

Safety Notes

> **All of the unknown liquids are flammable, and some of them are caustic or have hazardous vapors. Avoid contact with the liquids, do not breathe their vapors, and keep them away from flames and hot surfaces.**
> **2,4-Dinitrophenylhydrazine is harmful if absorbed through the skin and it will dye your hands yellow. Avoid contact with the DNPH reagent.**
> **The chromic acid reagent is corrosive, very toxic, and carcinogenic. Wear gloves and avoid contact with the reagent.**

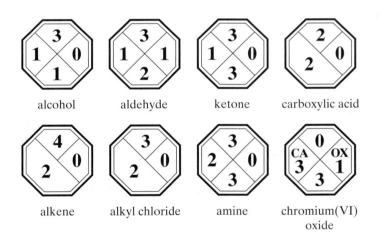

alcohol   aldehyde   ketone   carboxylic acid

alkene   alkyl chloride   amine   chromium(VI)
                                                        oxide

Take Care! Avoid contact with the unknown and do not breathe its vapors.

Stop and Think: Compare your boiling point with those obtained by other students in your group. What can you say about your compound's intermolecular forces?

Stop and Think: Compare your density with those obtained by other students in your group. Does your compound contain any "heavy" atoms such as O, N, or Cl?

A. *Physical Properties of the Unknowns*
Obtain a sample of an unknown organic liquid from your instructor and record its identification number in your laboratory notebook. Purify the liquid by simple distillation [OP-27] using a water-cooled condenser and an accurate thermometer. If the liquid boils over a range of 2°C or less, you can use the median distillation temperature as its boiling point [OP-31]. Otherwise, determine the boiling point of the purified liquid by a capillary-tube method.

Use a measuring pipet (**Take Care!** Do not pipet by mouth!) or automatic pipet to accurately measure [OP-5] 0.200 mL of the purified unknown into a tared vial. Stopper the vial and weigh it to the nearest milligram on an accurate balance; then calculate the density of your unknown.

Use a measuring pipet or a syringe to measure 0.10 mL of the liquid into a 13 × 100-mm test tube. Add 0.10 mL of water, stopper the tube, and shake it vigorously for a few seconds. If two liquid layers separate on standing, or if the mixture is cloudy or contains undissolved droplets, add another 3 mL of water and shake it again. If the unknown dissolves, save the solution for the litmus test in part **B**.

B. *Chemical Tests*

Except for the litmus test, which is performed using an aqueous solution of the unknown, all of the following tests should be carried out with the undiluted unknown liquid. Keep a careful record of your observations in your lab notebook.

Litmus Test.   If the unknown is miscible or soluble in water, test its *aqueous solution* with red and blue litmus paper.

DNPH Test.   Add 1 drop of the unknown to 1 mL of the DNPH reagent in a small test tube. Stopper and shake the test tube, then let the mixture stand for 15 minutes.

Chromic Acid Test.   Dissolve 1 drop of the unknown in 1 mL of reagent-grade acetone in a test tube. Add 1 drop of the chromic acid reagent, shake the mixture, and observe it for at least one minute, noting the time required for any positive test.

Potassium Permanganate Test.   Dissolve 1 drop of the unknown in 2 mL of 95% ethanol, and add 10 drops of 0.1 *M* potassium permanganate while shaking.

Beilstein's Test.   Make a small loop in the end of a copper wire. Heat the loop to redness in a burner flame. (Do this under a hood or in an area away from the unknowns or other flammable liquids.) Let it cool, then dip the loop into a small amount of your unknown liquid, and then hold it in the lower outside part of the flame.

C. *Infrared Spectrum*

Record an IR spectrum [OP-36] of your unknown, using the neat liquid, or obtain the spectrum from your instructor.

Decide which functional group is present in your compound. If your instructor assigns Exercise 1, report the chemical family, boiling point, density, and solubility of your unknown to your coworkers and obtain the same information from them. If your instructor assigns Exercise 2, make copies of your IR spectrum for yourself and your coworkers and obtain copies of their spectra in return (alternatively, you can examine their spectra and record the relevant spectral data). Turn in the original spectrum to your instructor.

Stop and Think: If your compound forms a separate layer, does it float or sink? Why?

Take Care!   Wear gloves and avoid contact with the chromic acid reagent.

Waste Disposal: Put wastes for the DNPH, chromic acid, and potassium permanganate tests in designated waste containers.

Stop and Think: At this point, which functional group do you think your compound contains?

Waste Disposal: Place the rest of your unknown in a waste container designated by your instructor.

Exercises

1   Using the data provided by your coworkers, group together compounds with similar physical properties. You should have one set of groups arranged according to boiling point, another set according to density, and a third set according to water solubility. Then explain the differences between the groups, specifying the type of intermolecular force involved, where appropriate. For example, you can group together compounds that have boiling points within 10°C of one another, listed from higher

to lower boiling point, and explain why the compounds in the first group have higher boiling points than those in the second group, etc.

2 Compare the IR spectra obtained by the members of your team, looking for significant similarities and differences. For each spectrum, point out any IR bands that are associated with a functional group and indicate what kind of chemical bond is responsible for each of these IR bands.

3 Describe and explain the possible effect on your results of the following experimental errors or variations. (a) Unknown to you, the automatic pipet you used to measure your liquid for the density determination was set at 250 μL. (b) Most of your liquid (an aldehyde) distilled around 75°C, but some of it distilled above 100°C, and you used all of the distillate for the subsequent tests. (c) You cleaned a test tube for the DNPH test by rinsing it with acetone, but failed to dry it completely.

4 Construct a flow diagram that could be used to efficiently classify compounds from the families in Table 14.1, using classification tests from this experiment. There should be two branches for each classification test, one leading to compounds that give a positive result and the other leading to compounds that give a negative result.

5 Draw structures for all possible constitutional isomers of your unknown having the same functional group (if there are any), and give their systematic names.

6 Compound X, having the molecular formula $C_4H_6O_2$, is believed to be one of the compounds in the margin. (a) Name the chemical family or families to which each compound belongs. (b) Draw a flow chart diagramming a procedure that could be used to determine the identity of compound X, using chemical tests described in Part IV of this book.

Margin structures:

1. $\overset{\displaystyle O}{\overset{\|}{CH_3CH_2C}}-\overset{\displaystyle O}{\overset{\|}{CH}}$

2. $\overset{\displaystyle O}{\overset{\|}{CH_3C}}-\overset{\displaystyle O}{\overset{\|}{CCH_3}}$

3. $\overset{\displaystyle OH}{\overset{|}{CH_2{=}CHCH}}-\overset{\displaystyle O}{\overset{\|}{CH}}$

4. $CH_3CH{=}CH\overset{\displaystyle O}{\overset{\|}{C}}-OH$

Other Things You Can Do

(Starred items require your instructor's permission.)

*1 If the reagents are available, you can use additional or alternative classification tests such as Tollens' test (C-23) for aldehydes, sodium iodide in acetone (C-22) for alkyl chlorides, and bromine (C-7) for alkenes.

*2 In Minilab 14, find out who else has the same compound that you have.

3 Write a research paper describing how hydrogen bonding can be detected and studied by spectrometric methods, using sources listed in the Bibliography or elsewhere.

Thin-Layer Chromatographic Analysis of Drug Components

Qualitative Analysis. Thin-Layer Chromatography.

Operations

OP-6   Making Transfers
OP-19   Thin-Layer Chromatography

Before You Begin

1 Read the experiment, read or review OP-19 carefully, and write a brief experimental plan.

Scenario

A patient who swallowed a large number of drug tablets was just wheeled into the emergency room of a local health clinic suffering from a drug overdose. The bottle that contained the drug is missing, but some unused tablets were left on a counter in the bathroom. The patient's spouse recalls seeing a bottle of an over-the-counter analgesic drug in the medicine chest, but doesn't remember what was on the label. Because there was no nearby toxicology lab in operation, the clinic rushed the tablets to your supervisor for analysis. Your assignment is to determine the identity of the drug so that emergency room physicians can apply the appropriate treatment.

Applying Scientific Methodology

This experiment requires you to identify the components of an analgesic drug tablet and then identify the tablet as one of the commercial drug preparations listed in Table 15.1. You will probably not be able to formulate a meaningful hypothesis until after you examine your thin-layer chromatography (TLC) plate under ultraviolet light. Then you can use another visualization method to provide evidence that may either support or disprove your tentative hypothesis.

Table 15.1   Composition of some analgesic drug preparations (in milligrams per tablet)

Drug name	Aspirin	Acetaminophen	Ibuprofen	Salicylamide	Caffeine
Advil			200		
Anacin	400				32
Aspirin*	325				
B.C. Tablets	325			95	16
Excedrin	250	250			65
Tylenol		325			

*5-grain tablet (1 grain = 64.8 mg)

Drugstore Chemicals

Druggists were once called *chemists*, and in Great Britain, they still are. Whenever you go to the drugstore to buy a bottle of aspirin, Advil, Tylenol, or another of the dozens of different analgesic (painkilling) drug preparations available, you are purchasing an organic chemical or a mixture of several chemicals. That shouldn't be surprising since all matter—including you—is composed of chemicals, where *chemical* is just another name for a substance (element or compound). Most of the materials we encounter in daily life are complex mixtures of substances, while many drugs are reasonably pure substances or mixtures of only a few substances. In fact, most analgesic drug preparations contain only one or two active ingredients, either aspirin, acetaminophen, ibuprofen, or some combination of these. The most popular combination is aspirin and acetaminophen, but salicylamide (a chemical relative of aspirin) is combined with aspirin in a few drug preparations. Caffeine is sometimes added to an analgesic preparation for its stimulant effect and because it interacts with some analgesics to enhance their pain-relieving effects.

aspirin   acetaminophen   ibuprofen

salicylamide   caffeine

A small amount of starch is generally added as a binder to hold the tablets together. Table 15.1 shows the composition (excluding the starch) of some representative analgesic preparations.

Most analgesic drugs do more than just kill pain. Aspirin, acetaminophen, and ibuprofen are also antipyretics, meaning that they reduce fever, while aspirin and ibuprofen are both nonsteroidal anti-inflammatory drugs (NSAIDs), meaning that they help reduce swelling and other symptoms of inflammation. Aspirin, in fact, has such a surprisingly wide range of benefits that it can be regarded as a true "wonder drug." In recent years it has been shown to reduce the incidence of heart disease, strokes, and certain cancers. It may also help prevent cataracts, reduce the occurrence of gallstones, and improve brain function in people who have suffered small strokes.

Although aspirin (acetylsalicylic acid) is the most widely used drug in the world, nobody understood how it worked until 1971, when the British pharmacologist John Vane showed that aspirin blocks the overproduction of natural biological substances called *prostaglandins* by deactivating a key enzyme required to manufacture them, prostaglandin synthase. Although prostaglandins are essential biological regulators, an oversupply of certain prostaglandins can promote the formation of blood clots that lead to heart attacks or strokes, while others trigger pain, fever, and inflammation.

Vane shared the 1982 Nobel prize in medicine as a result of his discovery.

prostaglandin E_2

Recent studies have shown that prostaglandin synthase functions like a factory assembly line; raw materials enter one end of a channel that passes through the enzyme and leave the other end as fully assembled prostaglandin molecules. Aspirin molecules sabotage this operation by blocking the channel, thereby preventing raw materials from getting by. Aspirin appears to inhibit cancers of the digestive system by a different mechanism, stimulating the production of cancer-fighting substances used by the body's immune system.

If the other analgesics can't compete with aspirin in versatility, they do have certain advantages over aspirin. Aspirin tends to promote bleeding, especially in the stomach, and it has been implicated in Reye's syndrome, a rare but often fatal disease that affects children. Acetaminophen is just as effective as aspirin at reducing pain but has fewer side effects at normal dosages. Overdoses of acetaminophen can cause serious liver damage, especially when taken in conjunction with alcohol. Ibuprofen is a powerful analgesic with about the same painkilling effect as aspirin at one-third the dosage. It is especially effective in treating arthritis and menstrual cramps, but it can cause problems in people with asthma, ulcers, high blood pressure, or kidney, liver, or heart disease.

Understanding the Experiment

In this experiment, you will use TLC to identify the components of the analgesic drug that your instructor assigns. Your drug may contain from one to three of these substances: acetaminophen, aspirin, caffeine, ibuprofen, and salicylamide.

You will prepare a solution of the drug by dissolving part of a tablet in 1:1 ethanol/dichloromethane, then spot a TLC plate with this solution along with standard solutions of all of the active substances the drug tablet is likely to contain. The TLC plate can be developed with a solvent mixture such as ethyl acetate/acetic acid (200:1). By using two different methods to visualize the spots and comparing the R_f values of your unknown's spot(s) with those of the standards, you should be able to match each unknown spot with the spot of a known substance. Then you can refer to Table 15.1 to find out which commercial drug preparation contains the components you have identified.

Caffeine is only a minor component of a drug such as Excedrin, so its spot might be too faint to see clearly in some cases. In this event, you can usually arrive at the correct commercial drug by comparing the remaining components with those in the table. For example, Excedrin is the only drug listed that contains both aspirin and acetaminophen, and B.C. is the only one that contains salicylamide.

Directions

Safety Notes

dichloromethane

Take Care! Avoid contact with the developing solvent and do not breathe its vapors.

Take Care! Avoid contact with the solvent mixture and do not breathe its vapors.

You can use a scrap piece of TLC plate to practice your spotting technique.

Waste Disposal: Place the developing solvent and unused solutions in designated waste containers.

Observe and Note: Record in your lab notebook any characteristics of the spots that might help you identify them.

Dichloromethane may be harmful if ingested, inhaled, or absorbed through the skin. There is a possibility that prolonged inhalation of dichloromethane may cause cancer. Minimize contact with the ethanol/dichloromethane solvent mixture and the standard solutions and do not breathe their vapors.
Avoid contact with the developing solvent and do not breathe its vapors.

Preparation of the Developing Chamber. Unless your instructor indicates otherwise, use ethyl acetate/acetic acid (200:1) as the developing solvent. Fill an appropriate developing chamber (see OP-19) containing a paper wick to a depth of about 5 mm with the solvent, cover it, and set it aside under a hood while you prepare the TLC plate.

Preparation and Development of the TLC Plate. Read or review OP-19 carefully before you begin. Obtain a quarter tablet of the unknown analgesic drug and grind it to a powder with a flat-bottomed stirring rod or flat-bladed spatula. (If the tablet is coated, remove as much of the coating as you can before grinding.) Transfer the powder to a test tube, and add 2.5 mL of 1:1 ethanol/dichloromethane. Thoroughly crush and mix the solid in the solvent with a stirring rod to dissolve as much of the solid as possible. Use a filter-tip pipet to transfer [OP-6] the solution to a small vial, leaving the solid behind, and cap the vial.

Spot the unknown solution and all of the standard solutions provided on a silica gel TLC plate [OP-19] having a fluorescent indicator, and label the spots. To avoid cross contamination, use a different micropipet (or other spotting device) for each solution. If possible, the unknown should be spotted in different concentrations (from 1–3 applications) at two or more locations along the starting line. Develop the plate *under the hood* in the developing chamber you prepared. Do not disturb the developing chamber until the plate is developed. Be sure to mark the solvent front before the plate dries.

Visualization and Analysis. When the TLC plate is dry, observe the spots under short-wavelength (254 nm) ultraviolet light, outline them with a pencil, and mark the center of greatest intensity for each spot. (**Take Care!** Do not look directly at the light source.) Then visualize the spots by using iodine vapor or another visualizing reagent selected by your instructor. Calculate the R_f value of each spot, and identify the active ingredients of your unknown by comparison with the R_f values and visualization results for the known spots.

Use Table 15.1 to determine the commercial name of your analgesic drug preparation. (Your tablet may be a generic equivalent of one of those listed.) Turn in your TLC plate with your report.

Waste Disposal: Place the solution of your unknown in an appropriate waste container.

Exercises

1 Suppose you carry out a TLC separation of acetaminophen and phenacetin (see Experiment 2) on silica gel using a nonpolar developing solvent. Which should have the higher R_f value, and why?

2 Assuming that the drug components whose spots you identified dissolved completely when you stirred the tablet with 1:1 ethanol/dichloromethane, what was the solid that remained behind?

3 Describe and explain the possible effect on your results of the following experimental errors. (a) You allowed your chromatogram to develop too long, and you couldn't find the solvent front. (b) The lab assistant who prepared the developing solvent mistakenly used aqueous ammonia in place of acetic acid. (c) You used an open-ended melting-point capillary rather than a Drummond microcap to apply the spots. (d) You marked the starting line with a ballpoint pen.

4 Describe one or more simple chemical tests that would distinguish acetaminophen from ibuprofen. Refer to Part IV of this book if necessary.

5 Another analgesic drug, Aleve, contains the sodium salt of naproxen as its active ingredient. (a) Look up the systematic name and structure of naproxen. (b) The physiological effects of naproxen are very similar to those of a drug component you studied in this experiment. Which one do you think it is, and why?

Other Things You Can Do

(Starred items require your instructor's permission.)

*1 Carry out the TLC analysis of felt-tip-pen inks as described in Minilab 11.

*2 Carry out a quantitative or qualitative analysis of an analgesic drug mixture using high-performance liquid chromatography. See OP-35 and *J. Chem. Educ.* **1983**, *60*, pages 163 and 1000.

3 Write a research paper describing some of the methods used to test for drugs and their metabolites in body fluids, starting with sources listed in the Bibliography. For example, you might tell how the presence of proscribed drugs can be detected in professional athletes or racehorses.

EXPERIMENT **16**   Separation of an Alkane Clathrate

Reactions of Alkanes. Clathrates. Infrared Spectrometry.

Operations

OP-10   Mixing
OP-13   Vacuum Filtration
OP-15   Extraction
OP-16   Evaporation
OP-22   Drying Liquids
OP-23   Drying Solids
OP-36   Infrared Spectrometry

Before You Begin

1 Read the experiment, read or review the operations as necessary, and write a brief experimental plan.
2 Calculate the mass of 1.00 mmol of hexadecane and the mass of 20 mmol of urea.
3 Use Equation **1** to help you estimate the theoretical yield of a hexadecane–urea clathrate.

Scenario

The Petit Prix Racing Group mixes its own auto racing fuel by combining clean-burning high-octane hydrocarbons such as 2,2,4-trimethylpentane (also called "isooctane") with other racing-fuel components such as methanol and nitromethane. An inexperienced employee, Rusty Tappet, was mixing the last batch of fuel when he accidentally added a can of diesel fuel intended for his employer's Mercedes-Benz to a barrel containing 2,2,4-trimethylpentane and methanol. Diesel fuel is not clean-burning, and its hydrocarbons have extremely low octane numbers, so the contents of the barrel are now worthless as racing fuel. But the racing group can't afford to waste expensive fuel, so it has contacted your supervisor to see whether your Institute's consulting chemists can develop a method for separating the diesel fuel components from the mixture.

Urea molecules are known to form tunnel-like channels in certain solvents by hydrogen bonding with one another. Your supervisor thinks it should be possible to trap the straight-chain hydrocarbon molecules of diesel fuel inside such channels while excluding the bulkier 2,2,4-trimethylpentane molecules. You will test this hypothesis by combining a mixture of hexadecane (a major component of diesel fuel), 2,2,4-trimethylpentane, and methanol with urea to see if the urea will remove the hexadecane from the mixture. If it does, you are to determine the ratio of urea molecules to hexadecane molecules (the host/guest ratio) in the urea-hexadecane complex so that the minimum amount of urea needed for the separation can be estimated.

$$CH_3CH_2CH_2CH_2CH_2CH_2CH_2CH_2CH_2CH_2CH_2CH_2CH_2CH_2CH_2CH_3$$

hexadecane

$$CH_3$$
$$|$$
$$CH_3CCH_2CHCH_3$$
$$|\quad\ \ |$$
$$CH_3\ \ CH_3$$

2,2,4-trimethylpentane

Applying Scientific Methodology

There are two problems in this experiment, so you should formulate two working hypotheses after reading it. You will test one hypothesis by recording the infrared (IR) spectrum of any hydrocarbon that you isolate, and the other by calculating the host/guest ratio from the masses of urea and hexadecane in the complex.

Clathrates and Fuel Quality

In 1976, scientists drilling into the ocean floor near Central America brought up a core sample that contained softball-sized lumps of an icelike solid that sizzled like frying bacon and emitted liquid droplets and a flammable gas, leaving nothing behind but a puddle of water. The icelike solid was a gas hydrate, consisting mostly of single methane molecules trapped inside a cage made up of water molecules. Methane hydrate consists of 20 water molecules arranged to form a nearly spherical 12-sided geometric figure (a dodecahedron), with a free methane molecule floating inside its watery cage.

Recent discoveries of enormous methane hydrate deposits beneath ocean floors have attracted attention because they represent both an opportunity and a potential threat. It is estimated that about 20 quadrillion (2×10^{16}) cubic meters of methane are trapped in such hydrates; this is energetically equivalent to approximately twice the earth's coal, oil, and gas reserves combined. If the oceans could somehow be "mined" for methane hydrates, they would represent a huge energy reserve. But methane is also an efficient greenhouse gas, and some scientists believe that global warming or geological disturbances could cause hydrate deposits to decompose, releasing huge quantities of methane into the atmosphere. This could cause a runaway global-warming effect that would drastically alter the earth's weather patterns.

Methane hydrate is an example of a *clathrate*, a complex formed when molecules of one compound are enclosed in cavities within other molecules or crystal lattices. In 1941, a German chemist discovered that urea combines with straight-chain alkanes having seven or more carbon atoms to form a crystalline clathrate, but does not combine with most branched alkanes. This property of urea provides a means of improving fuel quality.

The quality of a motor fuel is measured by its octane number, which compares its antiknock properties to those of the highly branched alkane 2,2,4-trimethylpentane (octane number = 100). Long straight-chain alkanes have very low octane numbers, so treating a gasoline-grade petroleum fraction with urea to remove such alkanes increases its octane number. Jet fuels are also improved by treatment with urea because straight-chain alkanes tend to freeze

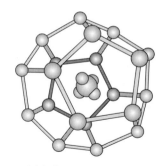

Figure 16.1 Structure of a methane hydrate

$$O$$
$$\|$$
$$H_2NCNH_2$$
urea

$$CH_3 \quad CH_3 \qquad\qquad CH_3$$
$$| \qquad | \qquad\qquad\qquad |$$
$$CH_3CCH_2CCH_2CHCH_2CCH_3$$
$$| \qquad | \qquad | \qquad\quad |$$
$$CH_3 \quad CH_3 \quad CH_3 \quad CH_3$$

HMN

sooner than branched alkanes of the same carbon number. Unless the straight-chain alkanes are removed, wax crystals could form in jet fuel at high altitude, where the temperature can drop to $-60°C$.

The quality of a diesel fuel, on the other hand, is measured by its *cetane number*, which compares its ignition properties to those of hexadecane (also called cetane). The cetane number of a fuel is established by matching the fuel's performance to that of a mixture containing hexadecane, whose cetane number is 100, and 2,2,4,4,6,8,8-heptamethylnonane (HMN), whose cetane number is 15. For example, a diesel fuel that has the same ignition properties as a mixture containing 40% hexadecane and 60% HMN is assigned a cetane number of

$$100(0.40) + 15(0.60) = 49$$

Most diesel fuels marketed in the United States have cetane numbers between 40 and 65.

Understanding the Experiment

In this experiment, you will combine a solution of urea in methanol with a mixture of hexadecane and 2,2,4-trimethylpentane to see if the urea will form a clathrate with one of the alkanes. Many organic compounds—not just alkanes—combine with urea in solution to form crystalline clathrates in which six or more molecules of urea, the *host* compound, form a tubular channel large enough to hold a molecule of the *guest* compound. The channel's walls are formed by interpenetrating spirals of urea molecules that are hydrogen bonded to one another. The guest molecule is not covalently bonded to the host molecules but is held inside the channel by relatively weak intramolecular forces. Whether a compound qualifies as a guest depends mainly on the size and shape of its molecules. The diameter of the channel formed by urea molecules is about 0.52 nm, which is large enough to hold straight-chain molecules but not most branched molecules. In addition to alkanes, guest compounds can include straight-chain primary alcohols, carboxylic acids, and esters that have terminal functional groups and at least seven carbon atoms.

The longer the guest molecule, the more urea molecules are needed to surround it. The number of urea molecules per guest molecule (the host/guest ratio) can be estimated using Equation **1**, where n is the number of carbon atoms in the guest molecule.

$$\text{host/guest ratio for urea clathrates} \cong 1.5 + 0.65n \qquad \textbf{(1)}$$

For example, about eight urea molecules $[1.5 + (0.65 \times 10)]$ are sufficient to confine a molecule of decane, which has 10 carbon atoms.

Adding water to a urea clathrate causes it to decompose, releasing the guest compound. If urea forms a clathrate with a hydrocarbon in your mixture of hexadecane and 2,2,4-trimethylpentane, you should be able to identify the guest alkane from its IR spectrum. *Stretching* vibrations of C—H bonds in methyl (CH_3) groups produce IR bands around 2960 cm^{-1} and 2870 cm^{-1}. The corresponding methylene (CH_2) bands are near 2925 cm^{-1} and 2850 cm^{-1}. Often the methyl and methylene bands overlap so that only two or three peaks are observed. Methyl groups also show asymmetrical and symmetrical *bending* vibrations near 1450 cm^{-1} and 1375 cm^{-1}. If there are two or three methyl groups on the same carbon atom, the symmetrical bending band is split into two or more closely spaced peaks near 1385 cm^{-1} and 1370 cm^{-1}. Bending vibrations of methylene groups may also give rise to bands near 1465 cm^{-1}, between 1350 cm^{-1} and 1150 cm^{-1} (often weak), and around 720 cm^{-1}. The intensity of the 720 cm^{-1} band (called the methylene rocking band) increases in proportion to the number of adjacent methylene groups, so a band in this region is characteristic of unbranched long-chain alkanes. The terms used to describe certain bond vibrations—scissoring, twisting, wagging, rocking—remind us that molecules are dynamic entities and not the static particles that molecular models might suggest (see Figure G16, OP-36). For more information about bond vibrations and IR spectra, see "Interpretation of Infrared Spectra" in OP-36.

Key Concept: Molecules are in constant translational, rotational, and vibrational motion.

From the masses of the clathrate and the recovered guest alkane, you can determine the host/guest ratio—the number of urea molecules that surround each alkane molecule. For this result to be valid, material losses must be kept to a minimum and the masses of the clathrate and recovered guest

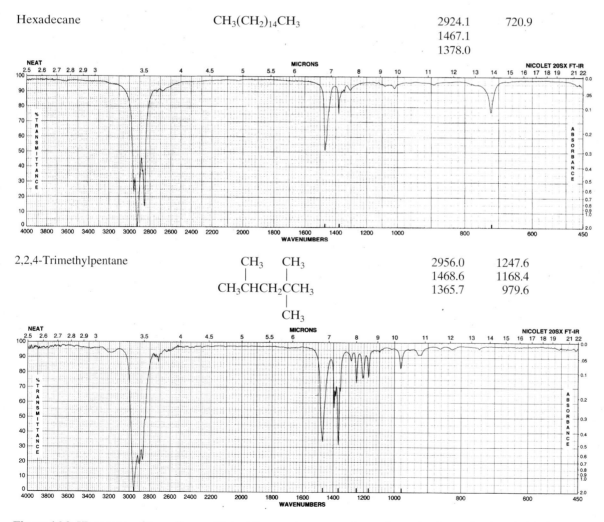

Figure 16.2 IR spectra of two alkanes. *Note:* The exact wave numbers of certain absorption bands (designated by "tick" markers at the base of the spectrum) will be listed above the IR spectra reproduced in this book.

alkane must be measured as accurately as possible. It is also essential that the clathrate be completely dry and the alkane be entirely free of solvent.

Reactions and Properties

Table 16.1 Physical properties

	M.W.	bp	mp	d	n_D^{20}
hexadecane	226.4	287	18	0.773	1.4345
2,2,4-trimethylpentane	114.2	99	−107	0.692	1.3915
methanol	32.0	65	−98	0.792	1.3292
urea	60.06		135	1.323	

Note: bp and mp are in °C, density is in g/mL; refractive index is for 20°C.

$$\underset{\text{alkane}}{C_nH_{(2n+2)}} + \underset{\text{urea}}{n \; H_2N\overset{\overset{\displaystyle O}{\|}}{C}NH_2} \longrightarrow \underset{\text{alkane-urea clathrate}}{C_nH_{(2n+2)} \cdot [H_2N\overset{\overset{\displaystyle O}{\|}}{C}NH_2]_n}$$

Directions

Safety Notes

> **Hexadecane and 2,2,4-trimethylpentane are flammable; keep them away from flames and hot surfaces.**
> **Methanol is flammable and harmful if ingested, inhaled, or absorbed through the skin. Avoid contact with the liquid and do not breathe its vapors. Dichloromethane may be harmful if ingested, inhaled, or absorbed through the skin. There is a possibility that prolonged inhalation of dichloromethane may cause cancer. Minimize contact with the liquid and do not breathe its vapors.**

2,2,4-trimethyl-   methanol   dichloromethane
pentane

Take Care! Avoid contact with methanol and do not breathe its vapors.

Stop and Think: What do you think is in the solid?

Reaction.   Accurately weigh 1.00 mmol of hexadecane in a small screw-cap vial, and add 0.50 mL of 2,2,4-trimethylpentane. In a small Erlenmeyer flask, combine 20 mmol of urea with 4.0 mL of methanol and add a stir bar. Warm this mixture to 55–60°C while stirring [OP-10] until all of the urea has dissolved; do not heat the methanol to boiling. While the solution is still warm, add the mixture of alkanes, using about 1 mL of methanol for the transfer. Stir until a white solid begins to separate, set the solution aside, and let it cool slowly to room temperature. Then cool it in an ice/water bath for 10 minutes or more until crystallization is complete.

Separation.   Collect the clathrate (the solid product) by vacuum filtration [OP-13] and wash it on the filter with ice-cold methanol. Dry [OP-23]
the solid to constant mass at room temperature (heating will cause it to decompose). If the clathrate cannot be left overnight or longer to dry, it should
be air dried thoroughly on the filter and left to dry in a well-ventilated location, such as a hood. Weigh the dry product accurately.

Transfer the clathrate quantitatively to a 15-mL centrifuge tube, add 2.5
mL of warm water, and heat the mixture in a boiling-water bath as you stir it
with a stirring rod until the crystals dissolve. Cool the mixture in an ice/water
bath; the alkane may solidify on cooling. Extract [OP-15] the guest alkane
with two separate 2-mL portions of dichloromethane. Dry [OP-22] the
combined extracts over anhydrous calcium chloride. Use a filter-tip pipet to
transfer the liquid to a tared 5-mL conical vial. Evaporate [OP-16] the solvent *completely* and measure the mass of the guest alkane accurately.

Analysis.   Record the IR spectrum [OP-36] of the guest alkane, and use it to
identify the alkane. Calculate the percent recovery and the host/guest ratio
from your experimental results. Turn in the IR spectrum along with the alkane.

Waste Disposal: Place the filtrate
in an appropriate solvent recovery
container.

Observe and Note: What happens
as the clathrate dissolves?

Take Care! Avoid contact with dichloromethane and do not breathe
its vapors.

Exercises

1 Estimate the host/guest ratio for hexadecane using Equation 1, and
   compare that value with your experimental value. Try to account for
   any significant differences.
2 Compare the IR spectrum of hexadecane with that of 2,2,4-
   trimethylpentane and point out some bands that might help you distinguish straight-chain alkanes from branched ones.
3 Describe and explain the possible effect on your results (including your
   host/guest ratio) of the following experimental errors. (a) The lab assistant accidentally put octane in the "isooctane" (2,2,4-trimethylpentane)
   bottle. (b) The lab assistant put hexane in the hexadecane bottle. (c)
   You dried the clathrate in a 90°C oven. (d) The clathrate crystals were
   not dry when you weighed them.
4 Rusty Tappet emptied a 5.0-gallon can of diesel fuel into a barrel containing 25 gallons of methanol and 2,2,4-trimethylpentane. Using your
   host/guest ratio, estimate the minimum mass of urea needed to remove
   the diesel fuel. You can approximate the composition of the diesel fuel
   by assuming that it is pure hexadecane.
5 Following the format in Appendix V, construct a flow diagram for this
   experiment.
6 What is the cetane number of a fuel that has the same ignition properties as a mixture containing 20% hexadecane and 80% HMN? Would it
   make a good diesel fuel?
7 Thiourea (H_2NCSNH_2) forms tubular clathrates that are similar in
   structure to those formed by urea. Combining a mixture of hexadecane
   and 2,2,4-trimethylpentane with thiourea produces a crystalline solid
   that decomposes in water to yield 2,2,4-trimethylpentane, but no hexadecane. Propose an explanation for this result.
8 Outline a possible synthesis of 2,2,4,4,6,8,8-heptamethylnonane (HMN)
   starting with isobutylene.

Other Things You Can Do

(Starred items require your instructor's permission.)

*1 Identify the guest alkane by measuring its refractive index [OP-32] instead of (or as well as) its IR spectrum.

*2 Recover the other alkane by adding a large amount of water to the filtrate and separating the alkane layer. After it has been dried with calcium chloride, the alkane can be characterized by obtaining its IR spectrum and refractive index.

3 Write a research paper about the use of urea and thiourea in the manufacture of barbiturates, starting with sources listed in the Bibliography. Describe some of the therapeutic uses and side effects of these drugs.

Isomers and Isomerization Reactions

Reactions of Alkenes. Isomerization Reactions. Isomerism.

Operations

OP-10   Mixing
OP-13   Vacuum Filtration
OP-23   Drying Solids
OP-30   Melting Point

Before You Begin

1   Read the experiment, read or review the operations as necessary, and write an experimental plan.
2   Calculate the mass of 2.20 mmol of potassium cyanate and the theoretical yield of phenylurea.
3   Calculate the mass and volume of 1.20 mmol of dimethyl maleate and the theoretical yield of dimethyl fumarate.

Scenario

A roving science historian, Dr. Perry Celsus, travels around the country delivering lectures on the history of chemistry at scientific meetings and educational institutions. Dr. Celsus would like to enliven his lectures by demonstrating some of the pivotal experiments in organic chemistry that led to major advances in the field. These experiments include

- Friedrich Wöhler's isomerization of ammonium cyanate to urea
- The conversion of maleic acid to its geometric isomer, fumaric acid

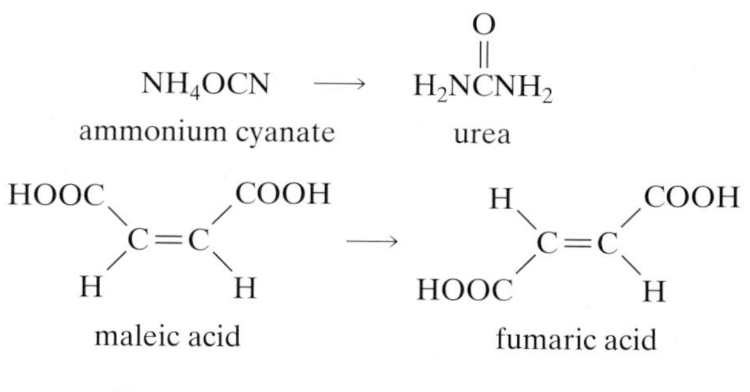

Pivotal experiments in organic chemistry

Wöhler's experiment helped demolish the vitalistic theory of organic chemistry, and the isomerization of maleic acid helped J. H. van't Hoff develop the concept of geometric (*cis-trans*) isomerism. These experiments are not suitable for demonstrations because of the time and conditions required, so Dr. Celsus has asked your Institute for help in developing updated versions of the ammonium cyanate and fumaric acid experiments that still demonstrate the scientific principles involved. Your supervisor thinks it may be possible to reenact these nineteenth century experiments in modern form by using phenylammonium cyanate in place of ammonium cyanate and dimethyl maleate in place of maleic acid. Your assignment is to find out whether these two substances do, in fact, yield constitutional and geometric isomers analogous to the ones in the original experiments.

Applying Scientific Methodology

Since you will be performing two separate experiments, you will need to develop two working hypotheses, then test them by measuring the melting points of your products.

Isomerism in the History of Chemistry

Early in the nineteenth century, most scientists believed that plants and animals possessed some "vital force" that made it possible for them to convert inorganic substances into organic compounds. Since this vital force was presumably absent in inanimate objects, it was assumed that organic compounds could not possibly be synthesized from inorganic substances in the laboratory. Vitalism received a serious blow in 1828, when Friedrich Wöhler mixed cyanic acid (HOCN) with ammonia, expecting to obtain ammonium cyanate, and came up instead with urea, a product of protein metabolism that is excreted in urine. This "lucky accident" was the first known synthesis of an organic compound from an inorganic one to take place outside a living organism. As Wöhler put it in a letter to another eminent chemist, J. J. Berzelius, "I can make urea without the aid of kidneys, either man or dog!" Wöhler's synthesis of urea suggested that there is no essential difference between inorganic and organic compounds, and it paved the way for the flowering of organic chemistry in the last half of the nineteenth century.

Ammonium cyanate and urea share the same molecular formula, CH_4N_2O, and are therefore *isomers*. At one time, scientists believed that no two compounds could be formed out of the same set of atoms. There is, after all, only one kind of H_2SO_4 and one kind of NaCl. So when two independent investigators reported the same elemental composition for cyanic acid (HOCN) and fulminic acid (HONC), Berzelius thought one of them had made a mistake. After further investigation, he became convinced that a given set of atoms can indeed combine in different ways to form compounds that have different properties, and he coined the term *isomerism* to describe this phenomenon.

Isomers such as ammonium cyanate and urea, which differ in the way their atoms are attached to one another, are called *constitutional isomers* or *structural isomers*. Isomers that differ only in the way their atoms are arranged in space are called *stereoisomers*. *Geometric isomers*, also called *cis-trans* isomers, are stereoisomers that differ from one another because of restricted rotation about the bonds connecting two or more atoms. Maleic acid and fumaric acid are geometric isomers because restricted rotation about their carbon-carbon double bonds prevents their interconversion under ordinary conditions. A geometric isomer that has certain groups other than hydrogen on the *same* side of the double bond is called a *cis* isomer, and one with those

Key Concept: The properties of a substance arise from its chemical structure, not from its source.

The term isomer *is derived from Greek words meaning "equal parts," referring to the fact that isomers have the same number and kind of atoms.*

groups on *opposite* sides is a *trans* isomer. Maleic acid is thus *cis*-2-butene-dioic acid and fumaric acid is *trans*-2-butenedioic acid.

Understanding the Experiment

Friedrich Wöhler prepared urea by several different methods. One of them involved mixing ammonium chloride and silver cyanate and then evaporating the resulting ammonium cyanate solution to dryness. When heated, ammonium cyanate decomposes to ammonia and cyanic acid, which combine to form urea.

$$NH_4Cl + AgOCN \longrightarrow \underset{\substack{\text{ammonium} \\ \text{cyanate}}}{NH_4OCN} + AgCl$$

$$NH_4OCN \xrightarrow{\text{heat}} NH_3 + \underset{\substack{\text{cyanic} \\ \text{acid}}}{HOC\equiv N} \longrightarrow \underset{\text{urea}}{H_2N\overset{\displaystyle O}{\overset{\|}{C}}NH_2}$$

You will attempt to isomerize the phenyl derivative of ammonium cyanate by a similar reaction. As shown in the "Reactions and Properties" section, aniline reacts with hydrochloric acid to form the organic salt phenylammonium chloride. By combining an aqueous solution of phenylammonium chloride with potassium cyanate, you can prepare a solution containing phenylammonium cyanate. Then you will see whether heating this solution yields a different product. If it does, a melting-point determination should tell you whether you have synthesized phenylurea, a constitutional isomer of phenylammonium cyanate.

Sometimes one of a pair of geometric isomers can be converted to the other under conditions that cause temporary cleavage of a pi bond. When maleic acid is heated with a trace of bromine in the presence of light, some of the bromine molecules break apart into bromine atoms. A bromine atom then adds to one end of the double bond of a maleic acid molecule, breaking the pi bond. This allows the molecule to rotate freely about the remaining sigma bond. If the bromine atom is ejected when the COOH groups are opposite one another, the pi bond is regenerated to yield a molecule of fumaric acid.

Mechanism of the isomerization of maleic acid

Maleic acid and fumaric acid are both solids, so this reaction must be run in solution, and the fumaric acid cannot be separated from the starting material by filtration. But dimethyl maleate (the methyl ester of maleic acid) is a liquid while dimethyl fumarate is a solid, so if dimethyl maleate can be induced to isomerize to dimethyl fumarate, the liquid *cis* isomer should change to a solid *trans* isomer that can be separated by filtration. If molecular model sets are available, you can simulate the isomerization of dimethyl maleate after building a molecular model of it.

Reactions and Properties

A   PhNH$_2$ + HCl $\longrightarrow$ PhNH$_3$Cl
      aniline   phenylammonium
                      chloride

PhNH$_3$Cl + KOCN $\longrightarrow$ PhNH$_3$OCN + KCl
                         phenylammonium
                          cyanate

$$\text{PhNH}_3\text{OCN} \longrightarrow \text{PhNH}\overset{\displaystyle O}{\overset{\|}{\text{C}}}\text{NH}_2$$
                         phenylurea

B   CH$_3$OOC, COOCH$_3$ $\xrightarrow{\text{Br}_2}$ CH$_3$OOC, H
               C=C   C=C
          H   H   H   COOCH$_3$
       dimethyl maleate   dimethyl fumarate

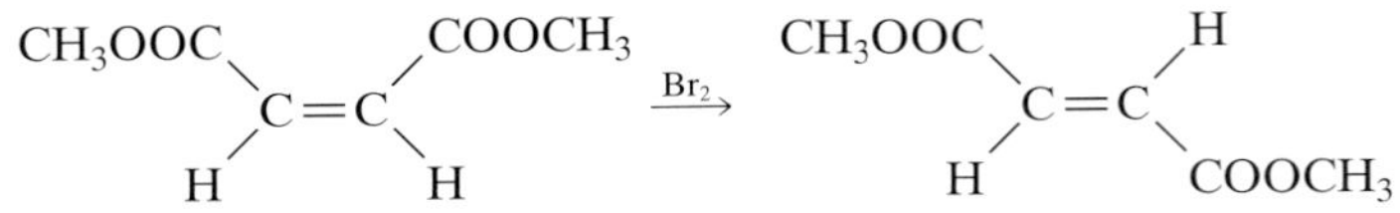

Table 17.1   Physical properties

	M.W.	mp	bp	d
aniline	93.1	−6	184	1.022
potassium cyanate	81.1	800 d		
phenylurea	136.2	147	238	
dimethyl maleate	144.1	8	205	1.151
dimethyl fumarate	144.1	104	193	

Note: mp and bp are in °C, d is in g/mL.

Directions

A. *Isomerization of Phenylammonium Cyanate*

Safety Notes

> **Aniline is poisonous and may be carcinogenic. It can cause serious injury or death if swallowed, inhaled, or absorbed through the skin. Wear protective gloves and dispense under a fume hood; avoid contact, and do not breathe its vapors.**

aniline

Take Care! Wear gloves, avoid contact with aniline and do not breathe its vapors.

Under the hood, prepare a solution of phenylammonium chloride by combining 1.0 mL of 3 *M* hydrochloric acid with 1.0 mL of water in a small beaker and adding 0.20 mL (~2.2 mmol) of recently distilled aniline. Dissolve 2.20 mmol of potassium cyanate in 2.0 mL of water in a small test tube and mix this solution with the phenylammonium chloride solution to form phenylammonium cyanate. Stir [OP-10] for 2–3 minutes, then add 2.0 mL of water and heat the solution to 80°C; any solid should dissolve during this time. Let the solution cool slowly to room temperature while you proceed with part **B**, then cool it further in an ice/water bath. Collect any product by vacuum filtration [OP-13] and dry [OP-23] it thoroughly. Weigh the dry product and measure its melting point [OP-30].

If the solution is dark colored, add a small amount of pelletized Norit, stir it for 2 minutes at 80°C, and then use a preheated filter-tip pipet to transfer it to another container for crystallization.

Observe and Note: What happened as the solution cooled?

Waste Disposal: Unless your instructor directs otherwise, pour the filtrate down the drain.

B. *Isomerization of Dimethyl Maleate*

> **Bromine is toxic and corrosive, and its vapors are very harmful. Dichloromethane may be harmful if ingested, inhaled, or absorbed through the skin. Avoid contact with the bromine/dichloromethane solution and do not breathe its vapors.**

Safety Notes

bromine

Take Care!   Avoid contact with the bromine solution and do not breathe its vapors.

Measure 1.20 mmol of dimethyl maleate into a small test tube, and mix in a drop of 1 *M* bromine in dichloromethane. Set the test tube in a small beaker that is half full of boiling water and place the beaker about 15 cm away from an unfrosted 75- or 100-watt light bulb. Keep the water boiling for 10 minutes or more with the light switched on, then remove the test tube and cool it in ice water until crystallization is complete. Use a stirring rod to break up the product into small particles (be careful not to punch a hole through the test tube). Collect the product by vacuum filtration [OP-13], wash it on the filter with a small amount of cold 95% ethanol, and dry it. Measure the mass and melting point [OP-30] of the product.

Waste Disposal: Pour the filtrate down the drain.

If molecular-model kits are available, construct a molecular model of dimethyl maleate and show that it cannot be converted to dimethyl fumarate without breaking bonds. Using a model of Br_2, simulate the course of the reaction you carried out in this experiment, referring to the mechanism for the isomerization of maleic acid.

Exercises

1 (a) Estimate the percentage of your dimethyl maleate that was converted to dimethyl fumarate. (b) Would it be possible to prepare dimethyl maleate in good yield from dimethyl fumarate (using bromine and light) under experimental conditions that permitted recovery of the dimethyl maleate? Why or why not?

2 Methylenemalonic acid, $CH_2{=}C(COOH)_2$, is an isomer of maleic acid. (a) What kind of isomers are methylenemalonic acid and maleic acid, constitutional or geometric? (b) Would methylenemalonic acid yield a geometric isomer under the conditions of this experiment (part **B**)? If so, give the structure of the product; if not, explain why.

3 Describe and explain the possible effect on your results of the following experimental errors or variations. (a) Your instructor was out of aniline, so *p*-toluidine (*p*-methylaniline) was substituted. (b) It was a dark and

stormy day, and the lights were out in the laboratory (but the gas was on). (c) You misread the label on a bottle of dimethyl malonate and used it instead of dimethyl maleate.

4 (a) Explain why dimethyl maleate has a higher boiling point than dimethyl fumarate. (b) Explain why dimethyl fumarate has a higher melting point than dimethyl maleate.

5 Propose a mechanism for the isomerization of phenylammonium cyanate to phenylurea.

Other Things You Can Do

(Starred items require your instructor's permission.)

*1 Investigate the structures and properties of isomers as described in Minilab 15.

*2 Make models for as many isomers having the molecular formula C_3H_6O as you can, and identify any pairs of geometric isomers. You should also be able to construct models for pairs of molecules that, like your two hands, are nonsuperposable mirror images of one another. These are called mirror-image isomers, or enantiomers.

3 Starting with sources listed in the Bibliography, write a research paper about the manufacture of urea and its use in the production of such commercial products as urethane plastics, urea-formaldehyde resins, barbiturates, and jet fuel.

Structures and Properties of Stereoisomers

Isomerization Reactions. Stereoisomerism. Infrared Spectrometry. Gas Chromatography.

Operations

OP-7   Heating
OP-10   Mixing
OP-15   Extraction
OP-16   Evaporation
OP-22   Drying Liquids
OP-31   Boiling Point
OP-32   Refractive Index
OP-33   Optical Rotation
OP-34   Gas Chromatography
OP-36   Infrared Spectrometry

Before You Begin

1 Read the experiment, read or review the operations as necessary, and write an experimental plan.
2 Before beginning this experiment, you should know how to draw stereochemical structural formulas, how to designate configurations at stereocenters, and how to classify stereoisomers as enantiomers and diastereomers. Review the sections of your lecture textbook on stereochemistry if you need help in any of these areas.

Scenario

The new-age herbalist Basil Wormwood explores the composition and uses of natural products from plants. For example, he isolates and analyzes essential oils from various plants using techniques he learned in his college organic chemistry course. While pursuing his studies, he observed several phenomena that puzzled him, so he has contacted the Institute to see if its consulting chemists could explain them.

- While Basil was away, his assistant accidentally spilled some vinegar in a batch of peppermint oil. When he returned, he removed the vinegar by extraction, but the aroma of the oil had changed markedly.
- After isolating the major component of caraway oil (from caraway seeds) and the major component of spearmint oil (from spearmint leaves), Basil discovered that these substances, whose odors are very different, have virtually identical physical and chemical properties.
- When examined under a microscope, the tartaric acid crystals he received from a supplier looked different than the crystals of natural tartaric acid that he had used before.

caraway plant   spearmint plant

Your assignment, then, is threefold:

- To find out what happens when menthone from peppermint oil is treated with acid.
- To find out how the major components of caraway and spearmint oils differ, if they do.
- To find out whether the tartaric acid provided is natural L-(+)-tartaric acid or an "unnatural" tartaric acid stereoisomer.

Applying Scientific Methodology

You should find enough information in the experiment to formulate working hypotheses for the first two problems. You will need to gather some experimental evidence before you can develop a meaningful hypothesis for the third problem.

Alice Through the Looking Glass: Dissymmetry and Life

In Lewis Carroll's *Through the Looking Glass*, Alice, contemplating the world she views in her mirror, wonders aloud to her cat whether looking-glass milk would be good to drink. Such an inquiry might seem naive, even to a cat. If you hold a glass of milk in front of a mirror, the milk and its reflection look exactly alike, so why should "mirror-image milk" be any different than the milk we ordinarily drink? You may be surprised to learn that mirror-image milk would not be very good to drink; it would be quite indigestible, and its taste would be bitter and unpleasant. To understand why, let's take an imaginary trip to a looking-glass world.

Imagine yourself on board the spaceship *Icarus* bound for the planetary system of Barnard's star. Arriving on the surface of its earthlike third planet, called *Arret* by the inhabitants, your landing party is invited to a feast by some friendly Arretians. You are served their standard banquet fare: roast *krop*, overdone *iloc'corb*, crusty *yawarac* seed rolls, and an aromatic *t'nim* tea. You had been looking forward to a change from the monotonous space diet, but you soon lose your appetite. Most of the food tastes flat or bitter, the seeds on the roll have a minty flavor, and the tea smells like caraway seeds! After politely declining a second helping of *krop*, you return to your shuttlecraft with the rest of the landing party. Before long, you and the rest of the crew are suffering from indigestion, causing a serious breach in interplanetary relations.

Earthly organisms are composed mostly of *chiral* molecules, such as the D-monosaccharides in complex carbohydrates and the L-amino acids that make up proteins and enzymes. It is conceivable that somewhere in the universe there exists a mirror-image planet, otherwise similar to Earth, where carbohydrates are composed of L-monosaccharides and proteins of D-amino acids, and the configurations of other chiral compounds are reversed as well. On such a planet, many foods would taste different from their earthly counterparts, and we earthlings would find them indigestible. The chiral compounds on that planet would still be optically active; that is, they would rotate plane-polarized light, as ours do. But they would rotate the light in the opposite direction.

One of the great unsolved mysteries of life concerns the origin of optically active compounds. Living systems are composed of dissymmetric (chiral)

A proposed expedition to Barnard's star was described in New Scientist **1974**, *63*, 522.

Key Concept: *A chiral compound is one exhibiting "handedness." Like a glove, a chiral molecule lacks a plane of symmetry and cannot be superposed on its mirror image.*

molecules, and molecular dissymmetry is necessary for life as we know it, but how did such dissymmetry come about? Was there some "molecular Adam," a single dissymmetric molecule that gave rise to all molecular dissymmetry on Earth? Or is molecular dissymmetry a phenomenon that appeared independently at many different locations as a consequence of some kind of fundamental dissymmetry in the universe? The discovery (in 1997) that some amino acids in an Australian meteorite contain an excess of the L-enantiomer appears to support the hypothesis that some natural process favors one enantiomer over another. But convincing proof of such a hypothesis is hard to come by, and the answer may never be fully known.

The basic molecules of life—proteins, carbohydrates, nucleic acids, and enzymes—are chiral and are built up of smaller units that are also chiral. A strand of deoxyribonucleic acid (DNA), for example, consists of two long chains, each having a backbone of linked D-2-deoxyribose molecules twisted into a right-handed double helix (see Figure 18.1). DNA and ribonucleic acid (RNA) regulate the synthesis of proteins from L-amino acids, which are combined in specific sequences inside cellular structures called ribosomes. Some of these proteins make up the enzymes that assist in the digestion of carbohydrates, yielding D-glucose to be used by the body for fuel. If life somewhere else in the universe were based on a mirror-image DNA made up of L-2-deoxyribose and twisted into a left-handed double helix, then protein synthesis could utilize only D-amino acids and the corresponding enzymes could digest only L-carbohydrates. In other words, if the configuration of one link in the chain of life (as we know it) is reversed, all the rest must be reversed as well.

If we could, by some magical contrivance, pass through the looking glass as Alice did, all of the people, plants, and other organic matter in the looking-glass world would presumably be constructed of these mirror-image molecules. We could not survive in such a world (though Alice did, in Lewis Carroll's imagination) because digestion, metabolism, reproduction, and other life processes involving chiral molecules would be inhibited or prevented entirely.

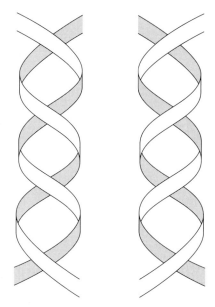

Figure 18.1 Left-handed and right-handed double helixes

Understanding the Experiment

Peppermint oil is a mixture of several optically active components, including (−)-menthol and the corresponding ketone, (−)-menthone. When an aldehyde or ketone has a stereocenter next to its carbonyl group, the configuration at the stereocenter may change under acidic or basic conditions, as illustrated by the following mechanism for the isomerization of (R)-3-methyl-2-butanone to its enantiomer.

Mechanism of the acid-catalyzed isomerization of (R)-3-methyl-2-butanone to (S)-3-methyl-2-butanone

In part **A** of this experiment, you will see whether treating (−)-menthone from peppermint oil with acid can bring about a similar isomerization. Note that (−)-menthone has two stereocenters (see "Reactions and Properties"), so changing the configuration at just one of them will yield a diastereomer of (−)-menthone, not its enantiomer. If the configuration at the carbon next to the carbonyl group changes, the product will be (+)-isomenthone. You will analyze the product by gas chromatography to determine whether or not isomerization has taken place and, if so, what percentage of the (−)-menthone has isomerized.

Enantiomers have identical physical properties except the direction in which they rotate plane-polarized light. If one enantiomer of menthone, for example, rotates polarized light in a clockwise (+) direction, the other enantiomer will rotate polarized light by the same amount in a counterclockwise (−) direction. Enantiomers may also differ in some chemical properties, such as the way in which they interact with chiral substrates. These differences are particularly important in biochemical systems. For example, some kinds of olfactory receptors are apparently chiral, so the (+)- and (−)-enantiomers of a compound, when they interact with such receptors, may smell different. Carvone, a ketone found in the essential oils of both caraway seeds and the spearmint plant, is an example of a chiral compound whose enantiomers have markedly different odors.

Experiments that proved the odor differences between carvone enantiomers are described in Science **1971**, *172*, 1043.

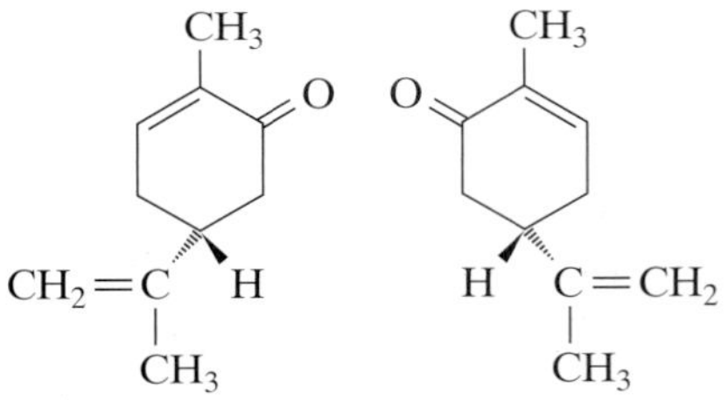

(R)-(−)-carvone   (S)-(+)-carvone

In part **B** of this experiment, you and your coworkers will compare some physical properties and the infrared (IR) spectra of the carvones from both spearmint and caraway oils, as well as their odors. You will also measure their optical rotations to find out whether the carvones are the same or different and, if they are different, which enantiomer is present in each oil.

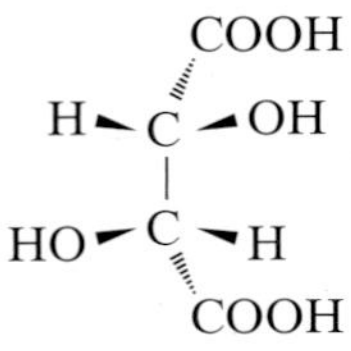

Tartaric acid can be produced from potassium hydrogen tartrate (cream of tartar), a by-product of winemaking. This natural tartaric acid is the L-(+)-form illustrated here. Since tartaric acid has two stereocenters, several other stereoisomers are possible. In part **C** of this experiment, you will measure the optical rotation of the tartaric acid provided to determine whether it is the natural form and, if it is not, what its structure is.

If molecular models are available, you can use them in part **D** of this experiment to explore the stereochemical relationships among these compounds.

L-(+)-tartaric acid

Reactions and Properties

CH$_3$ [structure] $\underset{}{\overset{\text{acid}}{\rightleftharpoons}}$ CH$_3$ [structure]

CH CH
CH$_3$ CH$_3$ CH$_3$ CH$_3$
(−)-menthone   (+)-isomenthone

Table 18.1 Physical properties

	M.W.	d	$[\alpha]$
(R)-carvone	150.2	0.96	−62°
(S)-carvone	150.2	0.96	+62°
(−)-menthone	154.2	0.895	−30°
(+)-isomenthone	154.2	0.900	+92°

Note: Densities are in g/mL.

Directions

Acetic acid causes chemical burns that can seriously damage skin and eyes; its vapors are highly irritating to the eyes and respiratory tract. Wear gloves, dispense under a hood; avoid contact and do not breathe its vapors. Diethyl ether is extremely flammable and may be harmful if inhaled. Do not breathe its vapors, and keep it away from flames and hot surfaces.

Safety Notes

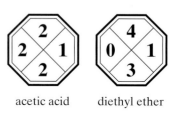

acetic acid   diethyl ether

Take Care! Wear gloves, do not breathe the vapors of acetic acid.

A. *Isomerization of (−)-Menthone*

Under the hood, mix 0.20 mL of (−)-menthone with 1.0 mL of glacial acetic acid and 1.0 mL of 1 M HCl in a 5-mL conical vial. Attach an air condenser and drop in a magnetic stirring device. Heat the mixture under reflux [OP-7], while stirring [OP-10], for 30 minutes, then let it cool to room temperature. Transfer the mixture to a centrifuge tube, using a small amount of water for the transfer. Add enough 4 M NaOH to neutralize the solution to red litmus. Cool it to room temperature and extract [OP-15] it with two 3-mL portions of solvent-grade diethyl ether. Dry [OP-22] the combined ether extracts with anhydrous magnesium sulfate or sodium sulfate, then evaporate [OP-16] the ether. Obtain gas chromatograms [OP-34] of (−)-menthone and of your product on an open tubular 10% Carbowax column or another column specified by your instructor. Calculate the percentage of (−)-menthone that isomerized to (+)-isomenthone, assuming that the GC detector response factors for the menthone isomers are equal.

Take Care! Keep the ether away from flames and hot surfaces.

B. *Properties of Carvones from Spearmint and Caraway Oils*

Students should work in pairs, with each student using about 1 mL of carvone from a different oil. Compare the odors of the two carvones. Measure the boiling point [OP-31] and refractive index [OP-32] of your carvone. Record its IR spectrum [OP-36] or obtain the spectrum from your instructor. Weigh the remainder of your carvone accurately and use a 10-mL volumetric flask to prepare a solution of it in 95% ethanol. Measure the optical rotation [OP-33] of this solution and of a blank in 1-dm polarimeter cells and calculate its specific rotation. Based on your results, decide which carvone enantiomer you have. Compare the properties you measured with those of your lab partner's carvone, and compare their IR spectra as well.

Waste Disposal: Unless your instructor indicates otherwise, place the solution in a designated waste container.

C. *Identification of a Tartaric Acid Stereoisomer*

Using about 1.0 g of solute per 10 mL of solution (measure accurately), prepare an aqueous solution of the tartaric acid stereoisomer provided, then measure its optical rotation [OP-33] and the optical rotation of a blank in 1-dm polarimeter cells and calculate its specific rotation. Using the fact that the specific rotation of L-(+)-tartaric acid is +12°, predict the specific rotations of the other stereoisomers, and deduce the structure of your tartaric acid (assume that it is not a racemic mixture).

Waste Disposal: Unless your instructor indicates otherwise, place the tartaric acid solution in a designated waste container.

D. *Stereochemical Exercises with Molecular Models*

Construct a molecular model of (−)-menthone in its most stable chair conformation. Show how it can be converted to a model of (+)-isomenthone by interchanging atoms or groups. Use the models to confirm that the two compounds are diastereomers, not enantiomers, and decide which one is more stable. Determine the configuration (*R* or *S*) at each stereocenter of each isomer. Construct a molecular model of your carvone, and compare it with your lab partner's molecular model to confirm that the two models represent enantiomers. Determine the configuration at the stereocenter of your model. Construct a molecular model of your tartaric acid stereoisomer, and determine the configuration at each stereocenter. Provide stereochemical drawings representing the structures of all the models you constructed, giving the configurations at their stereocenters.

Exercises

1 Explain any similarities or differences in the properties of the carvone enantiomers and interpret your IR spectrum as completely as you can.

2 (a) Draw chair-form structures for the most stable conformations of (−)-menthone and (+)-isomenthone and decide which diastereomer should be more stable. (b) Is the composition of your isomerization mixture consistent with this conclusion? Explain.

3 Describe and explain the possible effect on your results of the following experimental errors or variations. (a) In part **A**, you misread the label on a bottle of (−)-menthol and used it instead of (−)-menthone. (b) In part **B**, you mixed your carvone with an equal amount of your partner's carvone before measuring the optical rotation. (c) Your tartaric acid was

the (−)-stereoisomer, but in measuring its optical rotation, you read the scale when the analyzer was rotated 180° from the correct setting.

4 (a) Propose a mechanism for the isomerization of (−)-menthone to (+)-isomenthone in the presence of acid. (b) Explain why the acid-catalyzed isomerization of (−)-menthone doesn't yield its enantiomer, (+)-menthone.

5 A synthetic form of tartaric acid has a specific rotation of 0°, but its melting point is very different from that of *meso*-tartaric acid, which is also optically inactive. Explain, using stereochemical drawings.

6 The steroid cholic acid is said to have 2,047 possible stereoisomers. Indicate each stereogenic carbon atom on the cholic acid molecule with an asterisk, and perform a calculation to verify this isomer number.

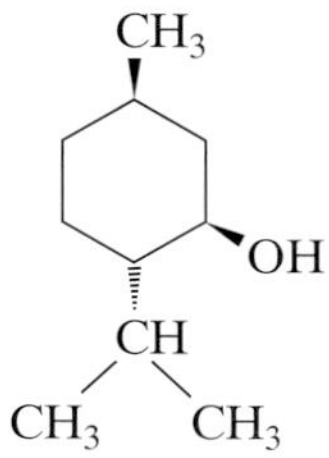

(−)-menthol

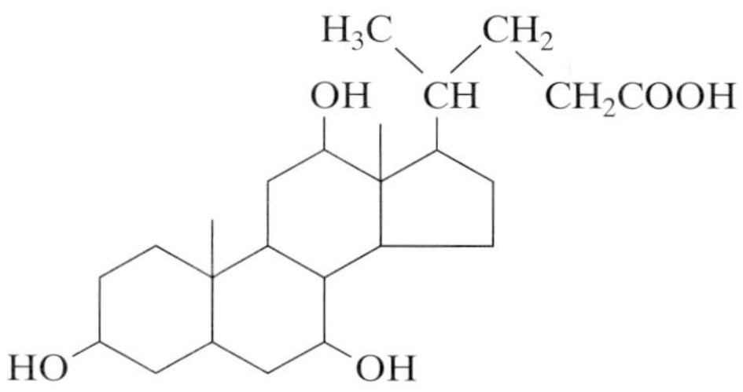

cholic acid

Other Things You Can Do

(Starred items require your instructor's permission.)

*1 You can isolate your carvone enantiomer for part **B** from spearmint or caraway oil by standard-scale column chromatography. Following the procedure in OP-18 for slurry-packing a microburet, pack a 25-mL buret with silica gel, using high-boiling petroleum ether. Introduce about 2 g of the essential oil onto the top of the column and elute it with the following solvents: 25 mL of high-boiling petroleum ether; 50 mL of 10% dichloromethane/petroleum ether; 25 mL of 20% dichloromethane/petroleum ether; and 125 mL of 50% dichloromethane/petroleum ether. Put the first 100 mL of eluate in a designated solvent recovery container and collect the rest in 25-mL fraction collectors. Evaporate the fractions and measure the refractive indexes of the residues. Use the purest carvone fractions ($n_D \cong 1.499$ at 20°C) in part **B**.

2 Read the articles beginning on pages 448 and 455 of *J. Chem. Educ.* **1972**, *49* and write a research paper about the origin and consequences of optical activity. Look for additional material from more recent literature.

3 Starting with sources listed in the Bibliography, write a research paper about chiral drugs, describing differences in the physiological properties of enantiomers of the same drugs.

Bridgehead Reactivity in an S$_N$1 Solvolysis Reaction

Reactions of Alkyl Halides. Nucleophilic Substitution. Carbocations. Reaction Kinetics.

Operations

OP-5   Measuring Volume
OP-7   Heating
OP-10   Mixing

Before You Begin

Read the experiment, read or review the operations, and write a brief experimental plan.

Scenario

Bridgehead Balms, Inc., is a small pharmaceutical company seeking to develop new drugs, and more economical routes to existing drugs, based on the adamantane ring system. One such drug, amantadine, is an antiviral agent that is effective against rubella (German measles) and some influenza viruses.

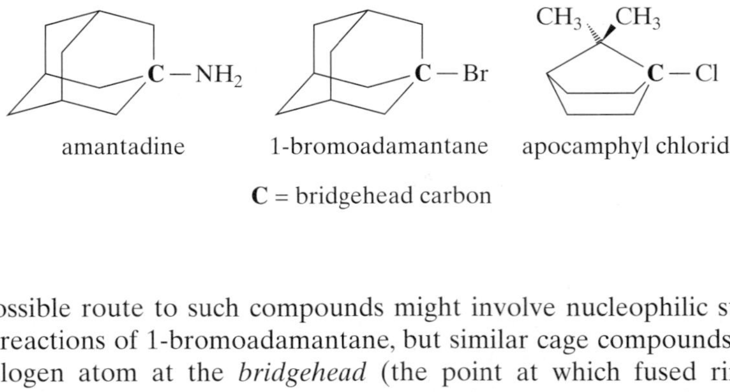

amantadine   1-bromoadamantane   apocamphyl chloride

C = bridgehead carbon

A possible route to such compounds might involve nucleophilic substitution reactions of 1-bromoadamantane, but similar cage compounds having a halogen atom at the *bridgehead* (the point at which fused rings are joined), such as apocamphyl chloride, are quite resistant to such reactions. Bridgehead Balms has commissioned your institute to study the reactivity of 1-bromoadamantane (1) to determine whether it will undergo nucleophilic substitution reactions with hydroxylic solvents and if so, (2) to see how its reactivity compares with that of a comparable open-chain tertiary halide, 2-bromo-2-methylpropane (*t*-butyl bromide). To compare the reactivities of these compounds quantitatively, you will need to measure the rate constants for their reactions.

Applying Scientific Methodology

After reading the experiment, you should be able to formulate working hypotheses based on each of the problems implied in the Scenario. You will need to complete the rate measurements to obtain a quantitative result relating to the second problem, but you should be able to predict which tertiary halide should react faster before you begin.

A Gem Among Molecules

Adamantane, whose name is derived from a Greek word for diamond, has molecules of elegant symmetry that can be regarded as fragments of a tetrahedral diamond lattice. Molecular models of this unique molecule reveal that it consists of four interlocking chair-form cyclohexane rings, arranged somewhat like the four planes of a tetrahedron. A space-filling model is nearly spherical, and this molecular shape results in a particularly stable crystal lattice that is responsible for adamantane's unusually high melting point of 268°C.

Like the spherical "soccer ball" structure of buckminsterfullerene (an unusual allotrope of carbon), the extraordinary structure of adamantane is intriguing to chemists because adamantyl systems have properties that make them almost ideal for the study of certain chemical phenomena. The rigid adamantane skeleton results in a system of known geometry with unstrained, tetrahedral bond angles; the cyclohexane rings making up the adamantane molecule come together at four points, forming a bridgehead at each junction; and the cagelike structure prevents certain kinds of interactions and reaction mechanisms, thereby simplifying the analysis of reaction parameters.

In 1939, P. D. Bartlett and L. H. Knox discovered that another bridged compound, apocamphyl chloride, is surprisingly inert to reagents that usually bring about nucleophilic substitution reactions. For a nucleophile to attack from the back side of its bridgehead carbon, as required for an S$_N$2 reaction, the nucleophile would have to somehow get inside a cagelike molecule of apocamphyl chloride, which is highly unlikely. In an S$_N$1 reaction, the leaving group must leave, forming an intermediate carbocation, before the nucleophile attacks. Carbocations prefer a planar geometry, as illustrated below for the *t*-butyl carbocation—the intermediate in the S$_N$1 reactions of 2-bromo-2-methylpropane. But attaining a planar geometry at the bridgehead carbon is impossible in the rigid apocamphyl system.

Unlike the highly strained cage of apocamphyl chloride, a molecule of 1-bromoadamantane has normal tetrahedral bond angles with no bond-angle strain. In a 1-adamantyl carbocation, the bridgehead carbon can flatten out a bit to attain a bond angle of 113°, which is somewhere between the tetrahedral angle of the adamantane ring and the 120° angle of the *t*-butyl carbocation.

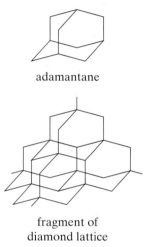

adamantane

fragment of
diamond lattice

You can read the paper by Bartlett and Knox in J. Amer. Chem. Soc. **1939**, *61*, 3184.

t-butyl carbocation   1-adamantyl carbocation

In an S$_N$1 reaction, a more stable carbocation forms faster than one that is less stable. Therefore, any factor that decreases the stability of a carbocation

Key Concept: *The more stable a reactive intermediate is, the faster it will form.*

tends to decrease the reactivity of the substrate from which the carbocation forms. You will be testing this principle as you compare the reactivities of 1-bromoadamantane and 2-bromo-2-methylpropane.

Understanding the Experiment

When Bartlett and Knox showed that bridgehead compounds such as apo-camphyl chloride are quite unreactive with nucleophiles, they suggested that studies of bridgehead reactivity might yield valuable information about reaction mechanisms and the geometry of transition states. That suggestion was taken up by numerous investigators; particularly fruitful results have been obtained in the study of *solvolysis reactions*—nucleophilic substitution reactions in which the solvent acts as the nucleophile. In this experiment, you will measure the first-order rate constants for the solvolysis reactions of 1-bromoadamantane and 2-bromo-2-methylpropane with an ethanol-water mixed solvent.

The solvolysis of 2-bromo-2-methylpropane and similar halides by hydroxylic solvents, such as water and ethanol, is believed to proceed by an S_N1 mechanism involving the formation of a carbocation intermediate.

1. $t\text{-BuBr} \longrightarrow t\text{-Bu}^+ + \text{Br}^-$ (slow)

2. $t\text{-Bu}^+ + \text{SOH} \longrightarrow t\text{-Bu}\overset{\text{H}}{\underset{\oplus}{\text{OS}}}$ (fast)

(SOH = hydroxylic solvent)

3. $t\text{-Bu}\overset{\text{H}}{\underset{\oplus}{\text{OS}}} + \text{Base:} \longrightarrow t\text{-BuOS} + \text{Base: H}^+$ (fast)

t-Butyl bromide solvolysis mechanism

Dissociation of the alkyl bromide is the rate-determining step, so the reaction is first order with the rate equation

$$\frac{-d[\text{RBr}]}{dt} = k[\text{RBr}]$$

Role of solvent in displacement reactions

solvent as nucleophile   solvent as electrophile

Although the solvent does not appear in the rate equation, it can affect the reaction rate by assisting in the formation of the carbocation. Some solvent molecules may help push the leaving group off from the rear (except for most bridgehead compounds), while others pull it off from the front. Thus the rate of an S_N1 solvolysis reaction should depend on both the polarity of the solvent and the stability of the carbocation formed in the first step; the more polar the solvent and the more stable the carbocation, the faster is the reaction.

The rate constant for a first-order reaction can be calculated from the following integrated rate law, where c_o is the initial concentration of the substrate and c is its concentration at time t:

$$\ln \frac{c}{c_o} = -kt \qquad \textbf{(1)}$$

A solvolysis reaction of an alkyl bromide with a hydroxylic solvent (SOH, where S = H or R) will produce hydrogen bromide according to the general equation

$$R\!-\!Br + SOH \rightarrow R\!-\!OS + HBr$$

Therefore, the concentration (c) in Equation **1** can be evaluated indirectly by measuring the amount of HBr that forms during the reaction. You will do this by adding measured portions of a KOH solution from a buret and recording the time it takes for the HBr evolved to neutralize the added KOH, as shown by the color change of an indicator. At any given time during the reaction, the concentration of the alkyl bromide (c) will equal its initial concentration (c_o) minus the concentration of the HBr that has formed by then (c_{HBr}). The concentration of HBr is proportional to the volume of KOH needed to neutralize it (V), and the initial concentration of the alkyl bromide is proportional to the total volume of KOH solution needed to neutralize all of the HBr produced (V_∞). Thus we can derive an expression for the concentration factor (c/c_o) from Equation **1** in terms of the volume of KOH solution added.

$$\frac{c}{c_o} = \frac{(c_o - c_{HBr})}{c_o} = 1 - \frac{c_{HBr}}{c_o} = 1 - \frac{V}{V_\infty}$$

The solvolysis of 1-bromoadamantane will be studied in 40% (by volume) aqueous ethanol, in which both water and ethanol act as nucleophiles. Because the reaction of 2-bromo-2-methylpropane in 40% ethanol is difficult to measure accurately, its solvolysis reaction will be conducted in 80% ethanol instead. To make a meaningful comparison between the two reactions, you will have to estimate the rate constant for 2-bromo-2-methylpropane in 40% ethanol using Equation **2** (the Winstein-Grunwald equation), which relates the rate of a solvolysis reaction to the ionizing power of the solvent.

$$\ln \frac{k}{k_o} = mY \qquad\qquad (2)$$

In this equation, k_o is the rate constant for a reaction in the reference solvent, 80% ethanol, and k is the rate constant in the actual reaction solvent (40% ethanol in this experiment) at the same temperature. Y is a measure of the reaction solvent's ionizing power, and m measures the sensitivity of the substrate to changes in ionizing power. The value of m for 2-bromo-2-methylpropane is 0.94, and the value of Y for 40% ethanol is 2.20.

For each kinetic run, you will prepare the appropriate reaction solvent by combining 95% ethanol and water in such proportions that the solvent will be 40% or 80% aqueous ethanol after you have added the alkyl bromide. It is important to measure the solvents accurately because an error in solvent composition can markedly affect the solvolysis rate. You will then add some bromthymol blue indicator and the alkyl bromide to the reaction solvent, followed by a measured portion of potassium hydroxide in the appropriate solvent. The indicator should turn blue as each portion of KOH solution is added, changing to green when enough HBr is produced by the solvolysis reaction to neutralize the added KOH. As the solution becomes more acidic, its color fades to yellow. Each portion of KOH solution will consume the HBr produced by the reaction of approximately 5% of the alkyl halide; to obtain sufficient data for the rate calculations, you should continue the run until the reaction is at least 50% complete, which requires 10 portions or more of the KOH solution. After the last portion of KOH has been added, you will heat the reaction flask gently to bring the reaction to completion, and then titrate the solution with more KOH to determine V_∞.

Reactions and Properties

$$RBr + H_2O \rightarrow ROH + HBr \quad and$$

$$RBr + CH_3CH_2OH \rightarrow ROCH_2CH_3 + HBr$$

$$(R = 1\text{-adamantyl or } t\text{-butyl})$$

$$HBr + KOH \rightarrow H_2O + KBr$$

Table 19.1 Physical properties

	M.W.	mp	bp	d
1-bromoadamantane	215.1	118		
2-bromo-2-methylpropane	137.0	−16	73	1.221

Note: mp and bp are in °C, density is in g/mL.

Directions

The alkyl halides must be protected from moisture; be sure that your glassware is clean and dry. With your instructor's permission, work in pairs, with one student recording the times and the other adding the KOH solution.

Safety Notes

2-Bromo-2-methylpropane is harmful if inhaled or absorbed through the skin, and it may be carcinogenic. Avoid contact and do not breathe its vapors.

A. *Solvolysis of 1-Bromoadamantane in 40% Ethanol*

Preparation of Solutions.   Prepare an indicator blank by measuring 25 mL of a pH 6.9 buffer into a 50-mL Erlenmeyer flask and adding 4 drops of bromthymol blue indicator solution (the blank should be green). Set this flask aside while you prepare the reaction mixture.

Measuring [OP-5] as accurately as possible, combine 10.0 mL of 95% ethanol and 14.5 mL of water in a 50-mL Erlenmeyer flask. Add 4 drops of bromthymol blue solution and swirl to mix. Clamp this reaction flask to a ring stand and lower it into a water bath containing room-temperature (20–25°C) water set on a magnetic stirrer. The water temperature should remain nearly constant throughout a kinetic run. Adjust the water level so that the bath is about two-thirds full, and measure the water temperature. Now fill a clean, dry 10-mL buret with 0.0050 *M* KOH in 40% ethanol, and record the initial buret reading accurately. Obtain a timer or a watch that measures seconds, if you don't already have one.

Take Care! Be sure to use the right KOH solution.

Observe and Note: Take note of the color changes throughout the kinetic runs and try to explain them.

Kinetic Run.   Drop in a stir bar and begin stirring [OP-10] the reaction mixture. Add 0.50 mL of a *freshly prepared* 0.10 *M* solution of

1-bromoadamantane in absolute ethanol to the reaction flask and *immediately* record the time of addition to the nearest second (or start the timer). Without delay, add 0.50 mL of the KOH/40%-ethanol solution to the reaction mixture and record the buret reading. (If the solution does not turn blue, add more KOH until it does, and then record the buret reading.) Place the indicator blank near the reaction mixture and record the time (to the nearest second) when the solution changes to the same shade of green as the blank. (The color change will be easier to see if you set the flasks on a sheet of white paper.) Within a minute of the color change, add another 0.50-mL portion of the KOH solution, and record the time when the solution again turns from blue to green. Repeat the addition of 0.50-mL portions of KOH, recording the buret reading and the time of the color change after each addition, until at least 10 portions have been added.

The 1-bromoadamantane solution can be prepared by dissolving 0.11 g of 1-bromoadamantane in 5.0 mL of absolute ethanol and then shared among up to five students.

Determination of V_∞.   After the last color change, seal the flask with a square of Parafilm and heat [OP-7] the reaction mixture in a 60°C water bath for 30 minutes (alternatively, let it stand overnight or longer at room temperature). Cool the solution to room temperature, and then titrate it with the 0.0050 M KOH solution to the green end point. If the green color fades to yellow after standing a few minutes, heat the flask in the water bath for about 10 minutes longer. After cooling, again titrate the solution to the green end point. Subtract your initial buret reading (recorded before you started the kinetic run) from the final reading to get V_∞. For each run, compute the time (t) of each color change in seconds, measured from the time of addition of the alkyl bromide ($t = 0$). Calculate $\ln(1 - V/V_\infty)$ for each t value, where V is the total volume of KOH that has been added up to that time. Using good graph paper, plot $\ln(1 - V/V_\infty)$ versus t and determine the value of k (in s^{-1}) for the alkyl bromide from the slope of the line. (Alternatively, you may use a calculator or computer that has a linear regression program to determine the least-squares slope from your data.)

You do not have to add exactly 0.50 mL of the KOH solution each time, but you must record the exact buret reading after each addition.

Observe and Note: Does the time between color changes increase, decrease, or stay the same as the reaction proceeds?

Waste Disposal: Unless your instructor directs otherwise, the reaction mixtures and unused KOH solutions can be poured down the drain.

B. *Solvolysis of 2-Bromo-2-methylpropane in 80% Ethanol*

Preparation of Solutions.   Accurately measure [OP-5] 21.0 mL of 95% ethanol and 4.0 mL of water into a 50-mL Erlenmeyer flask. Add 4 drops of bromthymol blue solution, swirl to mix, and support the flask in the room-temperature water bath as described in part **A**. Fill a clean, dry 10-mL buret with 0.10 M KOH in 80% ethanol and record the initial buret reading accurately.

Take Care! Be sure to use the right KOH solution.

Kinetic Run.   Drop in a stir bar and begin stirring [OP-10] the reaction mixture. Use an automatic pipet or a *dry* measuring pipet to measure [OP-5] 0.10 mL of 2-bromo-2-methylpropane into the reaction flask. *Immediately* record the time of addition (or start your timer). Carry out the kinetic run by the same procedure you followed in part **A**, using at least ten 0.50-mL portions of the 0.10 M KOH solution.

Take Care! Avoid contact with the alkyl halide and do not breathe its vapors or pipet it by mouth.

Stop and Think: Why does the time between buret readings change as the reaction proceeds?

Determination of V_∞.   After the last color change, seal the flask with Parafilm and heat [OP-7] the reaction mixture in a water bath at 60°C for about 10 minutes. Then titrate it with the 0.10 M solution of KOH in 80% ethanol to the green end point, and calculate V_∞ as in part **A**. Carry out the calculations described in part **A** to determine the rate constant

Waste Disposal: Unless your instructor directs otherwise, the reaction mixtures and unused KOH solutions can be poured down the drain.

for 2-bromo-2-methylpropane in 80% ethanol (k_o), and then use Equation **2** to estimate its rate constant in 40% ethanol. Calculate the relative solvolysis rate for 1-bromoadamantane by dividing its rate constant by the rate constant for 2-bromo-2-methylpropane in 40% ethanol.

Exercises

1 (a) Which tertiary carbocation is more stable, 1-adamantyl or *t*-butyl? Explain why it is more stable and tell how your experimental results support this conclusion. (b) Which solvent promotes the formation of a carbocation more effectively, 40% ethanol or 80% ethanol? Explain.

2 Write a mechanism for the solvolysis reaction of 1-bromoadamantane with ethanol.

3 Describe and explain the possible effect on your results of the following experimental errors or variations. (a) The pH of the buffer for the blank was 3.9 rather than 6.9. (b) You ran both reactions in absolute ethanol. (c) You used norbornyl bromide (1-bromobicyclo[2.2.1]heptane) in place of 1-bromoadamantane. (d) Your pipet was wet when you used it to measure the 2-bromo-2-methylpropane.

4 Hydroxide ion is a stronger nucleophile than either water or ethanol, yet the addition of KOH during the kinetic runs in this experiment has virtually no effect on the reaction rates. Explain.

5 (a) During its solvolysis reaction, some 2-bromo-2-methylpropane molecules lose HBr by an E1 reaction to form an alkene. Would you expect this to affect the measured reaction rate? Why or why not? (b) None of the 1-bromoadamantane undergoes elimination during its solvolysis reaction. Explain.

6 (a) Outline a synthesis of amantadine from 1-bromoadamantane. (b) Outline a synthesis of the antiviral agent rimantadine [$RCH(NH_2)$ CH_3, where R = 1-adamantyl] from 1-bromoadamantane using an organocuprate.

7 (a) Using your experimental rate constant, calculate the time it should take for 90% of your 1-bromoadamantane to react at room temperature. (b) How long should it take for 90% of the 2-bromo-2-methylpropane to react in 40% ethanol under the same conditions?

Other Things You Can Do

(Starred items require your instructor's permission.)

*1 Collect data for this experiment using a pH probe with a computer interface as described in *J. Chem. Educ.* **1991**, *68*, 609.

*2 Measure the relative reactivities of different alkyl halides as described in Minilab 16.

*3 Carry out an S_N1 reaction of trityl bromide with ethanol and isolate the product as described in Minilab 17.

4 Write a research paper about medical uses of adamantane derivatives, starting with sources listed in the Bibliography and the article in *J. Chem. Educ.* **1973**, *50*, 780.

Reaction of Iodoethane with Sodium Saccharin, an Ambident Nucleophile

Nucleophilic Substitution. NMR Spectrometry. Carboxylic Acid Derivatives. Heterocyclic Compounds.

Operations

OP-35   High-Performance Liquid Chromatography (optional)
OP-13   Vacuum Filtration
OP-23   Drying Solids
OP-30   Melting Point
OP-37   Nuclear Magnetic Resonance Spectrometry

Before You Begin

1 If you will be doing the optional HPLC analysis, read OP-35. Read the experiment, read or review the other operations, and write a brief experimental plan.
2 Calculate the mass of 2.00 mmol of sodium saccharin and the theoretical yield of ethylsaccharin.

Scenario

Saccharin is a nonnutritive sweetener, meaning that it is not metabolized by the body to produce energy. But saccharin is usually mixed with fructose or other Calorie-laden sweeteners to mask its bitter aftertaste, giving the mixture about half as many Calories as sucrose and thus making it less attractive as a sugar substitute. Dulcinea Petty IV directs a product development team at Sweet Nothings Ltd., which manufactures saccharin. She has learned that substances with N—H bonds often have bitter tastes, so she wonders if converting the N—H bond of saccharin to an N—C bond by alkylating it will mask the bitter taste and thus yield a better sweetener. Saccharin is converted to its more nucleophilic sodium salt prior to alkylation, but resonance structures of the salt reveal that it is an ambident nucleophile; that is, it has two potentially nucleophilic atoms, the nitrogen atom and an oxygen atom.

A nutritional Calorie is 1000 times as large as a scientific calorie.

saccharin   resonance structures of sodium saccharin

Before their quest for a better sweetener can be pursued, Sweet Nothings needs to know whether or not alkylation will occur mainly on the nitrogen atom. Your assignment is to carry out the alkylation of sodium saccharin with iodoethane and analyze the product mixture to determine the structure of the major product.

Applying Scientific Methodology

You should be able to formulate a working hypothesis (or at least a reasonable guess) after reading the experiment and then test it by either NMR or HPLC analysis of your product. You will not taste the product, but you can ask your instructor about its taste after completing the experiment.

Saccharin, an Accidental Sweetener

One rule that most chemists follow scrupulously is to never, *ever*, taste anything they make in the laboratory. A chemist should not even eat or drink anything while working in the lab because of possible contamination by toxic chemicals. During the nineteenth century, however, chemists were not so fastidious. It was a common practice to perform a "taste test" on any new chemical, sometimes with unfortunate results; but occasionally an accidental or deliberate tasting paid off with a new discovery.

Ira Remsen, a Johns Hopkins University chemistry professor, studied chemistry under a protégé of the "father of organic chemistry," Friedrich Wöhler, and became the most famous American chemist of the nineteenth century. In 1878, a German student working in Remsen's research group, Constantin Fahlberg, prepared some white crystals of a previously unknown compound from *o*-toluenesulfonamide. He later ate a piece of bread and was astonished to find that it tasted intensely sweet. It didn't take Fahlberg long to trace the sweet taste to the new compound he had just handled, which he named saccharin after the Latin word for sugar, *saccharum*.

Saccharin is about 500 times sweeter than sucrose (common table sugar). Its sweetness came as a surprise, because no one was looking for a synthetic sweetener at the time—most scientists believed that only natural compounds could be sweet. Fahlberg recognized the commercial possibilities of a nonfattening sweetener, so he applied for a patent and began to manufacture saccharin. Despite its somewhat bitter aftertaste, saccharin was the most popular artificial sweetener during most of the twentieth century, outselling other synthetic sweeteners such as dulcin (from the Latin *dulcis*, meaning sweet), which was discovered just six years after saccharin.

Concerns about the safety of saccharin cropped up from time to time, inspiring Theodore Roosevelt to proclaim, "anyone who says saccharin is injurious to health is an idiot!" Roosevelt, who liked to sweeten his chewing tobacco with saccharin, was no authority on the safety of commercial products, but his words must have reassured many Americans about saccharin. Then, in a Canadian study carried out in 1977, some rats developed bladder tumors when fed a diet containing 5% saccharin. Although the rats' diet was equivalent to a human consuming about 1,000 cans of diet soda per day, saccharin was promptly removed from the GRAS (generally recognized as safe) list and later banned in the United States. Reacting to protests by diabetics and overweight Americans, for whom consuming sugar was a far greater health risk than the remote possibility of saccharin-induced cancer, Congress suspended the ban in 1979, but foods containing saccharin were still required to carry a warning label. Saccharin was finally removed from the U.S. government's list of suspected carcinogens in 1999 in response to evidence that the rat-bladder tumors arose from mechanisms that are not relevant to humans.

Because of the cancer scare and competition from aspartame (Nutra-Sweet), saccharin use has declined sharply in recent years. Lacking the bitter aftertaste of saccharin, aspartame has become our most popular artificial sweetener, but it may face some tough competition before long. A French sweetener called superaspartame is 300 times sweeter than aspartame and—unlike aspartame—can be used in baking and frying. The natural sweetener thaumatin, which is extracted from the west African ketemfe plant, is reported to be nearly 100,000 times sweeter than sucrose, making it the sweetest natural substance ever discovered. It is also (like aspartame) a flavor enhancer, so it has been used to persuade farm animals to eat more—pigs gain up to 10% more weight when thaumatin is added to their feed.

Understanding the Experiment

In this experiment, you will carry out the reaction of sodium saccharin with iodoethane in the solvent N,N-dimethylformamide (DMF). This is a nucleophilic substitution reaction in which the nucleophilic atom can be either nitrogen or oxygen and the leaving group is iodide ion (I^-). The rate of a nucleophilic substitution reaction can be very sensitive to the solvent used. Polar protic solvents (solvents capable of hydrogen bonding), such as water and ethanol, form bulky solvation shells around a charged nucleophile, reducing its nucleophilic strength. Polar aprotic solvents, such as DMF, do not solvate the nucleophile strongly, leaving it free to attack the substrate. Thus they accelerate the rates of many substitution reactions, particularly S_N2 reactions, in which the strength of the nucleophile has a large effect on the reaction rate.

As shown in the "Reactions and Properties" section, nucleophilic attack by nitrogen yields N-ethylsaccharin, while nucleophilic attack by oxygen yields O-ethylsaccharin. Thus your product will be either N-ethylsaccharin, O-ethylsaccharin, or a mixture of the two, depending on whether saccharin's nitrogen atom or oxygen atom (or both) acts as the nucleophilic atom. Predicting the most likely product is not easy, because several competing factors may come into play. N-Ethylsaccharin is more stable than O-ethylsaccharin, so it should be the major (or only) product if the reaction reaches thermal equilibrium. But the oxygen atom of sodium saccharin has a higher partial negative charge than the nitrogen atom because oxygen is more electronegative than nitrogen, so a reaction involving oxygen as the nucleophile should occur faster than one involving nitrogen. For example, the reaction of potassium saccharin with 2-bromopropane in DMF yields mainly O-isopropylsaccharin.

You can determine the identity or composition of your product by using proton nuclear magnetic resonance (^{1}H NMR) spectrometry. An oxygen atom has a stronger deshielding effect on nearby protons than does a nitrogen atom, so the signal for the methylene protons (highlighted) of an $-OCH_2CH_3$ group will appear farther downfield ($\delta \approx 4.7$ ppm) than the corresponding signal for an $-NCH_2CH_3$ group ($\delta \approx 3.9$ ppm). Because the methylene protons have three methyl protons as neighbors, their signal in either case will be a quartet. If your product is either N-ethylsaccharin or O-ethylsaccharin, you can identify it from the chemical shift of its methylene quartet. If it is a mixture, you can measure the integrated signal areas for

DMF

Key Concept: Solvation reduces the strength of a nucleophile and therefore decreases the rates of its nucleophilic substitution reactions.

both quartets, calculate the percentages of *N*-ethylsaccharin and *O*-ethylsaccharin present, and decide which one is the major product.

At your instructor's option, you can analyze your product using high-performance liquid chromatography (HPLC) in addition to or instead of NMR spectrometry.

Reactions and Properties

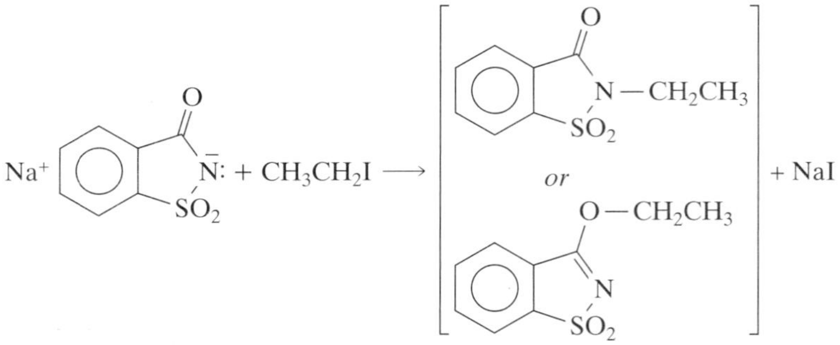

$$\text{Na}^+ \left[\begin{array}{c} \overset{O}{\underset{SO_2}{\bigcirc\!\!\!N:}} \end{array}\right] + CH_3CH_2I \longrightarrow \left[\begin{array}{c} \overset{O}{\underset{SO_2}{\bigcirc\!\!\!N-CH_2CH_3}} \\ or \\ \overset{O-CH_2CH_3}{\underset{SO_2}{\bigcirc\!\!\!N}} \end{array}\right] + \text{NaI}$$

Table 20.1 Physical properties

	M.W.	bp	mp	*d*
sodium saccharin	205.2			
iodoethane	156.0	72		1.950
N,*N*-dimethylformamide	73.1	153		0.945
N-ethylsaccharin	211.2		95	
O-ethylsaccharin	211.2		211	

Note: bp and mp are in °C, density is in g/mL.

Directions

These procedures are adapted from an article published in the *Journal of Chemical Education* **1990**, *67*, 611.

Safety Notes

N,*N*-dimethylformamide

Take Care! Wear gloves, avoid contact with DMF and iodoethane, and do not breathe their vapors.

> **Iodoethane severely irritates the eyes, skin, and respiratory tract, and it may be carcinogenic. Wear gloves, avoid contact, and do not breathe its vapors. *N*,*N*-Dimethylformamide (DMF) is harmful by inhalation, ingestion, and absorption through the skin. Avoid contact and do not breathe its vapors. Deuterochloroform is harmful if inhaled, ingested, or absorbed through the skin, and it may be carcinogenic. Avoid contact and do not breathe its vapors.**

Reaction.   *Carry out the reaction under the hood.* Weigh 2.00 mmol of sodium saccharin and add it to 1.0 mL of *N*,*N*-dimethyl formamide in a 25-mL Erlenmeyer flask. Heat the mixture in an 80°C water bath with swirling until the solid dissolves, and then add 0.16 mL (~2.0 mmol) of iodoethane. Cover the mouth of the flask with Parafilm and heat the mixture in the water bath with occasional swirling for 10 minutes, keeping the water temperature close to 80°C.

Separation.   Let the reaction mixture cool to room temperature, add 8.0 mL of water, and shake the stoppered flask until any liquid residue that forms has solidified. Cool the flask in an ice/water bath and break up the solid with a flat-bladed microspatula until it is finely divided. Collect the solid by vacuum filtration [OP-13], washing it twice with ice-cold water. Do *not* taste the product!

Analysis.   Dry [OP-23] the product and measure its mass and melting-point range [OP-30]. Obtain an integrated ^{1}H NMR spectrum [OP-37] of the product in deuterochloroform. (**Take Care!** Avoid contact with $CDCl_3$ and do not breathe its vapors.) (At your instructor's discretion, you can analyze the product by HPLC [OP-35]. Your instructor will demonstrate the operation of the instrument.) If the product is a single compound, deduce its identity. If it is a mixture, calculate its percentage composition and decide whether *N*-ethylsaccharin or *O*-ethylsaccharin is the major product.

Waste Disposal: Put the filtrate in a designated solvent recovery container.

Stop and Think: Is the product a single compound or a mixture? How can you tell?

Waste Disposal: Put the deuterochloroform solution in a designated solvent recovery container.

Exercises

1   (a) Assuming that the reaction was S_N2, which atom appears to be more nucleophilic, N or O? (b) Write a mechanism showing the transition state of the reaction that led to your major product.

2   Describe and explain the possible effect on your results of the following experimental errors or variations. (a) The reagent bottle labeled "sodium saccharin" contained saccharin instead. (b) You used water as the reaction solvent rather than DMF. (c) You heated the reaction mixture for 3 hours under reflux.

3   Following the format in Appendix V, construct a flow diagram for this experiment.

4   Most compounds containing N—H bonds are basic, but saccharin is acidic. Explain why, using resonance structures.

5   Outline a synthesis of saccharin from *o*-toluenesulfonamide.

6   One objection raised to the use of aspartame is that it decomposes in the presence of moisture to produce phenylalanine, which must be avoided by people who have the genetic condition phenylketonuria, and methanol, which can have an effect on mental behavior. Write an equation for a hydrolysis reaction of aspartame that yields both of these products.

aspartame

Other Things You Can Do

(Starred items require your instructor's permission.)

*__*1__   Add some aqueous sodium bicarbonate to a solution of saccharin (not sodium saccharin) and explain the result, writing an equation for the reaction.

*__*2__   Carry out an S_N1 reaction of trityl bromide with ethanol as described in Minilab 17.

3   Write a research paper about artificial sweeteners, starting with sources listed in the Bibliography.

Dehydration of Methylcyclohexanols

EXPERIMENT **21**

Reactions of Alcohols. Preparation of Alkenes. Elimination Reactions. Carbocations. Regioselectivity.

Operations

OP-10   Mixing
OP-21   Washing Liquids
OP-22   Drying Liquids
OP-27   Simple Distillation
OP-34   Gas Chromatography

Before You Begin

1 Read the experiment, read or review the operations as necessary, and write an experimental plan.
2 Calculate the mass and volume of 20.0 mmol of 2- and 4-methylcyclohexanol and the theoretical yield of methylcyclohexenes from each alcohol.

Scenario

(Unlike most other Scenarios in this book, this one describes a real chemical mystery, which is documented in the *Journal of Chemical Education*, **1994**, *71*, 440. Not even the names of the characters have been changed.)

For many years, the dehydration of 2-methylcyclohexanol to a mixture of alkenes has been carried out in college organic chemistry labs to demonstrate the application of Zaitzev's rule and the occurrence of the E1 mechanism in alcohol dehydration reactions. In 1994, David Todd (then a chemistry professor at Pomona College) carried out this reaction and was distilling the product alkenes from the reaction mixture when he received an invitation to lunch with the chemistry department secretary, Evelyn Jacoby. Dr. Todd stopped the distillation, which was about half done, and saved the distillate in its receiver. Upon returning from lunch he decided to collect the rest of the distillate in a new receiver, giving him two separate fractions. He then worked up both fractions and analyzed them by gas chromatography. Much to his surprise, the second fraction contained a markedly lower percentage of the expected product, 1-methylcyclohexene, than did the first. Because his decision to replace the receiver with a new one was a direct result of the secretary's invitation, Professor Todd named this unexpected result the "Evelyn Effect."

Although several mechanistic hypotheses have been proposed to explain the Evelyn Effect, it is by no means certain that any of them are correct. Your project group's assignment is to verify the existence of the Evelyn Effect for the dehydration of 2-methylcyclohexanol and to see if a similar effect exists for 4-methylcyclohexanol. You may then want to speculate about some possible causes of the Evelyn Effect.

Applying Scientific Methodology

As in the previous experiments, you need to state the problem as a question, formulate a working hypothesis, follow the course of action described in the Directions, gather and evaluate evidence, test your hypothesis, arrive at a conclusion, and report your findings.

Zaitzev's Rule and the Evelyn Effect

More than a century ago at the University of Kazan, Vladimir Vasilevich Markovnikov and Alexander Zaitzev were investigating a chemical reaction both forward and backward: Markovnikov was adding hydrogen iodide to alkenes to prepare alkyl iodides, and Zaitzev was removing hydrogen iodide from alkyl iodides to prepare alkenes. Markovnikov discovered that hydrogen iodide adds to propene to form mainly 2-iodopropane.

$$CH_3CH{=}CH_2 + HI \longrightarrow CH_3\overset{\displaystyle I}{\overset{\displaystyle |}{C}}HCH_3$$

From this and other results, Markovnikov formulated his well-known rule, which can be expressed as follows for a hydrogen-containing species represented by HZ:

Markovnikov's rule: When HZ adds to the carbon-carbon double bond of an unsymmetrical alkene, hydrogen adds preferentially to the carbon atom that already has more hydrogens.

In the meantime, Zaitzev learned that dehydrohalogenation of 2-iodobutane by alcoholic potassium hydroxide yields mainly 2-butene.

$$CH_3CH_2\overset{\displaystyle I}{\overset{\displaystyle |}{C}}HCH_3 \xrightarrow{\text{KOH}} CH_3CH{=}CHCH_3$$

He proposed an analogous rule for elimination reactions.

Zaitzev's rule: When HZ is removed from a species to form an alkene, hydrogen is lost preferentially from the carbon atom that has fewer hydrogens.

Markovnikov's and Zaitzev's rules together can be paraphrased by the well-known socioeconomic maxim "The rich get richer and the poor get poorer."

These examples show that organic reactions can be selective, favoring some products and not others—Zaitzev's reaction might have yielded as much 1-butene as 2-butene, but it did not. When a reaction *could* produce two or more different structural isomers but in fact yields mainly one of them, the reaction is said to be *regioselective*. Zaitzev's rule works because, in most cases, it predicts the formation of the most stable alkene. 2-Butene was the major product of Zaitzev's reaction not because hydrogen-poor carbon atoms have some innate tendency to lose the hydrogens they have but because 2-butene is more stable than 1-butene.

Although generalizations such as Zaitzev's rule can help us predict the products of many organic reactions, organic chemistry remains an empirical science—we cannot be certain that a rule that is valid for one system under a given set of conditions will apply equally well under different circumstances.

Chemists must study each system experimentally to see if it behaves in the expected manner, and if it doesn't, try to find out why.

For example, neomenthyl chloride undergoes dehydrohalogenation to yield a product mixture that consists of mostly alkene **A**, the one predicted by Zaitzev's rule. But menthyl chloride, which differs only in the geometry of the C—Cl bond, yields 100% of alkene **B** and none of the Zaitzev product. It also reacts much more slowly than neomenthyl chloride.

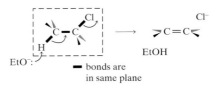

E2 mechanism for elimination of HCl

This result can be explained by assuming that the reaction occurs by an E2 mechanism, requiring that the H and Cl atoms being eliminated lie in the same plane and on opposite sides of the C—C bond separating them; this is called *anti*-periplanar geometry.

Neomenthyl chloride, in its most stable ring conformation, has the desired *anti*-periplanar geometry for formation of either **A** or **B**. Since **A** is the more stable alkene, it is the major product.

In its more stable conformation (with all large groups equatorial), menthyl chloride does not have the geometry necessary to form either product. In its less stable conformation, the *anti*-periplanar geometry needed to form product **A** cannot be attained because the isopropyl group rather than a hydrogen atom is *anti* to the chlorine atom. This conformation is suitable for the formation of product **B**, however, so it is the only product isolated.

The reaction is much slower than the reaction of neomenthyl chloride because only a small percentage of the menthyl chloride molecules are in the less stable conformation at any time. This example shows that, whenever a substrate yields the less stable alkene as a major product of an elimination reaction, there may be some stereochemical constraints inhibiting the formation of the Zaitzev product.

Key Concept: *For certain kinds of reactions, stereochemical constraints may lead to unexpected results.*

The acid-catalyzed dehydration of alcohols, which usually follows Zaitzev's rule, is generally believed to occur by an E1 (elimination, unimolecular) mechanism involving protonation of the hydroxyl group, loss of water to form a carbocation intermediate, and then loss of a proton.

Note that there are *no* stereochemical constraints in an E1 reaction because the leaving group leaves before the proton is lost. Thus in the E1 dehydration of an alcohol, H and OH do not need to be *anti*-periplanar or in any other particular orientation in order for elimination to occur.

Unlike E2 reactions, E1 reactions may involve rearrangements in which the initial carbocation rearranges to a more stable carbocation before it loses H⁺. A carbocation rearrangement may involve a *hydride shift*, during which a hydrogen next to the positively charged carbon moves to that carbon, taking its bonding electron pair along with it. Such rearrangements lead to alkenes whose double bond does not involve the carbon atom that was originally bonded to the hydroxyl group. Postulating such a rearrangement can explain the formation of 2-methyl-2-butene in the dehydration of 2-methyl-1-butanol, for example.

This brings us to the Evelyn Effect. When Professor Todd carried out the dehydration of 2-methylcyclohexanol, he obtained the following mixture of alkenes.

This reaction is performed by distilling the alkenes as they are formed, and the distillate typically contains 75–80% of product **A**, the product predicted by Zaitzev's rule. But when the distillate is collected in separate fractions and the fractions are analyzed separately, the first 10% of the distillate contains about 93% **A**, while the final distillate contains as little as 55% **A**. The remaining product in each case is mostly **B**, with only a trace of **C**. There is a clue to the origin of the Evelyn effect in the catalog of the Aldrich Chemical Company, where the 2-methylcyclohexanol used by Professor Todd is described as a mixture of *cis* and *trans* isomers. In fact, it is a nearly equimolar mixture of the two isomers. Previous researchers had reported that the *cis* isomer reacts much faster than the *trans* isomer, so Todd reasoned that the initial product mixture, containing mostly **A**, formed mainly by dehydration of the *cis* isomer while the final product mixture, with comparable amounts of **A** and **B**, formed mainly by dehydration of the *trans* isomer.

Since the *trans* isomer yields an unexpectedly large percentage of the less stable alkene, it appears that this reaction, like the E2 dehydrohalogenation of menthyl chloride, has some stereochemical constraints. The occurrence of E2 elimination from a protonated alcohol could explain a reduction in the amount of the expected product **A**, since elimination involving an *anti*-periplanar geometry can yield only product **B** and not product **A**.

It could also explain the lower reactivity of the *trans* alcohol, which can achieve the *anti*-periplanar geometry only in its less stable *diaxial* conformation. However, it does *not* explain why the *trans* alcohol yields any **A** at all, nor does it explain the existence of a small amount of methylenecyclohexane (product **C**) in the product mixture. Product **C** might be obtained by an E1 mechanism involving a carbocation rearrangement, but not by an ordinary E2 mechanism.

Does the reaction proceed by both E1 and E2 mechanisms? That possibility, raised by Todd, has been questioned by two other researchers, John J. Cawley and Patrick E. Lindner. Cawley and Lindner proposed an "E2-like" mechanism involving bridged ions (*J. Chem. Educ.* **1997,** *74,* 102), but it remains to be seen whether their mechanism will gain general acceptance. A mechanism is, after all, a scientific hypothesis about processes that we can't observe directly—the things that molecules do as they redistribute their atoms and change into new molecules—and may be revised or rejected as new evidence comes to light.

The Evelyn Effect illustrates how science often works. For decades, the results of alcohol dehydration reactions were adequately explained by the E1 hypothesis; no other explanation seemed necessary. Then a chance observation showed the inadequacy of the accepted hypothesis. A different hypothesis—that both E1 and E2 mechanisms are involved—was proposed and contested, followed by another hypothesis, and so on. The road to scientific discovery is a rocky one, and there may be many detours along the way, but every failed hypothesis yields new information, new ideas, and often new applications. Science is not simply a body of established facts and theories; the facts and theories of science are always subject to further inquiry that may disprove or modify them. Science is a dynamic *process* by which knowledge is acquired, ideas are debated, theories are proposed, and new ways of doing things are discovered.

Understanding the Experiment

In this experiment, you and your coworkers will carry out the dehydration of 2-methylcyclohexanol and 4-methylcyclohexanol by heating the alcohols in the presence of phosphoric acid. Both alcohols will be mixtures of *cis* and *trans* isomers, so either one or both may exhibit an Evelyn Effect.

Dehydration of a secondary alcohol proceeds readily with about half a mole of phosphoric acid for every mole of the alcohol. By protonating an alcohol, the acid catalyst converts the poor leaving group $-OH$ to a much better leaving group, $-OH_2^+$.

Elimination of H^+ and H_2O from the protonated alcohol yields an alkene, with the unprotonated alcohol serving as the reaction solvent.

According to Le Châtelier's principle, removing a product from a chemical system at equilibrium shifts the equilibrium in the direction favoring the formation of the products. You will carry out the dehydration reaction in a distillation apparatus so that the products (water and alkene) will continuously distill from the reaction mixture into a Hickman still as they are formed. Their removal will shift the equilibrium to the right and thus increase the yield of alkene.

The upward-pointing arrows in the equation indicate that the products are vaporized under the reaction conditions, not that they are gases at room temperature.

$$\text{alcohol} \rightleftharpoons \text{alkene} \uparrow + \text{water} \uparrow$$

If the reaction mixture is heated to a temperature above the boiling points of the product alkenes but below that of the alcohol, most of the unreacted alcohol will remain in the reaction flask while the alkenes and water collect in the well of the Hickman still. You will collect the distillate in two fractions of approximately equal volume. When the reaction is over, the residue in the reaction flask may begin to foam and emit white vapors. You should separate the apparatus from the heat source at this time because overheating the residue may form a black tar and generate toxic fumes.

After washing and drying the organic layer of each fraction, you will analyze the fractions by gas chromatography. If you started with 2-methylcyclohexanol, your gas chromatograms may show peaks for both 1- and 3-methylcyclohexene (the methylenecyclohexene peak will be resolved only if you use a capillary column). If you started with 4-methylcyclohexanol, you may obtain only 4-methylcyclohexene or a mixture of products including 3-methylcyclohexene and 1-methylcyclohexene. From the relative areas of your peaks, you can estimate the percentage composition of the product mixture in each fraction. A packed column may not separate 3- and 4-methylcyclohexene, so in that case you should calculate their combined percentage.

Reactions and Properties

Table 21.1 Physical properties

	M.W.	bp	*d*
2-methylcyclohexanol*	114.2	166	0.930
4-methylcyclohexanol*	114.2	173	0.914
1-methylcyclohexene	96.2	110	0.813
3-methylcyclohexene	96.2	104	0.801
4-methylcyclohexene	96.2	102	0.799
phosphoric acid (85%)	98.0		1.70

*Mixture of *cis* and *trans* isomers.

Note: bp is in °C, density is in g/mL. The molecular weight given for phosphoric acid is for the pure acid; 85% phosphoric acid is about 14.7 *M*.

Directions

Students can work in pairs, with each student using one of the two methyl-cyclohexanols.

Safety Notes

Reaction.   Accurately weigh 20.0 mmol of 2-methylcyclohexanol *or* 4-methylcyclohexanol into a 10-mL round-bottom flask. Mix in 0.70 mL of 85% phosphoric acid and add a stir bar. Clamp the reaction vessel to a ring stand over a suitable heat source and assemble an apparatus for simple distillation [OP-27] using a Hickman still, water-cooled condenser, and thermometer. Weigh and number two clean conical vials.

Start the stirrer [OP-10] and begin heating the reactants. Control the heating rate so that the product mixture collects *slowly* in the well of the Hickman still. Record the still-head temperature after distillation begins and observe it at intervals during the reaction. Using a Pasteur pipet, transfer distillate frequently to the first conical vial until it is filled with liquid to the 1.0 mL mark, then begin transferring distillate to the second vial. As the alkene volume in that vial approaches 1 mL, monitor the still-head temperature continually. Remove the apparatus from the heat source when you observe a marked temperature drop at the still head, which may be accompanied by foaming and dense white fumes in the reaction flask.

Separation.   For each separate fraction, wash [OP-21] the distillate with 1.5 mL of 5% aqueous sodium bicarbonate and carefully remove the aqueous (lower) layer with a Pasteur pipet. Dry [OP-22] each alkene mixture separately with anhydrous calcium chloride. Measure the mass of each alkene fraction and calculate the total mass of alkenes.

Take Care! Avoid contact with the acid and alcohol and do not breathe their vapors.

The water that codistills with the alkene reduces its boiling temperature, so the still-head temperature will be lower than the expected boiling point of the product.

Waste Disposal: Place the residue from the reaction flask into a designated waste container.

Stop and Think: What is the purpose of the sodium bicarbonate?

Waste Disposal: The aqueous layers can be poured down the drain.

Analysis.   Analyze both fractions by gas chromatography [OP-34] as directed by your instructor. Measure the area and retention time of each peak on your gas chromatograms. Identify each peak by comparison with a chromatogram provided by your instructor or by spiking your product mixture with an authentic sample of 1-methylcyclohexene and obtaining a chromatogram of the resulting mixture. Assuming that the detector response factors for the alkenes are equal, calculate the percentage composition of each fraction and obtain the same data from a coworker who started with the other alcohol. Decide whether either or both alcohols exhibit an Evelyn Effect.

Exercises

1 (a) Which kind of mechanism can better account for the product mixture obtained from the dehydration of *cis*- and *trans*-4-methylcyclohexanol: E1, E2, or a combination of the two? (Keep in mind that the actual mechanism may be none of these.) (b) Based on your answer, write detailed mechanisms explaining the formation of all of the observed products.

2 Following the format in Appendix V, construct a flow diagram for this experiment.

3 Describe and explain the possible effect on your results of the following experimental errors or variations. (a) You forgot to add the phosphoric acid. (b) You collected all of the distillate in one container rather than in two. (c) The 2-methylcyclohexanol you used was the pure *trans* isomer rather than a mixture of isomers.

4 In "Understanding the Experiment," 1-methylcyclohexene and 3-methylcyclohexene were mentioned as possible products of the dehydration of 2-methylcyclohexanol. Why was 2-methylcyclohexene not mentioned as a possible product?

5 In an E2 reaction, a base removes H^+ as the leaving group leaves the substrate. Draw structures of at least three possible bases (they may be weak bases) present in the reaction mixture for the reaction of 2-methylcyclohexanol.

6 (a) Predict the major alkene product that would result from dehydrating each of the following alcohols, with no carbocation rearrangements.

(b) In each case, predict the most stable dehydration product that could result after a single carbocation rearrangement.

Other Things You Can Do

(Starred items require your instructor's permission.)

*1 Prepare and test a gaseous alkene as described in Minilab 18.
*2 Test your product mixtures with bromine or potassium permanganate solution, and interpret the results. (See classification tests C-7 and C-19 in the "Classification" section of Part IV.)
*3 Dehydrate another alcohol, such as cyclohexanol, 3-methylcyclohexanol, or 4-methyl-2-pentanol, by the same procedure, adjusting the distillation temperature for the alkene or alkene mixture anticipated. Analyze the product mixtures by gas chromatography and interpret the results.
 4 The methylcyclohexanols used in this experiment are synthesized by catalytic hydrogenation of the corresponding cresols (methylphenols). Write a research paper about the production and uses of cresols, starting with sources listed in the Bibliography.

o-cresol   m-cresol   p-cresol

Synthesis of 7,7-Dichloronorcarane Using a Phase-Transfer Catalyst

Reactions of Alkenes. Preparation of Alicyclic Compounds. Alkyl Halides. Carbenes. Phase-Transfer Catalysis.

Operations

OP-9   Temperature Monitoring
OP-10 Mixing
OP-14 Centrifugation
OP-15 Extraction
OP-16 Evaporation
OP-22 Drying Liquids
OP-27 Simple Distillation
OP-36 Infrared Spectrometry

Before You Begin

1 Read the experiment, read or review the operations as necessary, and write an experimental plan.
2 Calculate the mass and volume of 20.0 mmol of cyclohexene and the theoretical yield of 7,7-dichloronorcarane.

Scenario

Erewhon is an imaginary kingdom described in the 1872 novel of the same name by Samuel Butler.

The somewhat backward kingdom of Erewhon has just modernized its air force, which previously consisted of 24 supercharged Sopwith Camels. But the F-14 Tomcats the Erewhonians just purchased don't perform well on aviation fuel designed for Camels, so they have commissioned your Institute to help them develop a high-energy jet fuel.

Hydrocarbons with high bond-angle strain have unusually high energies, so when burned, they release more energy than unstrained hydrocarbons. Cyclopropane has the highest bond-angle strain of any monocyclic hydrocarbon, and the strain increases when cyclopropane rings are fused onto other rings, as illustrated by the following examples.

1   **2**   **3**

Compounds **1** and **2** were developed by the Monsanto Research Corporation as part of a U.S. Air Force program to develop new jet fuels with high heats of combustion. Compound **3** (tricyclo[4.1.0.0^{2,7}]heptane), with two cyclopropane rings fused side by side onto a cyclohexane ring, has an extremely high bond-angle strain, which should make it a very energetic aircraft fuel. Your supervisor thinks it should be possible to prepare compound **3** by the following reaction of 7,7-dichloronorcarane, which should itself have some bond-angle strain. This reaction proceeds through the carbene intermediate shown here.

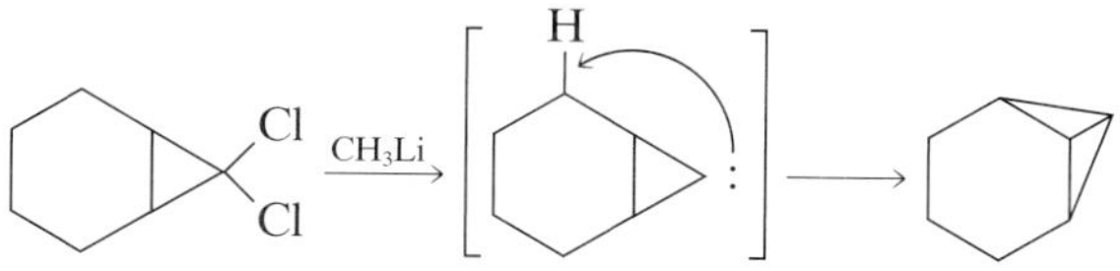

7,7-dichloronorcarane   carbene   **3**
intermediate

Before this idea can be tested, you will need to prepare some 7,7-dichloronorcarane, which can be synthesized by the addition of dichlorocarbene (:CCl$_2$) to cyclohexene. The synthesis involves a two-phase reaction mixture consisting of aqueous NaOH and an organic phase containing chloroform and cyclohexene. Such reactions tends to be very slow because the reactants are separated by the phase boundary, but some two-phase reactions can be accelerated by phase-transfer catalysts. Your assignment is to see whether 7,7-dichloronorcarane can be prepared from cyclohexene using a phase-transfer catalyst and to find out whether or not it shows evidence of bond-angle strain.

Applying Scientific Methodology

As in the previous experiments, you need to state each problem as a question, formulate working hypotheses, follow the course of action described in the Directions, gather and evaluate evidence, test your hypotheses, arrive at a conclusion, and report your findings. You should be able to test one or both hypotheses by obtaining and analyzing an infrared spectrum of the product.

The Ubiquitous Triangle

It often appears that nature loves the hexagon, since so many natural compounds contain six-membered rings in their molecules. This is not surprising, since both the aromatic benzene ring and the unstrained cyclohexane ring are unusually stable compared to other possible ring structures. But it is surprising to find numerous examples of the triangle, in the form of the highly strained cyclopropane ring, in everything from arborvitae to water molds.

Henry David Thoreau reported that northwoods lumbermen of the nineteenth century were accustomed to drinking "a quart of arborvitae, to make (them) strong and mighty." An extract of the leaves of the arborvitae (white cedar) tree was thought to impart strength and prevent illness, particularly rheumatism. One of the main constituents of the oil from arborvitae and other *Thuja* species is the bicyclic terpenoid thujone, which contains a three-membered ring fused onto a cyclopentane ring. Thujone is also found in wormwood (*Artemisia absinthium*), an ingredient in the addictive nineteenth-century drink absinthe. It has been speculated that thujone was responsible for the artist Vincent van Gogh's mental illness. Van Gogh was an absinthe

hexagon

triangle

Euell Gibbons, an authority on edible wild plants, tasted arborvitae tea and declared that he "would almost prefer rheumatism."

thujone

sirenin

illudin-S

drinker who once cut off part of his left ear—inspiring him to paint his "Self-Portrait with Pipe and Bandaged Ear"—and eventually committed suicide.

The sirens of Greek mythology—beautiful female creatures who lured sailors to destruction with their singing—inspired the name for sirenin, a sperm attractant produced by the female gametes of a water mold, *Allomyces javanicus*. Like the compound you will synthesize in this experiment, sirenin has a three-membered ring fused onto a six-membered ring.

One of the more startling sights in nature is a pumpkin-colored mushroom that glows in the dark around Halloween. This is the poisonous jack-o'-lantern fungus, *Clitocybe illudens*, which is sometimes mistaken for the edible chanterelle mushroom, with unpleasant consequences. It contains an antibiotic substance, illudin-S, which has a cyclopropane ring attached at its apex to a six-membered ring.

Pyrethrin is a natural biodegradable insecticide, nontoxic to humans, obtained from the flowers of a daisylike plant, *Chrysanthemum cineariae-folium*. It is made up of a mixture of esters, such as cinerin I, that contain a cyclopropane ring in the carboxylic acid portion. A number of *pyrethroids*, synthetic analogs of the natural pyrethrins, have been developed in an effort to find relatively safe but effective insecticides to replace environmentally unsound "hard" pesticides such as DDT. One of the most powerful of these insecticides is decamethrin, which is about 60 times more lethal to houseflies than parathion and over 600 times more effective against certain mosquitoes than DDT.

cinerin I

decamethrin

Understanding the Experiment

Key Concept: A phase-transfer catalyst speeds up a reaction by bringing together reactants that would otherwise remain in separate phases.

In this experiment, you will generate the highly reactive intermediate dichlorocarbene by the reaction of chloroform with a strong aqueous solution of sodium hydroxide. (Chloroform is toxic and is suspected of causing cancer in humans, so you should wear gloves and use a hood while you are working with it.) The carbene should then combine with cyclohexene to form 7,7-dichloronorcarane. But sodium hydroxide is insoluble in the organic phase and both chloroform and cyclohexene are insoluble in water, so the reactants would ordinarily have a hard time getting together. To circumvent this difficulty, you will use a phase-transfer catalyst to "escort" reactant molecules across the phase boundary.

The principles of phase-transfer catalysis (PTC) can be explained by reference to the nucleophilic substitution reaction of an alkyl halide with sodium cyanide to form a nitrile. If a high-molecular-weight halide, such as 1-chlorooctane, is heated with aqueous sodium cyanide, the reaction is exceedingly slow. Cyanide ions stay in the aqueous layer and alkyl halide molecules stay in the organic layer, so the reactants only meet infrequently at the phase boundary. If, instead of sodium cyanide, a quaternary ammonium (Q) salt such as tetrabutylammonium cyanide is used, the reaction proceeds quite readily and gives a high yield of product.

Reaction of 1-chlorooctane with sodium cyanide ($R = n\text{-}C_8H_{17}$):

$$RCl + Na^+CN^- \longrightarrow RCN + Na^+Cl^-$$

Improved reaction with quaternary ammonium cyanide:

$$RCl + Q^+CN^- \longrightarrow Q^+Cl^- + RCN$$

$$[Q^+ = (CH_3CH_2CH_2CH_2)_4N^+]$$

There are several reasons for this enhanced reactivity of the cyanide. First, and most importantly, the 16 carbon atoms of the cation make it soluble in the organic phase, and where the cation goes, the anion must follow. Second, the cyanide ion is more reactive in the organic phase than it would have been in the aqueous phase because it is not solvated by water molecules (a solvent shell would shield it from the alkyl halide and decrease its reactivity). Finally, the bulky alkyl groups around the positive nitrogen of the cation decrease the attractive forces between cation and anion and allow the cyanide ion more freedom to attack the alkyl halide. So substituting a quaternary ammonium ion for the sodium ion in the cyanide salt allows the desired reaction to proceed at a much higher rate. Its main drawback is that the quaternary ammonium cyanide is much more expensive than sodium cyanide.

The PTC technique gets around the high cost of quaternary ammonium salts by recycling them after each reaction step. If the quaternary ammonium cation in tetrabutylammonium chloride (represented in the previous equation by Q^+Cl^-) can be made to pick up some more cyanide to react with the alkyl halide, it will function as a true catalyst, accelerating the reaction without being used up. All that is needed is a reservoir of cyanide ions and a small amount of quaternary ammonium salt to keep the reaction going. The reservoir can be provided by an aqueous layer that contains sodium cyanide.

Figure 22.1 diagrams the process, which occurs as follows: A catalytic amount of Q^+Cl^- combines with cyanide ion in the aqueous phase, and the Q^+CN^- that forms crosses over to the organic layer. There it reacts with the alkyl halide to produce the nitrile (RCN) and form more Q^+Cl^-, which migrates across the interface, picks up some more cyanide, shuttles it back into the organic layer to react with the alkyl halide and form more product and Q^+Cl^-, and so on, until the alkyl halide or the cyanide ion is used up.

$$Q^+Cl^- + Na^+CN^- \longrightarrow Q^+CN^- + Na^+Cl^- \qquad \text{aqueous phase (reservoir)}$$
$$Q^+Cl^- + RCN \longleftarrow Q^+CN^- + RCl \qquad \text{organic phase}$$

Figure 22.1 Phase-transfer process for a nucleophilic substitution reaction

You will use the viscous liquid tricaprylmethylammonium chloride (also known as Aliquat 336) as the phase-transfer catalyst for the synthesis of 7,7-dichloronorcarane. In this catalyst, the quaternary cation contains a methyl group and three alkyl chains of 8 or 10 carbons attached to a nitrogen atom. The initial reaction mixture will consist of an organic phase containing cyclohexene and chloroform and an aqueous phase containing sodium hydroxide, with the catalyst distributed between both phases. The

$[CH_3(CH_2)_n]_3NCH_3^+Cl^-$
tricaprylmethylammonium chloride (Aliquat 336)
$n = 7$ or 9

sequence of events in the subsequent reaction is not entirely clear, but the following scenario seems reasonable. A hydroxide ion associated with the quaternary cation can remove a proton from chloroform at the phase boundary to produce CCl_3^-. This ionic species would ordinarily linger in the aqueous phase, but when paired with the quaternary cation it can invade the organic phase, in which it loses a chloride ion to form dichlorocarbene ($:CCl_2$). The highly reactive carbene then attacks a cyclohexene molecule in the organic phase to form the product, and the quaternary cation can return to the aqueous phase to repeat the process.

In the procedure used for this experiment, the phase-transfer catalyst promotes the formation of an emulsion in which the organic phase is dispersed throughout the aqueous medium in minute spherical clusters called *micelles*. This greatly increases the area of contact between the phases, which speeds up the reaction by increasing the rate at which reactant molecules cross the phase boundary. The formation of a thick emulsion, with color and texture similar to thick cream, is essential to the success of the reaction. If it does not form, the reaction temperature will not rise much above 40°C and you will recover little, if any, product. A small amount of cyclohexanol will help stabilize the emulsion, and cleaning your glassware thoroughly should remove impurities that might break up the emulsion or prevent its formation. The reaction is exothermic, so it may be necessary to cool the reaction mixture to keep the reactants from boiling away. When the reaction is nearly over, the temperature will start to drop spontaneously. After separation from the reaction mixture, the crude product is purified by simple distillation. A forerun containing cyclohexanol and unreacted starting materials may be collected below 170°C.

You will characterize your product by recording its infrared spectrum. A carbon-chlorine stretching band for an alkyl halide can occur anywhere between 550 and 850 cm^{-1}, but it tends to be at the higher end of the range when two or more chlorine atoms are bonded to the same carbon; thus chloroform ($CHCl_3$) has a strong C—Cl band near 750 cm^{-1} (see Figure 22.2). Bonds to strained ring carbons tend to vibrate at higher frequencies than normal, so a C—H bond on a cyclopropane ring may absorb near or above 3000 cm^{-1}.

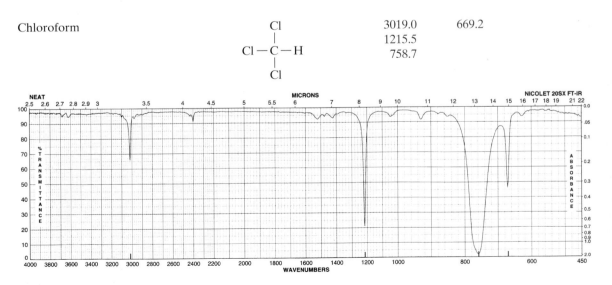

Figure 22.2 IR spectrum of chloroform

Reactions and Properties

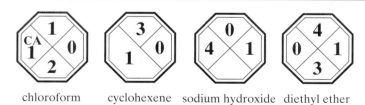

$$\text{cyclohexene} + CHCl_3 + NaOH \xrightarrow{\text{PTC}} \text{7,7-dichloronorcarane} + H_2O + NaCl$$

Table 22.1 Physical properties

	M.W.	bp	d
cyclohexene	82.15	83	0.810
chloroform	119.4	62	1.483
cyclohexanol	100.2	161	0.942
tricaprylmethylammonium chloride	404.2		0.884
7,7-dichloronorcarane	165.1	198	

Note: bp is in °C, density is in g/mL.

Directions

Chloroform is harmful if inhaled, swallowed, or absorbed through the skin. It is known to cause kidney and liver tumors when ingested by rats or mice and is a suspected carcinogen in humans. Wear gloves and work under a hood; avoid contact with the liquid, and do not breathe its vapors.
Cyclohexene is flammable and inhalation or skin absorption may be harmful. Avoid contact and do not breathe its vapors.
Sodium hydroxide is toxic and corrosive, and its concentrated solutions can cause severe damage to skin, eyes, and mucous membranes. Wear gloves and avoid contact with the NaOH solution.
Diethyl ether is extremely flammable and may be harmful if inhaled. Do not breathe its vapors, and keep it away from flames and hot surfaces.

Safety Notes

chloroform   cyclohexene   sodium hydroxide   diethyl ether

Reaction.   *Carry out this reaction under an efficient fume hood and wear gloves throughout.* Weigh approximately 20.0 mmol of cyclohexene into a *very clean* 25-mL Erlenmeyer flask and add 2.0 mL (~25 mmol) of chloroform, 0.20 mL of cyclohexanol, and 5 drops of tricaprylmethylammonium chloride. Have ready an ice/water bath large enough to accommodate the reaction flask. Drop in a 1-inch (~25 mm) stir bar and add 4.0 mL of 50% (by mass) aqueous sodium hydroxide with stirring. Set the stirrer [OP-10] to its maximum practical speed (the stir bar should rotate as rapidly as possible without spinning out of control) to whip the mixture into a frothy, cloudy liquid. After a few minutes, the mixture should form an opaque,

Take Care! Wear gloves, avoid contact with chloroform, and do not breathe its vapors.

Take Care! Wear gloves and avoid contact with the NaOH solution.

Stop and Think: Why is it important to maintain the emulsion?

creamy emulsion; at that point, monitor the temperature [OP-9] frequently. If the temperature exceeds 60°C, swirl the flask in an ice/water bath for a few seconds to bring the temperature down to 55°C, then resume stirring vigorously enough to maintain the emulsion. When the temperature drops spontaneously without external cooling, continue stirring until the temperature is 35°C or below. Then let the mixture stand until it is close to room temperature.

Separation. Transfer the reaction mixture to a 15-mL centrifuge tube (leave the stir bar behind), using about 1 mL of diethyl ether followed by 5 mL of saturated aqueous sodium chloride for the transfer. Shake gently to mix the layers and centrifuge [OP-14] the mixture for 2–3 minutes. Transfer the aqueous layer to a second centrifuge tube. If an opaque emulsion remains in the bottom of the first tube, transfer the overlying ether layer to a third centrifuge tube, stir the emulsion with 1.5 mL of diethyl ether, centrifuge the mixture, and combine the ether with that in the third tube. Extract [OP-15] the aqueous layer in the second tube with 5 mL of diethyl ether. Transfer the ether extract to the centrifuge tube containing the rest of the ether solution. Dry [OP-22] the ether solution with anhydrous calcium chloride, then evaporate [OP-16] the solvent.

Take Care! Remember that the aqueous layer contains caustic NaOH.

Waste Disposal: Pour all aqueous layers down the drain.

Purification and Analysis. Purify the residue by simple distillation [OP-27], collecting the liquid distilling in the 195–205°C range in a tared vial. Use an air condenser and insulate the neck of the Hickman still. Weigh the 7,7-dichloronorcarane and record its infrared spectrum [OP-36]. Interpret the spectrum as completely as you can and examine it for evidence of bond-angle strain.

Waste Disposal: Place any forerun and residue in a designated waste container.

Exercises

1 (a) Diagram the phase-transfer process for the reaction you carried out, using a format like that illustrated in Figure 22.1. (b) Write a mechanism for the reaction of cyclohexene with chloroform in the presence of sodium hydroxide and a phase-transfer catalyst, showing the role of the catalyst.

2 Give the systematic (IUPAC) name of 7,7-dichloronorcarane.

3 Describe and explain the possible effect on your results of the following experimental errors or variations. (a) You forgot to add the tricaprylmethylammonium chloride. (b) The lab assistant misplaced a decimal point and prepared 5.0% NaOH for this experiment. (c) Your stir bar was too small and no emulsion formed.

4 Following the format in Appendix V, construct a flow diagram for this experiment.

5 The carbon-carbon sigma bonds of cyclopropane rings exhibit some of the characteristics of pi bonds in other systems. For example, the cyclopropylmethyl cation undergoes reactions suggesting that it is best represented by the resonance structures shown on the following page. Explain the isomerization in acidic solution of chrysanthemyl alcohol to yomogi alcohol and artemisia alcohol by writing appropriate mechanisms.

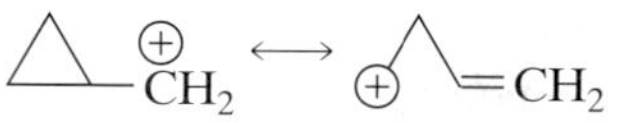
CH2
CH2

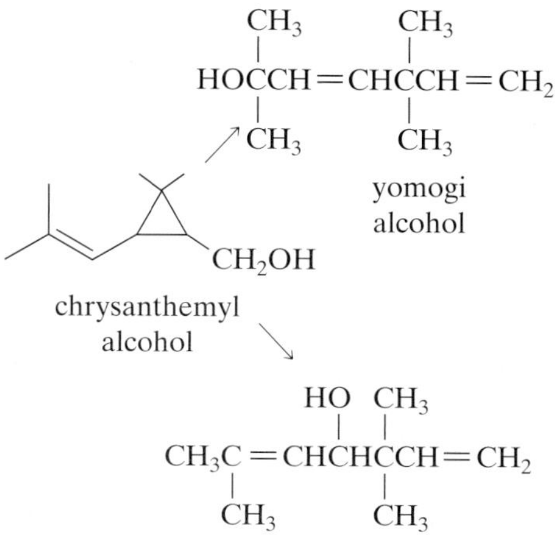
HOCCH=CHCCH=CH2
yomogi alcohol
CH2OH
chrysanthemyl alcohol
CH3C=CHCHCCH=CH2
artemisia alcohol

6 Outline syntheses of compounds **1** and **2** from the Scenario, starting with an appropriate alkene in each case.

Other Things You Can Do

(Starred items require your instructor's permission.)

*1 Assess the purity of your product by gas chromatography, or record its ^{1}H NMR spectrum in deuterochloroform. A general-purpose silicone oil/Chromosorb W column at ~110°C can be used for the chromatography.

*2 See what happens when you mix iodine with turpentine, whose components have strained rings, by carrying out Minilab 19.

3 Look up the article in *Tetrahedron Letters* **1975**, 3013, to find out which of the double bonds of limonene is attacked first by dichlorocarbene. Give the structure of the major product of this reaction and summarize the reaction conditions used, indicating how they differ from the conditions used in this experiment.

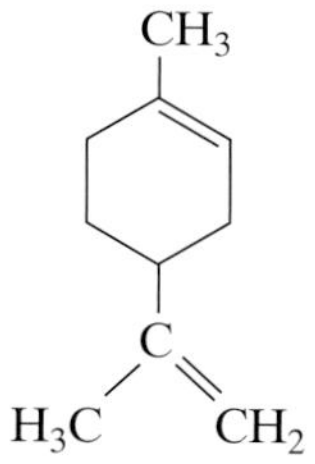
CH3
H3C
CH2

limonene

Stereochemistry of Bromine Addition to *trans*-Cinnamic Acid

EXPERIMENT 23

Reactions of Alkenes. Preparation of Alkyl Halides. Electrophilic Addition. Stereoselectivity.

Operations

OP-10   Mixing
OP-11   Addition of Reactants
OP-13   Vacuum Filtration
OP-23   Drying Solids
OP-25   Recrystallization
OP-30   Melting Point

Before You Begin

1 Read the experiment, read or review the operations as necessary, and write an experimental plan.
2 Calculate the mass of 1.00 mmol of *trans*-cinnamic acid and the theoretical yield of 2,3-dibromo-3-phenylpropanoic acid.

Scenario

The Bond Triplex is a chemical specialties company that supplies alkynes to order. Many of their alkynes are prepared by adding bromine to the corresponding alkenes and dehydrohalogenating the resulting dibromides. For example, they convert *trans*-cinnamic acid to 3-phenylpropynoic acid by way of 2,3-dibromo-3-phenylpropanoic acid.

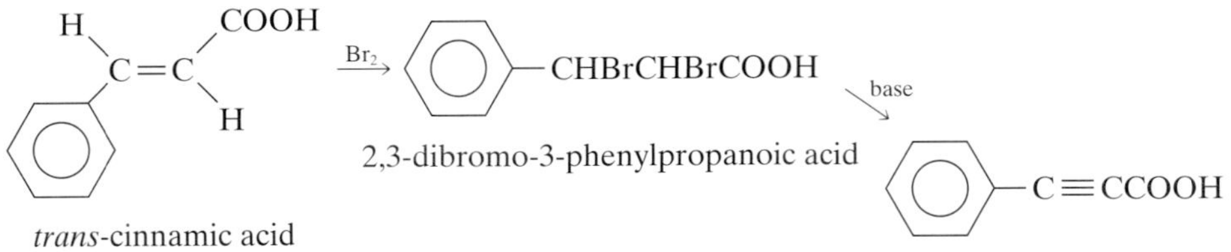

trans-cinnamic acid

2,3-dibromo-3-phenylpropanoic acid

3-phenylpropynoic acid

At a recent meeting of the three Bond partners (Sigmund, Bridget, and James) with the chemical engineers who developed their manufacturing processes, the engineers were compelled to admit that they weren't certain of the stereochemical structures of the intermediate dibromides. They had simply assumed that the bromine addition reactions proceeded by the well-established bromonium ion mechanism (see "Understanding the Experiment"), resulting in *anti* addition of bromine, and deduced the stereochemistry of the dibromides accordingly. But some of the alkenes the Bond Triplex is using have electron-donating substituents that could affect the mechanism of the addition reaction. For example, *trans*-anethole undergoes a significant amount of *syn* addition of bromine by an alternative mechanism.

Because knowing the stereochemistry of the intermediate dibromides could help the company's chemical engineers design a more efficient process for converting them to alkynes, Sigmund Bond has contacted your Institute for help in characterizing the intermediates. Your assignment is to carry out the bromination of *trans*-cinnamic acid, determine the stereochemical structure of the dibromide, and find out whether the reaction proceeds by the usual bromonium ion mechanism or by some other mechanism.

trans-anethole

Key Concept: *Changing the structure of the substrate may alter the mechanism of a reaction.*

Applying Scientific Methodology

Your hypothesis should include a prediction regarding the mechanism of the reaction. You will test your hypothesis by measuring the melting point of the product. This will reveal its identity, from which you can deduce the stereochemistry of the reaction.

The Cinnamic Acid Connection

Cinnamic acid and its close relatives, cinnamaldehyde and cinnamyl alcohol, are naturally occurring compounds that are important as flavoring and perfume ingredients and as sources for pharmaceuticals. Cinnamaldehyde, the major component of cinnamon oil, is used to flavor many foods and beverages and to contribute a spicy "oriental" note to perfumes. Cinnamic acid itself plays an important role in secondary plant metabolism. As an intermediate in the shikimic acid pathway for plant biosynthesis, cinnamic acid is involved in the formation of an enormous number of natural substances that (to give only a few examples) contribute structural strength to wood, give flavor to cloves, nutmeg, and sassafras, and produce many of the brilliant colors of nature—the flower pigments that attract insects for pollination, the vivid and delicate shades of a butterfly's wings, and the radiant colors of leaves in autumn.

In nature, cinnamic acid is formed by the enzymatic deamination (removal of ammonia) of the amino acid phenylalanine, which in turn is biosynthesized in a series of steps from shikimic acid.

cinnamic acid

cinnamaldehyde

coniferyl alcohol

myristicin

safrole

shikimic acid →(many steps)→ phenylalanine →(−NH₃)→ cinnamic acid

It can then be converted, by a wide variety of biosynthetic pathways, to coniferyl alcohol (a precursor of lignin) in sapwood, myristicin in nutmeg, safrole in sassafras bark, and flavonoids in a wide variety of plant structures.

The flavonoids are natural substances characterized by the 2-arylbenzopyran structure found in flavanone, which itself is biosynthesized from cinnamic acid by a process involving the linkage of three acetate residues to the carboxyl group of the acid.

acetates   cinnamic acid   flavanone

Flavonoids perform no single function in plants. Many are highly colored and attract insects for pollination or animals for seed dispersal; others help regulate seed germination and plant growth or protect plants from fungal and bacterial diseases. Certain flavonoids contribute the bitter taste to lemons and the bracing astringency of cocoa, tea, and beer. Some flavonoids in foods appear to act as antioxidants that may boost your immune system and help prevent cancer. Flavonoids and other derivatives of cinnamic acid provide much to delight the eye and stimulate the senses, and the world would be a duller place without them!

Understanding the Experiment

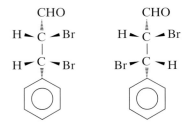

erythro-dibromide   *threo*-dibromide

erythrose   threose

In this experiment, you will carry out the addition of bromine to *trans*-cinnamic acid and identify the product from its melting point. The product, actually a mixture of enantiomers, could be either *erythro*-2,3-dibromo-3-phenylpropanoic acid [whose enantiomers have the (2R,3S) and (2S,3R) configurations], the *threo*-dibromide [(2R,3R) and (2S,3S)], or a mixture of the *erythro*- and *threo*-dibromides. The *erythro-threo* nomenclature is used to describe the configurations of compounds having two chiral centers but no plane of symmetry. It is based on the structures of the two simple sugars erythrose and threose.

The product you obtain will depend on the stereochemical course of the reaction. As explained in Experiment 21, a reaction is said to be regioselective if it might produce two or more structural isomers but in fact yields one of them preferentially. Similarly, a reaction is said to be *stereoselective* if it might produce two or more stereoisomers but in fact yields mainly (or entirely) one of them. For example, the electrophilic addition of bromine to cyclopentene is stereoselective because it yields *trans*-dibromocyclopentane and no *cis*-dibromocyclopentane, indicating that the components of Br_2 must add to opposite sides of the carbon-carbon double bond. This mode of addition is called *anti* addition, while addition of the components of a reagent to the same side of a double bond is called *syn* addition.

The following scenario has been proposed to explain the *anti* addition of bromine to cyclopentene. As a bromine molecule approaches perpendicular to the negatively charged pi cloud of the carbon-carbon double bond, its bonding electrons are repelled away from the bromine atom nearer the double bond, leaving it with a partial positive charge. As the positively charged bromine penetrates the pi cloud, a negative bromide ion breaks away from it, leaving a cyclic *bromonium ion*, in which the positive bromine is bonded to two carbon atoms. Backside attack on the bromonium ion by a bromide ion results in the observed *trans* product.

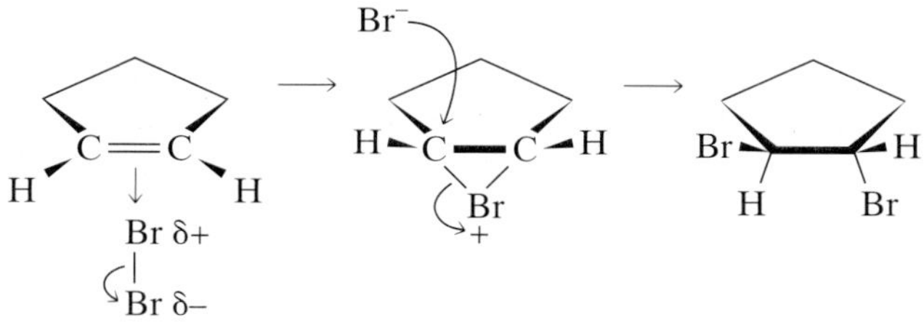

Other electrophilic addition mechanisms may lead to different stereo-chemical outcomes. In the addition of bromine to *trans*-anethole, conjugation with the ring stabilizes a carbocation intermediate that can be attacked on either side, leading to *syn* as well as *anti* addition.

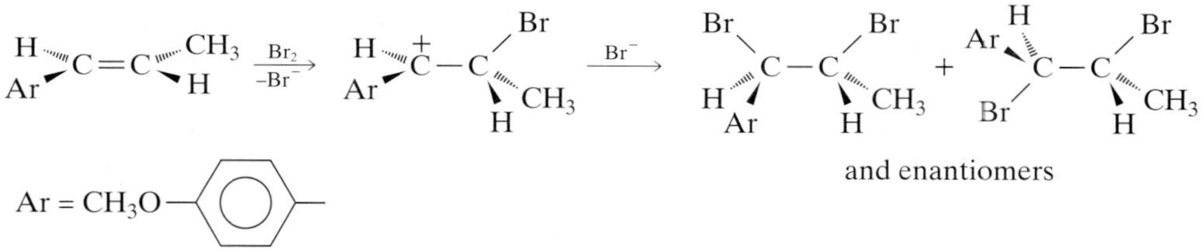

and enantiomers

$Ar = CH_3O-\bigcirc-$

In this particular reaction, about 35% of the product results from *syn* addition and 65% from *anti* addition. It is also conceivable that a concerted addition to the carbon-carbon double bond could occur, leading exclusively to *syn* addition, as illustrated here.

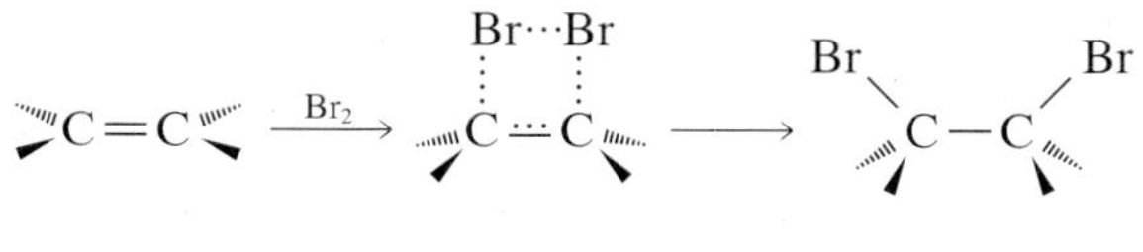

You will carry out the reaction by adding a solution of bromine in acetic acid to a solution of *trans*-cinnamic acid in the same solvent, and separate the resulting dibromide by vacuum filtration. Because the melting points of the *erythro*- and *threo*-dibromides differ by more than 100°C, the product can easily be identified from its melting point. A mixture of both products would melt over a broad range that should not coincide with the melting point of either pure dibromide. (Note that impurities may lower the melting point of your product somewhat, so a difference of a few degrees from the expected melting point does not indicate that both products are present.) From the identity of your product, you should be able to deduce whether the addition of bromine to *trans*-cinnamic acid involves *syn* or *anti* addition, or a mixture of the two. Molecular models will help you relate the configurations of the *syn* and *anti* addition products to the stereochemical structures shown for the *erythro*- and *threo*-dibromides.

Reaction and Properties

$$
\underset{\textit{trans-cinnamic acid}}{\ce{H\underset{|}{C}=C\underset{|}{H}}} \ce{^{COOH}} + Br_2 \xrightarrow{\text{HOAc}} \underset{\text{2,3-dibromo-3-phenylpropanoic acid}}{\ce{CHBrCHBrCOOH}}
$$

trans-cinnamic acid + Br₂ →(HOAc) 2,3-dibromo-3-phenylpropanoic acid

Table 23.1 Physical properties

	M.W.	mp	bp	d
trans-cinnamic acid	148.2	136		
bromine	159.8	−7	59	3.12
acetic acid	60.1	17	118	1.049
erythro-2,3-dibromo-3-phenylpropanoic acid	308.0	204		
threo-2,3-dibromo-3-phenylpropanoic acid	308.0	95		

Note: mp and bp are in °C, density is in g/mL.

Directions

A. *Preparation of 2,3-Dibromo-3-phenylpropanoic Acid*

Safety Notes

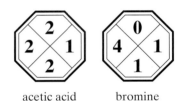

acetic acid   bromine

Take Care! Wear gloves, avoid contact with acetic acid and the bromine solution, and do not breathe their vapors.

Stop and Think: What is the purpose of the cyclohexene? What reaction is involved?

Waste Disposal: Unless your instructor indicates otherwise, flush the filtrate down the hood's drain with water.

> **Bromine is highly toxic and corrosive, and its vapors can damage the eyes and respiratory tract. Wear gloves when handling the bromine solution and dispense it under a hood; avoid contact and do not inhale its vapors. Acetic acid causes chemical burns that can seriously damage skin and eyes; its vapors are highly irritating to the eyes and respiratory tract. Wear gloves and dispense it under a hood; avoid contact and do not breathe its vapors.**

Reaction. *Carry out the reaction under the hood and wear gloves throughout.* Weigh 1.00 mmol of *trans*-cinnamic acid into a 3-mL conical vial and add 0.8 mL of glacial acetic acid and a stirring device. Attach an air condenser and clamp the apparatus to a ring stand, then stir [OP-10] this mixture in a 50°C water bath until the cinnamic acid has dissolved. Measure 1.0 mL of a 1.0 *M* solution of bromine in acetic acid into a conical vial and add [OP-11] this solution through the top of the condenser with a 9-inch Pasteur pipet. Heat the reaction mixture in the 50° hot water bath until the red-brown bromine color fades to light orange; then continue to heat the reaction mixture for 15 minutes. If the mixture becomes colorless (or nearly so) during this period, add more of the bromine/acetic acid solution dropwise until the color just persists. If the mixture has a distinct orange color at the end of the reaction period, add a drop or two of cyclohexene to turn it light yellow.

Separation. Cool the reaction mixture in ice/water bath for 10 minutes or more, scratching the sides of the vial to induce crystallization, if necessary. Collect the product by vacuum filtration [OP-13] and wash it on the filter with several portions of ice-cold water, until the acetic acid odor is hardly noticeable.

Purification and Analysis. Purify the product by recrystallization [OP-25] from 50% aqueous ethanol. Dry [OP-23] the 2,3-dibromo-3-phenylpropanoic acid and measure its melting point [OP-30], then decide whether you have prepared the *erythro* or *threo* isomer, or a mixture of the two.

Stop and Think: Was the result what you expected? If not, why not?

B. *Stereochemistry of Bromine Addition*
Construct a molecular model of *trans*-cinnamic acid (to simplify matters, use a colored ball to represent the phenyl group). Simulate the *syn* addition of bromine by removing one of the C=C connectors and inserting two orange bromine atoms, with connectors, into the vacant holes. (You may want to replace the remaining flexible connector by a rigid one.) Rotate around the carbon–carbon single bond that remains until the model corresponds to the stereochemical projection for either *threo*- or *erythro*-2,3-dibromo-3-phenyl-propanoic acid. Simulate the *anti* addition of bromine by removing the upper connector of the carbon–carbon double bond and moving one end of the lower connector from the hole it occupies to the vacant hole in the same carbon atom. Be careful not to rotate either carbon atom as you do so. Insert two bromine atoms, with connectors, into the vacant holes; then rotate the model as before, until it matches one of the stereochemical projections. Decide whether your product was formed by *syn* addition, *anti* addition, or a mixture of both. Write a mechanism that explains your results.

Exercises

1 (a) Write resonance structures showing how the aryl group of *trans*-anethole stabilizes the intermediate carbocation shown in "Understanding the Experiment." (b) Based on your results, explain any differences or similarities in the stereochemistry of the bromine addition reactions of *trans*-cinnamic acid and *trans*-anethole.

2 What product or products would you expect to obtain by the addition of bromine to *cis*-cinnamic acid, assuming that it reacts by the same mechanism as the *trans* acid?

3 (a) Write mechanisms showing why bromination of *trans*-cinnamic acid yields the product you obtained, as a racemic mixture of enantiomers. (b) Draw a stereochemical projection for each enantiomer and specify the configuration (*R* or *S*) at each stereocenter.

4 Describe and explain the possible effect on your results of the following experimental errors or variations. (a) The cinnamic acid you used was actually a mixture of *cis* and *trans* isomers. (b) You added only 0.5 mL of the bromine solution. (c) You misread the label on a bottle of cyclohexane and used it in place of cyclohexene.

5 Following the format in Appendix V, construct a flow diagram for this experiment.

6 Draw stereochemical projections for the products of bromine addition to maleic acid and fumaric acid (*cis* and *trans* HOOCCH=CHCOOH), assuming that bromine adds in the same way to these compounds as it does to cinnamic acid.

7 (a) Would you expect the product from this experiment to be optically active? Could it be resolved into optically active constituents? Explain. (b) Would the product of the bromination of fumaric acid (see Exercise 6) be optically active? Could it be resolved into optically active constituents? Explain.

Other Things You Can Do

(Starred projects require your instructor's permission.)

*1 Test some commercial products for unsaturation as described in Minilab 20.

*2 Synthesize 3-phenylpropynoic acid from your product by scaling down the procedure given in *J. Am. Chem. Soc.* **1942**, *64*, 2510. Find a suitable recrystallization solvent to use in place of carbon tetrachloride.

3 Write a research paper about the industrial preparation and commercial uses of cinnamic acid and its derivatives, starting with sources listed in the Bibliography.

Hydration of a Difunctional Alkyne

Reactions of Alkynes. Preparation of Carbonyl Compounds. Electrophilic Addition. Infrared Spectrometry.

Operations

OP-6   Heating
OP-15   Extraction
OP-16   Evaporation
OP-17   Steam Distillation
OP-21   Washing Liquids
OP-22   Drying Liquids
OP-27   Simple Distillation
OP-36   Infrared Spectrometry

Before You Begin

1 Read the experiment, read or review the operations, and write an experimental plan.
2 Calculate the mass and volume of 10.0 mmol of 2-methyl-3-butyn-2-ol and the theoretical yield of the product.

Scenario

Malthusian Solutions manufactures a line of birth control pills, called Malthusian Lozenges, which contain synthetic sex hormones as their active ingredients. Megestrol acetate, a sex hormone the company would like to incorporate in their pills, is one of the most powerful ovulation inhibitors known. Megestrol acetate is classified as an *acetoxyketo steroid* because it is a ketone with an acetoxy (CH_3COO) group on its five-membered ring (the D ring). A number of synthetic sex hormones such as norethynodrel are *alkynol steroids* having an ethynyl group ($HC \equiv C-$) and a hydroxyl function on the D ring.

Malthusian lozenges were introduced in Aldous Huxley's apocalyptic novel Brave New World.

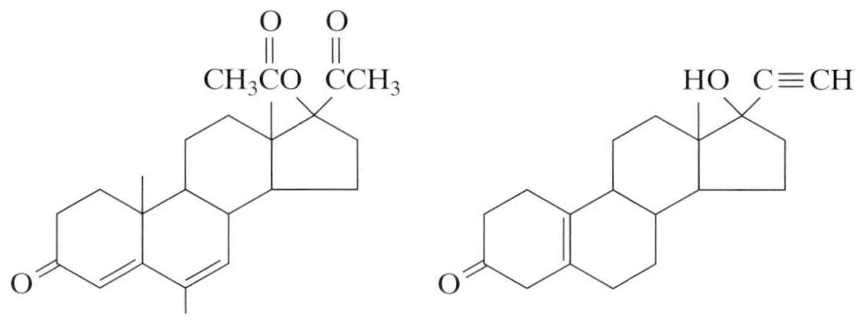

megestrol acetate   norethynodrel

Harry D. Stork, the product development director at Malthusian Solutions, thinks it should be possible to convert an alkynol steroid such as norethynodrel to an acetoxyketo steroid by hydrating the carbon-carbon triple bond and acetylating the hydroxyl group.

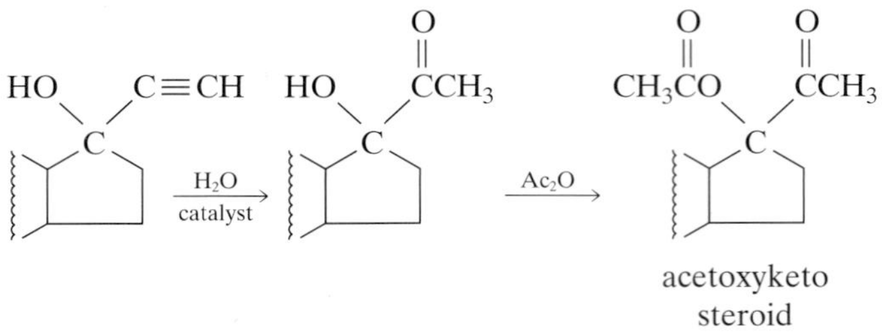

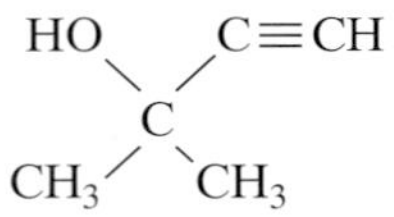

acetoxyketo
steroid

2-methyl-3-butyn-2-ol

If this reaction pathway is feasible, Malthusian Solutions might use it to produce megestrol acetate and related acetoxyketo steroids. But first they must establish that the alkyne hydration follows Markovnikov's rule and yields the desired carbonyl compound. Norethynodrel and other synthetic sex hormones are quite expensive, so you will test Stork's idea by carrying out the hydration reaction on a simpler "model compound," 2-methyl-3-butyn-2-ol.

Applying Scientific Methodology

Your hypothesis should include a prediction of the structure of the product, which may or may not be analogous to that of an acetoxyketo steroid. You will test your hypothesis using infrared (IR) spectroscopy.

From "The Pill" to Oblivon

Natural *estrogens* such as estrone and 17β-estradiol are responsible for promoting the development of secondary sex characteristics in females during puberty. Estrogens are also frequently prescribed to alleviate the mental and physical discomfort associated with menopause, and to prevent osteoporosis (weakening of bone structure) in older women. Natural *progestins* such as progesterone are necessary to maintain pregnancy in mammals.

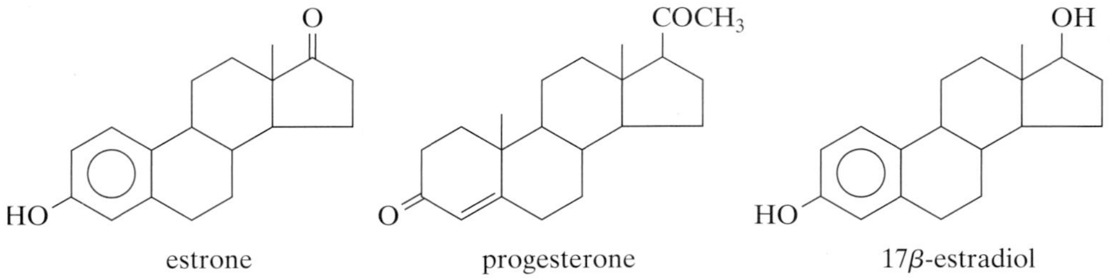

estrone   progesterone   17β-estradiol

Progestins have been used to treat menstrual disorders and uterine bleeding and to prevent miscarriages. Both types of hormones belong to an important class of natural products called *steroids*, which are characterized by a basic four-ring skeleton, the perhydrocyclopentanophenanthrene nucleus. The most widely known steroid, cholesterol, is often regarded as an undesirable ingredient in our food because of its role in cardiovascular disease, but it is also an essential precursor of sex hormones and other body regulators.

perhydrocyclopentanophenanthrene

cholesterol

Natural hormones cannot be taken orally because they are rapidly deactivated in the liver, so the search for synthetic hormones with similar activity began soon after the natural ones were isolated and characterized. One way to prevent the deactivation of a steroid such as 17β-estradiol is to stabilize the C-17 hydroxy group with an appropriate substituent. The ethynyl (HC$\equiv$C—) function seems to fill the bill nicely. For example, treating estrone with potassium acetylide in liquid ammonia yields ethinyl estradiol, a potent estrogen that can be taken orally.

ethinyl estradiol

One of the most intriguing properties of the natural estrogens and progestins is their ability to inhibit ovulation in females. During the 1940s it was a common practice to treat certain menstrual disorders with these hormones because preventing ovulation stops menstruation. Of course, preventing ovulation also prevents pregnancy, so the idea of using hormones for birth control undoubtedly occurred to some scientists. At that time there were no suitable sex hormones that could be taken orally, but by the early 1950s the picture had changed. The discovery that certain Mexican wild yams of the species *Dioscorea* contain the natural steroid diosgenin provided chemists and pharmaceutical manufacturers with a cheap, abundant starting material for the preparation of synthetic steroids. The first synthesis of a steroid oral contraceptive, 19-norprogesterone, was accomplished in 1951 by a research team at Syntex led by Carl Djerassi (see Bibliography, L13 for more about Djerassi's work and wide-ranging interests). Early in 1953 this and other synthetic steroids were evaluated for anti-ovulatory activity by Gregory Pincus and his colleagues. They soon discovered that the combination of a synthetic progestin with a small amount of synthetic estrogen provided the highest anti-ovulatory activity, and "the pill" was born soon afterward. The G. D. Searle Company marketed the first birth-control pill, Enovid, which contained 9.85 mg of the progestin norethynodrel and 0.15 mg of the estrogen mestranol. Modern oral contraceptives of the "mini-pill" type contain considerably smaller amounts of a progestin and no estrogen; although they are slightly less effective than the combination pills, they have fewer side effects. When used properly, most oral contraceptives are 99–100% effective in preventing pregnancy, and their impact on society has been enormous.

Alkynols having a hydroxy group adjacent to the triple bond (as in norethynodrel) are comparatively easy to synthesize commercially, and they are used in a number of pharmaceuticals. The 2-methyl-3-butyn-2-ol used as a model compound in this experiment is prepared by combining acetone and acetylene with an alkali metal in liquid ammonia.

$$CH_3-\overset{\overset{\displaystyle O}{\|}}{\underset{\underset{\displaystyle CH_3}{|}}{C}} + HC\equiv CH \xrightarrow{\text{Li, NH}_3} CH_3-\overset{\overset{\displaystyle OH}{|}}{\underset{\underset{\displaystyle CH_3}{|}}{C}}-C\equiv CH$$

Preparation of 2-methyl-3-butyn-2-ol

$$CH_3CH_2-\overset{\overset{\displaystyle OH}{|}}{\underset{\underset{\displaystyle CH_3}{|}}{C}}-C\equiv CH$$

methylparafynol
(Oblivon)

A similar alkynol, named methylparafynol, has been used in sleeping pills under the trade name Oblivon. Apparently the tertiary alcohol portion of the molecule causes methylparafynol to act as sedative or depressant, while the acetylenic group gives it hypnotic (sleep-producing) properties.

Understanding the Experiment

In this experiment you will carry out the hydration of 2-methyl-3-butyn-2-ol and identify the product by IR spectrometry. Alkyne hydration, the electrophilic addition of water to a carbon-carbon triple bond, is generally accomplished by heating the alkyne with water in the presence of an acid and a mercury(II) salt. As shown in the proposed reaction mechanism below, mercuric ion catalyzes the reaction by a process that involves its addition to the triple bond to form a pi complex (**1**). The pi complex is then attacked by water to give intermediate **2**, which loses a proton and is hydrolyzed to an enol (**3**). Isomerization of the enol to the corresponding keto tautomer yields the final product.

$$-C\equiv C- \xrightarrow{\text{HgSO}_4} \underset{\mathbf{1}}{\overset{\overset{\displaystyle Hg}{\overset{\displaystyle 2+}{\cdots}}}{C\!\equiv\!C}} \xrightarrow{\text{H}_2\text{O}} \underset{\mathbf{2}}{\overset{\displaystyle Hg^+}{C\!=\!C}}\overset{}{OH_2^+}$$

$$\xrightarrow[-Hg^{2+}]{-H^+ \;\; H_3O^+} \underset{\mathbf{3}}{\overset{\displaystyle H}{C\!=\!C}}\overset{}{OH} \xrightarrow{\text{tautomerization}} \overset{\overset{\displaystyle H}{|}}{\underset{\underset{\displaystyle H}{|}}{-C-}}C\overset{\displaystyle O}{\lVert}$$

See Experiment 21 for a discussion of regioselective reactions.

The hydration of alkenes is regioselective because the intermediate is a carbocation that can be stabilized by electron-donating alkyl substituents, resulting in Markovnikov addition. The above mechanism for hydration of alkynes does not involve a free carbocation, so it is not obvious that such a reaction should follow Markovnikov's rule, especially in the presence of an electron-withdrawing group such as the OH in 2-methyl-3-butyn-2-ol. If it does, the product from hydration of a terminal alkyne (one having a hydrogen on a triple-bonded carbon) should be a methyl ketone. Anti-Markovnikov addition of water to a terminal alkyne should yield an aldehyde.

$$RC{\equiv}CH \xrightarrow{\ H_2O\ }$$

$$\begin{array}{c} \overset{O}{\overset{\|}{RC}}-\overset{H}{\overset{|}{CH}} \\ | \\ H \end{array}$$ Markovnikov addition

$$\begin{array}{c} \overset{H}{\overset{|}{RC}}-\overset{O}{\overset{\|}{CH}} \\ | \\ H \end{array}$$ anti-Markovnikov addition

Adding 2-methyl-3-butyn-2-ol to an aqueous solution containing sulfuric acid and mercury(II) sulfate (the catalysts) will get the reaction started, and you will heat the reaction mixture under reflux to bring it closer to completion. The product will be separated from nonvolatile impurities in the reaction mixture by internal steam distillation with water. During the reaction, some mercury(II) ion may be reduced to metallic mercury, which will remain in the reaction flask. Because mercury vapor is very toxic, it is important to dispose of the residue properly.

Any acid that distills with the product will be neutralized by adding some hydrated potassium carbonate to the distillate. The potassium carbonate also salts out the product (see OP-15b) so that it can be removed more completely during the subsequent extraction with diethyl ether. Mercury and other heavy metals can form organometallic salts of terminal alkynes that may *explode* if heated to dryness. Because such salts do not form at a low pH, it is unlikely that your product will contain any, but you will wash the ether extract with dilute sulfuric acid as a precautionary measure. Using anhydrous potassium carbonate to dry the product removes any traces of wash acid that may remain. After evaporating the solvent, you will purify the crude product by simple distillation.

Infrared spectrometry is particularly useful for detecting functional groups in molecules, so it can readily confirm the conversion of one functional group to another in this synthesis. As you can see in Figure 24.1, the IR spectrum of the reactant is characterized by a strong, narrow ${\equiv}CH$ stretching band near 3300 cm^{-1} and a weak $C{\equiv}C$ stretching band near 2100 cm^{-1}. Because the broad $O-H$ band of the reactant occurs in the 3300–3400 cm^{-1} region, it partially obscures the sharper ${\equiv}C-H$ band. Aldehydes and ketones both give rise to strong carbonyl bands near 1700 cm^{-1}, but the CHO group of an aliphatic aldehyde produces sharp $C-H$ bands of moderate strength near 2720 cm^{-1} and 2840 cm^{-1}. (See OP-36 for additional information about interpretation of IR spectra.) By comparing the spectrum of your product with that of the reactant, you should be able to decide whether or not the desired product has been formed.

Reactions and Properties

$$\begin{array}{c} \overset{OH}{\overset{|}{CH_3C}}-C{\equiv}CH + H_2O \\ | \\ CH_3 \end{array} \xrightarrow[HgSO_4]{H_2SO_4} \begin{array}{c} \overset{OH}{\overset{|}{CH_3C}}-\overset{O}{\overset{\|}{CCH_3}} \\ | \\ CH_3 \end{array} \quad or \quad \begin{array}{c} \overset{OH}{\overset{|}{CH_3C}}-CH_2\overset{O}{\overset{\|}{CH}} \\ | \\ CH_3 \end{array}$$

2-methyl-3-butyn-2-ol   3-hydroxy-3-methyl-2-butanone   3-hydroxy-3-methylbutanal

Table 24.1 Physical properties

	M.W.	bp	d
2-methyl-3-butyn-2-ol	84.1	104	0.868
mercury(II) sulfate	296.7		
3-hydroxy-3-methyl-2-butanone	102.1		0.953
3-hydroxy-3-methylbutanal	102.1		

Note: bp is in °C, density is in g/mL.

2-methyl-3-butyn-2-ol

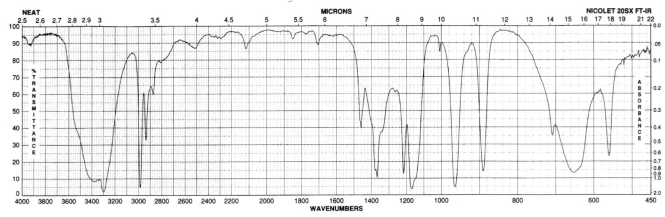

Figure 24.1 IR spectrum of 2-methyl-3-butyn-2-ol.

Directions

Safety Notes

Mercury(II) sulfate is very poisonous if inhaled or ingested. Do not breathe its dust, or allow it to contact skin or eyes. Wash hands thoroughly after handling the compound.

2-Methyl-3-butyn-2-ol and 3-hydroxy-3-methyl-2-butanone may be harmful by inhalation, ingestion, or skin absorption. Avoid contact with the reactant and product and do not breathe their vapors.

Mercury (formed during the reaction) emits toxic vapors whose concentration can build up to dangerous levels in poorly ventilated areas. Dispose of mercury-containing residues properly and clean up any spills immediately.

Diethyl ether is extremely flammable and may be harmful if inhaled. Do not breathe its vapors and keep it away from flames and hot surfaces.

If the experimental directions are not followed carefully, the reaction mixture could contain an unstable organomercury compound that could explode when strongly heated. Be sure to wash the product with 1*M* sulfuric acid and do not distill it to dryness.

Take Care! $HgSO_4$ is very poisonous; wear gloves and avoid contact or ingestion.

Reaction. Carefully weigh 50 mg of mercury(II) sulfate into a 10-mL round-bottom flask, and weigh 10.0 mmol of 2-methyl-3-butyn-2-ol into a conical vial. Add 4.0 mL of 3 *M* sulfuric acid and a magnetic stir bar to the flask, and attach an air condenser. Start the stirrer and stir until the mercury(II) sulfate dissolves. Add the 2-methyl-3-butyn-2-ol all at once through the condenser and continue to stir for 2–3 minutes. Heat the reaction mixture to boiling; then continue heating under gentle reflux [OP-6] for 30 minutes.

Observe and Note: Can you see any evidence for a reaction? Describe any observations in your lab notebook.

Separation.   Calibrate a 15-cm centrifuge tube by adding 7.5 mL of water and marking the level with a marking pen (discard the water). Assemble an apparatus for internal steam distillation [OP-17] by inserting a Hickman still between the round-bottom flask and the air condenser. Measure 5.0 mL of water into a small beaker. Distill the reaction mixture, adding water periodically to keep the flask about half full, until all of the water has been added. Continue distilling until about 7.5 mL of distillate has been collected.

Take Care! Foaming may occur during distillation. Do not distill to dryness.

Shake the distillate with 1.5 g of potassium carbonate sesquihydrate (or dihydrate) until the salt dissolves. Extract this liquid mixture [OP-15] with two successive 2.5-mL portions of diethyl ether. Wash [OP-21] the combined extracts with 2.5 mL of 1 M sulfuric acid in a centrifuge tube, stirring the liquid layers until any fizzing subsides before you stopper and shake the tube. Dry [OP-22] the organic layer with anhydrous potassium carbonate. Evaporate [OP-16] the solvent, in portions, from a 3-mL conical vial.

Waste Disposal: Place the residue from the boiling flask, which may contain a small droplet of mercury, into a designated waste container.

Take Care! Do not breathe the vapors of diethyl ether.

Purification and Analysis.   Purify your crude product by simple distillation [OP-27], collecting the liquid that distills around 138–142°C. Weigh the purified product; then record its IR spectrum [OP-36] or obtain the spectrum from your instructor. From its IR spectrum, deduce the name and structure of your product.

Take Care! Do not distill to dryness.

Exercises

1 Write a mechanism for the hydration of 2-methyl-3-butyn-2-ol that explains your results. Assume that the bridged ion **1** in the alkyne hydration mechanism (see "Understanding the Experiment") is asymmetric, with a larger partial positive charge on the alkyl-substituted carbon.

2 Interpret the IR spectrum of your product as completely as you can, and explain clearly how it supports your conclusion about the structure of the product.

3 Describe and explain the possible effect on your results of the following experimental errors or variations. (a) You added potassium carbonate sesquihydrate to the reaction flask (after reflux) rather than to the initial distillate, forgot to wash the ether solution with 1 M sulfuric acid, and distilled the final product to dryness. (b) You added concentrated sulfuric acid to the reaction mixture rather than 3 M sulfuric acid. (c) You forgot to add potassium carbonate sesquihydrate to the initial distillate.

4 (a) Identify each signal in the ^{1}H NMR spectrum (Figure 24.2) of the Markivnikov addition product, 3-hydroxy-3-methyl-2-butanone, by indicating the proton set responsible for it. (b) Sketch the ^{1}H NMR spectrum you would expect to obtain for the anti-Markovnikov addition product, 3-hydroxy-3-methylbutanal, showing approximate signal areas, multiplicities, and chemical shifts.

5 Following the format in Appendix V, construct a flow diagram for this experiment.

3-hydroxy-3-methyl-2-butanone

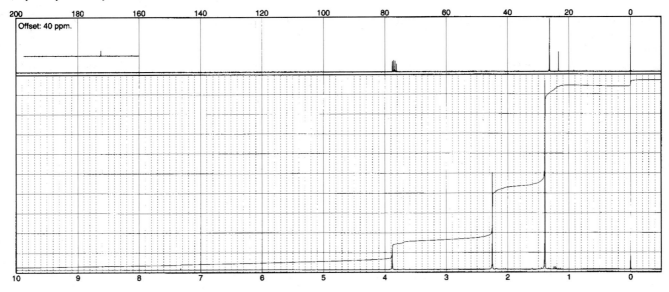

Figure 24.2 ^{1}H NMR spectrum of 3-hydroxy-3-methyl-2-butanone.

estrone methyl ether

↓

an acetoxyketo steroid

6   Draw structures for the products that you would expect to obtain from the hydration of acetylene (ethyne), 1-butyne, 2-butyne, and norethynodrel.

7   Using reactions discussed in this experiment, outline a synthetic pathway for the conversion of estrone methyl ether to the acetoxyketo steroid shown.

Other Things You Can Do

(Starred projects require your instructor's permission.)

*1   Test your reactant and product with bromine (test C-7 in Part IV) and potassium permanganate (test C-19 in part IV). Explain the results.

*2   To verify the direction of addition, prepare a semicarbazone derivative of the product (procedure D-4 in Part IV) and obtain its melting point. The semicarbazone of the Markovnikov product melts at 163°C; that of the anti-Markovnikov product melts at 222–223°C.

3   Oral contraceptives have been the subject of scientific and ethical controversy since Enovid was first marketed in 1960. Starting with sources listed in the Bibliography, write a research paper about recent developments in the field of contraception, discussing the impact of modern birth-control methods on society.

Preparation of Bromotriphenylmethane and the Trityl Free Radical

Reactions of Hydrocarbons. Preparation of Alkyl Halides. Free-Radical Substitution. Free Radicals.

Operations

OP-24 Drying and Trapping Gases
OP-6   Making Transfers
OP-7   Heating
OP-10   Mixing
OP-11   Addition of Reactants
OP-12   Gravity Filtration
OP-13   Vacuum Filtration
OP-16   Evaporation
OP-23   Drying Solids
OP-25   Recrystallization
OP-30   Melting Point

Before You Begin

1 Read the experiment and operation OP-24, read or review the other operations as necessary, and write an experimental plan.
2 Calculate the mass of 0.900 mmol of triphenylmethane and the theoretical yield of bromotriphenylmethane.

Scenario

Dr. Perry Celsus, the science historian who asked you to help develop a modern version of Wöhler's urea synthesis (Experiment 17), now wants you to recreate the experiment that led to the discovery of the first stable free radical by Moses Gomberg. In a landmark paper that was published in the 1900 *Journal of the American Chemical Society*, Gomberg described his preparation of the triphenylmethyl (trityl) free radical by the reaction of bromotriphenylmethane with various metals. Dr. Celsus has requested that you prepare some bromotriphenylmethane and find out whether it can, in fact, be converted by metallic zinc to the free radical that Gomberg described.

J. Am. Chem. Soc. **1900**, *22*, 757.

Applying Scientific Methodology

Your working hypothesis should deal with the likelihood that the free radical will be produced and detected by the procedure described in part **B** of the Directions. You will test the hypothesis by observing and interpreting the behavior of the reaction mixture obtained by treating bromotriphenyl-methane with zinc.

The Case of the Disappearing Dimer

In the mid-nineteenth century, many chemists were convinced that carbon could exist in a trivalent state as a free *radical*, where a radical is a group of atoms such as methyl (CH_3—) that generally exists only in combination with other atoms or groups, as in methyl bromide (CH_3Br). For example, it seemed reasonable to believe that if magnesium chloride could react with sodium metal to yield magnesium, then methyl halides should react with sodium to form methyl. But when Charles Wurtz added sodium to methyl iodide, he obtained not methyl but ethane in a reaction known today as the Wurtz reaction.

Reaction of methyl iodide and sodium

$$CH_3I + Na \longrightarrow CH_3 + NaI \qquad \text{(expected reaction)}$$
$$2CH_3I + Na \longrightarrow CH_3CH_3 + 2NaI \qquad \text{(actual reaction)}$$

After many similar attempts to make free radicals ended in failure, chemists began to doubt that they could exist at all. Then in 1900, Moses Gomberg, a young chemistry instructor at the University of Michigan, published a remarkable paper describing his discovery of the world's first stable free radical, triphenylmethyl.

Gomberg never meant to make a free radical. He was trying to synthesize hexaphenylethane to prove a point that, had he been successful, would be remembered today by only a handful of scientists. His initial attempts to prepare this compound using the Wurtz reaction were not successful, so he tried different metals, such as silver and zinc.

Gomberg's attempted synthesis of hexaphenylethane (Ph = phenyl, C_6H_5—):

$$2Ph_3CBr + Zn \rightarrow Ph_3C\text{---}CPh_3 + ZnBr_2 \qquad \text{(expected reaction)}$$

Each time, he obtained a snow-white solid that melted at 185°C and had the wrong molecular formula for hexaphenylethane. After repeated attempts to prepare this compound, it finally occurred to him that the product might be reacting with oxygen in the air and forming triphenylmethyl peroxide, $Ph_3COOCPh_3$. When Gomberg next ran the reaction, he was careful to exclude air from the reaction mixture, and he obtained a white solid that melted at 147°C and gave the correct analysis for hexaphenylethane. But this compound behaved very strangely for a hydrocarbon. It reacted in solution with air to form triphenylmethyl peroxide and it rapidly decolorized dilute halogen solutions—something no ordinary hydrocarbon would do. Gomberg eventually concluded that he had synthesized the world's first stable free radical, triphenylmethyl.

$$2Ph_3CBr + Zn \rightarrow 2Ph_3C\cdot + ZnBr_2$$

He had, in a way, reversed Wurtz's experiment; Wurtz tried to make a radical and obtained its dimer, while Gomberg tried to make the dimer of a radical and ended up with the radical itself. Although he didn't know it at the time, the dimer he finally made was not even the one he expected.

Gomberg believed that the colored triphenylmethyl (trityl) radical was in equilibrium with hexaphenylethane in solution.

Gomberg's proposed equilibrium

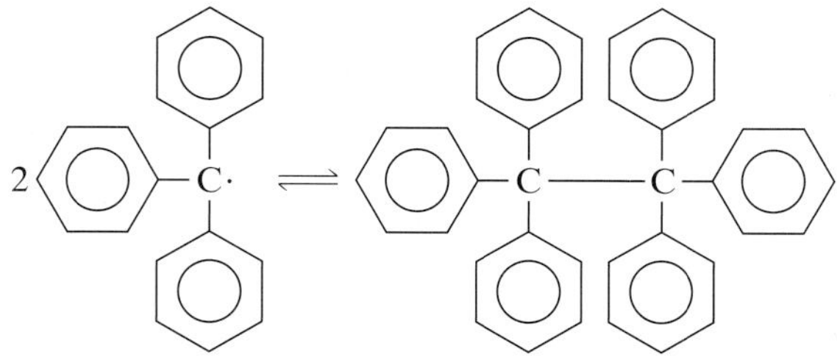

triphenylmethyl
(trityl) free
radical

hexaphenylethane (assumed)

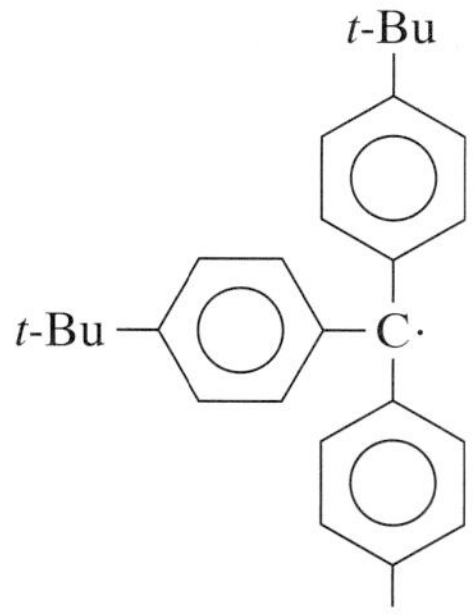

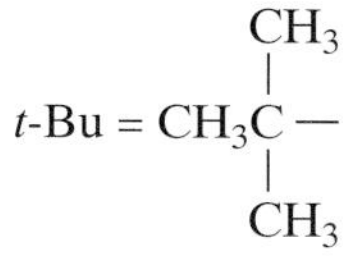

But, in 1968, a team of chemists from the Netherlands reported that they had prepared a similar free radical that would not dimerize. This new radical, *tris*(4-*t*-butylphenyl)methyl, had bulky *t*-butyl groups at each *para* position. Because the Dutch chemists could not explain why such substituents would prevent the formation of a hexaphenylethane-type dimer, they decided to prepare Gomberg's dimer and find out whether it had the structure he proposed. It didn't—NMR analysis showed that it actually has the following structure:

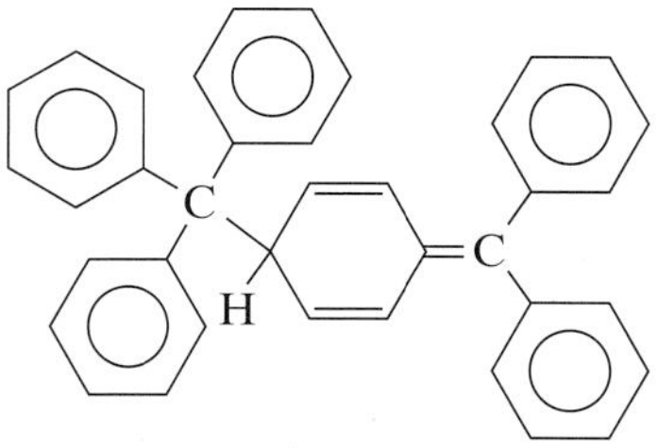

tris(4-*t*-butylphenyl)methyl

$$t\text{-Bu} = CH_3C \begin{matrix} CH_3 \\ | \\ | \\ CH_3 \end{matrix} —$$

dimer of triphenylmethyl

So Gomberg's dimer was not hexaphenylethane after all; that elusive hydrocarbon has probably never existed!

Understanding the Experiment

In this experiment, you will attempt to brominate triphenylmethane with elemental bromine using light to initiate the reaction. The bromination reaction proceeds by a chain mechanism similar to that for the chlorination of methane and other hydrocarbons. A source of atomic bromine is needed to initiate the chain reaction. In a typical alkane bromination, bromine atoms are produced by irradiating molecular bromine in solution.

$$Br_2 \xrightarrow{h\nu} 2Br\cdot$$

In the chain-propagating stage of the reaction, triphenylmethane should react with bromine atoms to produce trityl radicals. Thus you could actually be making this radical twice during the experiment: once as an intermediate in the

The symbol hv over a reaction arrow means that some kind of electromagnetic radiation, such as ultraviolet or visible light, is used to catalyze the reaction.

Key Concept: A chain reaction involves the initial formation of a reactive species that is regenerated after each cycle of product-forming steps and initiates the next cycle.

synthesis of bromotriphenylmethane, and later as a result of its reaction with zinc. Each trityl radical then reacts with a molecule of bromine, yielding a molecule of the product and another bromine atom that starts another cycle of chain propagating steps.

$$Ph_3C\text{---}H + Br\cdot \rightarrow Ph_3C\cdot + HBr$$

$$Ph_3C\cdot + Br_2 \rightarrow Ph_3C\text{---}Br + Br\cdot$$

You will carry out the reaction by adding a solution of bromine in dichloromethane to a solution of triphenylmethane in the same solvent while heating the reaction mixture under reflux and irradiating it with light to generate bromine atoms. The reaction evolves hydrogen bromide, and bromine vapors may also escape from your apparatus, so you must either work under a fume hood or use a gas trap. When the reaction is complete, some excess bromine may remain that will color the reaction mixture. It can be removed by adding a drop or so of cyclohexene.

Removal of excess bromine

cyclohexene   1,2-dibromocyclohexane (a liquid)

As always, you will want to minimize material losses by making nearly quantitative transfers and avoiding unnecessary transfers. You can reduce the number of transfers by evaporating the solvent directly from the reaction flask and recrystallizing the product in the same flask.

You will attempt to prepare the triphenylmethyl radical by treating bromotriphenylmethane with metallic zinc. Exposing the solution to air may provide evidence for the existence of a dynamic equilibrium between the radical and its dimer.

Reactions and Properties

triphenylmethane   triphenylmethyl bromide

$$B \quad 2 \left(\text{Ph}\right)_3\text{C-Br} + \text{Zn} \xrightarrow{\text{toluene}} 2 \left(\text{Ph}\right)_3\text{C·} + \text{ZnBr}_2$$

triphenylmethyl radical
(dimerizes in solution)

Table 25.1 Physical properties

	M.W.	mp	bp	d
triphenylmethane	244.3	94	359	
bromine	159.8	−7	59	3.12
dichloromethane	84.9	−97	40.5	1.326
bromotriphenylmethane	323.2	154		

Note: mp and bp are in °C, density is in g/mL.

Directions

A. *Preparation of Bromotriphenylmethane*

Safety Notes

bromine   dichloromethane   petroleum ether

Reaction. *All glassware that will be in contact with the reaction mixture or product must be dry.* Assemble an apparatus for heating under reflux [OP-7] using a 10-mL round-bottom flask and a water-cooled condenser.

Stop and Think: What will happen if your glassware is wet?

Construct a gas trap [OP-24c] and put it on top of the condenser. (The gas trap can be omitted if the reaction is carried out under a hood.) Clamp an unfrosted 100–200-watt light bulb within a few inches of the reaction flask (several students can use the same light). Weigh 0.900 mmol of triphenylmethane into the reaction flask, dissolve it in 2.0 ml of dichloromethane, and add a stir bar. Obtain 1.0 mL of *freshly prepared* 1.0 *M* bromine in dichloromethane in a capped vial. Turn on the light and magnetic stirrer [OP-10] and heat the mixture to a gentle reflux using a hot water bath [OP-7]. When the reaction mixture begins to boil, remove the gas trap (if one is used) momentarily and use a 9-inch Pasteur pipet to add [OP-11] about one-third of the bromine solution through the top of the condenser. Then replace the gas trap. Each time the color of the solution fades to light yellow-orange repeat the addition with another one-third of the bromine solution. Continue heating for 30 minutes after the last addition, or until the solution becomes light yellow, and then let the reaction mixture cool to room temperature. If the reaction mixture has a distinct orange or red-brown color, add 1–3 drops of cyclohexene (no more); the color should fade to yellow. Use 10% sodium thiosulfate solution (which reduces bromine to bromide ions) to rinse everything that contacted the bromine solution except the reaction flask.

Separation. Evaporate [OP-16] the dichloromethane from the reaction mixture. You can evaporate it directly from the reaction flask, under vacuum with stirring and gentle heating, using a setup like that shown in Figure C8c of OP-16.

Purification and Analysis. Recrystallize [OP-25] the crude bromotriphenylmethane from 4–5 mL of hexanes. The recrystallization mixture can be heated in the reaction flask with magnetic stirring, using an air condenser to prevent solvent loss. If the solution is quite dark, use a small amount of pelletized Norit to decolorize it. Separate the hot solution from the insoluble residue by transferring [OP-6] it to a small Erlenmeyer flask using a preheated filter-tip pipet. Allow the product to crystallize; then collect it by vacuum filtration [OP-13] and wash it on the filter with a small amount of cold low-boiling petroleum ether. Dry [OP-23] the bromotriphenylmethane and measure its mass and melting point [OP-30]. Save enough product for part **B** and turn in the rest to your instructor.

B. *Preparation and Reactions of the Trityl Free Radical*
At your instructor's option, this part can be carried out in small groups using either commercial trityl bromide or the purest product from each group.

Safety Notes

Toluene is flammable, and inhalation, ingestion, or skin absorption may be harmful. Avoid contact with the liquid and do not breathe its vapors.

toluene

Dissolve about 40 mg of *pure* bromotriphenylmethane in 1 mL of toluene in a 13 × 100-mm test tube. (**Take Care!** Avoid contact with toluene and do not inhale its vapors.) If the solution is significantly colored, add enough pelletized Norit to decolorize it. Add 0.10 g of 40-mesh zinc, stopper the tube immediately, and shake it for 10 minutes. Filter [OP-12] the mixture through a filtering pipet into another 13 × 100-mm test tube and observe the color. Without delay, bubble dry air through the solution until any color fades, stopper and shake the test tube, and let it stand for 3–4 minutes. Repeat this process until the solution remains colorless for at least a minute after stoppering. Describe any evidence you observed for the existence of the trityl radical and for a dynamic equilibrium involving its dimer. Give names and structures of all species and equations for all reactions for which you saw evidence.

Observe and Note: During this procedure, take careful notes and describe your observations in detail.

Waste Disposal: Place the solution in a designated waste container.

Exercises

1 (a) Show how Gomberg's dimer, whose true structure was given previously, can form by the combination of two trityl radicals. (b) Explain why *tris*(4-*t*-butylphenyl)methyl does not form an analogous dimer.

2 Describe and explain the possible effect on your results of the following experimental errors or variations. (a) You forgot to turn on the light. (b) The reaction flask for part **A** was wet. (c) In part **B**, you bubbled nitrogen, rather than air, into the reaction mixture.

3 Following the format in Appendix V, construct a flow diagram for the synthesis of bromotriphenylmethane.

4 Assuming a free-radical mechanism for the bromination of the following compounds, arrange them in order of their expected reactivity, starting with the most reactive.

5 Gomberg's dimer will decolorize a solution of iodine in dichloromethane. Explain why and write equations for all relevant reactions.

6 Write equations for three chain-terminating steps that can occur during the synthesis of bromotriphenylmethane.

Other Things You Can Do

(Starred projects require your instructor's permission.)

*1 Convert your bromotriphenylmethane to an ether as described in Minilab 17.

*2 Measure the bromination rates of some aromatic hydrocarbons as described in Minilab 21.

Chain-Growth Polymerization of Styrene and Methyl Methacrylate

Reactions of Alkenes. Preparation of Vinyl Polymers. Addition Polymerization. Free Radicals.

Operations

OP-7   Heating
OP-10 Mixing
OP-13 Vacuum Filtration
OP-23 Drying Solids
OP-27 Simple Distillation
OP-36 Infrared Spectrometry

Before You Begin

1 Read the experiment, read or review the operations as necessary, and write an experimental plan.
2 Calculate the mass of 4.00 mmol of styrene and the theoretical yield of polystyrene.

Scenario

The Consulting Chemists Institute has two clients this week, both seeking information about polymers and polymerization reactions. Polly von Ilek, an artisan who specializes in nature crafts, plans to embed natural objects such as leaves, insects, and flowers in a matrix of clear, recycled plastic. She wants you to develop a procedure for depolymerizing granulated poly(methyl methacrylate) and then repolymerizing the monomer under conditions that would allow objects to be embedded in the matrix before it hardens.

Mega Molecules, Inc., which manufactures styrofoam packing "peanuts" and other polymer-based products, has been manufacturing polystyrene by a process that involves benzene as an organic solvent. Because benzene is a known carcinogen, the company wants to phase it out and switch to the less toxic solvent toluene.

Your assignments are as follows:

- To see whether granulated poly(methyl methacrylate) can be depolymerized and then repolymerized to a clear plastic in which objects can be embedded.
- To find out whether the solution polymerization of styrene can be performed successfully in toluene.

Applying Scientific Methodology

In each part of this experiment the scientific problem is, in effect, to try something and see whether it works. If you can convert each starting material to a solid that has the properties of a plastic, that provides evidence that a polymerization reaction has occurred. At your instructor's option, you may be able to confirm the structure of your polystyrene by obtaining its infrared spectrum.

Chain-growth Polymers

The giant molecules we call polymers are immensely important to humankind—the meat, fruit, and vegetables we eat, the clothing we wear, and the wood we use for housing and furniture all consist partly or entirely of organic polymers. In fact, we are *made* of polymers—proteins in muscles, organs, blood cells, enzymes, and protoplasm; lipids in nerve sheaths, cell walls, and energy-storing fat tissues; and nucleic acids in the chromosomes that control our heredity. Compared to these natural polymers, synthetic polymers are newcomers on the scene. Polystyrene was first synthesized in 1839 (the year that Charles Goodyear learned how to vulcanize another polymer, natural rubber), but its properties were not appreciated at the time. The first commercially useful synthetic polymer did not appear until 1907, when Leo Baekeland synthesized Bakelite from phenol and formaldehyde. Bakelite is a good electrical insulator that is still used to make electrical plugs and switches. The synthesis of Nylon 66 and polyethylene in 1939, and the development of synthetic rubbers by German chemists during World War II, when Germany's supply of natural rubber was cut off, gave added impetus to the search for useful synthetic polymers. Today, it is possible to design "tailor-made" polymers having almost any desired combination of properties by using special catalysts (called Ziegler-Natta catalysts) to regulate the stereochemistry of a polymer, and by programming the sequence of monomers in copolymers, in which two or more different monomers are combined in various ways.

Synthetic polymers are often classified as addition or condensation polymers. *Addition polymerization* involves the combination of monomer units without eliminating any by-product molecules. The repeating unit of the polymer, therefore, has the same chemical constitution as the monomer, as illustrated in the margin for polyethylene. *Condensation polymerization* involves the combination of monomer units containing two or more reactive functional groups that lose a small molecule, such as a water or HCl molecule, as they combine. Thus the repeating unit of a condensation polymer does not have the same formula as the monomer (or monomers), as illustrated for Nylon 6 synthesized from 6-aminohexanoic acid. Note that Nylon 6 can also be formed by a ring-opening polymerization reaction from ω-caprolactam, so it is somewhat arbitrary to classify polymers according to the method used to prepare them.

The terms *chain-growth* and *step-growth* polymerization are used to describe two basic polymerization processes. Chain-growth polymerization is a process in which some monomer molecules are first activated by a polymerization initiator. Each polymer chain then grows very rapidly from the activated monomer by adding more monomer molecules to its reactive end—somewhat like adding beads to a string. Step-growth polymerization refers to a process in which each polymer chain grows from both ends through a series of individual reaction steps.

In a sense, polystyrene is a "natural" polymer since it is found in styrax from the sweet gum tree.

Formation of polyethylene by addition polymerization

$$n\,CH_2{=}CH_2 \longrightarrow {-}(CH_2CH_2)_{\overline{n}}$$

repeating unit

ethylene   polyethylene

(*n* is a large but indeterminate number.)

Formation of Nylon 6 by condensation polymerization

$$n\,H_2N(CH_2)_5\overset{\displaystyle O}{\overset{\|}{C}}OH \longrightarrow$$

6-aminohexanoic acid

$${-}(NH(CH_2)_5\overset{\displaystyle O}{\overset{\|}{C}})_{\overline{n}} + H_2O$$

Nylon 6

Formation of Nylon 6 by ring-opening polymerization

$$n\;\text{[caprolactam ring]} \longrightarrow {-}(NH(CH_2)_5\overset{\displaystyle O}{\overset{\|}{C}})_{\overline{n}}$$

Nylon 6

ω-caprolactam

The polymerization of styrene with a free-radical initiator is a typical example of chain-growth polymerization. The initiator (usually an organic peroxide) decomposes under the influence of heat or light to form free radicals, which add to styrene molecules according to Markovnikov's rule. Each activated styrene molecule then adds in similar fashion to another styrene molecule, leaving an unpaired electron at the end of the chain after each step. This process continues indefinitely until a reaction such as radical coupling or disproportionation occurs between chain ends (or with impurities) and deactivates the chain ends by forming stable products.

Key Concept: *Free-radical addition reactions obey Markovnikov's rule because radicals add in the direction that yields the most stable free-radical intermediate.*

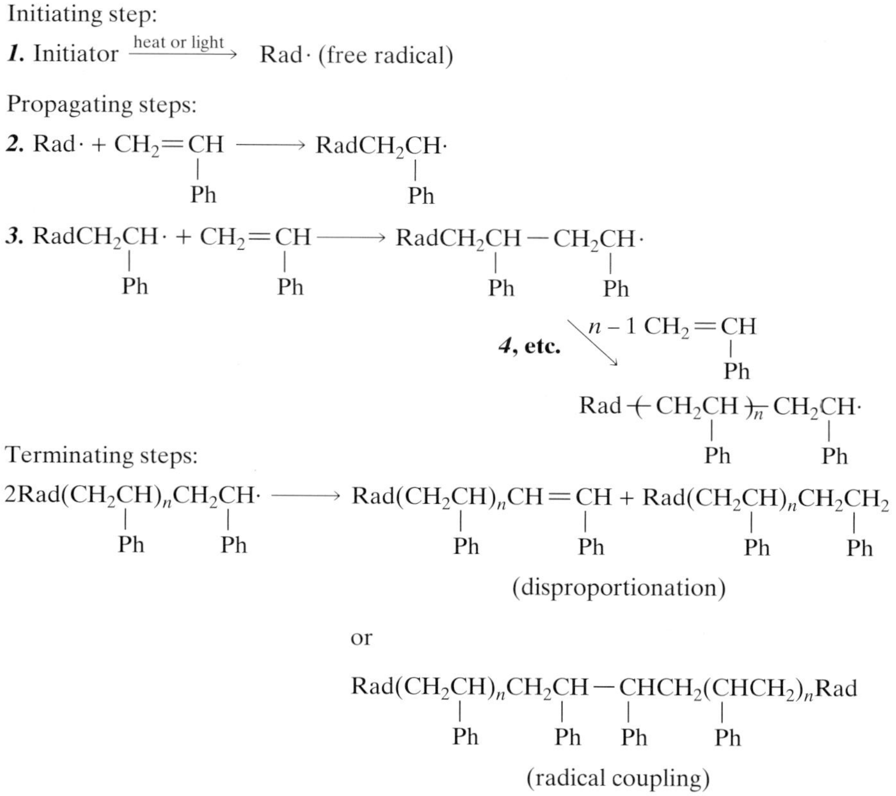

Initiating step:

1. Initiator $\xrightarrow{\text{heat or light}}$ Rad· (free radical)

Propagating steps:

2. Rad· + CH_2=CH $\longrightarrow$ RadCH$_2$CH·
 | |
 Ph Ph

3. RadCH$_2$CH· + CH_2=CH $\longrightarrow$ RadCH$_2$CH — CH$_2$CH·
 | | | |
 Ph Ph Ph Ph

$$4, \text{etc.} \searrow \quad n-1\ CH_2=CH$$
$$\underset{Ph}{|}$$

Rad$\left(\text{CH}_2\text{CH}\right)_nCH_2$CH·
 | |
 Ph Ph

Terminating steps:

2Rad(CH$_2$CH)$_n$CH$_2$CH· $\longrightarrow$ Rad(CH$_2$CH)$_n$CH=CH + Rad(CH$_2$CH)$_n$CH$_2$CH$_2$
 | | | | | |
 Ph Ph Ph Ph Ph Ph

(disproportionation)

or

Rad(CH$_2$CH)$_n$CH$_2$CH — CHCH$_2$(CHCH$_2$)$_n$Rad
 | | | |
 Ph Ph Ph Ph

(radical coupling)

Mechanism of chain-growth polymerization of polystyrene

Throughout a chain-growth polymerization, the bulk of the reaction mixture will consist of finished polymer molecules and unreacted monomers waiting to meet up with a reactive chain end. Because chain growth is so rapid once activation occurs, only one among many millions of molecules may be involved in the growth process at any given instant.

A number of different experimental methods are used for chain-growth polymerization, each with its advantages and disadvantages. *Bulk polymerization* is the simplest; it is carried out by adding a suitable initiator to the pure monomer and using heat or light to promote the reaction. The high heat of reaction makes bulk polymerization of vinyl monomers hard to control, and the method is seldom used commercially except for some polystyrene and poly(methyl methacrylate) products. Bulk polymerization can be used to preserve a botanical or zoological specimen (such as a desiccated flower or beetle) by suspending the specimen in a liquid monomer, which is then allowed to polymerize around it. Poly(methyl methacrylate), known by the trade names Lucite and Plexiglas, is often used for that purpose.

In *solution polymerization*, the monomer is dissolved in an organic solvent, which dilutes it enough to reduce the problem of heat generation. Because it may be difficult to remove the solvent entirely, solution polymerization is often used to make polymers that are commonly used in solution, such as acrylic finishes. *Suspension polymerization* is carried out by mechanically dispersing the monomer in water or a similar solvent, so that the polymer is obtained in the form of granular beads that can be easily isolated. *Emulsion polymerization* is similar to suspension polymerization in that the monomer is dispersed in a solvent, usually water. In this method, however, the initiator is dissolved in the aqueous phase, and the monomer is emulsified (broken up into tiny droplets) by a detergent or some other surfactant (surface-active agent). The resulting dispersion resembles a rubber latex and can be used in that form, but it is more often coagulated and isolated as a finely divided powder.

Understanding the Experiment

Heating benzoyl peroxide also forms benzoyloxy radicals $(PhCO_2 \cdot)$, *which may participate in the initiation process.*

In the first part of this experiment, you will attempt to carry out the solution polymerization of styrene in toluene using benzoyl peroxide as the polymerization initiator. Benzoyl peroxide decomposes on heating to yield phenyl radicals, which add to the double bonds of styrene molecules to start the polymerization process.

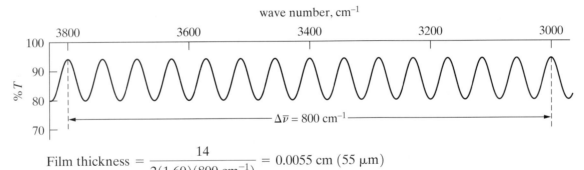

The resulting polystyrene precipitates as a gummy solid when methanol is added to the reaction mixture.

Polystyrene is, in effect, an aromatic hydrocarbon with a very long alkyl side chain, so its infrared spectrum should resemble that of an ordinary arene. The spectrum of a thin polymer film will often display a number of extraneous small peaks called *interference fringes*, which are caused by interference between beams of infrared radiation reflected from the film's surfaces. As illustrated in Figure 26.1, the thickness of such a film can be estimated by counting the number of fringes in a given wave number interval $\Delta \bar{\nu}$.

$$\text{Film thickness} = \frac{\text{number of fringes}}{2n(\Delta \bar{\nu})}$$

where

$$n = \text{refractive index} \ (\sim 1.60 \text{ for polystyrene})$$

$$\text{Film thickness} = \frac{14}{2(1.60)(800 \text{ cm}^{-1})} = 0.0055 \text{ cm} \ (55 \ \mu\text{m})$$

Figure 26.1 Determination of film thickness by counting interference fringes

In part **B**, you will heat some poly(methyl methacrylate) in a distillation apparatus to break it down into the monomer, methyl methacrylate. This monomer will be free of the polymerization inhibitors (usually aromatic phenols) that are added to commercial monomers to keep them from polymerizing in storage. You will add a small amount of a polymerization initiator, *t*-butyl peroxybenzoate, to the monomer and (if you like) suspend a small object of your choice in it. During the next lab period, you can see whether or not it has polymerized. With your instructor's permission, you can break the glass to recover the clear polymer (be sure to get permission, or you may be charged for the breakage!).

Reactions and Properties

A n ⬡—CH=CH$_2$ ⟶ —(CH—CH$_2$)$_n$—

styrene

polystyrene

B nCH$_2$=CCOOCH$_3$ ⟶ —(CH$_2$—C)$_n$—
$\quad\quad\quad$ | $\quad\quad\quad\quad\quad\quad\quad\quad$ |
$\quad\quad\quad$ CH$_3$ $\quad\quad\quad\quad\quad\quad\quad$ CH$_3$

$\quad\quad$ methyl methacrylate $\quad\quad$ poly(methyl methacrylate)

Table 26.1 Physical properties

	M.W.	mp	bp	*d*
styrene	104.2	−31	146	0.909
benzoyl peroxide	242.2	106		
methyl methacrylate	100.1	−48	100	0.944

Note: mp and bp are in °C, density is in g/mL.

Styrene

⬡—CH=CH$_2$

3026.8	1494.3	908.6
1630.0	1082.9	775.8
1575.6	991.1	697.1

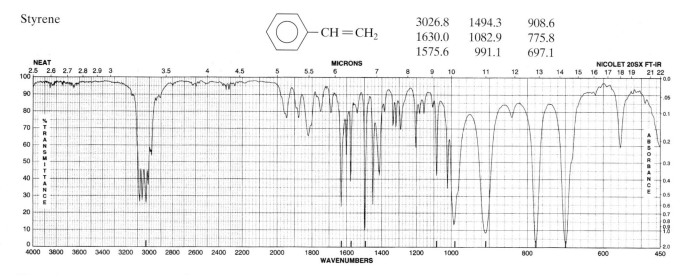

Figure 26.2 IR spectrum of styrene

Directions

A. *Solution Polymerization of Styrene*

Safety Notes

> Styrene has irritating vapors and inhalation, ingestion, or skin absorption may be harmful. Avoid contact with the liquid, do not breathe its vapors, and keep it away from flames.
> Toluene is flammable, and inhalation, ingestion, or skin absorption may be harmful. Avoid contact with the liquid and do not breathe its vapors.
> Benzoyl peroxide can explode due to shock or friction, and its decomposition can be catalyzed by metals. Do not allow it to contact any metal object and keep it away from other chemicals.
> 2-Butanone is flammable and ingestion, inhalation, or skin absorption may be harmful. Avoid contact, do not breathe its vapors, and keep it away from flames and hot surfaces.

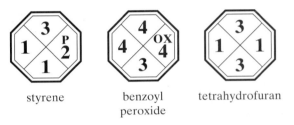

styrene   benzoyl   tetrahydrofuran
                    peroxide

Take Care! Avoid contact with styrene and toluene and do not breathe their vapors.

Take Care! Do not allow benzoyl peroxide to come in contact with a metal or another chemical. Clean up any spills as directed by your instructor.

Waste Disposal: Unless your instructor indicates otherwise, pour the wash methanol and filtrates down the drain.

Reaction.   Weigh 4.00 mmol of styrene into a 5-mL conical vial and add 2.0 mL of toluene. Use a *plastic* microscoop (see OP-6) to transfer approximately 20 mg of benzoyl peroxide to the reaction mixture. Drop in a stirring device [OP-10], add an air condenser, and heat the mixture in a boiling-water bath for *at least* one hour while stirring. (A longer reaction time will result in a higher yield.) Work on part **B** during the reaction period.

Separation and Purification.   Transfer the reaction mixture to a small beaker, using 5 mL of methanol for the transfer. Use a flat-bladed microspatula to stir the polystyrene in the methanol for 3–5 minutes, pressing the gummy solid against the sides of the beaker to squeeze out absorbed toluene. Remove the methanol with a Pasteur pipet, add 2 mL of fresh methanol, and repeat the stirring until the solid has hardened and is no longer sticky. Use additional 2-mL portions of methanol if necessary, removing most of the methanol after each treatment. Collect the polymer by vacuum filtration [OP-13], wash it on the filter with methanol, and let it air dry for a few minutes. Dry [OP-23] the product and weigh it.

Analysis.   Record the infrared spectrum [OP-36] of a polystyrene film provided by your instructor. Interpret the spectrum as completely as you can and calculate the thickness of the film. (Alternatively, try to prepare a thin film from your product as described in "Other Things You Can Do.")

B. *Bulk Polymerization of Methyl Methacrylate*
Your instructor may opt to omit the depolymerization procedure, in which case you can start at the polymerization step.

Methyl methacrylate is flammable and lachrymatory (tear producing) and can irritate the skin and eyes. Avoid contact, do not breathe its vapors, and keep it away from flames. Use protective gloves and a fume hood.

t-Butyl peroxybenzoate may react violently with strong oxidizing or reducing agents, or explode when strongly heated. Keep it away from combustible materials, heat, and flames.

Safety Notes

methyl
methacrylate

t-butyl
peroxybenzoate

Depolymerization of Poly(methyl methacrylate).   Assemble an apparatus for simple distillation [OP-27] using a 10-mL round-bottom flask. Weigh a clean, dry 10 × 75-mm test tube to collect the monomer. Place approximately 4.00 g of granulated low-molecular-weight poly(methyl methacrylate) and a boiling chip or two in the reaction flask and heat [OP-7] it at a medium-high power setting until the solid softens and starts to liquefy. Then reduce the heating rate so that the liquid distills slowly and collect the methyl methacrylate in the test tube. Stop the distillation when the residue in the flask begins to darken and become viscous. Clean all parts of the distillation apparatus thoroughly with acetone as soon as it has cooled. If any residue solidifies in the reaction flask, cover it with ethyl acetate and let it stand until the residue dissolves (this may takes several hours or days).

Waste Disposal: Place the residue and the wash liquid in a designated waste container.

Polymerization of Methyl Methacrylate.   Add two drops of *t*-butyl peroxybenzoate to the methyl methacrylate and shake gently to dissolve it and mix the solution. Heat the mixture in a boiling-water bath for 20 minutes or more. If you like, suspend a small object in the liquid by inserting the straight end of a hook (made from a piece of copper wire or part of a paper clip) into a cork of the appropriate size and hanging the object from a thread. Metal objects should first be coated with clear enamel. Stopper the test tube loosely with the cork (or seal it with Parafilm) and set it in direct sunlight until the next laboratory period or until the clear polymer has hardened. Measure the mass of the polymer. With your instructor's permission, wrap the test tube in a towel and *carefully* break it inside a cardboard box using a pestle or other hard object, then remove the polymer and place the glass in a sharps collector.

Take Care!   Keep t-butyl peroxybenzoate away from combustible materials and heat sources.

Exercises

1   Propose a mechanism for the polymerization of methyl methacrylate in the presence of *t*-butyl peroxybenzoate, which decomposes as shown here.

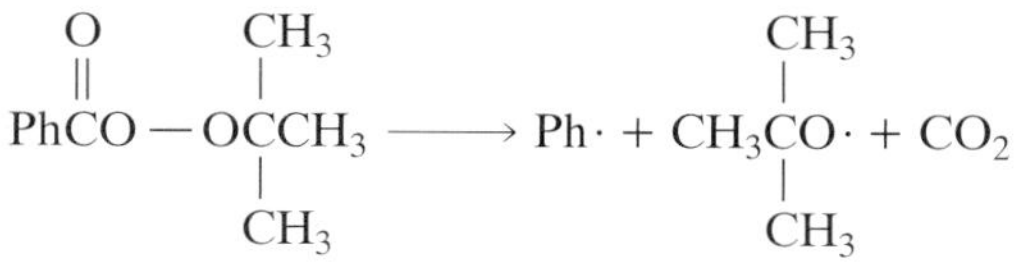

$$PhCO-OCCH_3 \longrightarrow Ph\cdot + CH_3CO\cdot + CO_2$$

2   Compare your polystyrene spectrum with the infrared spectrum of styrene in Figure 26.2, and try to account for any significant similarities and differences.

3 Describe and explain the possible effect on your results of the following experimental errors or variations. (a) In part **A**, you forgot to add the benzoyl peroxide. (b) So that it would dry faster, you washed the polystyrene with petroleum ether rather than methanol. (c) For the polymerization step of part **B**, you used commercial methyl methacrylate containing 50 ppm of 4-methoxyphenol.

4 Draw the structures of the monomers needed to prepare polymers having the following repeating units.

a. $-(CHClCHCl)-$ **b.** $-(CF_2CFCl)-$

c. $-(CH_2CH_2NH)-$ **d.** $-(CH_2C=CHCH_2)-$
$\quad\quad\quad\quad\quad\quad\quad\quad\quad\quad\quad\quad\quad |$
$\quad\quad\quad\quad\quad\quad\quad\quad\quad\quad\quad\quad CH_3$

e. $-(CH_2CH)-$ **f.** $-(CH_2CH-CH_2CH)-$
$\quad\quad\quad |$
$\quad\quad CH_2CH_3$

5 Show how the two monomers illustrated in the margin could combine to form a Diels-Alder addition polymer, and give the structure of the polymeric repeating unit.

6 Poly(ethylene glycol) can be prepared from either ethylene glycol or ethylene oxide. Write a balanced equation for each reaction.

7 Methyl methacrylate can be prepared commercially from acetone, hydrogen cyanide, and methanol. Propose a synthesis of methyl methacrylate from these starting materials, using any necessary inorganic reagents or solvents.

Other Things You Can Do

(Starred projects require your instructor's permission.)

*1 You can attempt to prepare a polystyrene film from your product by the following method. Dissolve about 50 mg of the product in 0.5 mL of 2-butanone. This solution can be shared by several students. Use a Pasteur pipet to transfer about 10 drops of the solution to a *very clean* microscope slide and tilt the slide to coat it evenly. Let the solvent evaporate completely under the hood, then carefully strip off the polymer film with a very sharp double-edged razor blade and mount it between salt plates in an infrared cell holder.

*2 Three students can work together on part **B**, one using the standard procedure, the second using no *t*-butyl peroxybenzoate, and the third using 40 mg of hydroquinone in place of the *t*-butyl peroxybenzoate. Compare and explain the results.

*3 Perform the "nylon rope trick" as described in Minilab 22.

4 Beginning with sources from the Bibliography, write a research paper on the use of Ziegler-Natta catalysts to synthesize stereoregular polymers and describe some properties and uses of these polymers.

$-(CH_2CH_2O)-$
poly(ethylene glycol)

OH OH   O
| | / \
CH_2CH_2 CH_2CH_2
ethylene   ethylene
glycol   oxide

Take Care!   Keep 2-butanone away from flames, avoid contact, and do not breathe its vapors.

Take Care!   Don't cut yourself!

Synthesis of Ethanol by Fermentation

EXPERIMENT 27

Reactions of Carbohydrates. Preparation of Alcohols. Biosynthesis of Organic Compounds.

Operations

OP-29   Fractional Distillation
OP-31   Boiling Point

Before You Begin

1   Read the experiment, read or review the operations as necessary, and write an experimental plan.
2   Calculate the mass of 5.00 mmol of sucrose and the theoretical mass and volume of ethanol that can be produced from that much sucrose.

Scenario

The Great Plains Farm Cooperative helps farmers market their crops to help ensure that they receive a fair price. Recently, a bumper crop of sugar beets resulted in a surplus of beet sugar and drastically lowered the price or the product. The Co-op has advised farmers to store their beets until the price rises; in the meantime it is exploring alternative uses for the surplus beets. One alternative would be to convert the beet sugar to ethanol for use as an energy source in gasahol and other commercial fuels.

The Co-op has sent you a quantity of beet sugar to see whether an ethanol conversion process would be feasible. To price the beet-sugar ethanol low enough to make it competitive with ethanol from corn and other sources, they require a product that is at least 180 proof. Your assignment is to ferment the sugar, distill it to concentrate the ethanol, and find out whether your product meets their specifications.

Applying Scientific Methodology

Whether or not you solve the problem in this experiment—that is, prepare ethanol that is 180 proof or higher—depends on your skill in performing the experiment.

The Chemistry of Brewing and Winemaking

The single-celled fungi called yeasts obtain the chemical energy they need to grow and reproduce by breaking down molecules of monosaccharides (simple sugars) such as glucose and fructose, and disaccharides (two-unit sugars), such as sucrose and maltose. To the yeast cells, ethyl alcohol is a useless by-product of this energy-generating process, but humans have developed a liking for that by-product as it exists in alcoholic beverages.

The fermentation of plant carbohydrates to produce alcohol is as old as civilization and probably began in prehistoric times. The first alcoholic beverage may have been a kind of mead (honey wine), which was probably discovered when some honey-sweetened water was forgotten and allowed to ferment. Early civilizations discovered that starchy grains such as barley, wheat, and millet could be treated to make them fermentable, leading to the brewing of beer and similar grain-based beverages. A poem dedicated to the Sumerian goddess of brewing contains a recipe for beer that dates back to about 2800 B.C. Since grains contain starch but no sugars that yeast cells can ferment, they are first allowed to germinate for several days in a process called *malting*. The dried, malted grain (usually barley) is soaked in warm water, allowing enzymes formed during the malting process to break down the starch into simple sugars. The liquid containing the sugars and flavor components, known as *wort*, is then separated from the solid residue. Additional cereals (rice, corn, or wheat) may be added to boost the wort's carbohydrate content, hops are included as a preservative and for their bitter, aromatic flavor, and the wort is boiled to sterilize and concentrate it. Brewing yeast is then added and the wort is allowed to ferment, during which the yeast cells transform the sugars into alcohol.

Grapevines grew wild in Central Asia and Western Europe long before humans appeared on Earth. By the sixth millennium B.C., grapes had been domesticated in Asia Minor, and the cultivation and use of grapes for food and wine spread from there to Egypt and Mesopotamia. Noah planted the first vineyard mentioned in the Bible; he was also the first Biblical character reported to have become intoxicated by drinking wine. Legend has it that Dionysus, the Greek god of wine, introduced wine drinking in Crete, and by Homer's time (ca. 700 B.C.) the ancient Greeks were making strong, sweet, thick wine that was always watered down before drinking. In Roman times, the Gauls, who occupied much of present-day France, produced the best wine in Europe. Their wines were popular in Rome until the emperor Domitian, to protect Italian winemakers from Gallic competition, demanded that half of the grape vines in Gaul be uprooted. Wine may have contributed to the decline and fall of Rome because a syrup used to preserve wines was boiled in lead-lined pots, and one of the symptoms of lead poisoning is mental confusion. After Rome fell to the barbarians, the art of winemaking was kept alive by monks in European monasteries, who cultivated grapes to supply their sacramental wine. A Benedictine monk, Dom Pérignon, is usually credited with the invention of champagne, and many of the best wines of France are still grown around former monasteries.

Even today, winemaking may be as much an art as a science. The first step is crushing ripe grapes into juice, called the *must*, which contains glucose and fructose along with smaller quantities of fruit acids (such as tartaric acid), tannins, and other components. To make white wine, the grape skins are removed before fermentation, so the wine contains little pigmentation or tannic material. Red wine musts are fermented in contact with red grape skins for most of the fermentation period. The longer the must is in contact with the skins, the deeper is its color and the more astringent or "tannic" its taste. Sulfur dioxide is added to prevent oxidation of flavor

In 1989, an American brewer reproduced this honey-sweetened beer following the ancient Sumerian recipe.

components and to inhibit the growth of bacteria and wild yeasts that could convert the must to vinegar. To balance their sweet and sour components, sugar is added to overly acidic musts, and tartaric acid is added to musts having insufficient acid.

Some grape skins are covered with a natural dusting of yeasts of the type needed for fermentation (saccharomycetes); with other grapes, a yeast culture must be added. The yeast not only converts grape sugars to alcohol, but also biosynthesizes flavor components—mainly esters and long-chain alcohols—that don't exist in the grapes themselves. A long, slow fermentation process at low temperature produces the most flavor components, giving the wine a more pronounced "nose" (aroma). Fermentation stops when nearly all of the sugar has been transformed. The yeast is then allowed to settle and the new wine is drawn off; this *racking* process is repeated several times as the wine ages. The wine may also be *fined*—that is, treated with a substance that coagulates suspended fine particles and carries them to the bottom. New wine has a harsh, raw flavor and a simple aroma, so it is aged in wooden casks to soften the flavor and enrich the aroma. A large number of chemical reactions take place during this stage, forming more flavor components and removing some components that have disagreeable characteristics. Many of these are oxidation reactions, made possible by pores in the wooden casks that allow air to enter the wine. The cask itself may contribute flavor as vanillin and other components of the wood leach into the wine. If oxidation proceeds too long it begins to degrade the wine, so after 6–24 months of aging, the wine is bottled to protect it from excessive oxidation. The chemical composition of a bottled wine continues to change, and a good wine may improve in flavor when aged in the bottle for several years, or even several decades.

Distilled beverages, such as whiskey, brandy, and vodka, are made by distilling beer, wine, or another fermented beverage to increase the alcohol content and concentrate the volatile flavor components. Thus brandy is distilled from wine, scotch from a special distiller's beer, rum from fermented molasses or sugar-cane juice, and vodka from fermented potatoes or grain. The distillate is usually aged to make the flavor richer and mellower.

Drinking alcoholic beverages of any kind can be both beneficial and harmful. Wine may enhance our enjoyment of a good meal, and a drink or two can help us relax and stimulate social interactions. In moderation, wine and other alcoholic beverages confer certain health benefits. Beer, wine, and liquor all raise the concentration of "good cholesterol" (HDL) in the blood, which may reduce the risk of heart attacks. Chemicals in wine may also boost the immune system, inhibit cancer, and reduce the risk of Alzheimer's disease.

Balanced against these beneficial effects of moderate drinking are the unquestionably harmful effects of excessive drinking, which include drunkenness, hangovers, alcoholism, cirrhosis of the liver, birth abnormalities, antisocial behavior, broken homes, wasted lives, and death. Alcohol is a drug—often a highly addictive one—that depresses the central nervous system. Its apparent stimulating influence, which is associated with a loss of inhibitions, is actually a result of alcohol's depressant effect on the brain centers that normally keep our social behavior in check. Because the destructive potential of beverage alcohol is so much greater than its apparent health benefit, most medical professionals agree that nondrinkers should not begin drinking alcohol for their health, and drinkers should limit their alcohol intake to one or two drinks a day.

Understanding the Experiment

Pasteur's salts contains 2.0 g potassium dihydrogen phosphate, 0.20 g calcium dihydrogen phosphate, 0.20 g magnesium sulfate, and 10 g ammonium tartrate per liter of water.

Enzymes are very efficient biochemical catalysts, produced by living cells, that facilitate the chemical processes necessary for life.

In this experiment, you will ferment an aqueous solution of beet sugar (sucrose) with commercial dry yeast (*Saccharomyces cerevisiae*) to obtain a dilute ethanol solution, which you will concentrate by fractional distillation. The fermentation mixture will contain Pasteur's salts, a mixture of nutrients that support the growth and reproduction of yeast. The fermentation takes time, so you should be prepared to get it started a week or so in advance. Carbon dioxide produced during the fermentation excludes air, which would otherwise oxidize the ethanol to acetic acid or other unwanted by-products. Thus a balloon is attached to the reaction flask to keep the carbon dioxide from escaping while fermentation is going on. When the reaction is over, it slowly leaks out, deflating the balloon.

Sucrose is a disaccharide having molecules that contain units of the two simple sugars glucose and fructose. The yeast you will use contains many *enzymes* that are necessary for the conversion of sucrose to ethanol. Invertase catalyzes the hydrolysis of sucrose to glucose and fructose. Then zymase, which is actually a group of at least 22 separate enzymes, catalyzes a series of steps in the very complex conversion of these simple sugars to ethanol and carbon dioxide.

$$4CH_3CH_2OH + 4CO_2 + energy$$

From a yeast cell's point of view, the important product of this process is energy. Ethanol is a waste product that is actually harmful to the yeast cells, which are deactivated after the alcohol concentration reaches 15% (v/v) or so.

Fermentation produces a number of by-products by various pathways. These include ethanal, glycerol, and fusel oil. Ethanal (acetaldehyde) is produced along the biosynthetic pathway that leads to ethanol, and some of it remains at the end of the fermentation. Glycerol improves a wine's "legs"—its ability to coat the inside of a wine glass—by increasing its surface tension. Fusel oil is a mixture of alcohols (including 1-propanol, 2-methyl-1-propanol, 2-methyl-2-butanol, and 3-methyl-1-butanol) that are responsible for the headaches that cheap wines often cause.

$$
\begin{array}{cc}
\underset{|}{OH} \quad \underset{|}{OH} \\
CH_2CHCH_2 \\
\underset{|}{OH} \\
\text{glycerol}
\end{array}
$$

$$
\underset{ethanal}{\overset{O}{\overset{\|}{CH_3CH}}}
$$

fusel oil:

$$
\underset{CH_3CH_2CH_2}{\overset{OH}{\overset{|}{}}} \qquad \underset{\underset{CH_3}{\overset{|}{}}CH_3CHCH_2}{\overset{OH}{\overset{|}{}}}
$$

$$
\underset{\underset{CH_3}{\overset{|}{}}CH_3CCH_2CH_3}{\overset{OH}{\overset{|}{}}} \qquad \underset{\underset{CH_3}{\overset{|}{}}CH_3CHCH_2CH_2}{\overset{OH}{\overset{|}{}}}
$$

To prepare alcohol for nonbeverage purposes, all of these by-products must be removed from the fermentation mixture, along with excess water and spent yeast cells. You will remove the yeast cells, which collect as a sediment at the bottom of the fermentation flask, by centrifugation. The remaining components, including most of the water, will be removed by fractional distillation. Acetaldehyde boils at 21°C, ethanol at 78°C, water at 100°C, the fusel oil alcohols at temperatures ranging from 97°C to 132°C and higher, and glycerol at 290°C. Thus any acetaldehyde present should escape as vapor, while most of the water and the other components should remain in the boiling flask. Water and ethanol form an *azeotrope*—a constant-boiling mixture—with a composition of 95.6% ethanol and 4.4% water (by mass), so it is impossible to obtain ethanol purer than 95.6 mass percent by distillation alone. Your percentage may be substantially lower because of such factors as the column-packing efficiency and the rate of distillation.

The *proof* of a commercial alcohol or alcoholic beverage is calculated by doubling its volume percent; that is, 1 degree of proof is equivalent to 0.5 volume percent. Thus an 80 proof bourbon contains 40% ethanol by volume. You can determine the volume and mass percent of ethanol in your product, and from that its proof and your percent yield, by measuring its density and using the data in Table 27.2.

Reactions and Properties

$$
\underset{\text{sucrose}}{C_{12}H_{22}O_{11}} + H_2O \longrightarrow \underset{\text{ethanol}}{4CH_3CH_2OH} + 4CO_2
$$

Table 27.1 Physical properties

	M.W.	mp	bp
sucrose	342.3	186d	
ethanol	46.1	−115	78.3

Note: mp and bp are in °C; d = decomposes at melting point.

Table 27.2 Density and composition of ethanol-water solutions at 20°C

Density at 20°C	Mass % ethanol	Volume % ethanol
0.914	50.0	57.8
0.903	55.0	62.8
0.891	60.0	67.7
0.879	65.0	72.4
0.868	70.0	76.9
0.856	75.0	81.3
0.844	80.0	85.5
0.831	85.0	89.5
0.818	90.0	93.3
0.804	95.0	96.8
0.789	100.0	100.0

Directions

Fermentation.   In a 50-mL Erlenmeyer flask, dissolve 5.00 mmol of sucrose in 12 mL of distilled water that has been warmed to about 30°C. Then add 1.5 mL of Pasteur's salts and 0.15 g of dry yeast. Mix the contents by shaking the flask, and then stretch the neck of a small balloon over the neck of the flask to seal it, leaving the rest of the balloon hanging down the side of the flask. Label the flask with your name and let the mixture stand in a 30–35°C oven or incubator for at least a week.

Ordinary bread-baking yeast will work, but you can also use brewer's yeast or wine yeast.
The balloon will fill with CO_2 as fermentation proceeds.

Separation.   At the end of the fermentation period, remove the balloon, transfer equal volumes of the liquid from the flask to two identical centrifuge tubes, and centrifuge them for 5 minutes or more. Decant the clear supernatant liquid from both centrifuge tubes into a 25-mL round-bottom flask (if you have one), leaving the residue behind. If you have only a 10-mL round-bottom flask, decant the liquid into a storage vial and (in the "Purification" step) distill it in three portions, filling the flask about half full of liquid each time.

Purification.   Assemble an apparatus for fractional distillation [OP-29] using stainless-steel sponge or another appropriate column packing. Insulate the column as directed by your instructor. Support a thermometer inside the Hickman still, making sure that it is positioned correctly (see Figure E9, OP-27). Distill the ethanol-water mixture slowly and collect all of the distillate that boils between 77°C and 82°C, periodically transferring it to a tared vial. If you are using a 10-mL boiling flask, stop the distillation each time the temperature reaches 82°C and discard any liquid that remains in the boiling flask, then add the next portion of liquid to the flask, and continue distilling as before, until all of the liquid has been distilled.

Waste Disposal: After the distillation, pour the contents of the boiling flask down the drain.

Analysis.   Measure the mass and boiling point [OP-31] of the distillate. With the temperature of this liquid as close to 20°C as possible, use an automatic pipet or other appropriate measuring device to accurately measure 0.200 mL of distillate into a tared screw-cap vial. Then weigh it and calculate its density. From the data in Table 27.2, prepare a graph of mass percent and volume percent ethanol (on the y-axis) versus density. Use the graph to determine the percentage composition and proof of your distillate. Calculate the mass of ethanol contained in the distillate (remember that it's not pure ethanol) and use it to calculate your percent yield.

Exercises

1 A strong vodka has a density of 0.909 g/mL. What is its proof? (Vodka is basically a solution of ethanol and water.)

2 Describe and explain the possible effect on your results of the following experimental errors or variations. (a) You forgot to add the Pasteur's salts. (b) During the fractional distillation, you continued to collect distillate until the temperature reached 98°C. (c) Unknown to you, someone reset the automatic pipet for the density determination to 190 μL.

3 Following the format in Appendix V, construct a flow diagram for the experiment.

4 Calculate the volume of carbon dioxide at 25°C and 1 atm that would be produced by complete fermentation of the amount of sucrose you used.

5 In an alternative procedure for the synthesis of ethanol, the carbon dioxide generated is bubbled into a calcium hydroxide solution, yielding calcium carbonate. (a) Write a balanced equation for the reaction of carbon dioxide with $Ca(OH)_2$. (b) Calculate the theoretical mass of calcium carbonate that could be produced by the CO_2 generated from the amount of sucrose you used (see Exercise 4).

6 A by-product of this fermentation is diethylacetal, $CH_3CH(OCH_2CH_3)_2$. Write a balanced equation for its formation from components of the fermentation mixture.

Other Things You Can Do

(Starred projects require your instructor's permission.)

*1 Estimate the ethanol concentration of a commercial vodka by the method used in this experiment.

2 Find out how absolute (100%) ethanol is manufactured and how ethanol is denatured to make it undrinkable.

3 Write a research paper about the chemistry and technology of wine-making, starting with sources listed in the Bibliography.

Reaction of Butanols
with Hydrobromic Acid

EXPERIMENT **28**

Reactions of Alcohols. Preparation of Alkyl Halides. Nucleophilic Aliphatic Substitution. Infrared Spectrometry.

Operations

OP-7   Heating
OP-10 Mixing
OP-21 Washing Liquids
OP-22 Drying Liquids
OP-24 Drying and Trapping Gases
OP-27 Simple Distillation
OP-31 Boiling Point
OP-36 Infrared Spectrometry

Before You Begin

1 Read the experiment, read or review the operations as necessary, and write an experimental plan.
2 Calculate the mass and volume of 18.0 mmol of 1-butanol and the theoretical yield of 1-bromobutane (both mass and volume). Repeat these calculations for 2-butanol and 2-bromobutane.

Scenario

The Bond Triplex (see Experiment 23) manufactures not only alkynes, but a variety of chemicals that can be prepared from ethyne (acetylene) and other alkynes. For example, they manufacture 1-butanol from ethyne by the following process:

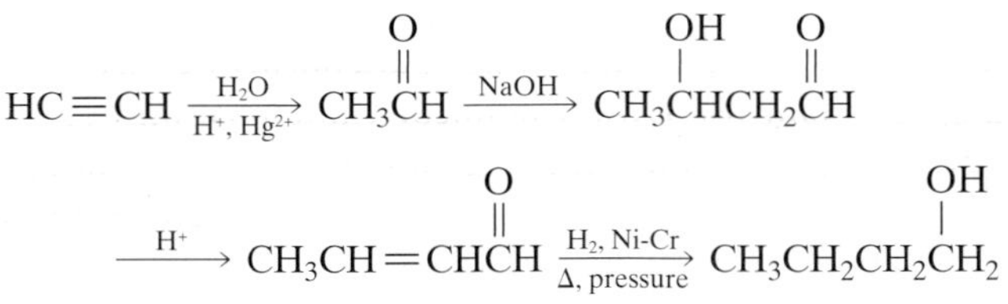

$$HC \equiv CH \xrightarrow[H^+, Hg^{2+}]{H_2O} CH_3\overset{\overset{\displaystyle O}{\|}}{C}H \xrightarrow{NaOH} CH_3\overset{\overset{\displaystyle OH}{|}}{C}HCH_2\overset{\overset{\displaystyle O}{\|}}{C}H$$

$$\xrightarrow{H^+} CH_3CH = CH\overset{\overset{\displaystyle O}{\|}}{C}H \xrightarrow[\Delta,\, pressure]{H_2,\, Ni\text{-}Cr} CH_3CH_2CH_2\overset{\overset{\displaystyle OH}{|}}{C}H_2$$

Now they want to manufacture 1-bromobutane and 2-bromobutane by treating the corresponding butanols with hydrogen bromide in the presence of a sulfuric acid catalyst. Economic considerations require that they use the amount of catalyst that will produce each alkyl bromide at the lowest cost per kilogram. They want your research team to carry out the syntheses of 1- and 2-bromobutane using different catalyst/substrate ratios, and then do a cost analysis to find out which ratio produces each product most economically.

Applying Scientific Methodology

Your working hypothesis should include a prediction of the effect on *each* reaction of increasing the catalyst/substrate ratio.

The Controversial Halides

The familiar expression "You can't get along with them and you can't get along without them" applies to many organic halides (organohalogen compounds). Organohalogens tend to be very stable. However, the stability that makes a particular organohalogen a useful commercial chemical can be a liability if the compound is released into the environment, where it may remain unchanged for decades, or even centuries. Most people first became aware that chlorinated organics might be a problem with the publication of Rachel Carson's book *Silent Spring*, in which she documented the effect of DDT on wildlife and the environment.

DDT

The polybrominated and polychlorinated biphenyls (PBBs and PCBs) have also caused serious environmental problems because they are toxic and believed to be carcinogenic and because they persist in the environment. PBBs are mixtures of various bromine derivatives of biphenyl, such as 2,2',4,4',5,5'-hexabromobiphenyl, a major component of fire-retardant mixtures such as Firemaster BP-6. In 1973, Firemaster BP-6 was accidentally added to cattle feed in Michigan, resulting in the illness of some dairy farmers and the death of many farm animals; it has since been withdrawn from the market. In past years, polychlorinated biphenyls were used in a number of commercial products, ranging from coolants for electrical transformers to plastic liners for baby bottles. In the United States, the manufacture, processing, and distribution of PCBs has been prohibited since 1979. Nevertheless, PCBs have persisted in the environment and may now be the most widespread pollutants on earth. They occur in dangerously high concentrations in some fish from the Great Lakes, and they have been detected in the tissues of many other organisms—even polar bears from the high Arctic.

2,2',4,4',5,5'-hexabromobiphenyl

Because of these and other instances of pollution, some environmental groups have called for a total ban on the manufacture of chlorine and chlorinated compounds, and the International Joint Commission, a U.S.–Canadian government-appointed group whose function is to preserve water quality in the Great Lakes, supports a gradual phasing out of these chemicals. People on opposite sides of this issue could hardly be farther apart. According to a spokesman for Dow Chemical Company, one of the largest producers and users of chlorine, "It is the single most important ingredient in [industrial] chemistry." Chlorine is used to manufacture an enormous number of consumer products, including plastics, pharmaceuticals, perfume bases, cosmetics, paper, medical devices, household adhesives, anesthetics, nonstick cookware, dry-cleaning solvents, refrigerants, photographic film, magnetic recording tape, floor coverings, wallpaper, food wraps, PVC pipe and fittings, electrical insulation, seat covers, baby strollers, compact discs, shoes, paints, solvents, brake fluid, food additives, and crop-protection chemicals. Reliable estimates are hard to come by, but chlorine probably accounts for tens of billions of dollars worth of commerce and one million or so jobs. Yet a Greenpeace research analyst has stated, "There are no uses of chlorine that we regard as safe."

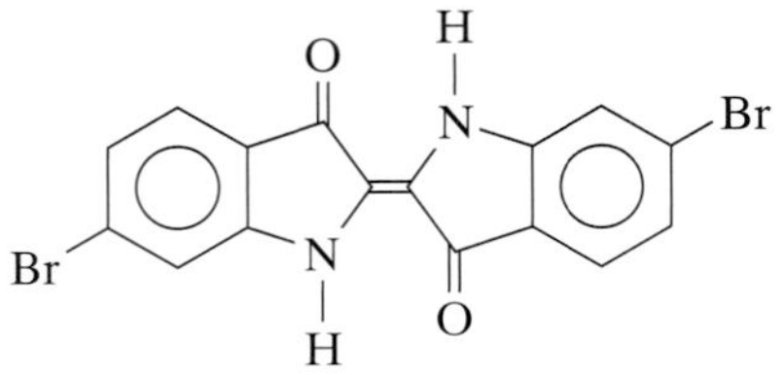

Tyrian purple (dibromoindigo)

Media reports on the issue tend to give the impression that all organohalogens are synthetic compounds that have never occurred naturally. In fact, there are a number of natural organohalogens. Tyrian purple is a bromine-containing dye derived from spiny carnivorous snails of the murex family. Because of its rarity—about 9000 snails are needed to produce one gram of the dye—Tyrian purple was a costly status symbol in the ancient world. Julius Caesar wore a purple toga as a sign of office, but a Roman senator was allowed only a purple stripe on his white toga.

Some seaweeds produce trichloromethane (chloroform, $CHCl_3$) and tetrachloromethane (carbon tetrachloride, CCl_4), compounds that are also produced as by-products of water chlorination. Bromomethane (methyl bromide, CH_3Br), an important insect fumigant, is produced in large quantities by ocean algae. These and other organohalogens are, in part, responsible for the sharp smell of seawater. As various forest fungi break down dead trees and other plant materials, they produce about 200 times as much natural chloromethane (methyl chloride, CH_3Cl) as humans produce synthetically. Volcanoes also produce chloromethane, along with dichloromethane (methylene chloride) and chlorofluorocarbons such as dichlorodifluoromethane (Freon 12). Our own bodies utilize organic iodides in the form of thyroid hormones, and our white blood cells oxidize chloride ions in the blood to molecular chlorine, which kills invading bacteria and other microorganisms.

Polychlorodibenzodioxins (PCDDs) and polychlorodibenzofurans (PCDFs) are considered to be among the most hazardous organohalogens.

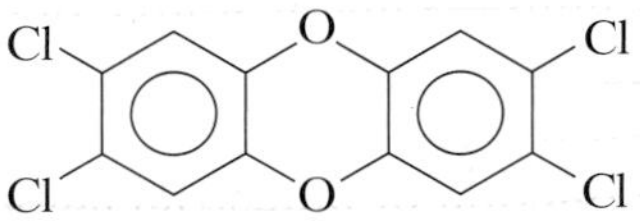

2,3,6,7-tetrachlorodibenzodioxin
(a PCDD)

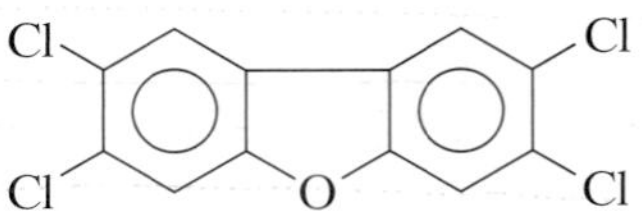

2,3,6,7-tetrachlorodibenzofuran
(a PCDF)

2,3,6,7-Tetrachlorodibenzodioxin, called "dioxin" by the news media, occurs as a by-product when the herbicide 2,4,5-trichlorophenoxyacetic acid (2,4,5-T) is manufactured from 2,4,5-trichlorophenol.

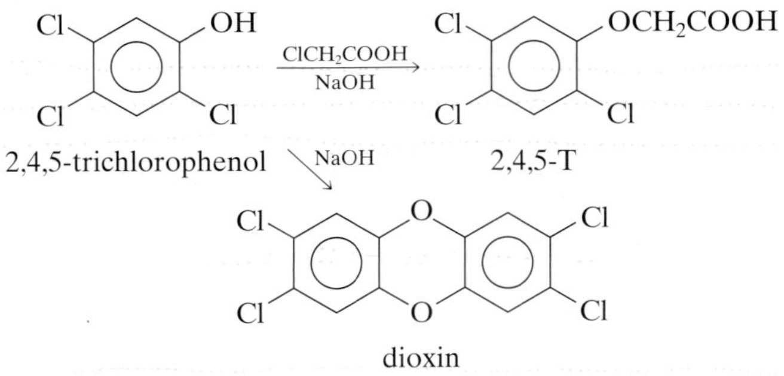

2,4,5-trichlorophenol   2,4,5-T

dioxin

2,4,5-T was the main constituent of Agent Orange, the Vietnam War defoliant believed to have caused health problems for many Vietnam veterans.

Thus 2,4,5-T is almost invariably contaminated with this chemical. Dioxin is regarded as one of the most toxic substances known, and some PCDFs may be even more toxic. Yet both PCDDs and PCDFs are produced by burning wood, which contains natural chlorides, and some researchers believe that forest fires and brush fires are the major source of dioxins in the environment. In addition, natural enzymes act on humic acids—phenolic acids formed by the decomposition of organic matter in the soil—to yield a number of chlorinated compounds, including chlorophenols that, like 2,4,5-trichlorophenol, could give rise to TCDDs. Dioxins have been identified in ancient marine sediments from as early as 6000 B.C., and they are even produced in garden compost piles.

This is not to say that synthetic organohalogens are not harmful because many of them are also produced naturally. Many naturally occurring substances are hazardous to our health, and the levels of certain organohalogens in humans, including PCDDs and PCDFs, are considerably higher now than they were in preindustrial times, when the only sources of these substances were natural. Environmental groups and health organizations are concerned that they may already be causing reproductive problems, such as declining sperm counts, and other health problems in humans.

Key Concept: *A naturally occurring compound is not safe because it is natural, nor is a synthetic compound harmful because it is synthetic. The physiological effect of any substance, natural or synthetic, depends on its molecular structure, not on its source.*

Most scientists agree that we should weigh the hazards and benefits of each organohalogen individually, rather than banning a whole family of compounds simply because it contains some bad actors. But those who support a chlorine ban believe we face a potential crisis that can't wait for a chemical-by-chemical study of all 15,000 or so chlorine-containing products that are now on the market. Meanwhile, measures are being taken to reduce the production of the most hazardous organohalogens. For example, the Environmental Protection Agency has proposed a regulation that would force pulp and paper mills—a major source of dioxin—to curtail their use of chlorine for bleaching. Nevertheless, the demand for many organohalogens has continued to grow, and there is little evidence to date of a slowdown in the production of industrial chlorine and its compounds.

Understanding the Experiment

In this experiment, you will convert either 1-butanol or 2-butanol to the corresponding alkyl bromide with HBr, using sulfuric acid as a catalyst. A *catalyst* is a substance that accelerates chemical reactions without being consumed in the process. For example, an alcohol cannot be converted to an alkyl bromide by sodium bromide, because hydroxide ion (OH^-) is too poor a leaving group to be displaced by bromide ion. A strong acid is needed to convert this poor leaving group to a better one, H_2O, by protonating it.

$$R\!-\!OH + H^+ \rightarrow R\!-\!OH_2^+$$

Sulfuric acid catalyzes the overall reaction by increasing the concentration of the protonated alcohol, which can then react with bromide ion by either an S_N1 or S_N2 mechanism to form an alkyl bromide. Primary alcohols tend to react by the direct substitution (S_N2) mechanism; secondary and tertiary alcohols are more likely to form an intermediate carbocation, which then combines with halide ion.

See your lecture course textbook for a more detailed discussion of nucleophilic substitution reactions.

S_N2 reaction of 1-butanol

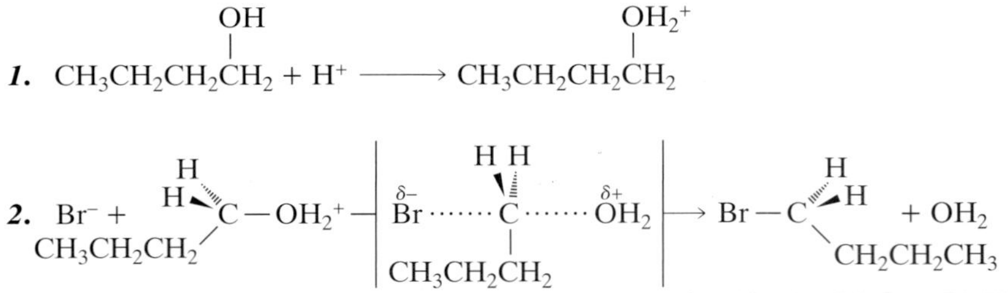

1.
$$\underset{\underset{CH_3CH_2CH_2CH_2}{|}}{OH} + H^+ \longrightarrow \underset{\underset{CH_3CH_2CH_2CH_2}{|}}{OH_2{}^+}$$

2.
$$Br^- + \underset{CH_3CH_2CH_2}{\overset{H}{\underset{}{\overset{}{C}}}} \diagdown \overset{H}{C} - OH_2{}^+ \longrightarrow \left[\overset{\delta-}{Br} \cdots\cdots \underset{\underset{CH_3CH_2CH_2}{|}}{\overset{\overset{H\ H}{|}}{C}} \cdots\cdots \overset{\delta+}{OH_2} \right] \rightarrow Br - \overset{H}{\underset{CH_2CH_2CH_3}{\overset{}{C}}} \diagup H + OH_2$$

S_N1 reaction of 2-butanol

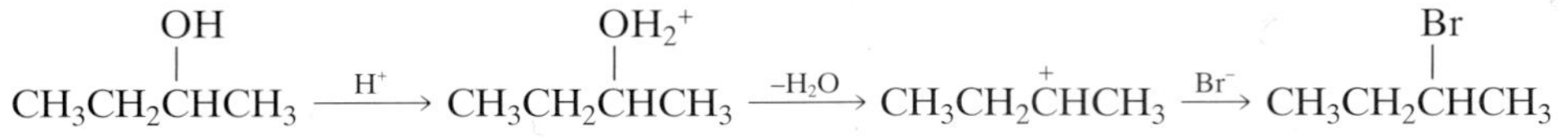

$$\underset{\underset{CH_3CH_2CHCH_3}{|}}{OH} \xrightarrow{H^+} \underset{\underset{CH_3CH_2CHCH_3}{|}}{OH_2{}^+} \xrightarrow{-H_2O} CH_3CH_2\overset{+}{C}HCH_3 \xrightarrow{Br^-} \underset{\underset{CH_3CH_2CHCH_3}{|}}{Br}$$

The results will be more reliable if average yields from several research groups are compared.

Since the reaction rate for both mechanisms depends on the concentration of the protonated alcohol, increasing the catalyst concentration should increase the rate. We might expect this to, in turn, increase the amount of alkyl bromide produced in a given reaction time, but it could increase the rates of side reactions as well. Accelerating such side reactions as E1 elimination, polymerization, and ether formation could actually reduce the product yield by converting the substrate into unwanted by-products.

By varying the sulfuric acid volume, your research team will be varying the ratio of catalyst to substrate from 0.25 to 1.00 for each of the two alcohols (see Table 28.1). Using the yields of the resulting alkyl bromides, you will carry out a cost analysis to determine which catalyst/substrate ratio will produce each product at the lowest cost per kilogram. Because variable product losses during the purification step may invalidate the results, the masses of the crude products will be used for the cost analyses.

You will carry out the reaction by heating the alcohol under reflux with a small excess of 48% hydrobromic acid and the assigned volume of sulfuric acid. Since sulfur dioxide and HBr fumes are evolved during the reaction, a hood or gas trap must be used to keep them from escaping into the laboratory. The organic layer containing the product will be separated from the reaction mixture after adding water. This layer ordinarily separates below the aqueous layer in a mixture with water, but with a low catalyst/substrate ratio there may be enough unreacted alcohol in the organic layer to make it separate above the aqueous layer. The organic layer will contain varying amounts of HBr, sulfuric acid, unreacted alcohol, and by-products; these impurities will be removed by washing with sulfuric acid, water, and aqueous sodium bicarbonate, followed by distillation. The alkyl bromides are less dense than sulfuric acid but denser than water, so the alkyl bromide layer will be on top after the first washing, but on the bottom in the next two.

You can confirm the conversion of reactant to product using infrared (IR) spectrometry. The IR spectra of alkyl bromides are characterized by a strong C—Br stretching band at low frequency, from 690 to 515 cm^{-1}. Since the range of some IR spectrometers does not extend below 600 cm^{-1}, the entire C—Br band is not observed in all cases.

Reactions and Properties

$$CH_3CH_2CH_2CH_2OH + HBr \xrightarrow{H_2SO_4} CH_3CH_2CH_2CH_2Br + H_2O$$

or

$$\underset{\displaystyle CH_3CH_2CHCH_3}{\overset{\displaystyle OH}{|}} + HBr \xrightarrow{H_2SO_4} \underset{\displaystyle CH_3CH_2CHCH_3}{\overset{\displaystyle Br}{|}} + H_2O$$

Table 28.1 Ratio of catalyst to alcohol in the eight runs

Run	Alcohol	ml H_2SO_4	Catalyst/substrate mole ratio
1	1-butanol	0.25	0.25
2	1-butanol	0.50	0.50
3	1-butanol	0.75	0.75
4	1-butanol	1.00	1.00
5	2-butanol	0.25	0.25
6	2-butanol	0.50	0.50
7	2-butanol	0.75	0.75
8	2-butanol	1.00	1.00

Note: 1 mL of concentrated (~96%) sulfuric acid contains about 18 mmol of the acid.

Table 28.2 Physical properties

	M.W.	bp	d
1-butanol	74.1	117	0.810
2-butanol	74.1	99.5	0.808
hydrobromic acid (48%)	80.9		1.49
sulfuric acid (~96%)	98.1		1.84
1-bromobutane	137.0	102	1.276
2-bromobutane	137.0	91	1.259

Note: bp is in °C, density is in g/mL.

1-Butanol $CH_3CH_2CH_2CH_2OH$

3335.4	1378.6	952.2
2959.7	1072.6	846.7
1466.0	1010.6	737.5

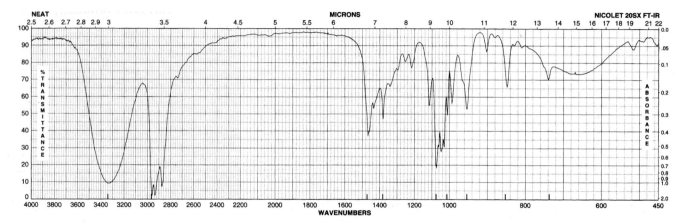

Figure 28.1 IR spectra of the starting materials

2-Butanol

$$CH_3CHCH_2CH_3$$
with OH group

3343.1	1374.6	991.0
2967.2	1299.9	912.6
1457.7	1108.9	668.4

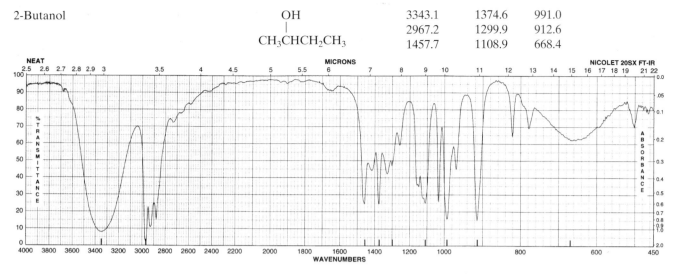

Figure 28.1 (*Continued*)

Directions

Safety Notes

sulfuric acid

butanols

bromobutanes

Take Care! Wear gloves, avoid contact with the acids, and do not breathe their vapors.

Hydrobromic acid is toxic and very corrosive and can cause very serious damage to the skin, eyes, and respiratory tract. Wear gloves, and dispense under a hood. Avoid contact with the acid and do not inhale its vapors. Sulfuric acid causes chemical burns that can seriously damage the skin and eyes. Wear gloves and avoid contact.
1-Butanol and 2-butanol are flammable and may cause eye or skin irritation. Avoid contact.
1-Bromobutane and 2-bromobutane are flammable and may be harmful if inhaled or absorbed through the skin. 2-Bromobutane is a suspected carcinogen. Avoid contact with the products and do not breathe their vapors.

You will be assigned one of the two alcohols and either 0.25, 0.50, 0.75, or 1.00 mL of sulfuric acid. Measure the quantities of all chemicals carefully for the cost analysis.

Reaction.   Assemble an apparatus for heating under reflux [OP-7] using a 10-mL round-bottom flask, a water-cooled condenser, and a gas trap [OP-24]. The gas trap can be omitted if you are performing the reaction under an efficient fume hood. Deliver 18.0 mmol of your assigned alcohol (1-butanol or 2-butanol), accurately weighed, into the round-bottom flask and drop in a stir bar. *Under the hood*, cool the flask in ice water and cautiously add 2.5 mL (22 mmol) of 48% aqueous hydrobromic acid while stirring. Use a measuring pipet or an automatic pipet to measure your assigned volume of concentrated sulfuric acid, and add it to the reaction mixture while stirring. Do not let either acid contact a plastic compression cap. Heat the mixture under reflux, with stirring [OP-10], for one hour from the time the solution starts to boil.

Separation.   Transfer the cooled reaction mixture to a 15-mL centrifuge tube using 5 mL of water for the transfer. Shake gently to mix the layers;

then allow the layers to separate completely, rubbing the inside of the centrifuge tube with a wooden applicator stick to facilitate separation, if necessary. There should be only two layers at this point; if there are three, continue shaking. Ordinarily the alkyl bromide is on the bottom and the aqueous layer on top, but occasionally the aqueous layer may be dense enough to end up on the bottom—in this case the bottom layer will be the larger one. (If you are not sure which layer to save, test each one by adding a drop of water to a drop of the layer.) After separating it cleanly from the aqueous layer, cautiously wash [OP-21] the organic layer with 2 mL of cold, concentrated sulfuric acid, shaking gently with occasional venting. Remove the lower H_2SO_4 layer, then wash the alkyl bromide layer with a 2-mL portion of water followed by a 2-mL portion of saturated aqueous sodium bicarbonate, saving the organic (lower) layer each time. (**Stop and Think:** What does the $NaHCO_3$ wash remove?) Dry [OP-22] the alkyl bromide with anhydrous calcium chloride. Transfer the crude product to a tared vial, weigh it, and report the yield to your instructor or coworkers.

Take Care! Possible violent reaction! Wear gloves and avoid contact with the acid.

Waste Disposal: Put the sulfuric acid into an acid wastes container and pour all aqueous layers down the drain.

Purification and Analysis.   Purify the alkyl bromide by simple distillation [OP-27]. Collect the product over a 3–4°C range centered around the expected boiling point. Measure the mass and boiling point [OP-31] of the purified product. At your instructor's option, obtain its IR spectrum [OP-36]. Using prices provided by your instructor, calculate the total cost of all the chemicals (except water) that you used in the "Reaction" phase of the experiment. Divide the cost in dollars by the mass of your crude product in kilograms and obtain the cost/mass values reported by other members of your team.

Exercises

1   Based on their structures, attempt to explain any differences in the optimum catalyst/substrate ratios for the two alcohols.

2   (a) Write structures for at least three by-products that might have formed during the reaction of 2-butanol. (b) Propose a mechanism for the formation of each by-product.

3   If you recorded the IR spectrum of your product, interpret the spectrum as completely as you can, and use the appropriate starting material spectrum in Figure 28.1 to show that the expected functional group conversion has taken place.

4   Following the format in Appendix V, construct a flow diagram for this experiment.

5   Describe and explain the possible effect on your results of the following experimental errors or variations. (a) After adding water to the reaction mixture, you kept the bottom layer and discarded the smaller top layer. (b) You omitted the sulfuric acid wash. (c) You used a boiling-water bath to heat a reaction mixture containing 1-butanol.

6   A gas trap for this experiment could be constructed using a test tube partly filled with 1 M sodium hydroxide. Write balanced equations showing how HBr and SO_2 would be consumed in such a gas trap.

7   Write a mechanism showing how 2,4,5-trichlorophenol is converted to 2,4,5-T.

Borohydride Reduction of Vanillin to Vanillyl Alcohol

EXPERIMENT **29**

Preparation of Alcohols. Reactions of Carbonyl Compounds. Reduction Reactions. Nucleophilic Addition.

Operations

The operations you use will depend on the procedure you develop.

Before You Begin

After reading the experiment, develop a procedure and an experimental plan for the sodium borohydride reduction of 2.50 mmol of vanillin to vanillyl alcohol. Calculate or estimate the quantities of reactants and other chemicals you will need, and calculate the theoretical yield of vanillyl alcohol. Describe the reaction conditions and tell how you intend to separate and purify the product. Your procedure should be clear and detailed enough so that anyone with sufficient background could carry it out successfully. (Alternatively, a group of students can work out a set of procedures in which some experimental parameters are varied to see how such variations alter the yield and purity of the product.)

Scenario

Pulpchem Inc., a subsidiary of a large paper company, produces useful chemicals from lignin and other by-products of paper production. Chemical treatment of lignin yields large quantities of vanillin, a white solid that is also responsible for the characteristic aroma of vanilla beans. Woody Aspin, Pulpchem's product development manager, has asked your Institute to develop a method for reducing vanillin to vanillyl alcohol, which shows promise as a starting material for the synthesis of synthetic drugs and flavoring ingredients. There is a procedure for the preparation of vanillyl alcohol in an obscure Swedish journal, *Acta Universitatis Lundensis*, which ceased publication in the 1960s. Unless you understand Swedish and can get your hands on a copy of the journal article, you will have to develop your own procedure for the synthesis.

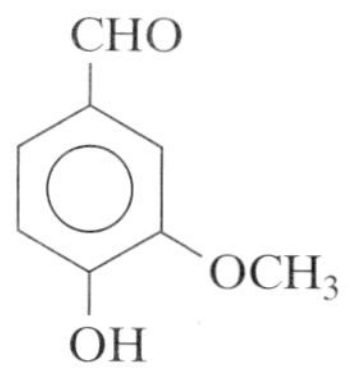

vanillin

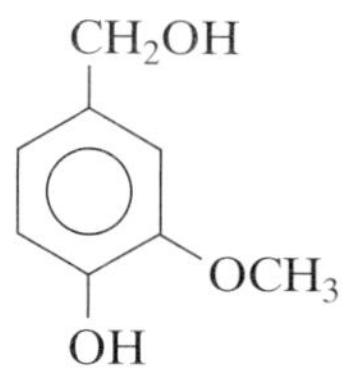

vanillyl alcohol

Applying Scientific Methodology

The problem, whether vanillin can be converted to vanillyl alcohol by a procedure that you have developed yourself, will be solved when (and if) you obtain and characterize the product. You will test your working hypothesis by measuring the product's melting point, and you can record its infrared (IR) spectrum for confirmation.

Fragrant and Fiery Aromatics

Chemists recognized at an early date that certain compounds obtained from natural sources showed a higher ratio of carbon to hydrogen than did typical aliphatic compounds. These compounds also had distinctive chemical properties. Because many of them came from such pleasant-smelling sources as the essential oils of cloves, sassafras, cinnamon, anise, bitter almonds, and vanilla, they were called *aromatic* compounds. The name stuck, although it is no longer associated with the odors of such compounds but with their structures and properties. Many aromatic compounds do live up to the original meaning of the term, however. Among the most interesting and important of these are vanillin and the *vanilloids*, structurally related compounds that may exhibit vanillin's characteristic ring-substitution pattern.

vanilla plant

In 1520, the Spanish conquistador Hernando Cortez was served an exotic drink by Montezuma II, emperor of the Aztecs, at his capital of Tenochtitlán on the site of modern Mexico City. Cortez enjoyed the drink, a combination of chocolate and vanilla, and it soon found its way back to Europe. The vanilla plant, a climbing orchid, *Vanilla planifolia*, was also shipped back to the Old World in the hope that it could be cultivated there. The transported plants grew well but, mysteriously, would not fruit. This mystery remained unsolved for more than 300 years, until someone discovered that the plant was pollinated by a native Mexican bee with an exceptionally long proboscis. A method of hand pollination was soon developed, allowing the cultivation of vanilla outside of Mexico.

Vanilla flavoring comes from the fruit of the vanilla plant, a long, narrow pod that, after curing, looks somewhat like a dark brown string bean. The principal component of vanilla flavoring, vanillin (3-methoxy-4-hydroxy-benzaldehyde), does not exist as such in the fresh vanilla bean, but is formed by the enzymatic breakdown of a glucoside during the curing process. Although the finest vanilla flavoring is still obtained from natural vanilla, synthetic vanillin is far less costly. It is used as a component of flavorings and perfumes and as a starting material for the synthesis of such drugs as L-dopa, which is used for treating Parkinson's disease.

At one time, synthetic vanillin was made mostly from isoeugenol, a naturally occurring phenol used widely as a perfume ingredient.

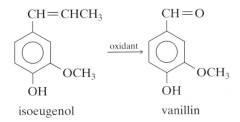

Most vanillin is now synthesized using lignin derived from wood pulp. Lignin is a complex polymer that gives rigidity to trees and other woody plants; after cellulose, it is the second most abundant organic material on earth. Many of the basic structural units of lignin are guiacylpropane units, and such units can be broken down chemically to yield vanillin.

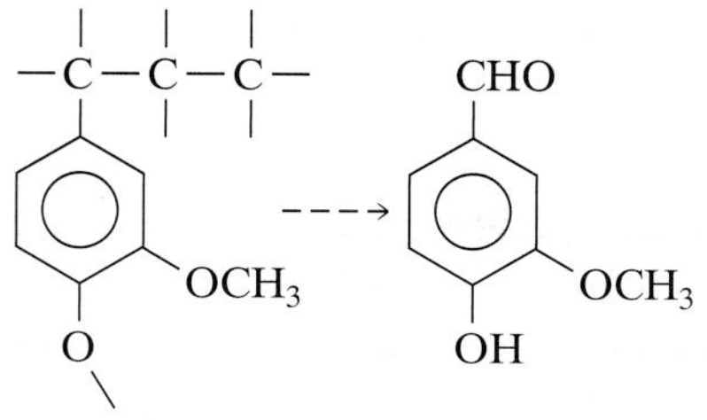

guiacylpropane unit

Safrole, a fragrant liquid derived from the roots and bark of the sassafras tree (*Sassafras albidum*), is structurally related to vanillin. Once widely used as a flavoring in root beer, toothpaste, and chewing gum, safrole can no longer be added to such products because of its toxic, irritant qualities and because it produces liver tumors in rats and mice. But black pepper, anise, nutmeg, and other spices contain minute amounts of safrole, so you probably consume a little of it every day. Safrole has also been used by illicit-drug manufacturers to synthesize the dangerous designer drug "Ecstasy" (methylenedioxymethamphetamine), which can cause mental confusion, kidney failure, and death. For this reason, the sale of safrole is now strictly controlled.

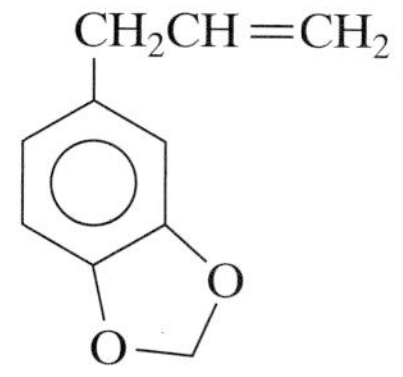

safrole

Black pepper gets its "bite" from piperine, which contains the methylenedioxy ($-OCH_2O-$) unit also found in safrole.

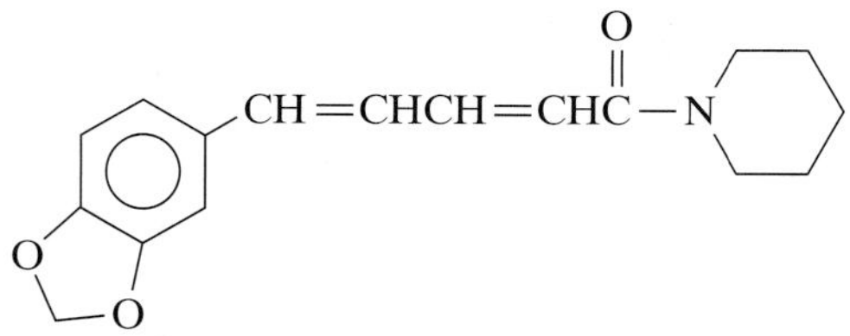

piperine (*trans* double bonds)

Capsaicin, piperine's much hotter cousin, has the 3-methoxy-4-hydroxy grouping characteristic of vanillin.

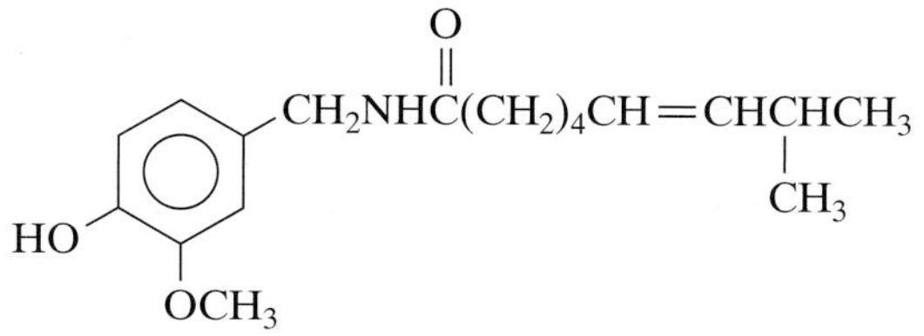

capsaicin

Capsaicin is a fiery component of many *capsicums*, pungent peppers such as cayennes, jalapeños, and the ultrahot habaneros. Capsicum peppers are used to make Tabasco Sauce, Jamaica Hell Fire and other hot sauces, as well

zingerone

Key Concept: *Developing a successful procedure for an organic synthesis requires a comprehensive knowledge of the characteristics of the reactants and a good understanding of the reaction.*

Reduction of a carbonyl group

as most commercial salsas. Another hot compound with the vanillin structural unit is zingerone, the pungent principle of ginger. The structural units that we associate with the pleasant flavor of vanillin contribute to a far different taste sensation in these fiery aromatics!

Understanding the Experiment

When organic chemists attempt to synthesize a new compound, they have no set procedure to follow. Instead, their options are to

- Adapt and modify an existing procedure for a similar synthesis
- Devise a procedure based on their knowledge of the substrate, reagent, product, and reaction involved
- Invent a new way of synthesizing the compound they wish to make

In this experiment, you will follow the second option. You will develop your own procedure for reducing vanillin to vanillyl alcohol based on information that you will find in this section about the reducing agent, sodium borohydride, and the reaction, reduction of a carbonyl compound to an alcohol, as well as information in the "Reactions and Properties" section about the substrate and the product.

Reduction Reactions.   In organic chemistry, *reduction* usually refers to a reaction that is accompanied by a gain of hydrogen atoms or a loss of oxygen atoms, or both. For example, a carbonyl compound is reduced to an alcohol when its carbonyl group gains two hydrogen atoms. The hydrogen is provided by an appropriate reducing agent.

Reducing Agents.   When lithium aluminum hydride ($LiAlH_4$) was introduced as a reducing agent in the late 1940s, it brought about a revolution in the preparation of alcohols by reduction. At that time, the two most popular reducing agents for carbonyl compounds were hydrogen and sodium metal. The greater simplicity and convenience of hydride reduction soon made it the preferred method for a broad spectrum of chemical reductions. Lithium aluminum hydride is a powerful reducing agent whose high reactivity is a disadvantage in some applications. Because it reacts violently with water and other hydroxylic solvents to release hydrogen gas, it can only be used in aprotic solvents, such as diethyl ether, under strictly anhydrous conditions. It is also expensive and somewhat hazardous to use—even grinding it in a mortar may cause a fire. By contrast, sodium borohydride ($NaBH_4$) is a much milder reducing agent that is comparatively safe to handle in its solid form. Unlike lithium aluminum hydride, it can even be used in aqueous or alcoholic solutions. $NaBH_4$ does decompose slowly in moist air, so its container should be tightly capped when it is not in use.

Sodium borohydride reductions involve nucleophilic addition of hydride ion ($H:^-$) to the carbonyl carbon, but apparently no free hydride ions are generated. Kinetic evidence suggests that one solvent molecule bonds to the boron atom while it is transferring hydride to the carbonyl compound, and another solvent molecule provides a proton to the carbonyl oxygen.

$$R{-}O{-}H + O{=}C + H{-}\overset{\overset{\displaystyle H}{|}}{B}{+}O{-}R \longrightarrow R{-}\overset{\ominus}{O} + H{-}O{-}\overset{|}{\underset{|}{C}}{-}H + B\overset{\ominus}{-}OR + H^{+}$$

(R = alkyl or H)

This process can continue until all of the hydride ions from BH_4^- have been used up.

Reaction Stoichiometry.   The overall stoichiometry of the borohydride reduction of a carbonyl compound (R′ = alkyl or H) is given by the following general equation:

$$\underset{\displaystyle 4R\overset{\overset{\displaystyle O}{\|}}{C}R'}{} + NaBH_4 + 4H_2O \longrightarrow \underset{\displaystyle 4R\overset{\overset{\displaystyle OH}{|}}{C}HR'}{} + H_3BO_3 + NaOH$$

In practice, it is best to use a 50–100% excess of sodium borohydride to compensate for any that reacts with the solvent or decomposes from other causes. Since the reaction is first order with respect to sodium borohydride (as well as the carbonyl compound), using an excess will also increase the reaction rate.

Reaction Solvents.   Sodium borohydride reductions are usually carried out in a dilute (~1 M) aqueous NaOH solution or in an alcohol, such as methanol, ethanol, or 2-propanol. The reagent is not stable at low pH, and even in a neutral aqueous solution it decomposes to the extent of about 4.5% per hour at 25°C. Acidic functional groups, such as COOH and the OH group of a phenol, may cause rapid decomposition of sodium borohydride. When a carbonyl compound having such a functional group is being reduced, enough 1 M NaOH should be used to neutralize the functional group *and* maintain a pH of 10 or higher. Sodium borohydride reacts slowly with alcohols, but ethanol and methanol are usually suitable solvents when there are no acidic functional groups and the reaction time is no more than 30 minutes at 25°C. For longer reaction times or reactions at higher temperatures, isopropyl alcohol is a better solvent, but it is more difficult to remove after the reaction is over.

Reaction Conditions.   In most reactions with sodium borohydride, the aldehyde or ketone is dissolved in the reaction solvent and a solution of sodium borohydride is added, with external cooling if necessary, at a rate slow enough to keep the reaction temperature below 25°C. Higher temperatures may decompose the hydride, and adding the carbonyl compound *to* the alkaline sodium borohydride solution may cause side reactions of base-sensitive substrates. The amount of solvent is not crucial, but enough should be used to completely dissolve the reactants and facilitate the workup of the reaction mixture. The solubility of sodium borohydride per 100 g of solvent is reported to be 55 g in water at 25°C, 16.4 g in methanol at 20°C, and 4.0 g in ethanol at 20°C.

The time required to complete the reaction depends on the reaction temperature and the reactivity of the substrate. Kinetic studies of borohydride reduction in isopropyl alcohol have shown that aldehydes are considerably more reactive than ketones and that aliphatic carbonyl compounds are more reactive than aromatic ones. Most reactions of aldehydes and aliphatic ketones are complete in 30 minutes at room temperature, but those of aromatic ketones or particularly hindered ketones may require more time or higher reaction temperatures. For example, the comparatively reactive ketone 4-*t*-butylcyclohexanone is completely reduced at room temperature in 20 minutes, but benzophenone is reduced by boiling it in isopropyl alcohol for 30 minutes.

Et   = ethyl
i-Pr = isopropyl
t-Bu = *t*-butyl
r.t.   = room temperature

Reaction conditions in borohydride reductions

Workup of the Reaction Mixture.   After the reaction is complete, the excess sodium borohydride is decomposed by acidifying the reaction mixture to pH 6 or below (slowly, while stirring) using dilute (~3 M) hydrochloric acid. Hydrogen gas is evolved as the excess sodium borohydride decomposes, so *there must be no flames in the vicinity*. Working under a hood is advisable. Addition of acid may also generate some diborane (B_2H_6), which can cause side reactions if other reducible groups, such as COOH, COOR, and C=C, are present.

Depending on the properties of the product and the reaction solvent used, the product can be separated from the reaction mixture by filtration, extraction, or partial evaporation of the solvent followed by extraction. If the product is a solid that crystallizes from the reaction mixture, it can be collected by vacuum filtration. The yield of the solid product can usually be increased by extracting the filtrate with diethyl ether or another suitable solvent, and then drying and evaporating the ether. Liquids or water-soluble products are generally separated from an aqueous reaction mixture by extraction with diethyl ether and recovered by evaporating the dried ether. If the reaction solvent is an alcohol, the reaction mixture is usually concentrated by evaporating most of the alcohol. Water is then added and the product is extracted with a suitable solvent.

Purification.   The product can be purified by any appropriate method, based on its physical state and properties. Vanillyl alcohol is reported to be soluble in hot and cold ethanol, hot and cold ether, and hot water. It is relatively insoluble in cold water, and it tends to form supersaturated solutions in water.

Reactions and Properties

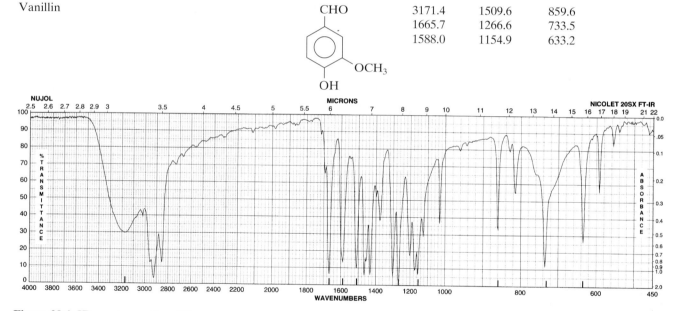

vanillin   + NaBH$_4$ + 4H$_2$O ⟶ 4   vanillyl alcohol   + H$_3$BO$_3$ + NaOH

Table 29.1 Physical properties

	M.W.	mp	bp
vanillin	152.2	79	285
sodium borohydride	37.83		400d
vanillyl alcohol	154.2	115	d

Note: mp and bp are in °C, density is in g/mL; d = decomposes.

Vanillin

3171.4	1509.6	859.6
1665.7	1266.6	733.5
1588.0	1154.9	633.2

Figure 29.1 IR spectrum of vanillin

Directions

Safety Notes

Sodium borohydride is toxic and corrosive, and it can react violently with concentrated acids, oxidizing agents, and other substances. Aqueous NaBH$_4$ solutions with pH values below 10.5 have been known to decompose violently, so be sure that your reaction mixture (if aqueous) is sufficiently alkaline. Avoid contact with NaBH$_4$, do not breathe its dust, and keep it away from other chemicals.

Hydrogen is generated when the reaction mixture is acidified, so be sure that there are no flames nearby during this step.

sodium borohydride

Develop your own procedure for this experiment and submit it to your instructor for approval. Carry out the synthesis in the laboratory, measure the yield and melting point of the purified product, and turn it in. At your instructor's option, you can obtain the IR spectrum of your product and compare it with the spectrum of the starting material to confirm that the expected functional-group conversion has occurred. Dispose of any waste solvents, etc., as directed by your instructor. Calculate your percent yield and compare your results with those of other students in your laboratory section.

Exercises

1 Write a mechanism for the reduction of vanillin by sodium borohydride under the reaction conditions you used.

2 Write a balanced equation for the decomposition of sodium borohydride in water to which HCl has been added.

3 Describe and explain the possible effect on your results of the following experimental errors or variations. (a) You used pure water as the reaction solvent. (b) You used 0.5 mL of 1 M NaOH for the $NaBH_4$ solution and 1.0 mL of 1 M NaOH for the vanillin solution. (c) Since you started with 2.5 mmol of vanillin, you assumed that a 100% excess of $NaBH_4$ was 5.0 mmol and used that amount.

4 Following the format in Appendix V, construct a flow diagram for the experiment.

5 Sodium borohydride is a strong base as well as a reducing agent. Explain why it is better to add the $NaBH_4$ solution *to* a base-sensitive substrate than to add the substrate to the $NaBH_4$ solution.

6 (a) Draw the structure of the product you would have isolated if you had used $NaBD_4$ in D_2O in this experiment, and write a mechanism explaining the result. (b) Give the product structure for the same reaction using $NaBD_4$ in H_2O. Assume that there is no hydrogen exchange between $NaBD_4$ and the solvent.

7 A careless student, Will Bobble, misread the label on a bottle of 10 M sodium hydroxide and used it (rather than 1 M NaOH) to dissolve his vanillin. Then he stored the solution and left the lab early to catch a ride home. During the next lab period he added $NaBH_4$ and finished the experiment. Although his product was a white solid, it melted over a broad temperature range that was much lower than the melting point of vanillyl alcohol. His instructor suggested washing the product with dilute sodium bicarbonate; when he did so, about half of the product dissolved and the remainder melted at 115°C. Explain what happened and write an equation for the reaction.

Other Things You Can Do

(Starred projects require your instructor's permission.)

*1 Carry out the following tests from Part IV on vanillin and on your product: 2,4-dinitrophenylhydrazine (Test C-11), ferric chloride (Test C-13), and Tollen's test (Test C-23). Interpret the results and comment on the purity of your product.

*2 Carry out a photoreduction reaction that uses isopropyl alcohol as the reducing agent, as described in Minilab 24.

*3 Reduce another carbonyl compound (such as benzaldehyde or camphor) with sodium borohydride after developing a detailed procedure, and turn in a pure sample of the resulting alcohol. Your instructor must approve the procedure before you start.

 4 Starting with sources listed in the Bibliography, write a research paper on the sources, nature, and uses of lignin, including a description of a commercial process for producing vanillin from the by-products of papermaking.

Synthesis of Triphenylmethanol and the Trityl Carbocation

Reactions of Carbonyl Compounds. Reactions of Organic Halides. Preparation of Alcohols. Nucleophilic Addition. Organometallic Compounds. Carbocations. Infrared Spectrometry.

Operations

OP-7   Heating
OP-10   Mixing
OP-11   Addition of Reactants
OP-13   Vacuum Filtration
OP-16   Evaporation
OP-21   Washing Liquids
OP-22   Drying Liquids
OP-23   Drying Solids
OP-24   Drying and Trapping Gases
OP-25   Recrystallization
OP-30   Melting Point
OP-36   Infrared Spectrometry (optional)

Before You Begin

1 Read the experiment, read or review the operations as necessary, and write an experimental plan.
2 Calculate the mass of 4.00 mmol of magnesium, the mass of 3.80 mmol of benzophenone, the mass and volume of 4.00 mmol of bromobenzene, and the theoretical yield of triphenylmethanol. Calculate the theoretical yield of trityl fluoborate from 0.100 g of triphenylmethanol.

Scenario

The Complimentary Colors Company manufactures synthetic dyes, including a number of triphenylmethane dyes. Gilda Lillie, product development director for the company, would like to develop some new colors to improve its market share in the dye industry. She has learned, for example, that having two *para*-dimethylamino groups on two of the three benzene rings in the parent structure yields a green dye (Malachite Green), while having three of them yields a violet dye (Crystal Violet).

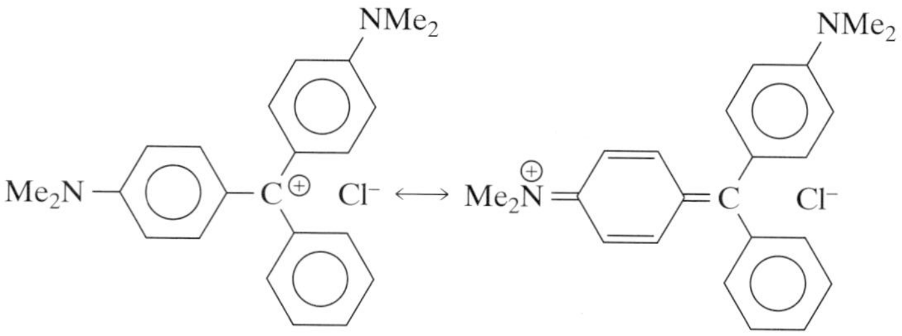

Two resonance structures for Malachite Green, a triphenylmethane dye

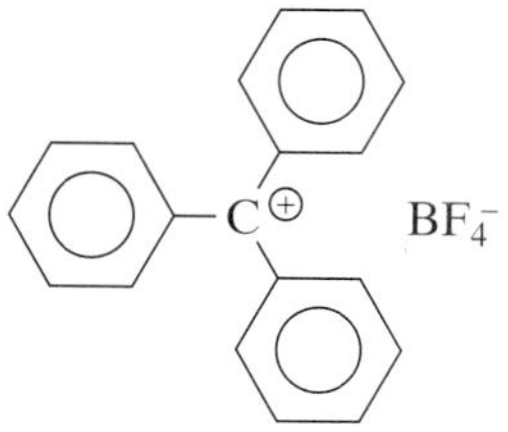

The company's technicians need to know how the kinds and positions of substituents on the parent triphenylmethyl ring structure affect the color of triphenylmethane dyes. For that they need a sample of the unsubstituted parent compound of these dyes, the triphenylmethyl (trityl) carbocation. Your assignment is to prepare triphenylmethanol and convert it to trityl fluoborate, a salt that contains the trityl carbocation.

Applying Scientific Methodology

The main scientific problem—determining the color of trityl fluoborate—will be solved when you prepare this substance.

The Colorful Career of the Triphenylmethanes

During his Easter vacation from the Royal College of Chemistry, 18-year-old William Perkin was trying to synthesize quinine when he came up with an unpromising black solid that had none of the properties of quinine. When Perkin dissolved it in water or alcohol, the solid yielded a purplish solution that could be used to dye cloth. Later named mauve, this was the first dye to be produced synthetically, and its preparation marked the birth of the synthetic-dye industry. The first triphenylmethane dye, fuchsin, was synthesized a few years later, and it was followed soon after by Malachite Green, Crystal Violet, and other triphenylmethane dyes. The race to develop commercially marketable dyes also stimulated research into the molecular basis of color. Why, for example, is Malachite Green green and its reduced form colorless?

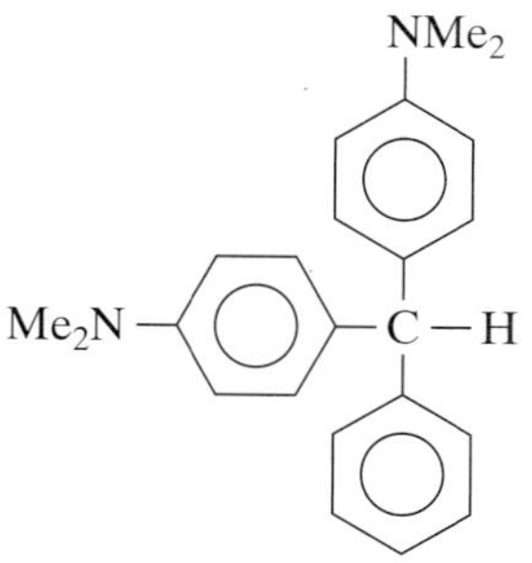

The *chromophore* of a compound is the part of its molecule over which electrons can be delocalized and which is responsible for its absorption of ultraviolet or visible light. Triphenylmethane dyes come in all colors of the rainbow, from the red of rosaniline through Malachite Green and Victoria Blue to Crystal Violet. The chromophores responsible for these colors appear to be nitrogen-substituted triphenylmethyl (trityl) cations, but they are not true carbocations because most of their positive charge is distributed to the nitrogen atoms, as in the iminium ion form of Malachite Green shown above. According to resonance theory, the iminium ion and carbocation forms are contributing structures of a resonance hybrid that has some characteristics of each. The electron delocalization suggested by such structures is responsible for their colors. As a rule, compounds with extensive chromophores that allow

triphenylmethyl (trityl) fluoborate

The synthesis of quinine was not accomplished until 1944—88 years after Perkin had attempted it.

Reduced form of malachite green

Key Concept: *When a substance cannot be represented satisfactorily by a single structural formula, its actual structure is regarded to be a composite of two or more contributing structures.*

One resonance structure of the triphenylmethyl cation

electron delocalization over many atoms tend to be colored; the longer the chromophore, the higher the wavelength of light they absorb. The reduced form of Malachite Green is not colored because the saturated carbon that connects the rings prevents electron delocalization over the three-ring system.

Although it is nearly a trillion times less stable than Crystal Violet, the triphenylmethyl cation is unusually stable for a carbocation. When protected from atmospheric moisture, trityl salts will keep almost indefinitely. The cation owes this unusual stability to delocalization of the positive charge about its three benzene rings. The cation is shaped somewhat like a propeller with the "blades" (benzene rings) pitched at a 32° angle, because steric interference between the *ortho* hydrogen atoms makes a planar configuration impossible.

Shape of triphenylmethyl cation

Understanding the Experiment

The year 1900 featured two milestones in organic chemistry: Moses Gomberg announced his discovery of the trityl free radical (see Experiment 25), and Victor Grignard reported the development of Grignard reagents. Just a year later, trityl carbocations were being prepared from triphenylmethanol, which is most easily synthesized using a Grignard reagent.

Grignard's original procedure for preparing a Grignard reagent was as follows: About a mole of magnesium metal was placed in a dry two-necked round-bottom flask fitted with a reflux condenser and addition funnel. A mole of the organic halide was dissolved in diethyl ether, and 50 mL of this solution was added to the magnesium. When a white turbidity appeared at the metal surface and effervescence began, more ether was added in portions, with cooling, followed by drop-by-drop addition of the remainder of the halide/ether solution. The reaction was brought to completion by heating under reflux in a hot water bath.

Although extensive studies of the reaction have led to some modifications of the reaction conditions, essentially the same method is used today for making most Grignard reagents. It is important that the reagents and apparatus be very dry, since water not only reacts with Grignard reagents but also inhibits their formation. In a study using butyl bromide, it was found that the *induction period* (the time between the combination of reactants and the start of a noticeable reaction) for forming the Grignard reagent was $7\frac{1}{2}$ minutes using sodium-dried diethyl ether, 20 minutes using commercial absolute diethyl ether, and 2 hours using diethyl ether half saturated with water. It is apparent that careful drying of the reaction apparatus and reagents saves time in the long run; it should increase the yield of the desired product as well.

The type and quantity of reagents and solvents used are also important. In order for magnesium to react readily, its oxide coating must be removed, so you will scrape a length of magnesium ribbon with a spatula and then cut it into small pieces. Magnesium can be activated further by placing it in a dry reaction flask, adding a small crystal or two of iodine, and using a heat gun to sublime iodine onto the surface of the magnesium; you may use this technique with your instructor's permission. Commercial anhydrous diethyl ether is suitable for most routine preparations, but the optimum quantity of ether depends on the kind of Grignard reagent being prepared. One study showed that the highest yields of phenylmagnesium bromide were obtained with 5 moles of diethyl ether per mole of bromobenzene.

In this experiment, you will prepare phenylmagnesium bromide by adding a solution of bromobenzene in dry diethyl ether to magnesium metal. A Grignard reaction can usually be started by rubbing or crushing the magnesium with a stirring rod while cupping the reaction flask in the palm of the hand to warm the ether. The onset of the reaction is signaled by the formation of small bubbles on the surface of the magnesium accompanied by the appearance of a cloudy precipitate. You can sometimes "jump-start" a balky Grignard reaction by adding a small amount of previously prepared Grignard reagent to the reaction mixture; consult your instructor for help.

Using a high concentration of bromobenzene helps to get the reaction started, but it can promote the formation of an undesirable by-product, biphenyl, through a side reaction on the metal's surface. Thus the bromobenzene solution is diluted by adding diethyl ether as soon as the reaction gets underway. Phenylmagnesium bromide reacts rapidly with water to form benzene and more slowly with oxygen to form a magnesium salt of phenol, so the reaction apparatus must be protected from moisture and the Grignard reagent should be used promptly after it is prepared.

When benzophenone is added to the Grignard reagent, a magnesium salt of triphenylmethanol precipitates from the reaction mixture, which usually turns pink during the addition. This salt is converted to triphenylmethanol by water, and dilute hydrochloric acid is added to dissolve the basic magnesium salts that form along with the triphenylmethanol. The ether solution containing triphenylmethanol is washed to remove impurities and the solvent is evaporated. The crude product is then *triturated* with hexanes or petroleum ether to remove biphenyl and purified by recrystallization.

Trituration of a solid involves rubbing or grinding it, usually in the presence of a solvent.

Carbocations can be prepared by mixing alcohols with a strong acid such as fluoboric acid (tetrafluoroboric acid); the fluoborate anion is a very weak nucleophile that doesn't react with the resulting carbocation. You will prepare trityl fluoborate by the reaction of triphenylmethanol with 48% fluoboric acid. The water in the aqueous fluoboric acid solution, as well as that produced during the carbocation-forming reaction, could prevent or reverse the reaction. Acetic anhydride is added to consume the water by the following reaction:

Preparation of carbocations using fluoboric acid

$$ROH + HBF_4 \rightleftharpoons ROH_2^+ + BF_4^-$$
$$\longrightarrow R^+BF_4^- + H_2O$$

$$\underset{\substack{\text{acetic}\\\text{anhydride}}}{CH_3\overset{O}{\overset{||}{C}}O\overset{O}{\overset{||}{C}}CH_3} + H_2O \longrightarrow \underset{\substack{\text{acetic}\\\text{acid}}}{2CH_3\overset{O}{\overset{||}{C}}OH}$$

Reactions and Properties

A bromobenzene + Mg $\xrightarrow{\text{ether}}$ phenylmagnesium bromide

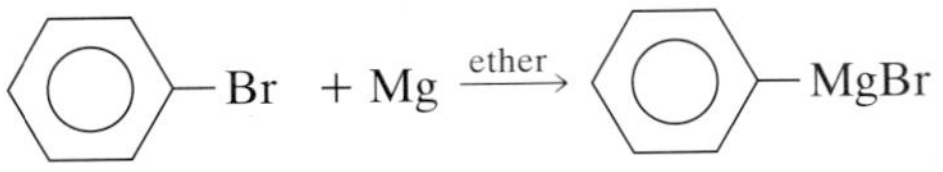

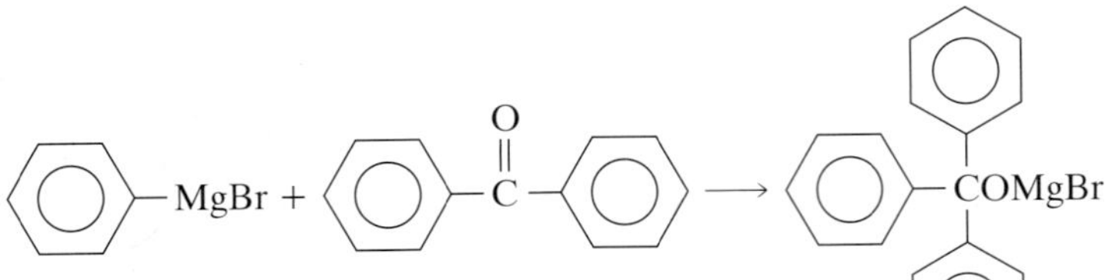

phenyl—MgBr + benzophenone $\longrightarrow$ triphenyl—COMgBr

triphenyl—COMgBr + H_2O $\xrightarrow{H^+}$ triphenyl—COH + Mg(OH)Br

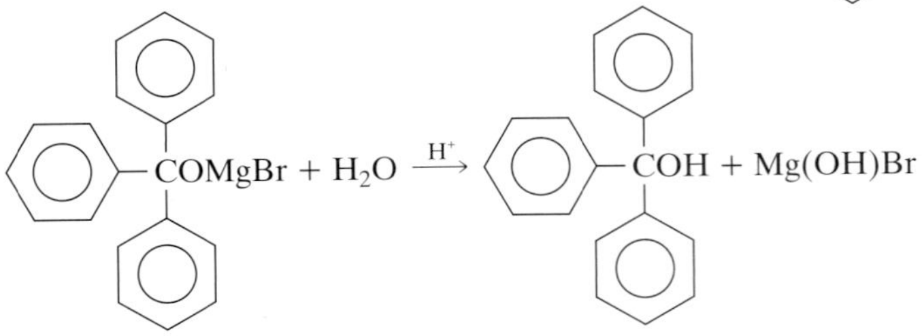

triphenylmethanol

B triphenyl—COH + HBF_4 $\xrightarrow{Ac_2O}$ triphenyl—$C^+BF_4^-$ + H_2O

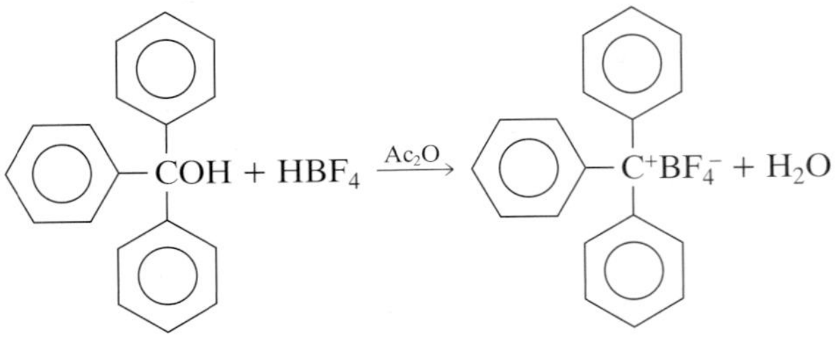

trityl fluoborate

Table 30.1 Physical properties

	M.W.	mp	bp	d
bromobenzene	157.0	−31	156	1.495
magnesium	24.3			
diethyl ether	74.1	−116	34.5	0.714
benzophenone	182.2	48	306	
triphenylmethanol	260.3	164	380	
fluoboric acid (48%)	87.8			1.41
acetic anhydride	102.1	−73	140	1.082
trityl fluoborate	330.1			

Note: mp and bp are in °C, density is in g/mL.

Bromobenzene

3064.8	1068.7	733.9
1578.4	999.5	683.8
1474.3	902.6	457.1

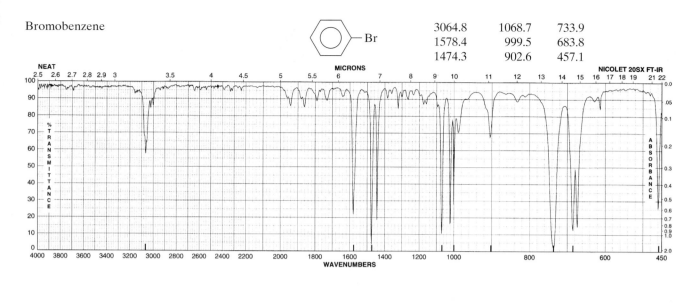

Benzophenone

1659.8	1277.4	698.2
1598.4	941.4	638.8
1447.3	763.4	407.3

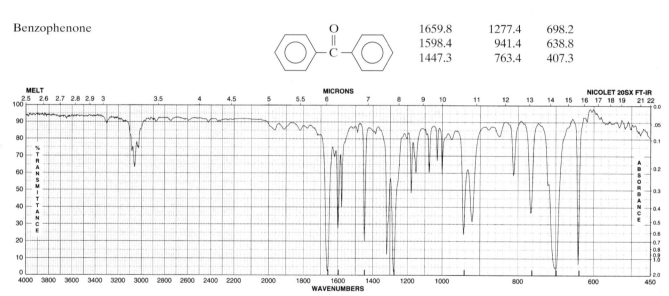

Figure 30.1 IR spectra of the starting materials

Directions

A. *Preparation of Triphenylmethanol*

If possible, clean and predry the glassware needed for the reaction before
the lab period to reduce the drying time.

Safety Notes

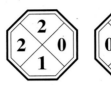

bromobenzene   diethyl ether

magnesium

Bromobenzene causes eye and skin irritation, and inhalation, ingestion, or skin absorption may be harmful. Avoid contact with the liquid and do not breathe its vapors.

Diethyl ether is extremely flammable and may be harmful if inhaled. Do not breathe its vapors and keep it away from flames and hot surfaces.

Magnesium can cause dangerous fires if ignited; keep it away from flames and hot surfaces.

Petroleum ether is extremely flammable and can be harmful if inhaled or absorbed through the skin. Avoid inhalation and prolonged contact, and keep it away from flames and hot surfaces.

Stop and Think: What might happen if the reaction apparatus isn't completely dry?

Take Care! Keep diethyl ether away from flames and hot surfaces.

If the reaction starts but then stops, add some ether as described here.

Reaction of Bromobenzene. *It is essential that all apparatus used during this reaction step be clean and scrupulously dried.* Clean the following items, place them in a labeled beaker, and dry them in a 110°C oven for at least 30 minutes: 10-mL round-bottom flask, Claisen adapter, water-cooled condenser, drying tube containing calcium chloride, 3-mL conical vial, 4-dram screw-cap vial, calibrated Pasteur pipet, 10-mL graduated cylinder, glass stirring rod, flat-bladed microspatula, and forceps. Keep the dried vials capped when not in use so they will stay dry. Also obtain a clean, dry syringe and all caps and cap liners that you will need for the apparatus; do not dry cap liners or plastic syringes in the oven. Unless it is already cut to length, cut a strip of magnesium ribbon to the length suggested by your instructor (usually ~5 cm). While holding it with a forceps, scrape both sides of the ribbon thoroughly with a microspatula to remove the oxide coating and reveal a shiny surface; avoid touching the ribbon with your fingers. Cut it into pieces 1–2 mm in length, weigh about 4.00 mmol of the magnesium pieces, and leave them in your round-bottom flask for the last 5 minutes of drying. As soon as the glassware is cool enough to handle, assemble an apparatus for addition under reflux [OP-11] using a water-cooled condenser and a clean, dry cap liner as a septum. Attach the drying tube [OP-24] to the top of the condenser before you start running water through the condenser.

Store about 6 mL of *anhydrous* diethyl ether in the dry 4-dram vial for later use. Weigh about 4.00 mmol of dry bromobenzene into the 3-mL conical vial and dissolve it in 1.0 mL of anhydrous diethyl ether from your storage vial. Use your syringe to add all of the bromobenzene solution to the reaction flask through the septum. Detach the flask momentarily, hold it in the palm of your hand to warm the ether, and *carefully* (don't punch a hole in the flask!) rub the magnesium pieces with the end of your stirring rod for at least 30 seconds. Then reattach the flask to the reaction apparatus. Observe the reaction mixture closely for evidence of a reaction, such as cloudiness and the evolution of bubbles from the magnesium surface. If the reaction does not begin within five minutes or so, detach the flask and rub the magnesium with your stirring rod as before; if the reaction still does not start, consult your instructor. Transfer 2.0 mL of diethyl ether from your storage vial to the conical vial that contained the bromobenzene. When the reaction mixture begins to boil quite vigorously without external heating, use your syringe to add the ether drop by drop at a rate just sufficient to keep the reaction mixture boiling. When all of the ether has been added, let the reaction continue until boiling has nearly stopped, then use a ~40°C water bath to heat the reaction mixture under gentle reflux [OP-7] for another 10–15 minutes. The reflux ring of condensing ether should be in the lower third of the condenser. If a significant amount of ether evaporates, reducing its volume in the reaction flask, replace it with fresh anhydrous ether. Do not stop at this point because the phenylmagnesium bromide solution will not keep for long.

Reaction of Phenylmagnesium Bromide.   Dissolve 3.80 mmol of benzophenone in 2.0 ml of anhydrous diethyl ether in your dry 3-mL conical vial. When the reaction mixture has cooled so that the ether is no longer boiling, remove the septum cap and drop a dry stir bar into the reaction flask, then replace the cap. Start the stirrer and use your syringe to add [OP-10] the benzophenone solution at a rate sufficient to keep the reaction mixture boiling gently without external heating. When the addition is complete, use 0.5 mL of anhydrous diethyl ether to rinse the vial that contained the benzophenone solution and add it to the reaction mixture. Use a warm water bath to heat the reaction mixture under gentle reflux for another 10 minutes; it may become so thick that the stirrer stops moving, in which case you can stop the stirring motor. (If you stop after the addition and allow the reaction mixture to stand overnight or longer, this heating period can be omitted.)

After the reaction mixture has cooled to room temperature, use your syringe to add 1.0 mL of water drop by drop to the reaction flask and wait for any reaction to subside. Add 2.0 mL of 5% (1.4 *M*) hydrochloric acid while stirring or shaking, then stir until the reaction subsides and most or all of the white solid has dissolved (some magnesium may remain undissolved). If undissolved white solid remains, detach the reaction flask from the apparatus and use a spatula to break up the solid, then shake the capped flask, adding 1–2 mL of solvent-grade (not anhydrous) diethyl ether or 5% HCl (or both) as needed to dissolve all of the solid. Transfer the reaction mixture to a 15-mL centrifuge tube, using a small amount of solvent-grade ether for the transfer; leave the stir bar and any undissolved magnesium behind. Measure the vertical length of the ether column in the centrifuge tube and, as necessary, add enough solvent-grade diethyl ether to make the column about 4 cm deep. Then shake gently to mix the layers thoroughly. (If an emulsion forms at this or any other stage, try using a wooden applicator stick to break it up and, if necessary, centrifuge the mixture for a few minutes.)

Separation.   Remove and discard the aqueous layer and cautiously wash [OP-21] the ether layer with 3 mL of aqueous 5% sodium bicarbonate. Then wash it with 3 mL of saturated aqueous sodium chloride. Dry [OP-22] the ether solution using anhydrous magnesium sulfate or sodium sulfate. Evaporate [OP-16] the ether to leave a solid residue. Add 2.0 mL of hexanes (or high-boiling petroleum ether) to the solid and use a flat-bladed microspatula to triturate (rub and grind) the solid in this solvent for 2–3 minutes. Collect the product by vacuum filtration [OP-13].

Purification and Analysis.   Recrystallize [OP-25] the crude triphenylmethanol from a 2:1 mixture of hexanes (or high-boiling petroleum ether) with absolute ethanol. Triphenylmethanol crystals form slowly, so allow 30 minutes or more for complete crystallization. Dry [OP-23] the purified

Stop and Think: What is the limiting reactant in this synthesis?

Observe and Note: What happens during the addition?

Take Care! A gas may be evolved; vent as necessary.

Waste Disposal: Pour the aqueous layers down the drain. Place the filtrate in a designated solvent recovery container.

Waste Disposal: Place the recrystallization filtrate in a designated solvent recovery container.

triphenylmethanol, weigh it, measure its melting point [OP-30], and turn it in to your instructor. If requested, obtain an infrared spectrum [OP-36] of the product.

B. *Preparation of Trityl Fluoborate*

Safety Notes

acetic anhydride   fluoboric
                         acid

Take Care! Wear gloves, avoid contact with acetic acid and fluoboric acid, and do not breathe their vapors.

Observe and Note: What color is the product?

Waste Disposal: Place the filtrate in a designated waste container.

> **Acetic anhydride can cause severe damage to skin and eyes, its vapors are very harmful if inhaled, and it reacts violently with water. Use gloves and a hood; avoid contact with the liquid, do not breathe its vapors, and keep it away from water.**
> **Fluoboric acid is poisonous, its solutions can cause severe damage to skin and eyes, and its vapors irritate the respiratory system. Use gloves and a hood; avoid contact with the acid solution, and do not breathe its vapors.**

Under the hood, mix 0.100 g of triphenylmethanol with 0.70 mL of acetic anhydride in a small, dry test tube. Carefully add 0.10 mL of 48% fluoboric acid and swirl to dissolve the solid. Stopper the test tube and let the mixture stand for about 15 minutes; then cool it in ice until crystallization is complete. Collect the product by vacuum filtration [OP-13], wash it on the filter with cold anhydrous diethyl ether, and let it air dry. Weigh the trityl fluoborate in a *dry* tared vial.

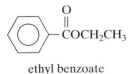

ethyl benzoate

Exercises

1 (a) Write a balanced equation for the coupling reaction of bromobenzene on the metal surface to form biphenyl. (b) Write balanced equations for the reactions of phenylmagnesium bromide and trityl fluoborate with water.

2 If you obtained the IR spectrum of triphenylmethanol, compare it with the spectra in Figure 30.1 and describe the evidence indicating that the expected reaction has taken place. Interpret your spectrum as completely as you can.

3 Describe and explain the possible effect on your results of the following experimental errors or variations. (a) You used solvent-grade (not anhydrous) diethyl ether for the reaction in part **A**. (b) After adding the bromobenzene solution in part **A**, you forgot to add anhydrous diethyl ether to the reaction mixture. (c) You used diethyl ether, rather than petroleum ether, to remove biphenyl from the crude triphenylmethanol.

4 Following the format in Appendix V, construct a flow diagram for the synthesis of triphenylmethanol (part **A**).

5 The reaction of phenylmagnesium bromide with benzophenone to form the salt of triphenylmethanol is an example of nucleophilic addition; its reaction with ethyl benzoate to yield the same product involves nucleophilic substitution followed by a nucleophilic addition step. Write reasonable mechanisms for both reactions.

6 Outline a synthetic pathway for preparing each of the following compounds, using the Grignard reaction and starting with benzene or toluene: (a) 1,1-diphenylethanol; (b) 1,2-diphenylethanol; (c) 2,2-diphenylethanol; (d) 2,3-diphenyl-2-butanol.

7 Excluding alternative Kekulé structures for the benzene rings, (a) draw all possible resonance structures for the trityl cation; (b) draw all possible resonance structures for Malachite Green.

8 One possible by-product from the triphenylmethanol synthesis is ethoxytriphenylmethane. Tell how and when it might form, giving an equation and a mechanism for the reaction.

Other Things You Can Do

(Starred projects require your instructor's permission.)

*1 Record the ultraviolet-visible spectrum (200–600 nm) of trityl fluoborate in dry acetone.

*2 Dissolve a small amount of trityl fluoborate in dry methanol, and record your observations. Dissolve about 0.1 g of trityl fluoborate in 1 mL of dry acetone; then add a solution of sodium iodide in dry acetone (0.1 g NaI in 1 mL acetone) drop by drop until no more changes are observed. Write balanced equations to explain your observations.

*3 Prepare the fluorescent dye fluorescein as described in Minilab 26.

4 Write a research paper about the structures, properties, and applications of Grignard reagents, starting with sources listed in the Bibliography.

Identification of a Conjugated Diene from Eucalyptus Oil

Reactions of Dienes. Preparation of Bicyclic Compounds. Cycloaddition. Infrared Spectrometry. Qualitative Analysis.

Operations

OP-7   Heating
OP-13   Vacuum Filtration
OP-23   Drying Solids
OP-25   Recrystallization
OP-30   Melting Point
OP-34   Gas Chromatography
OP-36   Infrared Spectrometry

Before You Begin

1 Read the experiment, read or review the operations as necessary, and write an experimental plan.
2 Be prepared to carry out the calculations described in part **A** of the Directions.
3 If time permits, record the gas chromatogram of the "eucalyptus oil" during a previous experiment.

maleic anhydride

Scenario

In 1927, Otto Diels and Kurt Alder treated a constituent of one kind of eucalyptus oil with maleic anhydride and obtained a new compound that they described as forming *"grosse glasglänzende Krystalle von ungewöhnlicher Schönheit"* (large lustrous crystals of unusual beauty). The reaction Diels and Alder used for this preparation was eventually named for them, and the lustrous crystals belonged to the Diels-Alder adduct of a natural diene. Gondwana Natural Products, Ltd., which obtains and markets useful products from Australian flora, is investigating the commercial possibilities of an essential oil from *Eucalyptus dives* that has been used to treat colds, as well as malaria and other fevers. The ultraviolet-visible spectrum of the eucalyptus oil suggests that it contains the same diene that Diels and Alder studied—one of the natural dienes described next. They want your Institute to identify the diene from a sample of the eucalyptus oil they have provided. Since only conjugated dienes undergo the Diels-Alder reaction, this reaction can be used to separate the diene from the eucalyptus oil, as well as identify it.

Applying Scientific Methodology

Since the problem involves the identity of an unknown diene, your working hypotheses can only be a guess, but you should be able to eliminate some of the possibilities from the list in the next section before you begin. Your hypothesis will be tested when you obtain the melting point of the adduct. Note that a triene having two conjugated double bonds can also qualify as the "diene" for a Diels-Alder reaction.

Dienes and Trienes in Nature

Dienes and trienes occur in the essential oils of a number of plants and contribute to the flavors and aromas of such plants. For example, limonene has a pleasant lemony odor that enhances the flavor of lemons, oranges, and other citrus fruits, while β-myrcene is responsible for much of the fragrance and flavor of bay leaves (*Myrcia acris*). β-Myrcene is also present in hops, verbena, and lemongrass oil. β-Ocimene was first isolated from the Javanese oil of basil (*Ocimium basilicum*) and is usually found in combination with *allo*-ocimene, which can be synthesized from α-pinene, the most abundant component of oil of turpentine. Both of the phellandrenes derive their name from the water fennel, *Phellandrium aquaticum*, but α-phellandrene apparently doesn't even occur in that plant; it was mistaken for its isomer, β-phellandrene, which does. α-Phellandrene *is* found in the oils of bitter fennel, ginger grass, cinnamon, and star anise, while β-phellandrene also occurs in lemon oil and Japanese peppermint oil. Another cyclic diene, α-terpinene, is obtained from the essential oils of cardamom, marjoram, and coriander.

limonene

$$CH_3C{=}CHCH_2CH_2CCH{=}CH_2$$

with CH_3 and CH_2 substituents

β-myrcene

$$CH_3C{=}CHCH_2CH{=}CCH{=}CH_2$$

with CH_3 and CH_3 substituents

β-ocimene

$$CH_3C{=}CHCH{=}CHC{=}CHCH_3$$

with CH_3 and CH_3 substituents

allo-ocimene

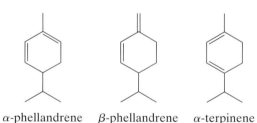

α-phellandrene β-phellandrene α-terpinene

Understanding the Experiment

Most conjugated dienes can form Diels-Alder adducts with maleic anhydride. Trienes such as β-myrcene may also form such adducts if at least two of their double bonds are conjugated. The adducts are usually crystalline solids that can be separated from the other components of an essential oil and used to identify the diene.

The Diels-Alder reaction is classified as a [4 + 2] cycloaddition reaction, because one reactant (the *diene*) contributes four carbon atoms and the other reactant (the *dienophile*) contributes two carbon atoms to the six-membered ring of the resulting cyclic compound (the *adduct*). As illustrated for the following reaction of 1,3-butadiene and ethene, the diene must be able to exist in an *s-cis* conformation, in which the carbon atoms bonding to the dienophile are on the same side of the C—C single bond.

Key Concept: *During a concerted cycloaddition reaction such as the Diels-Alder reaction, the new sigma bonds form simultaneously as electrons flow from the highest occupied pi molecular orbital of one reactant to the lowest unoccupied pi molecular orbital of the other.*

A 4 + 2 cycloaddition reaction

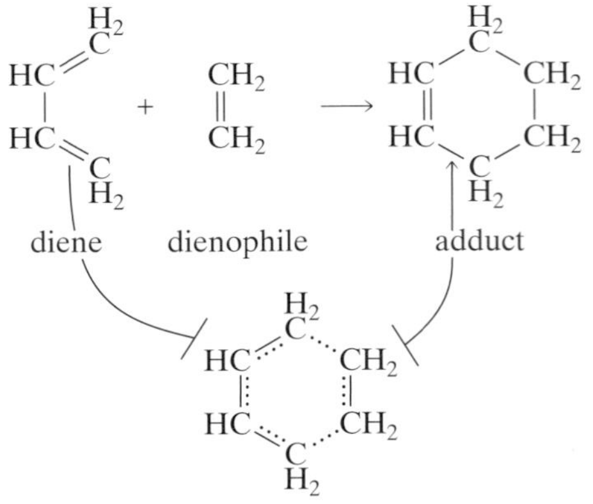

transition state

The dienophile must have either a double or a triple bond, often connected to one or more carbonyl groups or other electron-withdrawing groups.

The Diels-Alder reaction is stereoselective, usually yielding only one of several possible stereoisomers. For example, maleic acid could react with cyclopentadiene to yield either of two adducts, designated *exo* and *endo*. In fact, it yields entirely the *endo* adduct, in which the bulkier parts of the dienophile are closer to the developing carbon-carbon double bond. This orientation results from the fact that overlap between the pi electrons of the diene and those of the dienophile stabilizes the transition state leading to the adduct. Such overlap is possible only when the carbon-carbon double bonds of the diene are in close proximity to the carbonyl groups of the dienophile.

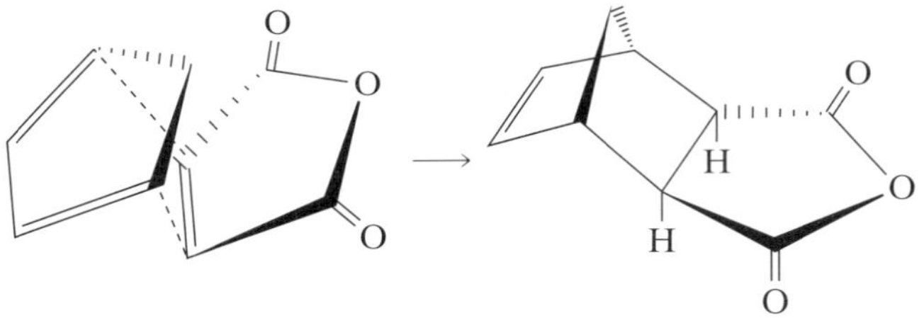

exo adduct

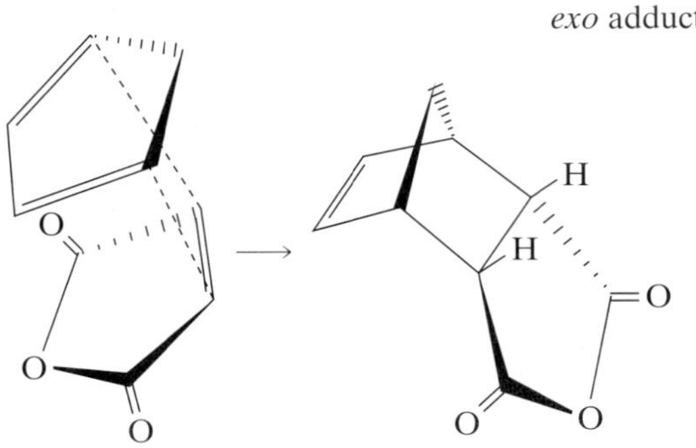

endo adduct

Sometimes more than one *endo* adduct is possible; in that event, the dienophile will tend to approach the diene from its less hindered side to give the more stable adduct.

In this experiment, you will prepare the Diels-Alder adduct of the un-known conjugated diene in "eucalyptus oil," separate the adduct, and iden-tify the diene from the melting point of its adduct. The unknown (whose molecular formula is $C_{10}H_{16}$) will be one of four conjugated dienes that were discussed in the previous section. Their names and the melting points of their adducts are listed in Table 31.2. You will determine the approximate percentage of diene in the eucalyptus oil from its gas chromatogram so that you can estimate the amount of maleic anhydride needed to react with the diene. Powdered maleic anhydride reacts quite rapidly with atmospheric moisture, so it is usually manufactured in the form of briquettes, which must be pulverized before use. If maleic anhydride is provided in powdered form, you should open its container only momentarily and replace the cap immediately after you have removed the amount needed. It is important to avoid using too much maleic anhydride in the reaction, because the excess can be difficult to remove from the product. Since both maleic anhydride and the adduct can be hydrolyzed by water, it is necessary to use dry glass-ware and to keep out moisture during the reaction and workup.

Hydrolysis of maleic anhydride

You will carry out the reaction by heating the reactants under reflux in diethyl ether. The adduct should precipitate from the reaction mixture as beautiful rectangular crystals; slow cooling may yield crystals several cen-timeters long. The adduct is then separated by vacuum filtration and puri-fied by recrystallization from methanol. Because the adduct may react with methanol to form a solvolysis product, you should avoid prolonged boiling during recrystallization. For the same reason, it is not a good idea to leave the crystallized adduct in methanol for more than a few hours.

Once you have identified the adduct, you should be able to deduce its structure with the help of molecular models. You can also characterize the adduct by recording its infrared (IR) spectrum. The IR spectra of anhy-drides show two carbonyl stretching bands that arise from symmetric and asymmetric stretching modes; maleic anhydride itself has $C{=}O$ bands near 1780 and 1850 cm^{-1}, as shown in Figure 31.1. The $C{-}CO{-}O{-}CO{-}C$ grouping can vibrate as a unit, causing additional bands that occur near 900 cm^{-1} and 1250 cm^{-1} for cyclic anhydrides.

Reaction and Properties

The actual structure of the adduct depends on the structure of the unknown diene.

Table 31.1 Physical properties

	M.W.	mp	bp	d
maleic anhydride	98.1	53	202	
diethyl ether	74.1	−116	34.5	0.714

Note: mp and bp are in °C, density is in g/mL.

Table 31.2 Melting points of maleic anhydride adducts of the possible dienes

Diene	mp of adduct
β-myrcene	33–34
allo-ocimene	83–84
α-phellandrene	126–127
α-terpinene	60–61

Maleic anhydride

3118.7	1240.1	840.6
1851.1	1058.1	695.9
1778.5	890.8	561.7

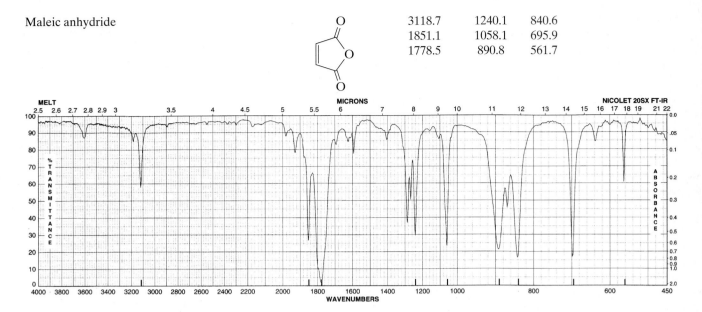

Figure 31.1 IR spectrum of maleic anhydride

Directions

A. *Preliminary Analysis and Calculations*

Obtain a gas chromatogram [OP-34] of the eucalyptus oil provided, using the column and conditions recommended by your instructor. Assuming that the unknown diene is responsible for the largest peak on the chromatogram and that peak areas are proportional to component masses, estimate the mass of the diene (molecular formula $C_{10}H_{16}$) in 1.00 g of the oil. Then calculate the mass of maleic anhydride needed to react with that much diene and the theoretical yield of the reaction.

B. *Preparation of the Adduct*

Safety Notes

> **Maleic anhydride is corrosive and can cause severe damage to the eyes, skin, and upper respiratory tract. Avoid contact with skin, eyes, or clothing, and do not breathe the dust. If you must pulverize maleic anhydride briquettes, wear gloves and work under the hood.**
>
> **Diethyl ether and petroleum ether are extremely flammable and may be harmful if inhaled. Do not breathe their vapors and keep them away from ignition sources.**
>
> **Methanol is harmful if inhaled or absorbed through the skin. Avoid contact and do not breathe its vapors.**

maleic anhydride   diethyl ether   petroleum ether   methanol

Reaction.   *All glassware must be dry.* In a clean, dry, 5-mL conical vial dissolve 1.00 g of the eucalyptus oil in 2.0 mL of anhydrous diethyl ether. Add the calculated amount of powdered maleic anhydride. Using a water-cooled condenser, heat the reaction mixture under gentle reflux [OP-7] on a hot water bath for 45 minutes. When the reaction period is over, remove the condenser and let the reaction mixture cool slowly to room temperature. If no crystals form by the time it reaches room temperature, dip the tip of a glass stirring rod into the reaction mixture, remove it long enough for the ether to evaporate, then reinsert it in the reaction mixture to induce crystallization. If that doesn't work, rub the rod tip against the inside walls of the vial. When crystallization seems nearly complete, cool the vial in an ice/water bath for a few minutes.

Take Care!   Keep diethyl ether away from flames and hot surfaces. Avoid contact with maleic anhydride and do not breathe its dust.

Separation.   Collect the adduct by vacuum filtration [OP-13], washing the crystals on the filter with 1.5 mL of cold, low-boiling petroleum ether.

Purification and Analysis.   Recrystallize [OP-25] the adduct from dry methanol, avoiding prolonged boiling. Dry [OP-23] the adduct and measure its mass and melting point [OP-30]. Deduce the identity of the eucalyptus oil diene from the melting point of its adduct. Record the IR spectrum [OP-36] of the adduct or obtain a spectrum from your instructor.

Take Care!   Keep petroleum ether away from flames and hot surfaces.

Waste Disposal: Place all filtrates in designated solvent recovery containers.

C. *Stereochemistry of the Adduct*
Construct molecular models for maleic anhydride and the diene. By moving their bonds around, find a way to connect them to make a model representing one form of the adduct. Disconnect and reconnect the diene and dienophile units until you have made models representing all possible structures for the adduct, and sketch stereochemical drawings of them for your report. Based on the description of the Diels-Alder reaction in "Understanding the Experiment," decide which is the most likely structure.

Exercises

1   If you obtained an IR spectrum of the adduct, interpret it as completely as you can. Compare the adduct's spectrum with that of maleic anhydride; point out and explain any significant similarities or differences.

2   Write a balanced equation for the reaction. Show the stereochemistry of the adduct and explain why it has that stereochemistry.

3   Which two dienes whose structures are shown in this experiment will not form a Diels-Alder adduct with maleic anhydride? Explain why in each case.

4   Describe and explain the possible effect on your results of the following experimental errors or variations. (a) You calculated the mass of maleic anhydride needed based on the total mass of the eucalyptus oil. (b) Your reaction vial was wet. (c) You dissolved the adduct in hot methanol and then stored the recrystallization solution until the next lab period.

5   Write the structure of the compound that would result if the adduct were heated too long in the recrystallization solvent. Write a balanced equation for its formation.

Dimerization of 1,3-butadiene

6 Following the format in Appendix V, construct a flow diagram for this experiment.

7 The side reaction most often encountered in Diels-Alder syntheses is dimerization or polymerization, in which the diene also acts as a dienophile. For example, butadiene can react with itself to yield 4-vinyl-cyclohexene as shown. Draw the structures of four possible Diels-Alder dimers of your diene.

Other Things You Can Do

(Starred projects require your instructor's permission.)

*1 Prepare the Diels-Alder adduct of an aromatic "diene" as described in Minilab 27.

2 A number of polychlorinated insecticides, such as dieldrin, aldrin, and chlordane, are synthesized using one or more Diels-Alder reactions. Starting with sources listed in the Bibliography, write a research paper about such insecticides, giving equations for their manufacture from cyclopentadiene and reporting on their uses and environmental effects.

Spectral Identification of Monoterpenoids

Infrared Spectrometry. Ultraviolet-Visible Spectrometry. Qualitative Analysis.

Operations

OP-36   Infrared Spectrometry
OP-37   Nuclear Magnetic Resonance Spectrometry (optional)
OP-38   Ultraviolet-Visible Spectrometry

Before You Begin

1   Read the experiment, read or review the operations as necessary, and write an experimental plan.

Scenario

Uncommon Scents, Inc., extracts essential oils from various plants and ships them to buyers around the world. Some of their buyers are aromatherapists from countries that classify aromatherapy oils as drugs and strictly regulate their contents. These countries require that the suppliers of essential oils report their major ingredients and list them on the label. Uncommon Scents has air freighted your Institute samples of the major components of one of their products, the essential oil of the motley marigold, *Calendula salmagundi*. Your project group's assignment is to identify the components using appropriate spectral methods and to determine what names should appear on the label.

Applying Scientific Methodology

This experiment can be performed as a group project in which each individual obtains the spectra for one of the components and the group members, working together, arrive at the identities of all of the components. You should be able to formulate a working hypothesis regarding the identity of your component after you record its spectra, but be prepared to have your hypothesis tested and either accepted or rejected by the other members of your group.

Monoterpenoids from Plants

Terpenes are hydrocarbons that can, in principle, be broken down into two or more isopentane units (also called isoprene units). Monoterpenes are composed of two such units, sesquiterpenes of three, diterpenes of four, etc. Analogous compounds that contain oxygen or other atoms in addition to carbon and hydrogen are called *terpenoids*.

See Experiment 12 for more information about terpenes and terpenoids.

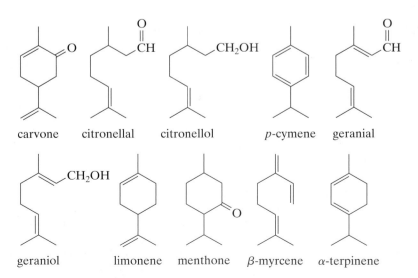

limonene (a monoterpene)

farnesol (a sesquiterpenoid)

Vitamin A (a diterpenoid)

Isopentane units in terpenes and terpenoids

Many terpenes and terpenoids are found in the essential oils of plants. Essential oils are usually obtained by steam-distilling the roots, bark, leaves, flowers, or other parts of plants. Their components are volatile, water-insoluble substances that often have pronounced aromas.

Structures of some monoterpenes and monoterpenoids obtained from essential oils are shown in Figure 32.1. Many of these compounds occur in a wide variety of plants. For example, limonene occurs in the oils of berg-amot, black pepper, cardamom, caraway, coriander, cypress, dill, eucalyp-tus, grapefruit, lemon, lime, neroli, and orange, among others. Carvone, which is obtained from caraway, coriander, dill, and peppermint oils, is used to flavor liqueurs and perfume soaps. Citronellal, the active ingredi-ent of insect-repelling citronella candles, occurs in lemon and lemongrass oils, as well as oil of citronella. Citronellol is also a constituent of citronella oil, but it is found in rose and geranium oils as well. The aromatic hydro-carbon *p*-cymene is a component of marjoram and oregano oils. Geranial and its geometric isomer neral often occur together in essential oils. The isomer mixture, called citral, is the major component of lemongrass oil and is also found in oils of lemon, orange, and verbena. Geraniol occurs in the

carvone   citronellal   citronellol   *p*-cymene   geranial

geraniol   limonene   menthone   β-myrcene   α-terpinene

Figure 32.1 Structures of monoterpenes and monoterpenoids from essential oils

oils of citronella, lemongrass, and roses, and is used widely in perfumery. Menthone, like the corresponding alcohol, menthol, is a constituent of peppermint oil, and it also occurs in the oils of pennyroyal and geranium. β-Myrcene is found in the oils of bay, juniper, and hops, and is an important intermediate in the manufacture of perfumes. α-Terpinene is a constituent of the oils from cardamom and marjoram.

You might think that the aromas of plant constituents should resemble the aromas of the plants in which they occur. In a few cases, this is so. For example, cinnamaldehyde—the major component of cinnamon bark—has an odor much like that of the spice. But most plants contain many different volatile components, and the odor of a single component may bear little or no resemblance to the overall odor of the plant or its essential oil. That aroma is due to the combined effect on your olfactory receptors of many different kinds of molecules. The aroma of a given compound may also depend on its stereochemistry. Thus one carvone enantiomer has an odor of spearmint while the other smells like caraway seeds.

cinnamaldehyde

Understanding the Experiment

In this experiment, you will use infrared (IR) and ultraviolet-visible (UV-VIS) spectrometry to identify the unknown compounds, which will be chosen either from those in Figure 32.1 or from a list provided by your instructor. With your instructor's permission, you may use ^{1}H NMR spectrometry as well.

The spectral interpretation section of OP-36 will help you interpret the IR spectra of the compounds assigned. The four functional groups found in the compounds of Figure 32.1, as well as the aromatic ring of *p*-cymene, are relatively easy to identify. Both aldehydes and ketones give rise to a strong carbonyl (C=O) band near 1700 cm^{-1}, and aliphatic aldehydes also show medium-intensity C—H stretching bands near 2720 and 2840 cm^{-1}. Primary alcohols are characterized mainly by their O—H and C—O stretching bands around 3300 and 1050 cm^{-1}. The carbon-carbon double bonds of alkenes usually give rise to =C—H stretching bands just above 3000 cm^{-1}, moderate to weak C=C stretching bands in the 1670–1640 cm^{-1} region (sometimes these are quite weak), and =C—H bending bands that are sensitive to substitution patterns in the 1000–650 cm^{-1} region. Aromatic rings are characterized by Ar—H stretching bands just above 3000 cm^{-1} and Ar—H bending bands in the 900–690 cm^{-1} region. The Ar—H bending patterns vary with the number and location of substituents on the benzene ring. Other structural features may affect the positions of certain IR absorption bands. For example, carbonyl groups that are conjugated with carbon-carbon double bonds or aromatic rings appear at lower wave numbers than usual. Thus while most saturated ketones have C=O bands near 1715 cm^{-1}, α,β-unsaturated ketones have C=O bands closer to 1670 cm^{-1}. Conjugation with a carbonyl group also affects both the position and intensity of the C=C stretching band; its intensity increases while its wave number is lowered by about 30 cm^{-1}. Unsymmetrical conjugated dienes may have two C=C stretching bands near 1650 cm^{-1} and 1600 cm^{-1}.

UV-VIS spectra of organic compounds generally contain only a few relatively broad absorption bands. Compounds that appear colored to the human eye absorb radiation in the visible range ($\sim$400–800 nm) and sometimes in the ultraviolet range (~200–400 nm) as well. Such compounds have extensive chromophores, usually with many conjugated double bonds or aromatic rings. Colorless compounds with less extensive conjugated systems absorb ultraviolet radiation above 200 nm. Compounds with unconjugated double bonds and no nonbonded electrons do not show significant absorption above 200 nm.

When a compound absorbs radiation in the UV-VIS region of the electromagnetic spectrum, its electrons undergo transitions from lower to higher energy levels. In a $\pi-\pi^*$ transition, pi (π) electrons in their ground-state energy levels (π molecular orbitals) jump to unoccupied higher energy levels (π^* molecular orbitals). The energy of the transition, and thus the wavelength of the resulting absorption band, depends on a number of factors, including the length of the conjugated system and the presence of certain substituents and structural features. Often, it is possible to estimate the λ_{max} value (wavelength of maximum absorption) of a conjugated compound's $\pi-\pi^*$ absorption band using a set of rules developed by chemistry Nobel laureate Robert B. Woodward and modified by Louis Fieser. Woodward-Fieser rules for conjugated dienes are given in Table 32.1.

Key Concept: *The wavelength of the UV-VIS radiation absorbed by a conjugated substance increases with the length of its conjugated system and the presence of substituents on the conjugated system.*

Table 32.1 Woodward-Fieser rules for $C{=}C{-}C{=}C$ systems

Structural feature	Wavelength or increment
base value for conjugated diene	214 nm
homoannular diene	+39 nm
double bond extending conjugation	+30 nm
alkyl substituent	+5 nm
exocyclic double bond	+5 nm

Note: Wavelengths are for ethanol solutions.

base value	214 nm	base value	214 nm
homoannular diene	+39 nm	exocyclic double bond	+5 nm
3 alkyl groups	+15 nm	2 alkyl groups	+10 nm
total	268 nm	total	229 nm

Examples illustrating Woodward–Fieser rules for dienes

An alkyl substituent can be either an open-chain group, such as methyl, or a ring residue—a carbon-containing group that is part of a ring. To be counted, the alkyl group must be attached directly to one of the carbons of the conjugated system. A homoannular diene is one in which both double bonds of the diene are in the same ring. An exocyclic double bond is one attached to a ring carbon from outside the ring. To apply the rules, start with the base value for the conjugated system and add wavelength increments for each of the designated structural features, as shown by the examples following Table 32.1.

Rules for α,β-unsaturated carbonyl compounds are given in Table 32.2, which is followed by examples. An α-alkyl group is on the carbon atom adjacent to the carbonyl group, and a β-alkyl group is on the second carbon from the carbonyl group.

Table 32.2 Woodward-Fieser rules for $C{=}C{-}C{=}O$ systems

Structural feature	Wavelength or increment
base value for conjugated ketone	215 nm
base value for conjugated aldehyde	210 nm
homoannular double bond extending conjugation	+69 nm
α-alkyl group	+10 nm
β-alkyl group	+12 nm
exocyclic double bond	+5 nm

Note: Wavelengths are for ethanol solutions.

base value, aldehyde	210 nm		base value, ketone	215 nm
α-alkyl group	+10 nm		exocyclic double bond	+5 nm
2 β-alkyl groups	+24 nm		α-alkyl group	+10 nm
total	244 nm		β-alkyl group	+12 nm
			total	242 nm

Examples illustrating Woodward-Fieser rules for α,β-unsaturated carbonyl compounds

Similar rules can be used to estimate the λ_{max} value for the strongest band of an aromatic compound. The benzene ring has a base value of 203.5 nm and adding a methyl substituent increases its λ_{max} by about 3 nm. An additional alkyl group increases λ_{max} by about 3 nm if it is *ortho* or *meta* to an existing substituent and by 10 nm if it is *para* to an existing substituent.

Directions

Unless your instructor indicates otherwise, you should work in groups. Each group can be provided with a selection of unknowns to be apportioned among its members, or the instructor can assign individual unknowns. Record the identification number of your unknown in your lab notebook as soon as you receive it. If necessary, the instructor will show you how to operate the instruments.

Citral, a mixture of geranial and its (Z)-isomer neral, may be substituted for geranial.

Safety Notes

> The monoterpenes and monoterpenoids may irritate the eyes and skin, and some are quite flammable. Minimize contact and keep them away from ignition sources.
> Deuterochloroform is harmful if inhaled or absorbed through the skin, and it is a suspected human carcinogen. Avoid contact with the liquid and do not breathe its vapors.

Infrared Spectrum.   Obtain an infrared (preferably FTIR) spectrum [OP-36] of your unknown compound as the neat liquid. Record accurate wave numbers for all significant absorption bands in the spectrum.

Ultraviolet Spectrum.   Using a microliter syringe, measure 1 μL of the unknown monoterpenoid and dissolve it in 25 mL of 95% ethanol in an Erlenmeyer flask. Using a 1-cm quartz or silica sample cell, 95% ethanol as the reference solvent, and a scanning UV-VIS spectrophotometer [OP-38], scan the ultraviolet spectrum of the solution over the 200–400 nm range. By adjusting the instrument's absorbance range or by diluting the sample with more 95% ethanol, obtain a spectrum in which the top of the highest absorption band is between the midpoint and top of the absorbance scale. Record the λ_{max} value for the strongest band and for any other significant bands.

Waste Disposal: Turn in your unknown when you are finished with it.

Take Care! Avoid contact with $CDCl_3$ and do not breathe its vapors.

¹H NMR Spectrum (optional).   Obtain a proton NMR spectrum [OP-37] of your unknown compound in deuterochloroform.

Identification of the Unknowns.   From its IR spectrum, decide what functional groups are present in your compound and deduce any other structural information you can, such as the existence of conjugation in a carbonyl compound. Based on this information decide which of the compounds in Figure 32.1 have structures that are consistent with your IR spectrum. From its UV spectrum, use the Woodward-Fieser rules to help you decide which structure is most consistent with both spectra. Use your ¹H NMR spectrum, if you obtained one, to confirm the structure. Share your spectrum with the other members of your group and examine their spectra until you come to a group consensus regarding the identities of all of the monoterpenes and monoterpenoids. Interpret your spectra as completely as you can and turn them in with your report.

Exercises

1 Describe and explain the possible effect on your results of the following experimental errors or variations. (a) You used a solution cell with a 0.1-mm spacer for the infrared spectrum. (b) You dissolved your unknown in petroleum ether for the UV analysis. (c) There was no deuterochloroform available for the NMR sample so you used ordinary chloroform instead.

2 Tell how infrared spectrometry can be used to differentiate the following compounds. In each case, indicate the significant infrared bands that will be observed and their approximate wave numbers. (a) 1-butanol,

2-butanol, and 2-methyl-2-butanol. (b) *ortho*-xylene, *meta*-xylene, and *para*-xylene (the xylenes are dimethylbenzenes). (c) butanal, 1-butanol, 2-butanone, butanoic acid, and butyl acetate.

3 Tell how UV-VIS spectrometry can be used to differentiate the compounds in each of the following groups. Calculate approximate λ_{max} values for the compounds when you can. (a) 1,4-hexadiene, 2,4-hexadiene, and 1,3-cyclohexadiene. (b) cyclohexanone, 2-cyclohexenone, and 2,4-cyclohexadienone. (c) the following compounds.

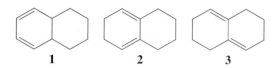

1   **2**   **3**

4 Classify each of the following (as a monoterpene, sesquiterpenoid, etc.) and show how each can be divided into isopentane units.

linalool   caryophyllene   β-selinene

5 Tell how you could use a spectrometric method to distinguish geranial from neral.

neral

Other Things You Can Do

(Starred projects require your instructor's permission.)

*1 Identify an unknown arene by NMR spectrometry as described in Minilab 28.
*2 Obtain and interpret the mass spectrum of a compound as described in Minilab 29.
3 Starting with sources listed in the Bibliography, write a research paper about cholesterol and its biological functions, including a discussion of the biosynthesis of cholesterol from the triterpene squalene.

Synthesis and Spectral Analysis of Aspirin

EXPERIMENT 33

Reactions of Phenols. Preparation of Esters. Nucleophilic Acyl Substitution. Infrared Spectrometry. NMR Spectrometry.

Operations

OP-7   Heating
OP-10   Mixing
OP-13   Vacuum Filtration
OP-23   Drying Solids
OP-25   Recrystallization
OP-36   Infrared Spectrometry
OP-37   Nuclear Magnetic Resonance Spectrometry

Before You Begin

1 Read the experiment, read or review the operations as necessary, and write an experimental plan.
2 Calculate the mass of 3.00 mmol of salicylic acid and the theoretical yield of aspirin.

Scenario

The Spectrum Publishing Co., a subsidiary of The Fulcourt Press, wants to produce a multimedia CD-ROM illustrating the principles of spectral interpretation for use with their new instrumental analysis textbook. The CD-ROM would, for example, use an animated molecular model of aspirin to simulate bond vibrations and their relationship to infrared (IR) spectral bands, and to illustrate nuclear magnetic transitions and how they give rise to ^1H and ^{13}C nuclear magnetic resonance (NMR) signals. To make the CD-ROM as accurate as possible, they have asked your Institute to perform an in-depth analysis of the IR and NMR spectra of aspirin. You can't use aspirin tablets because they contain impurities such as starch, which is used as a binder. Your assignment, therefore, is to prepare some pure aspirin, record its IR and NMR spectra, and propose assignments for its significant NMR signals and IR bands.

Applying Scientific Methodology

The initial scientific problem in this experiment is whether or not pure aspirin (free from unreacted starting material) can be prepared as described in the Directions. You will use a simple chemical method to test your initial hypothesis. As you study the spectra, you can develop hypotheses about the origins of various IR bands and NMR signals, then test your hypotheses by consulting sources of spectral data and studying the spectra of related compounds.

The Aspirin Saga

The more we learn about aspirin, the more it appears to be a true wonder drug. The family doctor who advises you to "take two aspirin and call me in the morning" knows that aspirin lowers fever, reduces inflammation, and relieves pain. In recent years, aspirin has been shown to reduce the incidence of heart disease, strokes, and certain cancers as well. It may also improve brain function in people who have suffered small strokes, help prevent cataracts, and reduce the occurrence of gallstones. But when it was first prepared about a century and a half ago, aspirin was considered so unremarkable that it was set aside and temporarily forgotten.

The aspirin saga begins with some relatives on its "family tree," a group of related compounds known as salicylates. Native Americans have known for centuries that willow bark can relieve pain and fight fever. The bark of the white willow and related species contains salicin, a natural salicylate chemically related to aspirin. Salicin can be broken down in the presence of water and an oxidant to form molecules of glucose (a simple sugar) and a sweet-smelling liquid named salicylaldehyde. German chemists discovered that salicylaldehyde obtained from another source, the meadowsweet plant (*Spiraea* species), reacts with strong alkali to yield a white solid on neutralization. The new compound was named *spirsäure* (*Spir*aea + *säure*, the German word for acid) by its discoverers, but was called salicylic acid by the English, who traced its lineage back to the white willow tree.

Salicylic acid can be synthesized from phenol and carbon dioxide using a method developed by the brilliant but irascible German chemist Hermann Kolbe. Phenol, also known as carbolic acid, has long been used as a surgical antiseptic. It is much too caustic to be taken internally, causing painful burns in the mouth and upper digestive tract. But Kolbe had an idea. What if the formation of salicylic acid from phenol and CO_2 were reversed inside the human body? Then a patient could swallow salicylic acid, which would break down inside the body to yield phenol, which would then (Kolbe hoped) kill the germs responsible for the patient's illness. He carried out tests that "proved" to his satisfaction that salicylic acid was indeed an effective germ killer, and soon recommended its use on patients suffering from a variety of bacterial diseases and infections. The first reports seemed promising—patients were still dying after salicylic acid treatments, but they felt much better while doing so! Before long, doctors began to suspect that the salicylic acid "cured" only those patients who would have survived without any medication.

Kolbe's idea was wrong; salicylic acid does not produce phenol in the body, and his tests were later found to be invalid. But like many scientific hypotheses that fail, this one led to important new discoveries. Although salicylic acid does not cure bacterial illnesses, it reduces fever and is a better pain reliever than salicin, so it was soon being prescribed for rheumatism, sciatica, headaches, and other painful conditions. But salicylic acid has a serious side effect. It irritates mucous membranes that line the mouth, esophagus, and stomach. One patient who suffered intensely from this side effect

$OC_6H_{11}O_5$
CH_2OH

salicin

CHO

salicylaldehyde

COOH

salicylic acid

$OCOCH_3$ other salicylates
COOH

aspirin

The aspirin family tree

Kolbe achieved scientific infamy with a vitriolic attack on J. H. van't Hoff, who helped develop the stereochemical theory of organic chemistry.

OH

phenol

was the father of Felix Hoffman, a chemist working for the Bayer division of a German pharmaceutical company. By a happy coincidence, Bayer was interested in finding a substitute for salicylic acid at about the same time Hoffman's father was experiencing its side effects, and this convergence of Hoffmann's research and personal interests provided the motivation that led to aspirin's rediscovery in 1893.

Hoffman knew that phenolic compounds were corrosive because of the presence of a free hydroxyl (OH) group on the benzene ring, and he reasoned that "masking" the OH with some easily removed substituent might provide the benefits of salicylic acid without the irritation. While studying some of the known derivatives of salicylic acid, Hoffman came across one in which the hydrogen of the OH group was replaced by an acetyl group. This was acetylsalicylic acid, which had been prepared some 40 years earlier by Charles Gerhardt. Tests showed that acetylsalicylic acid was superior to all known painkillers, in both its effectiveness against pain and fever and its freedom from serious side effects. Thus rediscovered, acetylsalicylic acid was introduced by Bayer in 1899 under the trade name "aspirin," and it soon became the world's most popular drug.

Understanding the Experiment

In this experiment, you will use acetic anhydride to convert salicylic acid to aspirin, and also to serve as a solvent for the reaction. Since acetic anhydride is very reactive, it will not be necessary to heat the reactants under reflux; warming them in a water bath is sufficient. When the reaction is complete, water is added to destroy the excess acetic anhydride, converting it to water-soluble acetic acid. Aspirin can be purified by recrystallization from an ethanol-water mixture. One possible impurity in the aspirin is salicylic acid itself. You will test both your crude and purified aspirin for the presence of salicylic acid with ferric chloride, which forms highly colored complexes with phenolic compounds. From the results, you may be able to tell whether the impurity (if there is any) resulted from incomplete reaction of the starting materials or was formed by hydrolysis of aspirin during the workup of the product.

Aspirin is both an ester and a carboxylic acid, so its IR spectrum shows characteristics of both kinds of compounds. The O—H stretching vibration of a free carboxylic acid molecule leads to a sharp band near 3520 cm^{-1}, but in the liquid and solid states the molecules are held together by strong hydrogen bonding, causing the O—H band to shift to lower frequency and spread out over much of the region between 3300 cm^{-1} and 2500 cm^{-1}. The IR spectrum of a typical carboxylic acid also features a C=O stretching band around 1725 cm^{-1}, an acyl C—O stretching band in the 1315–1280 cm^{-1} region, and an out-of-plane O—H bending band near 920 cm^{-1}. (The "C—O" bond vibrations for carboxylic acids and esters are actually coupled vibrations involving some adjacent atoms.) Conjugation with a benzene ring or a C=C bond moves a carbonyl band to a lower frequency. For example, the C=O band of benzoic acid occurs at 1688 cm^{-1}. The carbonyl stretching band of an ester usually occurs at higher frequency than that of a carboxylic acid, and the acyl C—O stretching band of an ester ranges from about 1240 cm^{-1} to 1140 cm^{-1}. The aromatic ring of an aryl ester increases the

aspirin

See Experiment 15 for more information about the properties of aspirin.

acetic anhydride

You can also recrystallize aspirin from ethyl acetate.

Aspirin decomposes at high temperatures, so its melting point is not a reliable indicator of its purity.

Stretching vibrations of carboxylic acids and esters

carboxylic acid

carboxylic ester

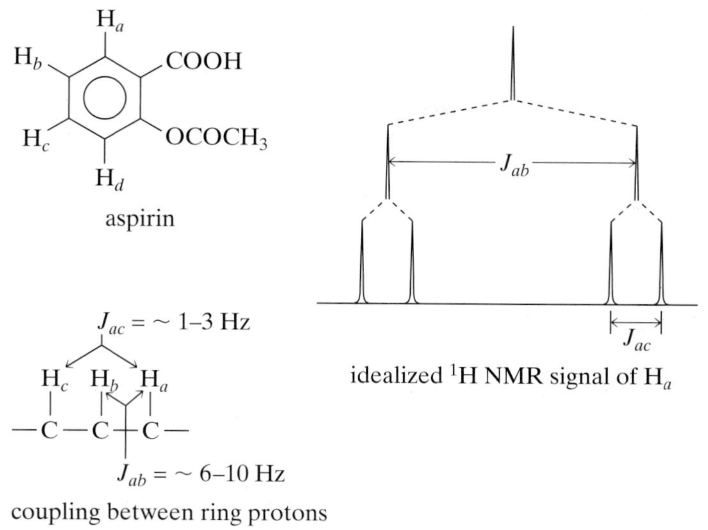

Figure 33.1 The effect of long-range coupling on an aromatic proton signal

C=O frequency and decreases the acyl C—O frequency of the ester function. For example, these bands occur at 1765 cm^{-1} and 1193 cm^{-1}, respectively, for phenyl acetate, compared to 1742 cm^{-1} and 1241 cm^{-1} for ethyl acetate.

The ^{1}H NMR spectrum of aspirin shows a complex pattern of aromatic proton signals characteristic of *ortho* substitution by groups of differing electronegativity. The —COOH substituent withdraws electrons from the ring, deshielding nearby ring protons, and the —OCOCH$_3$ substituent donates electrons by resonance, shielding the ring protons. This effect is particularly noticeable for proton H$_a$ in Figure 33.1, whose signal occurs well downfield of the rest because of its proximity to the —COOH group. Although H$_a$ has only one "nearest-neighbor" proton, its ^{1}H NMR signal has four peaks, because the delocalized pi cloud of the ring allows long-range coupling between nonadjacent protons. Thus the signal due to H$_a$ is split into a doublet by proton H$_b$, and each peak of that doublet is split into another closely spaced doublet by proton H$_c$, as illustrated in Figure 33.1. The long-range coupling constant for protons *meta* to one another is only about 1–3 Hz, compared to typical coupling constants for *ortho* protons of 6–10 Hz.

If you obtain a ^{13}C NMR spectrum of your aspirin, refer to OP-37b and your lecture textbook or another source to help you interpret it.

Key Concept: Anything that weakens a C=O bond, such as conjugation, decreases its vibrational frequency. Anything that strengthens the bond, such an electron-withdrawing group, increases its vibrational frequency (see Experiment 13).

Reactions and Properties

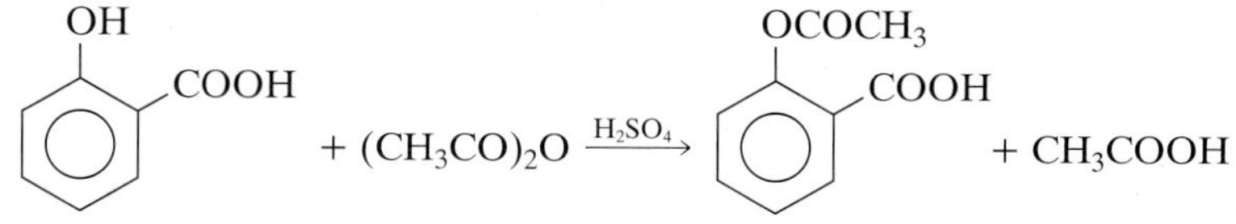

Table 33.1 Physical properties

	M.W.	mp	bp	d
salicylic acid	138.1	159		
acetic anhydride	102.1	−73	140	1.082
aspirin	180.2	135d		

Note: mp and bp are in °C, density is in g/mL.

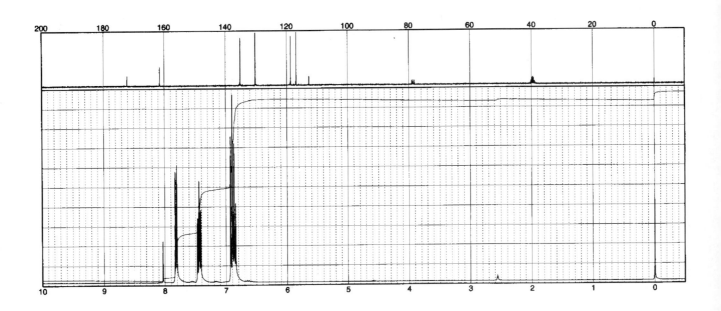

Salicylic acid

3237.9	1483.6	759.9
3004.3	1249.3	698.6
1659.8	1156.7	660.5

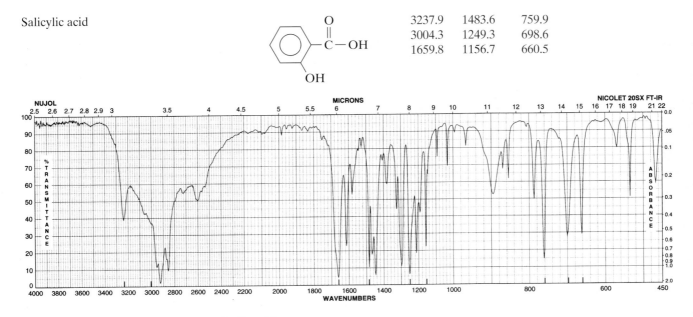

Figure 33.2 NMR and IR spectra of salicylic acid

Directions

Acetic anhydride can cause severe damage to skin and eyes, its vapors are very harmful if inhaled, and it reacts violently with water. Use gloves and a hood; avoid contact with the liquid, do not breathe its vapors, and keep it away from water.
Sulfuric acid causes chemical burns that can seriously damage skin and eyes. Wear gloves and avoid contact.
Deuterochloroform is harmful if inhaled or absorbed through the skin, and it is a suspected human carcinogen. Avoid contact with the liquid and do not breathe its vapors.

Safety Notes

acetic anhydride

sulfuric acid

Reaction.   *Under the hood*, add 1.0 mL of acetic anhydride to 3.00 mmol of salicylic acid in a dry 10-mL Erlenmeyer flask. Add 1 drop of concentrated sulfuric acid and a stir bar and start the stirrer [OP-10]. Heat [OP-7] the mixture in a 45–50°C water bath while stirring until the salicylic acid dissolves, then for about 5 minutes more.

Take Care! Wear gloves, avoid contact with acetic anhydride and sulfuric acid, and do not breathe their vapors.

Let the flask stand at room temperature until crystallization begins. If no product has precipitated when the solution is near room temperature, induce crystallization by scratching the wall of the flask at the surface of the solution with a glass stirring rod, or by adding a few seed crystals of pure aspirin. When a heavy precipitate has formed, stir in 6 mL of cold water and break up any lumps with a microspatula. Cool the mixture in an ice/water bath until crystallization is complete.

Separation.   Separate the aspirin from the reaction mixture by vacuum filtration [OP-13] and wash it with several 0.5 mL portions of ice-cold water, using a small amount of the wash water for transfer. Save about 10 mg of the crude product in a clean, labeled test tube for analysis.

Waste Disposal: Pour the filtrate down the drain.

Purification.   Recrystallize [OP-25] the aspirin from an ethanol-water mixture, using the following procedure to reduce the likelihood of hydrolysis. Dissolve the aspirin in the minimum volume of boiling 95% ethanol, and add another 6–8 drops of ethanol, estimating the total volume of ethanol used. Add twice that volume of warm (~60°C) water to the solution while it is still at the boiling point, and swirl to mix. If any precipitate forms, heat the solution gently until it is clear, but do not boil it. Let it cool slowly to room temperature, induce crystallization if necessary, and cool it in an ice/water bath until crystallization is complete. Collect the aspirin by vacuum filtration [OP-13], washing it on the filter with ice-cold water. Save another small sample of the aspirin in a clean, labeled test tube for analysis. Dry [OP-23] and weigh the remaining aspirin.

Stop and Think: What products would result from hydrolysis?

Waste Disposal: Pour all filtrates down the drain.

Analysis.   Place about 10 mg of salicylic acid in a clean, labeled test tube, then dissolve each reserved aspirin sample and the salicylic acid in 0.5 mL of 95% ethanol. The aspirin samples need not be completely dry. Add a drop of aqueous 2.5% ferric chloride to each test tube and record and explain your observations.

Record the IR spectrum [OP-36] of your aspirin as directed by your instructor. Identify as many IR bands as you can, indicating the kind of

Stop and Think: Which aspirin is the purest, and why?

bond vibration responsible for each. Record its ^1H NMR spectrum [OP-37] in deuterochloroform. If possible, use a sweep offset for the COOH signal and a scale expansion for the aromatic proton signals when you record the NMR spectrum. Locate the ^1H NMR signal for the proton designated H_a in Figure 33.1, determine the values of J_{ab} and J_{ac}, and assign as many other signals as you can. At your instructor's option, you can obtain and interpret a ^{13}C NMR spectrum of aspirin.

Exercises

1   Compare your IR and ^1H NMR spectra of aspirin with those of salicylic acid in Figure 33.2, and explain any significant similarities and differences.

2   (a) Write an equation for a reaction that might form salicylic acid during the workup of the product. (b) Tell how you could reduce or prevent the contamination of your product due to this reaction.

3   Describe and explain the possible effect on your results of the following experimental errors or variations. (a) The reagent bottle labeled "acetic anhydride" actually contained acetic acid. (b) The test tube used for analysis of the purified aspirin was rinsed with water and not completely dried, and you stored the aspirin sample for a week before testing it. (c) You boiled the recrystallization mixture after adding water.

4   Following the format in Appendix V, construct a flow diagram for the synthesis of aspirin.

5   Write a detailed mechanism for the reaction of salicylic acid with acetic anhydride, showing clearly the function of the catalyst, sulfuric acid.

6   A small bottle of 5-grain aspirin tablets holds 100 tablets, each containing 0.325 g of aspirin. Calculate the cost of the acetic anhydride and salicylic acid required to prepare the aspirin in such a bottle, assuming equimolar quantities of the reactants (the sulfuric acid, being a catalyst, can be recovered). You can look up their current prices in the *Chemical Marketing Reporter*, or in another source recommended by your instructor.

7   Which would you expect to be the stronger acid, aspirin or salicylic acid? Explain your answer.

Other Things You Can Do

(Starred projects require your instructor's permission.)

*1   Interpret the mass spectrum of a compound as described in Minilab 29.

*2   Test commercial aspirin tablets for salicylic acid by the ferric chloride test described. Try the test on freshly purchased aspirin and on aspirin that has been in use for some time. (Try to find some old aspirin that has a "vinegar" odor.) Test them for starch (often used as a binder) by boiling 2 mg of a ground-up tablet in 2 mL of water and adding a drop of a solution of iodine in potassium iodide. Starch forms a blue-violet complex with iodine.

3   Starting with sources listed in the Bibliography, write a research paper on some commercial uses for acetic anhydride other than in aspirin production.

Preparation of Nonbenzenoid Aromatic Compounds

EXPERIMENT **34**

Reactions of Unsaturated Hydrocarbons. Preparation of Nonbenzenoid Aromatic Compounds. Hydride-Transfer Reactions. Carbocations. Aromaticity. Infrared Spectrometry. NMR Spectrometry.

Operations

OP-7   Heating
OP-8   Cooling
OP-10   Mixing
OP-13   Vacuum Filtration
OP-23   Drying Solids
OP-24   Drying and Trapping Gases
OP-36   Infrared Spectrometry
OP-37   Nuclear Magnetic Resonance Spectrometry

Before You Begin

1   Read the experiment, read or review the operations as necessary, and write an experimental plan.
2   Calculate the mass and volume of 1.00 mmol of benzaldehyde and the theoretical yield of *meso*-tetraphenylporphin for part **A**. Calculate the mass of 1.20 mmol of triphenylmethanol and the theoretical yield of tropylium fluoborate for part **B**.

Scenario

In 1931, the German physicist and theoretical chemist Erich Hückel predicted that a seven-membered cyclic compound with six pi electrons would show aromatic properties. Hückel was unaware that just such a compound, tropylium bromide, had already been synthesized some 40 years before! Now Dr. Perry Celsus, the science historian (see Experiments 17 and 25), wants to reenact the discovery of this substance, the first nonbenzenoid aromatic compound. Dr. Celsus first wants to establish that the method used to prepare tropylium bromide did, in fact, yield the tropylium ion and that the ion is truly aromatic. To demonstrate this, he needs an authentic sample containing the tropylium ion, prepared by a different route, with which to compare it. Your assignment is to prepare tropylium fluoborate and look for evidence of aromaticity in its IR and NMR spectra. You will also prepare a sample of another nonbenzenoid aromatic compound, tetraphenylporphin, for an analytical chemist who is interested in using this compound to prepare colored metal ion complexes that might be used for spectrometric analysis of the ions.

Applying Scientific Methodology

The scientific problem regarding the tropylium ion should be apparent from the Scenario. After reading the experiment, you should be able to develop a working hypothesis, which you will test by spectral analysis.

Nonbenzenoid Aromatics: from Dracula to Dewar

In 1891, a team of German chemists was trying to determine the structure of atropine, an important alkaloid from the plant *Atropa belladonna*. One of them, G. Merling, had just brominated a degradation product from atropine, cycloheptatriene, and was distilling the liquid dibromide when a mass of yellow crystals collected in his distilling apparatus. He faithfully reported this observation in a German chemical journal, but it didn't appear to reveal anything about the structure of atropine, so it was filed away and forgotten. About 40 years later, Erich Hückel, a pioneer in applying the theories of quantum mechanics to organic molecules, formulated his well-known *Hückel rule*, which predicts that certain cyclic, unsaturated systems having $4n + 2$ pi electrons ($n = 0, 1, 2 \ldots$) will be aromatic. At that time, aromaticity was associated exclusively with benzene and related six-membered ring compounds, but Hückel predicted, for example, that a compound containing a cyclic $C_7H_7^+$ cation (which has six pi electrons delocalized around a seven-membered ring) would also show aromatic properties. When Merling's yellow salt was again prepared in 1954, it was identified as tropylium bromide, a nonbenzenoid aromatic compound of just the kind that Hückel had foretold.

Key Concept: To be aromatic, a molecule must have an uninterrupted circuit of 4n + 2 pi electrons distributed about a planar ring.

Merling's preparation of tropylium bromide

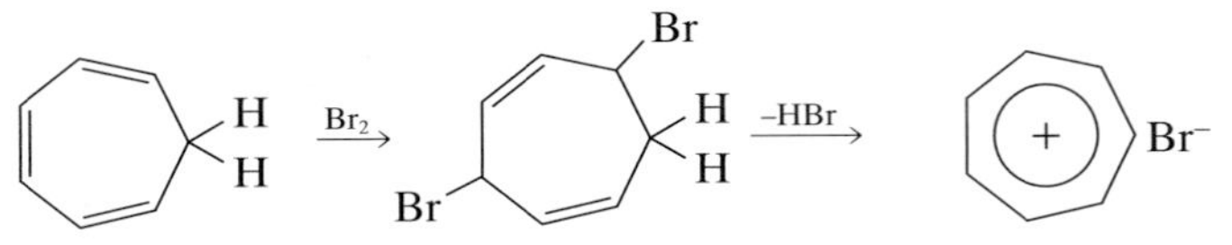

cycloheptatriene   tropylium bromide

Although their aromatic properties were unknown to scientists until a few decades ago, nonbenzenoid aromatic compounds have existed in nature for as long as green plants and animals have populated the earth. The most important of these compounds by far are the *porphyrins*, which play an important role in both photosynthesis and respiration. Porphyrins are constructed around the system of four linked pyrrole rings that constitutes the parent compound, porphin. The heavy line in the porphin molecule shown here traces an aromatic ring whose pi electrons number 18—one of the numbers that can be derived from Hückel's $4n + 2$ rule (when $n = 4$, $4n + 2 = 18$).

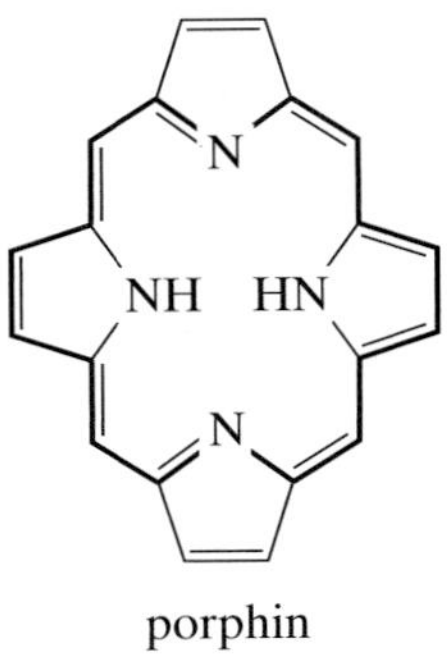

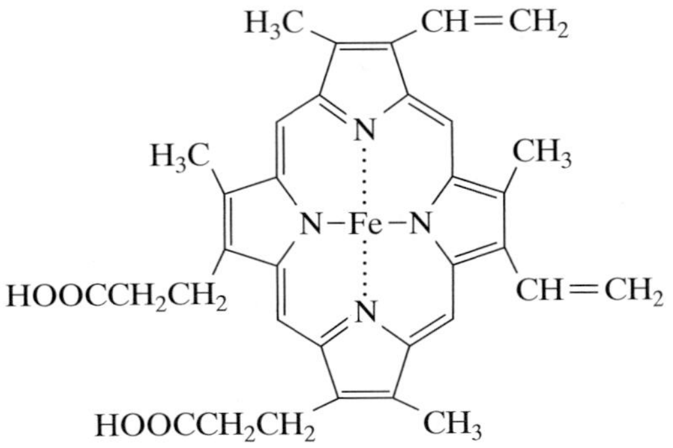

porphin

heme

Hemoglobin, which is a conjugated protein containing an iron-complexed porphyrin known as heme, shuttles oxygen through the bloodstream and keeps the respiratory process going. Hemoglobin itself is blue in color; when it picks up oxygen in the lungs, it is converted to bright red oxyhemoglobin, which gradually gives up its oxygen to the cells and is reduced to hemoglobin again before it is carried back to the lungs. Ordinarily, any excess porphyrins in the body are metabolized by the liver into iron-free substances, such as biliverdin and bilirubin, which collectively make up the bile pigments. A breakdown in this metabolic process leads to a disease known as porphyria, which often causes agonizing attacks that may resemble psychotic episodes. According to Raymond McNally, a history professor from Boston College, the original Count Dracula (also known as Vlad the Impaler for his habit of skewering people on stakes) may have suffered from porphyria. The medieval treatment for the disease was drinking blood, which Count Dracula obtained from serfs who farmed the valley below his castle. Victims of the disease may forage for food at night because they fear light, and they often have receding gums that make their teeth appear longer. All of these reported characteristics of the medieval Count Dracula were later attributed to Bram Stoker's fictional vampire of the same name.

Not long after Bela Lugosi turned Dracula into a Hollywood success story, the British chemist Michael J. S. Dewar was puzzling over a natural substance called stipitatic acid, which had been isolated from a culture of the mold *Penicillum stipitatum*. Because stipitatic acid showed aromatic properties, such as undergoing substitution (rather than addition) reactions with bromine, other scientists had assigned it a structure containing a benzene ring. Dewar disagreed and, with little experimental evidence to go on, he proposed a seven-membered ring structure and declared that the compound was one of a previously unknown class of aromatic compounds that he named tropolones. Dewar expected tropolone rings to show aromatic properties, because they can be represented, in part, by resonance structures in which six pi electrons are delocalized over the seven-membered ring. His guess turned out to be correct; before long, other investigators were proposing tropolone ring structures for compounds with similar properties. For example, a sample of an essential oil from the heartwood of western red cedar (*Thuja plicata*) had been set aside and nearly forgotten for 16 years while it slowly crystallized. The crystals yielded β-thujaplicin, an isopropyl derivative of tropolone, which is an effective fungicide and antibiotic. This compound and two other fungus-destroying thujaplicins are believed to be responsible for the great durability of red cedar, which was used by Native Americans to build canoes long before settlers discovered its value for making fence posts, shingles, and cedar chests.

Hemoglobin is an extremely complex protein whose structure differs slightly for different animal species; a typical empirical formula is $C_{738}H_{1166}O_{208}N_{203}S_2Fe$.

stipitatic acid

tropolone   resonance structure of tropolone

Stop and Think: What other resonance structures can be drawn for tropolone?

β-thujaplicin

Understanding the Experiment

Besides being the first representative of the class of nonbenzenoid aromatics, tropylium ion was probably the first stable carbocation to be synthesized. One way to prepare a stable carbocation is to remove a hydride ion from an appropriate hydrocarbon by a hydride-transfer reaction. Just as an acid can transfer a proton to a strong base, leaving behind a weaker (more stable) base, a hydrocarbon can transfer a hydride ion to a reactive carbocation, leaving behind a less reactive (more stable) carbocation. In this experiment, the reactive carbocation will be the trityl (triphenylmethyl) cation and the stable one will be the tropylium ion.

Proton transfer

$$HB_1 \; + \; B_2^- \; \longrightarrow$$
stronger base

$$HB_2 \; + \; B_1^-$$
weaker base

Hydride transfer

$$R_1H \; + \; R_2^+ \longrightarrow$$

less stable
carbocation

$$R_2H \; + \; R_1^+$$

more stable
carbocation

$$\text{Ph}_3C \oplus \; BF_4^-$$

trityl fluoborate

Stop and Think: What should the ¹H NMR spectrum look like if the product is not aromatic?

The hydride transfer takes place when trityl fluoborate reacts with the hydrocarbon cycloheptatriene, yielding tropylium fluoborate and triphenylmethane. Trityl fluoborate itself is prepared, as in Experiment 30, by treating the corresponding alcohol with aqueous fluoboric acid in acetic anhydride. In the procedure used here, the trityl fluoborate will not be isolated but will be treated with cycloheptatriene while still in solution. Tropylium fluoborate precipitates as the dark color of the trityl carbocation disappears; adding diethyl ether completes the precipitation.

Evidence for a compound's aromaticity can be obtained from its spectra. The infrared (IR) spectra of aromatic molecules show a number of characteristic bands. Aromatic C—H stretching vibrations generally occur between 3100 and 3000 cm^{-1}; C—H out-of-plane bending vibrations give rise to bands in the 900–650 cm^{-1} region; and skeletal vibrations involving carbon-carbon stretching within the ring result in one or more bands in the 1600–1400 cm^{-1} region. The IR spectra of symmetrical unsubstituted aromatic compounds are particularly simple, as shown by the IR spectrum of benzene in Figure 34.1.

The most unequivocal evidence for aromaticity is provided by NMR spectrometry. The circulation of electrons around an aromatic ring in a magnetic field results in a so-called *ring current* that deshields ring protons, moving their ¹H NMR signals well downfield (see Figure 34.2). As a rule, the larger the ring, the stronger is its ring current. The chemical shift of deshielded ring protons is 7.3 δ for the six-carbon aromatic compound benzene but 8.9 δ for the 18-carbon aromatic compound [18]annulene. The protons of a tropylium ion are also deshielded because of its ion charge, which reduces the electron density at each ring atom. The protons on most unsubstituted aromatic rings are equivalent, so the ¹H NMR spectra of unsubstituted aromatic compounds are very simple, usually consisting of one or two signals. Because tropylium fluoborate is very polar, you will need to use a polar solvent, the deuterated form of dimethyl sulfoxide (DMSO), to prepare the NMR sample. By comparing the IR and ¹H NMR spectra of your product with those of the starting material (see Figure 34.3), you should be able to determine whether tropylium fluoborate is or is not aromatic.

Benzene

3035.7	1478.8
1960.4	1035.8
1815.0	673.3

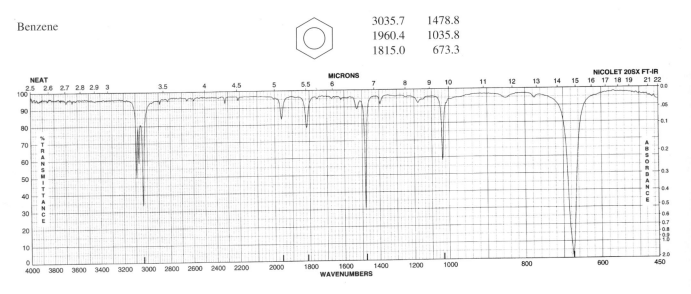

Figure 34.1 IR spectrum of benzene

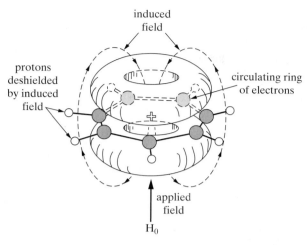

Figure 34.2 Ring current effect in the tropylium ion

You will also prepare a porphin derivative, *meso*-tetraphenylporphin, by a condensation reaction between benzaldehyde and pyrrole in boiling propanoic acid (propionic acid). The product precipitates from the cooled reaction mixture as lustrous purple crystals that are collected by vacuum filtration.

Reactions and Properties

A   4 pyrrole + 4 benzaldehyde + $\frac{3}{2}O_2$ $\xrightarrow{CH_3CH_2COOH}$ *meso*-tetraphenylporphin + $7H_2O$

pyrrole   benzaldehyde   *meso*-tetraphenylporphin

B $Ph_3COH + HBF_4 \xrightarrow{Ac_2O} Ph_3C^+BF_4^- + H_2O$ $Ph_3C^+BF_4^- +$ (cycloheptatriene) $\longrightarrow Ph_3CH +$ (tropylium $+$) BF_4^-

Table 34.1 Physical properties

	M.W.	mp	bp	*d*
benzaldehyde	106.1	−26	178	1.046
propanoic acid	74.1	−21.5	141	0.993
pyrrole	67.1		130–131	0.969
meso-tetraphenylporphin	614.8			
acetic anhydride	102.1	−73	140	1.082
cycloheptatriene	92.15	−80	117	0.887
fluoboric acid (48%)	87.8			1.400
triphenylmethanol	260.3	164	380	

Note: mp and bp are in °C, density is in g/mL.

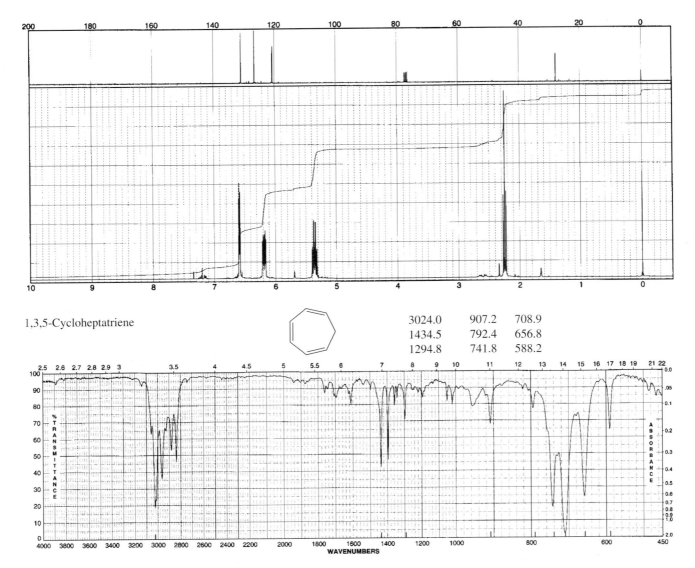

1,3,5-Cycloheptatriene

3024.0	907.2	708.9
1434.5	792.4	656.8
1294.8	741.8	588.2

Figure 34.3 NMR and IR spectra of 1,3,5-cycloheptatriene

Directions

A. *Preparation of* meso-*Tetraphenylporphin*

Safety Notes

propionic acid

Propanoic acid can burn the skin and eyes severely, and its vapors irritate the eyes and respiratory system. Wear gloves, avoid contact, and do not breathe its vapors.

Pyrrole is poisonous and may be very harmful if inhaled, ingested, or allowed to contact the skin or eyes. Wear gloves, avoid contact, and do not breathe its vapors.

Work under the hood and wear protective gloves until the product has been filtered and washed. Weigh 1.00 mmol of benzaldehyde into a 5-mL conical vial and add 3.0 mL of propanoic acid followed by 0.070 mL (~1 mmol) of freshly distilled pyrrole. Add acid-resistant boiling chips or a stirring device, attach a water-cooled condenser, and heat the mixture under reflux [OP-7] for 30 minutes or more. (If the reaction cannot be conducted under the hood, attach a gas trap [OP-24] to the top of the condenser.) During this time, begin part **B** of this experiment.

Cool the reaction mixture to room temperature and carefully collect the product by vacuum filtration [OP-13], using cold methanol for the transfer. Wash the solid on the filter with cold methanol, then wash it with several small portions of hot water until the odor of propanoic acid is gone. Dry [OP-23] the *meso*-tetraphenylporphin and weigh it.

B. *Preparation of Tropylium Fluoborate*

> **Fluoboric acid is poisonous, its solutions can cause severe damage to skin and eyes, and its vapors irritate the respiratory system. Use gloves and a hood; avoid contact with the acid solution, and do not breathe its vapors. Acetic anhydride can cause severe damage to skin and eyes, its vapors are very harmful if inhaled, and it reacts violently with water. Use gloves and a hood; avoid contact with the liquid, do not breathe its vapors, and keep it away from water.**
>
> **Cycloheptatriene is harmful if inhaled or absorbed through the skin; avoid contact and do not breathe its vapors.**
>
> **Diethyl ether is extremely flammable and may be harmful if inhaled. Do not breathe its vapors and keep it away from flames and hot surfaces.**
>
> **Tropylium fluoborate is corrosive and can damage the eyes, skin, and respiratory tract; avoid contact with the solid and do not breathe its dust.**

Reaction. *Carry out the reaction under the hood and wear protective gloves.* Carefully measure 3.0 mL of acetic anhydride into a 25-mL Erlenmeyer flask and cool [OP-8] the flask in an ice/water bath. Add slowly, with magnetic stirring [OP-10], 0.20 mL of aqueous 48% fluoboric acid. With continued cooling and stirring, add 1.20 mmol of pure triphenylmethanol in small portions, producing a solution of trityl fluoborate. Add 0.15 mL (~1.4 mmol) of cycloheptatriene drop by drop with cooling and stirring. If the solution remains dark, add a few more drops of cycloheptatriene until the color disappears or fades to pale yellow. Then mix in 5.0 mL of anhydrous diethyl ether and let the mixture stand (without stirring) in the ice bath for at least 10 minutes, until precipitation is complete. (**Stop and Think:** Why is tropylium fluoborate insoluble in diethyl ether, although cycloheptatriene is soluble?)

Separation and Analysis. Collect the tropylium fluoborate by vacuum filtration [OP-13] and wash it on the filter with anhydrous diethyl ether. Let the product dry [OP-23] thoroughly and weigh it. Record the IR spectrum [OP-36] of tropylium fluoborate as directed by your instructor. Record its ^{1}H NMR spectrum [OP-37] in dimethyl sulfoxide-d_6. Interpret the spectra and decide whether or not tropylium fluoborate is aromatic. Turn in the spectra with your report.

If propanoic acid is not available, you can use glacial acetic acid and increase the reflux time to 1 hour.

Take Care! Avoid contact with propanoic acid and pyrrole and do not breathe their vapors.

Take Care! Wear gloves and use a hood.

Waste Disposal: Place the filtrate in a designated waste container.

Safety Notes

acetic anhydride

Take Care! Avoid contact with acetic anhydride and fluoboric acid and do not breathe their vapors.

Stop and Think: What evidence do you see for the formation of the trityl cation?

Take Care! Keep diethyl ether away from ignition sources and do not breathe its vapors.

Waste Disposal: Place the filtrate in a designated waste container.

Your instructor may provide one or both spectra.

Exercises

1   Compare the spectra of the product with those of cycloheptatriene in Figure 34.3, and account for any significant similarities or differences.

2   Aqueous tropylium fluoborate has a K_a of 1.8×10^{-5}, making it as strong an acid as acetic acid. Write a balanced equation for an equilibrium reaction that explains the acidity of tropylium fluoborate.

3   Describe and explain the possible effect on your results of the following experimental errors or variations. (a) In part **A**, you heated the reactants under reflux for 30 minutes in acetic acid rather than propanoic acid. (b) You used methanol in place of triphenylmethanol in part **B**. (c) The diethyl ether you used to precipitate tropylium fluoborate was not anhydrous.

4   (a) Trityl fluoborate from Experiment 30 could have been used to make tropylium fluoborate in this experiment. Write a balanced equation for a reaction that could take place if you accidentally spilled some water on the trityl fluoborate. Assume a 1:1 mole ratio of the reactants. (b) What mass of trityl fluoborate (M.W. = 330) would be decomposed by a complete reaction with one drop of water, according to your equation in (a)? (Assume that 1 drop $\cong$ 0.050 mL.) If you started with 0.40 g of trityl fluoborate, how much of it would you have left?

5   Following the format in Appendix V, construct a flow diagram for the synthesis in part **B**.

6   (a) Explain why tropylium bromide (C_7H_7Br) is aromatic but the corresponding "oxide" $[(C_7H_7)_2O]$ is not. Draw structures for both compounds, showing clearly the kind of bonding between each ring and a Br or an O atom. (b) What kind of compound is the "oxide"?

7   Predict which of the following compounds will show aromatic properties and explain why each one is or is not aromatic.

Other Things You Can Do

(Starred projects require your instructor's permission.)

*1 Record and compare the ultraviolet spectra, from 200–400 nm, of $\sim 10^{-4}$ solutions of tropylium fluoborate in 0.1 M HCl and of cycloheptatriene in absolute ethanol. You can also obtain and interpret a ^{13}C NMR spectra of tropylium fluoborate.

*2 Prepare ditropyl ether by dissolving about 0.2 g of tropylium fluoborate in 1 mL of water and slowly adding 0.8 mL of saturated aqueous sodium carbonate, while stirring. Extract the product with dichloromethane; then wash the dichloromethane solution with water, dry it, and evaporate the solvent without heating. Obtain the ^{1}H NMR spectrum of the product and decide whether or not it is aromatic.

 3 Write a research paper about the importance of porphyrins to life on earth, starting with sources listed in the Bibliography.

Mechanism of the Nitration of Arenes by Nitronium Fluoborate

EXPERIMENT 35

Reactions of Arenes. Preparation of Nitro Compounds. Electrophilic Aromatic Substitution. Reaction Rates. Reaction Mechanisms.

Operations

OP-15   Extraction
OP-16   Evaporation
OP-21   Washing Liquids
OP-22   Drying Liquids
OP-34   Gas Chromatography

Before You Begin

1   Read the experiment, read or review the operations as necessary, and write an experimental plan.

Scenario

(*Note:* Professor Olah is a real chemist whose research group carried out the work described here.)

The research group of George A. Olah, recipient of the 1994 Nobel prize in chemistry, developed a highly reactive nitrating reagent, nitronium fluoborate (NO_2BF_4), and used it in a study of the mechanisms of aromatic nitration reactions. During its investigation, the group found evidence suggesting that the initial intermediate formed in nitronium fluoborate reactions is different than the one formed in most other aromatic nitration reactions, which use a mixture of nitric and sulfuric acids. According to Olah, such an intermediate could be either a sigma complex, a symmetrical pi complex, or an oriented pi complex. Your supervisor has devised two experiments that may reveal the nature of the initial intermediate involved in NO_2BF_4 reactions: (1) nitration of a mixture of toluene and mesitylene and measurement of their relative reaction rates; and (2) nitration of *t*-butyl-benzene and measurement of the product mixture's *ortho/para* ratio. Your assignment is to carry out these nitration reactions and, based on Olah's hypothesis and your experimental results, propose a reasonable structure for the intermediate in question.

For another interpretation of Olah's results, see J.H. Ridd, Accounts of Chemical Research, **1971,** *4, 248.*

Applying Scientific Methodology

After reading the experiment, you can formulate a working hypothesis about the nature of the intermediate, which you and your coworkers will test by analyzing the product mixtures from the reactions described in the Scenario. Keep in mind that experiments cannot prove that a hypothesis is true but can only support or refute the hypothesis, so your conclusion may be incorrect or incomplete regardless of your experimental results. In fact, there is evidence that the initial intermediate may be different from any of those mentioned above, but testing that possibility is beyond the scope of this experiment.

Intermediates in Nitration Reactions

The electrophile in aromatic nitration is the nitronium ion (NO_2^+), which is usually generated by mixing nitric acid with sulfuric acid.

$$HNO_3 + 2H_2SO_4 \rightleftharpoons NO_2^+ + H_3O^+ + 2HSO_4^-$$

With this "mixed acid" reagent, the equilibrium concentration of NO_2^+ is so low that elevated temperatures and long reaction times are generally required for successful nitration. Nitronium fluoborate (NO_2BF_4) is a salt that ionizes to yield NO_2^+ at a much higher concentration, allowing rapid nitration of most aromatic compounds at room temperature or below.

The nitronium ion is an electron-hungry species—an *electrophile*—that can steal a pair of pi electrons from a benzene ring's aromatic sextet. In a general mechanism proposed for electrophilic aromatic substitution, such an electrophile attacks the aromatic ring to form a *sigma complex*, in which the electrophile (symbolized here by E^+) is connected to a ring carbon by a sigma bond.

The sigma complex (also called an *arenium ion*) then loses a proton to a basic species in the reaction mixture to yield the corresponding substituted benzene, as shown in the margin. The ring atoms *ortho* and *para* to the incoming electrophile bear most of the positive charge in the sigma complex. Thus electron-donating groups located on those atoms stabilize the complex, favoring the formation of *ortho*- and *para*-substituted products. For example, only 4% of the product obtained from the nitration of toluene in mixed acid is *meta*-nitrotoluene; the other 96% is a mixture of *ortho*- and *para*-nitrotoluene.

In some aromatic substitution reactions, the initial intermediate may be a pi complex, in which the electrophile is loosely bonded to the pi electrons of an aromatic ring. Such an intermediate would presumably rearrange to a sigma complex before giving rise to the product. A pi complex involving a nitronium ion and benzene might be pictured as shown, with the electrophile sitting atop a "doughnut" of pi electrons, equidistant from all of the ring atoms.

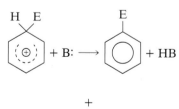

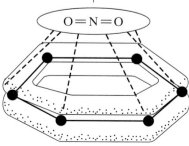

Pi complex involving nitronium ion and benzene

symmetrical
pi complex

oriented
pi complexes

BF_4^-

HBF$_4$ sigma complex
of mesitylene

Key Concept: *Anything that stabilizes a reactive intermediate will also stabilize the transition state leading to that intermediate, lowering the activation energy and speeding up the reaction.*

sigma complex
for *ortho* nitration
of *t*-butylbenzene

Table 35.1 *Ortho/para* ratios for the nitration of arenes in mixed acid

Arene	o/p ratio
toluene	1.57
ethylbenzene	0.93
isopropylbenzene	0.48
t-butylbenzene	0.22

The pi cloud of an alkyl-substituted benzene is "lumpier" than that of benzene, with bulges at the positions *ortho* and *para* to the alkyl group. An electrophile that spends more time near these regions of higher electron density will form an *oriented* pi complex rather than a symmetrical benzene-type complex. Both possibilities are illustrated in the margin.

Although alkyl groups alter the electron distribution in a pi cloud, they have only a small effect on its total electron density. Thus increasing the number of methyl groups on a benzene ring should not increase the stability of its pi complexes very much. On the other hand, the stabilities of sigma complexes are very sensitive to the electronic effects of substituents. Mesitylene, with three methyl groups, forms a sigma complex with HBF$_4$ that is nearly 300,000 times more stable than the corresponding sigma complex involving toluene; but the pi complex that mesitylene forms with HCl is only about twice as stable as the complex that toluene forms. As a rule, the more stable an intermediate is, the faster it will form. If the initial intermediate in nitration by nitronium fluoborate is a pi complex, the nitration rates for mesitylene and toluene should be within a factor of 10 or so of one another. But if the initial intermediate is a sigma complex, the nitration rate for mesitylene should be thousands of times greater than that for toluene.

When a substituent influences the outcome of a reaction as a result of its bulkiness (as opposed to its electronic effects), we say that a *steric effect* is operating. Steric effects in aromatic substitution reactions can be detected by measuring the ratios of *ortho* products to *para* products for the same reaction with different substituents. There are twice as many *ortho* as *para* hydrogens in a monosubstituted arene, suggesting that the ratio of *ortho*- to *para*-substituted products should be about 2:1 in the absence of steric effects. As shown in Table 35.1, *ortho/para* ratios for the nitration of arenes in mixed acid are considerably lower than 2:1, especially for very bulky substituents such as *t*-butyl. Apparently, a bulky alkyl group hinders substitution at the *ortho* position by crowding the attacking electrophile in the transition state that leads to the sigma complex.

The steric requirements of pi complexes have not been as thoroughly studied as those of sigma complexes. If an incoming electrophile approaches the ring from the top to form a symmetrical pi complex, there should be little if any steric effect, even with a bulky substituent such as the *t*-butyl group. On the other hand, the formation of an oriented pi complex should be markedly influenced by steric factors, giving product ratios comparable to those observed for mixed-acid nitrations.

Understanding the Experiment

Nitronium fluoborate is prepared by treating nitric acid with hydrofluoric acid and boron trifluoride:

$$HNO_3 + HF + 2BF_3 \longrightarrow NO_2BF_4 + BF_3 \cdot H_2O$$

The stable crystalline salt ionizes in polar solvents to provide "ready-made" nitronium ions:

$$NO_2BF_4 \longrightarrow NO_2^+ + BF_4^-$$

In this experiment, you will nitrate a mixture of mesitylene and toluene with nitronium fluoborate and analyze the product mixture by gas chromatography to determine their relative reaction rates. You will also nitrate *t*-butylbenzene and measure the *ortho/para* ratio of the product. From the results, you should be able to decide whether the initial intermediate is more likely to be a sigma complex, a symmetrical pi complex, or an oriented pi complex.

You will measure the relative rates of nitration for mesitylene and toluene by carrying out a competitive nitration reaction in which equimolar quantities of the two arenes compete for a limited amount of nitronium fluoborate. The arene that competes most successfully will form the most product, so the relative rates for the two arenes should be proportional to the relative amounts of nitroarene they produce:

$$\frac{\text{reaction rate for mesitylene}}{\text{reaction rate for toluene}} = \frac{\text{moles of nitromesitylene}}{\text{moles of nitrotoluenes}}$$

For a meaningful rate comparison, you must determine the relative rates per reaction site; otherwise, toluene, with five ring hydrogens, will have a statistical advantage over mesitylene, with only three. The rate per reaction site is proportional to the number of moles of product divided by the number of reaction sites, so the relative reactivity of a mesitylene site is given by this equation:

$$\frac{\text{reactivity of mesitylene site}}{\text{reactivity of toluene site}} = \frac{\text{moles of nitromesitylene}/3}{\text{moles of nitrotoluenes}/5}$$

Since the area of a peak on a gas chromatogram is proportional to the mass of the component that produces it, you can estimate the relative number of moles of each product by dividing its peak area by its molecular weight. (The relative masses of different components may not be in the exact ratio of their peak areas, but they will be close enough for the purposes of this experiment.) The peak areas for the three nitrotoluenes should be combined for this calculation.

The nitration reactions will be carried out at room temperature by adding nitronium fluoborate in sulfolane to an excess of each arene in the same solvent. The excess reactant prevents the formation of di- and trinitrated products, which would skew the results. Sulfolane is an excellent solvent for the reaction, because it dissolves both the nitronium salt and the arene, thus providing a homogeneous reaction mixture. It is also miscible with water, making it easy to separate the products from the reaction mixture. When water and diethyl ether are added to the reaction mixture, sulfolane and fluoboric acid (a by-product of the reaction) end up in the water layer and the aromatic compounds are extracted into the ether layer. Evaporation of the ether leaves a mixture of unreacted arenes and nitrated products.

The reaction mixtures will be analyzed by gas chromatography on a column that separates aromatic compounds in order of their boiling points. Unreacted arenes should elute from the column first, followed by *ortho*-, *meta*-, and *para*-nitroarenes, in that order. The nitromesitylene peak should appear later than all of the nitrotoluene peaks. If you failed to evaporate the diethyl ether completely, you will see an initial ether peak, and there may be additional peaks due to impurities in the commercial arenes used as starting materials. Compare your chromatograms with chromatograms of those arenes if such peaks make interpretation difficult.

sulfolane

Reactions and Properties

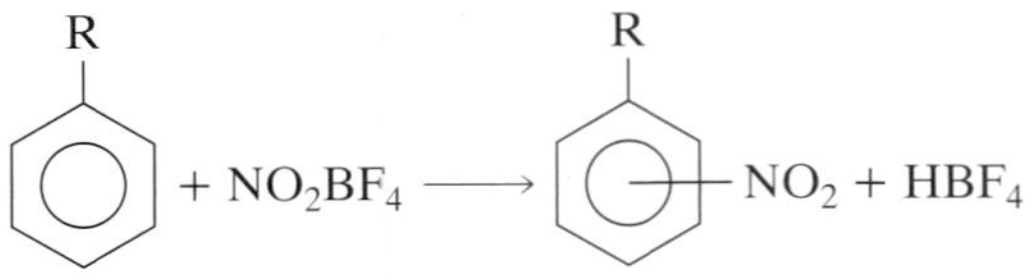

R = methyl or *t*-butyl

(NO$_2$ is predominantly *ortho* and *para*)

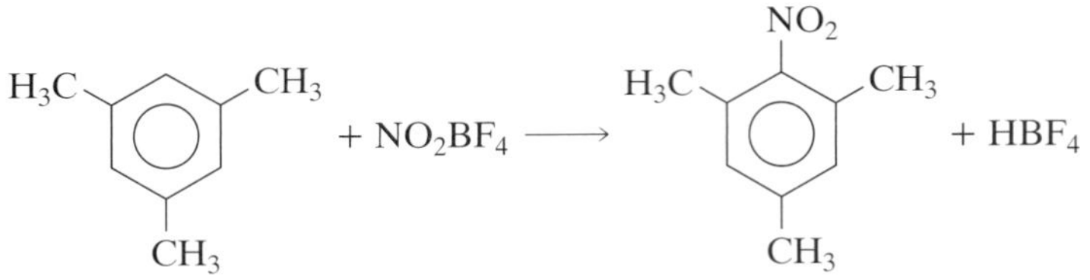

mesitylene   nitromesitylene

Table 35.2 Physical properties

	M.W.	mp	bp	*d*
toluene	92.1	−95	111	0.867
t-butylbenzene	134.2	−58	169	0.867
mesitylene	120.2	−45	165	0.862
nitronium fluoborate	132.8			
sulfolane	120.2	28	285	1.260
o-nitrotoluene	137.1	10	222	1.163
m-nitrotoluene	137.1	16	233	1.157
p-nitrotoluene	137.1	55	238	1.104
nitromesitylene	165.2	44	255	

Note: mp and bp are in °C, density is in g/mL.

Directions

If desired, students can work in pairs, with each student being responsible for one nitration reaction.

Safety Notes

The nitronium fluoborate solution is toxic and corrosive. Wear gloves and avoid contact.
The aromatic hydrocarbons are flammable, and inhalation or skin absorption may be harmful. Avoid contact and inhalation, and keep them away from flames.
Diethyl ether is extremely flammable and may be harmful if inhaled. Do not breathe its vapors and keep it away from flames and hot surfaces.
Aromatic nitro compounds are toxic and may cause serious harm if inhaled or absorbed through the skin. The reaction mixtures will also contain fluoboric acid and hydrofluoric acid, which cause very serious burns and have harmful vapors. Avoid contact with the reaction mixtures and products, and do not breathe their vapors.

toluene   diethyl ether   *p*-nitrotoluene

Reactions.   *Carry out the reactions under a hood, and wear protective gloves. All glassware must be clean and dry.* Clean, dry, and label two 15-mL centrifuge tubes with screw caps. Measure 0.60 mL of an equimolar mixture of mesitylene and toluene into the first centrifuge tube and add 1.0 mL of sulfolane. Measure 0.50 mL of dry *t*-butylbenzene into the second centrifuge tube and add 1.0 mL of sulfolane. Cap both tubes and shake gently to mix the contents. Slowly add 1.0 mL of 0.5 *M* nitronium fluoborate in sulfolane to each centrifuge tube, swirling to mix the reactants. Again cap both tubes and shake gently. Then let the tubes stand at room temperature for 10 minutes, with occasional shaking.

Take Care!   Do not breathe the vapors of the arenes or their solutions, and keep them away from flames. Wear gloves and avoid contact with nitronium fluoborate.

Separation.   Carry out the following procedure with each reaction mixture. Add 2 mL of solvent-grade diethyl ether and 3 mL of water to the centrifuge tube, then cap it and shake to extract [OP-15] the products and unreacted arene into the ether layer. Carefully remove the aqueous layer with a Pasteur pipet. Wash [OP-21] the ether layer with 2 mL of water. Then remove the aqueous layer, dry [OP-22] the ether solution with anhydrous calcium chloride, and decant it into a small screw-cap vial. Under the hood, carefully evaporate [OP-16] the diethyl ether using a stream of dry air or nitrogen.

Take Care! Keep diethyl ether away from ignition sources and do not breathe its vapors. Wear gloves when shaking the tube and vent it often.

Waste Disposal: Place the aqueous layer, which contains hydrofluoric acid and fluoboric acid, in a designated waste container.

Analysis.   Obtain a gas chromatogram [OP-34] of each product mixture as directed by your instructor. Identify the peaks on the gas chromatograms and measure the peak areas for all of the nitrated products. Calculate the *ortho/para* ratio for the nitration of *t*-butylbenzene and the reactivity per reaction site of mesitylene relative to toluene. Decide whether the initial intermediate in these nitrations by nitronium fluoborate is more likely to be a sigma complex, a symmetrical pi complex, or an oriented pi complex. Turn in the gas chromatograms with your report.

Stop and Think: Is there any evidence of incomplete evaporation? What gives rise to the largest peak(s) on each chromatogram?

Exercises

1   Based on your results, propose a detailed mechanism for the reaction of *t*-butylbenzene with nitronium fluoborate to yield *p*-nitro-*t*-butylbenzene.

2   (a) It has been estimated that the *meta* product arising from direct nitration of *t*-butylbenzene by NO_2BF_4 makes up about 2.0% of the product mixture, the remaining *meta* product arising from isomerization of the *ortho* product. Use this estimate to calculate a more accurate value of your *ortho/para* nitration ratio for *t*-butylbenzene. (b) Propose a mechanism for the isomerization reaction, which is apparently promoted by the HBF_4 formed during the reaction.

3   Describe and explain the possible effect on your results of the following experimental errors or variations. (a) You rinsed your reaction tubes with water and didn't dry them completely. (b) You added toluene rather than *t*-butylbenzene to the second centrifuge tube. (c) When analyzing the gas chromatogram from the *t*-butylbenzene nitration, you misidentified the *t*-butylbenzene peak as the peak for the *ortho* product and assumed that the next two peaks were for the *meta* and *para* products.

4   Following the format in Appendix V, construct a flow diagram for the nitration of *t*-butylbenzene.

5   In a solution of bromine in acetic acid, mesitylene is brominated nearly 300 million times faster than benzene. Do you think a sigma complex or a pi complex is formed in the rate-determining step of this reaction? Explain.

6   Predict the major product or products of the mononitration of (a) ethyl benzoate, (b) phenyl acetate, (c) phenyl benzoate, (d) *m*-nitrotoluene, (e) *p*-methoxybenzaldehyde.

Other Things You Can Do

(Starred projects require your instructor's permission.)

*1   Carry out the mixed-acid nitration of napthalene by the procedure in Minilab 30.

2   Read the paper by George A. Olah and his coworkers that describes the nitration of arenes with nitronium fluoborate (*J. Am. Chem. Soc.* **1961**, *83*, 4571) and compare their results and conclusions with your own.

Friedel-Crafts
Acylation of Anisole

Reactions of Aromatic Ethers. Preparation of Carbonyl Compounds. Electrophilic Aromatic Substitution. Infrared Spectrometry.

Operations

OP-7   Heating
OP-10 Mixing
OP-11 Addition of Reactants
OP-13 Vacuum Filtration
OP-16 Evaporation
OP-21 Washing Liquids
OP-22 Drying Liquids
OP-23 Drying Solids
OP-24 Drying and Trapping Gases
OP-30 Melting Point
OP-36 Infrared Spectrometry

Before You Begin

1 Read the experiment, read or review the operations as necessary, and write an experimental plan.
2 Calculate the mass and volume of 2.50 mmol of anisole and the theoretical yield of methoxyacetophenone.

Scenario

The purchasing agent for the Olfactory Factory mistakenly ordered 500 kilograms of anisole rather than 500 pounds, so the company needs to find some way to use up the excess anisole by converting it to perfume ingredients. One possibility is to prepare *p*-methoxyacetophenone, also known as crataeon, which occurs naturally in hawthorn blossoms (*Crataegus* spp.). The company's chemical technicians think it should be possible to synthesize *p*-methoxyacetophenone from anisole by a Friedel-Crafts reaction, but they are concerned that the reaction may yield the wrong isomer or a mixture of isomers that will be difficult to separate. According to their business manager, the Olfactory Factory cannot sell the product at a competitive price if they have to invest in expensive separation equipment. Your assignment is to see whether or not the Friedel-Crafts acetylation of anisole yields mainly *p*-methoxyacetophenone, another isomer, or a mixture of isomers.

hawthorn blossom

crataeon

Applying Scientific Methodology

After reading the experiment, you should be able to develop a working hypothesis related to the problem, which you will test by obtaining the melting point and infrared (IR) spectrum of the product.

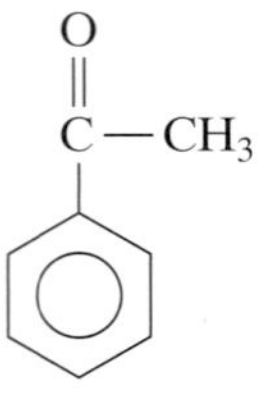

acetophenone

benzophenone

Friedel, Crafts, and Phenones

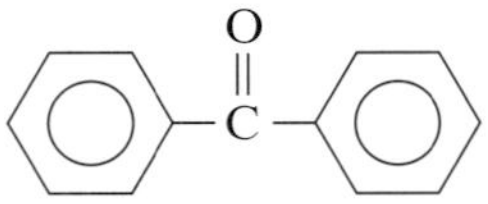

Aromatic ketones that have the carbonyl group adjacent to the benzene ring are called *phenones*. The simplest member of this group is acetophenone, a pleasant-smelling liquid that has been used to impart an odor of orange blossoms to perfumes, and has also been used as a sleep-producing drug under the generic name hypnone. Charles Friedel first prepared acetophenone in 1857 by distilling a mixture of calcium benzoate and calcium acetate. Another fragrant phenone, benzophenone, is a white solid with a geranium-like odor that has been used as a fixative for perfumes and a starting material for the manufacture of drugs and insecticides.

The natural and synthetic musks are powerfully odoriferous substances that supply the long-lasting, musky "end note" characteristic of some perfumes, deodorants, and aftershave lotions. The large-ring ketone called muscone (3-methylcyclopentadecanone) is the major constituent of natural musk, which is a secretion from the musk pod of the male musk deer. Muscone is very costly and its use threatens the existence of the deer, so it has been almost entirely replaced by synthetic musks. Among these are the phenones known as musk ketone and Celestolide.

Natural and synthetic musks

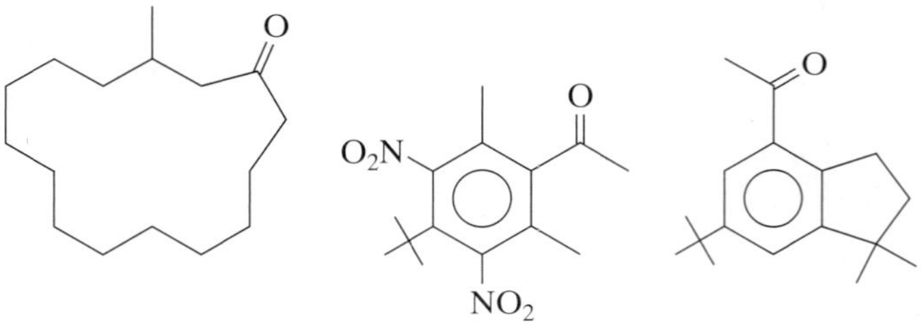

muscone   musk ketone   Celestolide

A Friedel-Crafts reaction involves the catalytic alkylation or acylation of an aromatic ring.

Musk ketone is prepared from *m*-xylene by two Friedel-Crafts reactions—alkylation with *t*-butyl chloride and acylation with acetyl chloride—followed by nitration of the aromatic ring.

Synthesis of musk ketone

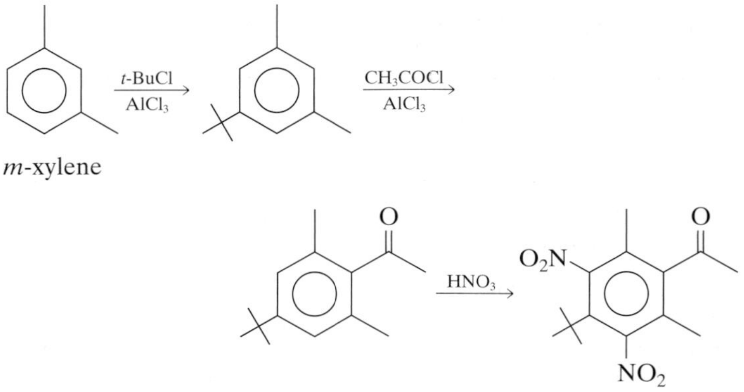

m-xylene

Like the synthetic musks, most phenones can be prepared by a Friedel-Crafts reaction of an aromatic compound with an appropriate acylating agent. Theodor Zincke first prepared benzophenone by heating benzoyl chloride with a metal in benzene. This was essentially a Friedel-Crafts reaction, but Zincke didn't know it because that reaction had not been discovered yet! In fact, the Friedel-Crafts reaction might well have been named the "Zincke reaction" if Theodor Zincke had understood the significance of an experiment that failed. In 1869, Zincke tried to synthesize 3-phenylpropanoic acid by combining benzyl chloride and chloroacetic acid in the presence of metallic silver—a variation of the Wurtz reaction. While carrying out the reaction with benzene as the solvent, Zincke observed, to his surprise, that a great deal of hydrogen chloride was evolved and that the major product was diphenylmethane instead of the expected carboxylic acid.

See Experiment 25 for the story of another Wurtz reaction that failed, with momentous consequences.

Zincke's attempted synthesis

$$PhCH_2Cl + ClCH_2COOH \xrightarrow[\text{benzene}]{Ag} PhCH_2CH_2COOH$$

The "Zincke reaction"

$$PhCH_2Cl + PhH \text{ (benzene)} \xrightarrow{Ag} PhCH_2Ph + HCl$$

About four years later, a Frenchman named Charles Friedel was watching a student in Wurtz's laboratory perform a "Zincke reaction" using (appropriately) powdered zinc as the catalyst. When the reaction suddenly became violent, Friedel helped the student separate the solution from the zinc powder, thinking that removing the catalyst would moderate the reaction. To the astonishment of both, the reaction was just as violent in the absence of zinc. Although there is no record of his thought processes after this event, Friedel must have recognized its significance. In 1877, he and his collaborator, an American named Charles Mason Crafts, published a paper that marked the inception of the Friedel-Crafts reaction as one of the most important synthetic procedures in the history of organic chemistry. Friedel and Crafts' major discovery was a simple one—it is a chloride of the metal, and not the metal itself, that catalyzes the reaction of organic halides with aromatic compounds. For example, during Zincke's attempted synthesis of 3-phenylpropanoic acid, traces of silver chloride had formed as a result of oxidation of the metal. Friedel and Crafts found that anhydrous aluminum chloride was the most effective catalyst of those then available. It is still the catalyst of choice for most Friedel-Crafts reactions.

Understanding the Experiment

The Friedel-Crafts reaction is not a single reaction type, although the term has most often been applied to alkylations and acylations of aromatic compounds using aluminum chloride (or another Lewis acid catalyst) and a suitable alkylating or acylating agent. A typical Friedel-Crafts acylation reaction uses a carboxylic acid chloride as the acylating agent and anhydrous aluminum chloride as the catalyst. As illustrated in Figure 36.1 for the reaction of benzene with an acyl chloride, aluminum chloride (a Lewis acid) removes a leaving group from the acylating agent, forming an acylium ion. The acylium ion then attacks the benzene ring to form an arenium ion, which loses a proton to yield the product and regenerate the catalyst.

Key Concept: During an electrophilic addition reaction, the aromatic sextet of the benzene ring acts as an electron source, contributing a pair of pi electrons to the electrophile to form the arenium ion intermediate.

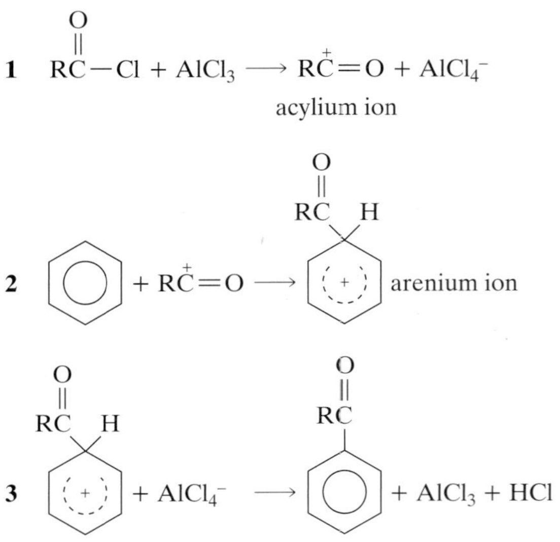

$$1 \quad \underset{\text{O}}{\overset{\text{O}}{\|}} \quad$$

1    $RC-Cl + AlCl_3 \longrightarrow R\overset{+}{C}=O + AlCl_4^-$
             acylium ion

2   (benzene) $+ R\overset{+}{C}=O \longrightarrow$ (arenium ion)   arenium ion

3   (arenium ion) $+ AlCl_4^- \longrightarrow$ (product) $+ AlCl_3 + HCl$

Figure 36.1 Mechanism for the Friedel-Crafts acylation of benzene

When the acyl group is acetyl (CH_3CO), acetic anhydride is often used as the acylating agent rather than acetyl chloride. The anhydride is safer to work with and it usually provides better yields and a simpler workup. More catalyst is needed with acetic anhydride, however, because some of the aluminum chloride forms complexes with the acetic acid produced during the reaction, making it ineffective as a catalyst. As a rule, 2–3 moles of $AlCl_3$ are used per mole of acetic anhydride.

In this experiment, you will use acetic anhydride as the acylating agent and dichloromethane as the reaction solvent. The reaction is highly exothermic, so it will be carried out by adding acetic anhydride slowly to the other reactants, then heating under reflux to complete the reaction. Pouring the product into ice water will decompose the aluminum chloride complex of the product and transfer inorganic salts to the aqueous phase. The product is obtained by evaporating the organic solvent. The IR spectrum of the product can be obtained by the method for melts described in OP-36, or by some other method.

In principle, acylation of a monosubstituted benzene can yield any or all of three different disubstituted products. From the melting point and IR spectrum of your product, you should be able to determine whether it is predominantly a single compound or a mixture of isomers and, if it is a single compound, to establish its identity. Disubstituted benzenes can be distinguished by the location of their out-of-plane C—H bending bands, which occur at frequencies (expressed in wavenumbers) below 850 cm^{-1}. The frequency of such a band decreases with the number of adjacent hydrogens on the ring, as shown in Table 36.1. Thus a *para*-disubstituted benzene, with its two sets of two adjacent hydrogens, should have an absorption band in the 840–810 cm^{-1} region; *meta* compounds, with three adjacent ring hydrogens, absorb in the 810–750 cm^{-1} region; and *ortho* compounds, with four adjacent ring hydrogens, absorb in the 770–735 cm^{-1} region. Absorption by the isolated hydrogen of a *meta* compound is usually very weak, and its frequency may vary. Monosubstituted and *meta*-disubstituted benzenes have an additional band in the 710–680 cm^{-1} region, which arises from a vibration involving the entire benzene ring.

Aluminum chloride complex with acyl compound

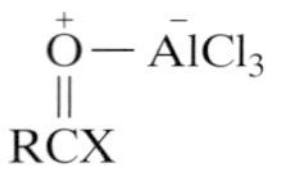

$$\overset{+}{\underset{\|}{\text{O}}}-\bar{\text{A}}\text{lCl}_3$$
$$\underset{}{\text{RCX}}$$

∴ need more catalyst when using acetic anhydride

Table 36.1 Frequencies of C—H out-of-plane bending bands in aromatic hydrocarbons

No. of adjacent hydrogens	Frequency range, cm^{-1}
1	900–860 (weak)
2	840–810
3	810–750
4	770–735
5	770–730

Possible products from the Friedel–Crafts acylation of anisole

o-methoxyacetophenone   m-methoxyacetophenone

p-methoxyacetophenone

Your product will be an ether as well as a ketone, so its IR spectrum will contain bands characteristic of both functional groups. The carbonyl band of a phenone usually appears in the 1685–1665 cm^{-1} region, and a weak carbonyl overtone band may be observed at twice the frequency of the fundamental band. Aryl alkyl ethers display an asymmetrical C—O—C stretching band at 1275–1200 cm^{-1} and a symmetrical C—O—C band near 1075–1020 cm^{-1}. These ether bands can be seen in the spectrum of anisole in Figure 36.2.

Reactions and Properties

anisole   acetic   ?-methoxy-
             anhydride   acetophenone

Table 36.2 Physical properties

	M.W.	mp	bp	d
anisole	108.2	−38	155	0.996
acetic anhydride	102.1	−73	140	1.082
aluminum chloride	133.3	193	subl	
dichloromethane	84.9	−95	40	1.327
o-methoxyacetophenone	150.2		245	1.090
m-methoxyacetophenone	150.2		240	1.034
p-methoxyacetophenone	150.2	39	258	1.082^{41}

Note: mp and bp are in °C, density is in g/mL (subl = sublimes).

Anisole

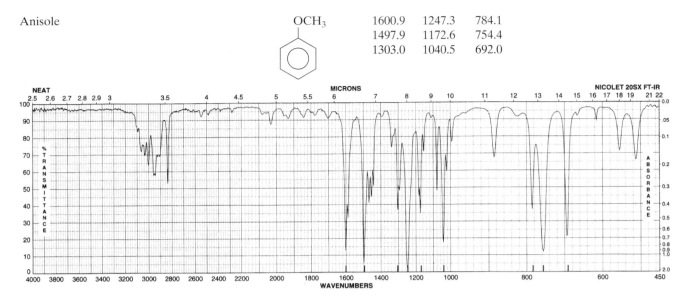

1600.9	1247.3	784.1
1497.9	1172.6	754.4
1303.0	1040.5	692.0

Figure 36.2 IR spectrum of anisole

Directions

Safety Notes

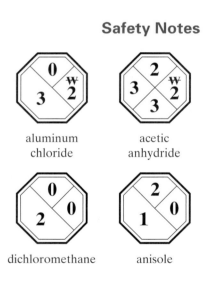

aluminum chloride

acetic anhydride

dichloromethane

anisole

Aluminum chloride reacts with atmospheric moisture and violently with water, generating HCl vapors. It can cause painful burns on moist skin and eyes, and inhaling the dust or vapors can damage the respiratory tract. Weigh it under a hood, wear gloves and safety goggles, avoid contact, do not inhale dust or vapors, and keep it away from water.

Acetic anhydride can cause severe damage to skin and eyes, its vapors are very harmful if inhaled, and it reacts violently with water. Use gloves and a hood; avoid contact with the liquid, do not breathe its vapors, and keep it away from water.

Dichloromethane may be harmful if ingested, inhaled, or absorbed through the skin. There is a possibility that prolonged inhalation of dichloromethane may cause cancer. Minimize contact with the liquid and do not breathe its vapors.

Reaction.   Wear gloves and eye protection! Work under a hood, if possible. Anhydrous aluminum chloride is deactivated by water, so protect it from atmospheric moisture and be sure that all glassware is thoroughly dried (oven drying is recommended). Assemble an apparatus for addition [OP-11] under reflux using a 10-mL round-bottom flask and a water-cooled condenser. Prepare a combination drying tube-gas trap [OP-24] by inserting a glass-wool plug into a drying tube, adding a layer of anhydrous calcium chloride and a layer of pelletized Norit, and inserting another glass-wool plug (the Norit can be omitted if the reaction will be carried out under a hood). Attach the drying tube to the top of the condenser and clamp the apparatus securely to a ring stand, raising it high enough so that you can remove the boiling flask to add the reactants without disturbing the rest of the apparatus. *Under the hood*, carefully weigh 0.73 g (~5.5 mmol) of finely powdered anhydrous aluminum chloride into a *dry* screw-cap vial (don't let it tip over). Immediately cap the vial and the aluminum chloride container. Weigh 2.50 mmol of anisole into the reaction flask and add 2.5 mL of dichloromethane and a stir bar. Have a beaker of cold water ready to cool the mixture if it begins to boil. Start the stirrer [OP-10], then cautiously add the aluminum chloride in small portions from a plastic weighing dish or through a "funnel" made of a square of glazed weighing paper folded into a cone. If necessary, wash any adherent aluminum chloride into the flask with a little dichloromethane, then reassemble the apparatus. *Under the hood*, measure 0.25 mL (~2.6 mmol) of acetic anhydride into a clean, dry syringe. Add the acetic anhydride drop by drop while stirring, so that the reaction mixture boils gently. When the addition is complete, use a hot water bath to heat the reaction mixture under gentle reflux [OP-7], while stirring, for 30 minutes. Rinse the syringe with water shortly after using it.

Take Care!   Wear gloves and goggles, avoid contact with $AlCl_3$ and dichloro-methane, and do not breathe their vapors.

Take Care!   Wear gloves, avoid contact with acetic anhydride, do not breathe its vapors, and keep it away from water.

Separation and Purification.   *Wear gloves and eye protection! Under the hood*, pour the warm reaction mixture *slowly* with vigorous manual stirring onto about 3 g of finely cracked ice in a beaker. Use a small amount of ice water to rinse any residue out of the flask into the beaker. Transfer the reaction mixture to a centrifuge tube, add 2 mL of dichloromethane, and stir gently to mix. Transfer the dichloromethane layer to another centrifuge tube and wash [OP-21] it with 2 mL of 3 *M* sodium hydroxide followed by 2 mL of saturated aqueous sodium chloride. Dry [OP-22] the dichloromethane layer with anhydrous magnesium sulfate or sodium sulfate and collect it in a 5-mL conical vial. Evaporate [OP-16] the dichloromethane under a stream of dry air or nitrogen. Remove the vial and let it stand, scratching with a stirring rod and cooling in ice water if necessary, until the product solidifies. Break up the solid with a flat-bladed microspatula, then add 1 mL of ice-cold low-boiling petroleum ether, and triturate (crush and rub) the product thoroughly in the solvent. Collect the product by vacuum filtration [OP-13] and dry it [OP-23].

Take Care!   Splattering may occur.

Stop and Think: Which is the dichloromethane layer? What does NaOH remove from the dichloromethane solution?

Analysis.   Weigh the dry product and measure its melting point [OP-30]. Record its IR spectrum [OP-36], or obtain a spectrum from your instructor. Interpret the spectrum as completely as you can and turn it in with your report.

Waste Disposal: Pour all aqueous layers down the drain. Place the petroleum ether in a designated solvent recovery container.

Exercises

1 If your product was a single compound, explain why it was that compound rather than another isomer.
2 Write a mechanism for the Friedel-Crafts reaction of anisole with acetic anhydride.
3 Describe and explain the possible effect on your results of the following experimental errors or variations. (a) You used 2.5 mmol of aluminum chloride for the reaction. (b) You used acetic acid as the acylating agent. (c) You used acetophenone as the substrate.
4 (a) Write an equation for the reaction of aluminum chloride with a large excess of water. (b) Write equations for one or more reactions that would account for the production of HCl during the acylation reaction.

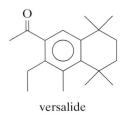

versalide

5 Following the format in Appendix V, construct a flow diagram for this experiment.

6 Explain why Friedel-Crafts reactions are usually carried out by adding the alkylating or acylating agent *to* the aromatic compound rather than vice versa.

7 2,5-Dichloro-2,5-dimethylhexane is an important starting material for the aroma chemicals called tetralin musks. Outline a synthesis of the tetralin musk versalide from this starting material and benzene, using any necessary inorganic or organic reagents.

Other Things You Can Do

(Starred projects require your instructor's permission.)

*1 Record the ^{1}H NMR spectrum of the product in deuterochloroform. Interpret it as completely as you can, assigning the signals from protons on the ring as well as those from protons on the side chains.

*2 Carry out some Friedel-Crafts reactions that yield colored products as described in Minilab 31.

3 Starting with sources listed in the Bibliography, write a research paper about the Friedel-Crafts reaction, including specific examples and industrial applications.

Determination of the Structure of a Natural Product in Anise Oil

EXPERIMENT **37**

Reactions of Alkenylbenzenes. Preparation of Carboxylic Acids. Side-Chain Oxidation. Structure Determination. Infrared Spectrometry.

Operations

OP-7   Heating
OP-10   Mixing
OP-13   Vacuum Filtration
OP-23   Drying Solids
OP-25   Recrystallization
OP-30   Melting Point
OP-36   Infrared Spectrometry

Before You Begin

Read the experiment, read or review the operations as necessary, and write an experimental plan.

Scenario

Basil Wormwood, the new-age herbalist with a chemistry degree, has another puzzle for you (see Experiment 18 for his previous puzzles). He obtained some Chinese star anise from an oriental-foods wholesaler, steam-distilled its essential oil, and isolated the major component of the oil. Not knowing its identity, he tentatively named this compound anisene. He sent it off to a chemical analyst for elemental analysis, and from the results found its molecular formula to be $C_{10}H_{12}O$. He also carried out some experiments (described next) showing that the compound contains a methoxyl group and a three-carbon side chain on a benzene ring, but he doesn't know the identity of the side chain or where it is located with respect to the methoxyl group. He has just shipped a sample of the compound to your supervisor, hoping that your Institute's consulting chemists can solve this structure puzzle. Your supervisor thinks that the position of the side chain can be determined by oxidizing it to a COOH group, and that its structure can be determined by infrared (IR) analysis. Your assignment is to determine the complete structure of anisene.

Applying Scientific Methodology

You will have to carry out some experimental work before you can propose a hypothesis about the structure of anisene.

star anise seed clusters

Key Concept: Chemists determine the structures of organic molecules by breaking them down into smaller fragments and identifying the fragments, probing them with different kinds of electromagnetic radiation and interpreting the resulting signals, or both.

The Structure Puzzle—Taking Molecules Apart and Putting Them Back Together

Chinese star anise (*Illicium verum*) is a small evergreen tree of the magnolia family. When its dried, star-shaped seed clusters are ground up and steam-distilled, they yield an oily liquid with a strong odor of licorice. Anise oil (from star anise and other spices) or its synthetic equivalent is widely used as a flavoring for licorice, cough drops, chewing gum, and liqueurs such as ouzo and anisette. In this experiment, you will use both classical and modern methods of structural analysis to determine the complete structure of its major component, which we will call "anisene" (not its real name).

Today, when a chemist can run an NMR spectrum or a mass spectrum of an organic compound and often determine its structure in a matter of minutes, it is hard to imagine how much time and effort were once required to determine the structures of even the simpler natural products. In a classical structure determination, the molecular formula of a compound is first obtained by elemental analysis and molecular-weight measurement. Then the compound is degraded (broken down) into smaller structural units that are isolated and, if possible, identified. Finding how the smaller units fit together to form the original molecule is an intellectual challenge that might be compared to putting together a jigsaw puzzle with some pieces missing, others that don't belong, and still others that have been chewed up by the family dog and are no longer recognizable. Finally, when enough information has been gathered to suggest a possible structure, that structure must usually be proven by an independent synthesis in which the compound is built up again, from known compounds, by reactions whose outcome can be reliably predicted.

In many cases, classical structure determinations involved the efforts of dozens or even hundreds of chemists over many decades, with the generation of much irrelevant or misleading information and many synthetic dead ends. The advent of modern spectrometric methods has simplified the process enormously by providing detailed structural information that was not readily available to the chemists of earlier times.

Understanding the Experiment

In this experiment, you will attempt to determine the structure of the major component of star anise oil, which has the molecular formula $C_{10}H_{12}O$. Most open-chain saturated organic compounds (except those containing nitrogen, phosphorus, or halogen atoms) have $2n + 2$ hydrogen atoms for every n carbon atoms. If anisene were such a compound it would have $2(10) + 2 = 22$ hydrogen atoms, but since it has only 12, it is said to be "deficient" by 10 hydrogens. Every ring or pi bond in a molecule represents a deficit of two hydrogens. That is, an open-chain compound must lose two hydrogen atoms to form a ring, and a saturated compound must lose two hydrogens to form a pi bond (or a pi-bond equivalent in the Kekulé structure of an aromatic ring). Thus its deficiency of 10 hydrogens indicates that there must be a total of five rings and/or pi bonds in an anisene molecule; this is called its *index of hydrogen deficiency* (IHD). The IHD of a compound having n carbon atoms and x hydrogen atoms can be calculated using the following formula:

$$IHD = \frac{(2n + 2) - x}{2}$$

Catalytic hydrogenation of anisene under high pressure yields a saturated compound with the formula $C_{10}H_{20}O$. The gain of eight hydrogens indicates that anisene has four pi bonds, so it must contain only one ring. A high carbon/hydrogen ratio often indicates an aromatic structure, and we can account for the ring and three pi bonds by assuming that anisene contains a benzene ring.

Heating anisene with hydriodic acid yields a phenol with the formula C_9H_9OH and a volatile compound identified as methyl iodide. This reaction is used to test for certain ether functions. Methyl ethers yield methyl iodide, and the formation of a phenol indicates that anisene is an aryl methyl ether, whose formula we write as $C_9H_9OCH_3$ in the following equation for the reaction:

$$C_9H_9OCH_3 + HI \longrightarrow C_9H_9OH + CH_3I$$

At this point, we know that anisene contains a methoxyl ($-OCH_3$) group and a benzene ring, which accounts for seven carbon atoms and three pi bonds. That leaves three more carbons and one pi bond to be accounted for. This remaining fragment could be a three-carbon unsaturated side chain, whose formula can be determined by subtracting the fragments already identified from the molecular formula of anisene:

molecular formula	$C_{10}H_{12}O$
disubstituted benzene ring	$-C_6H_4$
methoxyl group	$-CH_3O$
side chain	$\overline{C_3H_5}$

Now we can write a partial structure for anisene, shown in the margin. All that remains is to determine the structure of the unsaturated side chain and its location on the benzene ring.

Potassium permanganate is capable of oxidizing most aliphatic side chains all the way down to the benzylic carbon atom, leaving a COOH group where the side chain was originally located. Oxidizing anisene should yield one of three possible methoxybenzoic acids, whose melting points are given in Table 37.3. By identifying the oxidation product as one of these three, you will establish the position of anisene's side chain.

Although aqueous potassium permanganate is a powerful oxidizing agent, it reacts slowly with water-insoluble organic compounds because $KMnO_4$ is essentially insoluble in the organic phase. In 1974, Herriot and Picker added a quaternary ammonium salt to a stirred heterogeneous mixture of aqueous $KMnO_4$ and benzene, causing permanganate ions to dissolve in the organic layer and form "purple benzene." The quaternary salt acted as a phase-transfer catalyst, escorting the permanganate ions across the phase boundary into the organic phase (see Experiment 22 for a discussion of phase-transfer catalysis). When an oxidizable organic compound is dissolved in purple benzene, it reacts much more rapidly and under milder conditions than it would with aqueous $KMnO_4$.

Benzene is toxic and can cause leukemia in humans, so you will use a simplified procedure in which anisene and a phase-transfer catalyst are combined directly with aqueous potassium permanganate. Thus anisene

Another possibility, that anisene has two side chains, is explored in Exercise 3.

partial structure of anisene

a methoxybenzoic acid

itself will be the organic phase of the two-phase system, and no organic solvent is needed. Because you require only enough product for a melting point, you will start with only a few drops of anisene. Excess potassium permanganate is used because some of it may decompose during the reaction. As the reaction proceeds, permanganate ion is reduced to manganese dioxide, which forms a fine brown precipitate that is difficult to filter and wash. Fortunately, this precipitate can be dissolved during the workup by acidifying the solution and adding sodium bisulfite, which reduces manganese dioxide (and any unreacted permanganate ion) to soluble manganese(II) sulfate.

Removal of manganese dioxide

$$MnO_2 + NaHSO_3 + H^+ \longrightarrow MnSO_4 + H_2O + Na^+$$

The methoxybenzoic acid can then be separated by vacuum filtration and purified by recrystallization from water.

Carbon–carbon double bonds give rise to characteristic $=C-H$ out-of-plane bending bands in the 1000–650 cm^{-1} region of an IR spectrum. The wave numbers of these bands can reveal the number and location of substituents on the carbon–carbon double bond, as shown in Table 37.1. There are four possible structures for an unsaturated C_3H_5 side chain, corresponding to the four structure types in the table. From the wave number(s) of anisene's $=C-H$ bending band(s), you should be able to deduce the structure of the side chain. But first you must locate the right absorption bands, which is more easily said than done, because aromatic $C-H$ bonds give rise to strong bands in the same region, as shown in Table 37.2.

Table 37.1 Out-of-plane bending vibrations of vinylic $C-H$ bonds

Structure type	Frequency range, cm^{-1}
$RCH=CH_2$	995–985 and 915–905
$RCH=CHR$ (cis)	730–665
$RCH=CHR$ (trans)	980–960
$R_2C=CH_2$	895–885

R = alkyl or aryl.

Table 37.2 Out-of-plane bending vibrations of aromatic $C-H$ bonds

Ring substitution	Frequency range, cm^{-1}
ortho	770–735
meta	810–750 and 710–690
para	840–810

Once you learn the position of the side chain on anisene's benzene ring, you should be able to locate any bands due to aromatic $C-H$ bonds in its infrared spectrum, which will help you pick out one or more vinylic $C-H$ bands from the remaining strong bands in the 1000–650 cm^{-1} region.

Reactions and Properties

$$OCH_3 \quad \xrightarrow{KMnO_4,\ Q^+} \quad OCH_3$$

(structure with C_3H_5 substituent) → (structure with COO^-K^+ substituent)

(structure with COO^-K^+) $+ \ HCl \ \longrightarrow$ (structure with $COOH$) $+ \ KCl$

$$Q^+ \approx (CH_3(CH_2)_7)_3\overset{\oplus}{N}CH_3$$

Table 37.3 Physical properties

	M.W.	mp	bp	d
potassium permanganate	158.0			
o-methoxybenzoic acid	152.2	101		
m-methoxybenzoic acid	152.2	110		
p-methoxybenzoic acid	152.2	185		
toluene	92.2	−95	111	0.867

Note: mp and bp are in °C; density is in g/mL.

Directions

Potassium permanganate can react violently with oxidizable materials; keep it away from other chemicals and combustibles.
Sodium bisulfite produces harmful vapors when it reacts with acids; do not breathe them.

Safety Notes

potassium
permanganate

Take Care!   Keep KMnO$_4$ away from oxidizable materials.

Reaction.   Obtain some anisene (or anise oil) from your instructor, or isolate anise oil from anise seeds as described in "Other Things You Can Do." Add 0.50 g of crystalline potassium permanganate and 2 drops of tricaprylmethylammonium chloride (Aliquat 336) to 10 mL of water in a 25-mL Erlenmeyer flask, and then drop in a stir bar. Heat [OP-7] the mixture in a boiling-water bath while stirring [OP-10] for 5 minutes or more to dissolve most of the KMnO$_4$. Add 5 drops of anisene or anise oil using a medicine dropper (not a Pasteur pipet), place a watch glass (convex side down) over the mouth of the flask to prevent evaporation, and heat the mixture in a boiling water bath with vigorous stirring for 15 minutes or more.

Separation.   Cool the reaction mixture to room temperature and transfer it to a small beaker. *Under the hood,* add 1 mL of 6 *M* hydrochloric acid and test the solution with blue litmus paper; if it is not acidic, add more HCl until it is. Add just enough solid sodium bisulfite, in small portions while

Take Care! Do not breathe the vapors that may be produced.

Stop and Think: What vapors may form, and how are they produced? What are the brown and white precipitates?

Waste Disposal: Place the filtrate in a designated waste container.

stirring or swirling, to reduce any excess permanganate and remove the brown manganese dioxide (0.5–1.0 g of NaHSO$_3$ should be sufficient). Test the solution with pH paper after each bisulfite addition and add 6 M HCl as needed to keep it acidic. When all of the brown precipitate has disappeared and only a white precipitate remains, again test the solution with pH paper. If the pH is higher than 2, add enough 6 M HCl to reduce it to 2. Separate the product from the reaction mixture by vacuum filtration [OP-13] and wash it on the filter with ice-cold water.

Purification and Analysis.   Recrystallize [OP-25] the product from boiling water and dry [OP-23] it to constant mass. Measure the melting point [OP-30] of the methoxybenzoic acid. Record the IR spectrum [OP-36] of anisene (*not* of the methoxybenzoic acid), or obtain its spectrum from your instructor. Deduce the location and structure of the side chain and draw the structure of anisene. Turn in the IR spectrum with your report.

Exercises

1 Derive a systematic name for anisene and find its common name in *The Merck Index* or another reference book.
2 (a) Write a balanced equation for the reaction of anisene with potassium permanganate, assuming that acetic acid is a by-product of the reaction. (*Note*: the reaction mixture is alkaline.) (b) Assuming that 5 drops of anisene is about 1.0 mmol, calculate the mass of potassium permanganate required to oxidize that much anisene and the percentage in excess that was actually used.
3 (a) The three carbon atoms of anisene's side chain might have formed two separate side chains rather than one. Give the structures of these side chains. (b) Give the structures of all of the dicarboxylic acids that could have resulted from complete side-chain oxidation of anisene had it contained these two side chains.
4 Describe and explain the possible effect on your results of the following experimental errors or variations. (a) You forgot to add the tricaprylammonium chloride. (b) The pH of the reaction mixture was 7 when you filtered it, and you obtained a brown solid. (c) You recorded the IR spectrum of the oxidation product rather than that of anisene itself.
5 Describe the probable role of the phase-transfer catalyst in this reaction, giving equations for the relevant reactions.
6 Following the format in Appendix V, construct a flow diagram for the synthesis of your methoxybenzoic acid.
7 (a) Draw the structure of the compound C$_{10}$H$_{20}$O that is obtained by the catalytic hydrogenation of anisene. (b) Draw the structure of the compound C$_9$H$_9$OH that is obtained when anisene is treated with hydriodic acid.
8 You could confirm the structure of anisene by synthesizing it from known starting materials. Outline a synthesis of anisene from benzene and alcohols that have four carbon atoms or fewer.

9 The structure in the margin has been proposed for coniferyl alcohol, which can be obtained by the hydrolysis of coniferin, a natural product found in the sap of conifer trees. Assuming that the structure of coniferyl alcohol had not been reported in the literature, describe how you would go about proving its structure. Indicate what chemical tests and degradations might be carried out, describing the expected results and conclusions. Summarize the information that could be derived from infrared analysis. Then show how the alcohol could be synthesized from readily available starting materials.

proposed structure for coniferyl alcohol

Other Things You Can Do

(Starred projects require your instructor's permission.)

*1 Isolate anise oil from anise seeds (or star anise) as follows. Weigh out 1 g of fresh anise seeds and grind them finely using a spice grinder or a mortar and pestle. Isolate the anise oil by steam distillation and extraction of the distillate with dichloromethane, as described for clove oil in Experiment 10. Dry the dichloromethane solution and evaporate the solvent completely. You can obtain a gas chromatogram of the oil and estimate the percentage of anisene it contains.

*2 Use air as an oxidizing agent to convert fluorene to fluorenone as described in Minilab 32.

3 Starting with sources listed in the Bibliography, write a research paper on the use of chemical methods for structure determination of natural products, illustrating it with examples of actual structure determinations.

Identification of an Oxygen-Containing Organic Compound

EXPERIMENT 38

Reactions of Aldehydes and Ketones. Reactions of Alcohols. Infrared Spectrometry. Qualitative Analysis.

Operations

OP-23   Drying Solids
OP-25   Recrystallization
OP-27   Simple Distillation
OP-30   Melting Point
OP-31   Boiling Point
OP-36   Infrared Spectrometry
OP-37   Nuclear Magnetic Resonance Spectrometry (optional)

Before You Begin

1 Read the experiment, read or review the operations as necessary, and write an experimental plan.
2 Read Part IV, Qualitative Organic Analysis, except for the sections headed "Directions."

Scenario

You can find a description of Arkham in the *Dictionary of Imaginary Places, Expanded Edition* (Harcourt Brace Jovanovich, 1987).

Dr. Keziah Armitage, professor of medieval metaphysics at Miskatonic University in Arkham, Massachusetts, was exploring an abandoned and nearly forgotten room in the basement of the metaphysics building when she came across a grime-encrusted bottle containing an unknown liquid. Its label had long since decomposed to dust, but the liquid in the bottle was clear and colorless. Curious about its origin, she sent the bottle and its contents to your Institute for analysis. She thinks it might be a potion used in unmentionable rites practiced by her ancestor, Keziah Mason, a witch whose trial scandalized Arkham in 1692. Your supervisor thinks it is more likely to be an alcohol or carbonyl compound misplaced by an absent-minded alchemy professor and then forgotten. Your assignment is to find out what family the mysterious liquid belongs to and then identify it.

Applying Scientific Methodology

As you carry out the experiment, you should propose provisional hypotheses about the nature and identity of your unknown. You will test—and perhaps reject or revise—your hypotheses as you gather additional experimental evidence. Your conclusion should, if possible, be consistent with all of the experimental evidence you obtain. If any evidence is not consistent with your conclusion, you should attempt to explain why.

The Chemist as Detective

Qualitative organic analysis, the process of identifying unknown organic compounds, can be compared to the approach used by a detective in identifying the perpetrator of a crime. The detective first looks for clues that help to characterize the criminal and indicate the most productive areas of investigation. Once a list of possible suspects has been assembled, the detective can evaluate the evidence already acquired and gather additional evidence to help narrow the list of suspects, focusing the investigation on the most likely suspects. Finally, the detective must evaluate all of the evidence, come to a conclusion regarding the identity of the perpetrator, and organize the facts of the case in such a way as to convince a jury that the accused is, in fact, guilty of the crime.

Key Concept: *Identifying an unknown organic compound involves matching the physical and chemical properties of the unknown with those of some known compound.*

In carrying out the identification of an organic compound, you, like the detective, should be constantly on the lookout for clues to its identity. Chemical and spectral data should allow you to confine your search to a particular chemical family. Additional physical and chemical evidence will help you narrow down the list of "suspects" and focus your attention on a few of the most probable compounds. Finally, the preparation of one or more derivatives should lead you to a definite conclusion and provide you with sufficient evidence to convince the "jury" (your instructor) that your compound is, in fact, what you believe it to be.

As in the solution of any other problem, you must first ask yourself the right questions before you can arrive at the correct answer. Some important questions to be answered regarding an unknown compound are these: (1) Is it pure? (2) What functional group(s) does it contain? (3) Are there any other significant structural features that might aid in its identification? Each bit of evidence you obtain should, if interpreted correctly, help reveal the answer to one or more of these questions. All of them combined should provide you with an answer to the ultimate question, "What is it?"

A detective trying to solve a case will almost invariably come upon clues that lead nowhere or, even worse, to false conclusions. The same is true in chemical problem solving, so it is important to keep an open mind throughout your investigation and to avoid jumping to conclusions before all the evidence is in. You may formulate tentative assumptions based on your initial observations (for example, "It turns chromic anhydride reagent green, so it may be an alcohol"), but you should be ready to revise or discard such assumptions if they are not supported by subsequent observations (for example, "Its IR spectrum has a $C=O$ stretching band but no $O-H$ band, so it may be an aldehyde instead").

Chemical and physical evidence can be misleading for a variety of reasons:

- Some compounds of a given family may undergo an atypical reaction with a given reagent and yield either a false positive or a false negative result.
- Some reagents give positive tests with more than one functional group.
- Impurities may complicate or invalidate a test.
- Spectral bands may occur outside the expected frequency ranges or may be incorrectly assigned.

Because of these and other possible sources of error, it is best not to rely on a single piece of evidence in formulating a conclusion. For example, the classification of an unknown as a secondary alcohol can be convincingly established by a positive chromic anhydride test, a slow reaction with Lucas's reagent, *and* an infrared (IR) band in the 1100 cm^{-1} region, but not by any one of these alone.

Understanding the Experiment

This section provides a general discussion of most of the procedures you will follow to identify your unknown. See the appropriate sections in Part IV for more detailed information about the interpretation of test results. For information about the interpretation of spectra, see the corresponding operations.

Throughout this experiment, you should have your lab notebook handy to record the data you collect and all of your observations as you make them. Keeping meticulous records can often mean the difference between the successful identification of a compound and a failure that could prove very costly, insofar as it affects your lab grade.

Since an unknown liquid may be impure, it should be purified by distillation before any chemical tests or spectra are run. The median distillation temperature should also give you a good estimate of its boiling point. Solids can be purified by recrystallization, but unless your instructor indicates otherwise, you can assume that an unknown solid is pure enough to use without purification. It is very important to measure the boiling point or melting point of your unknown as accurately as you can, since your list of possibilities will be based on the value you obtain. If your measured boiling point or melting point is inaccurate, your list of possibilities may not even contain the name of your unknown compound, and identifying it correctly may then be impossible.

A preliminary examination of your unknown may provide some clues that will help you identify it. For example, observing that a compound is a liquid at room temperature eliminates most compounds with reported melting points of 30°C or higher. The ignition behavior of a substance can provide clues about its structure; many oxygen-containing compounds burn with a blue flame, but those with a high molecular weight may exhibit a clean yellow flame and those with aromatic rings a sooty yellow flame. The solubility behavior of your unknown compound in water can also tell you something about its structure. Most alcohols and carbonyl compounds containing up to four carbon atoms are soluble and most with six carbons or more are relatively insoluble.

To find out what family your compound belongs to, you will carry out several classification tests and record its IR spectrum. For the procedures for the classification tests and directions for their interpretation, see "Classification Tests" in Part IV. The tests you will use include the following:

- The 2,4-dinitrophenylhydrazine (DNPH) test, which is positive for both aldehydes and ketones
- The chromic acid test, which is positive for 1° and 2° alcohols and for aldehydes, but which gives a faster reaction with alcohols
- The Tollens test, which is positive only for aldehydes

IR bands that you should look for include a strong, broad O—H stretching band near 3300 cm^{-1}, a strong carbonyl (C=O) stretching band near 1700 cm^{-1}, and one or two weak-to-moderate bands in the 2700–2850 cm^{-1}

region, which arise from C—H stretching vibrations involving the carbonyl carbon of an aldehyde.

Additional structural information can be obtained both from chemical tests and from your infrared spectrum. The iodoform test is positive for methyl carbinols (alcohols having a CH_3 group on the carbon that holds the OH) and methyl ketones. The bromine test can show whether your compound contains any carbon–carbon double or triple bonds. And Lucas' test can tell you whether an alcohol is primary, secondary, or tertiary, *if* the alcohol's boiling point is below 150°C (the test is invalid for most alcohols with higher boiling points). The IR spectrum of an alcohol may also help you decide what kind of alcohol it is. The C—O bands of most open-chain primary, secondary, and tertiary alcohols occur near 1050 cm^{-1}, 1110 cm^{-1}, and 1175 cm^{-1}, respectively. The wave number of the C—O band is about 25–50 cm^{-1} lower (to the right) for cyclic alcohols and alcohols having aromatic rings or C=C groups on the carbinol carbon. Other structural features that can be detected from IR spectra include aromatic rings and carbon–carbon double or triple bonds. (See OP-36 for more detailed information on the interpretation of IR spectra.)

NMR spectra can provide a great deal of information about the structure of a molecule. If your instructor allows you to obtain the ^1H or ^{13}C NMR spectrum of your unknown, see OP-37 or your textbook for information about NMR spectral interpretation.

After you carry out the classification tests and interpret your spectra, you should have a short list of possible compounds. Keep in mind, however, that classification tests and spectral interpretations are subject to error, so later you may want to reconsider some of the compounds you eliminated to arrive at your list. To decide which of the compounds on your list is the correct one, you will need to prepare a derivative. A derivative preparation is a small-scale chemical synthesis in which the unknown compound is converted to a different compound, a solid whose melting point may indicate the identity of the unknown. As for any synthesis, the formation of by-products, incomplete purification, and insufficient drying can lower the melting point of the product. This makes it important to follow the directions carefully and to be certain that the product is completely dry before you measure its melting point. It is also important to select the right derivative. Depending on the reagents available, you can prepare a *p*-nitrobenzoate, 3,5-dinitrobenzoate, α-naphthylurethane, or phenylurethane if you have an alcohol; and a 2,4-dinitrophenylhydrazone, semicarbazone, or oxime if you have an aldehyde or ketone. But a derivative may be unsuitable because its melting point is too low or is not listed for some of the compounds on your list. Derivatives with melting points of 60° or below are often hard to purify, because they tend to melt to an oil in the hot recrystallization solvent. You should also avoid derivatives whose melting points (for the compounds on your list) are too close together. For example, the semicarbazones of 3-methyl-2-butanone and 2-pentanone melt at 113°C and 112°C, while their 2,4-dinitrophenylhydrazones melt at 124°C and 143°C, making the second derivative a better choice for distinguishing between these ketones. If you have difficulty preparing a certain derivative, or if the derivative you prepare does not eliminate all of the possibilities but one, you should prepare a second derivative.

When you think you have gathered enough evidence to identify your unknown with some certainty, you are free to write down your conclusion.

But keep in mind that your evidence should be sufficient to convince your instructor—and yourself—that your conclusion is correct. Thus you should go back over the evidence and make sure that it all points to the same conclusion. If some evidence is not consistent with that conclusion—for example, if a chemical test does not give the result expected for a compound with the proposed structure—be prepared to either reevaluate your conclusion or explain the inconsistency.

Reactions and Properties

General equations for classification test reactions are given in the "Classification Tests" section of Part IV. General equations for derivative preparations are given in the "Preparation of Derivatives" section of Part IV. The properties of the different classes of compounds and their derivatives are given in Appendix VI.

Directions

Safety Notes

> **You should consider your unknown compound to be flammable and harmful by inhalation, ingestion, and skin absorption. Minimize your contact with the unknown and do not breathe its vapors.**
> **Safety information for chemicals used in classification tests and derivative preparations are included with the corresponding procedures in Part IV.**

Take Care! Minimize contact with the unknown and do not breathe its vapors.

Preliminary Work.   Obtain an unknown compound from your instructor and record its identification number in your laboratory notebook. If the unknown is a liquid, purify it by simple distillation [OP-27] and record its distillation boiling range and median boiling temperature. Then measure the boiling point [OP-31] of the pure liquid using a capillary-tube method. If it is a solid, measure its melting point [OP-30]. Describe the physical state, general appearance, and any other notable characteristics of the purified compound in your lab notebook. Carry out an ignition test (see "Ignition Test" in Part IV), and test the solubility of the unknown in water (see "Solubility Tests" in Part IV).

Functional-Class Determination.   Test the unknown with 2,4-dinitrophenylhydrazine reagent (DNPH, classification test C-11), chromic acid reagent (C-9), and Tollens' reagent (C-23), or carry out other classification tests suggested by your instructor. Record the IR spectrum [OP-36] of your unknown. Use it and the results of the classification tests to decide whether your unknown is an alcohol, aldehyde, or ketone.

Detection of Structural Features.   In your lab notebook, list all compounds from the appropriate table in Appendix VI that have melting or boiling points within ±10°C of your observed value, recording their melting or boiling points and the melting points of their derivatives. At your instructor's option, show him or her your list; the instructor may approve the list if it includes your unknown or suggest additional work if it doesn't.

Write the structure of every compound on your list and consider whether additional classification tests, such as the bromine test (C-7), iodoform test (C-16), or Lucas test (C-17), will help you eliminate any compounds from the list. Also look for evidence from your observations and your IR spectrum that suggest specific structural features, such as aromatic rings, conjugation with double bonds, or the structural class (1°, 2°, or 3°) of an alcohol. With your instructor's permission, you can obtain and interpret the ^{1}H or ^{13}C NMR spectrum [OP-37] of your unknown as well. At this point, prepare a short list of compounds by eliminating the least likely possibilities.

Preparation of a Derivative.   Using procedure D-1, D-2, D-3, or D-4, (see "Preparation of Derivatives" in Part IV) prepare a suitable derivative of your unknown. Purify the derivative by recrystallization [OP-25] as described in the appropriate procedure, dry [OP-23] it thoroughly, and obtain its melting point [OP-30]. Deduce the identity of your unknown from the derivative melting point and all other relevant evidence.

Exercises

1 Interpret the spectrum or spectra you obtained as completely as you can.

2 (a) Write balanced equations for the reactions involved in all of the classification tests for which you obtained a positive result. (b) Write balanced equations for the reaction(s) involved in your derivative preparation(s).

3 Describe and explain the possible effect on your results of the following experimental errors or variations. (a) The test tube you used to carry out a DNPH test had just been rinsed with acetone. (b) The watch glass you used for the ignition test had previously been used to weigh sodium sulfate. (c) You performed Lucas' test on a compound that had a boiling point of 175°C. (d) Your derivative formed an oil when you heated it in the recrystallization solvent, but the oil solidified on cooling, so you used it to obtain a melting point.

4 Construct a flow diagram showing the process you followed to identify your unknown.

5 The unknown assigned to a student was an aldehyde, but about half of the sample distilled around 65°C and the rest of it distilled near 155°. (a) What was the aldehyde? (b) What else was in the sample, and why? Write a balanced equation for its formation.

6 An unknown liquid is water soluble and reacts with chromic acid within 2 seconds. It dissolves in the Lucas reagent, but the solution remains clear for 30 minutes. The iodoform test yields a yellow precipitate. Give the name and structure of the unknown.

7 An unknown liquid with a boiling range of 179–181°C is insoluble in water, gives a blue-green suspension with chromic acid, and immediately forms a separate layer when shaken with the Lucas reagent. Its IR spectrum contains bands at 3060 cm^{-1}, 2805 cm^{-1}, 2730 cm^{-1}, 1705 cm^{-1}, 745 cm^{-1}, and 690 cm^{-1}. Give the name and structure of the unknown.

8   Write mechanisms for the following reactions, which are used in chemical tests and derivative preparations. (a) The reaction of butanal with 2,4-dinitrophenylhydrazine reagent. (b) The preparation of the 3,5-dinitrobenzoate of 1-butanol. (c) The iodoform reaction of 2-butanone. (d) The reaction of 2-methyl-2-butanol with the Lucas reagent.

Other Things You Can Do

(Starred projects require your instructor's permission.)

*1 Observe the effect of aqueous potassium permanganate on different classes of alcohols as described in Minilab 25.
*2 Use ^{1}H NMR to identify an unknown arene as described in Minilab 28.
 3 Starting with sources listed in the Bibliography, write a research paper about the use of gas chromatography-mass spectrometry (GC-MS) to identify illicit drug samples and trace them back to their sources.

Wittig Synthesis
of 1,4-Diphenyl-1,3-butadiene

EXPERIMENT **39**

Reactions of Carbonyl Compounds. Preparation of Dienes. Nucleophilic Addition. Wittig Reaction. Ylides.

Operations

OP-7   Heating
OP-10   Mixing
OP-13   Vacuum Filtration
OP-15   Extraction
OP-16   Evaporation
OP-23   Drying Solids
OP-25   Recrystallization
OP-30   Melting Point

Before You Begin

1   Read the experiment, read or review the operations as necessary, and write an experimental plan.
2   Calculate the mass and volume of 1.00 mmol of *trans*-cinnamaldehyde and the theoretical yield of 1,4-diphenyl-1,3-butadiene.

Scenario

Marvelous Molecules Incorporated has experienced a lower than expected demand for its product line of aldehydes, so it has a large inventory of cinnamaldehyde that it would like to reduce. Penny Wise, MMI's peripatetic marketing executive, has learned that there is some demand for novel dienes such as 1,4-diphenyl-1,3-butadiene, which can be converted to interesting products such as *p*-terphenyl through Diels-Alder reactions.

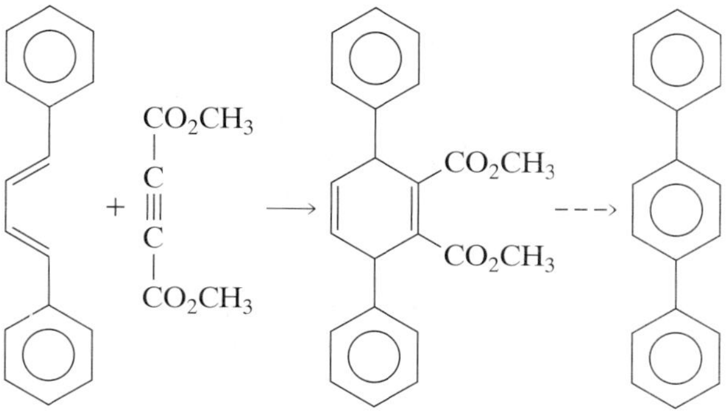

(*E,E*)-1,4-diphenyl-1,3-butadiene   *p*-terphenyl
(*s-cis* conformation)

A staff chemist has informed her that it should be possible to prepare 1,4-diphenyl-1,3-butadiene from cinnamaldehyde by a procedure known as the Wittig synthesis, but its value as a Diels-Alder diene will depend on its stereochemistry. The (*E,E*) diene can easily attain the *s-cis* conformation needed to form a Diels-Alder adduct, but the (*E,Z*) diene is less likely to do so. Your assignment is to find out whether or not you can prepare 1,4-diphenyl-1,3-butadiene from *trans*-cinnamaldehyde and whether or not the major product has the desired (*E,E*) stereochemistry.

Applying Scientific Methodology

The scientific problems in this experiment are implicit in the Scenario. After reading the experiment, you should be able to develop a working hypothesis, which you will test by obtaining the melting point of the major product.

Bark Spices and Cinnamaldehyde

cinnamon stick

Despite claims by wild foods advocate Euell Gibbons that the inner bark of the slippery elm and other trees is edible and nutritious, tree bark is not, as a rule, a popular source of food products. The most notable exceptions to the rule are certain trees of the genus *Cinnamomum*, which provide the spices cinnamon and cassia. True cinnamon is obtained from *Cinnamomum zeylanicum*, a tree that grows in Sri Lanka (formerly Ceylon) and southern India. Cinnamon is obtained by peeling the bark from the cut branches of the cinnamon tree and then scraping off any wood and the outer layers of bark. The thin strips of inner bark are dried in the sun, forming rolled-up "quills" that are up to a meter in length. The quills are generally cut into shorter lengths to produce cinnamon sticks or ground up to make powdered cinnamon. Cinnamon bark from Sri Lanka contains 1–2% of an aldehyde-rich essential oil of which about 70% is cinnamaldehyde (3-phenyl-2-propenal). Other components of cinnamon oil include benzaldehyde, *p*-isopropylbenzaldehyde, 3-phenylpropanal, nonanal, and 2-furaldehyde, along with such nonaldehyde ingredients as 2-heptanone, caryophyllene, and various other flavor components. Cinnamon leaves, surprisingly, contain no cinnamaldehyde but are rich in eugenol, the main flavor ingredient of cloves.

Nearly all of the "cinnamon" consumed in the United States is actually cassia, which is obtained from the Chinese cassia tree, *Cinnamomum cassia*, and several related species grown in southeast Asia. Chinese cassia oil is 80–95% cinnamaldehyde, but its strong, spicy-sweet flavor is quite different from that of true cinnamon oil, which is less sweet but more complex and fragrant, with citrus overtones.

Synthetic cinnamaldehyde, which is mainly the *trans* isomer, is prepared by an aldol-condensation reaction of benzaldehyde and ethanal (acetaldehyde).

$$
\text{C}_6\text{H}_5\!-\!\overset{\overset{\text{O}}{\|}}{\text{CH}} + \text{CH}_3\overset{\overset{\text{O}}{\|}}{\text{CH}} \xrightarrow{\text{NaOH}} \underset{\text{C}_6\text{H}_5}{\overset{\text{H}}{\diagdown}}\text{C}\!=\!\text{C}\underset{\text{H}}{\overset{\overset{\overset{\text{O}}{\|}}{\text{CH}}}{\diagup}} + \text{H}_2\text{O}
$$

Cinnamaldehyde, as well as natural cinnamon and cassia oils, is used to flavor candies, chewing gum, and baked goods, and as an ingredient in perfumes.

Understanding the Experiment

The German chemist Georg Wittig developed the Wittig reaction in 1954, but it took 25 years before he received full recognition for originating one of the most synthetically useful reactions in organic chemistry. In 1979, Wittig shared the Nobel prize in chemistry with another well-known synthetic organic chemist, Herbert C. Brown of the United States.

Like the aldol condensation (discussed in Experiment 40), the Wittig reaction is used to construct larger molecules from smaller ones, connecting the components of the smaller molecules with carbon-carbon double bonds. Both reactions involve the attack of a nucleophilic carbon atom, stabilized by a neighboring electron-withdrawing group, on the carbonyl carbon atom of an aldehyde or ketone, as shown by the following examples.

aldol condensation

$$\underset{\underset{H}{|}}{\overset{O}{\underset{\|}{RC}}}\overset{}{\text{:CH}_2\text{CH}}\quad\text{an enolate ion}\longrightarrow \underset{\underset{H}{|}}{\overset{\overset{-}{O}\text{:}}{\underset{|}{RC}}}-\overset{O}{\underset{\|}{\text{CH}_2\text{CH}}}\xrightarrow{H^+}\underset{\underset{H}{|}}{\overset{OH}{\underset{|}{RC}}}-\overset{O}{\underset{\|}{\text{CH}_2\text{CH}}}$$

Wittig reaction

$$\underset{\underset{H}{|}}{\overset{O}{\underset{\|}{RC}}}\overset{}{\text{:CH}_2}-\overset{+}{\text{PPh}_3}\quad\text{a phosphorus ylide}\longrightarrow \underset{\underset{H}{|}}{\overset{\overset{-}{O}\text{:}}{\underset{|}{RC}}}-\text{CH}_2-\overset{+}{\text{PPh}_3}$$

In the aldol condensation, the nucleophile is an enolate ion, which is stabilized by resonance involving the carbonyl group. In the Wittig reaction, the nucleophile is a phosphorus ylide, which is stabilized by resonance involving a triphenylphosphonium group. In both reactions, the carbon–carbon double bond then forms by elimination (in several steps) of a molecular species; H_2O in the case of the aldol condensation and $Ph_3P{=}O$ in the case of the Wittig reaction.

Key Concept: Carbon is ordinarily not nucleophilic because it has no unshared electron pair, but a nucleophilic carbon atom can be formed by transfer of a proton from a C—H bond to a strong base. This transfer is facilitated by the presence of some group that can stabilize the resulting species.

aldol condensation

$$\underset{\underset{H}{|}}{\overset{\overset{HO\ \ H\ \ O}{|\ \ \ |\ \ \|}}{RC}}-\text{CHCH}\xrightarrow{-H_2O}\text{RCH}{=}\text{CHCH}$$

Wittig reaction

$$\underset{\underset{H}{|}}{\overset{\overset{\ \ \ \overset{-}{O}\text{:}\ \ \overset{+}{\text{PPh}_3}}{\ \ \ |\ \ \ |}}{RC}}-\text{CH}_2\overset{-Ph_3P=O}{\dashrightarrow}\text{RCH}{=}\text{CH}_2$$

During a Wittig reaction, the Ph_3P group needed to stabilize the nucleophilic carbon atom is lost along with the carbonyl oxygen. Thus the Wittig reaction, unlike the aldol condensation, can be used for the synthesis of unsaturated hydrocarbons that have no additional functional groups.

The key intermediate in a Wittig synthesis is the resonance-stabilized phosphorus ylide, which is typically prepared by the reaction of triphenylphosphine with an alkyl halide to yield a phosphonium salt, followed by treatment with a strong base such as butyllithium.

See your lecture textbook for a more complete description and mechanism of the Wittig reaction.

$$Ph_3P + CH_3I \longrightarrow [Ph_3\overset{+}{P}{-}CH_3]I^- \xrightarrow{C_4H_9Li} [Ph_3\overset{+}{P}{-}\overset{=}{C}H_2 \longleftrightarrow Ph_3P{=}CH_2]$$

In this experiment, the phosphonium salt will be benzyltriphenylphos-phonium chloride, which is prepared by the reaction of triphenylphosphine with benzyl chloride. This is an S_N2 reaction in which triphenylphosphine (the nucleophile) displaces chloride ion (the leaving group) from benzyl chloride (the substrate).

$$Ph_3P + ClCH_2 \text{—} \bigcirc \longrightarrow [Ph_3\overset{+}{P} \text{—} CH_2 \text{—} \bigcirc] Cl^-$$

benzyltriphenylphosphonium chloride

The phosphonium salt may be provided, or you may have to synthesize it as described in part **A** of the Directions. The reaction is carried out in the high-boiling solvent *p*-cymene (1-isopropyl-4-methylbenzene) over a 2-hour re-action period.

Formation of the ylide is facilitated by the presence of the phenyl group, which helps stabilize it, so concentrated sodium hydroxide is used rather than a stronger base such as butyllithium.

$$[Ph_3\overset{+}{P} \text{—} CH_2 \text{—} \bigcirc] Cl^- \xrightarrow{\text{NaOH}} Ph_3\overset{+}{P} \text{—} \overset{..}{\overset{-}{C}}H \text{—} \bigcirc$$

You will carry out the ylide-forming reaction in the presence of cin-namaldehyde, which reacts with the ylide to form the product, 1,4-diphenyl-1,3-butadiene.

$$\bigcirc \text{—} CH \text{=} CH\overset{\overset{\displaystyle O}{\parallel}}{C}H + Ph_3\overset{+}{P} \text{—} \overset{..}{\overset{-}{C}}H \text{—} \bigcirc$$

cinnamaldehyde

$$\bigcirc \text{—} CH \text{=} CHCH \text{=} CH \text{—} \bigcirc + Ph_3P \text{=} O$$

1,4-diphenyl-1,3-butadiene

This reaction takes place at room temperature in a two-phase mixture, with dichloromethane as the organic phase. Most of the major product will re-main in the organic phase; the rest is extracted from the aqueous phase with additional dichloromethane. The crude product is triturated with aqueous 60% ethanol to remove triphenylphosphine oxide and a minor product of the reaction (which can be isolated as described in "Other Things You Can Do"), then purified further by recrystallization from 95% ethanol. The flat crystals tend to stick to glass surfaces and are hard to scrape off, so you should avoid any unnecessary transfers.

You will be using the *trans* isomer of cinnamaldehyde, so the geometry of one double bond of the major product will also be *trans*. The geometry of the second double bond—the one formed in the reaction—could be either *cis* or *trans*, allowing for the formation of two possible products, (*E*,*E*)- and (*E*,*Z*)-1,4-diphenyl-1,3-butadiene.

(*E*,*E*)-1,4-diphenyl-1,3-butadiene   (*E*,*Z*)-1,4-diphenyl-1,3-butadiene

A melting point measurement should tell you whether or not you obtained the desired (*E*,*E*) isomer. Ultraviolet spectrometry can also be used to distinguish geometric isomers, as described in "Other Things You Can Do."

Reactions and Properties

A Ph$_3$P + PhCH$_2$Cl $\longrightarrow$ [Ph$_3\overset{+}{\text{P}}$—CH$_2$Ph] Cl$^-$

  benzyl chloride   benzyltriphenylphosphonium chloride

B [Ph$_3\overset{+}{\text{P}}$—CH$_2$Ph] Cl$^-$ + NaOH $\longrightarrow$ Ph$_3\overset{+}{\text{P}}$—$\overset{..}{\overset{-}{\text{C}}}$HPh + H$_2$O + NaCl

  trans-cinnamaldehyde   (*E*,*E*)- or (*E*,*Z*)-1,4-diphenyl-1,3-butadiene

Table 39.1 Physical properties

	M.W.	mp	bp	*d*
benzyl chloride	126.6	−43	179	1.100
triphenylphosphine	262.3	80.5	377	
benzyltriphenylphosphonium				
chloride	388.9			
p-cymene	134.2	−68	177	0.857
trans-cinnamaldehyde	132.2	−7.5	246	1.050
dichloromethane	84.9	−97	40.5	1.326
(*E*,*E*)-1,4-diphenyl-1,3-butadiene	206.3	153		
(*E*,*Z*)-1,4-diphenyl-1,3-butadiene	206.3	88		

Note: mp and bp are in °C, density is in g/mL.

Directions

A. *Preparation of Benzyltriphenylphosphonium Chloride*
This step can be omitted if commercial or previously prepared benzyltriphenylphosphonium chloride is available.

benzyl chloride   *p*-cymene

Benzyl chloride is toxic when inhaled and highly irritating to the skin and eyes. It is a suspected carcinogen and an experimental teratogen (a substance that may harm a developing fetus). It may react violently with oxidants. Wear gloves and use a hood; avoid contact, do not breathe its vapors, and keep it away from other chemicals.

Triphenylphosphine is toxic when ingested and may be harmful if inhaled; do not breathe its vapors.

p-Cymene is flammable and irritates the skin. Avoid contact and keep it away from flames.

Take Care! Wear gloves, avoid contact with the reactants and solvent, and do not breathe their vapors.

Stop and Think: Why does the product precipitate from this reaction mixture?

Waste Disposal: Place the *p*-cymene in a designated solvent recovery container.

Reaction.   *Under the hood*, combine 0.30 mL of benzyl chloride with 0.90 g of triphenylphosphine and 4.5 mL of *p*-cymene in a 10-mL round-bottom flask. Add a stir bar [OP-10], attach an air condenser, and heat the reaction mixture under reflux [OP-7] for 2 hours or more, stirring to prevent bumping as the solid product forms. Let the mixture cool to room temperature, then cool it in an ice/water bath for 15 minutes and collect the product by vacuum filtration [OP-13], washing it with 2 mL of cold, low-boiling petroleum ether to remove the *p*-cymene. Dry [OP-23] the benzyltriphenylphosphonium chloride and measure its mass.

B. *Preparation of 1,4-Diphenyl-1,3-butadiene*

If you obtained less than 0.39 g of benzyltriphenylphosphonium chloride from part **A**, scale down all quantities of reactants and solvents proportionately.

dichloromethane   sodium hydroxide

trans-Cinnamaldehyde is a skin irritant; avoid contact.

Dichloromethane may be harmful if ingested, inhaled, or absorbed through the skin. There is a possibility that prolonged inhalation of dichloromethane may cause cancer. Minimize contact with the liquid and do not breathe its vapors.

Sodium hydroxide is toxic and corrosive, causing severe damage to skin, eyes, and mucous membranes. Wear gloves and avoid contact with the NaOH solution.

Take Care! Avoid contact with dichloromethane and do not breathe its vapors. Wear gloves and avoid contact with the NaOH solution.

Waste Disposal: Flush the aqueous layer down the drain with plenty of water.

Reaction.   In a clean, dry 3-mL conical vial, combine 1.00 mmol of pure *trans*-cinnamaldehyde, 1.0 mL of dichloromethane, and 0.39 g of benzyltriphenylphosphonium chloride. Drop in a stirring device; then add 0.50 mL of 50% aqueous (~19 M) sodium hydroxide while stirring [OP-10]. Attach an air condenser, clamp the apparatus to a ring stand, and stir the mixture vigorously at room temperature for 30 minutes.

Separation.   Transfer the reaction mixture to a centrifuge tube, using 2.0 mL of dichloromethane followed by 1.5 mL of water for the transfer. Shake gently but thoroughly to extract [OP-15] the product into the dichloromethane layer. Dry [OP-23] the dichloromethane layer with anhydrous magnesium sulfate or sodium sulfate, and collect it in a clean, dry, 5-mL conical vial.

Evaporate [OP-16] the dichloromethane until only an oily or solid residue remains. If the residue is an oil, it should solidify as it cools; if necessary, cool it in an ice/water bath.

Purification and Analysis.   Add 3.5 mL of aqueous 60% (by volume) ethanol to the solidified residue and triturate (rub and crush) the solid thoroughly with a flat-bladed microspatula. This procedure removes triphenylphosphine oxide and other impurities. Collect the solid by vacuum filtration [OP-13], washing it on the filter with several small portions of ice-cold 60% ethanol. Recrystallize [OP-25] the crude 1,4-diphenyl-1,3-butadiene from 95% ethanol. Then dry it [OP-23] and measure its mass and melting point [OP-30]. Decide whether your product is the (*E,E*) or the (*E,Z*) isomer.

Waste Disposal: At your instructor's option, save the filtrate for further work. Otherwise, dispose of it as requested.

Exercises

1 Why do you think the product you obtained, rather than another isomer, is the major product of the Wittig reaction?

2 Of the three 1,4-diphenyl-1,3-butadiene isomers, (*E,E*), (*E,Z*), and (*Z,Z*), which would be most suitable as a diene in the Diels–Alder reaction? Which would be least suitable? Explain. (b) Why is (*Z,Z*)-1,4-diphenyl-1,3-butadiene not a likely product of the reaction you carried out?

3 Describe and explain the possible effect on your results of the following experimental errors or variations. (a) The lab assistant, thinking chlorobenzene and benzyl chloride were different names for the same compound, put chlorobenzene in the benzyl chloride bottle. (b) There was no *p*-cymene available, so you used toluene instead. (c) You didn't have a magnetic stirrer, so in part **B** you shook the reaction mixture for a few seconds every minute or so.

4 Following the format in Appendix V, construct a flow diagram for the synthesis of 1,4-diphenyl-1,3-butadiene in part **B**.

5 Outline a Wittig synthesis of each of the following compounds from an appropriate alkyl halide and carbonyl compound.

a.

b.

c.

Other Things You Can Do

(Starred projects require your instructor's permission.)

*1 Isolate and characterize another isomer of 1,4-diphenyl-1,3-butadiene from the filtrate saved from the first vacuum filtration in part **B**. First transfer the filtrate to a centrifuge tube (or conical vial) and withdraw any insoluble oil using a Pasteur pipet. Dissolve the oil in 1.0 mL of dichloromethane (more if necessary), dry the dichloromethane solution, and evaporate the solvent. Keep this substance in the dark, as it is converted to yet another isomer by light. Dissolve a small crystal of the substance in 10 mL of hexanes and obtain its UV spectrum [OP-38] between 200 and 400 nm, diluting the solution with more hexanes as necessary to keep the peaks on scale. Add a small crystal of iodine and illuminate the solution with a 100-watt light bulb for 10 minutes or more, then obtain its UV spectrum in the same region. Compare both spectra with the UV spectrum of your original isomer from part **B**, recorded in the same way, and explain your results.

*2 Carry out a synthesis that involves C═N bond-forming reactions, as described in Minilab 33.

3 Starting with sources listed in the Bibliography, write a research paper about the mechanism, stereochemistry, and applications of the Wittig reaction.

Effect of Reaction Conditions on the Condensation of Furfural with Cyclopentanone

EXPERIMENT 40

Reactions of Carbonyl Compounds. Preparation of α,β-Unsaturated Carbonyl Compounds. Nucleophilic Addition. Condensation Reactions. Carbanions. NMR Spectrometry.

Operations

OP-8   Cooling
OP-10   Mixing
OP-13   Vacuum Filtration
OP-14   Centrifugation
OP-15   Extraction
OP-16   Evaporation
OP-21   Washing Liquids
OP-22   Drying Liquids
OP-23   Drying Solids
OP-25   Recrystallization
OP-28   Vacuum Distillation
OP-30   Melting Point
OP-37   Nuclear Magnetic Resonance Spectrometry

Before You Begin

1 Read the experiment, read or review the operations as necessary, and write an experimental plan. Read OP-28 carefully if you have not performed a vacuum distillation before.
2 For part **A**, calculate the mass and volume of 10.0 mmol of cyclopentanone and 10.0 mmol of furfural.
3 For part **B**, calculate the mass and volume of 2.50 mmol of cyclopentanone.

Scenario

Grits 'n Groats, a breakfast-cereal manufacturer, produces huge quantities of oat hulls and corncobs while processing cereal grains. In the past, the company has simply disposed of such by-products, but the rising cost of waste disposal has convinced the board of directors that Grits 'n Groats should find a way to profit from them rather than paying to get rid of them. Oat hulls, corncobs, and other agricultural by-products can be processed to yield furfural, an aldehyde with an aromatic furan ring. While searching the chemical literature for references to furfural, project leader Farina Millet came across a paper in the *Journal of Organic Chemistry* describing a Claisen-Schmidt reaction (a type of aldol condensation) between cyclopentanone and furfural. Under one set of conditions, the reaction yields a low-melting yellow solid, but under a different set of conditions it produces a high-melting golden-orange solid. Dr. Millet needs some samples of these compounds and their structures so that she can explore their potential for commercial development. Your assignment is to synthesize both compounds and identify them by using your chemical intuition, with some help from NMR spectrometry.

Applying Scientific Methodology

After reading the experiment and comparing the reaction conditions, you should be able to formulate a tentative hypothesis about the structures of the products and predict the kind of ^{1}H NMR spectrum each product should have. You will test your hypothesis when you obtain the actual NMR spectra of the products.

From Oats to Furfural

furfural

Furfural, also known as 2-furaldehyde, is the most important member of the furan series of aromatic compounds. The furan ring is aromatic because it has six pi electrons (two from the oxygen atom) distributed about a five-membered ring. The aromatic sextet of a furan ring is less stable than that of a benzene ring, so furans undergo such reactions as electrophilic addition, cycloaddition, and cleavage more readily than the corresponding benzene compounds.

Furfural can be prepared in large quantities by treating such agricultural by-products as bran, oat hulls, corncobs, and peanut shells with dilute acids. These materials contain polysaccharides known as pentosans that are hydrolyzed to pentoses (simple five-carbon sugars) under acidic conditions. The pentoses are then converted to furfural by acid-catalyzed dehydration.

Conversion of a pentose to furfural. (This equation represents the overall process, not the reaction mechanism.)

a pentose   furfural

Figure 40.1 Aromatic furan ring

The first commercial process for the manufacture of furfural was developed in 1922 by the Quaker Oats Company, which was trying to convert oat hulls into better cattle feed at the time. Instead, it came up with a valuable commercial product that can be made cheaply on a large scale. Furfural is used in the purification of lubricating oils, the extractive distillation of 1,3-butadiene (used in the manufacture of rubber), the synthesis of phenolic resins, and the manufacture of a large number of chemical intermediates.

Furfural behaves like a typical aromatic aldehyde in many of its reactions. It can be oxidized to the corresponding carboxylic acid and reduced to the corresponding alcohol and, like benzaldehyde, it undergoes the Cannizzaro reaction and the benzoin condensation. It also reacts with compounds that have active α-hydrogen atoms such as aldehydes and ketones, to yield condensation products. For example, the reaction of furfural with acetone yields furfurylideneacetone by a Claisen-Schmidt reaction, and its condensation with acetic anhydride and sodium acetate forms furylacrylic acid by a Perkin reaction.

furylacrylic acid

Claisen-Schmidt reaction of furfural and acetone

furfurylideneacetone

Understanding the Experiment

The Claisen-Schmidt reaction is a kind of crossed aldol condensation between an aromatic aldehyde and an aliphatic aldehyde or ketone that yields an α,β-unsaturated aldehyde or ketone. As illustrated for the reaction of benzaldehyde and acetone, the aliphatic carbonyl compound loses a proton to form an enolate ion, which attacks the carbonyl carbon of the aromatic aldehyde to yield (after protonation) a β-hydroxy carbonyl compound. This intermediate is generally not isolated but undergoes base-catalyzed dehydration by an E1cb mechanism. The base removes a proton from the α-carbon to form another enolate ion, which loses an OH$^-$ ion to yield the unsaturated product.

In one of the procedures referred to in the Scenario, equimolar amounts of cyclopentanone and furfural are dissolved in diethyl ether and stirred with an aqueous solution of dilute sodium hydroxide. Furfural tends to oxidize and turn dark brown in storage, so furfural from a previously opened bottle should be distilled before use. Because the organic reactants are not very soluble in water, they tend to stay in the ether phase, necessitating vigorous stirring to mix the layers. Extraction of the reaction mixture with diethyl ether yields an impure yellow liquid that is distilled under vacuum to produce a liquid product that should crystallize to a low-melting yellow solid, product **A**. If desired, product **A** can be purified further by mixed-solvent recrystallization from methanol and water, but this may not be practical if the yield of distillate is low.

Mechanism of a Claisen-Schmidt condensation

When cyclopentanone in diethyl ether is stirred with excess furfural in the presence of an effective phase-transfer catalyst such as tricaprylmethyl-ammonium chloride (Aliquat 336), golden-orange crystals of product **B** begin to crystallize from the reaction mixture almost immediately. The exothermic reaction generates enough heat to vaporize the ether; thus it is necessary to cool the reactants in an ice/water bath before and during the reaction period. Product **B** is separated by filtration and purified by recrystallization from 2-butanone. The role played by the phase-transfer catalyst in aldol-type condensations is not entirely clear, but the catalyst may carry hydroxide ions into the organic phase where they can generate enolate ions, which then react with furfural molecules. (See Experiment 22 for further information about phase-transfer catalysis.)

From the structures of the reactants and your knowledge of aldol-type condensation reactions, you should be able to propose likely structures for the two products. The ^{1}H NMR spectra will show clearly which product you have actually prepared in each case. In interpreting the spectra, you should pay particular attention to the number and multiplicity of signals produced by the methylene protons of the cyclopentanone ring. Comparing your spectra with the ^{1}H NMR spectra of the reactants in Figure 40.2 should help you assign the signals in your spectra to specific proton sets.

Although the stereochemistry of the products has not been reported in the literature, most reactions of this type yield the (E) isomers, as in the condensation of benzaldehyde with 4,4-dimethyl-1-tetralone.

4,4-dimethyl-
1-tetralone

(E)-2-benzal-4,4-dimethyl-
1-tetralone

The nearby carbonyl group has a deshielding effect on the vinylic proton of the (E) isomer, raising its chemical shift to 7.7 ppm, compared to a value of 6.6 ppm for the (Z) isomer.

Reactions and Properties

cyclopentanone   furfural   (equation not balanced)

Table 40.1 Physical properties

	M.W.	mp	bp	d
furfural	96.1	−39	162	1.159
cyclopentanone	84.1	−51	131	0.949
2-butanone	72.1	−86	80	0.805
tricaprylmethylammonium chloride	404.2			0.884
product **A**	162.2	60.5	154^{15}	
product **B**	240.3	162		

Note: mp and bp are in °C, superscripts indicate pressure in torr; density is in g/mL.

Cyclopentanone

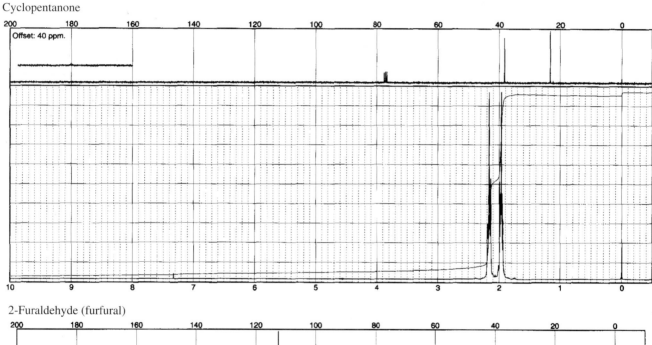

2-Furaldehyde (furfural)

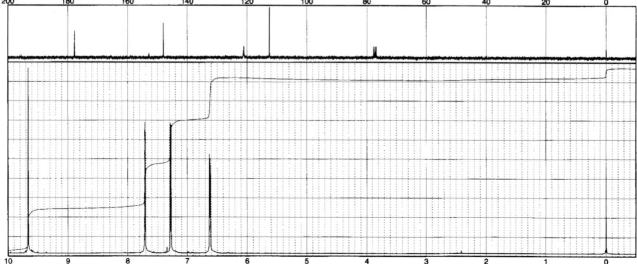

Figure 40.2 NMR spectra of the starting materials

Directions

A. *Claisen-Schmidt Reaction of Cyclopentanone and Furfural*

Take Care! Keep diethyl ether away from ignition sources and do not breathe its vapors. Wear gloves, avoid contact with furfural, and do not breathe its vapors.

Observe and Note: What happens during the reaction period?

Stop and Think: What could this solid be?

Waste Disposal: Pour all aqueous layers down the drain.

Reaction. In a 25-mL Erlenmeyer flask, dissolve 10.0 mmol of cyclopentanone in 6.0 mL of solvent-grade diethyl ether. Add 9.0 mL of aqueous 0.10 M sodium hydroxide and a stir bar. Cool [OP-8] the mixture to 5°C in an ice/water bath and add 10.0 mmol of freshly distilled furfural while stirring. Seal the flask with Parafilm and stir [OP-10] the reaction mixture vigorously in a cold water bath (10–15°C) for 45 minutes. Parafilm dissolves in ether, so don't let it contact the reaction mixture. Replace any ether that evaporates during the reaction.

Separation. Obtain and label three 15-mL centrifuge tubes. Cool a clean filter flask in an ice/water bath; then filter the reaction mixture quickly by vacuum filtration [OP-13] through a Hirsch funnel and save the filtrate, which contains product **A**. Turn the vacuum off immediately to keep the ether from evaporating. Wash any solid on the filter with 2 mL of diethyl ether, combining the wash liquid with the filtrate. Save this solid and weigh it when it is dry. If any solid remains in the filtrate, it will be removed during subsequent operations.

   Transfer all of the filtrate to the first two centrifuge tubes, alternating from one to the other, so that both contain about the same volume of each layer. Use a small amount of diethyl ether for the transfer. Centrifuge [OP-14] the contents of both tubes until the layers are cleanly separated. If necessary, add enough additional ether to provide a 2-cm layer of ether in each centrifuge tube and shake gently to mix the layers thoroughly. Transfer both aqueous (lower) layers to the third centrifuge tube. Extract [OP-15] this aqueous solution with 3 mL of diethyl ether, remove the aqueous layer, and combine all of the ether layers in one centrifuge tube. Wash [OP-21] the ether solution with two separate 2-mL portions of saturated aqueous sodium chloride. Dry [OP-22] it with anhydrous sodium sulfate or magnesium sulfate and evaporate [OP-16] the ether (in portions, if necessary) in a 5-mL conical vial until the volume of the liquid residue remains essentially constant.

Purification and Analysis.   Assemble an apparatus for vacuum distillation [OP-28] using a Hickman still and a 10-mL round-bottom flask and have your instructor approve it. Make sure that a safety shield or hood sash is between you and the apparatus, then purify the residue by vacuum distillation. Any residual ether and unreacted starting materials should distill below 100°C and should be removed before the main fraction begins to distill. Before allowing the Hickman still to cool, transfer the liquid distillate (the main fraction) to a tared vial or another suitable container and cool it in ice water, if necessary, until it completely solidifies. (At your instructor's request, you can purify the product further by mixed solvent recrystallization [OP-25] from methanol and water.) Dry [OP-23] it at room temperature. Measure the mass and melting point [OP-30] of product **A**. Record its ^{1}H NMR spectrum [OP-37] in deuterochloroform or obtain a spectrum from your instructor. Interpret the NMR spectrum as completely as you can, deduce the structure of product **A**, and name it.

B. *Claisen-Schmidt Reaction Using a Phase-Transfer Catalyst*

> **2-Butanone is flammable and ingestion, inhalation, or skin absorption may be harmful. Avoid contact, do not breathe its vapors, and keep it away from flames and hot surfaces.**
> **See part A for safety notes about cyclopentanone, diethyl ether, furfural, and deuterochloroform.**

Reaction.   Weigh 2.50 mmol of cyclopentanone into a 25-mL Erlenmeyer flask and dissolve it in 3.0 mL of diethyl ether. Add 3.0 mL of aqueous 0.10 *M* sodium hydroxide, 2 drops of tricaprylmethylammonium chloride (or 50 mg of a suitable solid phase-transfer catalyst), and a stir bar. Cool [OP-8] the mixture in an ice/water bath for at least 5 minutes and add 0.50 mL of freshly distilled furfural while stirring [OP-10]. Seal the flask with Parafilm and stir it vigorously at room temperature for 15 minutes, occasionally swirling it in the ice/water bath to reduce pressure buildup from vaporizing ether. Then let the reaction mixture stand at room temperature, with continued stirring, for 10 minutes more.

Separation.   Collect the product by vacuum filtration [OP-13], wash it with two portions of diethyl ether, and air dry it on the filter.

Purification and Analysis.   Recrystallize [OP-25] product **B** from 2-butanone, using about 6 mL or less of the solvent. Wash the product on the filter with diethyl ether, dry it [OP-23], and weigh it. (**Waste Disposal:** Place the filtrate in a designated solvent recovery container.) Measure the melting point [OP-30] of product **B**. Record its ^{1}H NMR spectrum [OP-37] in deuterochloroform, or obtain a spectrum from your instructor. Interpret the NMR spectrum as completely as you can, deduce the structure of product **B**, and name it.

Take Care! A vacuum-distillation apparatus may implode if any of its components are cracked or otherwise damaged. Follow the precautions described in OP-28 and the Safety Notes.

Stop and Think: At about what temperature should the product distill?

Waste Disposal: Place any low-boiling forerun in an appropriate waste container.

Take Care! Avoid contact with CDCl$_3$ and do not breathe its vapors.

Safety Notes

2-butanone

Take Care! Keep diethyl ether away from ignition sources and do not breathe its vapors. Wear gloves, avoid contact with furfural, and do not breathe its vapors.

Observe and Note: Compare your observations during this reaction with your observations during the first reaction.

Waste Disposal: Place the filtrate in a designated solvent recovery container.

Take Care! Do not breathe the vapors of 2-butanone and keep it away from ignition sources.

Take Care! Avoid contact with CDCl$_3$ and do not breathe its vapors.

Exercises

1 Discuss the effect of reaction conditions on the outcome of the Claisen-Schmidt reaction, telling what reaction conditions promote the formation of each product and why.

2 (a) What is the probable identity of the solid that was filtered from the reaction mixture in part **A**? How could you have confirmed its identity? (b) What percentage of the cyclopentanone that you started with in part **A** was converted to condensation products? (This is not the same as the percentage yield of **A**.)

3 Write balanced equations and detailed mechanisms for the formation of both products, **A** and **B**.

4 Describe and explain the possible effect on your results of the following experimental errors or variations. (a) You used 10 *M* NaOH rather than 0.10 *M* NaOH in part **A**. (b) In part **A**, you rinsed the reaction flask with acetone and didn't dry it completely. (c) You left out the tricaprylmethylammonium chloride in part **B**. (d) You used 10 mmol of cyclopentanone in part **B** as well as part **A**.

5 Diagram a possible phase-transfer process for the formation of product **B**, using the format illustrated in Experiment 22.

6 (a) Following the format in Appendix V, construct a flow diagram for the synthesis in part **A**. (b) Construct a flow diagram for the synthesis in part **B**.

7 From your ^{1}H NMR spectra, is it more likely that your products are (*Z*) or (*E*) stereoisomers? Explain your answer.

8 An error-prone student, Mel A. Droyt, forgot to add the furfural in part **A**, but he recovered a small amount of liquid that distilled at 139–142°C at 20 torr and did not solidify on cooling. The ^{1}H NMR spectrum of the liquid showed no signals from vinylic or hydroxylic protons. Propose a structure for this product and write a mechanism for its formation.

Other Things You Can Do

(Starred projects require your instructor's permission.)

*1 As a group project, carry out part **B** using different phase-transfer catalysts and compare the crude yields to find out which catalysts are most effective. Suggested catalysts are tetrabutylammonium bromide, tetrabutylphosphonium bromide, cetyltrimethylammonium bromide, and 1-hexadecylpyridinium chloride.

*2 Prepare some other aldol condensation products as described in Minilab 34.

3 Starting with sources listed in the Bibliography, write a research paper about the production and uses of furan, furfural, and some derivatives of these compounds.

Haloform Oxidation of 4'-Methoxyacetophenone

Reactions of Methyl Ketones. Preparation of Carboxylic Acids. Oxidation. Haloform Reaction. Infrared Spectrometry.

Operations

OP-8   Cooling
OP-10 Mixing
OP-13 Vacuum Filtration
OP-21 Washing Liquids
OP-23 Drying Solids
OP-24 Drying and Trapping Gases (optional)
OP-25 Recrystallization
OP-30 Melting Point
OP-36 Infrared Spectrometry

Before You Begin

1 Read the experiment, read or review the operations as necessary, and write an experimental plan.
2 Calculate the mass of 1.00 mmol of 4'-methoxyacetophenone and the theoretical yield of 4-methoxybenzoic acid.

Scenario

The Olfactory Factory, which previously commissioned your Institute to prepare 4'-methoxyacetophenone from anisole (see Experiment 36), has begun to manufacture this compound, but the demand for it has not met their expectations. Accordingly, they would like to explore the possibility of using it to manufacture other useful aroma chemicals. One possibility would be to convert it to 4-methoxybenzoic acid (*p*-anisic acid), which could be used to make pleasant-smelling esters. Now the Olfactory Factory has asked the Institute to come up with a safe and inexpensive way of carrying out the conversion of 4'-methoxyacetophenone to 4-methoxybenzoic acid. The conversion of the —$COCH_3$ group of a methyl ketone to a —COOH group can be accomplished by a haloform reaction. In a typical procedure for this reaction, the methyl ketone is stirred with bromine in aqueous sodium hydroxide for several hours, but your supervisor believes the reaction can be carried out more quickly and safely using ordinary laundry bleach and a phase-transfer catalyst. Your assignment is to see whether or not 4'-methoxyacetophenone can be converted to 4-methoxybenzoic acid by laundry bleach in the presence of a phase-transfer catalyst.

4′-methoxyacetophenone → 4-methoxybenzoic acid

Proposed synthesis of 4-methoxybenzoic acid

Applying Scientific Methodology

Develop a working hypothesis related to the problem posed in the Scenario. You will test your hypothesis by obtaining a melting point and an IR spectrum of the product—if you obtain a product.

Haloforms

The haloforms—chloroform, bromoform, and iodoform—are trihalomethanes that received their trivial names from the fact that they are **halo**gen compounds that can be hydrolyzed to **form**ic acid, HCO_2H.

$$CHCl_3 \qquad CHBr_3 \qquad CHI_3$$
chloroform   bromoform   iodoform

They can all be prepared by a *haloform reaction,* in which acetone is treated with an appropriate hypohalite salt in basic solution.

$$CH_3\overset{O}{\overset{\|}{C}}CH_3 \xrightarrow[\text{OH}^-]{\text{OX}^-} CH_3\overset{O}{\overset{\|}{C}}O^- + CHX_3 \quad (X = Cl, Br, I)$$

Chloroform is also prepared commercially by the controlled chlorination of methane.

$$CH_4 + 3Cl_2 \xrightarrow{h\nu} CHCl_3 + 3HCl$$

In 1846, the British obstetrician James Young Simpson became the first physician to use general anesthesia in childbirth. This practice was criticized as "unnatural" at the time because the book of Genesis reports that, as God cast Adam and Eve out of the Garden of Eden He said to Eve, "...in pain will you bring forth children." Most of Simpson's critics were silenced when he used chloroform to deliver Queen Victoria's seventh child. Although Simpson preferred to use chloroform as an anesthetic, its toxicity allows little margin for error; an inhaled chloroform concentration of about 1.5% is necessary to maintain anesthesia for surgery, but a concentration of 2% causes respiratory arrest. For this reason chloroform was eventually replaced by safer anesthetics such as diethyl ether. Chloroform has also been used as an industrial solvent, a dry-cleaning agent, a fire extinguisher, and a starting material for the synthesis of chlorofluorocarbons (CFCs), but its toxicity and apparent carcinogenic properties, as well as the current phaseout of ozone-depleting CFCs, have reduced its use considerably in recent years.

When surface water is purified by treatment with chlorine, very small amounts of chloroform (averaging about 20 parts per billion) and even smaller amounts of other trihalomethanes are produced. These trihalomethanes result from the reaction of chlorine with natural organic compounds such as humic acids, which are produced from decaying vegetable matter. No one knows for certain whether the small amounts of trihalomethanes in chlorinated water represent a serious health threat, but the U.S. Environmental Protection Agency (EPA) has established a maximum contaminant level (MCL) of 100 ppb for total trihalomethanes in drinking water.

Bromoform is a very dense liquid (d = 2.90 g/mL) that has been used to separate low-density minerals from mineral mixtures. For example, quartz, calcite, halite, and feldspar all float on bromoform because they have densities lower than 2.9 g/mL.

Iodoform, a yellow solid with a disagreeable "medicinal" odor, was once used as an antiseptic to keep wounds from becoming infected, but it has been replaced for that purpose by safer and less odorous materials. It is the product of the well-known iodoform test for carbonyl compounds that contain the —COCH$_3$ group and alcohols that can be oxidized to such carbonyl compounds. The test is carried out by heating the unknown compound in an alkaline iodine solution.

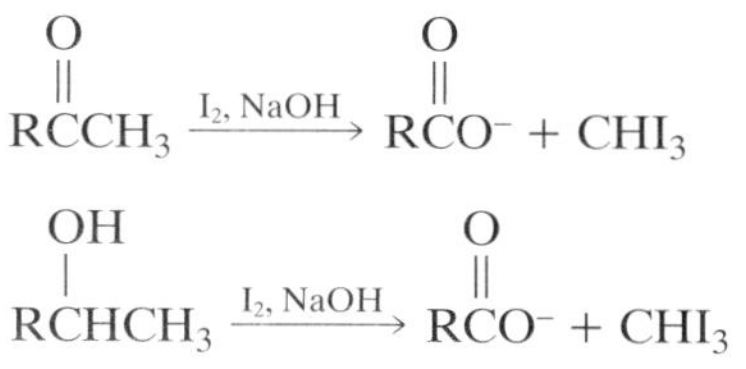

The MCL for a substance is a legally enforceable national standard set by the EPA to protect the public from hazardous materials in water.

$$\underset{RCCH_3}{\overset{O}{\|}} \xrightarrow{I_2,\ NaOH} \underset{RCO^-}{\overset{O}{\|}} + CHI_3$$

$$\underset{RCHCH_3}{\overset{OH}{|}} \xrightarrow{I_2,\ NaOH} \underset{RCO^-}{\overset{O}{\|}} + CHI_3$$

Iodoform test reactions

Understanding the Experiment

In this experiment, you will attempt to convert 4'-methoxyacetophenone to 4-methoxybenzoic acid using laundry bleach, which contains about 5.25% sodium hypochlorite (NaOCl) in water. Under the conditions of the reaction, molecular chlorine and hydroxide ion are produced according to the following reaction equation.

$$OCl^- + Cl^- + H_2O \rightleftharpoons Cl_2 + 2OH^-$$

The haloform reaction is, in effect, two separate reactions that occur under the same conditions. The first is a base-catalyzed halogenation reaction, proceeding via enolate ions, that replaces all three hydrogen atoms on the methyl group by chlorine atoms.

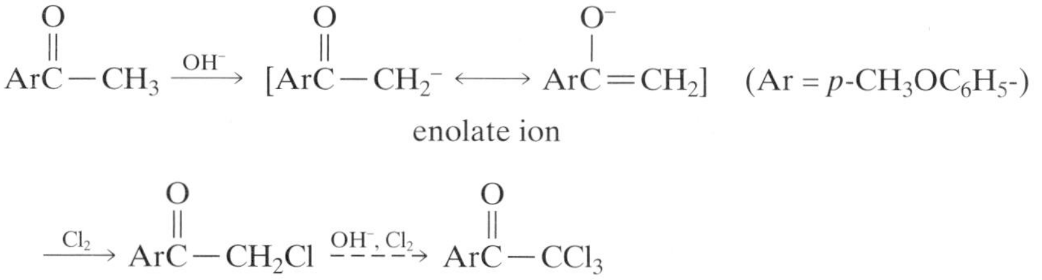

$$\underset{ArC-CH_3}{\overset{O}{\|}} \xrightarrow{OH^-} [\underset{ArC-CH_2^-}{\overset{O}{\|}} \longleftrightarrow \underset{ArC=CH_2}{\overset{O^-}{|}}] \quad (Ar = p\text{-}CH_3OC_6H_5\text{-})$$

enolate ion

$$\xrightarrow{Cl_2} \underset{ArC-CH_2Cl}{\overset{O}{\|}} \xrightarrow{OH^-,\ Cl_2} \underset{ArC-CCl_3}{\overset{O}{\|}}$$

The second involves a nucleophilic acyl substitution reaction in which hydroxide ion is the nucleophile and CCl$_3^-$ is the leaving group.

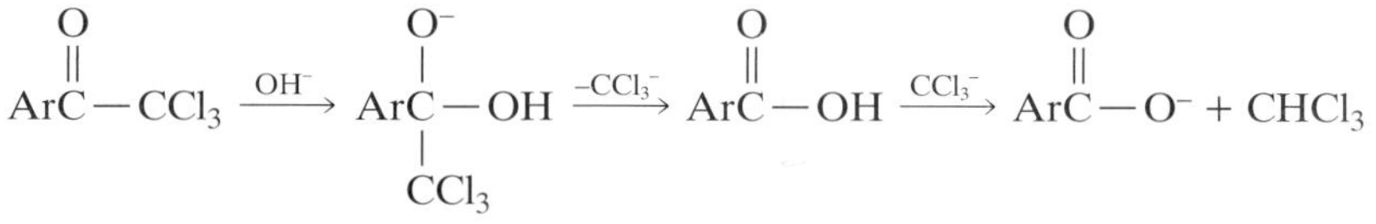

$$\underset{ArC-CCl_3}{\overset{O}{\|}} \xrightarrow{OH^-} \underset{\underset{CCl_3}{|}}{\overset{O^-}{\underset{ArC-OH}{|}}} \xrightarrow{-CCl_3^-} \underset{ArC-OH}{\overset{O}{\|}} \xrightarrow{CCl_3^-} \underset{ArC-O^-}{\overset{O}{\|}} + CHCl_3$$

Carbanions are ordinarily very poor leaving groups, but the three chlorine atoms on CCl$_3^-$ stabilize its negative charge, allowing its displacement by OH$^-$. The last step of this reaction sequence is an acid-base reaction that yields the haloform, in this case chloroform, and the salt of the carboxylic acid. Acidification of the reaction mixture then yields the free carboxylic acid.

Key Concept: Charge delocalization stabilizes a charged species, so a carbanion is stabilized by electron-withdrawing groups that can spread out its negative charge.

Chloroform is toxic and some of its vapors may escape from the reaction mixture, so you should carry out the reaction under a hood or use a gas trap. 4′-Methoxyacetophenone is a low-melting solid that should melt to an oily liquid shortly after the reaction begins. Since the liquid is insoluble in the aqueous hypochlorite solution, it is important to keep the reaction mixture well stirred. A phase-transfer catalyst, tricaprylmethylammonium chloride, is used to help ionic species involved in the reaction cross the phase boundary. During the reaction, you should test the reaction mixture occasionally with starch-iodide paper to make sure that sodium hypochlorite is still present. Sodium hypochlorite tends to bleach out the blue-black color that signifies a positive test, but you should see a fringe of color around the bleached spot. Any excess NaOCl left at the end of the reaction period is consumed by treatment with acetone, and the aqueous solution is washed with diethyl ether to remove chloroform and any unreacted starting material. The product is then precipitated with dilute hydrochloric acid and recrystallized from aqueous ethanol.

If your product is 4-methoxybenzoic acid, you should be able to identify the C=O, acyl C—O, and O—H stretching bands and an O—H bending band arising from the carboxyl group in its IR spectrum. Additional C—O stretching and Ar—H bending bands from the ether function and benzene ring should be observed as well. See OP-36 for additional information about the interpretation of IR spectra.

Reactions and Properties

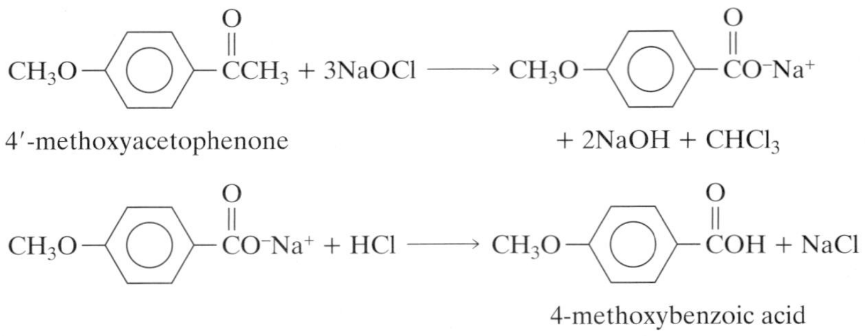

4′-methoxyacetophenone

+ 2NaOH + CHCl$_3$

4-methoxybenzoic acid

Table 41.1 Physical properties

	M.W.	mp	bp	d
4′-methoxyacetophenone	150.2	38	154^{26}	
4-methoxybenzoic acid	152.2	185	280	
chloroform	119.4	−63.5	62	1.484

Note: mp and bp are in °C, density is in g/mL.

Directions

The reaction should be performed under a hood to keep chloroform and chlorine vapors out of the lab. If this is not possible, it should be carried out in a reflux apparatus fitted with a gas trap [OP-24].

4-Methoxybenzoic acid is a mild sensitizer that may cause contact der-matitis. Minimize contact with the product.
Aqueous sodium hypochlorite can irritate the skin, eyes, and respiratory tract. Avoid contact and do not breathe its vapors.
The reaction mixture may evolve some chlorine gas, which irritates the eyes and respiratory tract, and chloroform, which is harmful if inhaled and is a suspected human carcinogen. Do not breathe the vapors from the reaction mixture.

Safety Notes

chloroform

Reaction.   Accurately weigh 1.00 mmol of 4′-methoxyacetophenone into a small Erlenmeyer flask. *Under the hood,* add 5.0 mL (~3.7 mmol) of aqueous 5.25% sodium hypochlorite (household bleach) and 1 drop of tri-caprylmethylammonium chloride, then drop in a magnetic stir bar. Stir [OP-10] the reaction mixture *vigorously* for at least 1 hour in a 60°C water bath. Test the reaction mixture with starch-iodide paper after 10 minutes and more often thereafter. If at any time the test is negative (no immediate blue-black color), add more 5.25% NaOCl solution in small portions until the test is positive. At the end of the reaction period, there should be little if any oily liquid left on the surface of the reaction mixture.

Take Care! Avoid contact with the bleach, and do not breathe its vapors.

Observe and Note: What evidence do you see for a reaction?

Separation.   Test the reaction mixture with starch-iodide paper and add acetone a few drops at a time with swirling until the test is negative (no blue-black color). When the reaction mixture has cooled to room tempera-ture, transfer it to a centrifuge tube and wash [OP-21] it with two 1-mL portions of solvent-grade diethyl ether. Be sure to save the aqueous layer, which contains the product. Add enough 3 M hydrochloric acid (about 1 mL or more) to the aqueous layer with stirring to bring the pH of the solution down to 2. Cool [OP-8] the reaction mixture in an ice/water bath for 5 min-utes or more. Collect the product by vacuum filtration [OP-13], washing it on the filter with several portions of cold water.

Stop and Think: What does the acetone do?

Stop and Think: What is the ether removing from the reaction mixture?

Waste Disposal: Place the wash ether in the appropriate solvent re-covery container. Pour all filtrates down the drain.

Purification and Analysis.   Purify the product by recrystallization [OP-25] from a small amount of aqueous 60% ethanol. Dry [OP-23] it to constant mass and measure its mass and melting point [OP-30]. Record an IR spec-trum [OP-36] of the product or obtain a spectrum from your instructor. In-terpret the spectrum as completely as you can and turn it in with your report.

Exercises

1 Write a balanced equation for the reaction undergone by acetone during the separation step.
2 Ketones can be chlorinated under acidic as well as basic conditions. Could the haloform reaction be carried out successfully in the presence of an acidic catalyst such as HCl? Why or why not?
3 Describe and explain the possible effect on your results of the follow-ing experimental errors or variations. (a) The laundry bleach you used was from an old bottle and contained only 3.5% NaOCl. (b) You didn't

wash the reaction mixture with diethyl ether. (c) You used benzophenone in place of 4′-methoxyacetophenone.

4 Following the format in Appendix V, construct a flow diagram for this experiment.

5 Which of the following compounds should give positive iodoform tests?

6 During the haloform reaction, each successive hydrogen atom of the methyl group is replaced by chlorine more rapidly than the previous one. Explain why.

7 (a) The time-averaged permitted emission limit (PEL) for chloroform in air set by the Occupational Safety and Health Administration (OSHA) is 2 ppm by volume. What is the minimum mass of chloroform that would have to vaporize at 25°C and 1 atm in a laboratory having a volume of 300 m^3 in order to attain a concentration in air of 2.0 ppm? (b) If each student in this laboratory treats 1.0 mmol of 4′-methoxyacetophenone with laundry bleach and allows all of the chloroform produced to vaporize into the lab, what is the minimum number of students it would take to exceed the PEL?

Other Things You Can Do

(Starred projects require your instructor's permission.)

*1 Observe a spontaneous reaction of benzaldehyde and identify the product, as described in Minilab 35.

*2 Prepare iodoform by the following procedure. Dissolve 0.75 g of potassium iodide in 5 mL of water in a small Erlenmeyer flask and add 0.2 mL of acetone. Then add, in small portions with stirring or shaking, 15 mL of bleach solution (5.25% NaOCl). Let the mixture stand for 5 minutes with stirring or shaking, cool it in ice, and collect the product by vacuum filtration. Dry the iodoform and obtain its melting point. (It can be crystallized from methanol if desired.)

3 Starting with sources listed in the Bibliography, write a research paper about general anesthetics. Describe the applications and physiological effects of some modern general anesthetics, giving their chemical structures and their advantages and disadvantages.

Electronic Effect
of a *para*-Iodo Substituent

*Reactions of Aromatic Amines. Preparation of Aryl Halides. Aromatic Nucle-
ophilic Substitution. Carboxylic Acids. Diazonium Salts. Linear Free-Energy
Relationships.*

Operations

OP-7   Heating
OP-8   Cooling
OP-10   Mixing
OP-13   Vacuum Filtration
OP-23   Drying Solids
OP-25   Recrystallization

Before You Begin

1 Read the experiment, read or review the operations as necessary, and write
   an experimental plan.
2 Calculate the masses of 2.00 mmol of *p*-aminobenzoic acid, 2.00 mmol of
   sodium nitrite, and 3.0 mmol of potassium iodide. Calculate the theoretical
   yield of *p*-iodobenzoic acid.
3 For each acid listed in Table 42.1, calculate the mass of 0.20 mmol of the
   acid.

Scenario

Si Starr, an energetic professor of metaphysical chemistry at Miskatonic
University, is using molecular orbital theory to predict the electronic effects
of various substituents on such reactions as the ionization of aromatic carboxylic acids. Substituent effects can be expressed quantitatively by empirical parameters called sigma values. To test the validity of his results,
Professor Starr wants to compare his calculated sigma values with experimentally measured sigma values. Sigma values have been determined for
most of the common substituents, but Starr has been unable to locate reliable sigma values for the iodo substituent ($-I$) on a benzene ring. Your assignment is to prepare *p*-iodobenzoic acid, measure its pK_a value, and use
that to determine the sigma value for a *para*-iodo substituent.

Applying Scientific Methodology

After reading the experiment, you can propose a hypothesis about the electronic effect of a *p*-iodo substituent. The sigma value you obtain will show
whether or not your hypothesis is correct.

Linear Free-Energy Relationships

Overlap of lone-pair electrons with pi system in anisole

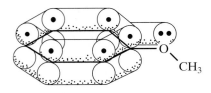

In general chemistry, you learned that electrons in a covalent bond tend to migrate toward the more electronegative atom, building up its electron density at the expense of the less electronegative atom. The situation is not always so simple in conjugated organic compounds, where the electrons of unshared pairs or in pi bonds sometimes migrate *away* from a more electronegative atom into the "electron sink" of a delocalized pi-electron system. For example, lone-pair electrons from an oxygen-containing substituent such as methoxyl ($—OCH_3$) tend to migrate toward and overlap with the pi-electron cloud of a benzene ring, as illustrated for anisole (methoxybenzene). The net electronic effect of a substituent is indicated by an empirical (experimentally measured) parameter called its *sigma value*. Electron-withdrawing groups have positive sigma values and electron-donating groups have negative ones. For example, the measured sigma value of $—OCH_3$ is -0.27, indicating that it has a net electron-donating effect at the *para* position of a benzene ring, even though oxygen is considerably more electronegative than carbon.

Consider a general substituent Z at the *para* position of a substituted benzoic acid. If Z is more electronegative than carbon, it will tend to withdraw electrons through the sigma-bond framework of the benzene ring by an *inductive effect*, and this should increase the strength of the acid by stabilizing its conjugate base. Conversely, a substituent that is more electropositive than carbon can donate electrons inductively and weaken an acid. But inductive effects decrease rapidly with distance, and because a *para* substituent is remote from the reaction site (the COOH group), its inductive effect may be quite small.

Substituents that have unpaired electrons and are adjacent to a conjugated system tend to donate electrons by a *resonance effect*. Electron donation by resonance can cause a buildup of electron density at locations within the ring and on conjugated substituents, as illustrated by the following resonance structures for the conjugate base of a substituted benzoic acid:

Conjugate base is destabilized by electron donation
transmitted through the pi-electron system.

Key Concept: *Stabilization of the conjugate base of an acid strengthens the acid; destabilization of its conjugate base weakens the acid.*

In this example, the conjugate base is destabilized by a concentration of negative charge near the reaction site. Destabilization of the conjugate base shifts the ionization equilibrium to the left and reduces the strength of the acid. Resonance effects are transmitted freely throughout a conjugated system and do not decrease significantly with distance. They are most important when the substituent is *ortho* or *para* to the ring carbon nearest the reaction site. Substituents such as the methoxyl group can donate electrons by resonance but withdraw them inductively, and thus may either weaken or strengthen an acid, depending on the relative importance of the two effects. Thus, a *meta*-methoxyl substituent (for which the resonance effect is weak) increases the acidity of a substituted benzoic acid by an inductive effect, while a *para*-methoxyl substituent decreases the acidity of such an acid by resonance.

The pK_a of an acid is defined as the negative logarithm of its ionization constant:

$$pK_a = -\log K_a$$

For example, the pK_a of phosphorous acid, which has a K_a value of 1.0×10^{-2}, is 2.00. At a given temperature, the pK value for any equilibrium reaction is directly proportional to its standard free-energy change, $\Delta G°$. This can be shown by using the definition of pK_a to rewrite the thermodynamic equation that relates $\Delta G°$ to the equilibrium constant:

$$\Delta G° = -(2.303 \cdot RT) \log K = (2.303 \cdot RT) \, pK$$

The value of $\Delta G°$ for a reaction is a measure of its tendency to go to completion, so pK must measure the same tendency. Negative pK_a and $\Delta G°$ values are observed for strong acids that dissociate completely, or nearly so. Most organic acids have positive pK_a values because they are only partly dissociated in ionizing solvents; a high positive pK_a value indicates a weak acid, and a low one indicates a stronger acid.

Suppose we compare the pK_a values for a pair of acids that differ in only one respect—that one acid has a substituent at a site where the other has only a hydrogen atom. Since each acid's pK_a is a measure of its acid strength, the difference between their pK_a values must measure the substituent's effect on acid strength. For example, in water at 25°C benzoic acid has a pK_a value of 4.19, while *p*-nitrobenzoic acid has a pK_a value of 3.41. The difference between these pK_a values, $4.19 - 3.41 = 0.78$, is a quantitative measure of the *para*-nitro substituent's acid-strengthening effect. Similarly, *p*-toluic acid (p-CH$_3$C$_6$H$_4$COOH) has a pK_a value of 4.36; subtracting this value from the pK_a for benzoic acid gives a difference of -0.17, indicating that the methyl substituent has a small acid-weakening effect. The difference between the pK_a value for the unsubstituted acid and that for the substituted acid, which we shall call ΔpK, is a measure of the substituent's ability to donate or withdraw electrons.

$$\Delta pK = pK_a \text{ of unsubstituted acid} - pK_a \text{ of substituted acid} \qquad \textbf{(1)}$$

Because substituents that give negative ΔpK values weaken acids, they must be electron donors, whereas those that give positive ΔpK values are electron acceptors. The numerical value of ΔpK in either case measures the magnitude of the substituent's effect.

The ionization of benzoic acid in water at 25°C has been selected as the reference reaction for measuring substituent effects. *Substituent constants*, symbolized by the Greek letter sigma (σ) are, in effect, adjusted ΔpK values for substituted benzoic acids.

$$\sigma = \Delta pK \text{ (benzoic acids)} = pK_a \text{ (bz)} - pK_a \text{ (Zbz)} \qquad \textbf{(2)}$$

(bz = benzoic acid; Zbz = substituted benzoic acid)

benzoic acid

p-nitrobenzoic acid

Don't confuse a negative ΔpK value, which is associated with an acid-weakening substituent, with a negative pK_a value, which indicates a strong acid.

phenylacetic acid

p-nitrophenylacetic acid

Table 42.1 in the "Reactions and Properties" section lists the reported sigma values for some common *para* substituents.

The effect of a substituent will usually change if the substrate or the reaction type is changed, or even if the reaction conditions are varied. For example, the pK_a values for phenylacetic acid and *p*-nitrophenylacetic acid in water are 4.28 and 3.85, respectively; thus ΔpK for the *para*-nitro group on this substrate is 0.43, which is considerably less than its ΔpK of 0.78 in the benzoic acid system. We can use the ratio of these ΔpK values to estimate the relative effect of a substituent on the ionization of the substituted phenylacetic acid (keeping in mind that $\sigma \approx \Delta pK$ for the benzoic acid system):

$$\frac{\Delta pK(\text{phenylacetic acids})}{\sigma} = \frac{0.43}{0.78} = 0.55$$

So the nitro group has only about half the acid-strengthening effect on phenylacetic acid that it does on benzoic acid. This is to be expected, because the substituent is farther from the reaction site in phenylacetic acid. Comparing the effects of a large number of substituents on the same reaction yields an averaged $\Delta pK/\sigma$ value of 0.49 for the ionization of phenylacetic acid. This value, called the *reaction constant*, is symbolized by the Greek letter rho (ρ). Rearranging the equality $\rho = \Delta pK/\sigma$ leads to the *Hammett equation*,

$$\Delta pK = \rho\sigma \qquad\qquad\qquad \textbf{(3)}$$

which, for organic acids, can also be written as

$$\log \frac{K_a(\text{substituted acid})}{K_a(\text{unsubstituted acid})} = \rho\sigma$$

According to the Hammett equation, the effect that any substituent (Z) will have on the pK of a particular reaction can be estimated by multiplying the sigma value for Z (the parameter that measures the electronic effect of Z on a reference reaction) by the rho value for the reaction in question (the parameter that measures the sensitivity of the reaction to the electronic effects of substituents). Because of the proportionality between pK and $\Delta G°$ mentioned previously, the Hammett equation is called a *linear free-energy relationship*.

By definition, the reaction constant (rho value) for the ionization of benzoic acids in water is 1; reactions that are more sensitive to substituent effects than the reference reaction have rho values greater than 1, and reactions that are less sensitive have rho values between 0 and 1. Negative rho values are observed for reactions in which electron donors increase the extent of reaction and electron acceptors decrease it, rather than the other way around. The rho value for a particular reaction may change if the solvent or other reaction conditions are changed. Just as a car's power is needed more on a steep hill than on the level, substituents have the greatest effect when a reaction is most difficult and the least effect when it is easiest.

Understanding the Experiment

In this experiment, you will prepare *p*-iodobenzoic acid from *p*-aminobenzoic acid, which is also known as PABA. Measuring the pK_a value of your product and other acids (from Table 42.1) in 95% ethanol will then allow you to determine the σ value for *p*-iodobenzoic acid and the ρ value for the reaction. An aryl iodide can be prepared by treating the corresponding aromatic amine with nitrous acid in the presence of a mineral acid such as HCl, then heating the resulting diazonium salt with aqueous potassium iodide.

$$ArNH_2 + HONO + HCl \rightarrow ArN_2^+ \, Cl^- + 2H_2O$$

$$ArN_2^+ \, Cl^- + KI \rightarrow ArI + N_2 + KCl$$

Although some displacement reactions of diazonium salts may involve free radical intermediates, others appear to be S_N1 reactions in which the diazonium salt loses nitrogen (as N_2) to form an intermediate aryl carbocation (Ar^+). A nucleophilic species then combines with the carbocation to form either the product or another intermediate that yields the product after a proton exchange.

The diazonium salt of a primary aromatic amine can be prepared by dissolving or suspending the amine in an aqueous acid such as dilute HCl, cooling the solution to 5°C or below, and adding aqueous sodium nitrite. Sodium nitrite reacts with some of the acid to generate nitrous acid according to the equation

$$NaNO_2 + HCl \rightarrow HONO + NaCl$$

The diazonium salt solution must be kept cold because diazonium salts can react with the solvent at elevated temperatures. When this solution is heated with a solution of potassium iodide, *p*-iodobenzoic acid separates as a fine precipitate that may take some time to filter.

To determine the pK_a of *p*-iodobenzoic acid or another acid in 95% ethanol, you will first dissolve some of the acid in 95% ethanol to obtain an approximately 0.1 M solution. You will measure two equal portions of this solution and add just enough dilute NaOH to one of the portions to convert all of the acid it contains to the conjugate base, according to the following reaction equation.

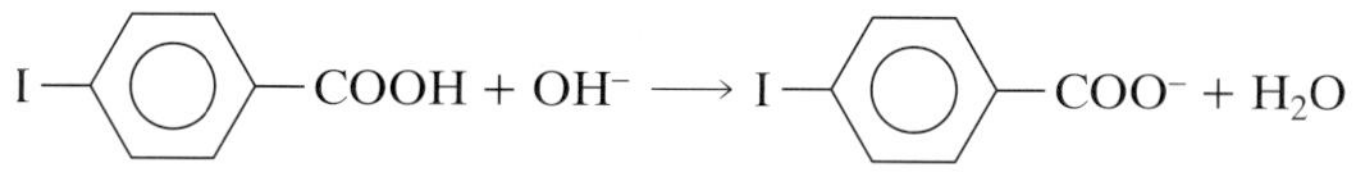

$$I-\!\!\left\langle\bigcirc\right\rangle\!\!-COOH + OH^- \longrightarrow I-\!\!\left\langle\bigcirc\right\rangle\!\!-COO^- + H_2O$$

You will then combine this neutralized portion with the second portion containing the unneutralized acid, giving a solution that contains equimolar amounts of the acid and its conjugate base. After preparing solutions of *p*-iodobenzoic acid and most or all of the acids listed in Table 42.1, you and your coworkers will measure the pH of each solution. Equation **4**, where n_{HA} = moles of acid and n_{A^-} = moles of conjugate base, relates the pH of a solution to the pK_a of the acid.

$$pK_a = pH + \log \frac{n_{HA}}{n_{A^-}} \qquad \textbf{(4)}$$

When n_{HA} and n_{A^-} are equal, the logarithmic term equals zero, so $pK_a = pH$. Thus you can determine the pK_a and ΔpK values of the acids from the measured pH values of their solutions. Graphing the ΔpK values for the acids against the sigma values for the substituents yields a *Hammett plot* of Equation **3**, from which you should be able to determine the sigma value for the *para*-iodo substituent and the rho value for the ionization of benzoic acids in 95% ethanol.

Reactions and Properties

Table 42.1 Molecular weights and sigma values for *para*-substituted benzoic acids

Acid	M.W.	Substituent (Z)	σ
p-aminobenzoic acid	137.1	$—NH_2$	−0.66
p-hydroxybenzoic acid	138.1	—OH	−0.37
p-anisic acid	152.2	$—OCH_3$	−0.27
p-toluic acid	136.2	$—CH_3$	−0.17
benzoic acid	122.1	none	0.00
p-chlorobenzoic acid	156.6	—Cl	0.23
terephthalic acid	166.1	—COOH	0.45
p-nitrobenzoic acid	167.1	$—NO_2$	0.78

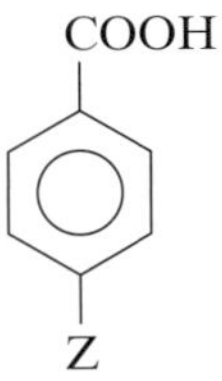

COOH
Z

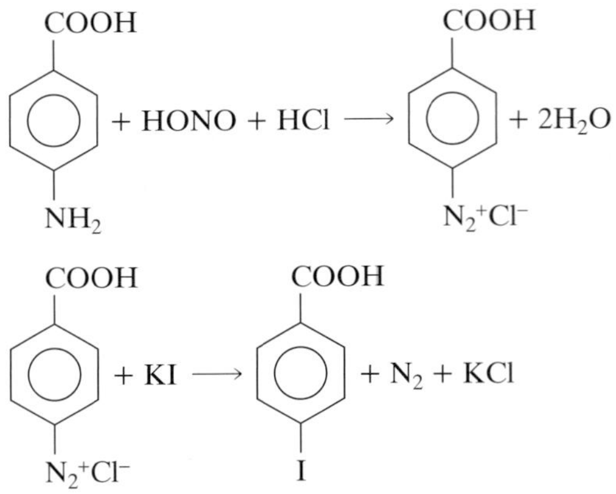

COOH + HONO + HCl $\longrightarrow$ COOH + $2H_2O$
NH₂ $\qquad$ N₂⁺Cl⁻

COOH + KI $\longrightarrow$ COOH + N_2 + KCl
N₂⁺Cl⁻ $\qquad$ I

Table 42.2 Physical properties

	M.W.	mp
p-aminobenzoic acid	137.1	189
p-iodobenzoic acid	248.0	270
potassium iodide	166.0	681
sodium nitrite	69.0	271

Note: M.W. = molecular weight, mp is in °C.

Directions

A. *Preparation of* p-*Iodobenzoic Acid*

benzoic acid

Safety Notes

> Sodium nitrite may cause a fire if mixed with combustible materials. Keep it away from such materials.
>
> Benzoic acid and the substituted benzoic acids in Table 42.1 may be harmful if inhaled, ingested, or absorbed through the skin. Minimize contact with the acids and do not breathe their dust.

Reaction.   Weigh 2.00 mmol of *p*-aminobenzoic acid (PABA) into a 10-mL Erlenmeyer flask, add 2.0 mL of 3 *M* hydrochloric acid and a stir bar, and warm the solution gently with stirring [OP-10] until the PABA dissolves. Dissolve 2.00 mmol of sodium nitrite in 2.0 mL of water in a small test tube. Cool [OP-8] both solutions in an ice/water or ice-salt bath for at least 10 minutes; their temperatures should be 5°C or below. To form the diazonium salt, add the sodium nitrite solution with stirring to the flask containing the PABA, adding it slowly enough so that the temperature remains below 10°C (leave the solution in the cold bath during the addition). Test the solution with starch-iodide paper and, if the test is positive, add just enough urea to give a negative test (no immediate blue-black color).

Dissolve 3.0 mmol of potassium iodide in 20 mL of water in a 100-mL beaker. Pour the diazonium salt solution into the potassium iodide solution with swirling (rinse the reaction flask contents into the beaker with a little water). Heat [OP-7] the reaction mixture in a boiling water bath for 10–15 minutes with occasional manual stirring. If the resulting foam threatens to overflow, poke it with your stirring rod to make it subside. Cool the reaction mixture in an ice/water bath for 5 minutes or more.

Separation and Purification.   Collect the crude *p*-iodobenzoic acid by vacuum filtration [OP-13], washing it on the filter with cold water. Some product may be in the foam, so transfer it to the funnel along with the liquid. Purify the product by recrystallization [OP-25] from aqueous 80% ethanol. If desired, use pelletized Norit to partly decolorize the dark solution. Dry [OP-23] the product to constant mass and measure its mass.

B. *Determination of* pK_a *Values*

This part can be performed by teams of three to four students. Each team should carry out pH measurements with *p*-iodobenzoic acid (preferably using the product with the lightest color) and all of the acids listed in Table 42.1 that are provided. For each acid, prepare an approximately 0.10 *M* solution in 95% ethanol by dissolving 0.20 mmol of the acid in 2.0 mL of 95% ethanol, with gentle heating if necessary. Prepare half-neutralized solutions of all of the acids by the following procedure. For each acid, transfer two accurately measured 0.500 mL portions of its solution to two labeled test tubes or two labeled wells in a well plate. Add a drop of phenolphthalein indicator to *one* of the portions, then neutralize *that* portion to a light pink end point by adding ~0.1 *M* NaOH in 95% ethanol drop by drop from a disposable narrow-tipped transfer pipet. Use a thin glass stirring rod to keep the solution well mixed during the titration. (If you go past the end point, discard the mixture and titrate a fresh portion of the same solution.) Combine the neutralized and unneutralized solutions in a labeled container that will accommodate the pH meter's probe, and mix them thoroughly with a stirring rod. Measure the pH values of all of the solutions you have prepared.

Report your pH values to your coworkers and obtain the pH values for the remaining acids from them. Tabulate the pK_a and ΔpK values for all of the acids and construct a Hammett plot. From your plot, determine the sigma value for the *para*-iodo substituent and the rho value for the ionization of benzoic acids in 95% ethanol. Decide whether an inductive or resonance effect of the *para*-iodo substituent can best account for your results.

Stop and Think: Why is it important to keep the temperature low?

Observe and Note: Describe what happens as the reaction proceeds.

Take Care! A vigorous reaction with considerable foaming will occur.

Stop and Think: What gas is responsible for the foaming?

Waste Disposal: The filtrates can be poured down the drain.

Waste Disposal: Dispose of the solutions as directed by your instructor.

Stop and Think: How is the pK_a of each acid related to the pH of its solution?

Exercises

1   Is the ionization of benzoic acids in 95% ethanol more or less sensitive to substituent effects than its ionization in water? Explain why.

2   Derive Equation **4** (see "Understanding the Experiment") from the equilibrium constant expression for the ionization of an acid, HA.

3   Describe and explain the possible effect on your results of the following experimental errors or variations. (a) You misread the label on a bottle of sodium nitrate and used it in place of the sodium nitrite. (b) When titrating the *p*-iodobenzoic acid solution, you added enough dilute NaOH to turn it dark red. (c) You misread the directions and neutralized both portions of the *p*-iodobenzoic acid solution before combining them for the pH measurement.

4   Following the format in Appendix V, construct a flow diagram for the synthesis of *p*-iodobenzoic acid.

5   (a) Bea Wilder heated the diazonium salt solution *before* adding potassium iodide and isolated a white solid with a melting point of 215°C. What did she make instead of *p*-iodobenzoic acid? (b) Propose a mechanism for the reaction.

6   Using resonance structures, explain why *m*-methoxybenzoic acid is more acidic than benzoic acid, while *p*-methoxybenzoic acid is less acidic than benzoic acid.

7   Predict the acid ionization constant (K_a) of *p*-trifluoromethylbenzoic acid in water and in 95% ethanol, given that a *para*-trifluoromethyl substituent ($-CF_3$) has a sigma value of 0.54.

8   (a) Would you expect the rho value for the ionization of *para*-substituted phenols (p-ZC_6H_4OH) in water to be greater than or less than 1? Explain. (b) Give examples of reactions that you would expect to have negative rho values.

Other Things You Can Do

(Starred projects require your instructor's permission.)

*1   As an alternative to the method described in part **B**, you can measure the pK_a of *p*-iodobenzoic acid (or the other acids) by carrying out a pH titration. Dissolve approximately 0.5 mmol of the acid in 10 mL of 95% ethanol. Titrate this solution with ~0.05 *M* KOH in 95% ethanol, adding about 0.5 mL at a time and measuring the pH after each addition. When the pH begins to rise rapidly, add the KOH in 0.2-mL increments and continue to titrate until the pH begins to level off. Plot pH versus volume of base, mark the inflection point of the curve (where it changes direction), and determine the volume of KOH solution added at that point. The pH reading at exactly half that volume equals the pK_a of the acid.

*2   Determine the relative strengths of some common organic acids and bases by carrying out Minilab 36.

3   Write a research paper about the natural sources, physiological functions, and commercial uses of *p*-aminobenzoic acid (PABA), starting with sources listed in the Bibliography.

Synthesis and Identification
of an Unknown Carboxylic Acid

EXPERIMENT **43**

Reactions of Acid Anhydrides. Reactions of Carbon-Carbon Double Bonds.
Preparation of Carboxylic Acids. Qualitative Analysis.

Operations

OP-10   Mixing
OP-12   Gravity Filtration
OP-13   Vacuum Filtration
OP-23   Drying Solids
OP-30   Melting Point

Before You Begin

1 Read the experiment, read or review the operations as necessary, and write
   an experimental plan.
2 Calculate the mass of 5.00 mmol of maleic anhydride and of 5.5 mmol of
   zinc.

Scenario

Tetrahydrofuran (THF) is manufactured by the catalytic hydrogenation of
maleic anhydride.

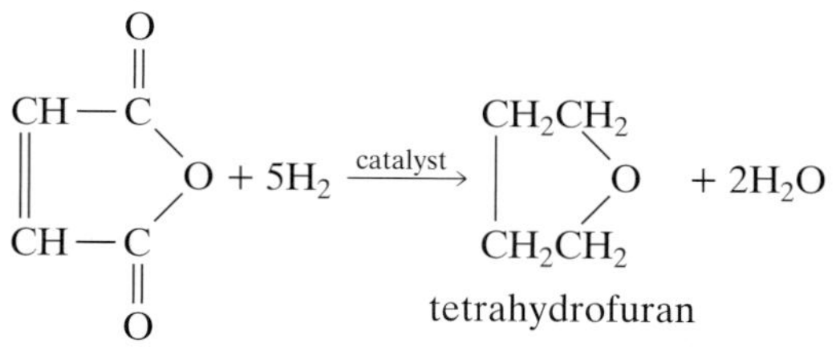

Reduction of maleic anhydride to tetrahydrofuran

Willy Hackett, an inexperienced graduate student at Miskatonic University,
was attempting to develop an improved method for preparing this useful
solvent as part of his thesis research project. He reasoned that if zinc and
hydrochloric acid were combined in the presence of maleic anhydride, they
would react to form hydrogen, which would then reduce the maleic anhydride to tetrahydrofuran.

$$Zn + 2HCl \longrightarrow H_2 + ZnCl_2$$

He thought this procedure would avoid most of the hazards associated with
the use of hydrogen gas, as well as the need for expensive apparatus and
catalysts. So he dissolved maleic acid in boiling water, stirred in some zinc

followed by concentrated HCl, and was pleased when the reaction mixture started fizzing vigorously, as expected. But when the reaction mixture cooled down, a white solid crystallized from solution. Since tetrahydrofuran is a low-boiling liquid, Willy realized that something had gone wrong.

Not wishing to admit to his research mentor that the experiment had failed, Willy has decided to study the reaction as part of his research project. But first he must find out what the white solid is, and whether his results can be reproduced by other chemists. Your supervisor learned of his predicament and has agreed to have the Consulting Chemists Institute work on the problem. Preliminary tests suggest that the white solid is a carboxylic acid. Your assignment is to see whether you can prepare the unknown acid by Hackett's method and, if so, to identify it.

Applying Scientific Methodology

After reading the experiment, you should be able to formulate a hypothesis about the structure of the unknown carboxylic acid by considering the structure of the starting material and the composition of the reaction mixture. You will test your hypothesis by determining the melting point and equivalent weight of the acid, which you will use to identify it.

Maleic Anhydride and Reid's Reaction

Maleic anhydride is manufactured by air oxidation of benzene in the presence of a vanadium oxide catalyst.

$$\text{benzene} + 4\tfrac{1}{2}O_2 \xrightarrow{V_2O_5,\ heat} \text{maleic anyhdride} + 2CO_2 + 2H_2O$$

maleic
anyhdride

maleic acid

fumaric acid

cinnamic acid

$$HOOCC \equiv CCOOH$$
butynedioic acid

Because both air and benzene are cheap and abundant, maleic anhydride is an attractive starting material for many other marketable chemicals. Besides being used in the production of tetrahydrofuran, maleic anhydride is used to manufacture agricultural chemicals, polyester resins, surface coatings, dye intermediates, pharmaceuticals, lubricant additives, and other commercially useful products. As noted in Experiment 31, it is also an excellent dienophile that forms Diels-Alder adducts with most conjugated dienes.

The chemical transformation described in the Scenario is an example of a little-known reaction that (unknown to Willy Hackett) was first reported by Dr. E. Emmett Reid. When Reid added maleic anhydride to a mixture of granular zinc and boiling water, he noticed an immediate reaction and the separation of a white precipitate. After acidifying the reaction mixture with hydrochloric acid, he recovered the product and identified it as a carboxylic acid. This result was not published until 1972, when it was mentioned—almost as an afterthought—in a "chemical autobiography" describing Reid's 76 years of research. Reid's reaction works only with maleic anhydride and certain other unsaturated carboxylic acids and anhydrides including maleic acid, fumaric acid, cinnamic acid, and butynedioic acid, whose structures are shown in the margin.

The reaction with cinnamic acid requires a zinc-mercury amalgam rather than pure zinc.

Understanding the Experiment

You will synthesize the unknown acid by adding granular zinc to a hot aqueous solution of maleic anhydride. Because powdered maleic anhydride reacts quite rapidly with atmospheric moisture, it is usually manufactured in the form of briquettes, which must be pulverized before use. If maleic anhydride is provided in powdered form, you should open its container only momentarily and replace the cap immediately after you have removed the amount needed. After the initial reaction is complete, concentrated hydrochloric acid is added to liberate the final product and dissolve any excess zinc. Because the product is somewhat soluble even in cold water, the solution should be concentrated to increase the yield. The product should require no further purification after it crystallizes from the cold solution.

The *equivalent weight* of an acid, sometimes called its neutralization equivalent, can be defined as the mass of the acid that provides a mole of protons (H^+) in a neutralization reaction. For example, 1 mole of acetic acid (CH_3COOH, M.W. = 60) and $\frac{1}{2}$ mole of oxalic acid (HOOCCOOH, M.W. = 90) both contain 1 mole of acidic hydrogens that can be removed as H^+ ions by a strong base. Thus, the equivalent weight of acetic acid is 60, but the equivalent weight of oxalic acid is 45—half its molecular weight. If the structure of an acid is known, its equivalent weight can be calculated by dividing its molecular weight by the number of acidic hydrogens per molecule.

Key Concept: An equivalent of a substance is the amount of that substance having a mass equal to its equivalent weight, in grams. One equivalent of an acid or base is the amount that will produce or react with one mole of protons, so x equivalents of any acid should neutralize x equivalents of any base.

The equivalent weight of an unknown acid can be measured by titrating an accurately weighed sample of the acid with a standardized solution of sodium hydroxide. The volume of the base required for the titration and its molar concentration are used to calculate the number of moles of base neutralized by the acid, which is equal to the number of moles of protons that the acid provided. For example, suppose 0.110 g of an unknown carboxylic acid with a melting point of 210°C was titrated with 11.6 mL of 0.114 *M* NaOH. Multiplying the molarity of the base by the volume used (in liters) gives the number of moles of base and protons. Dividing the mass of the acid by the number of moles of protons it contains then yields the equivalent weight of the acid, which is about 83.

Note that equivalent weight, like molecular weight, is dimensionless.

$$\text{moles of } H^+ = \text{moles of NaOH} = \frac{0.114 \text{ mol}}{1 \text{ L}} \times 0.0116 \text{ L}$$

$$= 1.32 \times 10^{-3} \text{ mol}$$

$$\text{equivalent weight} = \frac{\text{mass of acid}}{\text{moles of } H^+} = \frac{0.110}{1.32 \times 10^{-3}} = 83.2$$

The melting point of this unknown compound is too high for a carboxylic acid with a molecular weight near 83, so the unknown acid must have two or more carboxyl groups and a molecular weight that is some multiple of 83. From the data in Table 7, Appendix VI, we can deduce that the unknown is phthalic acid, whose molecular weight is 166.

The unknown acid you will synthesize in this experiment is also listed in Table 7 of Appendix VI, so you should be able to identify it after you have

phthalic acid

measured its equivalent weight and melting point. With your instructor's permission, you can obtain the IR or ^{1}H NMR spectrum of the acid to help confirm its identity.

Properties

Table 43.1 Physical properties

	M.W.	mp	bp
maleic anhydride	98.1	53	202
zinc	65.4	419	907d

Note: mp and bp are in °C, density is in g/mL; d = decomposes.

Directions

Safety Notes

maleic anhydride

hydrochloric acid

> Maleic anhydride is corrosive and toxic and can cause severe damage to the eyes, skin, and upper respiratory tract. Wear gloves, avoid contact, and do not breathe its dust. If you must powder maleic anhydride briquettes, do it under the hood and wear safety goggles and protective clothing.
> Hydrochloric acid is poisonous and corrosive. Contact or inhalation can cause severe damage to the eyes, skin, and respiratory tract. Wear gloves and dispense under a hood; avoid contact and do not breathe its vapors. The reaction of zinc with hydrochloric acid produces hydrogen gas, which is highly flammable. Keep the reaction mixture away from flames.
> The product may irritate the eyes, skin, or respiratory tract. Avoid contact and do not breathe its dust.

Take Care! Avoid contact with maleic anhydride and do not breathe its dust.

Observe and Note: What evidence is there that a reaction is occurring?

Take Care! Wear gloves, avoid contact with HCl, do not breathe its vapors.

Reaction.   Dissolve 5.00 mmol of maleic anhydride in 3.0 mL of water in a 10-mL Erlenmeyer flask by heating the water just to boiling, but don't let any water boil away. Remove the flask from the heat source and immediately add 5.5 mmol of 40-mesh zinc with swirling. Let the flask stand for 15 minutes with magnetic stirring [OP-10] or occasional swirling. *Under the hood*, slowly add 1.0 mL of concentrated HCl with stirring or swirling.

Separation.   When the zinc (or most of it) has dissolved, heat the mixture to boiling under the hood. Any white solid that formed during the reaction should dissolve. Filter [OP-12] the hot solution through a preheated filtering pipet into a small beaker. Rinse the filter with 0.5 mL of hot water and collect the water in the beaker. Boil the filtrate gently under the hood until the solution becomes *just* cloudy at the boiling point. Cover the beaker with a watch glass and set it aside to cool to room temperature. Then cool the beaker in an ice/water bath until crystallization is complete. Collect the product by vacuum filtration [OP-13] and wash it on the filter with cold acetone. Air dry the product on the filter for a few minutes; then dry [OP-23] it to constant mass.

Waste Disposal: Pour the filtrate down the drain.

Analysis. Measure the mass and melting point [OP-30] of the thoroughly dried product. Weigh about 0.10 g of the product to the nearest milligram and dissolve it in 10 mL of water in a 25-mL Erlenmeyer flask. Add a drop of phenolphthalein indicator and a magnetic stir bar and titrate with a standardized ~0.1 M NaOH solution to the light pink end point. Calculate the equivalent weight of your unknown acid and identify it by referring to Table 7 of Appendix VI. Write the structure of the product and a balanced equation for its formation, and give its systematic and common names.

Observe and Note: Record the exact concentration of the NaOH solution in your lab notebook.

Exercises

1 Estimate the amount of product that was lost in the filtrate, assuming that the volume of the reaction mixture was 1.0 mL when you collected the product by vacuum filtration. Compare this with the amount of product you would have lost if you had not boiled off most of the water. The solubility of the product is 6.8 g/100 mL at 20°C.

2 (a) What was wrong with the graduate student's idea (see the Scenario) that the reaction of maleic acid with zinc and aqueous HCl would yield tetrahydrofuran? (b) The actual synthesis that you carried out involves two separate reaction steps. Write a balanced equation for each step.

3 Describe and explain the possible effect on your results of the following experimental errors or variations. (a) You used phthalic anhydride (1,2-benzenedioic anhydride) instead of maleic anhydride. (b) After filtering the hot solution in the separation step, you forgot to boil down the filtrate. (c) You didn't record the concentration of the NaOH solution (which was 0.127 M), but assumed a concentration of exactly 0.1 M.

4 Following the format in Appendix V, construct a flow diagram for the synthesis you carried out in this experiment.

5 Propose structures for the products you would obtain by treating fumaric acid, cinnamic acid, and butynedioic acid with excess zinc and HCl. Write balanced equations for the reactions.

6 Suppose that 0.101 g of an unknown carboxylic acid is titrated using 18.1 mL of 0.106 M NaOH. What is the probable identity of the unknown if its melting point is 132 ± 5°C? (Use Table 7 in Appendix VI.)

7 Sketch the ^{1}H NMR spectrum you would expect from the product of this experiment, giving peak areas, multiplicities, and approximate chemical shifts for all signals.

8 Write balanced equations for the reactions of maleic anhydride with water, cyclohexanol, aniline, and isoprene (2-methyl-1,3-butadiene).

Other Things You Can Do

(Starred projects require your instructor's permission.)

*****1** Record the infrared spectrum of the product and compare it with the spectrum of the starting material in Figure 31.1 (Experiment 31). Interpret both spectra as completely as you can.

Preparation
of the Insect Repellent
N,N-Diethyl-*m*-toluamide

EXPERIMENT **44**

Reactions of Carboxylic Acids. Preparation of Amides. Nucleophilic Acyl Substitution. Acid Chlorides. Infrared Spectrometry.

Operations

OP-7   Heating
OP-8   Cooling
OP-10 Mixing
OP-11 Addition of Reactants
OP-15 Extraction
OP-16 Evaporation
OP-18 Column Chromatography
OP-21 Washing Liquids
OP-22 Drying Liquids
OP-24 Drying and Trapping Gases
OP-34 Gas Chromatography
OP-36 Infrared Spectrometry
OP-37 Nuclear Magnetic Resonance Spectrometry (optional)

Before You Begin

1 Read the experiment, read or review the operations as necessary, and write an experimental plan.
2 Calculate the mass of 3.0 mmol of *m*-toluic acid, the mass of 2.50 mmol of diethylamine hydrochloride, and the theoretical yield of *N,N*-diethyl-*m*-toluamide.

Scenario

Skeeters 'n Such manufactures insect repellents, including *Bug Off*, whose main active ingredient is *N,N*-diethyl-*m*-toluamide, better known as Deet. Its current process for the manufacture of Deet requires the use of diethyl ether as a solvent and a 2:1 molar ratio of nucleophile (diethylamine) to substrate. To reduce production costs, Skeeters 'n Such wants your Institute to develop a synthetic method that uses a cheaper and more environmentally friendly solvent, such as water, and a 1:1 or lower nucleophile:substrate ratio. Your supervisor thinks a procedural variation called the Schotten-Baumann reaction may accomplish both of these objectives, but is not sure whether the product will be pure enough to meet specifications. Your assignment is to see whether or not Deet can be synthesized in at least 95% purity using the Schotten-Baumann procedure.

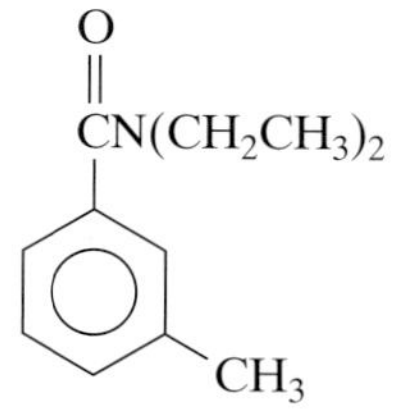

Deet
(*N,N*-diethyl-*m*-toluamide)

a hungry mosquito

Applying Scientific Methodology

You should develop a working hypothesis, based on the Scenario, which will be tested when you analyze the product by gas chromatography and infrared (IR) spectrometry.

Chemical Mosquito Evasion

Is there anyone on earth who hasn't cringed upon hearing the high-pitched whine of a hungry mosquito? Besides their capacity to torment us, these bloodthirsty insects have a well-earned reputation for spreading diseases, including malaria, yellow fever, West Nile disease, and viral encephalitis. Numerous methods of controlling mosquitoes have been developed, but the hardy creatures can withstand a variety of adverse conditions. Mosquitoes have been known to breed in the hot alkaline volcanic pools of Uganda, and even in a tank of hydrochloric acid! Since we are not likely to eradicate mosquitoes from the earth anytime soon, we must resign ourselves to living with them. This fate is made more tolerable by the availability of effective insect repellents. Among the best of these is Deet, which is the major ingredient of most commercial mosquito repellents manufactured in North America.

Mosquito "repellents" don't really repel mosquitoes in the same way that a disagreeable odor might repel a human from its source. Instead, they appear to jam the insect's sensors so that it can't find its victim. Warm objects generate convection currents in the air around them; warm living objects also emit carbon dioxide, which alerts a mosquito to the presence of a blood source and starts it on its flight. This flight is initially random, but when the insect encounters a warm, moist stream of air, it moves toward the source, which is generally a living object. Unless the object takes rapid evasive action, the mosquito follows the convection current until it makes contact—unless it gets squashed first.

When you (the intended victim) are protected by an effective insect repellent, the mosquito still knows you're around but is unable to find you. This is because the repellent prevents the insect's moisture sensors from responding normally to the high humidity of your convection current. Ordinarily, when a mosquito passes from warm, moist air into drier air, its moisture sensors send fewer signals to its central nervous system, causing it to turn back into the air stream. By blocking these sensors, the repellent reduces the signal frequency and convinces the mosquito that it is heading into drier rather than moister air; so it turns away before landing. Individual variations in diet, skin chemistry, or other factors such as the color of one's clothing (dark colors attract mosquitoes) may help explain why some unlucky individuals are eaten alive by mosquitoes and others escape unscathed. For example, thiamine (Vitamin B_1) taken orally is excreted through the skin, where it acts as an insect repellent. On the other hand, lactic acid and many other chemicals that emanate from human skin attract mosquitoes.

Scientists from the U. S. Department of Agriculture's research station have identified more than 340 different skin chemicals.

The molecular features that make a compound a good insect repellent are not well understood at this time. Repellents occur in nearly every chemical family and exhibit a large variety of molecular shapes, as shown in Figure 44.1. A number of *N,N*-disubstituted amides similar to Deet are represented, including the diethylamide of thujic acid, a constituent of the western red cedar that may be partly responsible for that tree's resistance to insect attack. Other effective repellents include esters such as dimethyl phthalate and diols such as 2-ethyl-1,3-hexanediol.

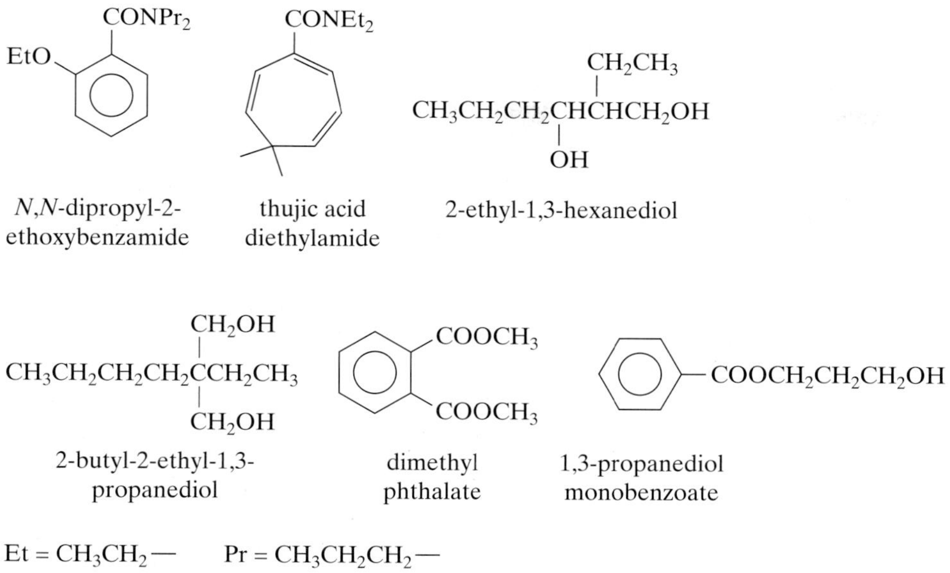

CONPr₂ structure with EtO group — *N,N*-dipropyl-2-ethoxybenzamide

CONEt₂ structure — thujic acid diethylamide

$CH_3CH_2CH_2CHCHCH_2OH$ with CH_2CH_3 and OH substituents — 2-ethyl-1,3-hexanediol

$CH_3CH_2CH_2CH_2CCH_2CH_3$ with CH_2OH groups — 2-butyl-2-ethyl-1,3-propanediol

dimethyl phthalate (COOCH₃, COOCH₃)

$-COOCH_2CH_2CH_2OH$ — 1,3-propanediol monobenzoate

$Et = CH_3CH_2-$ $Pr = CH_3CH_2CH_2-$

Figure 44.1 Some typical insect repellents

Recent lab tests with catnip oil, which contains two nepetalactone stereoisomers, have shown it to be ten times more effective than Deet as a mosquito repellent. Further tests are needed (perhaps with both mosquitoes and cats) to see whether the oil can be used safely and effectively by humans. When more is learned about the structural features that make a substance act as an insect repellent, it should be possible to develop even more effective and convenient repellents for long-term protection against insect bites.

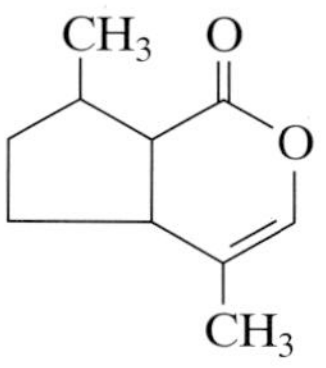

nepetalactone

Understanding the Experiment

Amides are usually prepared by treating a carboxylic acid derivative with ammonia or with a primary or secondary amine. This is a nucleophilic acyl substitution reaction in which ammonia or the amine acts as the nucleophile. Acid anhydrides and esters are sometimes used as the acid derivative, but acyl chlorides are the most useful for preparing the widest variety of amides.

General reactions for preparing amides

$$R-\overset{O}{\overset{\|}{C}}-Z + NH_3 \longrightarrow RCNH_2 + HZ$$

$$R-\overset{O}{\overset{\|}{C}}-Z + R'NH_2 \longrightarrow RCNHR' + HZ$$

$$R-\overset{O}{\overset{\|}{C}}-Z + R'-\underset{R''}{NH} \longrightarrow R\overset{O}{\overset{\|}{C}}N\underset{R''}{R'} + HZ$$

Z = Cl, OR, OCOR, etc.

Because of the high reactivity of acyl chlorides, the reactions are usually rapid and exothermic—so much so that, in many cases, the rate must be controlled by cooling or by using an appropriate solvent. When the reaction is carried out in an inert solvent such as diethyl ether, it is necessary to use at least a 2:1 mole ratio of amine (or ammonia) to the substrate because the reaction produces HCl, which reacts with one equivalent of amine (or ammonia) to form an ammonium salt. The salt is not nucleophilic, so it can't react with the acyl compound; it may also be difficult to separate from the product.

Key Concept: *A nucleophile is usually basic enough to accept a proton from an acid. A protonated nucleophile, having donated a lone pair, is either a much weaker nucleophile or is not nucleophilic at all.*

Example of reaction in inert solvent

$$\underset{\text{RCCl}}{\overset{\text{O}}{\underset{||}{}}} + 2R'NH_2 \longrightarrow \underset{\text{RCNHR}'}{\overset{\text{O}}{\underset{||}{}}} + R'NH_3^+Cl^-$$

The Schotten-Baumann reaction is a synthetic method that uses aqueous sodium hydroxide (or potassium hydroxide) as a solvent for the reactions of certain acyl chlorides. Some of the sodium hydroxide (NaOH) combines with the HCl produced during the reaction, making it unnecessary to add excess amine for that purpose. Some acyl chlorides, particularly low-molecular-weight aliphatic ones, hydrolyze rapidly in water to form carboxylic acids and thus are unsuitable candidates for the Schotten-Baumann reaction. However, aromatic and long-chain aliphatic acyl chlorides are nearly insoluble in water and therefore hydrolyze much more slowly. The small amount of acyl chloride lost by hydrolysis can be compensated for by using an excess of this reactant. The presence of aqueous NaOH in the reaction mixture also makes it possible to use an amine hydrochloride in place of the free amine, since the amine hydrochloride reacts with base to liberate the amine *in situ*. Because many amines are volatile, corrosive, and quite unpleasant to handle, the use of the comparatively well-behaved amine salt is a definite advantage.

The acyl chlorides used for preparing amides are generally made from the corresponding carboxylic acids. Several reagents can be used for this transformation, each with its advantages and disadvantages. Thionyl chloride ($SOCl_2$) is usually the reagent of choice because its inorganic reaction products (HCl and SO_2) are gases that can easily be removed from the acyl chloride, which can thus be used for further reactions without purification.

Schotten-Baumann method for preparing amides

$$\underset{\text{RCCl}}{\overset{\text{O}}{\underset{||}{}}} + R'NH_2 + NaOH \longrightarrow$$

$$\underset{\text{RCNHR}'}{\overset{\text{O}}{\underset{||}{}}} + NaCl + H_2O$$

Reaction of an amine hydrochloride with a base

$$RNH_3^+Cl^- + NaOH \longrightarrow$$

$$RNH_2 + NaCl + H_2O$$

Methods for preparing acyl chlorides

$$\underset{\text{RCOH}}{\overset{\text{O}}{\underset{||}{}}} + PCl_5 \longrightarrow \underset{\text{RCCl}}{\overset{\text{O}}{\underset{||}{}}} + POCl_3 + HCl$$

$$3\underset{\text{RCOH}}{\overset{\text{O}}{\underset{||}{}}} + 2PCl_3 \longrightarrow 3\underset{\text{RCCl}}{\overset{\text{O}}{\underset{||}{}}} + 3HCl + P_2O_3$$

$$\underset{\text{RCOH}}{\overset{\text{O}}{\underset{||}{}}} + SOCl_2 \longrightarrow \underset{\text{RCCl}}{\overset{\text{O}}{\underset{||}{}}} + SO_2 + HCl$$

In this experiment, you will prepare *m*-toluoyl chloride by heating *m*-toluic acid with excess thionyl chloride, and then treat the acid chloride with diethylamine (from the hydrochloride) in aqueous NaOH to obtain *N,N*-diethyl-*m*-toluamide. The excess thionyl chloride from the first step is used up by reacting with sodium hydroxide in the second. Because corrosive gases are released, you will need to carry out the reaction under a fume hood or use a gas trap to keep them out of the atmosphere. Since the acid chloride is relatively insoluble in the aqueous reaction mixture, efficient mixing is necessary to provide adequate contact between the phases. The detergent sodium lauryl sulfate helps disperse the acyl chloride into smaller droplets, increasing the area of contact between phases and thus increasing the reaction rate.

After the product has been isolated by extraction with diethyl ether, it will be purified by column chromatography, and then analyzed by gas chromatography to assess its purity. By comparing the infrared spectrum of your product with those of the starting materials in Figure 44.2, you should find evidence indicating whether or not the expected reaction has taken place.

Reactions and Properties

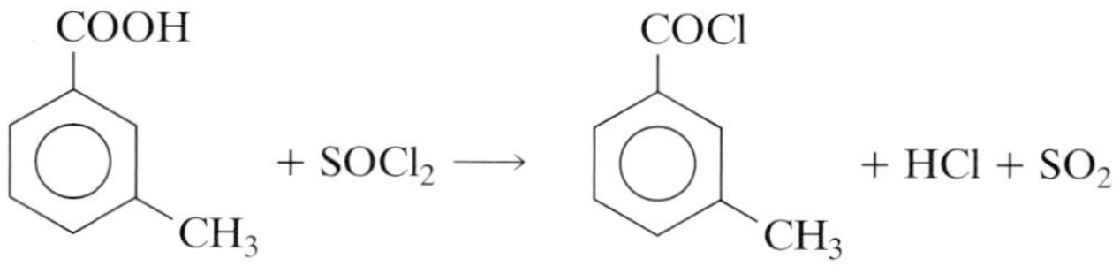

COOH $+ \; SOCl_2 \longrightarrow$ COCl $+ \; HCl + SO_2$

CH$_3$ CH$_3$

m-toluic acid   *m*-toluoyl chloride

$(CH_3CH_2)_2NH_2{}^+Cl^- + NaOH$

diethylamine
hydrochloride

$\longrightarrow (CH_3CH_2)_2NH + NaCl + H_2O$

diethylamine

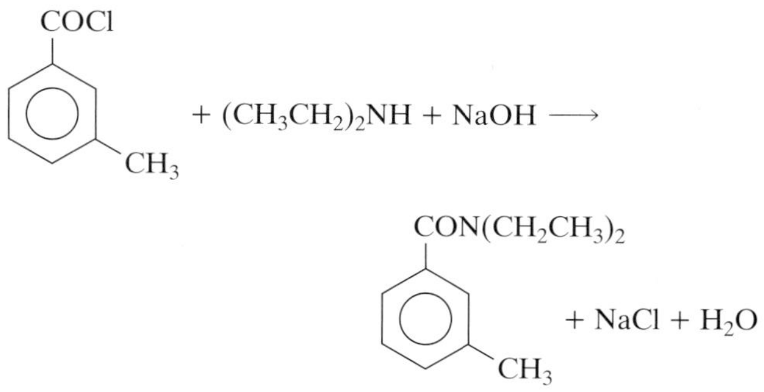

COCl

CH$_3$ $+ \; (CH_3CH_2)_2NH + NaOH \longrightarrow$

CON(CH$_2$CH$_3$)$_2$

CH$_3$ $+ \; NaCl + H_2O$

Deet

Table 44.1 Physical properties

	M.W.	mp	bp	d
m-toluic acid	136.2	113		
thionyl chloride	119.0		79^{746}	1.65
m-toluoyl chloride	154.6		86^5	1.173
diethylamine hydrochloride	109.6	227–230		
hexanes	86.2		~68–70	~0.67
N,N-diethyl-*m*-toluamide	191.3		160^{19}	0.996

Note: mp and bp are in °C, density is in g/mL, superscripts indicate pressure in torr.

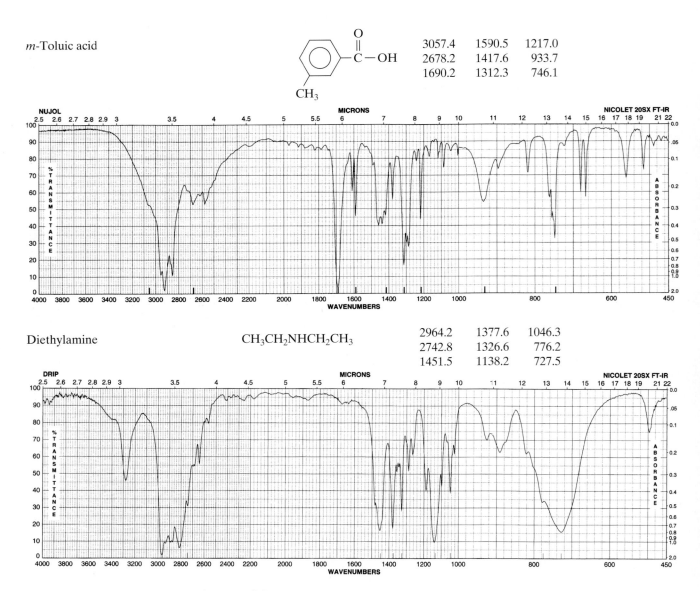

m-Toluic acid

3057.4	1590.5	1217.0
2678.2	1417.6	933.7
1690.2	1312.3	746.1

Diethylamine $CH_3CH_2NHCH_2CH_3$

2964.2	1377.6	1046.3
2742.8	1326.6	776.2
1451.5	1138.2	727.5

Figure 44.2 IR spectra of the starting materials

m-Toluic acid

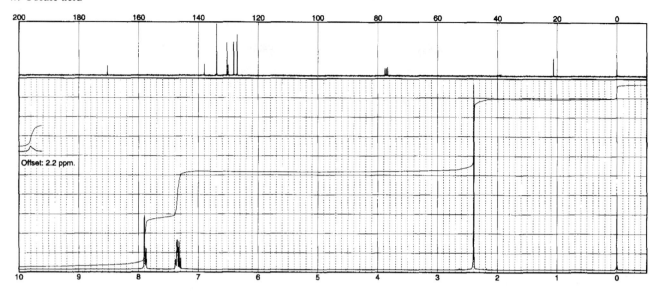

Diethylamine

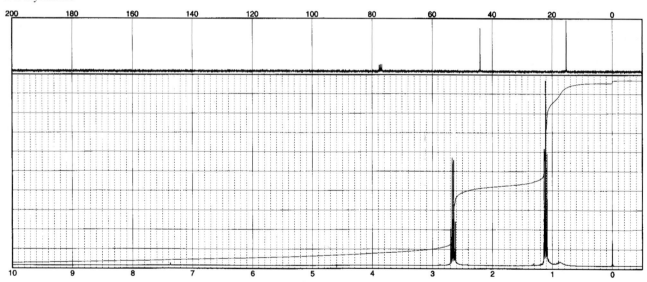

Figure 44.3 NMR spectra of the starting materials

Directions

Safety Notes

diethylamine

diethyl ether

> Thionyl chloride and *m*-toluoyl chloride are both corrosive and lachry-matory (tear inducing) and can damage the eyes, skin, and respiratory system. Thionyl chloride decomposes violently on contact with water to produce corrosive gases. Wear gloves, work under a hood, avoid contact with thionyl chloride and the reaction mixture, keep them away from water, and do not breathe their vapors.
> Diethylamine hydrochloride irritates the skin, eyes, and respiratory tract, so avoid contact and inhalation. In the reaction mixture it forms diethyl-amine, which is corrosive and has toxic vapors.
> Diethyl ether is extremely flammable and may be harmful if inhaled. Avoid breathing its vapors and keep it away from flames and hot sur-faces.
> Hexanes (a mixture of isomeric 6-carbon alkanes) is very flammable and may irritate the eyes and respiratory tract. Avoid breathing its vapors and keep it away from flames and hot surfaces.

Preparation of m-Toluoyl Chloride.   All glassware must be thoroughly dried for this step. The reaction should be carried out under a hood, if pos-sible. Assemble an apparatus for microscale addition [OP-11] under reflux by inserting a Claisen adapter into a clean, *dry* 10-mL round-bottom flask. Then cap the adapter's straight arm using an unperforated cap liner as a septum, and attach a water-cooled condenser to its bent arm. If you cannot perform the experiment under a fume hood, equip the condenser with a gas trap [OP-24]. Be sure the apparatus is securely clamped so that you can re-move and reattach the reaction flask safely. Remove the reaction flask and measure 3.0 mmol of *m*-toluic acid into it, then drop in a stir bar. *Under the hood*, add 0.26 mL (~3.6 mmol) of thionyl chloride, then reassemble the reaction apparatus. Heat the reaction mixture gently under reflux [OP-7] with stirring [OP-10] for at least 20 minutes. Cool the reaction mixture in an ice/water bath while you prepare the diethylamine solution in the next step.

Take Care! Wear gloves, avoid contact with thionyl chloride, and do not breathe its vapors.

Reaction.   Measure 3.5 mL of 3.0 *M* NaOH into a 5-mL conical vial and cool [OP-8] the vial in an ice/water bath for 5 minutes or more. *Under the hood*, slowly add 2.50 mmol of diethylamine hydrochloride to the NaOH solution with manual stirring or swirling. The amine salt will be converted to diethylamine as it is added. Then mix in about 10 mg (0.010 g) of sodium lauryl sulfate, cap the vial, and shake it gently to mix the contents. Fill a sy-ringe with the diethylamine solution and insert it through the cap liner on the reaction apparatus. (Don't puncture an unperforated cap liner without your instructor's permission; you can replace it with a perforated liner hav-ing a hole that the syringe needle fits tightly.) Leaving the reaction flask containing *m*-toluoyl chloride in its ice/water bath with the condenser water and stirrer running, add 0.5 mL portions of the diethylamine solution at 1-minute intervals until it has all been added. Rinse the syringe with water immediately after use (if the needle is corroded, discard it in a sharps con-tainer). Test the pH of the solution with pH paper; if it is not strongly basic, add more 3.0 *M* NaOH until it is. Then heat the reaction mixture using a

Take Care! Wear gloves, avoid con-tact with diethylamine hydrochlo-ride, and do not breathe its vapors.

Stop and Think: What is the pur-pose of the sodium lauryl sulfate?

boiling-water bath, with vigorous stirring, for 15 minutes or more. At this point, the acid chloride odor should be gone, and the reaction mixture should still be basic.

Separation.   Allow the reaction mixture to cool to room temperature, and then transfer it to a centrifuge tube. Extract [OP-15] the reaction mixture with three successive 2.0-mL portions of solvent-grade diethyl ether, using the first portion of ether for the transfer; then combine the extracts. Wash [OP-21] the combined extracts with 3.0 mL of 1 *M* HCl, and then with 3.0 mL of saturated aqueous sodium chloride. Dry [OP-22] the ether solution using anhydrous magnesium sulfate or sodium sulfate. Evaporate [OP-16] the ether, leaving crude Deet as an oily residue.

Take Care! Keep diethyl ether away from ignition sources.

Waste Disposal: Pour all aqueous layers down the drain.

Purification.   Pack a microscale chromatography column [OP-18] (made using a clean, *dry* Pasteur pipet) with 1.5–2.0 g of Brockman grade III activated alumina. Obtain about 5 mL of hexanes, a *dry*, tared 5-mL conical vial, and a 3-mL conical vial. Set the 3-mL vial under the column. Add 1 mL of hexanes to the column and let it drain until the liquid surface just reaches the top of the adsorbent (or sand) layer, then add the impure Deet to the column immediately. Use a few drops of hexanes to rinse the vial that contained the Deet and add the rinse liquid to the column. When the liquid surface again reaches the adsorbent layer, fill the column nearly to the top with hexanes. Continue to add hexanes during elution to keep the eluent level reasonably constant. Collect about 1 mL of eluate in the 3-mL conical vial, collect the next 4 mL of eluate in the 5-mL vial, and then replace it with the 3-mL vial until the column has drained dry. Evaporate [OP-16] the solvent from the solution in the 5-mL vial.

Take Care! Keep hexanes away from ignition sources.

Analysis.   Measure the mass of the product. Analyze it by gas chromatography [OP-34] and estimate its purity as mass percent Deet. Record the IR spectrum [OP-36] of the neat liquid or obtain a spectrum from your instructor. Compare it with the IR spectra of the reactants in Figure 44.2 and describe the evidence indicating that the expected reaction has taken place. At your instructor's request, record its ^{1}H NMR spectrum [OP-37] as well.

Waste Disposal: Put any recovered hexanes in the appropriate solvent recovery container.

Exercises

1   If you recorded the product's ^{1}H NMR spectrum, compare it with the spectra of the reactants in Figure 44.3 and try to account for any similarities or differences. Interpret your spectrum as completely as you can.

2   Write balanced equations for the reactions of aqueous NaOH with excess *m*-toluoyl chloride and with excess thionyl chloride.

3   Describe and explain the possible effect on your results of the following experimental errors or variations. (a) The reaction flask you used for the preparation of *m*-toluoyl chloride was wet. (b) You combined the diethylamine hydrochloride with 3.5 mL of 1.0 *M* NaOH rather than 3.0 *M* NaOH. (c) You didn't stir the reaction mixture during the reaction of diethylamine with the acid chloride.

4   Following the format in Appendix V, construct a flow diagram for the synthesis of Deet.

—NCOCH$_3$
CH$_2$CH$_2$CH$_2$CH$_3$

N-butylacetanilide

5 Write reasonable mechanisms for (a) the reaction of *m*-toluic acid with thionyl chloride, and (b) the reaction of *m*-toluoyl chloride with diethylamine.

6 The insect repellent *N*-butylacetanilide is used to treat clothing for fleas and ticks. Outline a synthesis of this compound, starting with aniline.

7 (a) Calculate the total volume of 3.0 *M* NaOH required for the preparation, including the amount used up in combining with excess reactants (see Exercise 2). (b) Calculate the percentage excess of NaOH that was actually used.

8 A student misread the label on a reagent bottle and used a 50% NaOH solution in this experiment instead of the 3.0 *M* (11%) solution. Very little Deet was obtained from the ether layer after extraction, but acidification of the aqueous layer yielded a white solid that melted at 112°C. Identify the solid and write balanced equations showing how it formed.

9 Hair is made up of the protein keratin, which is a polyamide containing many *α*-amino acid residues. Hair is often responsible for clogged drains, which can be unclogged by drain cleaners that contain lye (sodium hydroxide). Explain how such drain cleaners remove hair.

Other Things You Can Do

(Starred projects require your instructor's permission.)

*1 Test the effectiveness of your product as a mosquito repellent. Deet is a mild irritant and your product might contain harmful impurities, so do *not* apply it directly to your skin. Instead, prepare a 15% solution of the Deet in isopropyl alcohol and use it to saturate a piece of absorbent cheesecloth. Saturate an identical piece of cheesecloth with pure isopropyl alcohol to use as a control. When the treated cloths are dry, take them to a mosquito-infested area, drape one over each arm, and see which arm is targeted by more mosquitoes.

*2 Prepare another amide, benzamide, as described in Minilab 38.

3 Deet repels insects, but there are other chemicals that attract them. Starting with sources listed in the Bibliography, write a research paper about some of the pheromones that function as insect attractants and tell how they can be used for insect control.

Synthesis of 2-Acetylcyclohexanone by the Stork Reaction

Reactions of Carbonyl Compounds. Preparation of Dicarbonyl Compounds. Nucleophilic Acyl Substitution. Enamines. Infrared Spectrometry.

Operations:

OP-7   Heating
OP-15   Extraction
OP-16   Evaporation
OP-22   Drying Liquids
OP-27   Simple Distillation
OP-28   Vacuum Distillation
OP-36   Infrared Spectrometry

Before You Begin

1 Read the experiment, read "Water Separation" in OP-27c, read or review the other operations as necessary, and write an experimental plan.
2 Calculate the mass and volume of 10.0 mmol of cyclohexanone, 13.0 mmol of morpholine, 12.0 mmol of triethylamine, and 11.0 mmol of acetic anhydride. Calculate the theoretical yield of 2-acetylcyclohexanone.

Scenario

In a beehive there is only one queen bee, the fertile female that produces all of the eggs from which succeeding generations of bees are propagated. Scientists have discovered that queen pheromone, a "chemical messenger" produced by the queen honeybee, is responsible for her preeminence. Queen pheromone not only keeps the worker bees from rearing new queens, it also inhibits the development of their ovaries and attracts male bees to the queen for mating. Midge Miller, a Miskatonic University natural products chemist, believes she has isolated the major active ingredient of queen pheromone. To support Dr. Miller's case, the substance must be prepared in the laboratory and tested to see whether it produces the same effects on honeybees as natural queen pheromone. That requires some synthetic chemists, so the university has contacted your Institute for assistance. Your supervisor believes it should be possible to synthesize this substance, (2*E*)-9-oxo-2-decenoic acid, from 7-oxooctanoic acid, which can be prepared by the alkaline hydrolysis of 2-acetylcyclohexanone.

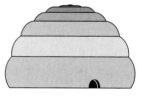

beehive

Proposed synthetic route to the queen pheromone component from 2-acetylcyclohexanone

First you will need to make some 2-acetylcyclohexanone. It has been synthesized by direct acetylation of cyclohexanone in the presence of boron trifluoride, but the yield is low, so your supervisor developed an alternative procedure based on the Stork enamine reaction. Your assignment is to see whether or not 2-acetylcyclohexanone can be prepared using the Stork reaction.

Applying Scientific Methodology

The scientific problem described in the Scenario involves a straightforward preparation, but another problem discussed in the experiment deals with the structure of 2-acetylcyclohexanone. After reading the experiment, you should develop a hypothesis regarding the structure of this compound, and then test your hypothesis by obtaining its infrared (IR) spectrum.

Enamines and the Stork Reaction

In 1954, Gilbert Stork and his co-workers sent a letter to the editor of the *Journal of the American Chemical Society* that began with the sentence, "We have discovered a new method for the alkylation and acylation of ketones." Their brief communication caused a stir in the world of organic chemistry because it announced a synthetic tool of exceptional utility, which came to be known as the Stork enamine reaction.

Before the development of the Stork reaction, carbonyl compounds were alkylated directly by a reaction in which an alpha proton is removed by strong base to form an enolate ion. An enolate ion can act as a carbon nucleophile toward an alkyl halide, displacing the halide ion and attaching itself to the alkyl group. But multiple alkylation can occur, with a second alkyl group usually ending up on the same alpha carbon atom as the first, as illustrated for the direct alkylation of cyclohexanone by bromomethane (methyl bromide).The need for a strongly basic catalyst such as sodium methoxide can also result in side reactions, such as aldol condensations. For these reasons, the yields of most direct alkylation reactions are quite low.

Direct alkylation of cyclohexanone

Tautomeric equilibria of enols and enamines

$$-\underset{|}{C}=\underset{|}{C}-OH \rightleftharpoons -\underset{|}{\overset{H}{\underset{|}{C}}}-\underset{|}{C}=O$$

enol   keto

$$-\underset{|}{C}=\underset{|}{C}-NH \rightleftharpoons -\underset{|}{\overset{H}{\underset{|}{C}}}-\underset{|}{C}=N$$

enamine   imine

In the Stork enamine reaction, a carbonyl compound is converted to a tertiary enamine before it is alkylated or acylated. Enamines are 1-aminoalkenes, the nitrogen analogs of enols. Just as enols tend to isomerize to the corresponding ketones, enamines with N—H bonds are in equilibrium with the corresponding imines. However, tertiary enamines, which have no N—H bonds, cannot tautomerize to imines. In the following example of the Stork reaction, cyclohexanone is first condensed with the secondary amine pyrrolidine to yield the tertiary enamine 1-pyrrolidinocyclohexene (**1**). This enamine is then alkylated under neutral reaction conditions to yield a monoalkylated iminium salt, which is hydrolyzed to the corresponding α-alkyl ketone, as illustrated for the preparation of 2-methylcyclohexanone. Since one of the enamine's two resonance structures has a negative charge on the alpha carbon atom, the alkylation step can be viewed as a nucleophilic substitution reaction with a carbanion acting as the nucleophile.

Formation of an enamine

Alkylation and hydrolysis of the enamine

Enamines thus provide an indirect method for preparing α-alkyl derivatives of carbonyl compounds, in which the enamine functional group both activates the molecule and prevents unwanted side reactions.

Enamines can also be acylated by acid chlorides and anhydrides, yielding α-acyl derivatives of carbonyl compounds. This provides a unique method for lengthening a carbon chain by six atoms. For example, acetyl chloride, which is prepared from acetic acid, reacts readily with 1-morpholinocyclohexene (**2**) to yield, after hydrolysis of the acylated enamine, 2-acetylcyclohexanone. Treatment of this diketone with strong base results in ring cleavage to 7-oxooctanoic acid, the precursor of the queen pheromone component described in the Scenario.

Enamine synthesis of octanoic acid

7-Oxooctanoic acid can also be reduced, by a Wolff-Kishner reaction, to octanoic acid; in this way, a two-carbon acid, acetic acid, is converted to an eight-carbon acid by way of its acyl chloride. Other carboxylic acids can undergo the same chain-lengthening process by way of their acyl chlorides or anhydrides.

Understanding the Experiment

Because this experiment involves two reflux periods and a room-temperature reaction extending until the next lab period, it is important that you plan your lab time carefully. You should be ready to start the room-temperature reaction by the end of the first lab period, if at all possible.

Although pyrrolidine reacts faster with ketones than morpholine, morpholine enamines give higher yields in acylation reactions, so you will carry out the reaction of morpholine with cyclohexanone to form the enamine 1-morpholinocyclohexene. The formation of a ketone enamine usually requires a catalyst such as *p*-toluenesulfonic acid. The reaction is carried out under reflux, using toluene as a solvent. To prevent premature hydrolysis of the enamine, water that forms during the reaction is distilled as a toluene-water azeotrope into the well of a Hickman still, from which toluene overflows into the reaction flask. This process, known as water separation, is described in OP-27c. Some morpholine may distill along with the water, so a small excess of morpholine will be used. After the reaction period is over, the toluene and any excess morpholine will be removed by simple distillation.

The acylation of 1-morpholinocyclohexene will be carried out, using acetic anhydride as the acylating agent, by simply combining the reactants and allowing the solution to stand for several days. Triethylamine is added to neutralize the acetic acid evolved during the reaction. The acylated enamine is then hydrolyzed by heating it with water under reflux, and the product is purified by vacuum distillation.

You can analyze the product by recording its IR spectrum; the IR spectra of cyclohexanone and acetic acid are shown in Figure 45.1 for comparison. Most β-diketones exist partly or predominantly in an enolic form, as illustrated for 2,4-pentanedione.

p-toluenesulfonic acid

Keto and enol forms for 2,4-pentanedione

$$CH_3CCH_2CCH_3 \rightleftharpoons CH_3C{=}CHCCH_3$$

keto form   enol form

The IR spectrum of 2-acetylcyclohexanone should tell you whether this compound exists mainly in the diketo form or in the enolic form. The carbonyl band of the diketo tautomer appears in the usual frequency range for a ketone, but as a doublet with two closely spaced peaks. The IR spectrum of the enol tautomer of a diketone contains the expected $O{-}H$ and $C{=}O$ stretching bands, but hydrogen bonding between the enolic OH group and the carbonyl oxygen alters their appearance and locations considerably. The $O{-}H$ band broadens and moves into the $3200{-}2500$ cm^{-1} region; sometimes this band is so broad and low that it is hard to identify. At the same time, the carbonyl band broadens and moves from its normal position (~ 1715 cm^{-1}) to the $1650{-}1580$ cm^{-1} region.

Reactions and Properties

A. cyclohexanone + morpholine $\xrightarrow{\text{TsOH}}$ 1-morpholinocyclohexene + H_2O

B. 1-morpholinocyclohexene + $(CH_3C)_2O$ + $(CH_3CH_2)_3N$

acetic anhydride

→ acylated enamine + $CH_3CO^-[(CH_3CH_2)_3NH]^+$

acylated enamine + H_2O → 2-acetylcyclohexanone + morpholine

Table 45.1 Physical properties

	M. W.	mp	bp	d
cyclohexanone	98.2	−16	156	0.948
morpholine	87.1	−5	128	1.000
toluene	92.15	−95	111	0.867
p-toluenesulfonic acid	172.2	105	140[20]	
1-morpholinocyclohexene	167.3		119[10]	
acetic anhydride	102.1	−73	140	1.082
triethylamine	101.2	−115	89	0.728
2-acetylcyclohexanone	140.2		112[18]	1.078

Note: mp and bp are in °C; density is in g/mL.

Cyclohexanone

2937.8	1311.2	908.4
1714.0	1221.7	749.8
1449.5	1118.8	489.7

Acetic acid

$$CH_3\overset{\overset{\displaystyle O}{\|}}{C}OH$$

3275.1	1714.5	1294.1
3043.0	1412.9	628.5
2631.9	1360.8	480.8

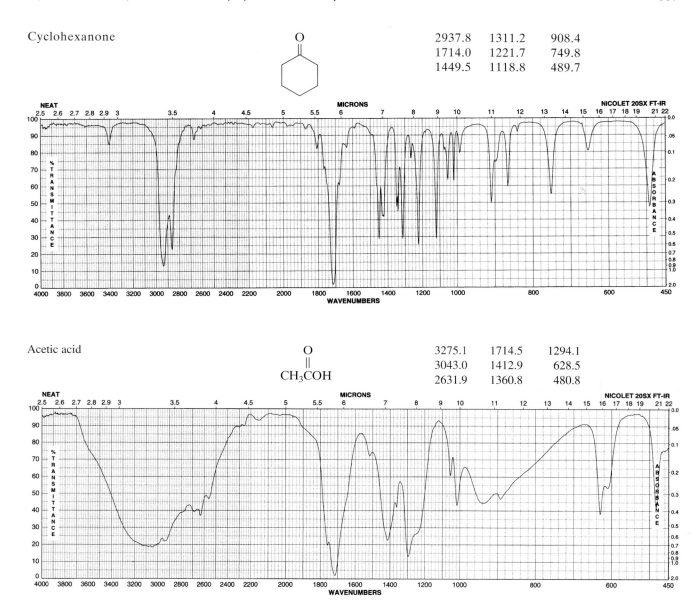

Figure 45.1 IR spectra of cyclohexanone and acetic acid

Directions

A. *Preparation of 1-Morpholinocyclohexene*

> **Toluene is flammable, and inhalation, ingestion, or skin absorption may be harmful. Avoid contact and do not breathe vapors.**
> **Morpholine is flammable and can cause severe damage to the skin, eyes, and respiratory tract. Wear gloves, dispense from a hood; avoid contact and do not breathe vapors.**
> ***p*-Toluenesulfonic acid can cause serious damage to the skin, eyes, and respiratory tract. Avoid contact and do not breathe dust.**

toluene	morpholine	*p*-toluenesulfonic acid	cyclohexanone

Take Care! Wear gloves, avoid contact with morpholine, toluene, and *p*-toluenesulfonic acid, and do not breathe their vapors or dust.

Measure 10.0 mmol of cyclohexanone, 13.0 mmol of morpholine, and 10 mg of *p*-toluenesulfonic acid into a 10-mL round-bottom flask. Add 4.0 mL of toluene, then drop in a boiling chip or stir bar. Assemble an apparatus for simple distillation [OP-27] using a Hickman still equipped with an air condenser as a water separator (see OP-27c). Boil the reaction mixture gently for at least an hour, allowing the liquid that collects in the well to overflow into the reaction flask (do not collect any distillate at this time). When the reaction period is over, remove the contents of the well and distill to remove the toluene and excess morpholine; the temperature should rise above 110°C during the distillation. Discontinue the distillation when no more liquid condenses in the still and the temperature drops below 90°C. If the 1-morpholinocyclohexene (in the reaction flask) cannot be used the same day, keep it in a tightly stoppered container in the refrigerator.

Stop and Think: What else is in the well besides toluene, and where did it come from?

Waste Disposal: Place the distillate in an appropriate waste container.

B. *Preparation of 2-Acetylcyclohexanone*

acetic anhydride	triethylamine

> **Acetic anhydride can cause severe damage to skin and eyes. Its vapors are very harmful if inhaled, and it reacts violently with water. Use gloves and a hood. Avoid contact with the liquid, do not breathe its vapors, and keep it away from water.**
> **Triethylamine is toxic, flammable, corrosive, and lachrymatory and can cause severe damage to the eyes, skin, and respiratory tract. Use gloves and a hood. Avoid contact with the liquid, do not breathe its vapors, and keep it away from flames or hot surfaces.**
> **Diethyl ether is extremely flammable and may be harmful if inhaled. Do not breathe its vapors and keep it away from flames and hot surfaces.**
> **A vacuum distillation apparatus may implode if any of its components are cracked or otherwise damaged. Inspect the parts for damage and have your instructor check your apparatus. Work behind a hood sash or safety shield while the apparatus is under vacuum.**

Reaction.   *Under the hood*, cool the reaction flask containing the 1-morpholinocyclohexene in an ice/water bath and add 12.0 mmol of triethylamine and 11.0 mmol of acetic anhydride. Cap the flask tightly and let it stand until the next laboratory period. (If your instructor requests that you turn in the round-bottom flask, transfer the reaction mixture to an appropriate storage container and seal it tightly.)

Add 3.0 mL of distilled water to the reaction mixture in the round-bottom flask, then drop in a boiling chip or stir bar and attach an air condenser. Heat the mixture under reflux [OP-7] for 20 minutes to hydrolyze the acylated enamine.

Take Care! Wear gloves, avoid contact with triethylamine and acetic anhydride, and do not breathe their vapors.

Stop and Think: What is the purpose of the triethylamine?

Separation.   Transfer the reaction mixture to a 15-mL centrifuge tube and extract [OP-15] it with two 2.5-mL portions of solvent-grade diethyl ether. Wash the combined ether extracts with 2.5 mL of saturated sodium chloride solution. Dry [OP-22] the ether layer with anhydrous sodium sulfate and evaporate [OP-16] the ether.

Waste Disposal: Pour the aqueous layer down the drain.

Purification and Analysis.   Purify the product by vacuum distillation [OP-28]. Make sure that a safety shield or hood sash is between you and the apparatus. 2-Acetylcyclohexanone should distill above 110°C under moderate vacuum; discard any forerun that distills below either that temperature or the boiling temperature estimated from a manometer reading. Weigh the product and record its IR spectrum [OP-36], or obtain a spectrum from your instructor.

Take Care! See the Safety Notes regarding vacuum distillations.

Exercises

1 From its IR spectrum, decide whether 2-acetylcyclohexanol exists mostly in the diketo form or in an enolic form, and if the latter, decide which enolic structure (of two possibilities) it has.

2 Why is it necessary to separate water from the reaction mixture in part **A**? Write an equation for the reaction that would occur if it weren't removed.

3 Describe and explain the possible effect on your results of the following experimental errors or variations. (a) In part **A**, you omitted the Hickman still from the reaction apparatus. (b) You used equimolar amounts of cyclohexanone and morpholine in part **A**. (c) You omitted the triethylamine in part **B**.

4 Following the format in Appendix V, construct a flow diagram for this experiment.

5 Acylation of enamines differs from their alkylation reaction in that an acylated iminium salt loses a proton to form an enamine before hydrolysis. Propose detailed mechanisms for the acylation of 1-morpholinocyclohexene by acetic anhydride and for the hydrolysis of the resulting enamine.

6 Propose a mechanism for the cleavage of 2-acetylcyclohexanone to the salt of 7-oxononanoic acid in aqueous KOH.

7 (a) Propose a synthesis for 9-oxo-2-decenoic acid from 7-oxononanoic acid. (b) Do you think your synthesis would yield the (2*E*) or the (2*Z*) isomer as the major product? Explain.

8 Outline enamine syntheses for the following compounds, starting with cyclohexanone and using any other needed reagents.

(a) [structure: cyclohexanone with $CH_2CCH_2CH_3$ substituent bearing carbonyl O]   (b) [bicyclic structure with CH_3 group and =O]   (c) $HOC(CH_2)_5C(CH_2)_5COH$ (with three C=O groups)

Other Things You Can Do

(Starred projects require your instructor's permission.)

*1 Record the ^1H NMR spectrum of 2-acetylcyclohexanone in deuterochloroform and interpret it as completely as you can. Use the sweep offset control to locate any enolic proton signal, which should occur around 15δ. This signal may be quite broad as a result of proton exchange.

*2 Assess the purity of your 2-acetylcyclohexanone by gas chromatography.

 3 Write a research paper about the pheromones that control behavior among social insects such as ants and honeybees, starting with sources listed in the Bibliography.

Synthesis of Dimedone and Measurement of Its Tautomeric Equilibrium Constant

EXPERIMENT **46**

Reactions of α,β-Unsaturated Carbonyl Compounds. Preparation of Dicarbonyl Compounds. Nucleophilic Addition to Carbon-Carbon Double Bonds. Condensation Reactions. Decarboxylation. Reaction Equilibria. NMR Spectrometry.

Operations

OP-7   Heating
OP-13   Vacuum Filtration
OP-16   Evaporation
OP-23   Drying Solids
OP-24   Drying and Trapping Gases
OP-25   Recrystallization
OP-30   Melting Point
OP-37   Nuclear Magnetic Resonance Spectrometry

Before You Begin

1 Read the experiment, read or review the operations as necessary, and write an experimental plan.
2 Calculate the mass and volume of 5.00 mmol of dimethyl malonate and the theoretical yield of dimedone.

Scenario

Harry Lingo, a punctilious chemistry editor for the Fulcourt Press, is faced with the dilemma of how to deal with compounds that exist in two tautomeric forms. In a soon-to-be published encyclopedia of organic compounds, he wants to list such a compound under the name corresponding to the major species present. For example, acetylacetone, which goes by the IUPAC name 2,4-pentanedione, contains four times more enol than keto tautomer, so he plans to list this compound under the name 4-hydroxy-3-penten-2-one.

Key Concept: The enolic form of a β-dicarbonyl compound is stabilized by conjugation of its C=C bond with a carbonyl group and by hydrogen bonding between the enolic OH and the carbonyl oxygen.

$$
\underset{\text{20\% keto}}{CH_3\overset{O}{\overset{\|}{C}}CH_2\overset{O}{\overset{\|}{C}}CH_3} \rightleftharpoons \underset{\text{80\% enol}}{CH_3\overset{O}{\overset{\|}{C}}CH=\overset{OH}{\overset{|}{C}}CH_3}
$$

acetylacetone

The organic compound known by the common name dimedone is represented as a β-diketone in most textbooks of organic chemistry. But many β-diketones contain a substantial amount of enolic tautomer, so Mr. Lingo has asked your Institute for help in determining the most appropriate structure and name for dimedone. Your assignment is to prepare dimedone,

Keto-enol equilibrium for dimedone

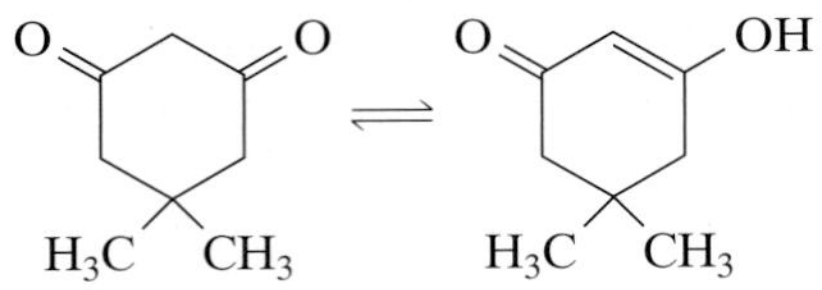

β-diketone   enolic
                      ketone

measure its tautomeric equilibrium constant to see whether the predominant species present is the β-diketone or the enolic ketone, and give this species an appropriate IUPAC name.

Applying Scientific Methodology

Your working hypothesis should include an "educated guess" about the structure of dimedone. You will test your hypothesis by analyzing its ^{1}H NMR spectrum.

Tautomers and Life

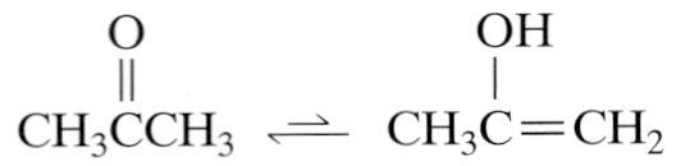

Enols have long been known as unstable intermediates in reactions involving carbonyl compounds, such as the bromination of acetone. Although acetone exists overwhelmingly in the keto form, it is the minute amount of the enolic form present that actually reacts with bromine under acidic conditions. Other enols are considerably more stable than that of acetone. For example, ethyl 3-oxobutanoate (ethyl acetoacetate) contains about 7.5% enol at equilibrium, and 2,4-pentanedione (acetylacetone) contains about 80% enol. Certain natural compounds such as Vitamin C, an enediol, exist almost entirely in the enolic form.

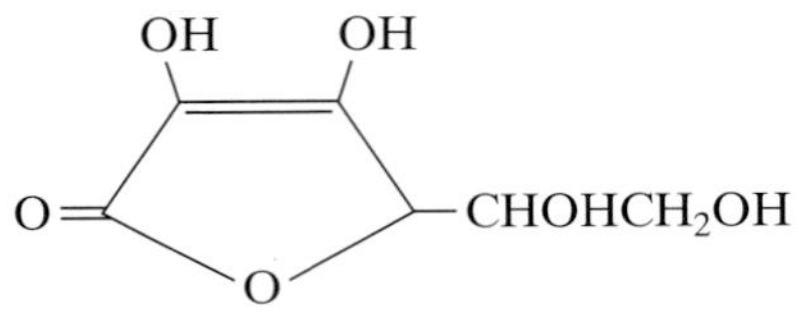

~100% keto   0.00025% enol

acetone

The Kekulé structures of phenols are enol-like, suggesting that some phenols may exist in keto forms as well. For example, phlorglucinol (1,3,5-trihydroxybenzene) forms carbonyl-type derivatives with hydroxylamine and similar derivatizing agents, which indicates that its phenolic form may exist in equilibrium with a triketone.

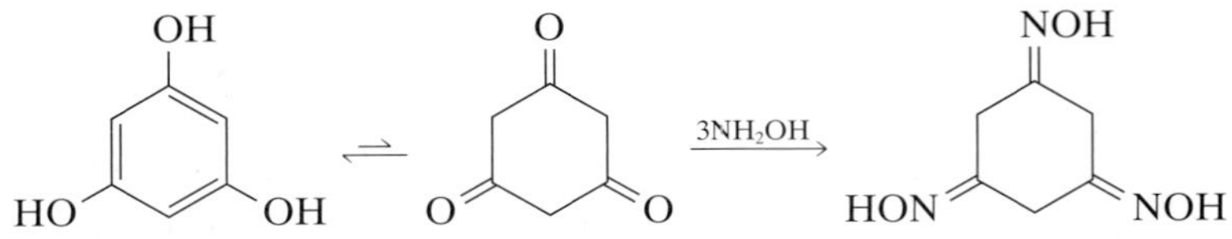

Vitamin C
(ascorbic acid)

phlorglucinol   phlorglucinol
                                       oxime

A molecule's preference for one of several possible tautomeric structures might appear to be a matter of interest only to a few chemists, but in fact it plays a crucial role in living systems. Ordinarily, the keto form of a phenolic compound is the less stable tautomer by far because its formation is accompanied by a loss of resonance energy. In the early 1950s, however, chemists discovered that certain biological amines (bases) such as guanine and thymine exist mainly in the keto forms at the pH of physiological systems.

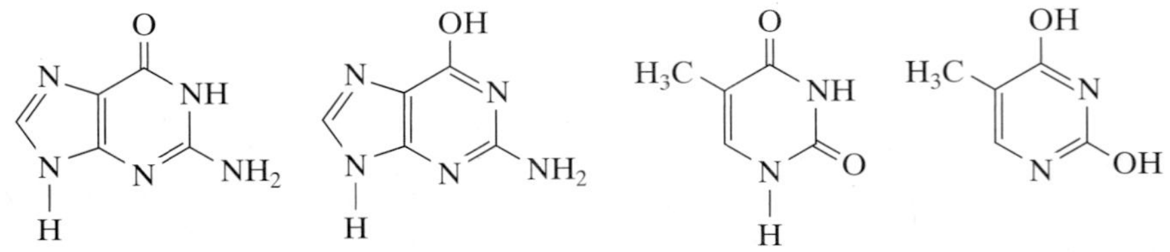

keto   enol   keto   enol

guanine   thymine

This discovery provided Francis Crick and James D. Watson with the key to the structure of DNA, and thus to the genetic code. The four bases guanine, thymine, adenine, and cytosine, which are attached to the polyester backbone of a nucleic acid molecule, are responsible for both the transmission of genetic traits (by DNA) and the synthesis of proteins (by RNA) in living beings. Only the keto forms of guanine and thymine allow for the formation of adenine-thymine and guanine-cytosine base pairs, and such base pairing is essential for nucleic acid molecules to function. If these two bases existed only in enolic forms rather than keto forms, life as we know it might be impossible!

Watson's book The Double Helix *[Bibliography, L77] gives a fascinating account of the discovery of the DNA structure.*

Understanding the Experiment

In this experiment, you will synthesize dimedone starting with the α,β-unsaturated ketone 4-methyl-3-penten-2-one, commonly known as mesityl oxide. This synthesis involves four distinct reactions that take place in sequence (see "Reactions and Properties"). The first reaction is a *Michael addition* of dimethyl malonate to mesityl oxide. Combining sodium methoxide with dimethyl malonate generates an enolate ion that undergoes nucleophilic addition to the carbon-carbon double bond of mesityl oxide, yielding a keto diester (**1**). In the presence of sodium methoxide, the methyl group alpha to the keto carbonyl group loses a proton, forming another enolate ion that attacks the carbonyl carbon of one of the ester functions and displaces its methoxyl group. The result of this Claisen-type *cyclization* reaction is a diketo ester (**2**) with a six-membered ring. Both of these reactions occur spontaneously in the same reaction mixture. Subsequent *hydrolysis* of the ester yields a carboxylic acid, which when heated undergoes *decarboxylation* to form dimedone.

A literature procedure in *Organic Syntheses* [Bibliography, B20] for the synthesis of dimedone requires the *in situ* preparation of sodium ethoxide by adding metallic sodium to absolute ethanol. Since any operation involving sodium can be hazardous, commercial sodium methoxide in methanol will be used instead. The enolate ion of dimethyl malonate, $[CH(COOCH_3)_2]^-$, forms immediately when sodium methoxide is added to dimethyl malonate. This species reacts rapidly and exothermically with mesityl oxide, which should therefore be added slowly to prevent side reactions. The Michael addition and the subsequent cyclization reaction should go virtually to completion during a 1-hour reflux period. The cyclic ester is then hydrolyzed with 3 M NaOH and the solution is acidified to form the carboxylic acid, which loses carbon dioxide on heating. The resulting dimedone is isolated by vacuum filtration and recrystallized from aqueous acetone.

One of the best ways to detect and analyze enol content is by ^{1}H NMR spectrometry. The enolic OH protons of α,β-diketones absorb far downfield, with chemical shift values in the 10–16 δ range. Enols also show vinylic proton (H—C=C) absorption in the usual range for these signals, about 4.5–6.0 δ. The keto form of an α,β-diketone can usually be recognized by the signal of the protons alpha to both carbonyl groups [H—C(C=O)$_2$]; these protons absorb near 3.5 δ. Since there are two such protons for every enolic vinyl proton in the enol form of dimedone, the equilibrium constant for dimedone's keto-enol equilibrium can be determined from the integrated areas of these two signals using the following formula:

$$K = \frac{[\text{enol}]}{[\text{keto}]} = \frac{2 \times \text{area of the } H-C{=}C \text{ signal}}{\text{area of the } H-C(C{=}O)_2 \text{ signal}}$$

Enols are stabilized by hydrogen-bonding interactions with other polar species, so the equilibrium constant might vary depending on the solvent used. The "Other Things You Can Do" section describes some additional studies you can carry out with your product.

Reactions and Properties

Note: All of the cyclohexanedione derivatives shown can also be represented by enolic forms.

Table 46.1 Physical properties

	M.W.	mp	bp	d
mesityl oxide	98.2	−52	129	0.858
dimethyl malonate	132.1	−62	181	1.154
sodium methoxide	54.0			
methanol	32.0	−94	65	0.791
dimedone	140.2	151		

Note: mp and bp are in °C; density is in g/mL. A 25% solution of sodium methoxide in methanol has a density of 0.945 g/mL.

Mesityl oxide (4-methyl-3-penten-2-one)

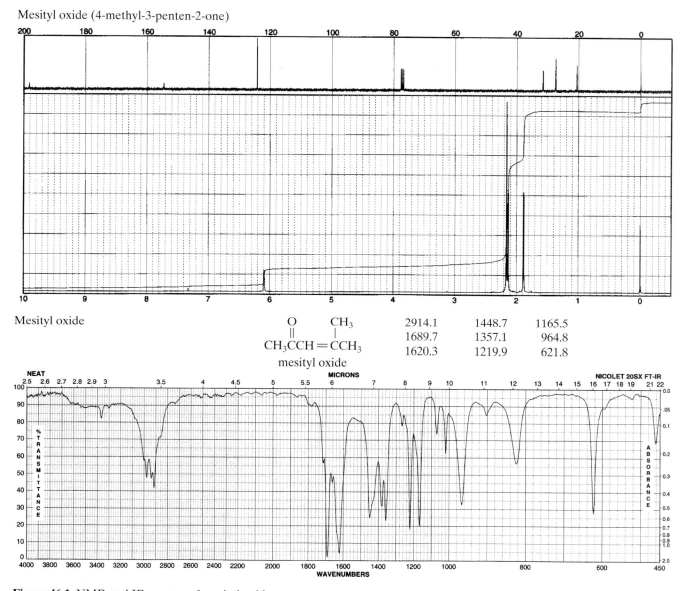

Mesityl oxide

O CH₃	

$$CH_3CCH = CCH_3$$

mesityl oxide

2914.1	1448.7	1165.5
1689.7	1357.1	964.8
1620.3	1219.9	621.8

Figure 46.1 NMR and IR spectra of mesityl oxide

Directions

The sodium methoxide/methanol solution is corrosive, toxic, and flammable. Contact or inhalation can cause severe damage to the skin, eyes, and respiratory tract. Wear gloves, avoid contact, do not inhale its vapors, and keep it away from flames.

Mesityl oxide is a lachrymator and a strong irritant; inhalation and skin absorption are harmful. Avoid contact and do not breathe its vapors. Mesityl oxide forms explosive peroxides on standing, so it should never be distilled to dryness.

Deuterochloroform is harmful if inhaled or absorbed through the skin, and it is a suspected human carcinogen. Avoid contact with the liquid and do not breathe its vapors.

Safety Notes

mesityl oxide

methanol

Reaction. *The reaction apparatus must be dry.* Weigh 5.00 mmol of dimethyl malonate into a 10-mL round-bottom flask and drop in a stir bar, then stir as you add 1.0 mL of methanol and 1.20 mL (~5.2 mmol) of a 25% solution of sodium methoxide in methanol. Attach a water-cooled condenser and use a hot water bath to heat [OP-7] the solution to boiling, with stirring. Any solid in the reaction mixture should dissolve (or nearly so) on heating. Remove the apparatus from the heat source and use a 9-inch Pasteur pipet to add 0.60 mL (~5.2 mmol) of recently distilled mesityl oxide drop by drop through the condenser over a period of 2–3 minutes, with stirring. Attach a drying tube [OP-24] containing calcium chloride to the top of the condenser and heat the reaction mixture under gentle reflux [OP-7] for one hour with stirring. Remove the condenser and evaporate [OP-16] the methanol directly from the round-bottom flask. Stop the evaporation when the solid residue is still moist but all standing liquid has been removed.

Add 4.0 mL of 3.0 M sodium hydroxide to the reaction flask, attach an air condenser, and heat the mixture under reflux [OP-7] for one hour, with stirring, to hydrolyze the intermediate ester. Measure 3.0 mL of 6.0 M hydrochloric acid into a 25-mL Erlenmeyer flask and, while swirling or stirring the HCl solution manually, transfer the warm reaction mixture to the Erlenmeyer flask. *Under the hood*, place a small watch glass over the mouth of the flask and boil the mixture gently, with stirring, for at least 10 minutes until no more gas is evolved. Add water if necessary to replace any that evaporates. Allow the reaction mixture to cool to room temperature, scratching to induce crystallization if necessary, and cool the flask in an ice/water bath until crystallization is complete.

Separation. Collect the product by vacuum filtration [OP-13] and let it air dry on the filter.

Purification and Analysis. Purify the air-dried dimedone by recrystallization [OP-25] from aqueous 50% acetone. Dry [OP-23] and weigh the dimedone and measure its melting point [OP-30] Record the ^{1}H NMR spectrum [OP-37] of dimedone in deuterochloroform, or obtain a spectrum from your instructor. On your NMR spectrum, identify the enol H—C=C and keto H—C(C=O)$_2$ signals. Calculate the keto-enol equilibrium constant, decide which species (if either) predominates in the solution, and give it an IUPAC name.

Exercises

1 (a) Compare the ^{1}H NMR spectrum of your product to that of mesityl oxide shown in Figure 46.1, and interpret both spectra as completely as you can. (b) Interpret the IR spectrum of mesityl oxide as completely as you can. Account for the wavenumber of the carbonyl band, comparing it with the usual value of ~1715 cm^{-1} for an aliphatic ketone.

2 (a) Write a mechanism for the Michael addition of dimethyl malonate to mesityl oxide in the presence of sodium methoxide/methanol. (b) Write a mechanism for the cyclization step in the synthesis of dimedone. (c) Write a mechanism for the decarboxylation step of this synthesis, showing the structure of the transition state.

3 Describe and explain the possible effect on your results of the following experimental errors or variations. (a) You neglected to evaporate methanol from the reaction mixture. (b) The mesityl oxide bottle was contaminated with water. (c) The bottle labeled "dimethyl malonate" actually contained dimethyl maleate.

4 Following the format in Appendix V, construct a flow diagram for this experiment.

5 What inexpensive starting material do you think is used to prepare mesityl oxide commercially? Outline the synthesis of mesityl oxide from this starting material.

6 A rather confused student, Ina Fogg, hydrolyzed the intermediate ester with 6 M HCl rather than 3 M NaOH and then boiled the reaction mixture with 50% NaOH instead of 6 M HCl. Filtration of the cooled solution yielded only a little dimedone. Finally realizing her mistake, she acidified the filtrate and a white solid precipitated, but its melting point was different from that of dimedone. What was this solid and how did it form?

7 Draw structures of some by-products that might be present in your product if you neglected to use a drying tube during the first step of the synthesis of dimedone. Write reaction pathways for their formation.

8 When heated with sodium ethoxide in ethanol, the diethyl esters of butanedioic acid and heptanedioic acid both yield products having six-membered rings. Write structures for both products and mechanisms for their formation.

9 Write a mechanism for the following reaction of dimedone with 3-buten-2-one.

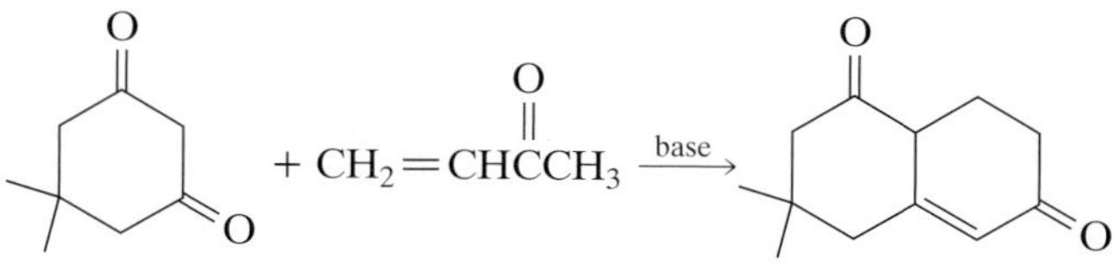

Other Things You Can Do

(Starred projects require your instructor's permission.)

*1 Record ^{1}H NMR spectra of dimedone in solvents other than deuterochloroform, such as DMSO-d_6, and compare the keto-enol equilibrium constants for the different solvents. Try to explain any differences.

*2 Record the infrared spectrum [OP-36] of dimedone using a KBr disc or Nujol mull and then using a chloroform solution. Look for absorption bands arising from the enolic form and the diketo form in both spectra, and compare the relative amounts of enol. Try to explain any differences.

*3 Use your product to prepare dimedone derivatives of benzaldehyde as described in Minilab 39.

4 Like the synthesis of dimedone from mesityl oxide, a Robinson annulation sequence involves a Michael addition followed by a ring-forming condensation reaction. Starting with sources listed in the Bibliography, write a research paper about the Robinson annulation, describing some of its applications in the synthesis of terpenoids and steroids.

Preparation of Para Red and Related Azo Dyes

EXPERIMENT 47

Reactions of Amines. Reactions of Diazonium Salts. Preparation of Azo Compounds. Electrophilic Aromatic Substitution.

Operations

OP-8   Cooling
OP-10   Mixing
OP-13   Vacuum Filtration
OP-23   Drying Solids

Before You Begin

1 Read the experiment, read or review the operations as necessary, and write an experimental plan.
2 Calculate the mass of 2.00 mmol of *p*-nitroaniline, the mass of 2.00 mmol of 2-naphthol, and the theoretical yield of Para Red. Be prepared to calculate the mass of 2.00 mmol of any other coupling component or diazo component and the theoretical yield of any other azo dye.

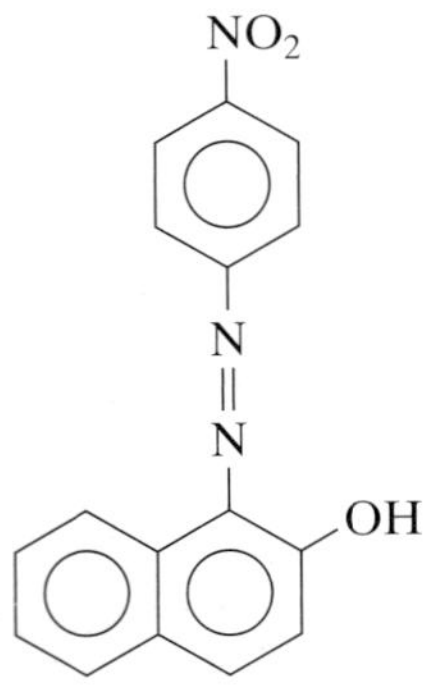

"American Flag Red"
(Para Red)

Scenario

The despotic King of Erewhon (see Experiment 22) has just been overthrown in a coup led by Sergeant Obmar, a soldier of fortune from California. In his honor, the provisional government has authorized a new flag, consisting of a navel orange on a field of blue. But the first flags they ordered were sent back because the orange was the wrong color—it looked more like a pink grapefruit. Acting President Gib Retsim has asked your Institute to come up with a dye that will produce just the right shade of orange. Your supervisor knows how to make "American Flag Red," an orange-red dye the color of the stripes in the United States flag, and thinks that by tinkering with its molecular structure, your project team should be able to come up with a suitable orange dye for the new flag.

Applying Scientific Methodology

Based on the information in the experiment, you should develop a hypothesis regarding a pair of reactants from Table 47.1 that you think might produce a dye the color of a navel orange. Alternatively, members of your project group can get together and select a different reactant pair for each member to work on.

Dyes and Serendipity

In the Persian fairy tale *The Three Princes of Serendip*, the title characters were forever discovering things they were not looking for at the time. Thus, *serendipity* is the aptitude for making happy discoveries by accident. The preparation of the first commercially important synthetic dye by William Henry Perkin in 1854 is a good example of a serendipitous discovery in science. During the nineteenth century, quinine was the only drug known to be effective against malaria, and it could be obtained only from the bark of the cinchona tree, which grew in South America. French chemists had isolated pure quinine from cinchona bark in 1820, but the inaccessibility of the tree made natural quinine very expensive. Perkin, then an 18-year-old graduate student working for the eminent German chemist August Wilhelm von Hofmann, realized that anyone who could make synthetic quinine might well become rich and famous. Perkin knew nothing about the molecular structure of quinine—structural organic chemistry was in its infancy in the mid-1800s—but he knew its molecular formula, $C_{20}H_{24}N_2O_2$. So he prepared some allyltoluidine ($C_{10}H_{13}N$), apparently thinking that two molecules of allyltoluidine plus three oxygen atoms minus a molecule of water would magically yield $C_{20}H_{24}N_2O_2$—quinine!

Perkin's idea

$$2 \text{ allyltoluidine} \xrightarrow[-H_2O]{+3[O],} \text{quinine}$$

allyltoluidine   quinine

Of course, Perkin had attempted the impossible—the molecular structure of allyltoluidine bears no resemblance to that of quinine, which was not synthesized until 1940. But he hopefully oxidized allyltoluidine with potassium dichromate and came up with a reddish-brown precipitate that he quickly realized was not quinine. Most chemists would have thrown out the stuff and started over, but it had properties that interested Perkin, so he decided to try the same reaction with a simpler base, aniline. This time he obtained a black precipitate that, when extracted by ethanol, formed a beautiful purple solution that impressed some of the local dyers. Perkin knew a good thing when he saw it, so he gave up his study of chemistry and went into the business of manufacturing "aniline purple," or mauve as the dye soon came to be known. Ironically, Perkin became rich and famous by *failing* to synthesize quinine. His dyestuffs plant was so successful that he was able to retire at the age of 36 and devote the rest of his life to pure research.

While Perkin was getting the synthetic-dye industry under way, other chemists were experimenting with aniline and many other compounds that can be extracted from coal tar. One of these was a brewery chemist named

aniline

mauve

Peter Griess, who took time off from the brewing of Alsopps' Pale Ale to discover the azo dyes. Undiscouraged by the fact that many of the diazo compounds he prepared had a tendency to explode, Griess did some fundamental research into the diazotization of aromatic amines and went on to discover the coupling reaction by which virtually all azo dyes are now synthesized. Aniline Yellow and Bismark Brown were synthesized in the 1860s, and the production and use of azo dyes grew rapidly thereafter.

Aniline Yellow

Bismark Brown

The advancement of organic chemistry as a science is due, in large part, to the discovery that chemists could prepare synthetic colors that were in many ways superior to the natural ones.

Understanding the Experiment

In this experiment, you will prepare the azo dye Para Red and at least one other azo dye, and use your products to dye cloth. Para Red, which is made from *para*-nitroaniline and 2-naphthol, was once called "American Flag Red" because it was used to dye the cloth used for the stripes in the American flag.

Most azo dyes can be prepared by combining a diazonium salt (the *diazo component*) with an activated aromatic compound (the *coupling component*). The preparation of the dye involves two stages, known as *diazotization* and *coupling*. In the diazotization stage, a primary aromatic amine reacts with nitrous acid (HONO) to form the diazonium salt. Nitrous acid is generated *in situ* from sodium nitrite and a mineral acid. The reaction is carried out at a low temperature because diazonium salts react with water to form phenols and other by-products at higher temperatures. In the coupling stage, the diazonium salt is added to a solution of the coupling component, which is usually a phenol or an aromatic amine. Phenols couple most readily in mildly alkaline solutions, whereas amines react best in acidic solutions. However, too low a pH will prevent an amine from reacting by causing protonation of the amino group, whereas too high a pH will cause the diazonium salt to change to a diazotate ion, which is incapable of coupling.

Diazotization

$$ArNH_2 \xrightarrow{\text{HONO}} Ar-N_2^+$$

diazonium salt

Coupling

$$Ar-N_2^+ + H-Ar' \xrightarrow{-H^+}$$

coupling component

$$Ar-N=N-Ar'$$

azo compound

(Ar′ must contain an activating group such as —OH or —NR$_2$.)

Low pH: $ArNR_2 \underset{}{\overset{H^+}{\rightleftharpoons}} ArNHR_2^+$

High pH: $ArN_2^+ \underset{}{\overset{OH^-}{\rightleftharpoons}} ArN=N-O^-$

diazotate ion

Formation of unreactive species at low and high pH

The coupling reaction is an electrophilic aromatic substitution reaction, with the diazonium salt acting as the electrophile. Because the diazo group, $-N_2^+$, is only weakly electrophilic, the coupling component must contain strongly activating groups, such as OH or NR_2, in order for coupling to occur.

To prepare the diazonium salt, a primary aromatic amine is dissolved in about 2.5 equivalents of dilute hydrochloric acid (or another suitable acid) and the solution is cooled to 5°C or below. Aqueous sodium nitrite is then added and the solution is tested for excess nitrous acid using starch-iodide paper. If the amine is insoluble, the diazotization reaction is carried out in suspension, with stirring.

The coupling reaction is carried out by adding the diazo compound, with cooling and stirring, to a solution of a coupling component in dilute acid or base. If the coupling component is a phenol, it is dissolved in about 2 equivalents of 1 *M* NaOH and cooled before adding the diazonium salt solution. If necessary, the pH can be adjusted after the addition to obtain a better yield of azo dye. If the coupling component is an amine, it is dissolved or suspended in 1 equivalent of 1 *M* HCl. After the diazo component is added and coupling is complete, the solution is neutralized to litmus paper by adding 3 *M* sodium carbonate. The dye is then cooled in an ice bath before filtering. The azo dye should be dried at room temperature, because it may decompose on heating.

Cloth can be dyed by several different processes. In the *direct process*, the dye is dissolved in water, the solution is heated, and the cloth is immersed in the hot solution. The dye molecules attach themselves to the cloth fibers by direct chemical interactions. In the *disperse process*, a water-insoluble dye is suspended in water and a small amount of a carrier substance is added. The carrier dissolves the dye and carries it into the fibers. With the *ingrain process*, a dye is synthesized on the cloth itself, usually by the combination of a diazonium salt with a coupling component. The cloth is immersed in a solution of one of the components, allowed to dry, and then immersed in a solution of the other component to develop the color. The comparatively small molecules of the separate components can diffuse into the spaces between the fibers; after they combine to form the dye, the larger dye molecules are trapped there. You will be using the ingrain method of dyeing in this experiment, but you can experiment with the other dyeing methods as described in "Other Things You Can Do."

The color of a dye depends on the wavelengths of visible light it absorbs. If a dye absorbs light at certain wavelengths, the color perceived by the human eye arises from the wavelengths that are not absorbed, and are therefore reflected to the eye. The color of each dye you prepare in this experiment should be some shade of yellow, orange, or red; a yellow color is associated with absorption of light of relatively short wavelengths, and a deep red color with absorption of light of longer wavelengths. The light-absorbing portion of a dye molecule, called a *chromophore*, is a conjugated system of delocalized pi electrons. The chromophoric system of Para Red, for example, includes the benzene and naphthalene rings, the two doubly bonded nitrogen atoms that connect them, and the unsaturated nitro group. In general, the more extended the chromophore, the longer the wavelength of the light it absorbs. Thus Para Red, with 21 atoms in its chromophore, absorbs light of considerably longer wavelengths than does Aniline Yellow, with only 14 atoms in its chromophore.

Key Concept: Increasing the length of a conjugated system decreases the energy separation between its pi molecular orbitals and therefore the ΔE of its $\pi-\pi^*$ electronic transitions. The wavelength of light absorbed during a transition is inversely proportional to the energy of the transition, so decreasing ΔE increases λ.

Certain saturated substituents called *auxochromes* can, in effect, extend a conjugated system by resonance. Auxochromes such as the OH, OCH_3, and NR_2 groups have one or more pairs of nonbonded electrons that they can share with a chromophore, thereby increasing the wavelength of the light it absorbs. Such groups exert the greatest effect if they are *ortho* or *para* to the $-N=N-$ group so that they can participate in resonance with the rest of the chromophore.

Reactions and Properties

Equations are given for the preparation of Para Red only.

Diazotization

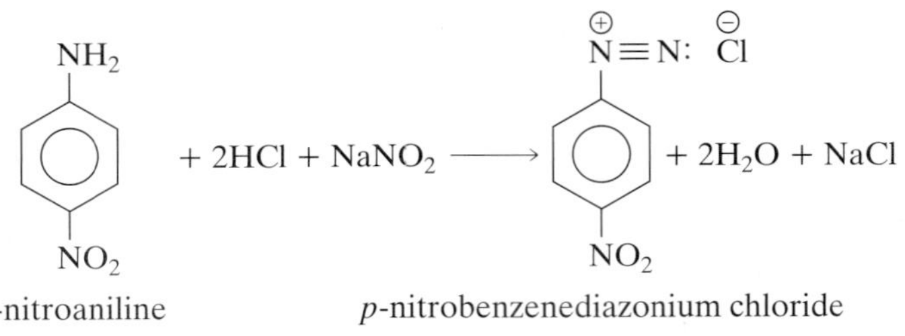

p-nitroaniline   p-nitrobenzenediazonium chloride

Coupling

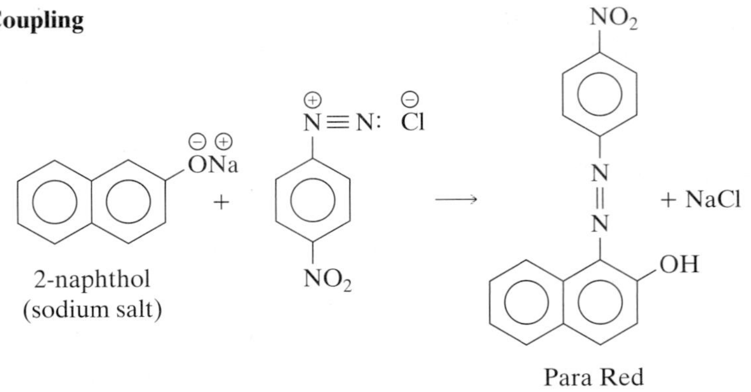

2-naphthol
(sodium salt)

Para Red

Table 47.1 Suggested diazo and coupling components

Diazo component	M.W.	Coupling component	M.W.
aniline	93.1	aniline	93.1
m-anisidine	123.2	*N*-methylaniline	107.2
m-nitroaniline	138.1	*N,N*-dirnethylaniline	121.2
m-toluidine	107.2	*m*-phenylenediamine	108.1
p-anisidine	123.2	phenol	94.1
p-nitroaniline	138.1	1-naphthol	144.2
p-toluidine	107.2	2-naphthol	144.2
		resorcinol	110.1

Note: Select one diazo component and one coupling component for each dye you prepare.

Directions

Each student should prepare Para Red and at least one other dye. With the instructor's permission, students can work in small groups, with each person preparing a different dye. Any diazo component from Table 47.1 can be used in combination with any coupling component from that table.

A. *Preparation of an Azo Dye*

> **Aromatic amines are very harmful if inhaled, ingested, or absorbed through the skin. Some aromatic amines are suspected carcinogens. Wear gloves and dispense under a hood. Avoid contact and do not inhale their vapors.**
> **Most phenols are harmful if inhaled, ingested, or absorbed through the skin, and some are very corrosive, causing severe irritation or damage to skin and eyes. Some phenols are suspected carcinogens. Wear gloves, avoid contact, and do not inhale their dust or vapors.**
> **Some azo dyes may be carcinogenic; wear gloves and avoid contact with the products.**

Safety Notes

aromatic amines   phenols

1. Diazotization of an Aromatic Amine.   Mix 2.00 mmol of the diazo component with 1.6 mL of 3 *M* HCl in a 10-mL beaker. If the diazo component doesn't dissolve completely, heat the solution gently, adding up to 2 mL of water to get most or all of it in solution. Cool [OP-8] this solution for 5 minutes or more in an ice/water or ice-salt bath with magnetic stirring [OP-10]. The amine salt may precipitate as you cool the solution, but it will diazotize satisfactorily if the reaction mixture is well stirred. Continue to stir as you add 2.0 mL of freshly prepared 1 *M* sodium nitrite, drop by drop, over a 2–3 minute period. Test the solution with starch-iodide paper; if necessary, add enough additional sodium nitrite, drop by drop, to give a positive test (blue-black color). As accurately as you can, divide the solution into two equal parts (labeled **d1** and **d2**), keeping both parts cold in an ice/water bath. Go to step *2a* if the coupling component is a phenol or to *2b* if it is an amine.

Take Care! Wear gloves, avoid contact with the amine and do not breathe its vapors.

Observe and Note: What evidence is there that a reaction is occurring?

2a. Coupling with a Phenol.   Dissolve or suspend 2.00 mmol of the phenol in 4.0 mL of 1 *M* NaOH (use 8.0 mL for resorcinol), and cool the solution in an ice/water bath. As accurately as you can, divide the solution into two equal parts (labeled **c1** and **c2**). Slowly add diazonium salt solution **d1** to coupling component solution **c1**, with manual stirring, and leave the mixture in the cold bath for 15 minutes or more. (Keep solution **d2** cold and save it and **c2** for part **B.**) If little or no colored solid appears, adjust the pH with dilute HCl or NaOH to induce coupling. Collect the azo dye by vacuum filtration [OP-13], washing it on the filter with water. Dry [OP-23] your azo dye at room temperature and weigh it.

Take Care! Wear gloves, avoid contact with the phenol, and do not breathe its dust or vapors.

Observe and Note: What evidence is there that a reaction is occurring?

Stop and Think: Why may adjusting the pH help induce coupling?

Take Care!   Wear gloves and avoid contact with the azo dye.

2b. Coupling with an Amine.   Dissolve or suspend 2.00 mmol of the aromatic amine in 2.0 mL of 1 *M* HCl (use 4.0 mL for *m*-phenylenediamine), and cool the solution in an ice/water bath. As accurately as you can, divide the solution into two equal parts (labeled **c1** and **c2**). Slowly add diazonium salt solution **d1** to coupling component solution **c1**, with manual stirring. (Keep solution **d2** cold and save it and **c2** for part **B.**) Neutralize

Take Care! Wear gloves, avoid contact with the amine, and do not breathe its vapors.

Observe and Note: What evidence is there that a reaction is occurring?

Stop and Think: What is the purpose of the neutralization step? How should it affect the color of the dye?

Take Care! Wear gloves and avoid contact with the azo dye.

Stop and Think: What is happening on the cloth to account for your observations?

the solution to litmus with 3 M sodium carbonate (add it slowly to minimize foaming) and leave the mixture in the cold bath for 15 minutes or more. Collect the azo dye by vacuum filtration [OP-13], washing it on the filter with water. Dry [OP-23] your azo dye at room temperature and weigh it.

B. *Dyeing a Cloth by the Ingrain Process*

Mix 8 mL of water into coupling component solution **c2** and soak a piece of clean white cloth in it for 2–3 minutes. Remove the cloth with forceps or a pair of stirring rods, blot it between paper towels to remove most of the water, and hang it up to dry. Mix 8 mL of ice-cold water into diazonium salt solution **d2**, insert the dry cloth, and agitate the solution with a stirring rod long enough to dye the cloth uniformly. If your coupling component was an aromatic amine, dip the cloth briefly into a small amount of 3 M sodium carbonate solution. Remove the cloth and dry it as before. If there is a navel orange handy, compare its color to that of your dyed cloth. Prepare a table describing the colors of the cloths dyed by the azo dyes prepared by your group or lab section.

Exercises

1   Discuss the effects of structural features (such as substituents and chromophore size) on the colors of the dyes prepared by your group.

2   Write a mechanism for the coupling reaction of *p*-nitrobenzenediazonium chloride with 2-naphthol.

3   Describe and explain the possible effect on your results of the following experimental errors or variations. (a) You forgot to cool the solution of your diazo component before adding aqueous sodium nitrite. (b) Your coupling component was a phenol, but you followed the procedure in *2b* to couple it. (c) You tried to dye a cloth by dipping it into a solution of 2-naphthol in 1 M NaOH, drying it, then dipping it into a solution of aniline in 1 M HCl.

4   Following the format in Appendix V, construct a flow diagram for your synthesis of Para Red in part **A**.

5   Mel A. Droyt was trying to prepare *p*-dimethylaminoazobenzene (Butter Yellow) by coupling 2.0 mmol of *N,N*-dimethylaniline with an equimolar amount of aniline. He first added 4.0 mL of 1 M sodium nitrite to the diazo component. Mixing this solution with the coupling component yielded some Butter Yellow along with a pale yellow oil. (a) What did he do wrong, and what was the yellow oil? (b) Write a balanced equation for its formation.

6   Bea Wilder was attempting to prepare chrysoidine by coupling 2.0 mmol of *m*-phenylenediamine with 2.0 mmol of aniline. To her surprise, she had to add 4.0 mL of 1 M sodium nitrite in the diazotization step before the solution turned starch-iodide paper blue. After pouring the diazonium salt solution into the solution of the other component, she recovered a dark-colored precipitate that was not chrysoidine. Next day the filtrate contained another precipitate that she identified as resorcinol. (a) What did she do wrong, and what was the structure of the azo dye she synthesized? (b) Write balanced equations for its synthesis and for the reaction that formed resorcinol.

chrysoidine

7 Why is it important to keep the temperature low during diazotization and coupling? Give the structure of the product that might form if the reaction mixture is heated during the diazotization of *p*-nitroaniline, and write an equation for its formation.

8 Explain why coupling of *p*-nitrobenzenediazonium chloride occurs mainly *para* to the —N(CH$_3$)$_2$ group of *N*,*N*-dimethylaniline but *ortho* to the —OH group of 2-naphthol.

9 Outline syntheses of Aniline Yellow and Bismark Brown (see "Dyes and Serendipity"), starting with benzene.

Other Things You Can Do

(Starred projects require your instructor's permission.)

*1 Use your dye for direct dyeing by suspending 0.10 g of the dye in 20 mL of hot water and acidifying the mixture with a few drops of concentrated sulfuric acid. Immerse pieces of wool or cotton cloth in the mixture for 5 minutes or more. Remove the cloth, rinse it with water, and let it dry. Adjusting the pH of the dyeing mixture with dilute HCl or NaOH may give better results in some cases.

Take Care! Wear gloves and avoid contact with H$_2$SO$_4$.

*2 Use your dye for disperse dyeing by suspending 0.10 g of the dye in 20 mL of hot water and stirring in 0.02 g of biphenyl (the carrier) and one small drop of liquid detergent. Then immerse a piece of cloth made of Dacron or another polyester in the mixture, and heat the solution with a boiling water bath for 15–20 minutes. Remove the cloth and let it dry.

*3 Dissolve 3–5 mg of an azo dye in 10 mL of 95% ethanol. If any solid remains undissolved, filter the solution. Record the ultraviolet–visible spectrum [OP-38] of the dye over the tungsten lamp range (~800–350 nm), diluting the solution with more ethanol if necessary. Compare the λ_{max} values of different dyes and try to explain some of the differences you observe. Note that such comparisons are meaningful only if the bands you are comparing arise from the same kind of electronic transition; such bands should be similar in appearance and intensity.

*4 In Minilab 40, see what happens when you diazotize anthranilic acid and heat the resulting solution.

5 Starting with sources listed in the Bibliography, write a research paper about the chemistry and uses of food colorings. Outline syntheses for some azo dyes that have been used as food colorings and discuss the controversy surrounding such dyes as FD&C Red No. 2.

Reaction of Phthalimide with Sodium Hypochlorite

Reactions of Amides and Imides. Nucleophilic Acyl Substitution. Functional Derivatives of Carboxylic Acids. Molecular Rearrangements.

Operations

OP-7   Heating
OP-8   Cooling
OP-9   Temperature Monitoring
OP-10 Mixing
OP-13 Vacuum Filtration
OP-23 Drying Solids
OP-25 Recrystallization
OP-30 Melting Point

Before You Begin

1 Read the experiment, read or review the operations as necessary, and write an experimental plan.
2 Calculate the mass of 5.00 mmol of phthalic anhydride, 2.50 mmol of urea, and 3.00 mmol of phthalimide. Calculate the theoretical yield of $C_7H_7NO_2$.

Scenario

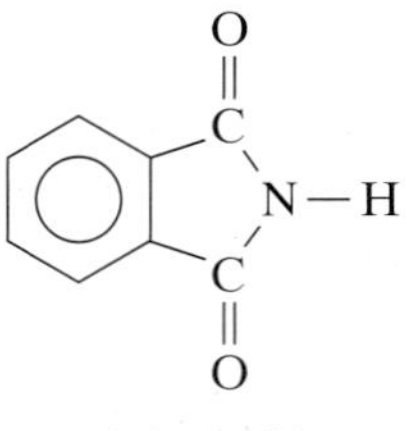

Otto Fökus, a nearsighted chemistry professor from Miskatonic University, was attempting to hydrolyze some phthalimide to phthalic acid when he mistook a bottle of chlorine bleach for a similar bottle containing aqueous sodium hydroxide (NaOH). By the time he took a close look at the label and realized his error, he had already added some bleach to the reaction mixture. Thinking he might still salvage the experiment, he then added the designated amount of sodium hydroxide and heated the reaction mixture, but the crystalline solid he obtained had the wrong melting point for phthalic acid. Curious about this unexpected result, Professor Fökus sent a sample of the product to a commercial laboratory for analysis. The analytical report showed that the product contained nitrogen, but no chlorine.

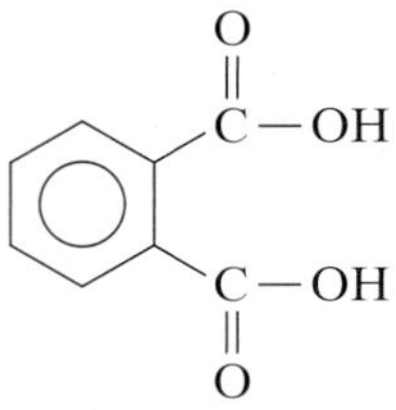

phthalimide

phthalic acid

Analysis of sample	
Carbon	61.30%
Hydrogen	5.15%
Nitrogen	10.23%
Chlorine	0.00%

Based on the product's elemental composition and an estimate of its molecular weight, the professor determined that its molecular formula is $C_7H_7NO_2$.

Unfortunately, 18 compounds with that formula were listed in the edition of the *CRC Handbook of Chemistry and Physics* that he consulted. Because of an inadequate science equipment budget, Professor Fökus doesn't have access to the instruments needed to determine the structure of this compound, so he has asked your Institute for help. Your supervisor suspects that the reaction of phthalimide with chlorine bleach (which contains sodium hypochlorite, NaOCl) may involve some kind of molecular rearrangement. Your assignment is to carry out the reaction of phthalimide with sodium hypochlorite and sodium hydroxide, and to identify the product from its melting point and by using your chemical intuition. If there is no phthalimide in the Institute's chemical stockroom, you may prepare it from a less expensive starting material, phthalic anhydride.

Applying Scientific Methodology

The following sections on the Curtius rearrangement and the reactions of imides contain some clues that should lead you to a reasonable hypothesis regarding the structure of $C_7H_7NO_2$, which is one of the 18 compounds listed in Table 48.1. You will test your hypothesis by measuring the melting point of the product.

The Curtius Rearrangement

Acyl azides are compounds containing the azide ($-N_3$) functional group on a carbonyl carbon. They can be prepared by treating acid chlorides with sodium azide (NaN_3). When an acyl azide is heated, it evolves gaseous nitrogen to yield an isocyanate by a reaction called the Curtius rearrangement. At one time, it was thought that the rearrangement involved an electron-deficient intermediate called a *nitrene*, as illustrated in the margin for the Curtius rearrangement of benzoyl azide. This reaction resembles a carbocation rearrangement, except that the migrating group moves to an electron-deficient nitrogen atom instead of to a carbon atom. In this example, migration of a phenyl group to the nitrogen atom would restore its missing pair of electrons and yield a stable product.

More recent studies of such rearrangements suggest that free nitrenes are probably *not* involved. The rearrangement step is thought to proceed by a concerted mechanism, as shown in the following general example, where Z is a good leaving group (such as $-N_3$) and R is an alkyl or aryl group:

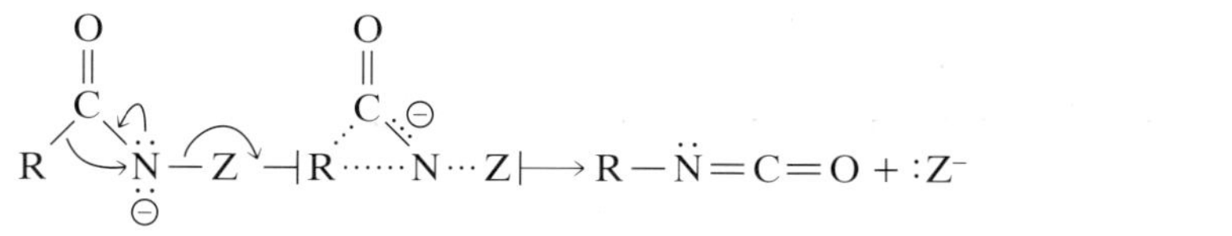

Thus the migration to a *potentially* electron-deficient nitrogen occurs as the leaving group is being ejected, not after it has already left to form a nitrene. Depending on the reaction conditions, the resulting isocyanate may be isolated as such, or it may react further with the solvent. If the Curtius rearrangement is run in an alcoholic solution, for example, the alcohol adds to the $C=N$ bond of the isocyanate to form a urethane. In water, a carbamic acid is formed at first, but it loses carbon dioxide spontaneously to yield an amine as the final product.

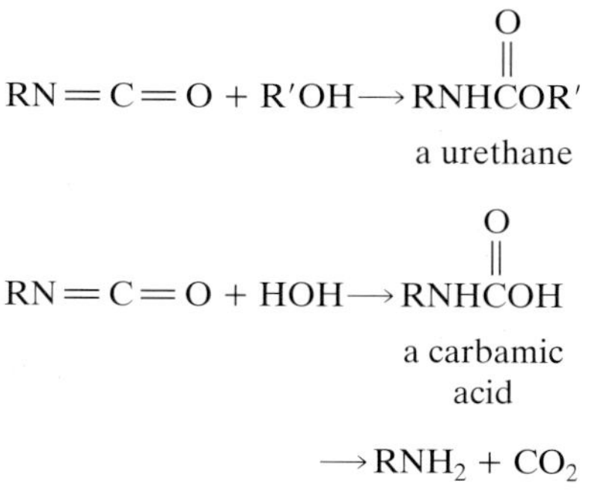

Nitrene mechanism for the Curtius rearrangement

benzoyl azide

a nitrene

phenyl isocyanate

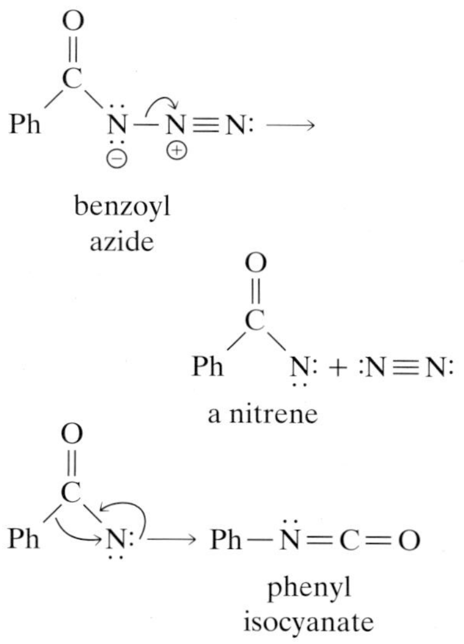

Key Concept: *A hydrogen atom, alkyl group, or aryl group may migrate to a neighboring electron-deficient site by moving its bonded electron pair to the new site via a bridged transition state.*

$$RN=C=O + R'OH \longrightarrow RNHCOR'$$

a urethane

$$RN=C=O + HOH \longrightarrow RNHCOH$$

a carbamic acid

$$\longrightarrow RNH_2 + CO_2$$

Reactions of isocyanates with hydroxylic solvents

Imides and Their Reactions

Imides are nitrogen analogs of carboxylic anhydrides, compounds containing a —CONRCO— functional group, where R is H, alkyl, or aryl. The chemistry of imides resembles that of amides and other acyl compounds. For example, imides can undergo nucleophilic acyl substitution reactions with good nucleophiles.

Nucleophilic acyl substitution with an imide (H$^+$ may be obtained from water or another solvent)

$$\underset{RCNHCR'}{\overset{O\quad O}{\overset{\|\quad\|}{}}} \xrightarrow{Nu:} \xrightarrow{H^+} \underset{RCNu}{\overset{O}{\overset{\|}{}}} + \underset{R'CNH_2}{\overset{O}{\overset{\|}{}}}$$

The amide formed in such a reaction may react further under appropriate conditions. Thus phthalimide is hydrolyzed easily by dilute aqueous sodium hydroxide to give the sodium salt of phthalamic acid, which yields phthalamic acid upon acidification. Heating phthalamic acid (or phthalimide itself) for 30 minutes or more with concentrated NaOH and acidifying the reaction mixture yields phthalic acid.

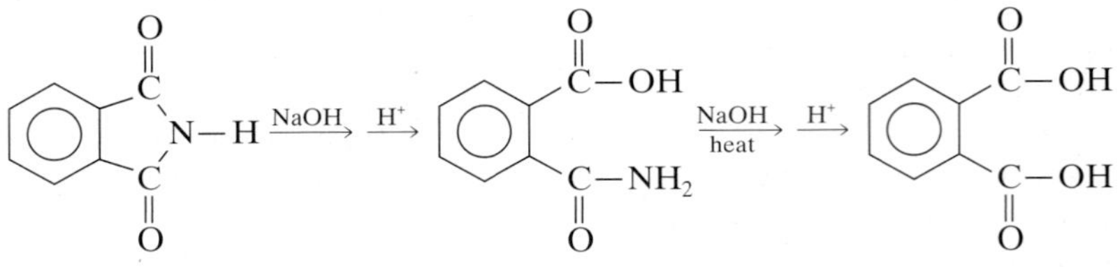

phthalimide   phthalamic acid   phthalic acid

Because of the two carbonyl groups adjacent to nitrogen, the N—H hydrogen of an imide is acidic enough to be removed by moderately strong bases. For example, phthalimide reacts with potassium hydroxide or potassium carbonate to form potassium phthalimide.

Reaction of phthalimide with potassium carbonate

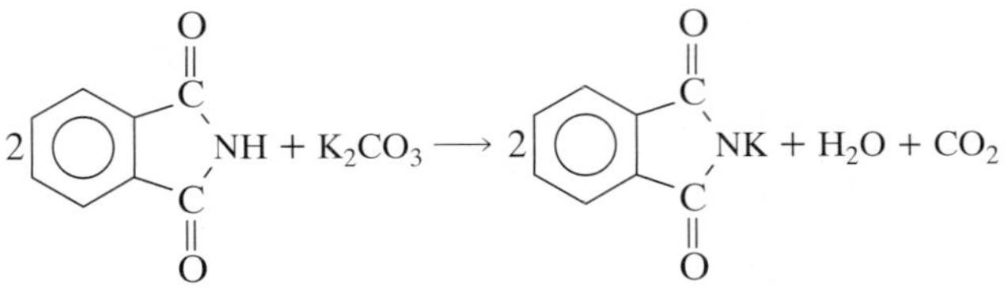

potassium
phthalimide

Imides and most other compounds having N—H bonds can be chlorinated by sodium hypochlorite. For example, NaOCl converts indole to N-chloroindole by the following reaction:

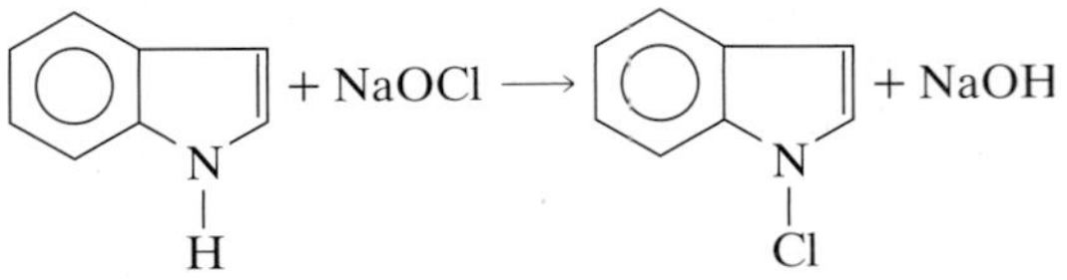

N-chloroindole

Understanding the Experiment

In this experiment, you will prepare phthalimide by heating phthalic anhydride with urea, unless the phthalimide is provided. Then you will treat phthalimide with sodium hypochlorite and sodium hydroxide and attempt to identify the product.

During the preparation of phthalimide, the reaction mixture expands to about three times its initial volume near the end of the reaction period due to the rapid evolution of carbon dioxide. You will carry out the reaction of phthalimide by heating it with a chlorine laundry bleach (such as Clorox or Javex) and aqueous sodium hydroxide. Most chlorine bleaches of this type contain about 5.25% NaOCl by mass, in water. When the reaction is complete, the resulting alkaline solution should contain the sodium salt of the product, which is precipitated with acetic acid after most of the sodium hydroxide has been neutralized with hydrochloric acid. Considerable foaming may occur during the acidification step, but it can be reduced by keeping the reactants cool and adding the acid slowly. The solid product is separated from the reaction mixture by vacuum filtration and purified by recrystallization from water.

The product will be one of the compounds listed in Table 48.1, some of whose structures are shown in Figure 48.1. (Structures of the remaining compounds can easily be deduced from the ones given.) After reading the experiment, you should be able to propose one or more reasonable structures for the product and a mechanism for its formation. The melting point of the product should then lead you to the correct structure. With your instructor's permission, you can also use the infrared (IR) spectrum of the product to help verify its structure.

Table 48.1 Compounds with the molecular formula $C_7H_7NO_2$

Compound	mp
2-hydroxybenzaldoxime*	63
3-hydroxybenzaldoxime	90
benzohydroxamic acid*	131–132
2-aminobenzoic acid*	146–147
3-aminobenzoic acid	174
4-aminobenzoic acid	188–189
2-hydroxybenzamide*	142
3-hydroxybenzamide	170.5
4-hydroxybenzamide	162
methyl 3-pyridinecarboxylate*	42–43
methyl 4-pyridinecarboxylate	8.5
1-methyl-3-pyridinecarboxylic acid*	218
2-hydroxy-5-nitrosotoluene	134–135
5-hydroxy-2-nitrosotoluene*	165
α-nitrotoluene	(bp 225–227)
2-nitrotoluene*	−10
3-nitrotoluene	16
4-nitrotoluene	54.5

*Structure is illustrated in Figure 48.1.

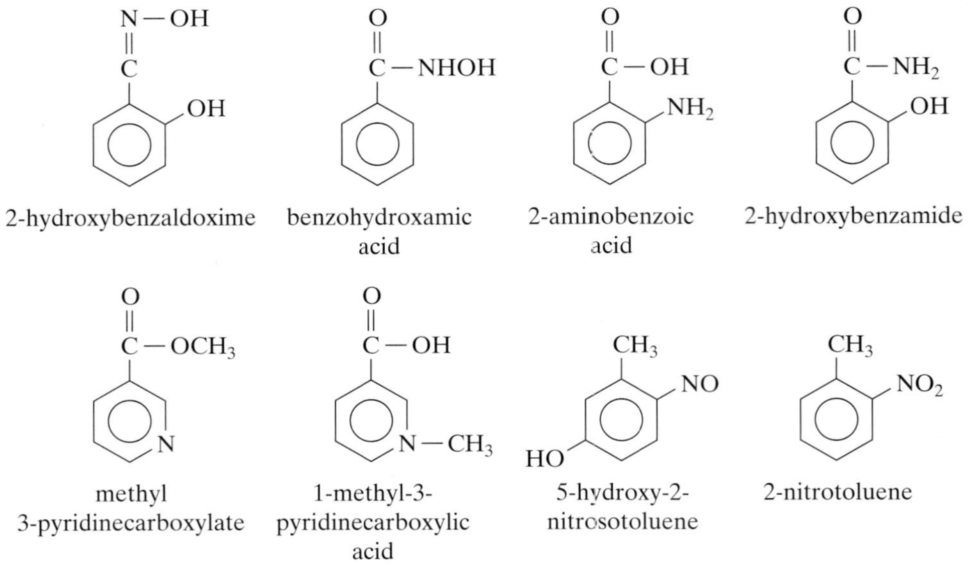
2-hydroxybenzaldoxime
benzohydroxamic acid
2-aminobenzoic acid
2-hydroxybenzamide
methyl 3-pyridinecarboxylate
1-methyl-3-pyridinecarboxylic acid
5-hydroxy-2-nitrosotoluene
2-nitrotoluene

Figure 48.1 Structures of some representative compounds from Table 48.1

Reactions and Properties

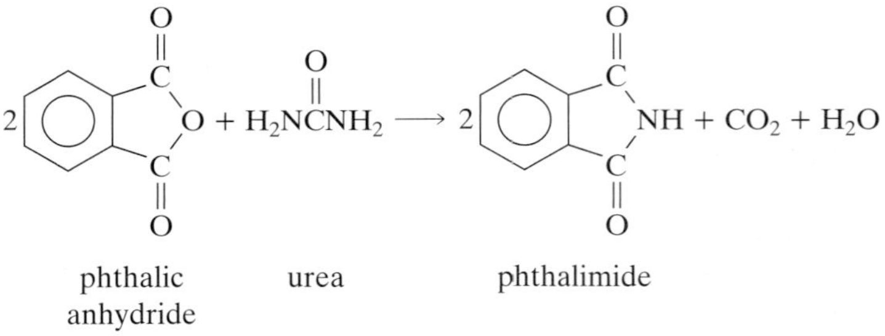
phthalic anhydride
urea
phthalimide

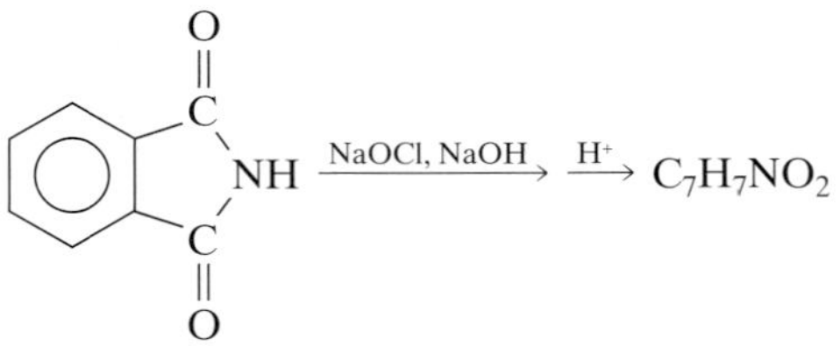
NaOCl, NaOH
H+
C7H7NO2

Table 48.2 Physical properties

	M.W.	mp
phthalic anhydride	148.1	132
urea	60.1	135
phthalimide	147.1	238
sodium hydroxide	40.0	322
sodium hypochlorite	74.4	

Note: mp is in °C, density is in g/mL. 5.25% sodium hypochlorite has a concentration of about 0.74 *M*.

Phthalimide

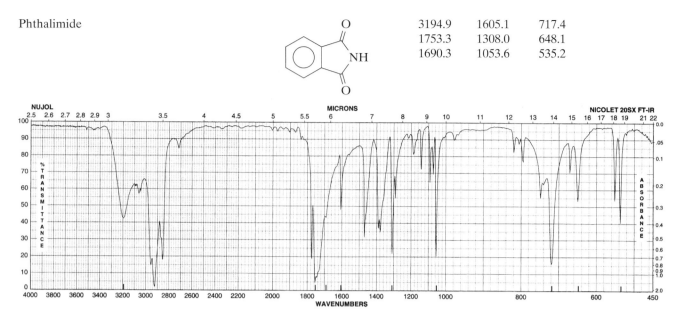

3194.9	1605.1	717.4
1753.3	1308.0	648.1
1690.3	1053.6	535.2

Figure 48.2 IR spectrum of phthalimide

Directions

Phthalic anhydride irritates the skin and eyes, and its dust can irritate the respiratory system. Avoid contact and do not breathe its dust.

Sodium hydroxide is toxic and corrosive, causing severe damage to skin, eyes, and mucous membranes. Wear gloves and avoid contact with the NaOH solution.

Hydrochloric acid is poisonous and corrosive; contact or inhalation can cause severe damage to the eyes, skin, and respiratory tract. Wear gloves and dispense under a hood. Avoid contact and do not breathe its vapors.

Acetic acid causes chemical burns that can seriously damage skin and eyes; its vapors are highly irritating to the eyes and respiratory tract. Wear gloves and dispense under a hood. Avoid contact and do not breathe its vapors.

Safety Notes

phthalic anhydride   sodium hydroxide

hydrochloric acid   acetic acid

Take Care! Avoid contact with phthalic anhydride and do not inhale its dust.

Stop and Think: What causes the frothing?

Preparation of Phthalimide. This part can be omitted if commercial phthalimide is provided. Intimately mix 5.00 mmol of pure phthalic anhydride with 2.50 mmol of urea in a 10-mL round-bottom flask and attach an air condenser. Heat [OP-7] the reaction flask in a heating block or sand bath that has been preheated to 150°C. Within a few minutes, the mixture should suddenly froth up and become nearly solid. At this point, turn off the heat, but leave the reaction flask in the block or sand bath for about 15 minutes. Add 1 mL of cold water and break up the solid with a spatula. Collect the product by vacuum filtration [OP-13], washing it with a small amount of cold water. Dry [OP-23] the phthalimide and measure its mass and melting point [OP-30]. Grind it to a powder with a flat-bladed microspatula.

Reaction. (If you obtain less than 3.0 mmol of phthalimide, scale down the quantities of other reactants and solvents proportionately.) Measure 4.5 mL (~3.3 mmol) of 5.25% sodium hypochlorite into a small flask or beaker, and then add 1.5 mL of 8 *M* (25%) NaOH. Cool [OP-8] this solution for at least 5 minutes in an ice-water or ice-salt bath. Weigh 3.00 mmol of finely powdered phthalimide in a 25-mL Erlenmeyer flask, drop in a stir bar, and add the NaOCl–NaOH solution with stirring [OP-10], using a small amount of water for the transfer. Add another 1.0 mL of the 8 *M* NaOH solution while stirring and monitor the temperature [OP-9] of the reaction mixture. After the temperature drops about 5°C below its highest point, heat the reaction mixture in an 80°C water bath, with stirring, for 10 minutes. Then let it stand at room temperature for about 10 minutes more.

Under the hood, cool the reaction mixture in an ice-water bath, then carefully stir in 1.4 mL of concentrated hydrochloric acid. Test the solution with pH paper and slowly add *just* enough additional concentrated HCl to bring its pH down to 10 (if the pH is already below 10, add enough 8 *M* NaOH dropwise to raise it to 10). Do not add too much HCl or the product will dissolve. Still under the hood, add 0.50 mL of glacial acetic acid *slowly*, with continuous stirring or swirling. (Don't let the reactants foam out the top of the flask.) Let the reaction mixture stand in the ice-water bath until precipitation is complete.

Separation. Collect the precipitate by vacuum filtration [OP-13], wash it on the filter with ice water until the odor of acetic acid is gone, and let it air dry on the filter.

Purification and Analysis. Purify the product by recrystallization [OP-25] from boiling water, using pelletized Norit as necessary to remove colored impurities. Dry [OP-23] the product at room temperature and measure its mass and melting point [OP-30]. Deduce the structure of the product, name it, and write a detailed mechanism that shows how phthalimide is converted to the product.

Exercises

1 What gas was responsible for the foaming when you acidified the reaction mixture? Write a balanced equation for the reaction that liberated the gas.

2 (a) May Bobble misread the directions and acidified the reaction mixture to a pH of 2 with HCl, and was surprised when there was no precipitate to filter. Explain what went wrong and write an equation for the reaction that caused the problem. (b) How could she have recovered the product and salvaged the experiment?

3 Describe and explain the possible effect (if any) on your results of the following experimental errors or variations. (a) The phthalimide wasn't dried completely and was allowed to sit for a week before it was used in the reaction. (b) The laundry bleach came from an old bottle and its NaOCl concentration was about 4%. (c) The lab assistant put out a bottle of Clorox 2 Ultra bleach rather than standard Clorox (check the labels at a local store). (d) You used 3.0 *M* acetic acid rather than glacial acetic acid and there was no foaming after you had added 0.50 mL of it.

4 Following the format in Appendix V, construct a flow diagram for this experiment (excluding the preparation of phthalimide).

5 (a) When 35 mg of the product from this experiment was mixed with 0.46 g of camphor, the melting point of the mixture was found to be 157°C. Calculate the approximate molecular weight of the product if the melting point of pure camphor is 179°C and its freezing-point depression constant (K_f) is $40°C \cdot kg \cdot mol^{-1}$. (b) Show how the molecular formula $C_7H_7NO_2$ can be derived using this result and data from the Scenario.

6 When 9-fluorenone hydrazone is treated with sodium nitrite in aqueous sulfuric acid, it rearranges to form phenanthridone. Propose a detailed mechanism that explains this reaction, showing the transition state for the rearrangement step. (*Hint:* What happens to amino groups in an acidified solution of sodium nitrite?)

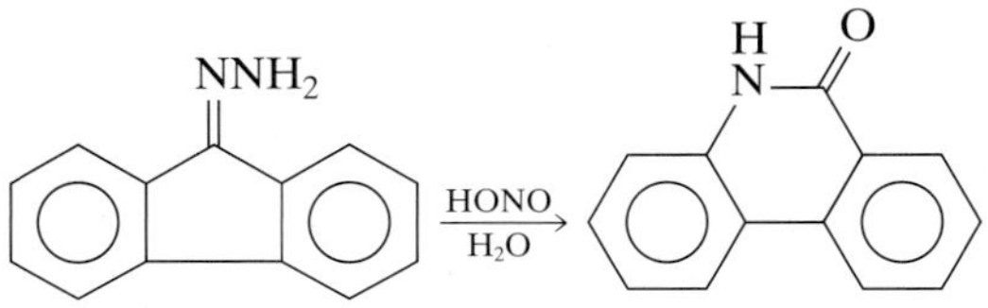

9-fluorenone
hydrazone

phenanthridone

7 (a) When a famous German chemist warmed a solution of *N*-bromoacetamide in base, he recognized an unmistakable pungent odor that slowly faded and was replaced by the strong ammonia-like odor of an escaping gas. What were the two substances that his nose told him were there? Write a mechanism explaining the formation of both. (b) Who was the chemist, and what reaction had he discovered?

Other Things You Can Do

(Starred projects require your instructor's permission.)

*1 Record the infrared spectrum [OP-36] of the product and use it to assist you in identifying the product. Compare its spectrum with that of phthalimide in Figure 48.2, and interpret it as completely as you can.

*2 Carry out a molecular rearrangement of benzophenone oxime as described in Minilab 41.

3 Starting with sources listed in the Bibliography, write a research paper about the use of potassium phthalimide in the synthesis of amines and amino acids. Include a discussion of the phthalimidomalonic ester method and the Gabriel synthesis, and illustrate synthetic routes to specific products using each method.

Identification
of an Unknown Amine

Reactions of Amines. Infrared Spectrometry. Qualitative Analysis.

Operations

OP-23   Drying Solids
OP-25   Recrystallization
OP-27   Simple Distillation
OP-30   Melting Point
OP-31   Boiling Point
OP-36   Infrared Spectrometry

Before You Begin

1 Read the experiment, read or review the operations as necessary, and write an experimental plan.
2 Read or review Part IV, Qualitative Organic Analysis, except for the sections headed "Directions."

Scenario

The city health department of Arkham, Massachusetts, has been deluged with reports of bad-tasting city water, which may have been responsible for several reported cases of illness. Arkham's municipal water supply comes from the Miskatonic River that winds through the town. A routine water analysis indicates the presence of a basic, nitrogen-containing compound, apparently an amine. The three most likely sources of the pollution, all up-stream of the town, are

- A dye factory that uses aniline and other aromatic amines as raw materials
- A chemical specialties company that synthesizes aliphatic amines for the manufacture of surfactants, corrosion inhibitors, and antioxidants
- An abandoned graveyard on Hangman's Hill overlooking the river

With some help from the alchemy faculty at nearby Miskatonic University, health department personnel have obtained a sample of the unknown amine, and the Arkham town council has now asked your Institute for help in identifying the source of this water contaminant. Your assignment is to identify the unknown amine and determine its probable source.

Applying Scientific Methodology

As you carry out the experiment, you should develop provisional hypotheses about the nature and identity of the unknown amine, which you will test—and perhaps reject or revise—as you gather additional experimental evidence. Your conclusion should, if possible, be consistent with all of the experimental evidence you obtain. If any evidence is not consistent with your conclusion, you should be able to explain why. Once you have identified the unknown, you should be able to deduce its probable source.

Biological Amines

Certain families of organic compounds, such as aldehydes and esters, tend to be associated with the pleasant aromas of fruits and perfumes. Amines, on the other hand, are more often associated with the unpleasant smells of body wastes, not-so-fresh fish, and decaying flesh. The amine family includes deadly poisons such as coniine, a component of the poison hemlock that killed Socrates, and dangerous drugs such as LSD, heroin, and methamphetamine. But the same family also includes some highly beneficial members that we couldn't get along without.

Some amines are produced by the enzyme-catalyzed breakdown of proteins and their component amino acids in decaying plant or animal material. For example, bacteria containing the enzymes called amino acid decarboxylases bring about the degradation of the amino acids ornithine and lysine to putrescine (1,4-butanediamine) and cadaverine (1,5-pentanediamine), respectively.

$$^+H_3NCH_2CH_2CH_2\underset{\underset{NH_2}{|}}{CH}\overset{\overset{O}{\|}}{C}O^- \xrightarrow[-CO_2]{enzyme} H_2NCH_2CH_2CH_2CH_2NH_2$$

ornithine   NH$_2$ putrescine

$$^+H_3NCH_2CH_2CH_2CH_2\underset{\underset{NH_2}{|}}{CH}\overset{\overset{O}{\|}}{C}O^- \xrightarrow[-CO_2]{enzyme} H_2NCH_2CH_2CH_2CH_2CH_2NH_2$$

lysine   NH$_2$ cadaverine

As their common names suggest, these amines are responsible for much of the objectionable odor of decaying flesh. The next homolog in this family of diamines, 1,6-hexanediamine, is more often associated with hosiery, camping gear, and outdoor clothing. Along with hexanedioic acid (adipic acid), it is a monomer used in the preparation of nylon 6,6, a commercially important polyamide.

$$n H_2NCH_2CH_2CH_2CH_2CH_2CH_2NH_2 + n HO\overset{\overset{O}{\|}}{C}CH_2CH_2CH_2CH_2\overset{\overset{O}{\|}}{C}OH \xrightarrow[-H_2O]{\Delta}$$

1,6-hexanediamine   hexanedioic acid

$$-[HNCH_2CH_2CH_2CH_2CH_2CH_2NH-\overset{\overset{O}{\|}}{C}CH_2CH_2CH_2CH_2\overset{\overset{O}{\|}}{C}]_n-$$

nylon 6,6

The family of amines called *phenethylamines*, whose parent compound is 2-phenylethylamine ($C_6H_5CH_2CH_2NH_2$), has many biologically active members, including body regulators such as epinephrine (adrenaline), useful drugs such as pseudoephedrine, and dangerous street drugs such as methamphetamine. Epinephrine is synthesized in the body as the end product of an important

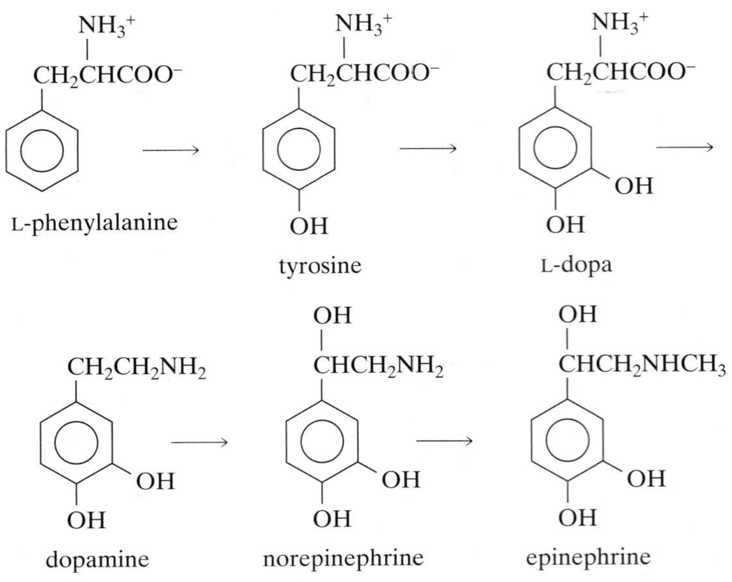

Figure 49.1 Biosynthetic pathway to epinephrine

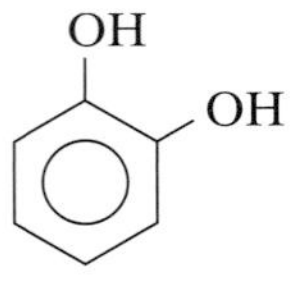

catechol

biosynthetic pathway that starts with the amino acid L-phenylalanine (see Figure 49.1). Along this pathway are two other amino acids, tyrosine and L-dihydroxyphenylalanine (L-dopa), and the three *catecholamines* dopamine, norepinephrine, and epinephrine. Catecholamines are so named because they can be regarded as derivatives of the phenol 1,2-dihydroxybenzene, whose common name is catechol. L-Dopa, which is converted to dopamine in brain tissue, is used to help reduce the characteristic tremors and other abnormal movements characteristic of Parkinson's disease. The disease is apparently associated with a deficiency of dopamine in the brain, causing an imbalance of the chemicals responsible for transmitting nerve impulses.

Norepinephrine is one of the body's major neurotransmitters and is responsible for the transmission of nerve impulses along the sympathetic (adrenergic) nervous system. Epinephrine, a hormone produced in the adrenal gland, increases the heart rate and blood pressure by constricting blood vessels, and also dilates bronchial passageways to permit free breathing. In this way epinephrine prepares the body for "fright, flight, or fight," giving rise to the adrenaline rush sometimes experienced by athletes and people in stressful situations. Both norepinephrine and epinephrine have been used therapeutically, the former to maintain blood pressure for persons in shock and the latter to treat allergies and stimulate the heart during heart attacks.

Another phenethylamine, ephedrine, occurs naturally in several species of leafless green-stemmed shrubs of the genus *Ephedra*. Twigs from the Chinese shrub ma huang (*Ephedra sinica*) have been used for more than 5000 years to treat asthma and a variety of other ailments. Like epinephrine, ephedrine can constrict the walls of blood vessels and dilate bronchial tubes. Such properties have led to the widespread use of ephedrine and its diastereomer pseudoephedrine as decongestants for people suffering from asthma, sinus congestion, allergies, and even the common cold.

(1*R*,2*S*)-ephedrine   (1*S*,2*S*)-pseudoephedrine

Ephedrine has also been used widely in dietary supplements that are claimed to help people lose weight, feel more energetic, and develop their muscles. However, overdoses of ephedrine can cause heart attacks, strokes, seizures, and sometimes death, so the U.S. Food and Drug Administration has limited the amount of ephedrine that can be included in any dietary supplement and banned its use in weight loss and body building products.

The illicit drug methamphetamine has a structure very similar to that of ephedrine but is considerably more dangerous. Before its addictive potential was recognized, methamphetamine—also known as "speed"—was prescribed as an appetite suppressant and to treat depression. It is a nervous system stimulant said to generate a feeling of confidence and mental alertness, but its side effects include hallucinations, paranoia, and death. Epinephrine and methamphetamine both appear to stimulate the release of norepinephrine in the body, thereby increasing the transmission of nerve impulses.

methamphetamine

Understanding the Experiment

This section provides a general discussion of most of the procedures you will follow to identify your unknown. For more detailed information about the interpretation of test results, see the appropriate sections in Part IV. For information about the interpretation of infrared spectra, see Operation 36. Locations of classification test procedures are given in Table 49.1.

Since an unknown liquid amine may be impure, you should purify it by distillation before you perform any chemical tests or record its spectra. Solid amines can be purified by recrystallization, but see your instructor first to find out if an unknown solid requires purification. Measure the boiling point or melting point of your unknown as accurately as you can, since your list of possibilities will be based on the value you obtain. A preliminary examination of your unknown may provide some clues that will help you identify it, as will its solubility behavior in water and 5% HCl.

Since tertiary amines are listed on a different table and have different derivatives than primary or secondary amines, it is important to classify your amine correctly. Many amines can be classified with reasonable accuracy using a simple color test, the quinhydrone test. Before you can interpret this test, you must know whether your amine is aromatic or aliphatic; aliphatic amines are basic enough to dissolve in a buffer solution having a pH of 5.5, while most aromatic amines are not. To confirm your tentative classification, you should carry out Hinsberg's test, which is based on the properties of the products formed when an amine reacts with an arenesulfonyl chloride in aqueous NaOH. The arenesulfonamide formed by a primary amine is soluble

Key Concept: Because tertiary amines have no N — H bonds, they don't undergo many of the reactions of primary and secondary amines, which require loss of H+ from N to form the desired product.

in the reagent and precipitates when HCl is added, while the arenesulfon-amide formed by a secondary amine is insoluble in the reagent and does not dissolve when HCl is added. Most tertiary amines leave an insoluble residue that dissolves when HCl is added.

You can also classify an amine and characterize it further from its infrared (IR) spectrum. A typical primary amine has a medium-intensity two-pronged N—H stretching band near 3350 cm^{-1}, a medium to strong N—H bending band near 1615 cm^{-1}, and a strong, broad N—H bending band in the vicinity of 800 cm^{-1}. A typical secondary amine has a single weak N—H stretching band near 3300 cm^{-1} and a strong, broad N—H bending band around 715 cm^{-1}. A tertiary amine has no N—H bonds, so its IR spectrum shows none of these bands, but it may have a recognizable C—N stretching band in the 1340–1020 cm^{-1} region. The position of the C—N band, which is present for primary and secondary amines as well, depends on the structure of the amine; for aromatic amines it is around 1340–1250 cm^{-1}, and for aliphatic amines around 1250–1020 cm^{-1}. The IR spectrum of an aromatic amine should also exhibit characteristic Ar—H stretching and bending bands, as described in the "Characteristic Infrared Bands" section of OP-36.

Your unknown amine will be one of the amines listed in Tables 5 and 6 in Appendix VI. After performing the classification tests and interpreting your IR spectrum, you should be able to prepare a short list of possibilities from the appropriate tables. You should then be able to identify your unknown by preparing a derivative whose melting point will distinguish it from all other amines with similar boiling or melting points. If possible, avoid preparing derivatives whose melting points for the compounds on your list are too low, too close together, or not listed for some of the possibilities. Keep in mind that incomplete purification and insufficient drying of the derivative can lower its melting point, so follow the directions for preparing the derivative carefully, and be sure that the product is completely dry before you measure its melting point. If you have difficulty preparing a certain derivative, or if the melting point of the derivative you prepare does not eliminate all of the possibilities but one, you should prepare a second derivative.

When you think you have gathered enough evidence to identify your unknown with some certainty, you are free to write down your conclusion. But keep in mind that your evidence should be sufficient to convince your instructor—and yourself—that your conclusion is correct.

Reactions and Properties

General equations for derivative preparations for amines are given under "Derivatives of Primary and Secondary Amines" and "Derivatives of Tertiary Amines" in Part IV. General equations for the classification tests are given under "Classification Tests" in Part IV, on the pages listed in Table 49.1.

Table 49.1 Classification tests for amines

No.	Test	Page
C-4	basicity test	525
C-15	Hinsberg's test	531
C-20	quinhydrone test	535

Directions

Safety Notes

Preliminary Work.   Obtain an unknown amine from your instructor and record its identification number in your laboratory notebook. If the unknown is a liquid, purify it by simple distillation [OP-27] and record its distillation boiling range and median boiling point. Then measure the boiling point [OP-31] of the purified liquid using a capillary-tube method. If it is a solid, purify it by recrystallization [OP-25] if necessary and measure its melting point [OP-30]. Describe the physical state, general appearance, and any other notable characteristics of the compound in your lab notebook. Carry out an ignition test as described under "Preliminary Work" on page 516.

Take Care! Wear gloves, avoid contact with the amine, and do not breathe its vapors.

Solubility Tests.   Test the solubility of the unknown in water as described under "Solubility Tests" on page 517. If it is soluble, test its aqueous solution with red litmus paper, then add 5% HCl dropwise to see whether its odor (if any) disappears. If it is insoluble in water, test its solubility in 5% HCl as described under "Solubility Tests" on page 517.

Stop and Think: Why might its odor disappear?

Classification of the Amine.   Refer to the section "Classification Tests" in Part IV for procedures. Classify the unknown amine as aliphatic or aromatic using the basicity test (classification test C-4). Then classify it as primary, secondary, or tertiary using the quinhydrone test (C-20) or Hinsberg's test (C-15), or both. Obtain an infrared spectrum [OP-36] of your unknown and use it to verify your classifications.

Detection of Structural Features.   In your lab notebook, list all compounds from the appropriate table (Table 5 or 6) in Appendix VI that have melting or boiling points within ±10°C of your observed value, and record their melting or boiling points and the melting points of the derivatives listed. At your instructor's option, show him or her your list; the instructor may approve the list if it includes your unknown or suggest additional work if it doesn't. Write the structure of every compound on your list and consider whether additional classification tests, such as Beilstein's test (C-5), would help you select the most likely possibilities. Inspect your IR spectrum to find out what you can about any structural features or secondary functional groups, such as aromatic rings, alkoxyl groups, or nitro groups. At this point, you should be able to prepare a short list of compounds by eliminating the least likely possibilities. Keep in mind, however, that classification tests and IR spectral bands are not infallible indicators of molecular structure, so it may be necessary later to reconsider some of the compounds that you eliminated from your short list.

Preparation of a Derivative.   Refer to the section "Preparation of Derivatives" in Part IV for procedures. Select the derivative that should best differentiate the compounds on your short list. If your amine is primary or secondary, you can prepare one or more of the following derivatives for which reagents are available: benzamide (derivative D-8), *p*-toluenesulfonamide (D-9), phenylthiourea (D-10), or picrate (D-12). If your amine is tertiary, you can prepare a methiodide (D-11) or picrate (D-12). Purify the derivative by recrystallization [OP-25] as described in the appropriate procedure, dry [OP-23] it thoroughly, and measure its melting point [OP-30]. Deduce the identity of your unknown from the derivative melting point and all other relevant evidence, and justify your conclusion based on the evidence. Then deduce the probable source of the amine based on information in the Scenario.

Exercises

1 Interpret the infrared spectrum you obtained as completely as you can.

2 (a) Write balanced equations for the reactions involved in all of the classification tests for which you obtained a positive result. (b) Write balanced equations for the reaction(s) involved in your derivative preparation(s).

3 Describe and explain the possible effect on your results of the following experimental errors or variations. (a) When you carried out Hinsberg's test, you inadvertently used 3 *M* HCl in place of 3 *M* NaOH. (b) You mistakenly classified a tertiary amine as secondary and tried to prepare a *p*-toluenesulfonate derivative. (c) While performing the basicity test on a water-insoluble unknown, you inadvertently used a sodium acetate solution rather than the acetate–acetic acid buffer.

4 Construct a flow diagram showing the process you followed to identify your unknown.

5 The basicity test differentiates aromatic and aliphatic amines based on their solubility in a pH 5.5 buffer. Given that K_b for aniline is 4.2×10^{-10} and K_b for cyclohexylamine is 5.0×10^{-4}, calculate the ratio of amine salt to dissolved amine for both compounds in such a buffer, and explain the difference in their solubility behavior.

6 An unknown liquid boiling around $185 \pm 5°C$ dissolves in 5% HCl but is insoluble in water and in a pH 5.5 buffer. Shaking the unknown with *p*-toluenesulfonyl chloride in aqueous NaOH produces a clear solution, which when acidified yields a white precipitate. Assuming that the unknown is listed in Appendix VI, give its name and draw its structure.

7 An unknown liquid boiling around $185 \pm 5°C$ dissolves in 5% HCl and in a pH 5.5 buffer but is insoluble in water. Shaking the unknown with *p*-toluenesulfonyl chloride in aqueous NaOH produces a white precipitate that doesn't dissolve in dilute HCl. Assuming that the unknown is listed in Appendix VI, give its name and draw its structure.

8 Write mechanisms for the following reactions, which are used in chemical tests and derivative preparations. (a) The reaction of aniline with benzoyl chloride in aqueous NaOH. (b) The formation of the methiodide of triethylamine. (c) The reaction of diethylamine with benzenesulfonyl chloride in aqueous NaOH.

Other Things You Can Do

(Starred projects require your instructor's permission.)

*1 Obtain and interpret the ^{1}H or ^{13}C NMR spectrum [OP-37] of your unknown.

  2 Starting with sources listed in the Bibliography, write a research paper describing the structures, biological functions, and therapeutic uses (if any) of some phenethylamines.

Preparation and Mass Spectrum of 2-Phenylindole

Reactions of Carbonyl Compounds. Preparation of Heterocyclic Amines. Fischer Indole Synthesis. Phenylhydrazones. Mass Spectrometry.

Operations

OP-39 Mass Spectrometry
OP-7   Heating
OP-9   Temperature Monitoring
OP-10   Mixing
OP-12   Gravity Filtration
OP-13   Vacuum Filtration
OP-23   Drying Solids
OP-25   Recrystallization
OP-26   Sublimation
OP-30   Melting Point

Before You Begin

1 Read the experiment and OP-39, read or review the operations as necessary, and write an experimental plan.
2 Calculate the mass and volume of 2.00 mmol of acetophenone and the theoretical yield of 2-phenylindole.

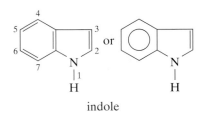

indole

Scenario

The indole family of heterocyclic amines has a shady reputation because many of its members, such as psilocybin and LSD, have mind-altering properties that have made them popular—but illegal—recreational drugs. But other indoles, such as tryptophan (an essential amino acid) and serotonin (a brain-regulating chemical), are necessary for normal body functioning. The Pharmstead Company is interested in developing new synthetic indole derivatives for use as legitimate drugs. To better focus their search for promising pharmaceuticals, they are conducting a preliminary study of the effect of ring substitution on physiological activity. Virtually all physiologically active natural indole derivatives have a substituent at the #3 position and none at the #2 position of the indole ring system, so the physiological effects of 2-substituted indoles have not been thoroughly explored. This may be because they *have* no significant physiological effects, but before coming to that conclusion, the drug development division at Pharmstead wants to obtain and test a representative sample of 2-substituted indoles whose structures are known with certainty. Your assignment is to prepare a sample of 2-phenylindole and confirm its structure by obtaining its mass spectrum.

Applying Scientific Methodology

After reading the experiment you should be able to predict some features of the mass spectrum of 2-phenylindole, including the relative intensities of the M + 1 and M + 2 peaks. Such predictions can be incorporated into a hypothesis, which will be tested when you obtain its mass spectrum.

Of Toads and Toadstools

The indoles constitute a large family of natural and synthetic compounds with extraordinary properties and functions. Indoles as a group display an ambivalent nature, as illustrated by the parent compound, indole, which has an odor of fine jasmine in dilute solutions, but a fecal smell when undiluted. 3-Methylindole is a major product of the digestive putrefaction of proteins and is mainly responsible for the repulsive odor of feces, giving rise to its common name, skatole—from scat, meaning animal droppings. However, skatole is also found in cabbage sprouts, tea, and even lilies!

In Mexico during the sixteenth century, Spanish conquistadors observed the Aztecs using some little brown mushrooms called *teonanacatl* ("flesh of the gods") in their religious ceremonies. According to the Spanish friar Bernardino de Sahagun, "They ate these little mushrooms with honey, and when they began to be excited by them, they began to dance, some singing, others weeping.... Some saw themselves dying in a vision and wept; others saw themselves being eaten by a wild beast; others imagined that they were capturing prisoners in battle, that they were rich, that they possessed many slaves, that they had committed adultery and were to have their heads crushed for the offense." Naturally the Spanish friars disapproved of these ceremonies and banned them. This only led to the formation of cults that consumed the mushrooms in secret, until ethnomycologist R. Gordon Wasson revealed the existence of such practices in the 1930s. The little brown mushrooms are from several different species of the genus *Psilocybe*. Their hallucinogenic constituents include psilocin and psilocybin, both of which are derivatives of tryptamine, the parent compound of many physiologically active indoles. A similar tryptamine derivative, bufotenin, is the active principle of cohoba snuff, which is inhaled by certain Indians of South America and the Caribbean area to produce hallucinations. Bufotenin has also been isolated from toads of the genus *Bufo* and "toadstools" such as the poisonous mushroom *Amanita porphyria*.

bufotenin

Other indole derivatives are more beneficial, such as the essential amino acid tryptophan, the plant-growth hormone 3-indoleacetic acid (also called heteroauxin), the beautiful dye indigo, and the bioregulator serotonin. Tryptophan is important as a source of serotonin and the B-vitamin nicotinamide, both of which are formed during its metabolism. Because the human body cannot biosynthesize aromatic compounds, tryptophan and the

magic mushrooms

skatole

psilocin

psilocybin

tryptamine

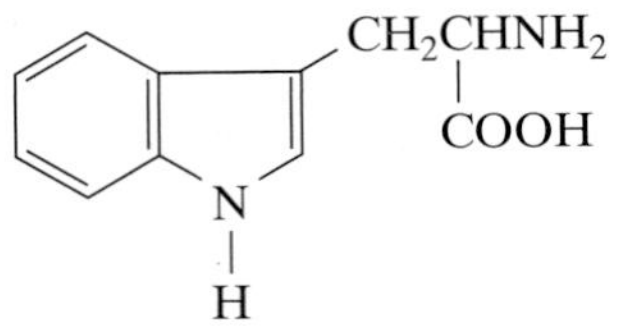

tryptophan

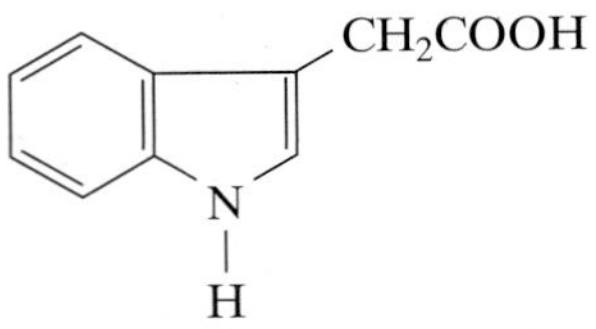

3-indoleacetic acid

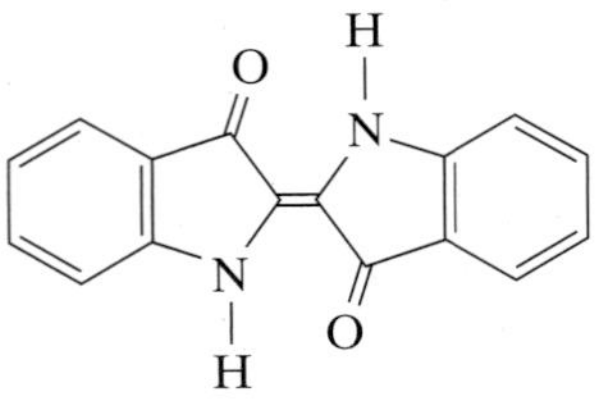

indigo

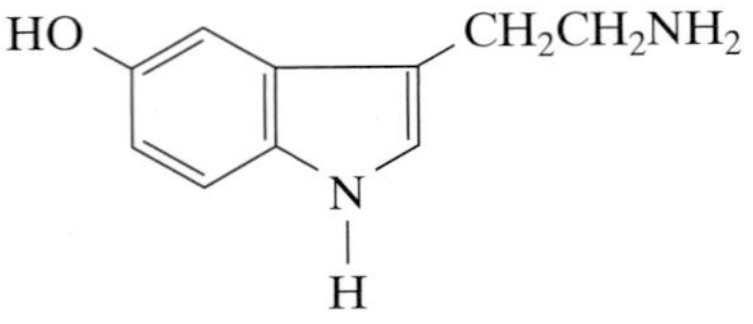

serotonin

other aromatic amino acids, phenylalanine and tyrosine, must be obtained from food. Indoleacetic acid promotes the enlargement of plant cells and is the principal natural growth regulator for many plants. Indigo was one of the first natural dyes to be prepared synthetically, and it is still used to dye blue jeans and other textiles. Indigo's structure was determined in 1883 by German organic chemist Adolph von Baeyer after 18 years of research. Although the function of serotonin in the human body is not fully understood, it appears to play an important role in mental processes. Some researchers see it as a mind stabilizer that helps to preserve sanity. The resemblance between serotonin and such mind-altering drugs as bufotenin is striking; some of the psychological activity of these drugs may arise from their interference with the action of serotonin.

Understanding the Experiment

The preparation of 2-phenylindole by the Fischer indole synthesis involves the acid-catalyzed cyclization of a phenylhydrazone with the loss of a molecule of ammonia.

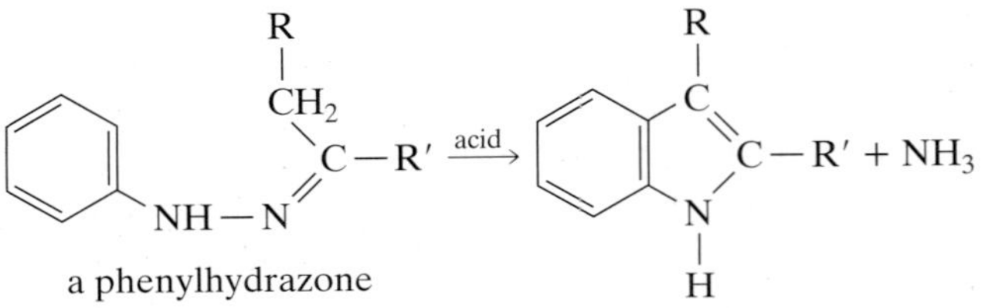

a phenylhydrazone

The phenylhydrazone can be prepared by combining phenylhydrazine or a substituted phenylhydrazine with an aldehyde or ketone having the structure RCH_2COR', where the R groups can be alkyl, aryl, or hydrogen. The generally accepted mechanism for this indolization reaction, proposed by Robinson and Robinson in 1918, includes the following steps:

1 A tautomeric shift of a proton from carbon to the β-nitrogen
2 Protonation of that nitrogen
3 A concerted electron shift that forms a C—C bond to the ring and breaks the N—N bond
4 Another tautomeric shift of a proton from carbon to nitrogen
5 Nucleophilic attack by nitrogen on a doubly bonded carbon atom, followed by the loss of a proton
6 Loss of ammonia to yield an aromatic pyrrole ring

Robinson mechanism

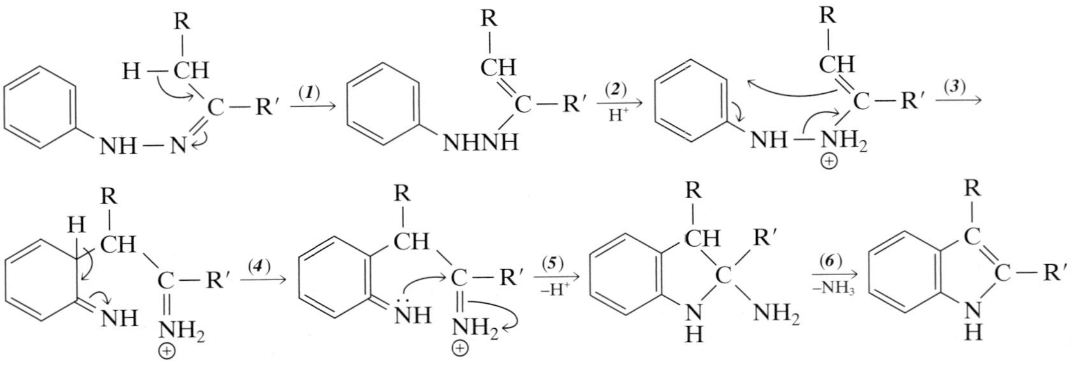

In this experiment, you will synthesize 2-phenylindole by first converting acetophenone to acetophenone phenylhydrazone, then heating acetophenone phenylhydrazone with the acid catalyst polyphosphoric acid. Because acetophenone phenylhydrazone is sensitive to heat and light, you should dry it at room temperature and store it in the dark. Its reaction in polyphosphoric acid is exothermic and requires only a few minutes to go to completion, after which the reaction mixture is poured into water to dissolve the acid catalyst and precipitate crude 2-phenylindole. You can purify 2-phenylindole by recrystallization from 95% ethanol/water mixed solvent, followed by vacuum sublimation if desired. Recrystallization is complicated by the fact that the crystals dissolve quite slowly in 95% ethanol, so you will need to use a condenser to keep the solvent from boiling away.

The mass spectrum of indole (Figure 50.1) is characterized by an intense molecular ion peak ($M \cdot ^{+}$, $m/e = 117$), which is also the base peak. The molecular ion is probably formed by loss of a nonbonded electron from the nitrogen atom. Most substituted indoles, like indole itself and other aromatic amines, have strong molecular ion peaks. Neutral fragments lost from the molecular ions of indoles include HCN and CH_2N, forming ions with m/e values of $M - 27$ and $M - 28$, respectively. The loss of CH_2N from 1-, 2-, and 3-phenylindoles leaves a $C_{13}H_9^{+}$ ion, which could be a fluorenyl cation. If so, the molecular ion must undergo considerable "scrambling" (intramolecular rearrangement) before it ejects CH_2N. Phenylindoles may lose the phenyl-substituted analogs of HCN (C_6H_5CN) and CH_2N (C_6H_5CHN) to form additional daughter ions. Because the molecular ion peak of 2-phenylindole is quite strong, its $M + 1$ and $M + 2$ peaks are also prominent, making it easy to compare their intensities with the expected values.

formation of the molecular ion

Key Concept: *The molecular weight of an amine, and thus the* m/e *value of its molecular ion peak, is usually an odd number.*
See OP-39 if you are not familiar with the principles and terminology of mass spectrometry.

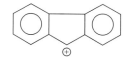

fluorenyl cation

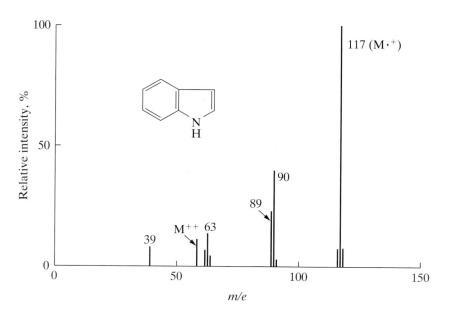

Figure 50.1 Mass spectrum of indole. (Reproduced from *Mass Spectrometry of Heterocyclic Compounds*, by Q. N. Porter and J. Baidas. Copyright © 1971 by John Wiley & Sons, Inc. This material is used by permission of John Wiley & Sons, Inc.)

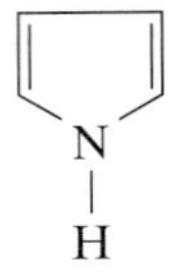

pyrrole

The mass spectra of some indoles also show "half-mass" peaks that result from the formation of ions that have a 2− charge. For example, the doubly charged molecular ion of indole itself gives an M^{++} peak with an *m/e* value of 58.5.

The infrared spectra of heteroaromatic compounds (see "Other Things You Can Do") are similar to those of analogous aromatic compounds, with some additional bands arising from the heteroatoms. Thus pyrroles and indoles, like other aromatic compounds, show ring-stretching bands in the $1650-1300$ cm^{-1} region and C—H out-of-plane bending bands in the $910-665$ cm^{-1} region. Most pyrrole rings are distinguished by a characteristically strong, broad C—H bending band near 740 cm^{-1}. Heteroaromatic amines with N—H bonds absorb in the $3500-3220$ cm^{-1} region. The N—H stretching band of indoles occurs near 3450 cm^{-1}.

Reactions and Properties

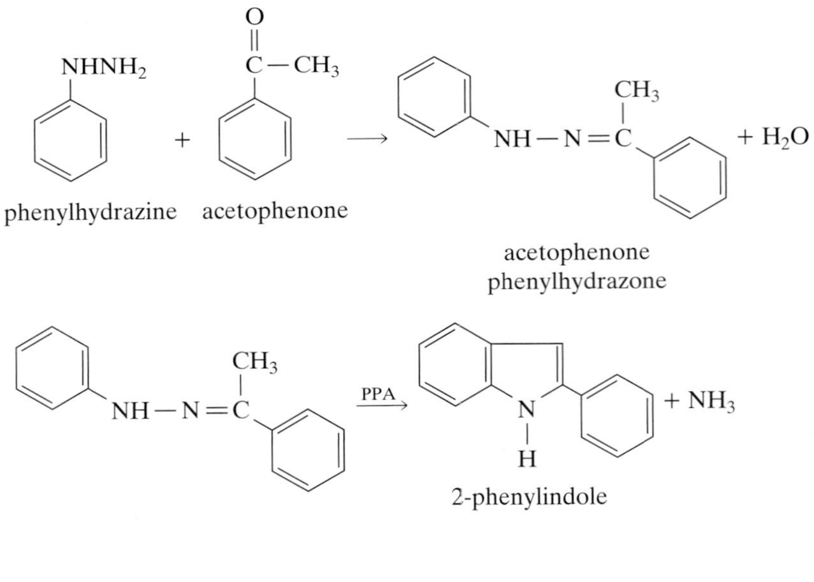

phenylhydrazine   acetophenone

acetophenone phenylhydrazone

2-phenylindole

Table 50.1 Physical properties

	M.W.	mp	bp	*d*
phenylhydrazine	108.15	20	243	1.099
acetophenone	120.2	20.5	202	1.028
acetophenone phenylhydrazone	210.35	106		
2-phenylindole	193.25	189	250^{10}	

Note: mp and bp are in °C, density is in g/mL.

Acetophenone

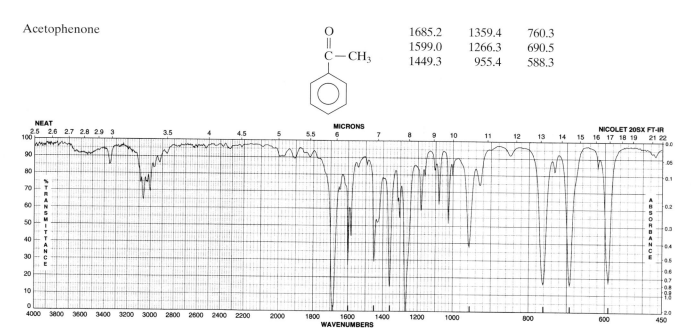

Phenylhydrazine

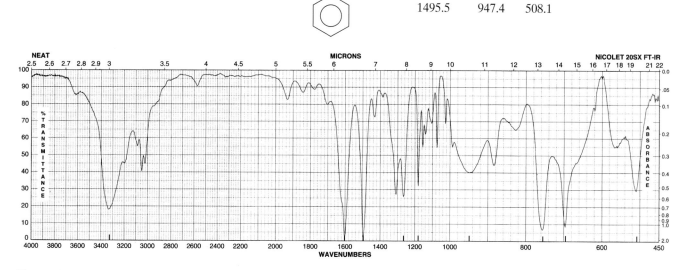

Figure 50.2 IR spectra of the starting materials

Directions

Phenylhydrazine is corrosive and very toxic if inhaled, ingested, or absorbed through the skin; contact may cause painful skin eruptions in sensitive individuals. It is also a suspected carcinogen. Use gloves, dispense under a hood, avoid contact, and do not breathe its vapors.
Acetic acid causes chemical burns that can seriously damage skin and eyes; its vapors are highly irritating to the eyes and respiratory tract. Dispense under a hood, avoid contact, and do not breathe its vapors.
Polyphosphoric acid can cause serious burns, particularly to the eyes, so avoid contact with skin and eyes.
Some indoles are suspected of causing cancer; minimize your contact with the product.

Take Care! Wear gloves, avoid contact with phenylhydrazine and acetic acid, and do not breathe their vapors.

Stop and Think: What is the purpose of the 1 *M* HCl?

Waste Disposal: Pour the filtrate down the drain.

Take Care! Avoid contact with the acid.

Observe and Note: What evidence is there that a chemical reaction is occurring?

Waste Disposal: Pour the filtrate down the drain.

Preparation of Acetophenone Phenylhydrazone. In a 3-mL conical vial, dissolve 2.00 mmol of acetophenone in 1.0 mL of 95% ethanol. Add 0.20 mL (~2 mmol) of freshly distilled phenylhydrazine followed by 1 small drop of glacial acetic acid (measured with a Pasteur pipet) and a stirring device. Then attach a condenser and start the stirrer [OP-10]. Heat the reactants under reflux [OP-7] with stirring for 15 minutes, using a boiling-water bath. Cool the mixture in an ice/water bath and scratch the sides of the vial if necessary to promote crystallization. When crystallization is complete, collect the acetophenone phenylhydrazone by vacuum filtration [OP-13]. Wash the product on the filter with 1 mL of 1 *M* hydrochloric acid followed by 1 mL of ice-cold 95% ethanol and let it air dry. If there isn't sufficient time to dry the product thoroughly, blot it as dry as possible between large filter papers (wear protective gloves!). Acetophenone phenylhydrazone should be used soon after it is prepared, or else stored in a cool, dark place.

Reaction. Measure 2.0 mL of polyphosphoric acid into a dry 5-mL conical vial. (Polyphosphoric acid can be warmed in a boiling-water bath to make it easier to transfer.) Place the vial in a sand bath or heating block. Clamp a thermometer [OP-9] in the liquid, near one wall of the vial but not touching it, to allow room for a microspatula or a thin glass stirring rod (don't use the thermometer as a stirring rod!). Heat the polyphosphoric acid to 50°C, then mix in the acetophenone phenylhydrazone with the stirring rod or microspatula. Heat [OP-7] the reaction vial, stirring until the solid dissolves, and monitor the temperature of the reaction mixture; when it reaches 100°C, raise the reaction vial out of the heat source temporarily. When the temperature drops to 90°C, lower the vial just far enough to keep the temperature in the 90–100°C range and heat it at that temperature for 10–15 minutes with occasional stirring.

Separation. Cautiously pour the warm reaction mixture into 5 mL of ice water in a small beaker, using more water for the transfer, and stir until all of the polyphosphoric acid has dissolved. Collect the crude 2-phenylindole by vacuum filtration [OP-13], wash it on the filter with cold methanol, and let it air-dry on the filter.

Purification and Analysis.   Recrystallize [OP-25] the product from 95% ethanol/water mixed solvent, using a 10-mL round-bottom flask fitted with an air condenser. Start with 3 mL of 95% ethanol and boil the mixture for about 10 minutes with magnetic stirring [OP-10]; if the solid doesn't dissolve completely, add more 95% ethanol in small portions, boiling for several minutes after each addition. When the product is completely dissolved, use Norit to partly decolorize the solution (unless your instructor suggests otherwise) and filter the hot solution by gravity [OP-12]. (Filtration should not be necessary if you don't use Norit.) With the solution boiling, add hot water to the cloud point and continue as described in OP-25b. Dry [OP-23] the 2-phenylindole at room temperature. At your instructor's request, purify the product (or part of it) further by vacuum sublimation [OP-26]. Measure the mass and melting point [OP-30] of the purified 2-phenylindole. Record a mass spectrum [OP-39] of the product, or obtain one from your instructor. Tabulate the *m/e* values and intensities of the molecular ion peak, the M + 1 and M + 2 peaks, and all other peaks whose intensities are 10% or more of the base peak intensity. Derive the molecular formula of 2-phenylindole from its structure, use the formula to calculate the expected intensities of the M + 1 and M + 2 peaks, and compare the experimental and theoretical values. Characterize as many of the other peaks as you can, in each instance giving the formula of the species lost from the molecular ion and the formula of the resulting daughter ion.

Exercises

1 (a) Rewrite the general mechanism given in the "Understanding the Experiment" section so that it applies to the synthesis of 2-phenylindole, and sketch the activated complexes for steps **3** and **5** of the Robison mechanism. (b) Write a mechanism for the reaction that forms acetophenone phenylhydrazone, showing the role of acetic acid.

2 Outline syntheses of 1-phenylindole and 3-phenylindole from appropriate starting materials.

3 Describe and explain the possible effect on your results of the following experimental errors or variations. (a) You rinsed the test tube used for the preparation of acetophenone phenylhydrazone with acetone and failed to remove all the acetone. (b) You used benzophenone rather than acetophenone. (c) You used phenylhydrazine hydrochloride ($C_6H_4NH_2NH_3{}^+Cl^-$) rather than phenylhydrazine.

4 Following the format in Appendix V, construct a flow diagram for this experiment.

5 Draw structures for the indoles that would be obtained from the phenylhydrazones of the following carbonyl compounds: (a) acetone; (b) phenylacetaldehyde (2-phenylethanal); (c) cyclopentanone; (d) pyruvic acid ($CH_3COCOOH$); (e) camphor.

6 When the phenylhydrazone of isobutyrophenone (2-methyl-1-phenyl-1-propanone) is heated to 150°C with polyphosphoric acid, it yields a mixture of the two compounds shown. Propose a mechanism that explains the formation of each product.

2-(3,4-dimethoxyphenyl) indole

7 The total synthesis of strychnine, accomplished by Robert B. Woodward in 1954, began with the preparation of 2-(3,4-dimethoxyphenyl)indole. Show how this compound can be prepared starting with catechol (1,2-dihydroxybenzene).

Other Things You Can Do

(Starred projects require your instructor's permission.)

*1 Record the IR spectrum of 2-phenylindole. Interpret this spectrum as completely as you can, compare it with the spectra of the starting materials shown in Figure 50.2, and describe the evidence suggesting that the expected reaction has taken place.
*2 Prepare some indigo and use it to dye cloth as described in Minilab 42.
 3 Starting with sources listed in the Bibliography, write a research paper about natural and synthetic plant-growth hormones (auxins), describing their sources and chemical structures and giving examples of their applications.

Nucleophilic Strength and Reactivity in S_NAr Reactions

Reactions of Aryl Halides. Nucleophilic Aromatic Substitution. Reaction Kinetics.

Operations

OP-5   Measuring Volume
OP-10   Mixing
OP-38   Ultraviolet-Visible Spectrometry

Before You Begin

Read the experiment, read OP-38b on colorimetry, read or review the other operations as necessary, and write an experimental plan.

Scenario

Organic chemists rely on handbooks and other sources of compiled chemical information to help them design their experiments. For example, a nucleophile's *nucleophilic constant*, which measures the nucleophile's ability to displace a leaving group from a specific kind of substrate, could help a chemist estimate how long it would take to carry out a given nucleophilic substitution reaction. Minnie Colfax, a chemical information specialist for the Fulcourt Press (see Experiment 46), is compiling a list of nucleophiles for the next edition of a chemistry reference book. She wants to list them in order of nucleophilic strength, as indicated by the values of their nucleophilic constants, but some of the nucleophilic constants she needs were measured many years ago by questionable methods and may not be accurate. Your project group's mission is to measure the nucleophilic constants for morpholine and piperidine in an S_NAr reaction, and from your results to determine which is the stronger nucleophile.

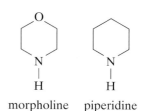

morpholine   piperidine

Applying Scientific Methodology

After reading the experiment, you should formulate a hypothesis predicting which of the two heterocyclic amines, morpholine or piperidine, should be the stronger nucleophile in S_NAr reactions. Your hypothesis will be tested when you determine and compare the nucleophilic constants of the two amines.

Nucleophilicity and the S_NAr Reaction

Nucleophilic aromatic substitution can occur by

- An S_N1 mechanism, as in some substitution reactions of aryldiazonium salts
- An elimination-addition mechanism involving an aryne intermediate
- A bimolecular addition-elimination route called the S_NAr reaction

In an S_NAr reaction, the nucleophile attacks an activated aromatic ring to form an intermediate complex, which then loses a leaving group to yield the product. The nature of that intermediate complex has long been the subject of speculation. The first evidence of its structure was obtained in 1898 when Jackson and Boos mixed picryl chloride (2,4,6-trinitrochlorobenzene) with sodium methoxide in methanol and isolated a red salt, which was converted by ethanol to another red salt. J. Meisenheimer prepared the second salt by two different methods, treating either 2,4,6-trinitroanisole with potassium ethoxide or 2,4,6-trinitrophenetole with potassium methoxide. He was then able to assign it the structure labeled *1* in the following reaction scheme:

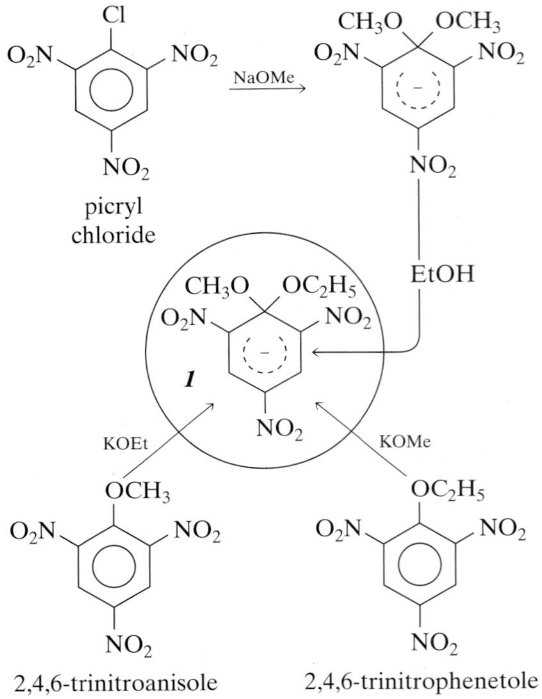

Negatively charged aromatic species analogous to structure *1* are called *Meisenheimer complexes*, and many such complexes have been prepared and characterized. A large body of experimental evidence indicates that the intermediates in S_NAr reactions are Meisenheimer complexes, explaining why only aromatic compounds that are activated by electron-withdrawing substituents, such as nitro groups, undergo S_NAr reactions easily. Such substituents remove excess electron density from the aromatic ring, stabilizing the intermediate complex and thus facilitating its formation.

The mechanism of most S_NAr reactions appears to be a simple two-step process involving the initial addition of the nucleophile to form a Meisenheimer complex, followed by the departure of the leaving group. The first step of the reaction is ordinarily rate limiting, so the nucleophilic strength of the reactant—its ability to donate its electron pair to the substrate and form a sigma bond—affects the reaction rate. Equation **1**, which is called the *Swain-Scott equation*, relates reaction rates to the strength of the nucleophile (n) and the sensitivity of the substrate to nucleophilic substitution (s).

General S_NAr mechanism

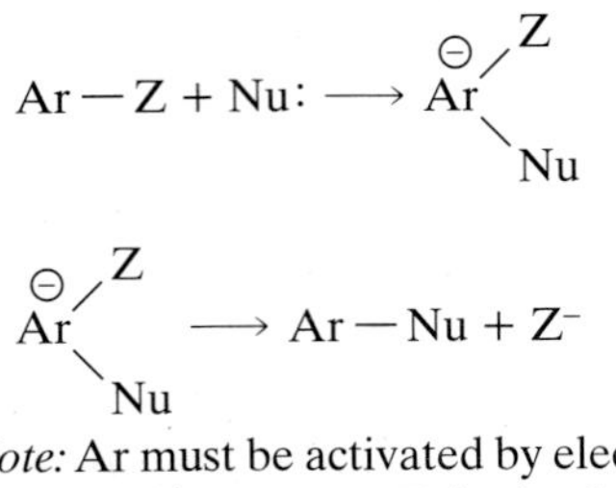

$$Ar-Z + Nu: \longrightarrow Ar\overset{\ominus}{\underset{Nu}{\overset{Z}{<}}}$$

$$\overset{\ominus}{Ar}\overset{Z}{\underset{Nu}{<}} \longrightarrow Ar-Nu + Z^-$$

Note: Ar must be activated by electron-withdrawing groups; Z is the leaving group and Nu: the nucleophile.

$$\log \frac{k}{k_0} = ns \qquad (1)$$

It has been applied widely to nucleophilic substitution reactions with aliphatic substrates, and some values of the nucleophilic constant n for reactions of various nucleophiles with methyl iodide are given in Table 51.1.

Attempts to use such correlations for $S_N Ar$ reactions have met with less success because changing the substrate or the solvent often changes the order of nucleophilic strength. However, an equation similar to Equation **1** can be useful in comparing the nucleophilic strengths of various reactants with reference to the same class of substrates. For this experiment, we define a nucleophilic constant (n_{Ar}) for the $S_N Ar$ reaction as follows:

$$n_{Ar} = \frac{1}{s} \log \frac{k}{k_0} \qquad (2)$$

The reaction of 2,4-dinitrochlorobenzene with ammonia in absolute ethanol, for which the rate constant (k_0) is 4.0×10^{-6} L mol^{-1} s^{-1} at 25°C, is used as the reference reaction for Equation **2**. For this substrate and nucleophile, s is 1 and n_{Ar} is zero by definition. The rate constant k is for a reaction involving the nucleophile whose n_{Ar} value is being determined.

Table 51.1 Nucleophilic constants for various nucleophiles

Nucleophile	n
CH_3OH	0.00
F^-	2.7
Cl^-	4.37
pyridine	5.23
NH_3	5.50
aniline	5.70
Br^-	5.79
CH_3O^-	6.29
$(CH_3CH_2)_3N$	6.66
$(CH_3CH_2)_2NH$	7.0
pyrrolidine	7.23
piperidine	7.30
I^-	7.42

Note: n values are measured relative to methanol, with methyl iodide as the substrate.

Understanding the Experiment

In this experiment, you and your coworkers will measure the rates of the reactions of piperidine and morpholine with 2,4-dinitrochlorobenzene and use the rate constants to calculate their nucleophilic constants, as defined in Equation **2**. The reaction of 2,4-dinitrochlorobenzene with an amine is second order in both substrate and nucleophile, and follows the general rate equation

$$\frac{dx}{dt} = k(S_0 - x)(N_0 - 2x) \qquad (3)$$

where x is the concentration of the product at time t, and S_0 and N_0 are the initial concentrations of substrate and nucleophile, respectively. The computations can be simplified considerably if the experiment is carried out with the initial concentration of nucleophile being just twice that of 2,4-dinitrochlorobenzene. Equation **3** then becomes $dx/dt = 2k(S_0 - x)^2$, and the integrated rate equation is

$$\frac{1}{(S_0 - x)} = 2kt + \frac{1}{S_0} \qquad (4)$$

By measuring the concentration of the product, x, at regular intervals during the reaction, you can calculate the term on the left side of Equation **4**. Graphing this expression versus time should then yield a straight line with slope $2k$.

The products of the $S_N Ar$ reactions are yellow-orange and absorb strongly in the visible region around 380 nm, so the concentration of each product will be determined indirectly by measuring the absorbance of 380-nm light by aliquots that are removed from the reaction mixture at various times. Since the absorbances are determined at a single wavelength, you can use a nonrecording spectrophotometer (colorimeter). Each aliquot must first be *quenched* by adding dilute acid to stop the reaction; this is done so that the product concentration will remain constant until you are ready to take the absorbance readings. The concentration term in Equation **4**, $S_0 - x$, is proportional to $A_\infty - A$, where A_∞ is the absorbance when the reaction is 100% complete, and A is the absorbance at time t. Therefore,

$$\frac{1}{(S_0 - x)} = \frac{q}{(A_\infty - A)} \tag{5}$$

where q is a proportionality constant. The "infinity" value of the absorbance A_∞ will be obtained by warming the reaction mixture to complete the reaction and measuring the absorbance of the resulting solution. The value of q can then be calculated from the relationship $q = A_\infty/S_0$. Note that S_0 is the initial substrate concentration in the *reaction mixture*, not in the stock solution you will use to prepare the reaction mixtures.

Because of their different reaction rates, the amines will be used in different initial concentrations so that both reactions will be about half complete after 30 minutes. The absorbance values of the quenched products are too high to measure, so you will have to dilute these solutions to obtain readings in a convenient range.

Reaction and Properties

2,4-dinitrochlorobenzene   piperidine (Y = CH$_2$) or morpholine (Y = O)   dinitrophenylated amine

The extra mole of amine combines with the HCl liberated during the reaction.

Table 51.2 Physical properties

	M.W.	mp	bp	d
2,4-dinitrochlorobenzene	202.6	53	315	
morpholine	87.1	−5	128	1.000
piperidine	85.2	−9	106	0.861

Note: mp and bp are in °C; density is in g/mL.

Directions

All glassware must be clean and dry. If equipment is limited, you may be asked to work in pairs or larger groups. Each student or group should have a timer or a watch that measures seconds. Stock solutions should be dispensed from burets or bottle-top dispensers.

2,4-Dinitrochlorobenzene is poisonous and skin contact may cause unpleasant and persistent dermatitis. Wear gloves and avoid contact with the 2,4-dinitrochlorobenzene solution.
Morpholine and piperidine are toxic and corrosive, capable of causing severe damage to the skin, eyes, and respiratory system. Avoid contact with their solutions.
Ethanol is very flammable, so keep ethanolic solutions away from flames and hot surfaces.

Safety Notes

2,4-dinitro-
chlorobenzene

morpholine   piperidine

A. *Reaction of 2,4-Dinitrochlorobenzene with Piperidine*
Label seven clean, dry 4-dram screw-cap vials (or 15-cm test tubes) from A1 to A7. Using a 10-mL volumetric pipet, accurately measure [OP-5] 10.0 mL of the quenching solution (0.5 M sulfuric acid in 50% ethanol) into each vial and cap the vials. Measure about 20 mL of absolute ethanol into a 25-mL volumetric flask; then accurately measure 2.00 mL of the 0.20 M stock solution of 2,4-dinitrochlorobenzene in ethanol and add it to the flask. Have your stopwatch or timer ready. Measure 2.00 mL of the 0.40 M piperidine/ethanol stock solution into the volumetric flask and start the timer (or record the starting time) when about half of the solution has drained from the pipet. Without delay, fill the flask to the mark with absolute ethanol, stopper and shake it, and pour the contents into a 50-mL Erlenmeyer flask (the reaction flask). Use a clean, dry 1-mL volumetric pipet to withdraw a 1.00-mL aliquot of the reaction mixture, and transfer it to vial A1 as you record the quenching time to the nearest second. Cap the vial and seal the reaction flask with Parafilm.

Stir or occasionally swirl [OP-10] the contents of the reaction flask during the reaction period. Approximately every 5 minutes, rinse the 1-mL volumetric pipet with the reaction mixture, withdraw a 1.00-mL aliquot, and transfer it to the next quenching vial, recording the quenching time accurately. Repeat this process until you have withdrawn and quenched a total of six aliquots. Then warm the reaction flask (still sealed with Parafilm) in a 50°C water bath for at least 2 hours, or let it stand for at least 48 hours at room temperature, to bring the reaction to completion. Pipet a final 1.00-mL aliquot into vial A7 for the A_∞ solution.

B. *Reaction of 2,4-Dinitrochlorobenzene with Morpholine*
Label seven clean, dry 4-dram screw-cap vials (or 15-cm test tubes) from B1 to B7. Using a 10-mL volumetric pipet, accurately measure [OP-5] 10.0 mL of the quenching solution (0.5 M sulfuric acid in 50% ethanol) into each vial and cap the vials. Measure about 17 mL of absolute ethanol into a 25-mL volumetric flask, then accurately measure 5.00 mL of the 0.20 M stock solution of 2,4-dinitrochlorobenzene in ethanol and add it to the flask. Have your stopwatch or timer ready. Measure 2.00 mL of the 1.0 M morpholine/ethanol stock solution into the volumetric flask, and start the timer (or record the starting time) when about half the solution has drained from the pipet. Without delay, fill the flask to the mark with absolute ethanol, stopper and shake it, and pour the contents into a 50-mL Erlenmeyer flask (the reaction flask). Use a clean, dry 1-mL volumetric pipet to withdraw a 1.00-mL aliquot of the reaction mixture and transfer it to vial B1 as you record the quenching time to the nearest second. Cap the vial and seal the reaction flask with Parafilm.

Take Care!   Do not pipet by mouth. Wear gloves and avoid contact with the stock solutions.

Stop and Think: Why does this stop the reaction?

Observe and Note: What evidence is there that a reaction is occurring?

Waste Disposal: Dispose of the remaining reaction mixture as directed by your instructor.

Take Care!   Do not pipet by mouth. Wear gloves and avoid contact with the stock solutions.

Observe and Note: What evidence is there that a reaction is occurring?

Waste Disposal: Dispose of the remaining reaction mixture as directed by your instructor.

Stop and Think: Why is this step necessary?

Waste Disposal: Dispose of all solutions as directed by your instructor.

Stir or occasionally swirl [OP-10] the contents of the reaction flask during the reaction period. Approximately every 5 minutes, rinse the 1-mL volumetric pipet with the reaction mixture, withdraw a 1.00-mL aliquot, and transfer it to the next quenching vial, recording the quenching time. Repeat this process until you have withdrawn and quenched a total of six aliquots. Then warm the reaction flask (still sealed with Parafilm) in a 50°C water bath for at least two hours, or let it stand for at least 48 hours at room temperature, to bring the reaction to completion. Pipet a final 1.00-mL aliquot into vial B7 for the A_∞ solution.

C. *Absorbance Measurements*
Obtain as many clean, dry 4-dram screw-cap vials (or 15-cm test tubes) as you have solutions to analyze and number them to correspond to the quenched solutions. Using a volumetric pipet, accurately measure [OP-5] 10.0 mL of 95% ethanol into each one. Then accurately measure 0.40 mL of each quenched solution into the corresponding vial, swirl to mix the contents, and cap the vial.

Zero the spectrophotometer [OP-38] at 380 nm and set the 100%-transmittance control using a cuvette filled with 95% ethanol. Make sure the sample cuvette is positioned correctly, then record the transmittance of each solution from each vial. Use the same instrument to analyze all of your solutions, including the A_∞ solution. Calculate the absorbance of each solution and use your data to compute $1/(S_0 - x)$ for each aliquot. For each amine, plot these values versus time and determine the slope of the line, or use a linear-regression program to calculate the slope. Calculate the second-order rate constant for each reaction and the n_{Ar} value for each amine. Arrange morpholine, piperidine, and ammonia ($n_{Ar} = 0$) in order according to their n_{Ar} values and try to explain their relative nucleophilicities.

Exercises

1 (a) Write a detailed mechanism for the reaction of piperidine with 2,4-dinitrochlorobenzene. (b) The rate constants for the reactions of piperidine with 2,4-dinitrochlorobenzene and with 2,4-dinitrobromobenzene are virtually identical. Identify the rate-determining step in your mechanism from (a) and explain your reasoning.

2 Explain why the reaction of 2,4-dinitrochlorobenzene with piperidine or morpholine is second order even though there are three reactant molecules in the overall equation for the reaction.

3 Describe and explain the possible effect on your results of the following experimental errors or variations. (a) The reaction flask you used in the kinetic runs was rinsed with water and not dried. (b) The stockroom was out of 2,4-dinitrochlorobenzene, so the lab assistant substituted chlorobenzene. (c) The lab assistant who prepared the stock solutions couldn't find any piperidine, so she used piperidine hydrochloride instead.

4 (a) Derive Equation **5** using Beer's law and evaluate the constant q in terms of Beer's law parameters. (b) Derive Equation **4** from Equation **3**.

5 A *spiro* species contains two rings that have one carbon atom in common. Outline a synthesis of the spiro Meisenheimer complex shown, starting with 1-chloro-2,4,6-trinitrobenzene.

6 (a) Calculate the concentrations of the dinitrophenylamines in the solutions used for the spectrophotometry infinity reading. (b) If you know the path length of the cuvette you used, calculate the molar absorptivities of these products at 380 nm.

7 Explain why quenching the reaction mixture with H_2SO_4 stops the S_NAr reaction, giving equations for any reactions involved.

a spiro Meisenheimer complex

Other Things You Can Do

(Starred projects require your instructor's permission.)

*1 Measure the reaction rates for one of the amines at different temperatures (0°C, 20°C, and 40°C, for example). Then plot ln k versus $1/T$ to determine the activation energy of the reaction, based on the following form of the Arrhenius equation:

$$\ln k = -\frac{E_{act}}{RT} + \ln A \qquad (R = 8.31 \text{ J mol}^{-1} \text{ K}^{-1})$$

*2 Carry out one of the reactions in other solvents, such as 2-propanol and aqueous ethanol, to measure the effect of solvent polarity on the rate constants. If you make up your own solutions, it is essential to wear protective gloves and to avoid contact with 2,4-dinitrochlorobenzene and the amines.

3 Read the paper by Bunnett, Garbisch, and Pruitt, *J. Am. Chem. Soc.* **1957**, *79*, 385. Then tell what is meant by the "element effect" in reactions of 1-substituted 2,4-dinitrobenzenes, and explain how it was used to elucidate the mechanism of the S_NAr reaction.

Structure of an Unknown D-Hexose

Reactions of Monosaccharides. Preparation of Alditols. Reduction Reactions. Structure Determination.

Operations

OP-7   Heating
OP-8   Cooling
OP-10   Mixing
OP-13   Vacuum Filtration
OP-23   Drying Solids
OP-25   Recrystallization
OP-30   Melting Point
OP-33   Optical Rotation

Before You Begin

1 Read the experiment, read or review the operations as necessary, and write an experimental plan.
2 Calculate the mass of 5.00 mmol of $C_6H_{12}O_6$ and the mass of 2.0 mmol of sodium borohydride ($NaBH_4$). Calculate the theoretical yield of the alditol from part **A**.

Scenario

S. A. Tucker, a food chemist working for the Global Food Research Foundation, collects and analyzes foods from around the world. From Hunza, an isolated state in a high valley of the Hindu Kush mountains, Tucker obtained a yogurt made from yak's milk that is said to contribute to the exceptional longevity of the valley's inhabitants. Now he wants to isolate and identify the major constituents of the yogurt to see whether any of them exhibit life-extending properties. He has already isolated a simple sugar, a D-hexose that has the same molecular formula as glucose, $C_6H_{12}O_6$, but is only half as sweet. Your assignment is to determine the structure of this unknown sugar.

Applying Scientific Methodology

After reading the experiment, you may be able to venture an "educated guess" about the identity of the unknown D-hexose and express it as a hypothesis. You should then be able to predict the results of each experimental procedure you will carry out and see whether your actual results agree with your predictions.

Sweet Molecules

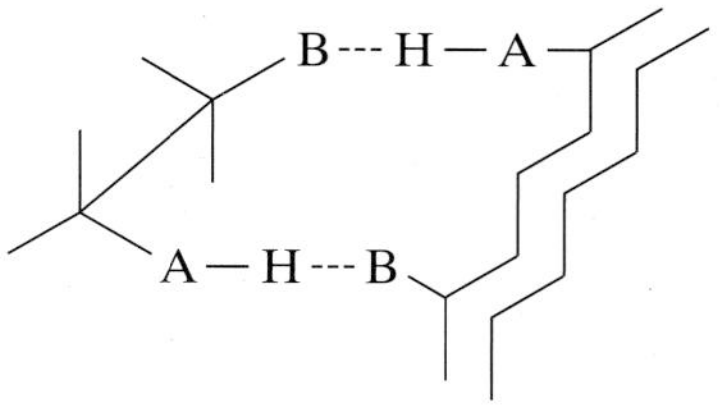

One of the most apparent properties of the sugars is their sweetness. Fructose is the sweetest known sugar—1.8 times sweeter (in its crystalline form) than sucrose—but some other substances are sweeter than any of the sugars. Cyclamates, such as sodium cyclohexylsulfamate, are approximately 30 times sweeter than sucrose, saccharin is nearly 500 times sweeter, and 1-*n*-propoxy-2-amino-4-nitrobenzene (P-4000) is estimated to be 4000 times sweeter.

A number of theories and models have been devised to explain the sweet taste sensation. According to the so-called AH,B theory, all sweet molecules contain an AH,B couple that is necessary to bind the compound to the taste bud receptor site. A and B are electronegative atoms, such as oxygen or nitrogen (but sometimes halogen or even carbon), that must be 0.25–0.40 nm apart in order to interact (by hydrogen bonding) with the receptor site, which is assumed to possess a similar AH,B couple. In sugars, AH and B are assumed to be an OH group and the oxygen atom of an adjacent OH group, respectively. For effective interaction, the OH groups should be in a *gauche* conformation, because in an *anti* conformation they would be too far apart to interact with the receptor site, and in a *syn* conformation they would tend to hydrogen bond intramolecularly rather than with the receptor site. Proposed AH and B sites for various sweet molecules are illustrated in Figure 52.1.

Another possible requirement for the sweet taste sensation is a hydrophobic or "greasy" site, X, on the sweet molecule. In the Keir tripartite

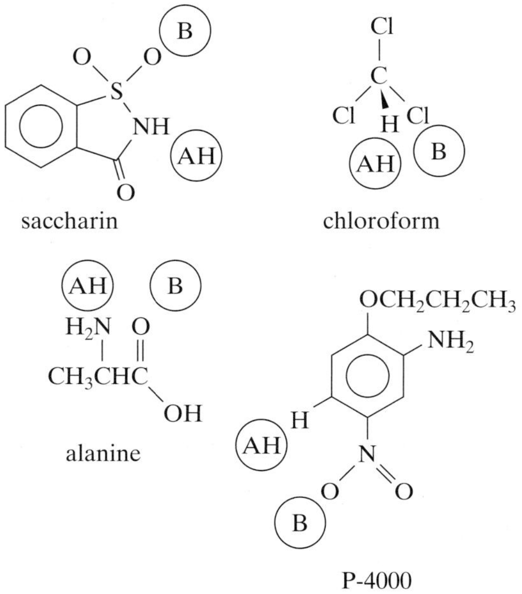

Interaction of a sweet molecule with the receptor site, according to the AH,B theory

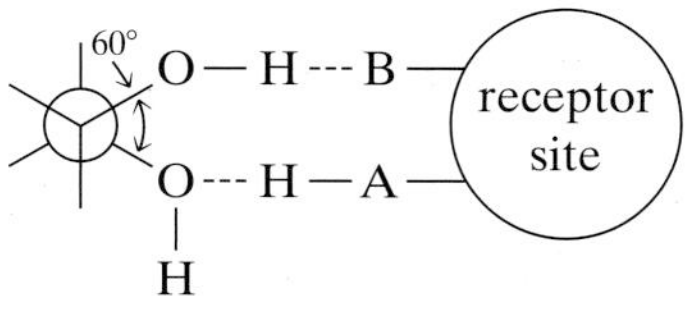

Gauche conformation in a sugar molecule

Figure 52.1 AH and B sites in sweet molecules

model, this site is estimated to be about 0.35 nm from A and 0.55 nm from B in the triangular grouping illustrated.

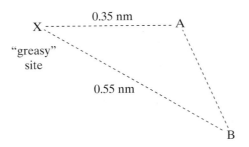

Keir tripartite model

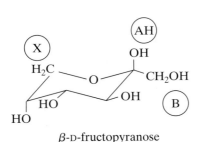

β-D-fructopyranose

See J. Chem. Educ. **1995,** *72,* 671.

In the crystalline form of fructose, β-D-fructopyranose, the AH, B, and X sites are believed to be the C-1 OH, the CH$_2$OH oxygen atom, and the ring methylene group, respectively. An even more complex model, the Tinti-Nofre model, postulates no fewer than eight sites, including the three of the Keir tripartite model. Not all of the sites are utilized by every sweet molecule, but it is assumed that the more sites a molecule occupies, the greater is its potential for sweetness.

Unfortunately, models proposed to explain the sweet taste sensation tend to be so general that they predict sweetness for many compounds that are actually tasteless or bitter, or so specific that only a few related molecules can satisfy their requirements for sweetness. Clearly, scientists do not yet have a complete understanding of the origin of the sweet taste sensation.

Understanding the Experiment

In this experiment, you will determine the structure of an unknown D-hexose by

- Using a simple chemical test that distinguishes aldohexoses from keto-hexoses
- Reducing the unknown with sodium borohydride and measuring the optical rotation of the product
- Converting the unknown to a phenylosazone and measuring its melting point

Both aldohexoses and ketohexoses are dehydrated in dilute acid to 5-(hydroxymethyl)-2-furaldehyde, which reacts with resorcinol to form a red condensation product.

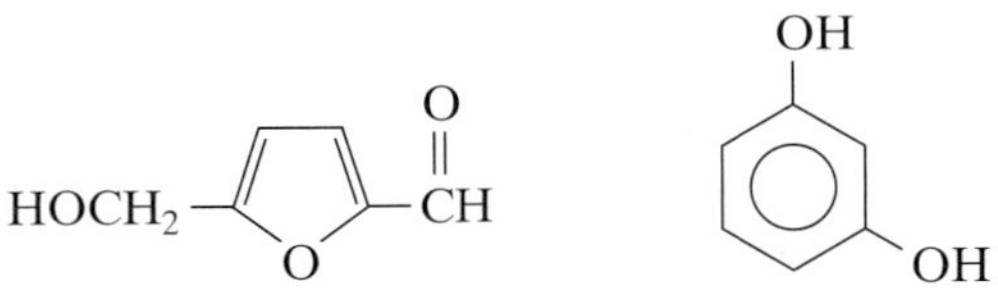

5-(hydroxymethyl)-2-furaldehyde   resorcinol

Ketohexoses dehydrate more rapidly than aldohexoses, making it possible to differentiate them using Seliwanoff's reagent, a solution of resorcinol in dilute HCl.

Reduction of a monosaccharide with sodium borohydride converts it to an alditol by reducing its carbonyl group. D-Glucose is converted to the alditol D-glucitol, which has no symmetry plane and is therefore optically active.

```
    CH=O                        CH₂OH
 H ─┼─ OH                    H ─┼─ OH
HO ─┼─ H      NaBH₄       HO ─┼─ H
 H ─┼─ OH     ─────→         H ─┼─ OH
 H ─┼─ OH                    H ─┼─ OH
    CH₂OH                       CH₂OH
  D-glucose                  D-glucitol
```

D-Allose is converted to D-allitol, which has a symmetry plane and is therefore optically inactive.

Key Concept: Compounds having a symmetry plane are achiral, meaning that they cannot exist in unique "left-handed" and "right-handed" forms. Achiral compounds do not rotate plane-polarized light, so they are optically inactive.

```
    CH=O                          CH₂OH
 H ─┼─ OH                      H ─┼─ OH
 H ─┼─ OH    NaBH₄            H ─┼─ OH
 H ─┼─ OH    ─────→          ─────────
 H ─┼─ OH    symmetry         H ─┼─ OH
    CH₂OH      plane           H ─┼─ OH
  D-allose                       CH₂OH
                               D-allitol
```

Thus the optical activity or inactivity of the alditol formed from reduction of a monosaccharide can yield structural information about the monosaccharide itself, thereby narrowing down the number of possible structures.

Converting monosaccharides to their phenylosazones in effect destroys any differences at the #1 and #2 carbon atoms of their molecules by changing both —CHOH—CHO and —CO—CH₂OH to the same structural unit:

```
CH=O          CH=NNHPh          CH₂OH
 |     ──→      |          ←──    |
CHOH           C=NNHPh           C=O
 |             |                  |
```

For example, D-glucose, its epimer D-mannose, and D-fructose are all converted to the same phenylosazone.

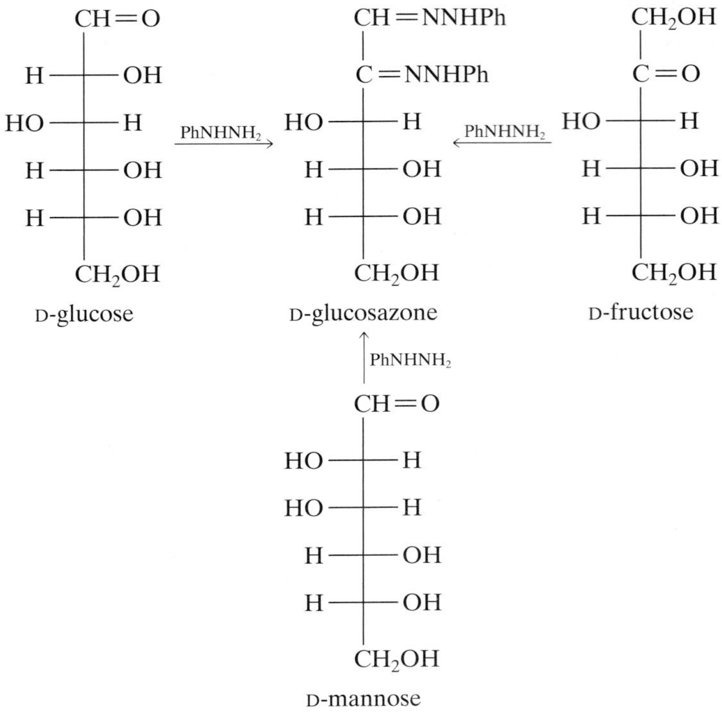

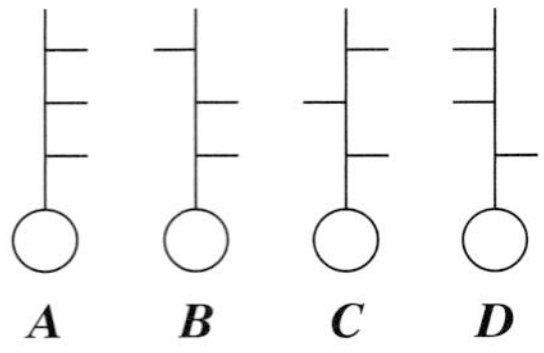

D-glucose   D-glucosazone   D-fructose

D-mannose

Figure 52.2 C-3 to C-6 structural units of D-hexoses

Only compounds that differ in structure from carbon #3 on down the chain will yield different phenylosazones. Among the D-hexoses, there are four C-3 to C-6 structural units that yield different phenylosazones; these are illustrated in Figure 52.2 using Rosanoff symbols.

Rosanoff symbols for functional groups

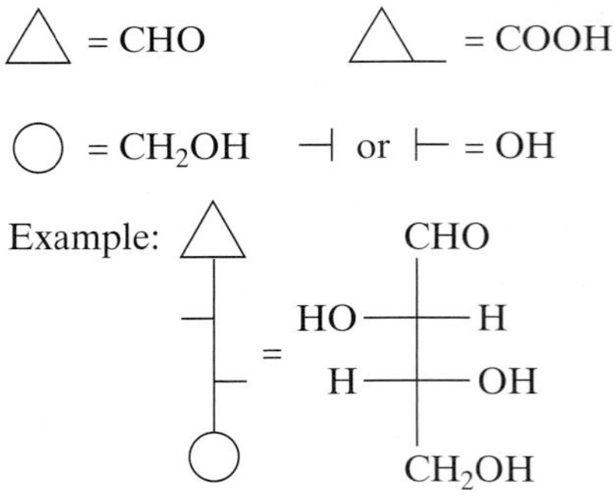

Thus D-glucose, D-mannose, and D-fructose yield the same phenylosazone because they all contain the same C-3 to C-6 structural unit, **B**.

In part **A** of this experiment, you will reduce your unknown D-hexose to an alditol by adding an excess of sodium borohydride in aqueous NaOH to an aqueous solution of the unknown. Like many sugars, the alditol tends to form supersaturated solutions and crystallizes slowly from solution. If your alditol refuses to crystallize after you cool the reaction mixture in ice water and rub the inside surface of its container with a stirring rod, you may be able to obtain a seed crystal from your instructor to help induce crystallization. Measuring the alditol's optical rotation in water will tell you whether it is optically active or optically inactive.

You will carry out Seliwanoff's test in part **B**. In part **C**, you will prepare the phenylosazone of the unknown monosaccharide by a simple test-tube reaction. Because phenylosazones decompose near the melting point, you will use a special technique to measure the melting point of your phenylosazone.

Reactions and Properties

See "Understanding the Experiment" for reactions of representative hexoses.

Table 52.1 Phenylosazone melting points

Structural unit	Phenylosazone mp
A	178
B	205
C	173
D	201

Note: melting points are in °C.

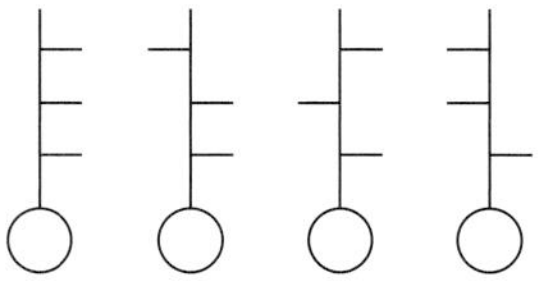

Directions

A. *Preparation of the Alditol*

> **Sodium borohydride is corrosive and can react violently with concentrated acids, oxidizing agents, and other chemicals. Aqueous sodium borohydride solutions with pH values below 10.5 have been known to decompose violently, so be sure that your reaction mixture is sufficiently alkaline. Keep NaBH₄ away from other chemicals; avoid contact and do not breathe its dust.**

Safety Notes

sodium borohydride

Reaction.   Dissolve 5.00 mmol of the unknown D-hexose in 3 mL of water in a 10-mL Erlenmeyer flask with magnetic stirring [OP-10] and gentle heating. Dissolve 2.0 mmol of sodium borohydride in 1.0 mL of 1 *M* NaOH in a small test tube. Cool [OP-8] both mixtures in an ice/water bath for about 5 minutes. Remove the beaker from the cold bath and add the cold sodium borohydride solution to the D-hexose solution drop by drop with stirring over a period of 3–5 minutes. When the addition is complete, let the

Take Care! Hydrogen is evolved, so keep flames away.

Stop and Think: What is the source of the hydrogen?

Waste Disposal: Pour the filtrate down the drain.

reaction mixture stand at room temperature, with stirring, for 20 minutes. *Under the hood*, add 6 *M* HCl drop by drop until foaming stops and the reaction mixture turns blue litmus paper red.

Separation and Purification.   Cool the reaction mixture in an ice/water bath and induce crystallization of the product, if necessary. Seal the flask with Parafilm and leave the reaction mixture in the cold bath while you complete the rest of the experiment. (For a better yield, leave it in a cool place overnight or longer.) Collect the product by vacuum filtration [OP-13] and wash it on the filter with two portions of cold 95% ethanol. Dry [OP-23] the product and measure its mass.

Analysis.   Prepare an aqueous solution containing about 0.2 g of the alditol (weighed to the nearest milligram) per 10 mL of solution. You may have to heat the mixture gently to dissolve the alditol. Measure the optical rotation [OP-33] of this solution in a 1-dm polarimeter cell, using pure water as the blank. If the optical rotation of the alditol is essentially equal to that of the blank (plus or minus 0.2°), assume that it is optically inactive; otherwise, calculate its specific rotation.

B. *Seliwanoff's Test*
Dissolve 10 mg of the D-hexose in 1.0 mL of water in a small test tube. Then add 2.0 mL of Seliwanoff's reagent, swirl to mix, and heat the test tube in a boiling-water bath for 2 minutes (no longer). Development of a deep red color within 2 minutes is a positive test for a ketose. Aldoses give a red color with additional heating or after standing for some time.

Take Care! The reagent is acidic, so avoid contact.

Waste Disposal: Dispose of the solution as directed by your instructor.

C. *Preparation of the Phenylosazone*

Phenylhydrazine hydrochloride is very toxic by ingestion, inhalation, and skin absorption, and it is a suspected carcinogen. Wear gloves, avoid contact, and do not breathe its dust.

Safety Notes

phenylhydrazine

Take Care! Wear gloves and avoid contact with phenylhydrazine hydrochloride.

Waste Disposal: Unless directed otherwise, pour the filtrates down the drain.

Dissolve 0.10 g of the unknown D-hexose in 2 mL of water in a test tube and stir in 0.20 g of phenylhydrazine hydrochloride, 0.30 g of sodium acetate trihydrate (or 0.18 g of anhydrous sodium acetate), and 0.20 mL of saturated aqueous sodium bisulfite. Seal the test tube with Parafilm and heat [OP-7] it in a boiling-water bath for 30 minutes, with occasional shaking. Add 3 mL of water and cool the reaction mixture in an ice/water bath. Collect the phenylosazone by vacuum filtration [OP-13], washing it on the filter with a little ice-cold methanol. Recrystallize [OP-25] the product from an ethanol/water mixture, dissolving it in the ethanol first. When you collect it by vacuum filtration, wash it on the filter with ice-cold methanol. Dry [OP-23] the phenylosazone thoroughly at room temperature. Pulverize enough of the dry solid for two melting-point tubes [OP-30] and measure the temperature, T_1, at which the first sample melts with rapid heating (a rise of 10–20°C per minute). After the melting-point apparatus has cooled below T_1, adjust it for a temperature rise of 3–6°C per minute, and insert the second sample just as the temperature reaches T_1. Record the temperature at which this sample becomes completely liquid as the melting point of the phenylosazone.

Deduce the structure of the D-hexose, draw its Haworth (flat-ring) projection, and use the formula index of *The Merck Index* or another reference work to find its common name.

Stop and Think: Which two C-3 to C-6 structural units might the D-hexose contain? Which one can you eliminate based on your results in part **A**?

Exercises

1 (a) Draw structures of all the monosaccharides that would yield the same alditol as your D-hexose. (b) Draw structures of all the monosaccharides that would yield the same phenylosazone as your D-hexose. (c) Draw structures of all the alditols that would be obtained by reduction of all possible D-hexoses, and indicate whether each one is optically active or optically inactive.

2 It is believed that the C-4 OH group and the C-3 oxygen atom of an aldopyranose, or the C-1 OH group and the CH_2OH oxygen atom of a ketopyranose, can interact as an AH,B couple with the sweet taste receptor site. Build molecular models of β-D-glucopyranose and the most stable pyranose ring form of the D-hexose that you identified. Use them to explain why your D-hexose is not as sweet as glucose.

3 Describe and explain the possible effect on your results of the following experimental errors or variations. (a) In part **A**, you dissolved the $NaBH_4$ in 1.0 *M* HCl rather than 1.0 *M* NaOH. (b) You had to leave the lab while the Seliwanoff test solution was in a boiling-water bath, and when you returned the test solution was red. (c) In part **C**, you rinsed the test tube with acetone and didn't dry it out completely.

4 Following the format in Appendix V, construct a flow diagram for the synthesis of the alditol in part **A**.

5 α-D-Mannose has a specific rotation of $+29.3°$ and β-D-mannose has a specific rotation of $-17.0°$. If either anomer is dissolved in water and allowed to stand until equilibrium is reached, the specific rotation of the equilibrium mixture is $+14.2°$. Calculate the percentage of each anomer in the equilibrium mixture.

6 (a) Draw the structures of the other D-hexoses that will yield the same phenylosazone as D-allose. (b) Draw the structures of the L-hexoses that will yield the same alditol as D-glucose.

Other Things You Can Do

(Starred projects require your instructor's permission.)

*1 Draw chair-form pyranose rings for both anomers of the D-hexose you identified and predict which one should be more stable. Measure the optical rotation of a solution containing about 0.4 g (accurately weighed) of the D-hexose in 10 mL of aqueous solution. *Under the hood*, mix 2 drops of concentrated aqueous ammonia into the solution, and again measure its optical rotation. If the optical rotation changes with time, wait until it

Take Care!   Avoid contact with aqueous ammonia and do not breathe its vapors.

equilibrates before taking a final reading, then calculate the specific rotation of the equilibrium mixture. Using *The Merck Index* or another reference book, look up the specific rotations in water of the α and β anomers of the D-hexose and calculate the percentage of each in the equilibrium mixture. Decide whether or not the calculated composition of the equilibrium mixture verifies your prediction.

*2 Carry out the reactions of some monosaccharides with phenols as described in Minilab 43.

 3 After referring to articles in *J. Chem. Educ.* **1995**, *72*, pages 671–683, write a research paper about sweet compounds and the theory of the sweet taste sensation.

Fatty Acid Content of Commercial Cooking Oils

Reactions and Preparation of Carboxylic Esters. Transesterification. Fats and Oils.

Operations

OP-7   Heating
OP-15   Extraction
OP-16   Evaporation
OP-22   Drying Liquids
OP-34   Gas Chromatography

Before You Begin

Read the experiment, read or review the operations as necessary, and write an experimental plan.

Scenario

The Consumers Advocate publishes *Caveat Emptor*, a periodical that rates consumer products and exposes inaccurate or misleading advertising (see Experiment 8). This organization is currently conducting a study of commercial vegetable oils to see whether their composition is consistent with advertising claims and to assess their relative dietary quality in terms of their effect on blood cholesterol. TCA's technical director, Patsy Haven, has asked your Institute for help in evaluating some of the vegetable oils sold in your geographic area. Your project group's assignment is to determine the fatty acid composition of each vegetable oil provided and to calculate the ratio of unsaturated to saturated fatty acids as a measure of their potential effect on cardiovascular health.

Applying Scientific Methodology

If you are given the names of the vegetable oils in advance of the lab period, try to predict their relative desirability with respect to cardiovascular health. Express your predictions as a hypothesis, to be tested when you obtain the results of your project group's analyses.

Fatty Acids and Health

Cooking fats and oils are glyceryl esters of long-chain carboxylic acids called *fatty acids.* An "oil" in this sense is essentially a fat that is liquid at room temperature. Fats and oils have been put to many uses throughout human history. Excavation of an Egyptian tomb more than 5000 years old yielded several earthenware vessels containing substances identified as palmitic acid and stearic acid, most likely formed by the breakdown of palm oil and beef or mutton tallow, which were placed there as provisions for the deceased.

$$CH_3(CH_2)_{14}COOH$$
palmitic acid

$$CH_3(CH_2)_{16}COOH$$
stearic acid

The Egyptians used olive oil as a lubricant when moving huge stones for their building projects, and a mixture of fat and lime as axle grease for their war chariots. They may also have originated the art of painting with oils or waxes, using pigments mixed with natural waxes to paint portraits on their mummy cases.

The Roman scholar Pliny the Elder described a process for making soap by boiling goat fat with wood ashes, then treating the pasty mass with sea water to harden it. But the art of making soap from animal fats must have originated much earlier; the Phoenicians of 600 B.C. were already trading soap to the Gauls, who later introduced it to Rome. Soap factories have been excavated at Pompeii, a Roman city that was buried under a blanket of volcanic ash during an eruption of Vesuvius. The same eruption killed Pliny, whose scientific curiosity impelled him to study the erupting volcano up close.

Modern research has revealed the importance of fats and oils in nutrition and is just beginning to clarify their role in human disease. Linoleic acid, for example, is an essential component of the human diet because it is not synthesized in the body. Polyunsaturated fatty acids (PFAs) such as linoleic acid are believed to be involved in the biosynthesis of the prostaglandins, which help control blood pressure and muscle contraction.

$$CH_3(CH_2)_4\overset{\overset{\displaystyle H}{|}}{C}=\overset{\overset{\displaystyle H}{|}}{C}CH_2\overset{\overset{\displaystyle H}{|}}{C}=\overset{\overset{\displaystyle H}{|}}{C}(CH_2)_7\overset{\overset{\displaystyle O}{\|}}{C}OH$$
linoleic acid

Polyunsaturated fatty acids also play a vital role in the functioning of biological membranes and have been implicated in the occurrence or prevention of such illnesses as atherosclerosis, cancer, and multiple sclerosis. Recent research suggests that PFAs of the $\omega3$ and $\omega6$ families are especially effective in preventing the buildup of cholesterol in arteries, which is a major cause of heart disease. The ω (omega) designation here refers to the distance of the last double bond from the methyl end of the fatty acid chain, as illustrated. Two $\omega3$ PFAs found in fish oils, eicosapentaenoic acid (EPA) and docosahexaenoic acid (DHA), appear to be even more beneficial than the PFAs in vegetable oils. Besides lowering serum cholesterol levels, they also reduce the tendency of blood to clot, and thus help prevent strokes as well as heart attacks.

With respect to cardiovascular health, the desirability of a fat or oil is related to the kinds and proportions of fatty acid residues that are incorporated into its triglyceryl ester molecules. Animal fats, coconut oil, and palm oil contain a relatively high proportion of saturated fatty acids (SFA), which raise blood cholesterol levels in humans and are believed to increase

$$CH_3CH_2CH=CH \ldots$$
$$\textcircled{$\omega1$}\textcircled{$\omega2$}\textcircled{$\omega3$}$$

the risk of heart disease. Vegetable oils and fish oils that are high in polyunsaturated fatty acids can have the opposite effect, lowering the amount of cholesterol in the blood by hastening its excretion. Monounsaturated fatty acids (MFAs), which occur in some vegetable oils, also help lower total cholesterol and are probably at least as healthful as the PFAs. Recent studies suggest that, unlike PFAs, MFAs lower total cholesterol without lowering high-density lipoprotein (HDL) levels. HDL is known as "good cholesterol" because it appears to reduce the risk of heart disease.

By contrast, *trans*-fatty acids raise total cholesterol levels and also lower HDL levels; thus they may be even more harmful to human health than saturated fatty acids. *trans*-Fatty acids, which form during the partial hydrogenation of vegetable oils, are identical to the natural mono- and polyunsaturated acids except for having *trans* double bonds where the natural acids have *cis* double bonds. They occur in most margarines and are used in a large number of processed food products.

Fatty acids are named and symbolized by several different systems. Most of the common fatty acids have trivial names that are still widely used, in part because of their simplicity compared to the systematic names. Linoleic acid, for example, is named *cis*-9-*cis*-12-octadecadienoic acid under the IUPAC system. Trivial names do not reveal the structures of the corresponding fatty acids, however, so shorthand notations are often used. For example, oleic acid can be represented by *c*-9-18:1, where *c* means *cis*, 9 refers to the position of the double bond (numbering from the COOH group), 18 is the total number of carbon atoms in the chain, and 1 is the number of double bonds. Linoleic acid is represented by *c,c*-9, 12-18:2 using this system. Since all of the acids to be studied in this experiment have *cis* double bonds, the *c* prefix will be omitted here.

Key Concept: The effect of a fatty acid on cardiovascular health is apparently related to its shape. Saturated fatty acids and trans-*fatty acids have more-or-less straight molecules, while natural monounsaturated and polyunsaturated fatty acids have molecules with kinks where their* cis *double bonds occur.*

$$CH_3(CH_2)_7 - \overset{\displaystyle H}{\underset{\displaystyle |}{C}} = \overset{\displaystyle H}{\underset{\displaystyle |}{C}} - (CH_2)_7COOH$$

oleic acid (9-18:1)

Understanding the Experiment

The cooking oils to be analyzed in this experiment are triglycerides; that is, esters of the trihydroxy alcohol glycerol containing three fatty acid residues. An example is tristearin, the stearic acid ester of glycerol. A mixture of fatty acid salts can be obtained by hydrolyzing a triglyceride in the presence of a base. This is the process used in soap formation, with the acid salts comprising the soap. Alternatively, a triglyceride can be transesterified with an alcohol such as methanol to yield a mixture of fatty acid methyl esters (FAME). Since the esters are much more suitable for analysis by gas chromatography than are the acid salts or the free acids themselves, the second method will be used in this experiment. The fatty acid composition of each cooking oil can then be derived from the percentages of the corresponding esters.

When methyl esters of saturated fatty acids are analyzed on a suitable gas chromatography column, their retention times (T_r) increase with the length of the carbon chain according to the relationship $\log T_r \propto$ number of carbons. The retention times of unsaturated fatty acid esters do not coincide with those of saturated esters. On a nonpolar liquid phase, their retention times are shorter than those of the saturated esters of the same chain length, whereas on a polar liquid phase, they are longer. The actual value of T_r depends on the number and positions of the double bonds. By comparing the retention times of a large number of such esters, a series of equivalent-chain-length (ECL) values

$$
\begin{array}{ll}
CH_3(CH_2)_{16}COOCH_2 & CH_2OH \\
\quad\quad\quad\quad\quad | & | \\
CH_3(CH_2)_{16}COOCH & CHOH \\
\quad\quad\quad\quad\quad | & | \\
CH_3(CH_2)_{16}COOCH_2 & CH_2OH
\end{array}
$$

a triglyceride,   glycerol
tristearin

$$CH_3(CH_2)_{16}COO^- Na^+$$

a soap, sodium stearate

$$CH_3(CH_2)_{16}COOCH_3$$

methyl ester of stearic acid
(methyl stearate)

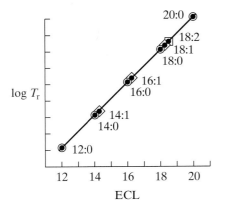

Figure 53.1 Plot of log T_r versus equivalent chain length for fatty acid methyl esters on a DEGS liquid phase

has been worked out for various esters on different liquid phases. The ECL for a *saturated* ester is the same as its actual chain length—for instance, 14.0 for myristic acid (as the methyl ester), 18.0 for stearic acid, etc. The ECL for an *unsaturated* ester on a polar liquid phase is greater than its actual chain length—for instance, 18.43 for methyl oleate (9-18:1) and 19.22 for methyl linoleate (9,12-18:2), both analyzed on a diethylene glycol succinate (DEGS) liquid phase. If the retention times for two or more saturated esters are known, a calibration curve can be constructed by plotting the logarithms of their retention times against their equivalent chain lengths, as shown in Figure 53.1.

To identify an unknown fatty acid ester, one can calculate the logarithm of its retention time, read its ECL value from the graph, and compare that value with the ECL values of known methyl esters (see Table 53.1).

Table 53.1 Shorthand notation and ECL values (on DEGS liquid phase) for representative fatty acids

Fatty acid	Shorthand notation	ECL	Type
lauric	12:0	12.00	SFA
myristic	14:0	14.00	SFA
myristoleic	9-14:1	14.71	MFA
palmitic	16:0	16.00	SFA
palmitoleic	9-16:1	16.55	MFA
stearic	18:0	18.00	SFA
oleic	9-18:1	18.43	MFA
linoleic	9,12-18:2	19.22	PFA
linolenic	9,12,15-18:3	20.12	PFA
behenic	22:0	22.00	SFA

Note: ECL values are for the corresponding methyl esters.

In order for ECL values to provide an accurate means of identifying the esters, the same type of column and carrier gas should be used in all cases. Even then, the experimental values may not correspond exactly to literature values because of differences in such factors as the kind of column support, the amount of liquid phase, the age of the column, and the experimental conditions. In addition, some peaks may overlap in complex mixtures, so that two or more columns must be used for their separation, and deviations from linearity in the log T_r plot may occur when short-chain fatty acids are analyzed. The relatively simple mixtures of fatty acids that you will encounter in this experiment should not present any serious experimental difficulties, however.

The transesterification of an oil to its methyl esters will be carried out by hydrolyzing its triglycerides in methanolic sodium hydroxide and esterifying the resulting fatty acids in methanol with a boron trifluoride catalyst. After the methyl esters have been isolated from the reaction mixture, they will be mixed with an internal standard, methyl heptadecanoate, and analyzed by gas chromatography. Heptadecanoic acid (17:0) does not occur naturally, so the 17:0 peak from its ester will not interfere with other component peaks. To minimize the effect of retention-time drift due to changes in operating parameters, relative retention values should be calculated by dividing the retention time for each peak by that of the 17:0 peak. Once the 17:0 peak has been located, those of the 16:0 and the 18:0 esters can be identified by the fact that the log T_r interval between their peaks and 17:0 is the same. (The 16:0 ester often gives the first strong peak on the chromatogram, making it easy to recognize.) Then a log T_r versus ECL plot can be prepared, from which the other peaks can be identified. Since the peak area for each component should be proportional to its mass, the percentage composition of the mixture can be determined from peak areas with good accuracy.

Assuming that saturated fatty acids have a negative effect on cardiovascular health, and that both monounsaturated and polyunsaturated fatty acids have a positive effect, you can use the ratio (MFA + PFA)/SFA as a rough indicator of the "healthfulness" of your cooking oil compared to the other oils analyzed in your laboratory section.

Reactions and Properties

$$\begin{matrix} R'COOCH_2 \\ | \\ R''COOCH \\ | \\ R'''COOCH_2 \end{matrix} + 3CH_3OH \xrightarrow[BF_3]{NaOH} \begin{matrix} R'COOCH_3 \\ + \\ R''COOCH_3 \\ + \\ R'''COOCH_3 \end{matrix} + \begin{matrix} H & H & H \\ O & O & O \\ | & | & | \\ CH_2CHCH_2 \end{matrix}$$

(R groups can be alike or different.)

Table 53.2 Physical properties

	M.W.	mp	bp	d
methanol	32.0	−94	65	0.791
boron trifluoride-methanol complex	131.9		59[4]	
boron trifluoride	67.8	−127	−100	

Note: mp and bp are in °C, density is in g/mL. The complex has the composition $BF_3 \cdot 2CH_3OH$.

Methyl stearate

$$CH_3(CH_2)_{16}\overset{\overset{\displaystyle O}{\|}}{C}OCH_3$$

2925.2	1360.7	1169.8
1744.9	1302.8	1117.3
1465.8	1247.5	721.3

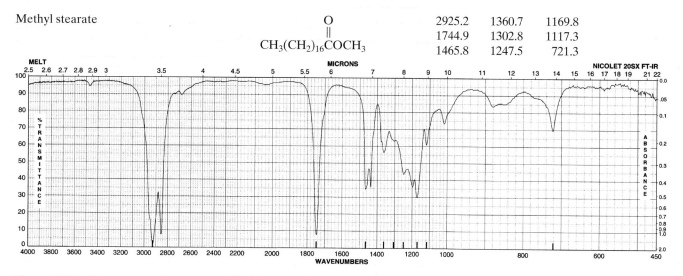

Figure 53.2A IR spectrum of a saturated 18-carbon methyl ester, methyl stearate

Methyl linolenate

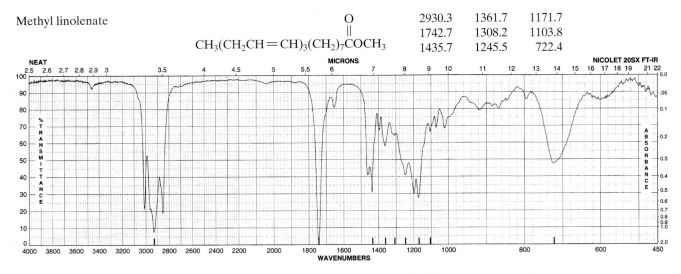

$$CH_3(CH_2CH = CH)_3(CH_2)_7COCH_3$$

2930.3	1361.7	1171.7
1742.7	1308.2	1103.8
1435.7	1245.5	722.4

Figure 53.2B IR spectrum of a polyunsaturated 18-carbon methyl ester, methyl linolenate

Directions

(Adapted from a procedure in the *Journal of Chemical Education* **1974**, *51*, 406, with permission.)

A selection of commercial cooking oils may be provided, or students may be asked to bring their own cooking oils. Different kinds of oils should be analyzed, such as canola, corn, olive, peanut, rice bran, safflower, soybean, and sunflower oils. All glassware and reagents should be dry for this experiment.

Safety Notes

The methanolic sodium hydroxide solution is flammable and toxic. Avoid contact, do not breathe its vapors, and keep it away from flames.
The boron trifluoride/methanol solution is flammable, corrosive, and toxic and can cause severe damage to the eyes, skin, and respiratory tract. Use gloves and a hood; avoid contact, and do not breathe its vapors.
Petroleum ether is extremely flammable. Keep it away from flames and hot surfaces.

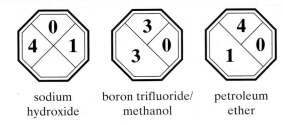

sodium   boron trifluoride/   petroleum
hydroxide   methanol   ether

Reaction.   In a clean, *dry* 15-cm screw-cap centrifuge tube, combine about 50 mg (0.050 g) of a commercial cooking oil with 1.7 mL of 0.5 *M* sodium hydroxide in methanol and add a boiling chip. Set the cap on top of the centrifuge tube loosely (do not screw it down) and use a hot water bath to heat [OP-7] the reactants at a gentle boil until the oil has completely dissolved, about 3–5 minutes. *Under the hood*, add 2.0 mL of a 12.5% (weight/volume) solution of boron trifluoride in methanol. Again cap the tube loosely and boil the mixture gently in a hot water bath for 2 minutes.

Take Care! Wear gloves, avoid contact with the methanolic NaOH and BF$_3$ solutions, and do not breathe their vapors.

Separation.   Add 4 mL of saturated aqueous sodium chloride and extract [OP-15] the reaction mixture with three 3-mL portions of low-boiling (~35–60°C) petroleum ether. Dry [OP-22] the petroleum ether solution with anhydrous sodium sulfate and evaporate [OP-16] the solvent completely.

Waste Disposal: Pour the aqueous layer down the drain.

Analysis.   Obtain a gas chromatogram [OP-34] of the fatty acid methyl ester mixture on a DEGS/Chromosorb W or similar column. Add a drop of a 50% solution of methyl heptadecanoate in dichloromethane to the remaining mixture, and obtain its gas chromatogram. Measure the areas of all peaks on the first gas chromatogram. Identify the 17:0 peak and measure the relative retention times of all peaks on the second gas chromatogram. Calculate relative retention times and log T_r values for all the components, and use these data to tentatively identify several of the methyl esters. For these esters, plot log T_r versus their ECL values and try to fit the rest of your data to the plot, redrawing it as necessary to get the best straight line. (If your data points don't fit the plot, you probably misidentified the esters. Make another guess and try again.) Identify the esters and calculate the total percentages of saturated, monounsaturated, and polyunsaturated fatty acids and the ratio of unsaturated to saturated fatty acids. Tabulate the class results and rank the different oils according to their expected effect on cardiovascular health.

Exercises

1   Write the structures of the fatty acids whose triglycerides were in the oil you analyzed and give their IUPAC names.

2   On the gas chromatogram of a mixture of methyl esters, methyl palmitate had a retention time of 150 seconds and methyl heptadecanoate had a retention time of 198 seconds. What is the probable identity of a methyl ester with a retention time of 363 seconds on the same chromatogram?

3   Describe and explain the possible effect on your results of the following experimental errors or variations: (a) The reaction test tube contained water. (b) You heated the oil with methanolic sodium hydroxide, but forgot to add the boron trifluoride/methanol solution. (c) You didn't evaporate all of the petroleum ether from your reaction mixture.

4   (a) Following the format in Appendix V, construct a flow diagram for this experiment. (b) Explain why there was no glycerol peak on the gas chromatogram of the methyl ester mixture, even though glycerol was a product of the hydrolysis.

5   Why was the Lewis acid boron trifluoride, rather than hydrochloric acid, used to catalyze the transesterification reaction?

6   Calculate the (MFA + PFA)/SFA ratio for palm oil, whose fatty-acid content is approximately 9% linoleic, 2% myristic, 40% oleic, 45% palmitic, and 4% stearic acid.

7   (a) Neat's-foot oil consists almost entirely of the triglycerides of oleic and palmitic acids. How many different triglycerides of these two acids can it contain? (b) Draw structures for the triglycerides that contain two oleic acid units and one palmitic acid unit.

8   Outline a synthesis of the detergent sodium lauryl sulfate from glyceryl trilaurate (trilaurin).

$$CH_3(CH_2)_{10}CH_2OSO_2O^-Na^+$$
sodium lauryl sulfate

Other Things You Can Do

(Starred projects require your instructor's permission.)

*1 Record an infrared spectrum [OP-36] of your methyl ester mixture and compare it with the spectra of the 18-carbon methyl esters in Figure 53.2. Try to account for significant similarities and differences in the spectra and point out any bands that show evidence of unsaturation.

*2 Extract the fat trimyristin from nutmeg as described in Minilab 44.

*3 Make some soap using a phase-transfer catalyst as described in Minilab 45.

4   Starting with sources listed in the Bibliography, write a research paper about the health implications of various types of fats and oils. Include a description of the HDL and LDL forms of cholesterol and a discussion of the role they play in heart disease.

Structure of an Unknown Dipeptide

Reactions of Peptides. Nucleophilic Aromatic Substitution. Amino Acids. Structure Determination.

Operations

OP-20 Paper Chromatography
OP-3   Using Glass Rod and Tubing
OP-15   Extraction
OP-16   Evaporation
OP-19   Thin-Layer Chromatography
OP-21   Washing Liquids

Before You Begin

Read the experiment and operation OP-20, read or review the other operations as necessary, and write an experimental plan.

Scenario

Snake venoms have been used in medicine as anticoagulants, which help dissolve blood clots or prevent their formation. Natural Nostrums, a pharmaceutical company that manufactures drugs based on natural substances, anticipates some medical uses for the venom of the black boomslang, a deadly African tree snake. Their native snake hunter recently had an unfortunate accident, so the supply of natural boomslang venom has dried up. Before they can prepare a synthetic version of the venom, which is a polypeptide, they need to determine the sequence of amino acids in its polypeptide chain. Ferdie Lance, the company's snake-venom expert, has used an enzyme called cathepsin C to break down the boomslang venom into dipeptide units. Now each dipeptide unit has to be identified so that Dr. Lance can reconstruct the amino-acid sequence in the polypeptide. Your project group's assignment is to determine the structures of the unknown dipeptides.

Applying Scientific Methodology

After you analyze the hydrolysis mixture from your assigned dipeptide using paper chromatography, you can formulate a tentative hypothesis about the structure of the dipeptide. Your hypothesis will be tested by TLC analysis of a hydrolysis mixture from the dinitrophenylated dipeptide.

Mushrooms, Black Mambas, and Memory Molecules

The naturally occurring polypeptides and proteins range in size from tripeptides such as glutathione to complex proteins having molecular weights on the order of 1 million; their variations in structure and function cover just as broad a range.

$$^+H_3NCHCH_2CH_2CNHCHCNHCH_2CO^-$$

glutathione

Some of them, such as the keratins of hair, horns, nails, and claws, act as biological building materials and have little or no biological activity. Others function as hormones, enzymes, antibiotics, or toxins, exerting a profound effect on biochemical reactions.

Bradykinin, sometimes known as the "pain molecule," is released by enzymatic cleavage of plasma glycoprotein whenever tissues are damaged.

Arg—Pro—Pro—Gly—Phe—Ser—Pro—Phe—Arg

bradykinin

It causes the sensation of pain by bonding to certain receptors on nerve endings. Peptides similar to bradykinin are present in wasp venom; other kinins act as hormones, stimulating a variety of physiological responses, such as the contraction or relaxation of smooth muscles and the dilation of blood vessel walls.

Oxytocin and vasopressin are both secreted by the posterior pituitary gland. Their structures are identical except for two amino acid residues, but their functions are entirely different. Vasopressin increases retention of water in the kidneys and is used as an antidiuretic in treating a form of diabetes that is characterized by excessive urine flow. Oxytocin intensifies uterine contractions during childbirth and is used clinically to induce labor. Both are cyclic compounds (cyclopeptides) that have a tripeptide side chain.

Poisonous mushrooms of the genera *Amanita* and *Galerina* contain cyclopeptides known as amatoxins and phallotoxins. These toxins are extremely potent; as little as one "death cap" mushroom, *Amanita phalloides*, can kill an adult. The Hollywood version of mushroom poisoning—in which the unlucky victim eats a serving of stewed mushrooms, turns pale, rises from the table clutching his throat, and immediately drops dead—is a myth. The victim may not even be aware that he or she has been poisoned for 8–24 hours after eating the mushrooms. Following a day or so of violent cramps, nausea, vomiting, and other unpleasant symptoms, the patient appears to recover and may even be sent home from the hospital. Death from kidney or liver failure often follows in several days. α-Amanitin and the other amatoxins are believed to attack the nuclei of liver and kidney cells, depleting their nuclear RNA and inhibiting the synthesis of more RNA so that protein synthesis stops and the cells die.

See Table 54.1 that follows for the names and abbreviations of some amino acids. The textbook for your lecture course should contain abbreviations for amino acids not listed in the table.

oxytocin

vasopressin

The Roman emperor Claudius died soon after eating mushrooms, but they were probably edible Amanita caesarea *mushrooms that had been laced with poison by his wife, Agrippina. She wanted him out of the way so that her son, Nero, would become emperor.*

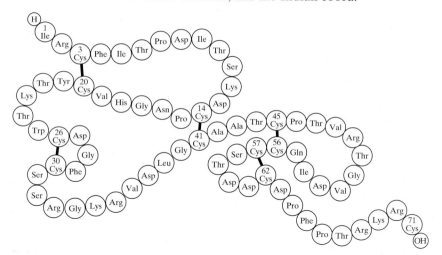

α-amanitin

Larger polypeptides, containing from 60 to 74 amino acid residues, are found in snake venoms such as those of the African black mamba, which can kill a mouse in less than 5 minutes, and the Indian cobra.

Indian cobra

Figure 54.1 Venom toxin of the Indian cobra (*Naja naja*) (Reprinted from *Biochemical and Biophysical Research Communications*, **1973**, 55, 435, by permission.)

Most of these venom toxins have long chains that are cross linked by four or more cystine disulfide bridges, as shown for the cobra's venom toxin in Figure 54.1.

One of the most exciting areas of polypeptide biochemistry is the study of "memory molecules." Researchers have discovered that an animal's learned behavior can be forgotten when a substance that interferes with peptide synthesis is injected into the animal's brain at a certain time. This suggests that peptide synthesis is involved in the consolidation of long-term memory. For example, a peptide called scotophobin was isolated from the brains of rats that had been conditioned to fear the dark.

Ser—Asp—Asn—Asn—Gln—Gln—Gly—Lys—Ser—Ala—Gln—Gln—Gly—Gly—TyrNH$_2$

scotophobin

(from the Greek *scotos*, dark; and *phobos*, fear)

When scotophobin was injected into unconditioned rats, they also became afraid of the dark. This raises the intriguing possibility that breaking the chemical memory code could enable chemists to synthesize the peptides corresponding to any kind of learning experience. Perhaps someday it will be possible to get an injection of Biology 101 or Philosophy 416 rather than absorbing the subject matter in the classroom!

Understanding the Experiment

The structures of long-chain polypeptides, such as those found in the venom of poisonous snakes, can be determined by breaking them down into shorter chains of amino acids and analyzing the fragments. For example, an enzyme called cathepsin C hydrolyzes a polypeptide into dipeptide units starting at the *N*-terminal amino acid residue (the one at the chain end terminated by an amino group), and the enzyme trypsin breaks a polypeptide chain after each lysine or arginine residue. By analyzing each fragment from these and other reactions, a biochemist can locate overlapping sequences of amino acids, from which the entire primary structure of the polypeptide can be deduced.

In this experiment, you will determine the structure of an unknown dipeptide that contains one or more of the amino acids listed in Table 54.1. To do so, you will first identify the amino acids your dipeptide contains by hydrolyzing it and analyzing the resulting amino acid mixture by paper chromatography. The dipeptides and amino acid derivatives you will be using are quite expensive; therefore, the experiment will be carried out on an ultramicro scale, using only a milligram of the dipeptide for each reaction. You will use Pasteur pipets for volume measurements and solvent extractions, and sealed capillary tubes for the hydrolysis reactions. It takes some practice to fill and seal a capillary tube properly, so it is recommended that you practice the technique with water before using the dipeptide and DNP-dipeptide solutions. The capillary method is very reliable when performed correctly, but there is always the chance that an improperly sealed tube will break or leak during the overnight hydrolysis period. To prevent the considerable loss of time that would result from a ruined sample, you will prepare at least three tubes for each hydrolysis reaction.

The paper chromatography of amino acids can be carried out in a multitude of developing solvents; no one solvent system is the best under all circumstances. The 2-propanol/formic acid/water (16 : 1 : 4) solvent system used in this procedure is suitable for most amino acids, but certain combinations of amino acids may be separated more readily with a different solvent system, such as 1-propanol/ammonia (7 : 3) or 2-butanone/propionic acid/water (15 : 5 : 6). The amino acids are colorless, so their spots will be made visible by spraying them with a solution of ninhydrin, which reacts with most amino acids (proline is an exception) to form a blue-violet product.

To work out the complete structure of your dipeptide, you must determine the sequence in which the amino acids appear in the dipeptide. This can be done using a dinitrophenylation procedure developed by Frederick Sanger, who won the chemistry Nobel Prize in 1956 for determining the structure of insulin and again in 1980 for his work on nucleic acid structures. 2,4-Dinitrofluorobenzene (DNFB) is an unusually reactive aryl halide that undergoes

Table 54.1 Names and abbreviations of selected amino acids

Name	Abbreviation
histidine	His
lysine	Lys
serine	Ser
aspartic acid	Asp
glycine	Gly
threonine	Thr
alanine	Ala
proline	Pro
tyrosine	Tyr
valine	Val
phenylalanine	Phe
leucine	Leu

The solvent ratios are volume ratios. Thus a 1-propanol/ammonia (7 : 3) solvent system can be prepared by mixing 70 mL of 1-propanol with 30 mL of concentrated aqueous ammonia.

ninhydrin

nucleophilic substitution reactions with unprotonated amino groups. Because most peptides exist as zwitterions in neutral solutions, sodium bicarbonate is used to raise the pH of the peptide solution enough to deprotonate the amino group, allowing it to react with DNFB. This reaction yields a dipeptide having a dinitrophenyl (DNP) substituent attached to the *N*-terminal amino acid residue, as shown in the "Reactions and Properties" section. In effect, the DNP group functions as a label for the *N*-terminal residue.

In the alkaline reaction mixture the DNP-substituted dipeptide exists in the anionic form, with a polar $-COO^-$ group. Extracting this solution with ether therefore removes the less polar DNFB, but leaves the DNP-dipeptide in the aqueous layer. Lowering the pH with hydrochloric acid then protonates the carboxyl group, allowing the DNP-dipeptide to be extracted from the aqueous layer by diethyl ether and isolated by evaporating the ether. It is important to remove the last traces of ether before hydrolysis, so the residue is dissolved in acetone, which is also evaporated. The DNP-dipeptide is then hydrolyzed in aqueous hydrochloric acid, yielding the free *C*-terminal amino acid (the one from the chain end terminated by a carboxyl group) and a dinitrophenyl derivative of the *N*-terminal amino acid.

DNP-amino acids are more easily separated on silica gel than on paper, so your product mixture will be analyzed by thin-layer chromatography using known DNP-amino acids as standards. Because the DNP derivatives are light sensitive, their chromatograms should be developed in a location away from strong light. The yellow spots formed by the DNP-amino acids are easy to locate, but they fade with time, so you should circle them in pencil soon after the TLC plate is developed. Dinitrophenol is a possible by-product of the dinitrophenylation procedure. However, its yellow spot is usually evident only when an alkaline developing solvent is used. If TLC analysis of your unknown produces two or more distinct spots, exposing the chromatogram to vapors of hydrochloric acid should bleach out the dinitrophenol spot.

Key Concept: The structures of amino acids and peptides are pH dependent. At low pH they exist as cations of the type $^+H_3N{\sim}COOH$, at high pH as anions of the type $H_2N{\sim}COO^-$, and at intermediate pH as zwitterions of the type $^+H_3N{\sim}COO^-$.

2,4-dinitrophenyl substituent
(DNP)

Reactions and Properties

A Dinitrophenylation of dipeptide

dipeptide

DNP-dipeptide

R_1, R_2 = amino acid side chains

B Hydrolysis of dipeptide and DNP-dipeptide

$$\underset{R_1}{\overset{O}{\overset{||}{^+H_3NCHC}}}-\underset{R_2}{\overset{O}{\overset{||}{NHCHCOH}}} + H_2O + HCl \longrightarrow \underset{R_1}{\overset{O}{\overset{||}{^+H_3NCHCOH}}} + \underset{R_2}{\overset{O}{\overset{||}{^+H_3NCHCOH}}} + Cl^-$$

N-terminal amino acid   C-terminal amino acid

$$O_2N-\underset{}{\overset{NO_2}{\bigcirc}}-\underset{R_1}{\overset{O}{\overset{||}{NHCHC}}}-\underset{R_2}{\overset{O}{\overset{||}{NHCHCOH}}} + H_2O + HCl \longrightarrow O_2N-\underset{}{\overset{NO_2}{\bigcirc}}-\underset{R_1}{\overset{O}{\overset{||}{NHCHCOH}}} + \underset{R_2}{\overset{O}{\overset{||}{^+H_3NCHCOH}}} + Cl^-$$

DNP-amino acid

Table 54.2 Physical properties

	M.W	mp	bp
2,4-dinitrofluorobenzene	186.1	27.5–30	178^{15}

Note: mp and bp are in °C.

Directions

Use a Pasteur pipet for all volume measurements; the pipet should deliver about 35–45 drops per milliliter. Organize your time efficiently so that you can have your hydrolysis tubes prepared and in the oven by the end of the first laboratory period.

A. *Dinitrophenylation of the Dipeptide*

Take Care! Wear gloves and avoid contact with the DNFB reagent.

Reaction. In a 3-mL conical vial, combine about 1 mg of your unknown dipeptide with 4 drops of distilled or deionized water, 1 drop of aqueous 4% sodium bicarbonate, and 8 drops of the DNFB reagent, a 5% solution of 2,4-dinitrofluorobenzene in ethanol. Cap the reaction vial securely and shake it occasionally over a 1-hour period. Every 10 minutes, check the pH by dipping the closed end of a melting-point capillary into the solution and touching it to a strip of narrow-range pH paper. As necessary, add a drop or two of 4% sodium bicarbonate solution to keep the pH between 8 and 9.

Stop and Think: Why is important to keep the pH in this range?

Separation. Add 10 drops of water and 10 drops of aqueous 4% sodium bicarbonate to the reaction mixture. Wash [OP-21] the reaction mixture two or three times with diethyl ether, each time using a volume of ether approximately equal to the total volume of solution. Carry out each washing operation as described here.

Take Care!   Keep diethyl ether well away from flames.

- Add the ether to the reaction vial.
- Draw both layers into a Pasteur pipet and eject the mixture forcefully into the reaction vial to mix the layers; repeat a half-dozen times or more.
- Cool the mixture in ice water and wait for any emulsion to settle.
- Remove the ether (top) layer with the Pasteur pipet.

Stop and Think: What is being removed by the ether?

The last ether layer should be colorless; wash the reaction mixture with more ether if it is not. Add enough 6 *M* hydrochloric acid (2 or 3 drops are usually sufficient) to the aqueous layer to bring its pH down to 1 or 2. (**Stop and Think:** Why is it necessary to acidify the solution?) Then extract [OP-15] the aqueous solution with two portions of diethyl ether by the same technique you used to wash it, and combine and save the extracts. Be careful not to include any of the aqueous layer when you withdraw the ether layers.

Take Care!   Avoid contact with the ether layers.

Waste Disposal: Place the wash ether layers in a designated solvent recovery container.

Waste Disposal: Pour the aqueous layer down the drain.

B. *Hydrolysis of the Dipeptide and DNP-Dipeptide*

Use a dry Pasteur pipet to distribute the ether solution equally among three wells on a porcelain spot plate and let the ether evaporate *completely* under the hood. Dissolve each residue in 2 drops of acetone and let the acetone evaporate under the hood. Then dissolve each residue in 4 drops of 6 *M* hydrochloric acid, using the closed end of a capillary melting point tube for stirring. Prepare at least three hydrolysis tubes by Method 1 if you are using capillary tubes that are sealed at one end, or by Method 2 if you are using open-ended capillary tubes.

Method 1.   Warm a capillary melting-point tube by holding it near its open end and moving it in and out of the "cool" part of a Bunsen burner flame for several seconds, then immediately insert the open end into the liquid in one of the wells. As the tube cools (blowing on it may help), it should begin to fill with liquid. When 2–3 cm of liquid is inside the tube, remove it from the liquid, and, holding the open end just above horizontal, cool the closed end under a cold water tap. Holding the tube upright near the open end, carefully strike it at a point below the liquid level with your fingernail until virtually all of the liquid is in the bottom of the tube. Capillary tubes are fragile, so don't tap the tube too hard or it will break. Holding the closed end so that your fingers just cover the liquid, rotate the tube as you move it rapidly in and out of the cool part of the burner flame for several seconds. (The tube should be inserted lengthwise into the flame so that most of the empty part is heated.) Immediately seal [OP-3] the open end by rotating it in the flame until that end closes. Because of the vacuum created by preheating the tube, its end should collapse and bend over as it seals.

Take Care!   Don't burn yourself!

Method 2.   Insert one end of an open capillary tube into one of the wells and draw in (by capillary action) about 2–3 cm of liquid. Hold the tube horizontally and tap it so that the liquid is near the middle, then seal one end by rotating it in the outermost edge of the flame. When that end has cooled, hold the open end up and strike the lower part of the tube with your fingernail until nearly all of the liquid is in the bottom, then seal the other end as described for Method 1.

Put the capillary tubes in a small test tube labeled with your name and "DNP-amino acid." Leave the test tube in a 100–110°C oven overnight or longer.

Dissolve about 1 mg of your unknown dipeptide in 10 drops of 6 *M* hydrochloric acid. Use this solution to prepare three or more capillary tubes by one of the previous methods. Put the capillary tubes in a small test tube labeled with your name and "amino acids." Leave the test tube in a 100–110°C oven overnight or longer.

C. *Separation and Analysis of the Amino Acids and the DNP-Amino Acid*

Safety Notes

> **Both chromatographic developing solvents contain flammable or toxic liquids with harmful vapors, and chloroform is a suspected carcinogen. Use a hood, avoid contact, do not breathe their vapors, and keep flames away.**

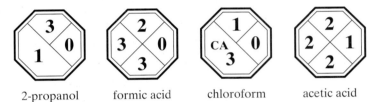

2-propanol   formic acid   chloroform   acetic acid

Take Care! Avoid contact with the developing solvents and do not breathe their vapors.

Remove the hydrolysis tubes from the oven and let them cool. *Under a hood*, prepare a large beaker (or another suitable developing chamber) for paper chromatography [OP-20] by adding enough of the 2-propanol/formic acid/water (16 : 1 : 4) solvent system to form a 1-cm layer on the bottom. Cover the beaker with plastic wrap and let the system equilibrate for at least 1 hour before using it. Prepare a developing chamber for the TLC analysis [OP-19] using enough of the chloroform/acetic acid/*t*-pentyl alcohol (23 : 1 : 10) solvent system to form a 5-mm layer on the bottom. Cover the developing chamber with plastic wrap and let the system equilibrate for at least 30 minutes before using it.

Use a sharp triangular file or another glass cutter to open one of the cooled amino acid capillary tubes just above the liquid level, and empty it into one well of a spot plate. Carefully evaporate [OP-16] the solvent in a dry air stream under the hood and dissolve the residue in 1 drop of water. Obtain a sheet of Whatman #1 chromatography paper that is at least 12 cm long (in the direction of development) and wide enough to accommodate the standard amino acid solutions as well as your unknown solution. Spot the paper in two (or more) places with the amino acid solution, then spot it with the amino acid standards. Develop the chromatogram under a hood in the 2-propanol/formic acid/water solvent system and let it dry under the hood. Then spray the paper lightly with the ninhydrin spray reagent. Develop the color by heating the paper in a 100–110°C oven for about 10 minutes. Measure the R_f values of all the spots.

Waste Disposal: Place the chromatography solvent in a designated solvent recovery container.

Open one of the DNP-amino acid capillary tubes and empty the contents into a conical vial, washing it down with 10–15 drops of water. Extract [OP-15] the aqueous solution twice with diethyl ether as described in part **A**, using a volume of ether equal to the volume of the solution for each extraction. Combine the extracts. Transfer the ether solution to a well on a

Waste Disposal: Unless you plan to confirm the identity of the *C*-terminal amino acid (see "Other Things You Can Do"), pour the aqueous layer down the drain.

spot plate and let the ether evaporate under the hood. Dissolve the residue in 3 drops of acetone and immediately use this solution to spot a silica gel TLC plate in two places. Spot the plate with the standard solutions of DNP-amino acids and develop it under a hood with the chloroform/acetic acid/ *t*-pentyl alcohol solvent system in a location away from strong light. Let the developed TLC plate dry under the hood. If your unknown solution forms more than one yellow spot, momentarily hold the chromatogram over an open bottle of concentrated HCl *under a hood*. Circle the yellow spots as soon as the developed plate is dry; you can use an ultraviolet lamp to make them more clearly visible. Measure the R_f values of all the spots. Tabulate the R_f values from your chromatograms and identify the two amino acids and the DNP-amino acid. Draw the structure of the unknown dipeptide and name it.

Waste Disposal: Put the chromatography solvent in a designated solvent recovery container. Place any unused capillary tubes in a sharps collector.

Take Care!   Do not look directly at the lamp.

Exercises

1 Write equations for the reactions undergone by your dipeptide during the dinitrophenylation and hydrolysis steps.

2 Tell whether the dipeptide you identified could be obtained by using cathepsin C to break down the polypeptide from the cobra venom illustrated in Figure 54.1, and explain your answer. The *N*-terminal amino acid is the one labeled "1."

3 Describe and explain the possible effect on your results of the following experimental errors or variations: (a) You failed to check the pH of the dinitrophenylation reaction mixture and it fell below 6. (b) You didn't add 6 *M* HCl to the dinitrophenylation reaction mixture before extracting it with diethyl ether. (c) You dissolved the DNP-dipeptide in 6 *M* NaOH rather than 6 *M* HCl before transferring it to the hydrolysis tubes.

4 Following the format in Appendix V, construct a flow diagram for the synthesis and hydrolysis of the DNP-dipeptide.

5 When you extracted the DNP-dipeptide hydrolysate with diethyl ether, why wasn't your *C*-terminal amino acid extracted along with your DNP-amino acid? Write equations explaining your answer.

6 (a) Propose a detailed mechanism for the reaction of 2,4-dinitrofluorobenzene with the dipeptide you identified. (b) With reference to the mechanism, explain why the reaction would not take place at low pH.

7 Will Bobble accidentally added aqueous sodium bisulfate instead of sodium bicarbonate to the dinitrophenylation reaction mixture just before the first ether extraction. His DNP-dipeptide hydrolysate produced no spots when analyzed by TLC. What happened to his product, and why?

8 Bea Wilder decided to save time by adding a large excess of sodium bicarbonate at the beginning of the dinitrophenylation reaction instead of adding smaller amounts during the reaction. When she developed her TLC plate with a solvent system containing butanol and ammonia, a large yellow spot showed up that didn't match any of the DNP-amino acid standards. What compound formed the yellow spot? Write an equation for its formation.

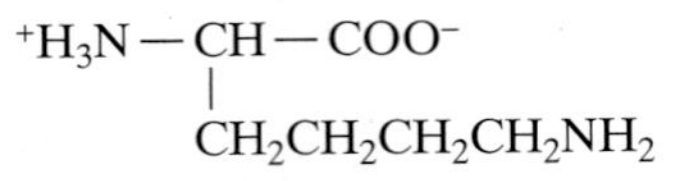

$$^+H_3N - CH - COO^-$$
$$|$$
$$CH_2CH_2CH_2CH_2NH_2$$

lysine

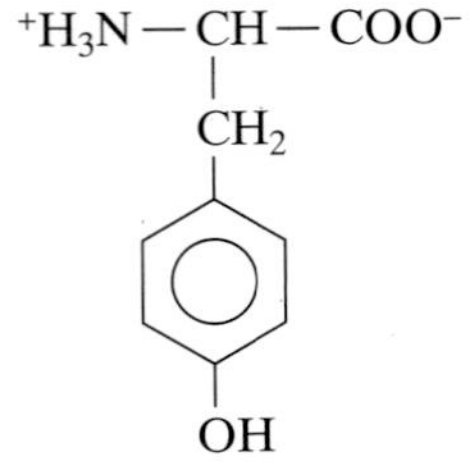

$$^+H_3N - CH - COO^-$$
$$|$$
$$CH_2$$

OH

tyrosine

9 (a) Some amino acids, such as lysine and tyrosine, yield dinitropheny-lated derivatives even when they are not at the end of a peptide chain. Explain, giving structures for the DNP derivatives. (b) Such derivatives do not usually interfere with the identification of the *N*-terminal DNP-amino acids since they are not extracted from the aqueous hydrolysis mixture at pH 1 by diethyl ether. Explain.

Other Things You Can Do

(Starred projects require your instructor's permission.)

*1 To confirm the identity of the *C*-terminal amino acid, evaporate the reserved aqueous layer from the DNP-dipeptide hydrolysate, add a drop of water to the residue, and identify the *C*-terminal amino acid by paper chromatography as described in the Directions.

*2 Isolate the protein casein from milk as described in Minilab 46.

3 Starting with sources listed in the Bibliography, write a research paper about primary structure determination of polypeptides and proteins, describing the applications of such reagents and enzymes as phenyl isothiocyanate, dansyl chloride, trypsin, chymotrypsin, cathepsin C, and cyanogen bromide.

Multistep Synthesis of Benzilic Acid from Benzaldehyde

Reactions of Carbonyl Compounds. Preparation of Carboxylic Acids. Nucleophilic Addition. Oxidation. Molecular Rearrangements.

Operations

OP-6   Making Transfers
OP-7   Heating
OP-10   Mixing
OP-12   Gravity Filtration
OP-13   Vacuum Filtration
OP-23   Drying Solids
OP-25   Recrystallization
OP-30   Melting Point
OP-36   Infrared Spectrometry

Before You Begin

1 Read the experiment, read or review the operations as necessary, and write an experimental plan.
2 Calculate the mass and volume of 15.0 mmol of benzaldehyde and the theoretical yields of benzoin, benzil, and benzilic acid expected from that much benzaldehyde.

Scenario

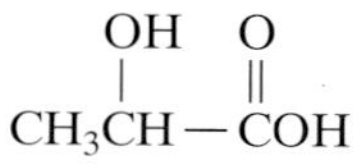

The α-hydroxy acids (AHAs) include lactic acid, which forms in milk as it sours and is also produced in muscles and blood after vigorous physical activity. Certain AHAs, such as glycolic acid ($HOCH_2COOH$), are used in anti-aging "wrinkle creams" that soften the skin and smooth out fine wrinkles and roughness. Golden Age Sundries is an organization that develops and markets various consumer goods for older people, and the success of AHAs in combating some effects of aging has led them to support research exploring the possible anti-aging benefits of other α-hydroxy acids. Your supervisor has just received a grant from them to investigate one of the AHAs, benzilic acid.

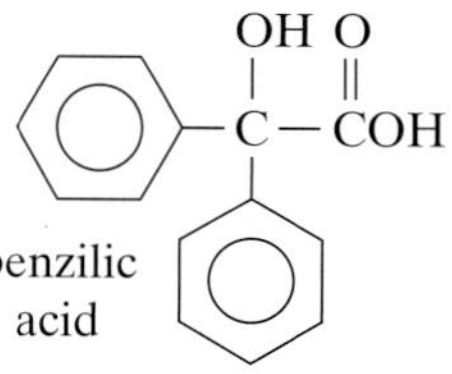

Benzilic acid can be prepared from an inexpensive starting material, benzaldehyde, by a three-step synthesis that involves a reaction called the benzoin condensation, an oxidation step, and a molecular rearrangement (see the "Reactions and Properties" section for the equations). The benzoin condensation has traditionally been carried out using cyanide ion as a catalyst, but because of the toxicity of cyanide your supervisor has decided to use Vitamin B_1 instead. Your assignment is to find out whether or not the vitamin is a suitable catalyst for the benzoin condensation and, if it is, to convert the benzoin you obtain to benzilic acid.

lactic acid

benzilic
acid

Applying Scientific Methodology

The scientific problem is outlined in the Scenario. You can solve it by the successful completion of the multistep synthesis described.

Justus von Liebig and the Bitter Almond Tree

Chemical warfare is generally thought of as being a recent and uniquely human invention, but, as with many such inventions, nature beat us to it. An otherwise unexceptional insect, the millipede *Apheloria corrugata*, discourages predators with a dose of poison gas powerful enough to kill a mouse. Many trees in the rose family, such as cherry, apple, peach, plum, and apricot trees, ensure their continued survival by protecting their seeds and foliage with cyanide-containing substances. There are cases on record of human fatalities from eating the seeds of these species; one man who considered apple seeds a delicacy died after eating a cup of them at one sitting. The chemical weapon in each of these examples is a natural cyanohydrin, mandelonitrile, which can be decomposed by enzymes or stomach acids into benzaldehyde and lethal hydrogen cyanide.

mandelonitrile   benzaldehyde   hydrogen cyanide

amygdalin

Laetrile

benzoin

In most cyanogenetic (cyanide-forming) plants, mandelonitrile is present as a carbohydrate derivative called a *glycoside*. The most common of these glycosides is amygdalin, which is an acetal of mandelonitrile and the carbohydrate gentiobiose. Amygdalin can be broken down enzymatically under certain conditions to yield mandelonitrile and its decomposition product, hydrogen cyanide. Closely related to amygdalin is the controversial cancer drug Laetrile, which allegedly kills malignant cells by releasing hydrogen cyanide (or possibly mandelonitrile) at the site of the malignancy. One of the most prolific sources of amygdalin is the bitter almond, which—unlike the sweet varieties used for human consumption—is grown for the oil that can be pressed from its seed kernels. The familiar "maraschino cherry" odor of almond oil comes from benzaldehyde formed by the breakdown of amygdalin. Both synthetic benzaldehyde and natural almond oil are used to flavor such food and beverage products as amaretto, cappucino, and marzipan.

In the early 1800s, distilled almond oil was treated with aqueous base to remove hydrogen cyanide and other impurities. This process yielded a small amount of a white solid that was subsequently identified as benzoin. Friedrich Wöhler and Justus von Liebig studied the reaction in 1832 and discovered that benzoin formation results from the catalytic action of sodium cyanide (formed when HCN from amygdalin reacted with the base) on benzaldehyde. Surprisingly, cyanide ion is almost the only substance that catalyzes this reaction, so the early discovery of the benzoin condensation depended on the coincidental production of both cyanide and benzaldehyde when bitter almond oil was washed with base.

Not long after his work on the benzoin condensation, von Liebig discovered yet another unusual reaction. Benzoin can be oxidized to the diketone benzil by a variety of oxidizing agents. When benzil is treated with potassium hydroxide, it rearranges to yield (upon acidification) benzilic acid. This reaction, called the benzilic acid rearrangement, is the oldest known molecular rearrangement, the prototype of a general class of rearrangements to electrophilic carbon atoms.

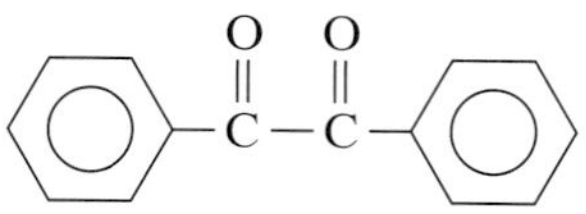

benzil

Understanding the Experiment

In this experiment, you will convert benzaldehyde to benzoin by the benzoin condensation, oxidize benzoin to benzil, and then convert benzil to benzilic acid. Because this is a multistep synthesis, you should try to keep material losses at a minimum in each step; otherwise, your overall yield will be quite low. It is essential that the benzaldehyde be pure because impure benzaldehyde contains benzoic acid, which inhibits the condensation reaction. Unless the benzaldehyde is from a previously unopened bottle or one that has been stored under nitrogen, it must be distilled before use.

For many years, it was believed that only cyanide ion could catalyze the benzoin condensation, but recently thiamine hydrochloride (Vitamin B_1) has proven to be an effective catalyst as well.

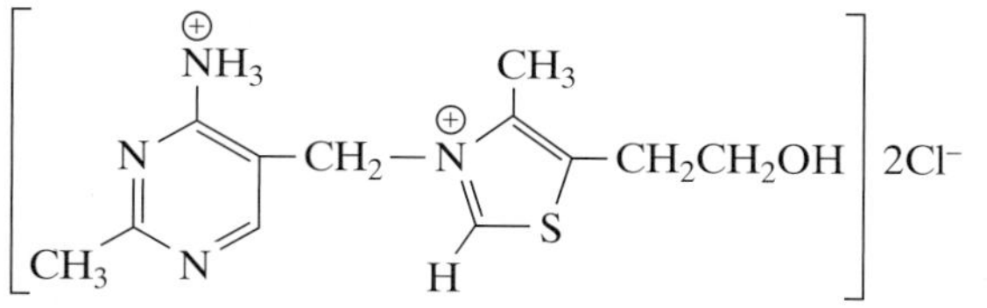

thiamine hydrochloride

It has the considerable advantage of being relatively hazard free, unlike the very poisonous cyanides. In the presence of sodium hydroxide, thiamine hydrochloride loses two protons to form a nucleophilic species that adds (with a proton) to the carbonyl group of benzaldehyde. The thiamine residue is sufficiently electron withdrawing to increase the acidity of the adjacent hydrogen atom, allowing its removal by the base to yield a resonance-stabilized carbanion, which undergoes nucleophilic addition to the carbonyl group of another benzaldehyde molecule. Loss of the thiamine residue from the product then yields benzoin.

Key Concept: Nucleophilic addition to an aldehyde or ketone involves nucleophilic attack at the carbonyl carbon followed by addition of a proton to the carbonyl oxygen.

Major steps in the mechanism of the thiamine-catalyzed benzoin condensation

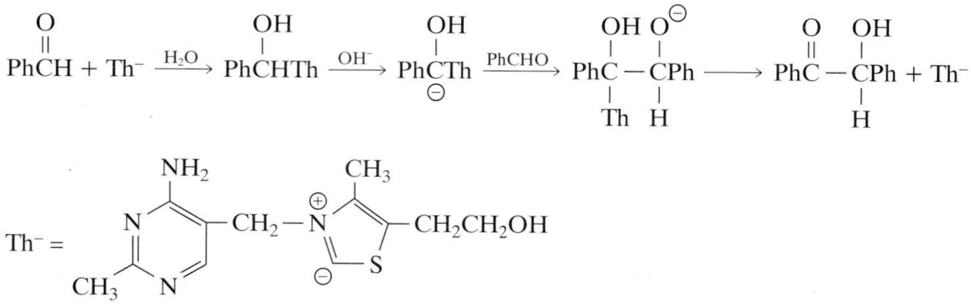

The benzoin condensation is carried out by heating a solution of benzaldehyde, sodium hydroxide, and thiamine hydrochloride in ethanol. To obtain a reasonable yield of benzoin, it is best to carry out the reaction for as long as possible (2–3 hours or more) during one lab period and let the reaction mixture stand until the next.

Benzoin can be converted to benzil by ammonium nitrate in the presence of a catalytic amount of copper(II) acetate. The benzoin is apparently oxidized by the direct action of copper(II) ions, which are reduced to copper(I) ions in the process. The ammonium nitrate then oxidizes the resulting copper(I) back to copper(II).

When treated with alcoholic potassium hydroxide, benzil undergoes the molecular rearrangement that was discovered by von Liebig. This rearrangement involves nucleophilic attack by hydroxide ion on one carbonyl carbon, followed by migration of the adjacent phenyl group to the other carbonyl carbon (see Exercise 2). The reaction initially yields a suspension of potassium benzilate, which should dissolve when the reaction mixture is heated. Benzilic acid is precipitated from the filtered solution by hydrochloric acid and purified by recrystallization from water.

You can gain additional experience in spectral interpretation by analyzing and comparing the infrared spectra of benzaldehyde and the products. These IR spectra should show bands from several different kinds of O—H, C—O, and C=O bonds, and in most of them you can observe aromatic C—H stretching vibrations with no interference from aliphatic C—H stretching bands.

Reactions and Properties

A   Benzoin condensation

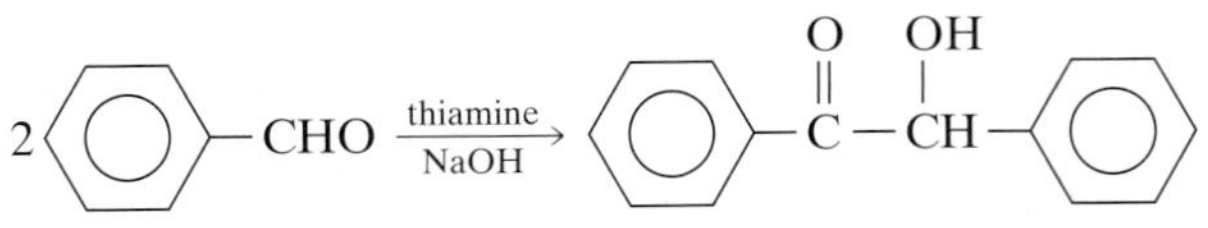

benzaldehyde   benzoin

B   Oxidation of benzoin

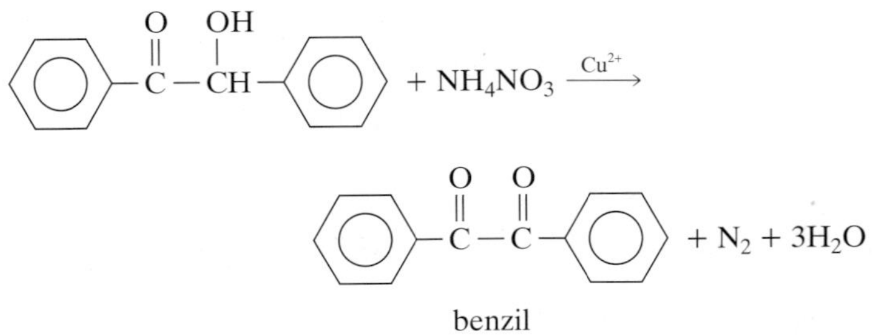

benzil

C   Benzilic acid rearrangement

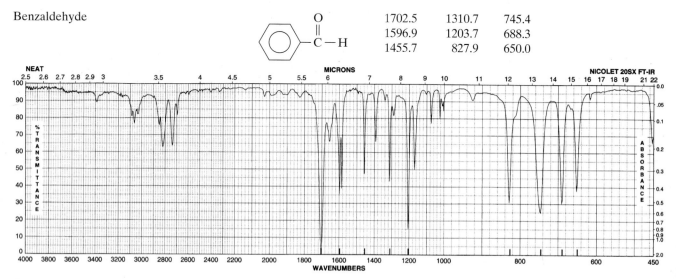

potassium benzilate

benzilic acid

Table 55.1 Physical properties

	M.W.	mp	bp	*d*
benzaldehyde	106.1	−26	178	1.042
thiamine hydrochloride	337.3	248 d		
benzoin	212.2	137		
benzil	210.2	95–6		
benzilic acid	228.2	151		
ammonium nitrate	80.0	170		
potassium hydroxide	56.1			

Note: mp and bp are in °C, density is in g/mL; d = decomposes. Potassium hydroxide pellets are about 85% KOH.

Benzaldehyde

1702.5	1310.7	745.4
1596.9	1203.7	688.3
1455.7	827.9	650.0

Figure 55.1 IR spectrum of benzaldehyde

Directions

A. *Preparation of Benzoin from Benzaldehyde*

The reaction should be started at the beginning of a lab period when a different experiment is being performed and allowed to proceed for as long as possible.

Safety Notes

methanol

Stop and Think: What impurity might benzaldehyde contain?

color change

Waste Disposal: Dispose of the filtrate as directed by your instructor.

Prod

mp, IR ↳ get prod

Safety Notes

ammonium   acetic acid
nitrate

Take Care! Avoid contact with acetic acid and do not breathe its vapors.

Stop and Think: What is the gas and where did it come from?

> Methanol is flammable and harmful if ingested, inhaled, or absorbed through the skin. Avoid contact with the liquid and do not breathe its vapors.

Reaction.   Dissolve 0.50 g of thiamine hydrochloride in 0.8 mL of water in a 25-mL Erlenmeyer flask. Add 5.0 mL of 95% ethanol and cool the solution in an ice/water bath. Add 1.0 mL of 3 *M* sodium hydroxide solution drop by drop, with stirring, over a 2-minute period. Test the pH of the mixture; if it is not strongly alkaline (9 or higher), add more 3 *M* NaOH dropwise until it is. Add 15.0 mmol of *pure* benzaldehyde to this solution. Seal the flask with Parafilm and heat [OP-7] the mixture in a 60° ± 5°C hot water bath until the end of the lab period. Then allow it to stand undisturbed in a dark place until the next lab period.

Separation and Analysis.   Scratch the inside of the flask to induce crystallization, if necessary, and then cool the reaction mixture in an ice/water bath until crystallization is complete. Collect the benzoin by vacuum filtration [OP-13] and wash it on the filter with cold water, then with two 1-mL portions of ice-cold methanol. Dry [OP-23] and weigh the product before proceeding to part **B**. Measure its melting point [OP-30] and record its IR spectrum [OP-36], or save enough product to do both later.

B. *Preparation of Benzil from Benzoin*

> Ammonium nitrate may cause a fire or explosion if it is heated strongly or allowed to contact combustible materials. Keep it away from other chemicals and combustibles.
> Acetic acid causes chemical burns that can seriously damage skin and eyes. Its vapors are highly irritating to the eyes and respiratory tract. Wear gloves and dispense under a hood; avoid contact and do not breathe its vapors.

Reaction.   Multiply the mass of dry benzoin by 0.5 and combine that mass of ammonium nitrate with 3.5 mL of aqueous 80% (volume/volume) acetic acid in a 10-mL round-bottom flask. Mix in the benzoin and 0.015 g of copper(II) acetate. Drop in a stir bar and attach a condenser. Heat the flask while stirring [OP-10] to start the reaction, which should be accompanied by evolution of a gas. When the gas evolution subsides, heat the solution to the boiling point, then heat it under gentle reflux [OP-7] for an hour or more.

Separation.   Cool the reaction mixture to 50°C. Transfer it, with manual stirring, to a small beaker containing 8 g of crushed ice, using a small amount of water for the transfer. Allow it to stand until the ice melts.

Then collect the benzil by vacuum filtration [OP-13] and wash it twice with cold water.

Purification and Analysis.   Recrystallize [OP-25] the benzil from 95% ethanol, washing the yellow crystals with cold aqueous 50% ethanol. Dry [OP-23] it at a low temperature (50°C or less), and measure its mass. Measure its melting point [OP-30] and record its IR spectrum [OP-36], or save enough product to do both later.

Waste Disposal: Pour all of the filtrate down the drain.

C. *Preparation of Benzilic Acid from Benzil*

> **The solution of potassium hydroxide in ethanol is flammable, toxic, and corrosive; it can cause severe damage to the eyes, skin, and respiratory tract. Wear gloves, avoid contact, and do not breathe its vapors. Hydrochloric acid is poisonous and corrosive. Contact or inhalation can cause severe damage to the eyes, skin, and respiratory tract. Wear gloves and dispense under a hood; avoid contact and do not breathe its vapors.**

Safety Notes

potassium   hydrochloric
hydroxide   acid

Reaction.   Multiply the mass (in grams) of dry benzil by 2.5 and measure out that *volume*, in milliliters, of 6 *M* potassium hydroxide. Then multiply the mass of benzil by 3.0 and measure out that *volume*, in milliliters, of 95% ethanol. Combine the benzil with these liquids in a round-bottom flask or conical vial, add a stirring device, and attach a condenser. Heat the mixture under gentle reflux [OP-7] with stirring [OP-10] for 15 minutes.

Separation.   Pour the *hot* reaction mixture, with manual stirring, into 10 mL of water in a small beaker and let the mixture stand for a few minutes. Then heat it to 50°C, with stirring [OP-10], to dissolve the potassium benzilate. There may be a colloidal suspension of unreacted benzil or by-products at this point, but most of the solid should dissolve. If it does not, add more warm water. Add about 50 mg of pelletized Norit and 25 mg of filtering aid (Celite), stir or swirl the mixture at 50°C for 2 minutes, and filter the warm solution by gravity [OP-12] into a suitable container. *Under the hood,* carefully add 1.5 mL of concentrated hydrochloric acid to 10 g of crushed ice in a beaker. Add 1.5 mL of the potassium benzilate solution to the cold HCl solution with manual stirring and scratch the sides of the beaker until crystals begin to form. Then add the rest of the solution slowly with continuous stirring. When the addition is complete, test the solution with pH paper. If the pH is higher than 2, add enough HCl to bring it down to 2. Cool the solution in ice water and collect the benzilic acid by vacuum filtration [OP-13], washing it with cold water.

Take Care!   Wear gloves, avoid contact with HCl, and do not breathe its vapors.

Purification and Analysis.   Purify the benzilic acid by recrystallization [OP-25] from boiling water in a small Erlenmeyer flask. If undissolved impurities remain, add enough water to prevent premature crystallization, heat the solution to boiling, and transfer [OP-6] it to a suitable container with a preheated filter-tip pipet. Dry [OP-23] and weigh the purified product. Measure the melting point [OP-30] of the benzilic acid and record its IR spectrum [OP-36], or obtain the spectrum from your instructor. Calculate the percent yield for each step of the synthesis and the overall percent yield of benzilic acid.

Waste Disposal: Pour the filtrate down the drain.

Waste Disposal: Pour the filtrate down the drain.

Exercises

1   Interpret the IR spectra of the products as completely as you can and describe the evidence suggesting that each reaction step has taken place as expected.

2   Propose a mechanism for the rearrangement of benzil to potassium benzilate in the presence of potassium hydroxide.

3   Describe and explain the possible effect on your results of the following experimental errors or variations: (a) The bottle of benzaldehyde for part **A** contained some white solid. (b) The label on the copper acetate bottle for part **B** read "cuprous acetate." (c) The pH of the solution in part **C** was 6 when you vacuum filtered it.

4   Following the format in Appendix V, construct a flow diagram for this experiment.

5   Write a detailed mechanism for the cyanide-catalyzed condensation of benzaldehyde, and explain the role played by the cyanide.

6   Write a synthetic pathway illustrating the preparation of Compound **1** from a suitable five-carbon starting material, using the reactions described in this experiment.

7   Benzil reacts with urea in the presence of sodium hydroxide to form the sodium salt of 5,5-diphenylhydantoin (Dilantin Sodium), a powerful anticonvulsant used to treat epilepsy. By analogy with the benzilic acid rearrangement, propose a mechanism for this reaction.

8   In the presence of sulfuric acid, benzilic acid reacts with acetone to form a compound with the molecular formula $C_{17}H_{16}O_3$. Propose a structure for this product and write a balanced equation for the reaction.

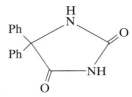

1

5,5-diphenylhydantoin

Other Things You Can Do

(Starred projects require your instructor's permission.)

*1   Test both your crude and purified benzil for the presence of unreacted benzoin by dissolving a few crystals of each in 1 mL of ethanol and adding a drop of dilute sodium hydroxide solution. Test a small amount of benzoin in the same way and compare the results. Shake the solutions with air, let them stand, and record your observations.

*2   Carry out another reaction of benzaldehyde as described in Minilab 35.

3   Starting with sources listed in the Bibliography, write a research paper about the use of α-hydroxy acids, Retin-A, and other skin care products to treat skin conditions, giving some of the advantages and disadvantages of each.

Using the Chemical Literature in an Organic Synthesis

Multistep Synthesis. Searching the Chemical Literature.

Before You Begin

1 Read the experiment and Appendix VII, "The Chemical Literature," and familiarize yourself with the layout of the Bibliography.

2 In your laboratory notebook, outline a synthetic pathway leading to the target compound from readily available starting materials or the starting material(s) recommended by your instructor. For the chemicals you will be using and the target compound, list information concerning their physical properties, purification, safety and health hazards, and disposal. If the synthesis requires laboratory techniques with which you are not familiar, find information about these techniques in the literature, describe them, and sketch any special apparatus required for them. Locate any published spectra or other data that will help you characterize the target compound, and list their sources.

3 In your laboratory notebook, write a complete description of your plan for the synthesis, including a detailed procedure for each synthetic step (with safety precautions) and a description of the methods you propose to use to analyze the purity and verify the structure of the product. The procedure should be designed to yield about 0.3–0.5 g of the final product, unless your instructor indicates otherwise. Your write-up should also include a concise description of the experimental methodology. Submit your laboratory notebook (or duplicate pages) to your instructor for approval well in advance of the scheduled lab period. You should also discuss your plans with the instructor to make sure that the necessary chemicals and equipment are available, that the procedure can be carried out safely under the existing conditions, and that you have a good understanding of the experiment and any problems that might arise.

Scenario

Your supervisor has a backlog of chemicals that need to be prepared for various clients, but doesn't have time to write up the procedures for you. You are on your own!

Applying Scientific Methodology

In the previous experiments, the scientific problem to be solved was specified and a course of action was laid out for you. For this experiment, you will have to operate like a research chemist—find a suitable scientific problem to work on, plan a course of action, work out detailed experimental procedures, carry out your experimental plan in the laboratory, and evaluate your results. Throughout, you will be expected to apply what you have already learned about organic synthesis in this course and what you can learn from the chemical literature.

Planning Your Synthesis

Your instructor will either suggest a target compound for you to synthesize or allow you to choose your own, subject to approval. Your instructor will also inform you whether you will work individually or in teams. If you are to select your own target compound, you should consult your instructor before coming to a final decision, since your plans may be either too ambitious or not challenging enough. You can search for ideas in chemical periodicals such as the *Journal of Chemical Education*, in other laboratory manuals, or in many of the sources described herein (see Appendix VII for more detailed information on these and other sources). Preferably, the target should be a compound that you have an interest in and really *want* to synthesize, such as an artificial flavor ingredient, a cosmetic ingredient, a perfume ingredient, an insect repellent, a plant-growth hormone, an analgesic drug, a pheromone, an artificial sweetener, a sunscreen, an antibiotic drug, a local anesthetic, a fluorescent dye, a chemoluminescent compound—the possibilities are limitless!

Once you have selected a target compound, you will have to outline a synthetic pathway for making it from some readily available starting material. To help you work out a synthetic pathway, you can consult your lecture textbook or refer to the Bibliography to find such sources as *Organic Synthesis: The Disconnection Approach* [B28] and *Principles of Organic Synthesis* [B18].

Sources from the Bibliography are referred to herein by category and number. For example, Organic Syntheses *[B20] is the twentieth entry under Category B,* Organic Reactions and Syntheses.

After you develop one or more promising synthetic pathways, you should learn more about the reactions required for the synthesis. *Modern Synthetic Reactions* [B12] and advanced organic chemistry texts by March [H5] and Carey-Sundberg [H1] describe a number of synthetic reactions, and also give literature references to specific synthetic procedures. *Organic Reactions* [B19] is a comprehensive source of information about specific reactions and includes some representative synthetic procedures, as well as numerous references to literature procedures. It describes the applications and limitations of many synthetic reactions and may help you tailor a synthetic procedure to fit the particular starting materials you will be working with. Many reactions and compound types are described in depth in review articles from such journals as *Chemical Reviews* and *Angewandte Chemie (International Edition in English)*.

Two excellent sources of detailed information about chemical reagents are Fieser's *Reagents for Organic Syntheses* [B9] and *Encyclopedia of Reagents for Organic Synthesis* [B21]. General reference books, such as *The Merck Index* [A10], *Lange's Handbook* [A5], the *CRC Handbook of Chemistry and Physics* [A14], the *Dictionary of Organic Compounds* [A11], and the *Aldrich Catalog* [A1], can be consulted for information about the physical properties of the reagents, as well as other chemicals (including solvents) involved in your synthesis.

The *Aldrich Catalog* also gives information about specific hazards associated with chemicals. More detailed information about chemical hazards can be found in *Sax's Dangerous Properties of Industrial Materials* [C5], the *Sigma-Aldrich Library of Chemical Safety Data* [C4], and other sources from Category C of the Bibliography. You should also obtain and read the Material Safety Data Sheets (MSDS) for any hazardous chemicals you will be working with. Information about safe laboratory practices may be found in *Prudent Practices in the Laboratory* [C7], which provides information about dealing with laboratory wastes as well.

There are many useful collections of detailed synthetic procedures. One of the best (and the first you may want to consult) is *Organic Syntheses* [B20], which provides a representative selection of carefully tested procedures. Other useful, if less comprehensive, sources include *Vogel's Textbook of Practical Organic Chemistry* [B29], *Organic Functional Group Preparations* [B25], and other works from Category B of the Bibliography, as well as a variety of basic and advanced organic chemistry lab textbooks. For more information about these and other useful sources on organic reactions and syntheses, as well as suggested literature searching strategies, read the three-part series "Information Sources for Organic Chemistry" [K7].

Most literature procedures involve the use of standard scale apparatus and relatively large quantities of chemicals, so you will probably have to scale down and otherwise revise such a procedure substantially before you can use it. For example, if a literature procedure calls for 12 g of acetophenone and you want to start with 0.60 grams, you should multiply the quantities of all reactants, reagents, catalysts, solvents, and other chemicals by the scaling factor 0.050, which is the ratio of 0.60 g to 12 g. You will also need to substitute appropriate microscale apparatus and techniques for the corresponding standard scale apparatus and techniques. For this purpose, you can compare the operation descriptions in a standard scale lab manual with the microscale versions of those operations in Part V of this book. Most scaled-down syntheses do not give as high a percent yield as the original procedure. For example, losing 20 mg of product during a transfer has a greater effect on the percent yield of a microscale experiment than it does when you have more product to work with. Other factors, such as differences in surface-contact area or the rate of heat transfer, may also cause differences in the outcome of a scaled-down experiment.

If you can't locate a suitable synthetic procedure for a specific synthetic conversion from the sources described previously, you may have to adapt a procedure for the same reaction of a closely related compound or search for a suitable procedure in the primary literature. Adapting a procedure is often just a matter of recalculating the masses of the reactants. For example, suppose you want to adapt a literature procedure for the $NaBH_4$ reduction of 3.00 g (30.5 mmol) of cyclohexanone (M.W. = 98.2) to the same reaction of 4-methylcyclohexanone. 4-Methylcyclohexanone has a molecular weight of 112.2, so the calculated mass of 30.5 mmol of this reactant is 3.43 g; this is the amount of 4-methylcyclohexanone that would be needed if all other quantities were kept the same. The quantities of all reactants, solvents, etc., can then be scaled down by multiplying them all by a scaling factor, as described previously. Such an adaptation may also require changes in such things as the amount or kind of recrystallization solvent required and the boiling range for a distillation, so it is important to identify changes in the properties of the reactants and products that necessitate procedural changes.

References to literature procedures for organic syntheses are given in many of the sources cited previously, such as *Organic Reactions* and the textbooks by March and Carey-Sundberg. *Theilheimer's Synthetic Methods of Organic Chemistry* [B30] is an extensive guide to synthetic procedures that is updated each year. Each entry gives a brief outline of the procedure and cites its source. Other useful, if less complete, sources include *Synthetic Organic Chemistry* [B32] (which covers many of the older reactions), *Comprehensive Organic Transformations* [B13], and *Compendium of Organic Synthetic Methods* [B11].

The most comprehensive guides to the chemical literature of organic chemistry are *Chemical Abstracts* (*CA*) [J3] and *Beilstein* [A3]. With some practice, you can use them to find information about every aspect of your synthesis. For example, to find out more about your target compound, you should first find its *CA* index name and registry number in the *Index Guide*, which appears with the collective indexes that are published periodically by *Chemical Abstracts*. Then you can conduct a search of *Chemical Abstracts*—from an on-line database, if one is available—to locate articles that refer to the target compound. Some of them should give procedures for its preparation. See Appendix VII for more information about the use of *CA*, *Beilstein*, and on-line searches.

If you locate a specific procedure for the synthesis of your target compound in the primary literature, you should not assume that it will be suitable for your purposes as written. It may call for unavailable or excessively hazardous chemicals or suggest techniques and apparatus that are not appropriate to your laboratory situation. Primary literature procedures tend to be rather condensed, so you will probably have to "fill in the blanks" from your own experience or by consulting other sources.

If you are unfamiliar with some of the laboratory techniques required for a synthesis, consult *Guide for the Perplexed Organic Experimentalist* [D2], the appropriate volume in Weissberger's *Technique of Organic Chemistry* [D7], *Encyclopedia of Separation Science* [D10], or other references from Category D and E of the Bibliography. If you need to purify a solvent or another chemical, see *Purification of Laboratory Chemicals* [D5] or another source from Category D.

Once you have synthesized your target compound (or what you think is your target compound), you will have to confirm its identity by measuring one or more physical properties, such as its melting point or boiling point, and obtaining and interpreting at least one kind of spectrum. Refer to the appropriate sources in Category F if you need information about spectral analysis and the interpretation of spectra. *Spectrometric Identification of Organic Compounds* [F19] is a good place to start. There are many sources of published spectra, such as the *Aldrich Library of FT-IR Spectra* [F17] and the *Aldrich Library of ^{13}C and ^{1}H FT-NMR Spectra* [F15]. Additional information about spectrometric and "wet" chemical methods for identifying organic compounds can be found in sources from Category G. You may also need to assess the purity of your product by gas chromatography, TLC, HPLC, or another method. Refer to Category E of the Bibliography for appropriate sources. For additional help in carrying out a literature search, consult Appendix VII and appropriate sources listed in Category K of the Bibliography.

Directions

Independent synthesis projects are intended for advanced or honors students who have mastered the major operations described in Part V. These directions only furnish general guidelines regarding your synthesis. You must provide the procedural details and have them approved by your instructor. Keep detailed notes of your work and observations as described in Appendix II.

Safety Notes

> **Except when you have reliable information to the contrary, assume that all chemicals you will use are flammable and are hazardous by ingestion, inhalation, and skin absorption. Wear gloves and use a hood whenever possible, avoid contact with the chemicals, do not breathe their dust or vapors, and keep them away from ignition sources and other chemicals that might react with them.**

If necessary, purify any starting materials, reagents, or solvents that appear to be insufficiently pure. Carry out the synthesis in the laboratory according to the procedure you have developed. Purify the product, weigh the thoroughly dried product, and calculate the percent yield. Measure appropriate physical constants of the product and assess its purity by an appropriate chromatographic method, if requested. Confirm the identity of the product by one or more spectrometric methods, or by preparing one or more derivatives, or both.

Report.   Write up your report as if it were a scientific paper being submitted to a professional publication such as the *Journal of Organic Chemistry*. Such papers are traditionally written in an impersonal, objective style using the passive voice (for instance, "the solution was stirred" rather than "I stirred the solution"). Your report should include the following items, unless your instructor directs otherwise:

1 A brief but descriptive title.
2 Your name(s) and affiliation.
3 A brief abstract that summarizes the principal results of the work.
4 An introductory section (usually untitled) that provides a concise statement of the purpose and possible applications of the work, supported by descriptions of related work from the literature. If your synthesis differs significantly from those reported in the literature, tell how and why.
5 A section titled "Experimental Methods" that gives enough detail about your materials and methods so that another experienced worker could repeat your work. Give current *Chemical Abstract* index names and registry numbers (see Appendix VII) for important starting materials and the product, and provide information about their purity if possible. Note any significant hazards and safety precautions in a separate paragraph labeled "Caution."
6 A section titled "Results and Discussion" (or two separate "Results" and "Discussion" sections) that summarizes the important experimental results and points out any special features, limitations, or implications of

your work. You should include any data (including spectral parameters) that will help justify your conclusion. You may also wish to suggest different approaches to the problem or areas that require further study.

7 A section titled "Conclusions" that states any conclusions you can draw from your work, based on the evidence presented.

8 A section titled "References" that gives complete citations for all literature sources referred to in the report.

Your instructor may suggest additions to or modifications of this list. *The ACS Style Guide* [L14] provides a more detailed discussion of these components, as well as helpful suggestions about writing style.

Exercises

1 Herbert C. Brown was a co-recipient (with Georg Wittig) of the 1979 chemistry Nobel prize for his work with reagents for organic syntheses, principally organoboranes. Give the structure of Brown's reagent 9-BBN (9-borabicyclo[3.3.1]nonane) and describe some of its applications and characteristics, giving equations where appropriate. Cite several papers reporting the use of 9-BBN.

2 The synthesis of dodecahedrane has been described as the "Mount Everest of alicyclic chemistry." (a) Give its *Chemical Abstracts* index name and registry number and draw its structure. (b) Locate the paper in which the synthesis of dodecahedrane was first reported. Give its title, the authors and their affiliation, and a standard literature citation showing where and when the paper appeared. (c) Summarize the salient points of the synthesis in your own words, specifying the starting material, the number of synthetic steps required, the yield of dodecahedrane, and the purification method. Tell how dodecahedrane was characterized, and report any spectral parameters. (d) Find a systematic name for dodecahedrane that is different from its *CA* index name. (e) Quote and explain the reference, in the paper, to an ancient Greek philosopher.

3 (a) Cite the paper in which the use of pyridinium chlorochromate to oxidize alcohols to carbonyl compounds was first reported. (b) Tell how pyridinium chlorochromate is prepared, describe a typical experimental procedure for oxidation of a primary alcohol to an aldehyde, and discuss the stoichiometry of the reaction. (c) Report on any hazards associated with the use of pyridinium chlorochromate, and describe safe disposal procedures for the reagent.

4 (a) Give a concise definition of the Knoevenagel condensation. (b) Describe typical experimental conditions for conducting a Knoevenagel condensation between an aldehyde and diethyl malonate. (c) Find and summarize a detailed procedure for the synthesis of ethyl coumarin-3-carboxylate using this reaction. (d) Describe the Doebner modification of the Knoevenagel condensation, giving at least one example.

5 (a) Describe any hazards associated with the use of thionyl chloride, and describe proper handling precautions and disposal procedures for this reagent. (b) Tell how thionyl chloride can be purified for use as a chemical reagent. (c) Cite a paper in which thionyl chloride was used to convert an amino acid to an ester in one step, and briefly describe the experimental conditions.

6 (a) Give the current *Chemical Abstracts* index name and registry number for (+)-camphor. (b) Draw a structure for (+)-camphor that shows its absolute configuration. (c) What is the melting point of the oxime of (+)-camphor? (d) Find an infrared spectrum for camphor and give the wave numbers of the major absorption bands. (e) Tell where you can find information about camphor in *Beilstein*. (f) Tell how most synthetic camphor is currently produced, giving equations for the reactions.

7 (a) Give the *CA* index name of the compound whose *CA* registry number is [5543-57-7]. (b) Give the trivial name of this compound and describe its major application. (c) Tell where information about this compound can be found in *Beilstein*.

8 For each of the following abbreviated names, provide the full name of the journal, any names under which it was previously published, the year in which it was first published (as volume 1, under a current or previous name), the language or languages in which it is published, and a nearby library that carries it. (a) *Acc. Chem. Res.* (b) *Helv. Chim. Acta.* (c) *Dokl. Akad. Nauk SSSR.* (d) *Chem. Ber.*

9 Find detailed synthetic procedures for the following compounds, briefly describe the experimental conditions, and write equations for the relevant reactions. (a) hexaphenylbenzene, (b) vanillic acid, (c) 1,2-cyclononadiene, (d) octadecanedioic acid.

10 For each of the following compounds, find and reproduce as many different kinds of published spectra (or spectral parameters) as you can. (a) mandelic acid, (b) resorcinol, (c) exaltone, (d) bourbonal, (e) testosterone.

A Research Project in Organic Chemistry

Research in Chemistry.

Before You Begin

1 Read the experiment and review the "Scientific Methodology" section of the Introduction. Review Appendix VII as necessary.
2 Write a research proposal based on one of the projects discussed in this experiment or on a project that you have developed yourself.
3 In your laboratory notebook, write a complete description of your research plans, including a detailed procedure for each experiment you intend to perform, with safety precautions. Submit your research proposal and lab notebook (or duplicate pages) to your research mentor.

Scenario

Your job with the Consulting Chemists Institute is nearly finished. Your supervisor, having a high regard for your abilities and potential, has encouraged you to consider making chemistry your career. To obtain employment as a professional chemist, you will probably have to earn a master's degree or a Ph.D. in chemistry. One of the requirements for your degree is the completion of an independent research project. You have decided to get a head start by carrying out an undergraduate research project under the direction of your supervisor or another faculty mentor. Now you must select a project to work on.

Applying Scientific Methodology

For this experiment, you will have to operate like a professional chemist—find a suitable scientific problem to work on, develop a hypothesis, plan a course of action, work out detailed experimental procedures, follow your experimental plan in the laboratory, evaluate your results, and arrive at a conclusion. Throughout, you will be expected to apply what you have already learned about organic chemistry in this course and what you can learn from the chemical literature.

Conducting Scientific Research

Conducting worthwhile research in chemistry requires a curious and logical mind, efficient work habits, superior laboratory skills, a good understanding of chemical concepts, and a great deal of perseverance. There is no single method to be followed in scientific research; different scientists often employ vastly different approaches. But the outline of scientific methodology in the Introduction to this book lists the important steps in a scientific research project.

- Define the problem
- Plan a course of action
- Gather evidence
- Evaluate the evidence
- Develop a hypothesis
- Test the hypothesis
- Reach a conclusion
- Report the results

First, you need to decide what problem to investigate. Some examples of possible research problems are discussed in the next section, but you may wish to select your own problem to work on, after consultation with your mentor. It should, above all, be some aspect of organic chemistry that interests you intensely; without such motivation it will be difficult for you to expend the time and effort required to complete a research project successfully. If possible, it should be in an area in which your mentor, or other chemists at your institution, have considerable interest and expertise. That will help you benefit from the guidance and enthusiasm they can provide. One way to come up with ideas for a research project in organic chemistry is to think about things that puzzled you during the lab or lecture course, such as unexplained lab observations, theories or mechanisms that don't account for certain experimental results, incomplete or inconsistent explanations of phenomena, and gaps in textbook descriptions of some aspect of organic chemistry. After consulting the chemical literature, you may find that there are already explanations for such phenomena—no textbook or course can cover the entire field of organic chemistry—but in the process of looking for explanations, you may find other problems worth exploring. You can also get ideas by looking through chemistry periodicals such as the *Journal of Chemical Education,* which occasionally contains articles describing potential research projects, and review journals or research reports, such as *Angewandte Chemie (International Edition in English), Chemical Reviews,* or *Annual Reports on the Progress of Chemistry, Section B* [J2]. When you do find an article dealing with a subject that interests you, be sure to check the citations in that article to locate more articles on the same subject. The important thing is to read and learn as much as you can about the areas that interest you; the more you learn, the easier it is to select a worthwhile problem.

Sources from the Bibliography are referred to herein by category and number. For example, Organic Syntheses *[B20] is the twentieth entry under Category B,* Organic Reactions and Syntheses.

Before you start working on any kind of scientific problem, you need to find out what research has already been carried out that is related to the problem. Therefore, you should do a thorough search of *Chemical Abstracts* [J3], and perhaps of other sources such as *Beilstein* [A3] as well. See Appendix VII for information about both of these references and about on-line searches of *Chemical Abstracts.* If, after a literature search, you learn that a problem you were planning to work on has already been "solved" satisfactorily, there is little point in working on the same problem. If only certain aspects of a problem have been investigated, you may decide to modify your objectives and concentrate on the areas in which original research can still be performed. And learning what has already been done on problems related to your own will help you fine-tune your objectives and plan your course of action.

The course of action you develop will depend on the kind of problem you select. If the problem requires a multistep synthesis, for example, you should read Experiment 56, "Using the Chemical Literature in an Organic Synthesis." In any case, you will need to consult the chemical literature to help you design your experimental approach and work out detailed procedures for all experimental work. Start by reading the appropriate sections of Appendix VII and consulting the Bibliography to find out about some of the sources that are available to you. You may also want to consult one or more of the sources in Category K of the Bibliography for more complete and detailed information about the chemical literature. If you will need special chemicals, supplies, or equipment, first check with your mentor or the stockroom manager to find out if they are available. If chemicals must be purchased, check *Chem Sources*, which lists the chemicals manufactured by hundreds of chemical companies, or consult the *Aldrich Catalog* [A1] or another chemical catalog. If supplies need to be purchased, check *The VWR Scientific Products Catalog*, the *Fisher Catalog*, or another chemical supplies catalog. Your plans and procedures should be written up in detail in your laboratory notebook, summarized on a form like the one in Figure 57.1, and submitted to your instructor for approval.

Other aspects of scientific methodology previously listed—gathering and evaluating evidence, developing and testing hypotheses, and arriving at a conclusion—are described in the *Scientific Methodology* section of this book.

Your plans, ideas, experimental work, significant observations, results, and other aspects of your research should be recorded faithfully in your laboratory notebook. Suggestions for maintaining and using a lab notebook effectively are covered in Appendix II and in *Writing the Laboratory Notebook* [L42].

Unless your instructor indicates otherwise, you should report your results as if they were to be published in a scientific journal—and they could be, since the results of undergraduate research projects are sometimes accepted for publication.

Suggested Research Projects

Many of these projects are open ended and can lead to further investigations not suggested in the outlines. Although some of them may not lead directly to new or unique contributions to the sum of chemical knowledge, others could yield hitherto unknown facts and new discoveries. Each project is outlined only briefly, so that you will have considerable latitude in developing and pursuing a course of action. Each project names one or more key references that will often have citations to other articles on the same subject.

1   Isolating and Testing a Natural Growth Inhibitor
Key Reference: *J. Chem. Educ.* **1977**, *54*, 156.

It has been speculated that juglone, a natural quinone that occurs in walnut trees, acts as a chemical defense agent for walnut seedlings, favoring their growth at the expense of competing species, particularly plants of the heath family (genus *Ericaceae*). You could isolate juglone or another substance that affects plant growth from a natural source and test its effect on the growth of various species or varieties of plants. Such substances could

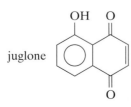

juglone

Name _____ Date submitted _____

Name of Project _____

Describe the scientific problem and discuss its significance.

State what you expect to accomplish and, in general terms, how you plan to accomplish it.

Describe, in approximate chronological order, specific experiments you expect to perform. In each case, give a rough estimate of the time you expect each experiment to take.

Describe any special chemicals, supplies, and equipment that will be needed. If any must be purchased, estimate their cost.

Approved by (mentor) _____ on (date) _____

Figure 57.1 Research project summary page

also be modified chemically to see what effect, if any, the modifications have on their growth-regulating properties.

2   Analysis of an Essential Oil

Key References: *J. Chem. Educ.* **1994**, *71*, A146. *J. Chem. Educ.* **1969**, *46*, 846.

Essential oils can be obtained from many plants by steam distillation, expression, and other isolation techniques. They can also be purchased in some health food stores and other stores that sell herbs and new-age supplies. You could obtain an essential oil and analyze it by gas chromatography to see if it contains a few major components that might be easily separated and identified. Its components could be separated by preparative gas chromatography, preparative HPLC, or another method. You may then be able to identify one or more components by IR, NMR, GC-MS, or some other spectrometric method.

3   Medicinal Components of Indigenous Wild Plants

Key References: S. Foster and J. A. Duke, *A Field Guide to Medicinal Plants: Eastern and Central North America* (Boston: Houghton Mifflin, 1990). R. Ikan. *Natural Products: A Laboratory Guide,* 2nd ed. (San Diego: Academic Press, 1991).

Many indigenous wild plants were used by Native Americans for medicine, and some still have medicinal uses. You could select a plant that appears to show significant pharmacological activity and attempt to isolate and identify one or more of its components. Keep in mind that the plant you select must be abundant enough to provide the large amounts of plant material that may be needed. In some cases, it may be possible to isolate major components by extraction or steam distillation, analyze them by gas chromatography, separate one or more components using basic laboratory operations, and identify them by IR or NMR spectrometry. But since most plants contain many diverse and complex constituents, it may be necessary to use advanced instrumental methods, such as high-performance liquid chromatography (HPLC) or gas chromatography-mass spectrometry (GC-MS), to separate and identify the components of a species you have selected. Unless you have access to such instruments, you may have to analyze the essential oils or extracts of several different species using gas chromatography or HPLC until you find one that is simple enough to work with.

4   The Effect of Molecular Modification on Odor

Key References: *J. Chem. Educ.* **1992**, *69*, A43. R. W. Moncrieff, *The Chemical Senses,* 3rd ed. (Cleveland, Ohio: CRC Press, 1967).

Relatively minor modifications in molecular structure can have drastic and often unpredictable effects on odor. Select an organic compound that has a discernible odor and two or more functional groups, such as *trans*-cinnamaldehyde or 4′-hydroxyacetophenone, and modify its molecules in various ways to find out how such modifications affect its odor.

trans-cinnamaldehyde   4′-hydroxyacetophenone

Note that the presence of several functional groups may cause complications in certain reactions unless one of them is protected. See *Protective Groups in Organic Synthesis* [B10] or a related source for information about the use of protective groups.

5   Gas Chromatographic Analysis of Gasoline
Key References: *J. Chem. Educ.* **1996**, *73*, 1056. *J. Chem. Educ.* **1976**, *53*, 51.

The compositions of different grades of gasoline and gasoline from different suppliers can vary considerably. Most gas stations label their gasoline accurately, but a few have been known to sell unleaded regular gasoline as premium gasoline, at premium prices. Some suppliers add certain oxygenates (oxygen-containing components) to their gasoline, such as methyl *t*-butyl ether (MTBE), which is a subject of environmental concern. You can analyze samples of gasoline from different sources by gas chromatography to look for oxygenates or for differences in their gas chromatographic hydrocarbon profile.

$$CH_3OCCH_3$$

MTBE

6   Stereochemistry of Addition Reactions
Key Reference: *J. Chem. Educ.* **1990**, *67*, 554.

Most bromine addition reactions appear to proceed through a bromonium ion intermediate that yields *anti* addition exclusively, but certain unsaturated substrates, such as *trans*-anethole, yield substantial amounts of *syn*-addition products. You can investigate the effect of substrate structure on stereochemistry by selecting one or more unsaturated compounds, carrying out a bromine addition reaction, and analyzing the product. You can explore the effect of reaction conditions on stereochemistry by using different brominating reagents, varying the solvent, or varying the reaction temperature. You might also extend your study to stereoselective reactions other than bromine addition.

7   Alkylation of a Bidentate Nucleophile
Key Reference: *J. Chem. Educ.* **1990**, *67*, 611.

Under different reaction conditions, the alkylation of sodium saccharin with ethyl iodide can yield varying proportions of *N*-ethylsaccharin and *O*-ethylsaccharin (see Experiment 20).

sodium saccharin   *N*-ethylsaccharin   *O*-ethylsaccharin

You can explore the effects of different factors, such as the metal cation, alkylating agent, substrate, reaction temperature, and use of a phase-transfer catalyst, on the product composition. You should analyze the product by HPLC, NMR, or some other instrumental method.

8   Migratory Aptitudes of Aryl Groups
Key Reference: *J. Chem. Educ.* **1971**, *48*, 257.

The pinacol rearrangement involves the migration of an alkyl or aryl group, presumably by means of a bridged intermediate. You can investigate the factors that contribute to migratory aptitude by preparing pinacols in

Bridged intermediate in the pinacol rearrangement

which either of two different groups (R and R′) can migrate and analyzing the products of their pinacol rearrangement reactions. By carrying out a reasonable number of reactions using carefully selected substrates, you should be able to arrange different groups in order of their migratory aptitude. The pinacols can be prepared from appropriate ketones and the products analyzed by NMR, gas chromatography, or another instrumental method.

9   Stability of Endocyclic and Exocyclic Double Bonds
Key Reference: *J. Chem. Educ.* **1973**, *50*, 372.

The stability of an alkene depends on the number of alkyl substituents attached to the double-bonded carbon atoms. In general, the more alkyl substituents there are, the more stable is the alkene. The stability of a cyclic alkene can also vary with the location of the double bond; in general, double bonds that are within the ring (*endo*) are more stable than those external to the ring but involving a ring carbon (*exo*). You can investigate the stability of *exo* double bonds in one or more ring systems by carrying out elimination reactions (such as dehydration or dehydrohalogenation) on *tertiary* substrates (**1**) that can yield both *endo* and *exo* double bonds without rearrangement. By varying certain features of the substrate, such as the ring size and the kinds of substituents on the starred carbon atom, you can explore the factors that affect double-bond stability. Alcohol substrates can be made by Grignard reactions with cyclic ketones; alkyl halides and other substrates can be made from the alcohols. The products can be analyzed by gas chromatography, HPLC, NMR, or another appropriate method.

Note: R, R′ can be alkyl, aryl, or hydrogen; Z is a leaving group such as OH or Br.

10   Synthesis and Activity of an Insect Pheromone
Key References: *J. Chem. Educ.* **1991**, *68*, 71. *J. Chem. Educ.* **1986**, *63*, 1014. *J. Chem. Educ.* **1984**, *61*, 927.

Insect pheromones are "chemical messengers" that attract other insects for mating, inform them of danger, help them find their way, and perform a variety of other functions. A number of pheromone syntheses are reported in the chemical literature. You can select a pheromone that interests you, synthesize it, and test its effect on the target insect (if you can obtain specimens). You may also be able to determine whether one stereoisomer is more effective than another or than a racemic mixture of the pheromone, and you can investigate the effect of impurities or molecular modifications on the activity of a pheromone.

11   Effect of Phase-Transfer Catalysts on a Reaction
Key References: E. V. Dehmlow and S. S. Dehmlow, *Phase Transfer Catalysis*, 3rd ed. (New York: VCH, 1993). W. P. Weber and G. W. Gokel, *Phase Transfer Catalysis in Organic Synthesis* (New York: Springer-Verlag, 1977).

Phase-transfer catalysts accelerate many two-phase reactions by helping reactants or intermediates cross the phase boundary and come into contact with one another. You should select an organic reaction that is carried out in two phases, preferably one for which phase-transfer catalysis has not been investigated extensively. Perform one or more examples of the reaction with and without a quaternary ammonium or phosphonium salt to see whether or not the onium salt catalyzes the reaction. If it does, you can use different onium salts and observe their effects on the reaction rate to determine their relative catalytic effectiveness. You can also vary substrates, solvents, and other factors to observe the effect of such changes.

12   Synthesis of a New Organic Compound
Key Reference: *J. Amer. Chem. Soc.* **1974**, *94*, 4024.

To prove the structure of α-pinene, Adolf von Baeyer started by oxidizing it to pinonic acid with potassium permanganate. This reaction has been shown to proceed in high yield using "purple benzene," a solution of potassium permanganate in benzene containing a crown ether.

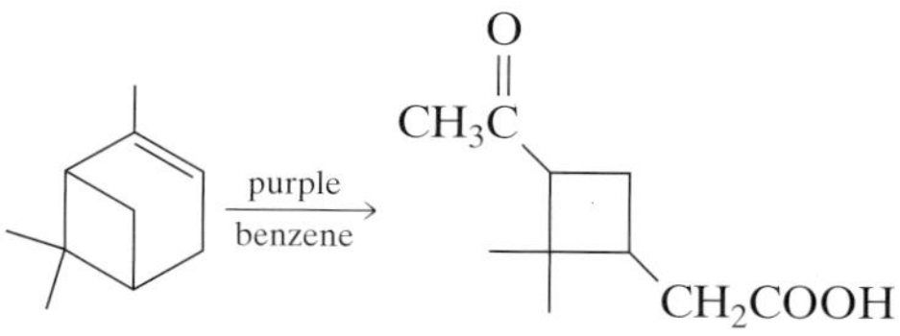

pinonic acid

Pinonic acid is a relatively uncommon compound containing two reactive functional groups. It should therefore be possible, using it as the starting material, to prepare a completely new organic compound—one that has never been reported in the chemical literature (you will need to conduct a thorough literature search to establish this). Benzene is hazardous and the crown ether needed for making purple benzene is quite expensive, so you should develop an alternative procedure for preparing pinonic acid. Note that some reagents may react with both of its functional groups unless one of them is protected. See *Protective Groups in Organic Synthesis* [B10] or a related source for information about the use of protective groups. Your new compound should not be a simple functional-group derivative (such as an ester or a phenylhydrazone) of pinonic acid; its preparation should require at least two synthetic steps. Once you have synthesized and purified the new compound, you will need to prove its structure by spectrometric or chemical methods, or both.

Directions

Research projects are intended for advanced or honors students who have mastered the major operations described in Part V of this book. These Directions only furnish general guidelines to help you complete your project. You must provide the procedural details and have them approved by your instructor. Keep detailed notes of your work and observations as described in Appendix II.

Except when you have reliable information to the contrary, assume that all chemicals you will use are flammable and are hazardous by ingestion, inhalation, and skin absorption. Wear gloves and use a hood whenever possible, avoid contact with the chemicals, do not breathe their dust or vapors, and keep them away from ignition sources and other chemicals that might react with them.

Safety Notes

Begin your research project by following the plan and procedures you have developed. Keep in mind that you are venturing into unknown territory, so your results may not turn out as you planned. If so, try to benefit from unexpected outcomes by considering what you might learn from them. You may then decide to revise your research objectives in order to investigate your findings. In any case, be flexible enough to follow new leads or even switch to a different research project if your original research plan seems unproductive. Meet with your mentor regularly to discuss your findings and receive guidance. Provide him or her with duplicate copies of your lab notebook pages. Continue to read everything you can find that relates to your research; this may help you overcome a difficulty that has hampered your research or suggest new avenues for you to explore.

Report.   Write up your report as if it were a scientific paper being submitted to a professional publication such as the *Journal of Organic Chemistry*. Such papers are traditionally written in an impersonal, objective style using the passive voice (for instance, "the solution was stirred" rather than "I stirred the solution"). Your report should include the following items, unless your instructor directs otherwise:

1 A brief but descriptive title.
2 Your name(s) and affiliation.
3 A brief abstract that summarizes the principal results of the work.
4 An introductory section (usually untitled) that provides a concise statement of the purpose and possible applications of the work, supported by descriptions of related work from the literature.
5 A section titled "Experimental Methods" that gives enough detail about your materials and methods so that another experienced worker could repeat your work. Give current *Chemical Abstract* index names and registry numbers (see Appendix VII) for important starting materials and the product, and provide information about their purity if possible. Note any significant hazards and safety precautions in a separate paragraph labeled "Caution."
6 A section titled "Results and Discussion" (or two separate "Results" and "Discussion" sections) that summarizes the important experimental results and points out any special features, limitations, or implications of your work. You should include any data (including spectral parameters) that will help justify your conclusion. You may also wish to suggest different approaches to the problem or areas that require further study.
7 A section titled "Conclusions" that states any conclusions you can draw from your work, based on the evidence presented.
8 A section titled "References" that gives complete citations for all literature sources referred to in the report.

Your instructor may suggest additions to or modifications of this list. *The ACS Style Guide* [L14] provides a more detailed discussion of these components, as well as helpful suggestions about writing style.

Minilabs

The minilabs are short, self-contained experiments that ordinarily take no more than an hour or two to complete. They can be used to supplement the experiments in Parts I and II and to fill in gaps that can occur when there is a long reaction time or when an experiment is completed well before the end of the lab period. You should submit a report for each minilab that includes your data, the percent yield when appropriate, and any other information specified in the minilab or by your instructor.

PART III

MINILAB 1   Making Useful Laboratory Items

Before You Begin: Read OP-3, "Using Glass Rod and Tubing."

A flat-bottomed stirring rod can come in handy when you are trying to dissolve lumps of a solid, pulverize crystals for a melting-point measurement, wash a solid on a Hirsch funnel, or stir a solution manually. You will need a boiling-point tube to carry out a semimicro boiling-point determination, and in preparing one, you will learn how to seal and fire-polish glass tubing. You can get some practice bending glass tubing by preparing a filter trap (also used as a solvent trap), which is used to prevent water from backing up into a filter flask from an aspirator and to recover volatile solvents that are being evaporated under vacuum.

Directions

Safety Notes

> Be careful not to burn or cut yourself while working with glass rods and tubing. See OP-3 for safe glassworking procedures.

Your instructor should demonstrate the correct glassworking techniques [OP-3]. Prepare as many of the following items as requested, and have your instructor inspect and approve the items when you are finished.

Making an acceptable flat-bottomed stirring rod may take practice. You can develop your technique using scrap pieces of glass rod.

Take Care! Don't burn yourself on the hot glass.

Flat-Bottomed Stirring Rod.   Cut a 20–25-cm length of soft glass rod having a diameter of 3–4 mm. Flatten one end by rotating that end in a burner flame until it is soft and incandescent and, without delay, hold the rod vertically and firmly press its hot end onto the metal base of a ring stand. The flat end should flare out to a diameter approximately twice that of the rod itself. Be sure to heat the rod only at its tip, as heating it farther up will cause it to bend rather than flatten. When the flat end has cooled, round off the other end in a burner flame.

Boiling-Point Tube.   (This part can be omitted if you will not be using the semimicroscale boiling-point determination method described in OP-31b.) Obtain a piece of 5-mm o.d. (outer diameter) soft glass tubing and carefully seal it at one end. When it has cooled, cut it to a length of 8–10 cm and fire-polish the open end. Test the tube as described in OP-3 to make sure it is sealed. Save it for use in Experiment 8 and later experiments.

Take Care! Hold the glass close to the stopper and protect your hands.

Filter Trap.   Using a 125–250-mL thick-walled bottle, 6–8-mm o.d. soft glass tubing, and a rubber stopper to fit, construct a filter trap like the one shown in Figure C2 of OP-13. One length of glass tubing should be about 10–15 cm long and the other should be about half that long. First, fire-polish both ends of each tube. Using a flame spreader, bend both pieces of tubing smoothly so that the shorter arm on each is approximately 3 cm long. Bore two holes of appropriate diameter in the rubber stopper (or use a 2-hole stopper provided) and insert the longer arm of each tube through each hole, using glycerol as a lubricant. When this assembly is inserted in the bottle, the lower end of the long tube should be 5 cm or so from the bottom; the short tube has only to extend through the stopper. Wrap the bottle with heavy-duty transparent plastic tape to prevent possible injury from implosion.

Extraction of Iodine by Dichloromethane

Before You Begin: Read the directions for liquid-liquid extraction in OP-15a.

Many extractions involve colorless solutions, making it impossible to observe the transfer of a solute from one layer to the other. In this minilab, the solute (iodine) is colored so you will be able to see and describe what is going on as the solute is extracted from an aqueous solution by an organic solvent. Iodine is nearly insoluble in water, so potassium iodide is added to the aqueous solution to increase its solubility. Because iodide ions combine chemically with I_2 in water, the iodine color in the aqueous and organic layers will be different (brown in one and violet in the other). You should interpret such a color in either layer as indicating the presence of iodine in that layer.

Directions

Record your observations carefully in your laboratory notebook, as you will have to explain them in your report.

Safety Notes

> **Dichloromethane may be harmful if ingested, inhaled, or absorbed through the skin. There is a possibility that prolonged inhalation of dichloromethane may cause cancer. Minimize contact with the liquid and do not breathe its vapors.**

Add 1.5 mL of 0.50 M iodine-potassium iodide solution to a 5-mL conical vial. Tilt the vial at an angle and use a Pasteur pipet to add 1.5 mL of dichloromethane down its side. The dichloromethane layer should slide underneath the colored aqueous layer with a minimum of mixing. Record the color of each layer and the intensity of the color (dark, very light, etc.). Cap the vial and shake it vigorously, with occasional venting, until the color of the aqueous layer does not change with further shaking. Again record the color of each layer and the intensity of the color. Let the mixture stand until there is a sharp interface (dividing line) between the liquid layers. Use a Pasteur pipet (preferably one with a filter tip) to transfer [OP-6] all of the dichloromethane layer to a small Erlenmeyer flask. Now add a fresh 2-mL portion of dichloromethane to the aqueous solution, shake the capped container as before, and record your observations.

Stop and Think: What happened that caused the color changes you observed?

Waste Disposal: Place the combined dichloromethane layers in a chlorinated-solvents waste container and pour the aqueous layer down the drain.

Describe the colors of the two liquid phases before mixing and before and after each extraction. Account for the color changes and for any differences in the color intensity of the layers after the first and second extractions.

Purification of an Unknown Compound by Recrystallization

Before You Begin: Read OP-25c, "Choosing a Recrystallization Solvent," and read or review the other operations as necessary.

When you are purifying a solid substance by recrystallization, the solid must be soluble in the boiling solvent but quite insoluble in the same solvent when it is cold. In this minilab, you will have to select a suitable solvent for the recrystallization of the unknown solid by carrying out some preliminary solubility tests in different solvents.

Directions

Safety Notes

The solvents and the unknown solid may be harmful if ingested, inhaled, or absorbed through the skin. Avoid contact with them and do not breathe their vapors.

Waste Disposal: Place the filtrate in an appropriate solvent recovery container (water and ethanol filtrates can be flushed down the drain).

Obtain an unknown solid from your instructor. Following the directions in OP-25c, "Choosing a Recrystallization Solvent," select a suitable recrystallization solvent by testing the unknown's solubility in hot and cold water, ethanol, hexane, 2-butanone, and any other solvents suggested by your instructor. Accurately weigh [OP-4] 0.2 g of the unknown solid and recrystallize [OP-25] it from the solvent you selected. Dry [OP-23] the purified solid to constant mass and weigh it. Measure the melting points [OP-30] of the impure and purified solids.

Developing and Testing a Hypothesis

Before You Begin: Read or review the "Scientific Methodology" section in the Introduction.

In this minilab, you will observe and attempt to explain the effect of a ferric chloride ($FeCl_3$) solution on solutions of seven organic compounds. Record your observations carefully and think about their significance. After noting what happens to the first three solutions, you will develop a hypothesis to account for your observations. You will then test your hypothesis by adding $FeCl_3$ to two more solutions, modify your original hypothesis as necessary to account for your observations, and repeat this procedure with two more solutions until you arrive at a hypothesis that accounts for all of your observations.

Directions

Safety Notes

Use a calibrated Pasteur pipet [OP-5] to add 0.5 mL of water and 0.5 mL of 95% ethanol to each of eight clean, numbered test tubes. Add one drop of (1) methyl salicylate to the first test tube, about 10 mg each of (2) salicylic acid, (3) aspirin, (4) acetaminophen, (5) phenacetin, and (6) menthol to test tubes two through six, and one drop of (7) benzyl alcohol to test tube number seven. Add two drops of 2.5% ferric chloride solution to test tube number eight, which will serve as a control.

methyl salicylate   salicylic acid   aspirin

acetaminophen   phenacetin   menthol   benzyl alcohol

Add two drops of 2.5% $FeCl_3$ to each of the first *three* test tubes; then **stop** and record your observations, comparing the contents of these test tubes to the control. Based on the structures of the compounds in these test tubes, formulate a hypothesis to account for your results.

Add two drops of 2.5% $FeCl_3$ to each of the next two test tubes, 4 and 5; then **stop** and record your observations. Do the results support your original hypothesis? If not, formulate a new hypothesis or revise your original one to explain them.

Add two drops of 2.5% $FeCl_3$ to each of the next two test tubes, 6 and 7. Do the results support your current hypothesis? If not, formulate a new hypothesis or revise your current one to explain them.

Report your observations in a table. Write down your hypothesis from each stage of the procedure, telling why it changed from one step to the next (if it did), and state your final conclusion.

Observe and Note: Which of the test tubes show evidence of a reaction?

Waste Disposal: Dispose of the solutions as directed by your instructor.

MINILAB 5   # Preparation of Acetate Esters

Before You Begin: Read or review the operations as necessary.

The pleasant aromas of many esters make them popular ingredients of flavorings and perfumes. In this minilab, you will prepare some acetate esters by treating the corresponding alcohols with acetyl chloride, and then compare their odors.

$$CH_3\overset{\displaystyle O}{\overset{\|}{C}}-Cl + ROH \longrightarrow CH_3\overset{\displaystyle O}{\overset{\|}{C}}-OR + HCl$$

acetyl chloride   acetate ester

Alternatively, your instructor may give you an unknown alcohol whose ester you will analyze by mass spectrometry as described in Minilab 29.

Directions

This minilab can be performed as a group project with each student responsible for preparing one ester.

Safety Notes

> **Acetyl chloride is very corrosive, and its vapors are irritating and toxic. It reacts violently with water and some alcohols. Use gloves and a hood, avoid contact, do not breathe its vapors, and keep it away from water.**

Take Care! Possible violent reactions. Wear gloves, avoid contact with acetyl chloride, and do not breathe its vapors.

Waste Disposal: Pour the aqueous layers down the drain. Dispose of the esters as directed by your instructor.

Obtain as many clean, *dry* 13 × 100-mm test tubes as there are alcohols, and number them consecutively. Add 10 drops or 0.3 mL of one of the following alcohols (or any other alcohols specified by your instructor) to each tube: ethanol, 1-propanol, 3-methyl-1-butanol, 1-octanol, benzyl alcohol. *Under the hood, slowly* add 10 drops or 0.3 mL of acetyl chloride to each of these reaction tubes (use of an automatic pipet is recommended). Let the solutions stand for 3 minutes or more. Carefully add 2 mL of water to each test tube and shake gently to mix. Extract [OP-15] the contents of each tube with 2 mL of diethyl ether. *Under the hood*, evaporate [OP-16] the ether from each tube using a stream of dry air or nitrogen. Do not apply heat or evaporate the contents to dryness. Transfer a drop of each ester to a spot plate and cautiously observe the odor of each ester. (Note whether the odor changes with time; if it does, the ether may not have evaporated completely.)

Describe and compare the odors of the esters as best you can. You may be able to associate an ester's odor with the odor of something familiar, such as a nail-polish remover. Write the formula and give the name of each ester you prepared.

Gas Chromatographic Analysis of Commercial Xylene

Before You Begin: Read or review OP-34, "Gas Chromatography."

Commercial xylene, sometimes called xylol, is actually a mixture of three isomeric xylenes (dimethylbenzenes). You may find some in your local hardware store, where it is sold for use as a solvent and thinner for oil-based paints and coatings, such as porch and deck enamels. It is also used for microscopy and in the manufacture of starting materials for the preparation of polyester fibers. You will be analyzing a commercial xylene mixture to determine the percentages of *ortho-*, *meta-*, and *para-*xylene it contains.

ortho-xylene   *meta*-xylene   *para*-xylene

Directions

Xylenes are flammable, toxic by inhalation and ingestion, and irritating to skin and eyes. Minimize contact with the xylene mixtures, and do not breathe their vapors.

Safety Notes

A standard mixture containing *ortho-*, *meta-*, and *para-*xylene will be provided. Record the percentage of each component from the label on the bottle. Obtain gas chromatograms [OP-34] of (1) the standard mixture and (2) a commercial xylene mixture as directed by your instructor. From the relative retention times and peak areas in the gas chromatogram of the standard mixture, identify the component corresponding to each peak and calculate its detector response factor. Identify each peak and measure all peak areas on the gas chromatogram of the commercial xylene. Then calculate the mass percent of each component in the commercial xylene.

Report your results in a table showing the retention time, peak area, detector response factor, and mass percent of each component. Show your calculations, and include your gas chromatograms with your report.

Isolation of Caffeine From No-Doz Tablets

Before You Begin: Read o1r review the operations as necessary.

No-Doz tablets are used by truck drivers, students studying for exams, and other people who need to stay awake at night. Their active

CH₃—N and CH₃ structure

caffeine

ingredient is caffeine, which—as a component of coffee, tea, cocoa, and many soft drinks—is probably the most frequently consumed drug in the world. Caffeine stimulates the cerebral cortex and, in small doses, can improve concentration and coordination. It also acts as a heart stimulant and makes skeletal muscles less susceptible to fatigue. As anyone with a bad case of "coffee nerves" can attest, caffeine also has its down side. Too much caffeine may cause headaches, heart irregularity, muscular trembling, irritability, and insomnia. In this minilab, you will use hot ethanol to extract caffeine from a No-Doz tablet.

Directions

Place a 200-mg No-Doz (or a generic equivalent) tablet in a 25-mL Erlenmeyer flask and crush it with a flat-bottomed stirring rod (see Minilab 1) or a flat-bladed spatula. Add 5.0 mL of 95% ethanol and boil the mixture gently for 3 minutes or more by swirling it in a boiling water bath [OP-7]. Replace any ethanol that evaporates. Filter the *hot* solution by gravity [OP-12] using a preheated filtering pipet. Wash the solid on the filter with 2.0 mL of hot 95% ethanol and combine the filtrates. Under the hood, concentrate [OP-16] the solution to a volume of 2 mL by heating it gently under a stream of dry air or nitrogen. If the solution is cloudy or contains a precipitate, heat it until it becomes clear. Let the solution cool at room temperature until a good crop of crystals is obtained, then cool [OP-8] it in an ice/water bath for 5 minutes. Collect the caffeine by vacuum filtration [OP-13], washing it twice with cold 50% ethanol. Dry [OP-23] it at room temperature and measure its mass and melting point [OP-30]. At your instructor's request you can purify the caffeine by sublimation [OP-26] and measure the melting point of the purified product. Calculate the percent recovery of caffeine from No-Doz tablets.

MINILAB **8**   A Missing-Label Puzzle

Before You Begin: After reading the minilab, devise a method for distinguishing the two dry-cleaning solvents.

Most dry-cleaning solvents are hydrocarbon mixtures or chlorinated hydrocarbons that, because of their low polarity, dissolve greasy stains in clothing. Many chlorinated solvents that were once used for dry cleaning, such as carbon tetrachloride, are no longer allowed because of their toxicity. Tetrachloroethene ($Cl_2C=CCl_2$), sold under such trade names as Perclene, is less toxic than most other chlorinated solvents. Mineral spirits, a mixture of petroleum hydrocarbons, is also used widely in dry cleaning.

For this minilab, you will assume that the labels have fallen off a bottle of tetrachloroethene and a bottle of mineral spirits, and you will devise a method of telling the solvents apart. Your method cannot be based on odor differences. Instruments such as balances or refractometers are not allowed. You can use only the equipment in your locker and basic lab utilities, plus reference books and anything else your instructor is willing to provide (check in advance if you're not sure what will be available). The simplest method may be the best—a few minutes of thought can often save hours of experimental work.

Directions

> **Tetrachloroethene is an eye and skin irritant and may be harmful by inhalation. It is also a suspected carcinogen. Avoid contact and do not breathe its vapors.**
> **Mineral spirits is very flammable and inhalation may be harmful. Do not breathe its vapors and keep flames away.**

In the laboratory, you will find two containers of "dry-cleaning solvent," labeled **A** and **B**. One of the solvents is tetrachloroethene and the other is mineral spirits. Assuming that the hydrocarbons in mineral spirits have properties similar to those of the alkanes in Experiment 8 (except for their boiling points), devise an experiment for finding out which solvent is in which container. Then carry out your experiment, report the identities of **A** and **B**, and tell how you arrived at your conclusion. Don't reveal your method or results to others—let them think for themselves, as you did.

Paper Chromatography of Dyes in Commercial Drink Mixes

MINILAB **9**

Before You Begin: Read OP-20, "Paper Chromatography."

Many *foods*, *drugs*, and *cosmetics* are colored with FD&C dyes to give them eye appeal or, in the case of foods, to give them the color we associate with the food in question. For example, the artificial flavors used to prepare a grape drink may be completely colorless, so dyes are added to give the drink a purple "grape" color. A number of former FD&C dyes have been removed from the market because they were found to be toxic or were suspected of causing cancer or birth defects. For example, FD&C Red No. 2, which was once the most widely used food dye, was banned in 1976 because it is a suspected carcinogen. At present, there are only eight FD&C dyes allowed for unrestricted food use, the four most popular being Blue No. 1, Red No. 40,

Yellow No. 5, and Yellow No. 6. In this minilab, you will analyze some powdered drink mixes (such as Kool-Aid) to see which FD&C dyes they contain. The four dyes mentioned can be identified by their colors and R_f values. Some foods also contain FD&C Green No. 3 and Violet No. 2, which can be identified by the colors of their spots alone.

Table M1 Colors and approximate R_f values for FD&C dyes

FD&C dye	Color	R_f
Yellow #5 (tartrazine)	bright yellow	0.31
Yellow #6 (sunset yellow)	orange	0.58
Red #40 (allura red)	bright red	0.62
Blue #1 (brilliant blue)	turquoise blue	0.67

Note: R_f values are for the solvent mixture 95% ethanol/ 1-butanol/2.0 M ammonia (1:1:1).

Directions

Ethanol and 1-butanol are flammable and 1-butanol may cause eye or skin irritation. Ammonia can cause severe skin and eye irritation. Avoid contact with the developing solvent, do not breathe its vapors, and keep flames away.

Take Care! Avoid contact with the developing solvent and do not breathe its vapors.

You will be provided with a selection of sugar-free drink mix flavors, each containing one or more of the food dyes listed in Table M1. *Under the hood*, place enough developing solvent (a 1:1:1 mixture of 95% ethanol, 1-butanol, and 2.0 M aqueous ammonia) in a suitable developing chamber (such as a 600-mL beaker) to cover its bottom to a depth of about 1 cm. Cover it with a lid or plastic wrap, swirl it to agitate the liquid, and let it sit to equilibrate. Label the wells of a spot plate to correspond to the drink mix flavors, using a one- or two-letter code. Place 0.2–0.3 g of each drink mix powder in a well of the spot plate; then add room-temperature water drop by drop with stirring to each well until the powder dissolves. Rinse your stirring rod before going on to the next spot plate.

Stop and Think: Why not use a pen?

Obtain an approximately 11 × 22-cm rectangle of Whatman #1 chromatography paper [OP-20] and use a pencil to draw a starting line along one of its long sides, about 1.5 cm from that side. Using a different end of a capillary micropipet (or toothpick) for each spot, spot the solutions 1.5–2.0 cm apart along the starting line. If space permits, apply two or three spots of varying concentration for each solution. Make each spot 2–3 mm in diameter and write its code letter underneath it with a pencil. Develop the chromatogram in the developing chamber and let it air dry (don't forget to mark the solvent front).

Waste Disposal: Place the used developing solvent in an appropriate solvent recovery container.

Prepare a table showing the color and your experimental R_f value for each spot in the chromatogram of each drink mix. Using information from Table M1, identify as many dyes in each drink mix as you can. Turn in your paper chromatogram with your report.

Gas Chromatographic Analysis
of an Essential Oil from Orange Peel

MINILAB 10

Before You Begin: Read OP-15c, "Liquid-Solid Extraction," and read or review the other operations as necessary.

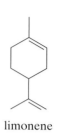

limonene

The orange tree originated in China and was first cultivated in the United States by Franciscan monks, in the part of Spanish North America that is now California. Orange oil is used medicinally and also to flavor foods, drinks, and confections. It is colored by β-carotene and contains a number of minor components, but its major component is limonene. Limonene is a ubiquitous terpene that occurs in a multitude of other essential oils, including the oils of lemon, caraway, and dill. Limonene exists in two enantiomeric (mirror image) forms, (R)-(+)-limonene and (S)-(−)-limonene, which have different odors.

In this minilab, you will isolate orange oil by extraction of grated orange peel with hexanes (a mixture of isomeric six-carbon alkanes) and analyze it by gas chromatography to determine the approximate percentage of limonene it contains. If your instructor has samples of (R)-(+)-limonene and (S)-(−)-limonene available, you can compare their odors to that of your extract to find out which enantiomer you have. Some people are unable to distinguish the enantiomers by odor; if you are one of them, ask a coworker for help. If a suitable preparative GC column is available, you can also isolate pure limonene and record its infrared spectrum as described in *J. Chem. Educ.* **1994**, *71*, A146.

Directions

Safety Notes

> **The hexanes mixture is flammable and its vapors may be harmful. Avoid inhalation and keep flames away.**

On a piece of wax paper or aluminum foil, grate the rind of an orange until you have about 1 g. Measure the mass of the grated orange rind and transfer it to a screw-cap centrifuge tube. Extract [OP-15c] the orange oil with two successive 4-mL portions of hexanes. Then dry [OP-22] the combined extracts through a drying column containing 0.5 g of anhydrous sodium sulfate. Use about 0.5 mL of hexanes to rinse the drying column, combining it with the extracts. *Under the hood*, remove the solvent by evaporation [OP-16] in a stream of dry air or nitrogen. Weigh the orange oil and, if necessary, resume evaporating until its mass is constant between weighings. Compare the odor of the orange oil to those of authentic samples of (R)-(+)-limonene and (S)-(−)-limonene, if available. Record a gas chromatogram [OP-34] of the orange oil.

Estimate the percentage of limonene in the orange oil by assuming the same detector response factor for all components, and calculate the percentage recovery of limonene from orange peel. If you can, identify the limonene in orange peel as (R)-(+)-limonene or (S)-(−)-limonene and draw a structural formula showing its stereochemistry.

Identification of an Unknown Felt-Tip-Pen Ink by TLC

MINILAB 11

This minilab is based on an article published in the *Journal of Chemical Education* **1983**, *60*, 232.

Before You Begin: Read or review OP-19, "Thin-Layer Chromatography."

Unlike food colorings, which are made from a very limited selection of approved dyes (see Minilab 9), pen inks are made from a wide variety of dyes, and the dye mixtures in pen inks from different manufacturers can vary greatly. These dyes can be separated by thin-layer chromatography (TLC), producing chromatograms with characteristic patterns of colored spots. For this reason, forensic scientists often use TLC to identify pen inks from documents associated with a criminal or civil case. For example, matching the ink from a pen in the suspect's possession with ink in the signature on a forged document provides evidence that may result in the suspect's conviction.

In this minilab, you will attempt to identify the ink from an unknown felt-tip pen by comparing its TLC pattern with those of known pen inks. The pens will have fine points, so they can be used to spot the TLC plate directly. You will develop the chromatogram using a 1-butanol/water/ethanol/acetic acid (120:40:20:1) mixed solvent or another suitable developing solvent.

Directions

Safety Notes

> The developing solvent is flammable and harmful by inhalation, ingestion, and skin absorption. Avoid contact, do not breathe its vapors, and keep flames away.

Under the hood, prepare a developing chamber using a 1-L beaker or another suitable container as described in OP-19, and add enough developing solvent to cover the bottom to a depth of about 5 mm. Cover the developing chamber with a lid or plastic food wrap, swirl it to agitate the liquid, and let it sit to equilibrate while you prepare the TLC sheet [OP-19]. Obtain a 10×10-cm silica gel TLC sheet and mark the starting line in pencil. Using the known and unknown felt-tip pens provided, carefully apply spots about 1.5 cm from each edge and 1 cm or more apart, labeling them in pencil. Spot the unknown in at least two places, applying the spots between known spots to make the inks easier to compare. Each spot should be no more than 2 mm in diameter, so practice your spotting technique on a scrap piece of TLC sheet first. Develop the TLC sheet in the pre-equilibrated developing chamber and let it air dry (don't forget to mark the solvent front).

Waste Disposal: Place the used developing solvent in an appropriate solvent recovery container.

For each pen ink, including the unknown, describe the pattern of colored spots you observed. Then identify the unknown pen ink by matching its pattern with that from one of the known pen inks. Turn in your chromatogram with your report.

Optical Rotation of Turpentine

Before You Begin: Read or review OP-33, "Optical Rotation."

North American turpentine contains $(+)$-α-pinene and $(-)$-β-pinene (see Experiment 12). In this minilab, you will measure the optical rotation of a commercial turpentine (or a simulated turpentine) to estimate the percentages of $(+)$-α- and $(-)$-β-pinene it contains.

Directions

> **Turpentine and its components are flammable and irritate the skin and eyes; inhalation of their vapors may be harmful. Minimize contact with the turpentine and do not breathe its vapors.**

Safety Notes

Using a 10-mL volumetric flask, prepare a solution containing about 1.0 g (accurately weighed) of turpentine in absolute ethanol and fill a 1-dm sample cell with the solution. Measure the optical rotation of the solution and of an absolute ethanol blank using an accurate polarimeter [OP-33]. Prepare solutions of $(+)$-α-pinene and $(-)$-β-pinene in the same way and measure their optical rotations.

Calculate the specific rotation of each solution. Assuming that it contains only $(+)$-α-pinene and $(-)$-β-pinene, calculate the percentage of each constituent in the turpentine.

Waste Disposal: Place the ethanol solutions in an appropriate solvent recovery container.

Stop and Think: What equation in OP-33 can you use to do this?

The Structures of Organic Molecules

Before You Begin: Familiarize yourself with the different kinds of formulas used to represent organic compounds and know how to name organic compounds by the IUPAC system.

In this minilab, you will study some molecular models to help you (1) better understand the structure and geometry of organic molecules and (2) learn how to write and interpret the different kinds of formulas that are used to represent them.

Directions

When you arrive, you will find numbered molecular models of representative organic compounds distributed around your organic chemistry laboratory. Start at one of the models and proceed around the laboratory until you have studied all of them. Unless your instructor indicates differently, assume that atoms are represented by balls of the following colors: carbon—black; hydrogen—white or yellow; oxygen—red; nitrogen—blue; chlorine—green; bromine—orange or brown; sulfur—yellow. The type of covalent bond connecting two atoms is indicated by the number of connectors; one for a single bond, two for a double bond, and three for a triple bond. For each model, write the following in your lab notebook, along with the number of the model:

- IUPAC name (the name may be provided in some cases)
- Empirical formula
- Molecular formula
- Condensed structural formula
- Expanded structural formula (Kekulé structure)
- Bond-line formula

Examples of these types of formulas are shown for 2-methylpropane.

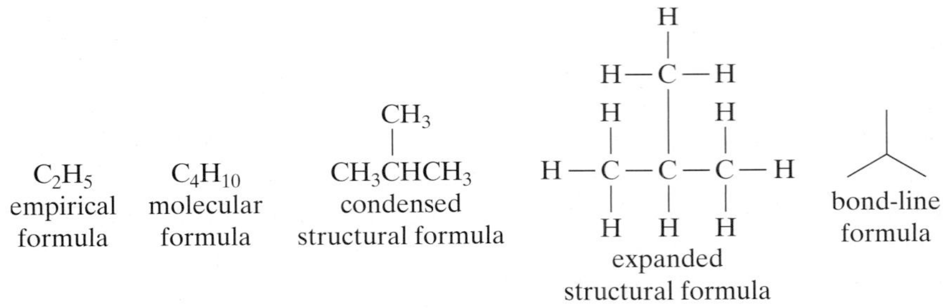

C_2H_5
empirical
formula

C_4H_{10}
molecular
formula

CH_3CHCH_3
condensed
structural formula

expanded
structural formula

bond-line
formula

formulas for 2-methylpropane

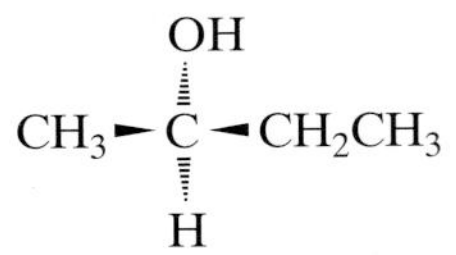

If molecular models of chiral compounds are provided, write their stereochemical formulas also, using either flying wedge, ball-and-stick, or Fischer projections. A flying wedge projection is illustrated in the margin for (R)-2-butanol. Include the (R) or (S) designation of any chiral compound in its name. Your instructor may request special formulas in some cases, such as chair-ring formulas for substituted cyclohexanes.

stereochemical
formula

MINILAB 14

Who Else Has My Compound?

This minilab is based on an article published in the *Journal of Chemical Education* **1995**, *72*, 1120.

Before You Begin: Develop an experimental plan based on the equipment and chemicals available to you, and read or review the appropriate operations as necessary.

Few, if any, scientists perform their work in isolation. Successful science requires cooperation and collaboration among scientists, who exchange information, ideas, and even materials. For example, suppose you isolated a natural product that was not previously reported in the chemical literature, but shortly afterward a Japanese chemist reported a natural product (isolated from a different source) whose spectra and properties appear to be the same as yours. You might then exchange data and samples with the Japanese chemist to establish whether your compounds were, in fact, the same.

In this minilab, you will receive an unknown organic compound, knowing that at least one other student in your laboratory has the same compound. You will be expected to learn enough about your compound so that, by sharing your findings with other students, you can find another unknown that appears to match your own. You and the student having that unknown should then confirm that your compounds are identical by, for example, having student X obtain data for his or her unknown that student Y has already obtained for the other unknown and vice versa. If more than two students share the same unknown, you may eventually identify yourself as part of a larger group of students. Since the outcome of the experiment depends on the efforts of all of the participants, your individual contribution is important for the success of the group.

Your instructor may advise you about the techniques you should use, or you may be asked to plan your own experimental strategy, given the chemicals and equipment available.

Directions

Safety Notes

Assume that the unknown compound may be flammable and harmful by inhalation, ingestion, and skin absorption. Avoid contact, do not breathe its vapors, and keep it away from flames and hot surfaces.

You will be issued an unknown compound with an identification number. Obtain as much information about your compound as you think is necessary to characterize it, given the equipment and chemicals available. Then approach other students with your findings (they may approach you first) to locate another student who appears to have the same compound you do. When you find one or more such students, perform any additional work necessary to confirm that your compounds are in fact identical.

Report the names and unknown numbers of other students whose unknowns are identical to your own and tell how you arrived at your conclusion. Report all of your findings in an appropriate format and turn in any spectra or chromatograms you obtained.

Isomers and Molecular Structure

MINILAB 15

Before You Begin: As necessary, review material from your lecture textbook about structural (constitutional) isomers and stereoisomers.

Just as zoology is the study of animals and botany the study of plants, organic chemistry can be regarded as the study of carbon-containing molecules.

Certain plants and animals are familiar to everyone because we see them all around us, but molecules are not. We can't *see* a molecule—the best we can do is to imagine what one looks like, and molecular models can help us do that. All models have their limitations. A plastic model of a Sopwith Camel is a far cry from the real thing, but it can show us what that WWI biplane looked like in three dimensions. A real molecule is obviously not made of colored balls, but a good molecular model can show the three-dimensional structure of a molecule far better than any drawing. In this minilab, you will use molecular models to help you see—in three dimensions—the structural and geometric differences between isomers.

Isomers are different compounds that share the same molecular formula. This means that the same set of atoms can be combined in different ways to form molecules with different structures. In this minilab, you will apply a "hands-on" approach to the study of isomers and molecular structure. You will be issued a set of atoms and bonds from a molecular-model kit, and will be expected to construct molecular models for as many isomers as you can in the time allowed. The most popular molecular models are the ball-and-stick type, in which balls of different colors represent different kinds of atoms and rigid or flexible connectors represent the bonds that hold the atoms together in molecules. Except for nitrogen, each ball is ordinarily drilled with a number of holes equal to the *normal covalence* of the corresponding atom—the number of covalent bonds formed by the neutral atom. Thus balls representing carbon atoms—colored black—are drilled with four holes arranged tetrahedrally, whereas those representing oxygen atoms have two holes, and those representing hydrogen atoms have one. Building models of all the isomers that have a given molecular formula is thus a matter of putting the appropriate colored balls together in all possible combinations, using just enough connectors to fill the holes. Often, a large number of isomers will share the same molecular formula, and it may take a considerable amount of ingenuity to find them all!

If the balls for nitrogen have four holes, use only three of them.

Directions

This minilab can be done individually or in groups. Your instructor will issue you or your group a set of atoms and will provide connectors, including flexible connectors for constructing strained and multiple bonds. Unless he or she indicates differently, assume that atoms are represented by balls of the colors given in Table M2.

Table M2 Colors of model-kit balls and normal covalences of atoms

Atom	Color of ball	Normal covalence
bromine	orange or brown	1
carbon	black	4
chlorine	green	1
hydrogen	white or yellow	1
nitrogen	blue	3
oxygen	red	2
sulfur	yellow	2

Construct molecular models for as many isomers as you can, based on the normal covalences in Table M2. In your lab notebook, neatly draw the structural formulas for the molecular models you constructed. Some of your molecular models may exhibit stereoisomerism; in that case, you can count each geometric (*cis-trans*) isomer or enantiomer as a different compound. Designate any geometric isomers as (*Z*) or (*E*) and any enantiomers as (*R*) or (*S*), and show their stereochemistry clearly. If your instructor requests, obtain another set of atoms and connectors and use it to prepare isomers as well.

Reactivities of Alkyl Halides in Nucleophilic Substitution Reactions

MINILAB **16**

Before You Begin: Draw the structures of all substrates you will be using in this reaction and classify them as 1°, 2°, 3°, aryl, benzylic, or some combination of these (such as 1° benzylic). Identify the nucleophile, substrate, and leaving group in the general equations for Reactions **1** and **2**.

In this minilab, you will compare the relative reactivities of different alkyl halides with two different reagents: sodium iodide in acetone and silver nitrate in ethanol. General equations for their reactions are

$$\textbf{1} \quad RX + NaI \xrightarrow{\text{acetone}} RI + NaX \qquad (X = Cl \text{ or } Br)$$

$$\textbf{2} \quad RX + AgNO_3 + EtOH \longrightarrow ROEt + AgX + HNO_3$$

(Additional organic products, such as alkenes, may be formed in reaction **2**.)

For each reagent, the occurrence of a reaction is indicated by the formation of a precipitate; sodium iodide (which is insoluble in acetone) in Reaction **1** and silver chloride or silver bromide in Reaction **2**. Your initial objective is to discover what kind of mechanism (S_N1 or S_N2) is involved in the reactions of each reagent. Based on your conclusions, you will predict the relative reactivities of other substrates with the reagents and test your predictions. In order to accomplish these objectives, you will have to think about the significance of your observations as you make them, so that your results from one set of reactivity measurements can lead to predictions about another set of measurements.

Directions

All of the alkyl bromides are toxic and have harmful vapors, and some of them are suspected carcinogens. Avoid contact and do not breathe their vapors.
Silver nitrate is toxic, causes skin irritation, and stains the skin black. Avoid contact with the silver nitrate reagent.

Safety Notes

A. *Reactions in Sodium Iodide/Acetone*

Directions for conducting Reaction **1** follow. Be sure to label the reaction tubes so you know which substrate each contains.

Carry out Reaction **1** using the following substrates: 2-bromobutane, 2-bromo-2-methylpropane, 1-bromobutane. Based on your results, decide whether the reaction in NaI/acetone is S_N1 or S_N2.

Predict the reactivities of the following substrates in NaI/acetone relative to the substrates in step 1: bromocyclohexane, 1-bromoadamantane. Then carry out Reaction **1** using these substrates.

Stop and Think: Are your results as you predicted? If not, how can you explain them?

Reaction 1. Obtain as many clean, dry 13×100-mm test tubes as there are alkyl halides to test. Measure 1 mL of 15% sodium iodide in acetone into each test tube. Add 2 drops of a different alkyl halide to each test tube, and stopper and shake the test tubes. (For a solid halide, use 0.1 g dissolved in a minimal volume of acetone). Observe them closely and record the time needed for any precipitate to form. After 5 minutes, put any test tubes that do *not* contain a precipitate into a 50°C water bath and leave them there for 6 minutes (loosen their stoppers first). Then cool the test tubes to room temperature and note the formation of any precipitate.

Waste Disposal: Place the sodium iodide test solutions in a designated waste container.

B. *Reactions in Silver Nitrate/Ethanol*

Directions for conducting Reaction **2** follow. Be sure to label the reaction tubes so you know which substrate each contains.

Carry out Reaction **2** using the following substrates: 2-bromobutane, 2-bromo-2-methylpropane, 1-bromobutane. Based on your results, decide whether the reaction in AgNO₃/ethanol is S_N1 or S_N2.

Predict the reactivities of the following substrates in AgNO₃/ethanol relative to the substrates in step 1: α-bromotoluene, bromobenzene, 1-bromoadamantane. Then carry out Reaction **2** using these substrates.

Stop and Think: Are your results as you predicted? If not, how can you explain them?

Reaction 2. Obtain as many clean, dry 13×100-mm test tubes as there are alkyl halides to test. Measure 2 mL of a 0.10 *M* solution of silver nitrate in ethanol into each test tube. Add one drop of a different alkyl halide to each test tube, and stopper and shake the test tubes. (For a solid halide, use 50 mg dissolved in a minimal volume of ethanol.) Observe them closely and record the time needed for any cloudiness or precipitate to form. After 5 minutes, gently boil any solutions that do *not* contain a precipitate in a hot water bath for 2–3 minutes and note the formation of any precipitate.

Waste Disposal: Place the silver nitrate test solutions in a designated waste container.

In your report, write balanced equations for the reactions undergone by each halide and write mechanisms for both of the reactions undergone by 2-bromobutane. Arrange the alkyl bromides in order of reactivity in each reaction, tell whether each reaction is S_N1 or S_N2, and explain your reasoning. If either reaction is S_N1, use your results to predict the relative stabilities of the carbocations formed by each substrate. If any substrates in the S_N1 reaction appear to be equally reactive (or unreactive), show their carbocations as being of approximately equal stability and try to learn their actual relative stabilities from your lecture textbook or elsewhere.

An S$_N$1 Reaction
of Bromotriphenylmethane

Before You Begin: Read or review the operations as necessary.

The traditional Williamson synthesis of ethers involves the reaction of an alkyl halide with the sodium salt of an alcohol or phenol:

$$RX + R'ONa \rightarrow ROR' + NaX$$

An alkoxide ion $(R'O^-)$ is strongly nucleophilic and displaces halide ion by an S$_N$2 mechanism. Because an alkoxide is also a strong base, most *tertiary* alkyl halides tend to undergo E2 elimination under these conditions, producing alkenes rather than ethers. Using a less nucleophilic alcohol $(R'OH)$ in place of the corresponding alkoxide prevents E2 elimination, but E1 elimination may then compete with S$_N$1 substitution to yield by-product alkenes along with the desired ether. Both E2 and E1 elimination are impossible for bromotriphenylmethane (trityl bromide), and the triphenylmethyl (trityl) carbocation is very stable, so this tertiary halide reacts readily with alcohols to form ethers but no by-product alkenes. In this minilab, you will heat bromotriphenylmethane with ethanol to form an ether, ethoxytriphenylmethane, by an S$_N$1 reaction.

Stop and Think: Why are these elimination reactions impossible?

Reaction of bromotriphenylmethane with ethanol

bromotriphenylmethane
(trityl bromide)

ethoxytriphenylmethane
(trityl ethyl ether)

Directions

Bromotriphenylmethane is harmful if inhaled or absorbed through the skin. Avoid contact and do not breathe its dust.
Since the reaction generates some gaseous hydrogen bromide, carry it out under a hood. Avoid contact with the reaction mixture and do not breathe its vapors.

Safety Notes

You can use purified bromotriphenylmethane from Experiment 25 in this minilab, scaling down the quantities if necessary. *Under the hood*, mix 0.10 g of bromotriphenylmethane with 1.0 mL of absolute ethanol in a small test tube or a Craig tube. Add a boiling chip and boil the mixture in a hot water bath [OP-7] until no more HBr is evolved (test by holding moist blue litmus paper over the mouth of the tube). Replace any ethanol that evaporates. Remove the boiling chip, let the reaction mixture cool to room temperature, induce crystallization if necessary, and cool [OP-8] it further using an ice/water bath. Then collect the ethoxytriphenylmethane by vacuum filtration [OP-13], or by centrifugation [OP-14] if you used a Craig tube. Dry [OP-23] the product and measure its mass and melting point [OP-30].

Write a detailed mechanism for the reaction and explain why E1 elimination, which often competes with S_N1 substitution, does not occur during this reaction.

MINILAB 18

Preparation and Properties of a Gaseous Alkene

Before You Begin: Predict the structure of the gaseous alkene that you will be preparing.

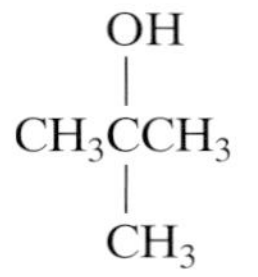

Alkenes can be prepared by elimination reactions in which a molecule of hydrogen halide (HX), halogen (X_2), or water is removed from a substrate. In this minilab you will be carrying out an acid-catalyzed elimination reaction of 2-methyl-2-propanol to form a gaseous alkene with the molecular formula C_4H_8. You should be able to predict the structure of this alkene knowing the structure of the substrate. You will test your alkene with bromine in dichloromethane (CH_2Cl_2 is the solvent) and with aqueous $KMnO_4$, looking for evidence of a reaction in each case. You will also test its flammability to see whether it reacts with atmospheric oxygen. You should then be able to explain your observations and write equations for the reactions involved. (You may have to consult your lecture textbook for help in some cases.)

OH
|
CH_3CCH_3
|
CH_3

2-methyl-2-propanol

Directions

With your instructor's permission, you can work in pairs.

Safety Notes

Sulfuric acid causes chemical burns that can seriously damage skin and eyes. Wear gloves and avoid contact.
Keep the gas you collect away from flames except when you are testing its flammability.

Fill four 18 × 150-mm test tubes with water, stopper them tightly, and invert them in a 1-L beaker or pneumatic trough that is about three-quarters filled with water. Remove the stoppers under water so that no water flows out of the test tubes, and leave them inverted in the beaker or trough. Measure 0.60 mL of 2-methyl-2-propanol (*t*-butyl alcohol) into a 3-mL conical vial, add 3 drops of concentrated sulfuric acid, swirl to mix, and drop in a boiling chip. (If necessary, warm the bottle of 2-methyl-2-propanol in a hot water bath until the alcohol flows freely.) Assemble a gas generator by attaching a plastic gas delivery tube (from your lab kit) to the conical vial with a compression cap, making sure that the O-ring is undamaged and the cap is screwed on securely.

Take Care!   Wear gloves and avoid contact with H_2SO_4.

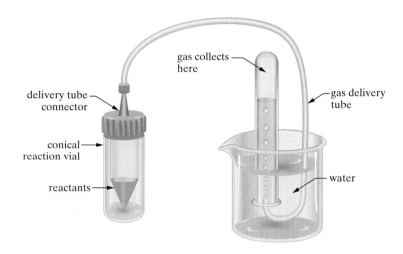

Figure M1 Apparatus for gas generation

Move one of the water-filled test tubes so that its mouth is over the outlet of the gas delivery tube. You may have to hold the end of the tube in place with a forceps or other device. Heat [OP-7] the reaction vial gently in a hot water bath until a steady stream of gas bubbles emerges from the delivery tube and fills the test tube. As each test tube fills with gas, quickly remove and stopper it and replace it with a water-filled test tube. Repeat this until all of the test tubes are filled. Remove the gas delivery tube from the water, and then remove the heat source and cool the reaction vessel in cold water to stop the generation of gas.

Take Care!   Keep flames away.

Don't stop heating until you have collected all of the gas samples and removed the gas delivery tube from the water. Otherwise, water will back up into the reaction vessel.

Discard the gas in the first test tube, which contains air. Add 5 drops of 0.10 *M* potassium permanganate to the second tube, then stopper and shake it. Add 5 drops of 1.0 *M* bromine in dichloromethane to the third tube, then stopper and shake it. Light one end of a wood splint with a match or a burner flame, and carefully lower the burning end into the fourth tube.

Observe and Note: What happens?

Waste Disposal: Dispose of the contents of the test tubes as directed by your instructor.

Write the structure of the gas and a balanced equation for its synthesis from 2-methyl-2-propanol. At your instructor's request, write a detailed mechanism for this reaction. Describe and explain your observations during the tests, and write a balanced equation for the chemical reaction that occurred in each of them.

MINILAB **19**   Addition of Iodine to α-Pinene

This minilab is based on an article published in the *Journal of Chemical Education* **1977**, *54*, 228.

Before You Begin: Review material in your lecture textbook about ring strain and carbocation rearrangements.

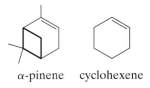

α-pinene   cyclohexene

In *All Creatures Great and Small,* the author, James Herriot, describes how veterinarians once disinfected wounds of cows, horses, and other large animals. The veterinarian would pack the wound with iodine, splash some turpentine on it, and stand back. The result was a violent reaction that forced gaseous iodine into the wound amid a cloud of purple vapor—accompanied by an equally violent reaction from the startled animal!

The major component of turpentine is α-pinene, an alkene having a carbon–carbon double bond in a six-membered ring and adjacent to the highly strained four-membered ring highlighted here. Although alkenes readily undergo addition reactions with chlorine and bromine, most of them are unreactive with iodine. If α-pinene is responsible for the reaction of turpentine with iodine, it must possess some special structural feature that makes it more reactive than most other alkenes. You will investigate the effect of treating α-pinene with iodine and compare it to the effect of treating a control compound, cyclohexene, with iodine. Then you will attempt to explain your observations by proposing a mechanism for the reaction and predicting its products.

Directions

The reaction may be carried out by your instructor, as a demonstration.

Safety Notes

α-**Pinene is flammable and irritates the skin and eyes; inhalation of its vapors may be harmful. Minimize contact and do not breathe its vapors. The reaction with iodine may be violent. Conduct it under a fume hood; wear safety goggles and adequate protective clothing.**

Take Care! Wear safety goggles and protective clothing

Waste Disposal: Dispose of the mixture in both test tubes as directed by your instructor.

Weigh 0.20 g of iodine into each of two 18 × 150-mm test tubes. *Under the hood,* add 0.40 mL of cyclohexene to the first test tube in one portion. Then add 0.40 mL of α-pinene to the second test tube in one portion and step back quickly. Record your observations in your lab notebook.

Write a mechanism that explains the exothermicity of the α-pinene reaction, assuming a carbocation as the initial intermediate. Write an equation for this reaction showing two likely products (neither is a vicinal diiodide). Tell how your mechanism explains your experimental observations.

Unsaturation in Commercial Products

Before You Begin: For as many of the commercial products as you can, try to predict whether or not the product should be unsaturated or contain unsaturated components.

A substance is *unsaturated* if it contains fewer hydrogen atoms than are possible based on its molecular structure. Compounds having multiple carbon–carbon bonds are unsaturated because they can add more hydrogen.

$$-\overset{|}{C}=\overset{|}{C}- \ + \ H_2 \ \xrightarrow{\text{catalyst}} \ -\overset{\overset{\displaystyle H}{|}}{\underset{|}{C}}-\overset{\overset{\displaystyle H}{|}}{\underset{|}{C}}-$$

Since bromine also adds to most carbon–carbon multiple bonds, a solution of bromine in carbon tetrachloride has often been used to test for unsaturation. If the compound tested is unsaturated, the red-brown color of the bromine fades as the bromine is consumed.

$$-\overset{|}{C}=\overset{|}{C}- \ + \ Br_2 \ \longrightarrow \ -\overset{\overset{\displaystyle Br}{|}}{\underset{|}{C}}-\overset{\overset{\displaystyle Br}{|}}{\underset{|}{C}}-$$

red-brown   colorless

Carbon tetrachloride is toxic and carcinogenic, so you will use a saturated solution of bromine in water instead. When this solution is shaken with a solution of an unsaturated compound dissolved in dichloromethane, bromine migrates to the dichloromethane layer, where it can react. Thus decolorization of the lower dichloromethane layer upon shaking indicates unsaturation, and the amount of bromine water it takes to produce a color in the lower layer is a measure of the degree of unsaturation.

Directions

Dichloromethane may be harmful if inhaled or absorbed through the skin, and it is a suspected carcinogen. Avoid contact and do not breathe its vapors.
Bromine is toxic and corrosive and its vapors are very harmful. Avoid contact with the saturated bromine water and do not breathe its vapors.

Safety Notes

Test as many of the following commercial products as possible (your instructor may add or remove some):

butter	canola oil
corn oil	dry-cleaning solvent
linseed oil	margarine
mineral oil	olive oil
paint thinner (mineral spirits)	rubber cement
rubbing alcohol	safflower oil
turpentine	vegetable shortening (Crisco, etc.)

Have ready enough clean 13 × 100-mm test tubes (with stoppers) to test each commercial product provided, and label each test tube appropriately. Using a calibrated Pasteur pipet [OP-5] to measure the solvent, dissolve 1 drop of each commercial product (or about 30 mg of a solid) in 0.5 mL of dichloromethane. *Under the hood,* use a clean, calibrated Pasteur pipet to add 0.5 mL of saturated bromine water to each test tube. Stopper each test tube and shake it vigorously for 10 seconds. Record the volume of bromine water added during this and all future additions. If the dichloromethane layer in a given test tube has a red-orange color, classify the commercial product it contains as saturated and set it aside. Add another 0.5 mL portion of bromine water to each of the remaining test tubes, stopper them, and shake as before. Repeat this process with any test tube whose dichloromethane layer is clear after shaking, until the dichloromethane layer of every test tube is a red-orange color. Record the total volume of bromine water you added to each test tube.

In your report, classify each commercial product as saturated or unsaturated and compare your results with your predictions. Considering the volume of bromine water you added to each test tube as a rough measure of the unsaturation of its contents, arrange the commercial products in order of degree of unsaturation. Note that there may be several compounds with about the same degree of unsaturation. Using sources such as those cited in the Bibliography, find out what kinds of unsaturated compounds may be present in each of the unsaturated commercial products. Draw structural formulas for representative components when you can.

Take Care! Avoid contact with dichloromethane and bromine water and do not breathe their vapors.

Waste Disposal: Place the dichloromethane layers in a designated solvent recovery container.

MINILAB **21**   Free-Radical Stability

Before You Begin: Try to predict the structures and relative stabilities of the free-radical intermediates that will form when each of the arenes reacts with bromine, and from that, predict the relative reactivities of the arenes.

The side chain of an arene such as toluene can undergo free-radical halogenation in the presence of light, as illustrated for the bromination of toluene. The rate of such a reaction depends on the stability of the intermediate free radical. As a rule, the more stable a free radical is, the faster it will form. In this minilab, you will determine the relative stabilities of the intermediates by measuring the relative reactivities of the corresponding substrates.

Directions

Safety Notes

Add 1.0 mL of dichloromethane to each of four clean, dry, labeled 13 × 100-mm test tubes. To each test tube, add 6 drops of a different aromatic hydrocarbon—toluene, ethylbenzene, isopropylbenzene, or *t*-butylbenzene—and swirl to mix. *Under the hood*, carefully add 6 drops of *freshly prepared* 1.0 *M* bromine in dichloromethane to each test tube, then stopper and shake it. Record the time of bromine addition to the nearest minute. Set the test tubes in a well-lighted location, such as a windowsill, and record the approximate time it takes for each solution to become colorless. Observe the solutions closely for the first 5 minutes or so and then at intervals during the lab period. If two or more solutions are not completely colorless by the end of the lab period, describe the relative intensity of their colors at that time.

Take Care!   Avoid contact with dichloromethane, the arenes, and the bromine solution; do not breathe their vapors.

Arrange the four hydrocarbons in order of their reactivity toward bromine, most reactive first, and compare your results with your predictions. Write a balanced equation for each reaction (assuming monobromination), and give the structure of the free-radical intermediate. Arrange the free radicals in order of stability and explain their relative stabilities. Write a complete mechanism for the bromination of toluene.

Waste Disposal: Place the bromine/ dichloromethane solutions in a designated solvent recovery container.

The Nylon Rope Trick   MINILAB **22**

Before You Begin: Review the reactions of acyl chlorides with amines in your lecture textbook.

In a chemical magic show, the "magician" will sometimes combine two immiscible solutions in a beaker and draw out a seemingly endless rope of nylon. This "nylon rope trick" is based on the fact that reactive molecules in two immiscible liquid layers may come together and form a product at the interface between the liquids. If the product is a polymer that—like nylon—forms strong fibers, it can be pulled from the interface like a rope. If the rope is drawn out slowly enough, the nylon that is removed is continually replenished at the interface until one of the reactants is used up.

In this minilab, the reactants are 1,6-hexanediamine (hexamethylenediamine) and decanedioyl chloride (sebacoyl chloride). The product, because it is made from a 6-carbon diamine and a 10-carbon acyl dichloride, is called nylon 6,10.

$$n\text{H} - \overset{\overset{\text{H}}{|}}{\text{N}}(\text{CH}_2)_6\overset{\overset{\text{H}}{|}}{\text{N}} - \text{H} + n\text{Cl} - \overset{\overset{\text{O}}{\|}}{\text{C}}(\text{CH}_2)_8\overset{\overset{\text{O}}{\|}}{\text{C}} - \text{Cl} \longrightarrow$$

1,6-hexanediamine   decanedioyl chloride

$$-[\text{NH}(\text{CH}_2)_6\text{NH} - \overset{\overset{\text{O}}{\|}}{\text{C}}(\text{CH}_2)_8\overset{\overset{\text{O}}{\|}}{\text{C}}]_n^- + n\text{HCl}$$

nylon 6,10

Directions

Safety Notes

> **Acid chlorides are corrosive and harmful by inhalation and skin absorption. Avoid contact with decanedioyl chloride and do not breathe its vapors.**
> **1,6-Hexanediamine is corrosive and harmful by inhalation. Avoid contact and do not breathe its vapors.**
> **Dichloromethane may be harmful if inhaled or absorbed through the skin, and it is a suspected carcinogen. Minimize contact with the liquid and do not breathe its vapors.**
> **Wear protective gloves throughout this experiment. Avoid touching the wet polymer with your hands; if you do, wash them immediately with soap and warm water.**

Take Care! Wear gloves, avoid contact with decanedioyl chloride, dichloromethane, and 1,6-hexanediamine; do not breathe their vapors.

Mix 10 mL of 95% ethanol with 10 mL of water in a 50-mL beaker and set it aside. *Under the hood*, dissolve 0.20 mL of pure decanedioyl chloride in 10 mL of dichloromethane in a 30-mL beaker. Combine 2.0 mL of 1.0 *M* sodium hydroxide with 4.0 mL of water in a small beaker, and then stir in 0.11 g of pure 1,6-hexanediamine until it dissolves. Tilt the beaker that contains the dichloromethane solution and slowly pour the 1,6-hexanediamine solution down its side, taking care not to mix the layers. Use a flat-bladed microspatula to free the polymer film from the side of the beaker if necessary; then use a piece of copper wire, bent at one end like a fishhook, to hook the film at its center. Wearing protective gloves, pull the film up slowly and continuously to form a strand of nylon, loop it around a cardboard tube (such as the core of a paper towel or toilet tissue roll), and rotate the tube to wind it out of the solution until no more nylon rope can be drawn out. Unwind the nylon into the aqueous ethanol in the 50-mL beaker and stir gently to wash it. Then decant the wash solvent, lay the polymer on a paper towel to dry (blotting it between two towels will reduce the drying time), and weigh it when it is dry [OP-23].

Waste Disposal: Place the liquids from the 30-mL beaker in a designated solvent recovery container. Pour the wash solvent down the drain.

Nucleophilic Substitution Rates of Alcohols

Before You Begin: Classify each alcohol as primary, secondary, or tertiary and try to predict their relative reactivities with the HCl-$ZnCl_2$ reagent.

In acidic solutions, alcohols can react with various nucleophiles by either an S_N1 or S_N2 mechanism. The S_N1 mechanism involves loss of water from the protonated alcohol to form a carbocation, which then combines with the nucleophile. The S_N2 reaction involves nucleophilic attack on the protonated alcohol to yield the product directly.

$$R-OH \xrightarrow{H^-} R-\overset{+}{\underset{\underset{H}{|}}{O}}H \begin{array}{c} \overset{S_N1}{\underset{-H_2O}{\nearrow}} R^+ \xrightarrow{Nu:^-} R-Nu \\ \underset{S_N2}{\overset{Nu:^-}{\searrow}} R-Nu \end{array}$$

In this minilab, you will carry out the reaction of an alcohol with hydrochloric acid in the presence of zinc chloride, a Lewis acid catalyst.

$$R-OH + HCL \xrightarrow{ZnCl_2} R-Cl + H_2O$$

The $ZnCl_2$ apparently coordinates with the oxygen atom of the alcohol, converting its $-OH$ (a very poor leaving group) to a better leaving group,

$$-\overset{+}{\underset{\underset{H}{|}}{O}}-\overset{-}{Zn}Cl_2$$

You will use your results to decide whether the reaction appears to proceed by an S_N1 or S_N2 mechanism.

Directions

Safety Notes

Label and number three clean, dry 13×100-mm test tubes. Measure 4 drops of (1) 1-butanol, (2) 2-butanol, and (3) 2-methyl-2-propanol into the corresponding test tubes. *Under the hood*, add 2.0 mL of the HCl-$ZnCl_2$ reagent to each test tube, stopper and shake each tube vigorously for 5 seconds, and then let them stand at room temperature. Look for any evidence of a reaction, recording your observations during the first 15 minutes or so and at intervals throughout the lab period.

Take Care! Wear gloves, avoid contact with the HCl-$ZnCl_2$ solution, do not breathe its vapors.

Waste Disposal: Put the contents of your test tubes in a designated waste container.

Compare your results with your predictions. Assuming that all of the alcohols react by the same mechanism, if they react, decide whether the reaction occurs by an S_N1 or S_N2 mechanism and explain how you arrived at your conclusion. Write balanced equations and mechanisms for any reactions for which you saw evidence.

MINILAB 24   Photoreduction of Benzophenone

Before You Begin: Read or review the operations as necessary, and propose a hypothesis regarding the outcome of the reaction.

Around 1900, an Italian scientist mixed benzophenone with isopropyl alcohol and left the mixture on a sunny rooftop in Bologna, Italy. When he returned, he discovered that a white solid had formed as a result of the action of sunlight on the reactants. This is a example of a *photochemical reaction*—a reaction that is initiated by visible or ultraviolet radiation.

When a benzophenone molecule absorbs light of an appropriate wavelength, one of two paired ground-state electrons on its oxygen atom jumps to an antibonding pi orbital without changing its spin state. This puts the molecule into a *singlet* excited state, in which both electrons have opposite spins. The singlet-state molecule can undergo another transition to a *triplet* excited state, in which both electrons have the same spin.

$$\uparrow\downarrow \quad \xrightarrow{\;h\nu\;} \quad \uparrow \quad \downarrow \qquad \longrightarrow \qquad \uparrow \quad \uparrow$$

ground   singlet   triplet
state   excited state   excited state

A triplet-state benzophenone molecule is very reactive and behaves like a diradical. It can even strip a hydrogen atom from a molecule of 2-propanol (a very unlikely reducing agent), which then loses a second hydrogen atom to ground-state benzophenone and is converted to a molecule of acetone.

$$\underset{\substack{|\\ Ph}}{\overset{\substack{O^*\\ ||}}{PhC}} + \underset{\substack{|\\ Ph}}{\overset{\substack{O\\ ||}}{PhC}} + \underset{}{\overset{\substack{OH\\ |}}{CH_3CHCH_3}} \longrightarrow 2\underset{\substack{|\\ Ph}}{\overset{\substack{OH\\ |}}{PhC\cdot}} + \underset{}{\overset{\substack{O\\ ||}}{CH_3CCH_3}}$$

benzophenone
(* = triplet state)

This leaves two identical diphenylhydroxymethyl radicals that might combine to yield a molecule of benzopinacol. Bases can catalyze cleavage of this product, so a drop of acetic acid is added to neutralize any base present in the reaction mixture. Another possible product of the reaction is

$$2\underset{\substack{|\\ Ph}}{\overset{\substack{OH\\ |}}{PhC\cdot}} \longrightarrow \underset{\substack{|\\ Ph}}{\overset{\substack{OH\\ |}}{PhC}}\!-\!\underset{\substack{|\\ Ph}}{\overset{\substack{OH\\ |}}{CPh}}$$

benzopinacol

benzhydrol, which can form if a diphenylhydroxymethyl radical abstracts a hydrogen atom from 2-propanol.

$$\underset{\text{PhCPh}}{\overset{\text{OH}}{|}} + \underset{\text{CH}_3\text{CHCH}_3}{\overset{\text{OH}}{|}} \longrightarrow \underset{\underset{\text{benzhydrol}}{\text{PhCHPh}}}{\overset{\text{OH}}{|}} + \underset{\text{CH}_3\text{CHCH}_3}{\overset{\text{OH}}{|}}$$

In this minilab, you will find out whether benzophenone will react with isopropyl alcohol in the presence of light and, if so, whether the product is benzhydrol or benzopinacol.

Directions

Safety Notes

Acetic acid causes chemical burns and its vapors are highly irritating to the eyes and respiratory tract. Dispense under a hood, avoid contact, and do not breathe its vapors.

Combine 0.40 g of benzophenone with 2.0 mL of 2-propanol (isopropyl alcohol) in a small test tube and warm [OP-7] the mixture gently with manual stirring to dissolve the benzophenone. *Under the hood,* add a small drop of glacial acetic acid, then stopper the test tube tightly and shake it. Place it on a windowsill where it will receive direct sunlight, if possible; otherwise, place it close to a strong artificial light source. After a week or more, collect the precipitate by vacuum filtration [OP-13], washing it on the filter with cold ethanol. Dry [OP-23] the product and measure its melting point [OP-30]. Leave the filtrate in sunlight for another week or so to see if any additional product forms. If it does, collect and dry it also; then weight the combined product.

Decide whether the solid you obtained is benzopinacol, benzhydrol, or unreacted benzophenone by looking up these compounds in *The Merck Index.* Write a balanced equation for the reaction.

Take Care! Avoid contact with acetic acid and do not breathe its vapors.

If the laboratory is very cold, some benzophenone may precipitate before it reacts.

Waste Disposal: Place the filtrate in a designated solvent recovery container.

Oxidation of Alcohols by Potassium Permanganate

MINILAB **25**

Before You Begin: Classify the alcohols as primary, secondary, and tertiary. Review the reactions of alcohols in your lecture textbook.

Primary and secondary alcohols can be converted to aldehydes and ketones by certain oxidizing agents. Tertiary alcohols are not converted directly to carbonyl compounds by oxidizing agents, because such a reaction would require cleavage of a carbon-carbon bond. In this minilab, you will treat several alcohols with neutral potassium permanganate, then with potassium permanganate acidified with HCl, and compare their reactivities. You will also observe the effect of adding sulfuric acid to any reaction mixture containing an alcohol that has not yet reacted.

$$\underset{\substack{|\\-\text{C}-\text{H}\\|}}{\overset{\text{O}-\text{H}}{|}} \xrightarrow[-2\text{H}]{\text{oxidizing agent}} \underset{|}{\overset{\text{O}}{\overset{\|}{-\text{C}}}}$$

Directions

Clean and number five small test tubes. Add 1 mL of aqueous 0.05% potassium permanganate to each test tube. Use test tube #1 as a control and add one drop of each of the following alcohols to the other test tubes: #2, methanol; #3, ethanol; #4, 2-propanol; #5, 2-methyl-2-propanol. Stopper and shake each test tube for 10 seconds and record your observations immediately after shaking, then after standing for 5 and 10 minutes, respectively. To any solution that shows little or no evidence of reaction (compared to the control) after 10 minutes, add 2 drops of 3.0 M HCl, shake, and observe at the same intervals as before. To any solution that shows little or no evidence of reaction 10 minutes after the HCl addition, add 2 drops of concentrated sulfuric acid, shake, and observe at the same intervals as before.

Stop and Think: What is the solid that forms?

Take Care! Avoid contact with H_2SO_4.
Waste Disposal: Place the potassium permanganate test solutions in a designated waste container.

Arrange the alcohols in order of reactivity toward $KMnO_4$ and describe the effect of substrate structure on reactivity. Based on alcohol reactions you have studied, explain any reaction of a tertiary alcohol for which you saw evidence. Draw the structure of the organic product of each reaction. Write a balanced equation for the reaction of methanol with acidic $KMnO_4$ (you can balance it using the half-reaction method taught in general chemistry).

MINILAB 26 # Preparation of a Fluorescent Dye

Before You Begin: Calculate the molar amounts of resorcinol and phthalic anhydride, decide which one is the limiting reactant, and calculate the theoretical yield of fluorescein.

Certain dyes and indicators, such as phenolphthalein and fluorescein, resemble the triphenylmethane dyes (see Experiment 30) in that their molecules contain three benzene rings attached to a central carbon atom. In this minilab, you will prepare fluorescein by a Lewis-acid catalyzed reaction of resorcinol (1,3-dihydroxybenzene) with phthalic anhydride.

You will convert the fluorescein to its basic form, disodium fluorescein, by dissolving it in dilute NaOH. Then you will irradiate the resulting solution with ultraviolet light and record your observations.

disodium fluorescein

Directions

As written, this minilab requires the use of a "shortie" (~150 mm) glass thermometer. If only standard-length thermometers are available, the quantities should be doubled.

> **Zinc chloride is toxic and corrosive, with harmful fumes. Resorcinol and phthalic anhydride are severe skin and eye irritants. Wear gloves and use a hood during this experiment. Avoid contact with zinc chloride, resorcinol, and phthalic anhydride, and do not breathe their fumes or dust. Before you light a burner, make sure that no flammable solvents are nearby.**

Safety Notes

Under the hood, measure 0.16 g of anhydrous zinc chloride into a 13 × 100-mm test tube. Heat [OP-7] it (preferably in an aluminum heating block) until the salt melts and no more bubbles of vapor are evolved. When you remove the test tube from the heat source, tilt and rotate it so that the zinc chloride coats its sides as it cools. Add 0.22 g of phthalic anhydride and 0.36 g of resorcinol, and heat the test tube cautiously until the temperature of the melt is 180°C; use a "shortie" thermometer, if available, to measure the temperature [OP-9]. Keep it near that temperature, stirring continuously with a stirring rod, until no more bubbles are evolved. Let the mixture cool for a minute or so; then add 2.5 mL of 2.0 *M* hydrochloric acid and heat the mixture to boiling with constant stirring. Continue to stir it at the boiling point for 2–3 minutes to dissolve zinc salts and unreacted starting materials, using the stirring rod (as needed) to break up the solid mass. Cool the mixture in an ice/water bath and collect the crude product by vacuum filtration [OP-13]. Transfer the air-dried solid to a 13 × 100-mm test tube and use a sturdy stirring rod, as needed, to crush any lumps and grind it to a powder. Add another 2.5-mL portion of 2.0 *M* HCl and boil the mixture with stirring as before. Cool it in an ice/water bath and collect the fluorescein by vacuum filtration. Dry [OP-23] the fluorescein and weigh it.

Dissolve about 10 mg (0.010 g) of fluorescein in 10 mL of 0.10 *M* sodium hydroxide. Irradiate the solution with an ultraviolet lamp in a darkened room. Add 2.0 *M* hydrochloric acid drop by drop until the appearance of the solution under the UV lamp changes markedly, and record your observations. Describe and explain your observations, and explain why the color of a fluorescein solution is different at high and low pH.

Take Care! Wear gloves, avoid contact with anhydrous zinc chloride, do not breathe its fumes.

Take Care! Avoid contact with phthalic acid and resorcinol. Do not use the thermometer as a stirring rod.

Fluorescein stains are very hard to remove, so keep it off your skin and clothing.

Waste Disposal: Pour the filtrates down the drain.

Take Care! Do not look directly at the UV light source.

Waste Disposal: Place the solution in a designated waste container.

MINILAB 27
Diels-Alder Reaction of Maleic Anhydride and Furan

Before You Begin: Read or review the operations as necessary. Referring to Experiment 31, propose a hypothesis regarding the stereochemistry of the adduct.

Some aromatic compounds behave like conjugated dienes when reacting with certain dienophiles. Furan is an aromatic heterocyclic compound that undergoes Diels-Alder reactions as if it existed as the diene shown in the margin (this is one of several resonance structures for furan). Furan forms a bicyclic adduct with maleic anhydride that can have either an *endo* or an *exo* stereochemistry.

furan   resonance structure
          for furan

furan   maleic   *endo* adduct   *exo* adduct
          anyhdride

The transition state leading to an *endo* adduct is stabilized by pi-electron overlap, causing the *endo* adduct to form faster. When a Diels-Alder reaction is under *kinetic control*, meaning that the product that forms faster is favored, the *endo* adduct is the major or sole product. However, most *endo* adducts are less stable than the corresponding *exo* adducts, so when a Diels-Alder reaction is under *thermodynamic control*, meaning that the more stable product is favored, the *exo* adduct should be the major or sole product.

In this minilab, you will prepare the Diels-Alder adduct of maleic anhydride and furan, measure its melting point, and decide whether the reaction is under kinetic or thermodynamic control. Since the reaction takes several days, you should prepare the reaction mixture during a lab period prior to the one in which you plan to isolate and analyze the product. The melting point of the *endo* adduct is 81°C and the melting point of the *exo* adduct is 114°C. The adduct tends to decompose near its melting point, so measure an approximate melting point (T_1) at a rapid rate, then obtain a more accurate melting point by raising the temperature slowly and inserting a melting point capillary when it reaches T_1.

Stop and Think: What products do you think the adduct forms when it decomposes?

Directions

Safety Notes

Maleic anhydride is corrosive and toxic and can cause severe damage to the eyes, skin, and upper respiratory tract. Wear gloves, avoid contact, and do not breathe its dust. If you must powder maleic anhydride

briquettes, do it under the hood and wear safety goggles and protective clothing.
Diethyl ether and hexanes are extremely flammable and may be harmful if inhaled. Do not breathe their vapors and keep them away from flames and hot surfaces.
Furan is flammable and harmful by inhalation, ingestion, and skin contact. Avoid contact, do not breathe its vapors, and keep it away from flames and hot surfaces.

Dissolve 0.24 g of finely divided maleic anhydride in 2.0 mL of anhydrous diethyl ether in a centrifuge tube or a small test tube by warming the mixture gently in a hot water bath [OP-7]. Replace any ether that evaporates. When the solution is cool, add 0.20 mL of furan (accurately measured). Cap or stopper the reaction tube tightly to prevent evaporation of ether, and leave it in a designated location for several days or until the next lab period. Collect the crystallized adduct by vacuum filtration [OP-13] and air-dry it on the funnel. Recrystallize [OP-25] the adduct from hexanes/ethyl acetate by heating it *just* to boiling in 1.0 mL of hexanes, then adding enough warm ethyl acetate (0.2–0.6 mL) to the mixture, with stirring, to dissolve the solid. After you collect the product by vacuum filtration, wash it on the filter with cold hexanes. Dry [OP-23] the adduct and measure its mass and melting point [OP-30].

Tell whether the adduct is *endo* or *exo* and propose an explanation for your result, writing equations for any relevant reactions. (*Hint:* Why do you think this adduct decomposes at its melting point while most other maleic anhydride adducts do not?)

Take Care! Wear gloves, avoid contact with maleic anhydride and furan, and do not breathe their dust or vapor. Keep flames away from ether.

Waste Disposal: Place the filtrate in the appropriate solvent recovery container.

Take Care! Keep flames away from the solvents.

Waste Disposal: Place the filtrate in the appropriate solvent recovery container.

Identification of an Unknown Arene by NMR Spectrometry

MINILAB 28

Before You Begin: Read or review OP-37, "Nuclear Magnetic Resonance Spectrometry." Be sure you understand how to interpret NMR spectra by reading the appropriate sections in OP-37 and, if necessary, relevant sections from your lecture textbook.

NMR spectrometry is one of the most valuable tools at the disposal of an organic chemist. A chemist skilled in spectral interpretation can often derive the molecular structure of a complex organic compound from its ^1H and ^{13}C NMR spectra alone. In this minilab, you will record the ^1H NMR spectrum of a relatively simple aromatic hydrocarbon (arene) or another type of compound selected by your instructor. At your instructor's option, you may also record its ^{13}C NMR spectrum. You will then use your spectrum (or spectra) to deduce the molecular structure of the unknown compound.

Directions

Your instructor may elect to assign you a compound with a different molecular formula than the one specified here.

Take Care! Avoid contact with $CDCl_3$ and do not breathe its vapors.

Waste Disposal: Place the deuterochloroform solution in a designated solvent recovery container. Place any remaining unknown in a designated waste container.

You will be issued a small amount of an unknown arene having the molecular formula $C_{10}H_{14}$. As directed by your instructor, record an integrated ^1H NMR spectrum [OP-37] of the compound in deuterochloroform ($CDCl_3$) using TMS as a reference compound. If requested, record its ^{13}C NMR spectrum as well, or obtain the spectrum from your instructor.

Letter the signals (a, b, c, etc.) from right to left and construct a table listing the chemical shift (δ), signal area, and multiplicity for each signal. For each signal, determine the number of protons giving rise to the signal, the number of neighboring protons (except for Ar**H** signals), and the proton type (aromatic, benzylic, etc.); include this information on your table. Use this information to deduce the structure of your unknown compound and explain, in detail, how you arrived at it. Turn in your table and NMR spectrum (or spectra) with your report.

MINILAB 29   Interpretation of a Mass Spectrum

Before You Begin: Read OP-39, "Mass Spectrometry." If you are to analyze an acetate ester of an unknown alcohol, also read the spectral interpretation rules provided by your instructor.

When an organic compound is injected into a mass spectrometer, its molecules are bombarded by a stream of high-energy electrons. When such an electron encounters a molecule, it can dislodge one of the molecule's electrons, producing a charged *molecular ion*, $M \cdot^+$. Each molecular ion can then fragment in a variety of ways, giving rise to an array of *daughter ions*. A mass spectrum of a compound is a record of all of the molecular ions and daughter ions arising from the fragmentation of its molecules. Each peak on a mass spectrum corresponds (usually) to one kind of ion; its location gives the mass/charge ratio of the ion (usually equal to its mass) and its intensity indicates the relative abundance of the ion.

In this minilab, you will obtain the mass spectrum of the major component of an unknown liquid using a gas chromatograph-mass spectrometer (GC-MS), or the mass spectrum of a pure liquid using a mass spectrometer. Following your instructor's directions and the procedure described here, you will then interpret your mass spectrum as completely as you can.

Directions

This minilab may be performed by teams of three or more students, each working with a different unknown.

Safety Notes

You will receive a liquid in a capped vial for analysis. Alternatively, you can use an acetate ester of an unknown alcohol prepared as described in Minilab 5. Your instructor will show you how to operate the mass spectrometer or GC-MS instrument [OP-39], or will operate it for you. Unless directed otherwise, inject a 2–3-μL head-space sample into the injection port and obtain a printout of the ester's mass spectrum.

Identify the base peak and molecular ion peak and determine the molecular weight of your compound. Analyze as many peaks as you can by assigning structures for the species that may be responsible for them, including (for example) species arising from α-cleavage and McLafferty rearrangements. At your instructor's request, determine the molecular formula of the compound and attempt to derive its structure, working with the other members of your team. Summarize your spectral data and interpretations in a table, and report and justify your conclusions. Write mechanisms for the formation of as many species as you can.

Nitration of Naphthalene   MINILAB **30**

Before You Begin: Read or review the operations as necessary. Review the mechanisms of electrophilic aromatic substitution reactions, as needed.

Polynuclear aromatic hydrocarbons such as naphthalene can be nitrated by the same methods as benzene derivatives, including the well-known "mixed acid" method that utilizes a mixture of nitric acid and sulfuric acid. The electrophile is the nitronium ion, NO_2^+, which is formed at low concentration by a reaction of the two acids. The mononitration of naphthalene could lead to either one of two products, 1-nitronaphthalene or 2-nitronaphthalene.

1-nitronaphthalene   2-nitronaphthalene
m.p. 61°C   m.p. 79°C

The major product of the reaction you will carry out is the kinetic product, the one that forms from the more stable carbocation intermediate (also called an arenium ion). By drawing all possible resonance structures for the carbocation intermediate leading to each of these products, you should be able to predict the major product. You will then carry out the nitration of naphthalene, identify the product from its melting point, and find out whether your prediction was correct. Because it is difficult to purify the product completely, your melting point may be somewhat lower than the listed value.

Directions

Safety Notes

If the mixed-acid nitrating solution has not been prepared, make it under the hood (wear gloves) by measuring 0.50 mL of concentrated nitric acid into a conical centrifuge tube or test tube, cooling it in ice, and cautiously mixing in 0.50 mL of concentrated sulfuric acid.

Take Care! Wear gloves while handling mixed acid, avoid contact, and do not breathe its vapors. Do not use your thermometer as a stirring rod.

Waste Disposal: Dispose of the liquid as directed by your instructor.

Waste Disposal: Pour the filtrate down the drain.

Take Care! Keep flames away from hexanes.

Waste Disposal: Place the filtrate and any remaining oil in a designated solvent recovery container.

> **Sulfuric acid and nitric acid can cause very serious burns and they react violently with water and other chemicals. Nitric acid produces toxic nitrogen dioxide fumes during the reaction. Use gloves and a hood, avoid contact with the acids, and do not breathe their vapors.**
> **The nitronaphthalene is a suspected carcinogen; avoid contact.**
> **The mixture of hexanes is very flammable; keep flames away.**

Measure 1.0 mL of the mixed acid (1:1 sulfuric acid and nitric acid) into a 3-mL conical vial. In small portions, add 0.50 g of finely divided naphthalene to the mixed acid, stirring or shaking [OP-10] after each addition and cooling [OP-8] as necessary to keep the temperature around 45–50°C. Then stir the reaction mixture in a 60°C water bath for 20 minutes. After the reaction mixture cools to room temperature, transfer it to a beaker containing 25 mL of ice-cold water while stirring. Wait for the yellow product to solidify, then remove the liquid with a filter-tip pipet [OP-6], leaving the solid behind. *Under the hood*, boil the solid with 10 mL of fresh water for 10 minutes or more. Then cool the mixture in ice and collect the product by vacuum filtration [OP-13]. Recrystallize [OP-25] about 0.10 g of the crude product by heating it under reflux [OP-7] for 5 minutes or more with 5.0 mL of hexanes and filtering [OP-12] the hot solution (leave any undissolved oil behind) with a preheated filtering pipet. When crystallization is complete, collect the product by vacuum filtration. Dry [OP-23] it and the unpurified product. Measure the total mass of the nitronaphthalene and the melting point [OP-30] of the purified product.

Deduce the structure of the product. Write a mechanism for the reaction, showing all resonance structures for the intermediate, and explain its orientation.

MINILAB 31

Preparation of Carbocations by the Friedel-Crafts Reaction

Before You Begin: Review the mechanism of Friedel-Crafts alkylation in your lecture textbook.

A simple chemical test for aromatic hydrocarbons can be carried out by pipetting a solution of the unknown compound in trichloromethane (chloroform) over some freshly sublimed aluminum chloride. If the un-known is aromatic, a characteristic color (other than light yellow) will ap-pear. The test is based on the Friedel-Crafts reaction of the aromatic hydrocarbon with trichloromethane, which, because it has three chlorine atoms, reacts with three molecules of the aromatic hydrocarbon to form a triarylmethane. The triarylmethane then loses a hydride ion to one of sever-al different carbocation intermediates, yielding a colored triarylmethyl car-bocation that is structurally related to the triphenylmethane dyes discussed in Experiment 30.

$$3ArH + CHCl_3 \xrightarrow{AlCl_3} Ar_3CH + 3HCl$$

$$Ar_3CH + R^+ \longrightarrow Ar_3C^+ + RH \qquad (R^+ = Ar_2CH^+, etc.)$$

In this minilab, you will carry out the Friedel-Crafts reaction on a test-tube scale with several aromatic hydrocarbons and observe the colors of the products. You will use only a few drops of trichloromethane for each reac-tion, but because it is a suspected carcinogen you must wear gloves and work under a hood.

Directions

At your instructor's option, this minilab can be carried out by teams of two or more students. Your instructor may add or remove some test compounds.

Safety Notes

> **Aluminum chloride reacts violently with water; skin and eye contact can cause painful burns, and inhaling its dust or vapors is harmful. Wear gloves, avoid contact, do not inhale dust or vapors, and keep it away from water.**
> **Trichloromethane is harmful if inhaled, ingested, or absorbed through the skin, and it is a suspected human carcinogen. Wear gloves and work under a hood. Avoid contact with the liquid and do not breathe its vapors.**
> **The aromatic hydrocarbons have harmful vapors; avoid inhalation.**

Label five or more clean, *dry* test tubes and add one of the following com-pounds—1 drop of a liquid or about 20 mg of a solid—to each test tube: toluene, limonene, naphthalene, anthracene, and biphenyl. *Under the hood*, add 8 drops of trichloromethane to each test tube. *Under the hood*, carefully measure about 0.10 g of anhydrous aluminum chloride into a clean, *dry* 13 × 100-mm test tube. Heat the test tube gently over a burner flame [OP-7] until the aluminum chloride forms a thin layer of sublimed solid up the sides of the tube. When it is cool, pipet the toluene solution down one side that is coated with the white solid. Repeat the test, apply-ing the remaining solutions to fresh areas of the sublimed solid. You should be able to use the same AlCl$_3$ layer to test several compounds; if necessary, prepare a fresh layer on another clean, dry test tube. Tabulate and explain your observations, telling which compounds are aromatic. Write equations and a detailed mechanism for the reaction of toluene with trichloromethane, assuming *para* substitution.

Take Care! Wear gloves, avoid contact with trichloromethane and aluminum chloride, and do not breathe their vapors.

Take Care! Be sure that there are no flammable solvents in the vicinity.

Observe and Note: What hap-pens? Describe your observations.

Waste Disposal: Rinse the test tubes that contain the trichlorome-thane solutions into a designated waste container.

Air Oxidation of Fluorene to 9-Fluorenone

MINILAB 32

Before You Begin: Read or review the operations as necessary.

Oxidizing the side chain of an arene usually requires a powerful oxidizing agent such as potassium permanganate or chromium(III) oxide in acetic anhydride, as illustrated for the following reactions of toluene:

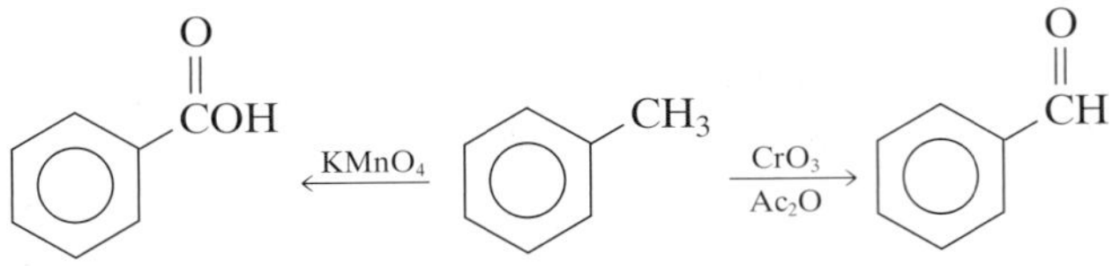

toluene

The methylene side chain of fluorene, however, can be oxidized to a carbonyl group by a much milder oxidizing agent, air.

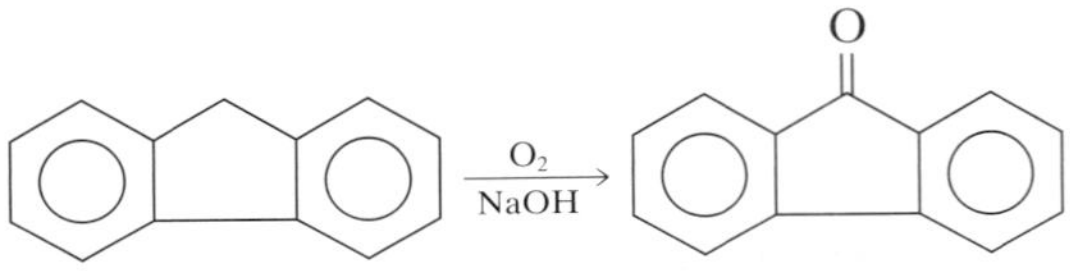

fluorene   9-fluorenone

This reaction, which yields the aromatic ketone 9-fluorenone, is ordinarily quite slow, but with vigorous stirring and a phase-transfer catalyst, it can be carried out in about an hour.

Directions

Safety Notes

Sodium hydroxide is toxic and corrosive, causing severe damage to skin, eyes, and mucous membranes. Wear gloves and avoid contact with the NaOH solution.

Take Care! Keep flames away from heptane. Wear gloves, and avoid contact with the NaOH solution.

Waste Disposal: Place the filtrates containing heptane and cyclohexane in a designated waste container.

Mix 0.40 g of fluorene with 4.0 mL of heptane in a 25-mL Erlenmeyer flask. Add 2.0 mL of 10 M (~30%) sodium hydroxide. Then add 5 drops of tricaprylmethylammonium chloride (Aliquat 336) and stir [OP-10] the mixture vigorously for an hour or more, using a 12-mm or longer magnetic stir bar. The stirring rate should be high enough to produce a froth on the surface of the reaction mixture. Transfer the reaction mixture to a 15-mL centrifuge tube, cool [OP-8] it in an ice/water bath, and carefully remove the aqueous (lower) layer with a Pasteur pipet. Collect the crude fluorenone by vacuum filtration [OP-13], washing it on the filter with a small amount of 1.0 M HCl, and then with water. Dry [OP-23] it at room temperature or in an oven at 60°C. Recrystallize [OP-25] the 9-fluorenone from cyclohexane and measure its mass and melting point [OP-30]. (**Take Care!** Keep flames away from cyclohexane.)

Tell why fluorene is oxidized by O_2 much more rapidly than is toluene, given that oxidation by O_2 involves free-radical intermediates.

A Nucleophilic Addition-Elimination Reaction of Benzil

MINILAB **33**

Before You Begin: Read or review the operations as necessary. Review the reactions of carbonyl compounds with ammonia derivatives in your lecture textbook.

Like the Wittig reaction (see Experiment 39), the reaction of an ammonia derivative with a carbonyl compound involves a nucleophilic addition step, followed by an elimination step that yields a double bond.

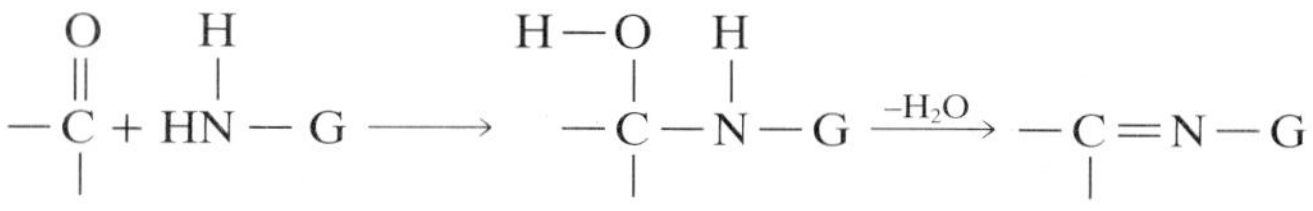

nucleophilic addition   elimination

G = R, Ar, NHAr, OH, etc.

Benzil has two carbonyl groups, so it can react with certain ammonia derivatives that have two NH_2 groups to form cyclic compounds. In this minilab, you will carry out the reaction of benzil with 1,2-benzenediamine (*o*-phenylenediamine) to prepare the heterocyclic amine 2,3-diphenylquinoxaline.

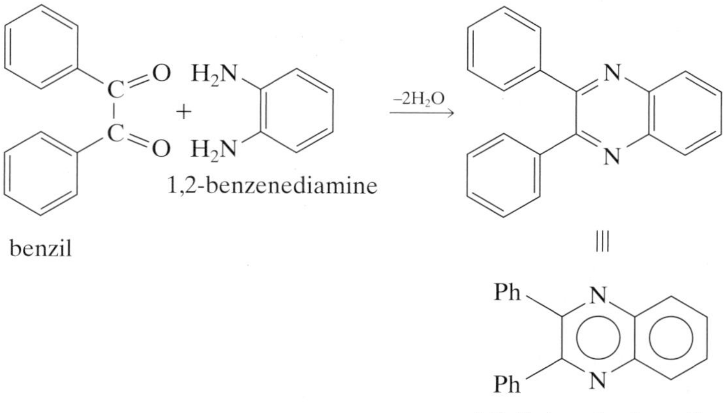

benzil   1,2-benzenediamine

2,3-diphenylquinoxaline

Directions

1,2-Benzenediamine can cause dermatitis and serious eye damage. Wear gloves and avoid contact.

Safety Notes

Dissolve 0.25 g of benzil in 1.0 mL of warm 95% ethanol in a 5-mL conical vial. Dissolve 0.13 g of 1,2-benzenediamine in 1.0 mL of 95% ethanol in a small test tube and add this solution to the benzil solution. Heat [OP-7] the reaction mixture in a 50°C water bath for 20–30 minutes. Add enough water dropwise, with manual or magnetic stirring [OP-10], to just saturate the warm solution (watch for a slight cloudiness that persists), then cool the reaction tube in an ice/water bath until crystallization is complete. Recover the product by vacuum filtration [OP-13]. Recrystallize [OP-25] the 2,3-diphenylquinoxaline from ethanol/water mixed solvent. Dry [OP-23] the purified product and measure its mass and melting point [OP-30].

Write a detailed mechanism for the reaction of benzil with 1,2-benzenediamine.

MINILAB 34

Preparation of Aldol Condensation Products

This minilab is based on an article published in the *Journal of Chemical Education* **1987**, *64*, 367.

Before You Begin: Read or review the operations as necessary. Review the mechanisms of aldol condensation reactions. Be prepared to calculate the mass and volume of 1.00 mmol of any ketone and 4.0 mmol of any aldehyde in Table M3.

Table M3 Aldehydes and ketones for the aldol condensation reactions

	Compound	Molecular weight	Density
aldehydes	benzaldehyde	106.1	1.045
	4-methylbenzaldehyde	120.2	1.019
	4-methoxybenzaldehyde	136.2	1.119
	trans-cinnamaldehyde	132.2	1.050
ketones	acetone	58.1	0.791
	cyclopentanone	84.1	0.951
	cyclohexanone	98.2	0.948
	4-methylcyclohexanone	112.2	0.916

Note: Density is in g/mL.

As described in Experiment 40, aromatic aldehydes react with aliphatic ketones in the presence of a base to form aldol-type condensation products. In the presence of excess aldehyde, each mole of ketone can condense with two moles of the aldehyde, as illustrated for the following reaction of benzaldehyde and acetone:

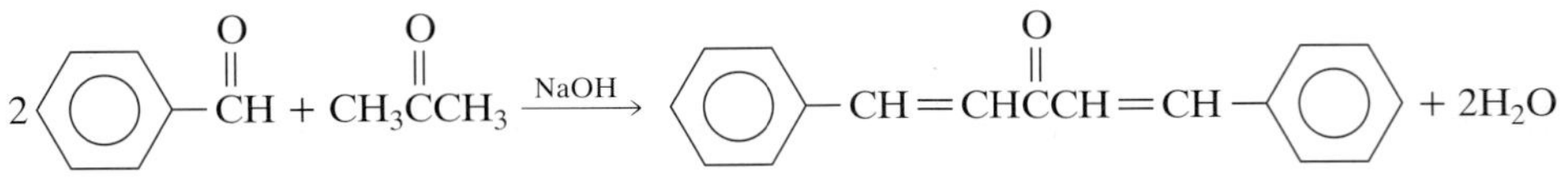

$$2\ C_6H_5{-}CHO + CH_3CCH_3 \xrightarrow{\text{NaOH}} C_6H_5{-}CH{=}CHCCH{=}CH{-}C_6H_5 + 2H_2O$$

dibenzalacetone

In this minilab, you will select your reactants from a list containing four aldehydes and four ketones, making it possible to synthesize up to 16 different aldol condensation products.

If you work in a group, the group members can select combinations of reactants that will enable them, by sharing information, to learn about the effect of structure on reactivity in aldol condensation reactions.

Directions

With your instructor's permission, you can work in groups of four to six or more students, with each student in a group using a different set of reactants.

Assume that the aldehydes and ketones are flammable and harmful by inhalation, ingestion, and skin absorption. Minimize contact and avoid inhaling their vapors.

Select one aldehyde and one ketone from Table M3, or from a list provided by your instructor. If you are working in a group, select a set of reactants that is not being used by any other member of the group. Accurately measure 1.00 mmol of your ketone into a 5-mL conical vial and dissolve it in 2.0 mL of 95% ethanol. Stir in 4.0 mmol of your aldehyde (it can be measured by volume) and 1.5 mL of 1.0 M NaOH. Let the reaction mixture stand at room temperature for 10 minutes with stirring [OP-10] (occasional manual stirring is sufficient). Record the time it takes for a precipitate to appear. If little or no precipitate has formed after 10 minutes, add a boiling chip and heat the reaction mixture in a boiling-water bath for an additional 10 minutes. Be careful that it does not boil over. Let the reaction mixture cool to room temperature, then cool it further in an ice-water bath. Collect the aldol condensation product by vacuum filtration [OP-13], washing it on the filter with 1 mL of cold 4% (v/v) acetic acid in 95% ethanol followed by 1 mL of cold 95% ethanol. (**Stop and Think:** What is the purpose of the acetic acid?) Dry [OP-23] the crude product and measure its mass and melting point [OP-30]. At your instructor's request, recrystallize [OP-25] some of the product from a suitable solvent before you measure its melting point. Either 95% ethanol, toluene, or 2-butanone may be a suitable recrystallization solvent.

Give the structure of your product and an equation and detailed mechanism for its formation. If you worked in a group, prepare a table giving the structure of each product and its mass, melting point, and percentage yield. Based on the yields and other information gathered from the members of your group, propose a hypothesis to account for the effect of aldehyde structure on reactivity in these reactions.

Safety Notes

Waste Disposal: Pour the filtrate down the drain.

Take Care! Avoid contact with toluene and 2-butanone, and do not breathe their vapors.

Waste Disposal: Place the filtrate from the recrystallization in a designated waste container.

A Spontaneous Reaction of Benzaldehyde

MINILAB **35**

Before You Begin: After reading the minilab, develop a hypothesis about the possible identity of the product. Then gather information and develop a plan of action that you will use to identify it.

In this minilab, you will observe a spontaneous reaction of benzaldehyde, isolate the product, and identify it by any means at your disposal. If you are not sure about the availability of certain equipment, test solutions, and so on, ask your instructor. As in many other experiments of this type, a little thought and preparation can save you hours of experimental work, so plan ahead.

Directions

Benzaldehyde and the product are mild irritants; avoid contact with them.

Put about 10 mL of clean dry sand into a small beaker. Stir in just enough benzaldehyde to moisten the sand—there should be no standing liquid on top. Bend a pipe cleaner into a "U" shape and push both ends into the sand so that the inverted "U" is vertical. Label the beaker with your name and set it aside until the next lab period. Scrape the solid off the pipe cleaner and transfer it to a test tube or small Erlenmeyer flask for purification. (If there is not much solid on the pipe cleaner, break up the sand with a spatula, stir it with 5 mL of boiling water, and filter the hot solution into a beaker. After crystallization begins, cool the solution and collect the product by vacuum filtration.) Recrystallize [OP-25] the crude product from boiling water. Dry [OP-23] the product and identify it by any means at your disposal.

Give the name and structure of your product, and write a balanced equation for its formation from benzaldehyde.

Waste Disposal: Pour the filtrate down the drain.

Stop and Think: What is there for the benzaldehyde to react with?

MINILAB 36

Acid-Base Strengths of Organic Compounds

Before You Begin: Review sections of your organic chemistry or general chemistry lecture texts about the effect of structure on acid-base strength and about writing resonance structures, and try to predict the relative acid-base strengths of the compounds listed.

In your general chemistry course, you learned some rules for predicting acid-base strengths. For example, the strength of an acid depends to some extent on the periodic-table location of the atom to which a proton is bonded; thus compounds with $H—O$ bonds tend to be more acidic than those with $H—N$ or $H—C$ bonds, and less acidic than those with $H—S$ bonds. Conversely, oxygen bases are less basic than comparable nitrogen bases and more basic than comparable sulfur bases. But most organic acids have $O—H$ bonds and most organic bases are nitrogen bases, so such rules are not very helpful in organic chemistry. To predict the strength of an organic acid, you must consider the stability of its conjugate base. Increasing the stability of an acid's conjugate base shifts its acid-base equilibrium to the right, increasing the strength of the acid.

Stop and Think: What are the general rules that lead to such predictions? Consult your general chemistry textbook if you don't remember.

$$H—A + H_2O \rightleftharpoons H_3O^+ + A^-$$
$$\text{acid} \qquad\qquad\qquad \text{conjugate base}$$

The conjugate base of an acid, if it has a negative charge (as most do), is stabilized by anything that tends to disperse (spread out) the charge. Thus electron-withdrawing groups and pi electron systems that allow delocalization of a negative charge by resonance can both stabilize a conjugate base, thereby increasing the acidity of the corresponding acid.

Uncharged bases, such as ammonia and its derivatives, may also be stabilized by resonance, but stabilizing the base and not its conjugate acid shifts the acid-base equilibrium to the left, *decreasing* the strength of the base.

$$B: + H_2O \rightleftharpoons B-H^+ + OH^-$$

$$\text{base} \qquad\qquad \text{conjugate acid}$$

In this minilab, you will be assessing the acid-base strengths of some organic compounds by measuring the pH values of their aqueous solutions. Before the lab period, you should try to predict the relative acid-base strengths of the compounds listed by, for example, guessing which compounds may be acidic and writing all possible resonance structures (if there are any) leading to dispersal of the negative charge of their conjugate bases.

Directions

Your instructor may add or remove some test compounds.

Some of the organic compounds are quite toxic, corrosive, or have other hazardous properties. Although the solutions are very dilute, minimize your contact with them.

Safety Notes

You will be provided with 0.10 *M* solutions in 50% ethanol of such organic compounds as benzenesulfonic acid, benzoic acid, benzyl alcohol, benzylamine, benzyltrimethylammonium hydroxide, *p*-cresol, and *p*-toluidine.

benzenesulfonic   benzoic acid   benzyl alcohol   benzylamine
acid

benzyltrimethylammonium   *p*-cresol   *p*-toluidine
hydroxide

Transfer 5 drops of each solution to a well of a clean spot plate (label the wells) and determine its pH using pH paper having a range from at least pH 1 to 12. (Alternatively, add a drop of universal indicator to each well, or measure enough of each solution into a micro well plate to use a pH meter with a mini probe.) Assuming that benzyl alcohol is neutral, decide whether each compound other than the alcohol is acidic or basic (in 50% ethanol, a neutral solution does *not* have a pH of 7).

Arrange the acidic compounds in order of acidity (most acidic first) and arrange the basic compounds in order of basicity (most basic first). Explain your results and compare them with your predictions, trying to account for any discrepancies. Write equations for the proton-transfer reactions of each of the acids and bases with water.

MINILAB 37   Hydrolysis Rates of Esters

Before You Begin: Review the mechanisms of acyl substitution reactions and try to predict the relative hydrolysis rates of the esters listed.

Under appropriate reaction conditions, an ester can be hydrolyzed to the corresponding carboxylic acid and alcohol.

$$\underset{\text{RC}-\text{OR}'}{\overset{\overset{\displaystyle O}{\|}}{}} + H_2O \rightleftharpoons \underset{\text{RC}-\text{OH}}{\overset{\overset{\displaystyle O}{\|}}{}} + R'OH$$

The pH of the reaction mixture decreases as the acidic component is formed, providing a means of monitoring the reaction rate. In this minilab, you will compare the hydrolysis rates of several esters by monitoring the pH values of their aqueous solutions as a function of time. Each reaction mixture will contain a small amount of NaOH, so the pH should be quite high at first and should decrease as the carboxylic acid forms (initially as the acid salt). Based on your knowledge of the mechanisms of nucleophilic acyl substitution reactions, you should be able to predict the relative hydrolysis rates of the esters. You can then compare your experimental results with your predictions.

Directions

Your instructor may add or remove some test compounds.

Safety Notes

The esters are flammable, and inhalation and skin contact may be harmful. Avoid contact, do not breathe their vapors, and keep flames away.

Place 1.0 mL of aqueous 50% ethanol, 1 drop of a universal indicator, and 1 drop of 1.0 M sodium hydroxide in each of four small, numbered test tubes. Stopper and shake the test tubes. Add 3 drops of one of the following esters to each of the four test tubes: ethyl acetate, ethyl benzoate, ethyl formate, and ethyl butyrate. Shake each test tube until the ester has dissolved and record the color of each solution. Heat [OP-7] the test tubes in a 50°C water bath for at least 30 minutes, recording the colors every 10 minutes. Use Table M4 (or a color chart provided) to estimate the pH of each solution at each time. Based on your results, arrange the esters in order of reactivity, most reactive first.

Compare your results with your predictions and try to account for any discrepancies. Propose a hypothesis that accounts for the reactivity order you observed.

Table M4 Colors and approximate pH values for Gramercy universal indicator solution

pH	Color
4	red
5	red-orange
5.5	orange
6	yellow-orange
6.5	yellow
7	yellow-green
7.5	green
8	dark green
8.5	blue-green
9	blue
9.5	violet
10	red-violet

Preparation of Benzamide

MINILAB **38**

Before You Begin: Read or review the operations as necessary.

The reaction of an acid chloride with ammonia is a traditional way of preparing primary amides. This is a nucleophilic acyl substitution reaction in which ammonia is the nucleophile. The reaction also produces HCl, which converts ammonia to ammonium ion.

$$\underset{\substack{|| \\ RC-Cl}}{O} + NH_3 \longrightarrow \underset{\substack{|| \\ RC-NH_2}}{O} + HCl$$

$$HCl + NH_3 \longrightarrow NH_4^+Cl^-$$

Since ammonium ion is not nucleophilic, it will not react with the acid chloride, so the HCl must be neutralized in order for the reaction to go to completion. In Experiment 44, this was done using aqueous NaOH in a procedure called the Schotten-Baumann reaction. In this minilab, it is accomplished by using excess ammonia to neutralize the HCl.

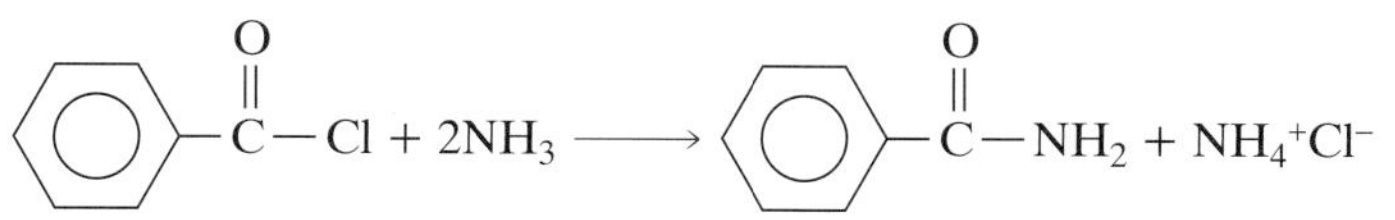

Directions

Both concentrated ammonia and benzoyl chloride can cause serious burns to skin and eyes, and their fumes irritate the eyes and respiratory system. Wear gloves, use them only under the hood, and do not breathe their vapors.

Safety Notes

Take Care! Possible violent reaction. Avoid contact with ammonia and benzoyl chloride, and do not breathe their vapors.

Waste Disposal: Pour the filtrates down the drain.

Under the hood, measure 0.50 mL of concentrated aqueous ammonia into a 3-mL conical vial and cool it well in an ice bath. *Wearing protective gloves, slowly* add 8 drops of benzoyl chloride to the cold ammonia with stirring [OP-10]. Let the solution stand in the ice bath for 5 minutes; then collect the benzamide by vacuum filtration [OP-13] and wash it on the filter with cold water. Recrystallize [OP-25] the benzamide from boiling water. Dry [OP-23] and weigh it, and measure its melting point [OP-30].

If you did Experiment 44, record an infrared spectrum [OP-36] of the benzamide, compare its IR spectrum with the spectrum of *N,N*-diethyl-*m*-toluamide, and discuss any significant similarities or differences. Write a mechanism for the reaction that formed benzamide.

MINILAB 39

Synthesis of Dimedone
Derivatives of Benzaldehyde

Before You Begin: Read or review the operations as necessary. Review the mechanisms of condensation and addition reactions of enolate ions.

Experiment 46 describes the preparation of dimedone (5,5-dimethyl-1,3-cyclohexanedione) by a synthesis that involves a Michael addition followed by a condensation reaction. In the presence of a base, dimedone itself reacts with aldehydes by a condensation reaction followed by a Michael addition, yielding a *bis*-dimedone derivative of the aldehyde. In the presence of an acid, the *bis*-dimedone derivative cyclizes to form an octahydroxanthenedione.

dimedone   benzaldehyde   *bis*-dimedone derivative   octahydroxanthenedione
                                                                        derivative

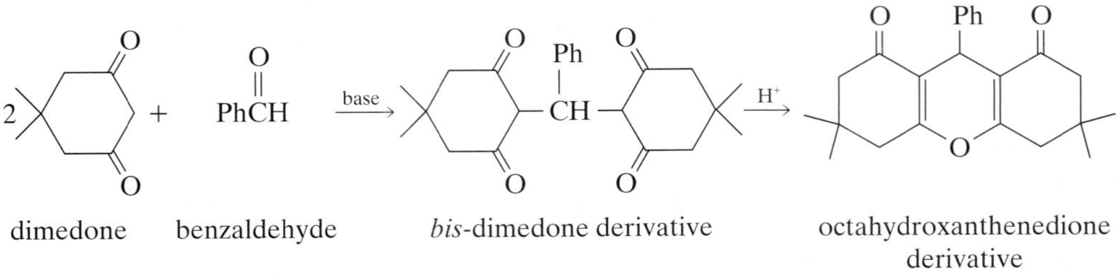

Both kinds of derivatives have been used to identify unknown aldehydes. In this minilab you will prepare the dimedone derivatives of benzaldehyde and test your mechanism-writing skills.

Directions

Safety Notes

Piperidine is very toxic and corrosive. Avoid contact and do not breathe its vapors.

Dissolve 0.10 g of benzaldehyde in 4.0 mL of 50% ethanol in a centrifuge tube or a 15 × 125-mm test tube. Stir in 0.30 g of dimedone and 1 drop of piperidine, and drop in a boiling chip. Boil [OP-7] the mixture gently for 5 minutes. If the hot solution is clear, add water drop by drop until it becomes cloudy. Cool [OP-8] the solution in an ice/water bath until crystallization is complete. Collect the *bis*-dimedone derivative by vacuum filtration [OP-13], wash it with a small amount of cold 50% ethanol, and then dry [OP-23] and weigh it.

Dissolve 0.10 g of the *bis*-dimedone derivative in 4.0 mL of absolute ethanol in a centrifuge tube or a 15 × 125-mm test tube with gentle heating. Stir in 1 mL of water and 2 drops of 6.0 *M* hydrochloric acid, drop in a boiling chip, and boil the mixture gently for 5 minutes. Add water drop by drop to the hot solution until it becomes cloudy, and then cool it until crystallization is complete. Collect the octahydroxanthene derivative by vacuum filtration [OP-13] and wash it with cold 50% ethanol. Then dry [OP-23] and weigh it. Measure the melting points [OP-30] of both derivatives.

Write balanced equations for the reactions and propose detailed mechanisms for the formation of both products. Remember that 1,3-diones such as dimedone can exist in both enolic and diketo forms.

Take Care! Avoid contact with piperidine and do not breathe its vapors.

Waste Disposal: Pour the filtrates down the drain.

A Diazonium Salt Reaction of 2-Aminobenzoic Acid

MINILAB **40**

Before You Begin: Read or review the operations as necessary, and try to predict the structure of the product.

Diazonium salts of aromatic amines are usually prepared at low temperatures to prevent unwanted side reactions. In this minilab, you will see what happens when a diazonium salt prepared from 2-aminobenzoic acid (anthranilic acid) is heated in solution. You should be able to deduce the structure of the product and propose a mechanism for its formation.

anthranilic acid
(2-aminobenzoic acid)

Directions

The product is a skin and eye irritant; avoid contact.

Safety Notes

In a 10-mL Erlenmeyer flask, add 0.27 g of 2-aminobenzoic acid (anthranilic acid) to 1.5 mL of 3.0 *M* HCl with swirling or magnetic stirring [OP-10]. If necessary, heat the solution gently and add up to 2 mL of water to get the amine salt to dissolve. Cool the solution to 5°C or below in an ice/water bath, then slowly add 2.0 mL of freshly prepared 1.0 *M* sodium nitrite solution with swirling or stirring. Test the solution with starch-iodide paper; if the test is negative, add just enough 1.0 *M* sodium nitrite to give a positive test. Drop in a boiling chip and boil [OP-7] the solution gently on a hot plate until no more gas is evolved. Cool [OP-8] the reaction mixture in an ice/water bath until crystallization is complete. Collect the product by vacuum filtration [OP-13] and purify it by recrystallization [OP-25] from boiling water. If desired, use a small amount of pelletized Norit to decolorize the

Stop and Think: What might the gas be?

Waste Disposal: Pour the filtrates down the drain.

recrystallization solution. Dry [OP-23] the product and measure its mass and melting point [OP-30]. Test the product with ferric chloride as described in classification test C-13 of Part IV.

Describe and interpret the results of the FeCl₃ test. Deduce the structure of the product and verify your conclusion by looking up its melting point in an appropriate reference book. Write balanced equations for the reactions involved and identify the gas that was evolved.

MINILAB 41   Beckmann Rearrangement of Benzophenone Oxime

Before You Begin: Read or review the operations as necessary. Review the mechanisms of molecular rearrangements involving nitrogen atoms.

The Beckmann rearrangement of an oxime involves a migration of an alkyl or aryl group to an electron-deficient nitrogen atom. The transition state for the rearrangement step resembles that for a Curtius-type rearrangement, described in Experiment 48.

$$
\underset{R}{\overset{\underset{|}{\overset{R'}{C}}}{\diagdown}}N-Z \longmapsto \left[R\cdots\overset{\underset{|}{\overset{R'}{C}}}{}\cdots N\cdots Z \right] \longrightarrow R-\overset{..}{N}=\overset{\oplus}{C}-R' + :Z^-
$$

In this minilab, you will prepare the oxime of benzophenone, carry out its Beckmann rearrangement to benzanilide, and work out a mechanism for the reaction.

benzophenone   benzophenone oxime   benzanilide

Directions

Take Care! Avoid contact with polyphosphoric acid.

Take Care! Wear gloves and avoid contact with hydroxylamine hydrochloride.

Measure 1.0 mL of polyphosphoric acid (PPA) into a 15 × 125-mm (or larger) test tube. You may need to warm the viscous acid before you can transfer it. Heat the mixture in a boiling-water bath for about 5 minutes. Then mix in 0.20 g of benzophenone and 0.24 g of hydroxylamine hydrochloride, using a stirring rod. Heat [OP-7] the reaction mixture in a boiling-water bath for a least 30 minutes, until frothing has stopped. Transfer the warm reaction mixture to a beaker containing about 8 g of crushed ice, using a small amount of ice water for the transfer. Let the mixture stand, with occasional stirring, until the ice has melted. Collect the product by vacuum filtration [OP-13], washing it on the filter with ice water. Purify the benzanilide by recrystallization [OP-25] from 95% ethanol. Dry [OP-21] it and measure its mass and melting point [OP-30].

Propose a detailed mechanism for the reaction, showing the transition state of the rearrangement step.

Dyeing with Indigo

Before You Begin: Read or review the operations as necessary.

The indigo plant, *Indigofera tinctoria*, is a shrub of the legume family that has been used to dye cloth since ancient times. Egyptian mummies were sometimes wrapped in cloth dyed blue with indigo. The leaves of the indigo plant contain a colorless natural product called indican, which is converted to indigo by a process that involves fermentation and oxidation. Indigo is an indole derivative that, in the presence of a reducing agent, changes to the colorless base-soluble compound leucoindigo.

indigo (blue)   leucoindigo (colorless)

When a piece of cloth is dipped into a solution of leucoindigo and exposed to air, indigo is regenerated, dyeing the cloth the color of blue jeans. Indigo fades with time, but the popularity of faded blue jeans has turned this apparent flaw into an advantage.

In this minilab, you will prepare synthetic indigo by a condensation reaction of 2-nitrobenzaldehyde with acetone in an alkaline solution, and then use it to dye a piece of cloth.

2-nitrobenzaldehyde $+ 2CH_3COOH + 2H_2O$

Directions

Safety Notes

Take Care! Avoid contact with 2-nitrobenzaldehyde.

Take Care! Wear gloves.

Dissolve 0.25 g of 2-nitrobenzaldehyde in 1.0 mL of acetone in a 3-mL conical vial. Add water drop by drop until the solution becomes cloudy; then add a drop or so of acetone to clear it up. Add 10 drops of 1.0 M sodium hydroxide slowly, with stirring [OP-10]. The solution should warm up and turn dark brown. Let the mixture stand for 15 minutes or more. Then cool [OP-8] it in an ice/water bath and collect the indigo by vacuum filtration [OP-13]. Wash the dye with a little ethanol, then with diethyl ether, and let it air-dry.

Transfer the indigo to an 18 × 150-mm test tube, add 75 mg of sodium dithionite, and grind the solids together with a stirring rod until they are well mixed. Add 5 mL of 1.0 M sodium hydroxide and heat [OP-7] the mixture in a boiling-water bath with stirring. When the indigo has dissolved, stopper and shake the tube until the indigo-blue color disappears (if the color persists, add a little more sodium dithionite). Immerse a small piece of cotton cloth in the solution with a stirring rod, stopper the test tube immediately, and shake it for about 30 seconds. Remove the cloth, blot it dry between paper towels, and hang it up to dry. Note what happens when the test tube is left open for a time and then stoppered and shaken. Describe and explain your observations during and after the dyeing step.

MINILAB 43

Reactions of Monosaccharides with Phenols

Before You Begin: Look up the structures of the monosaccharides you will be testing in this experiment and classify each as an aldopentose, aldohexose, or ketohexose.

Under acidic conditions, pentoses and hexoses undergo dehydration to furan derivatives. Aldopentoses and ketopentoses yield 2-furaldehyde (furfural); aldohexoses and ketohexoses yield 5-hydroxymethyl-2-furaldehyde.

Both of these aromatic aldehydes react with phenols under acidic conditions to form colored products. For example, 2-furaldehyde condenses with 1-naphthol to yield Compound **1**, which oxidizes to yield an intensely colored product, **2**.

Stop and Think: Why is Compound **2** colored although Compound **1** is not?

This reaction is the basis of the Molisch color test for pentoses and hexoses. In the presence of concentrated sulfuric acid and 1-naphthol, all pentoses yield Compound **2** and all hexoses yield an analogous derivative of 5-hydroxymethyl-2-furaldehyde. The phenols 1,3-dihydroxybenzene (resorcinol) and 3,5-dihydroxytoluene (orcinol) also give colored products similar to **2** with furaldehydes.

1,3-dihydroxybenzene   3,5-dihydroxytoluene

In this minilab, you will carry out the Molisch test on several monosaccharides and an unknown carbohydrate. You will test the same compounds with acidic solutions of the other two phenols to find out what results they give with different classes of monosaccharides. Based on your observations, you should then be able to classify your unknown.

Directions

Your instructor may add more compounds to be tested, or substitute some.

Sulfuric acid causes chemical burns that can seriously damage skin and eyes. Wear gloves and avoid contact.
The 3,5-dihydroxytoluene solution contains concentrated hydrochloric acid, which is poisonous and corrosive. Contact or inhalation can cause severe damage to the eyes, skin, and respiratory tract. Wear gloves and dispense under a hood, avoid contact, and do not breathe its vapors.
The phenols are toxic and corrosive, and some are suspected carcinogens or teratogens. Although their solutions are very dilute, avoid contact.

Safety Notes

Have ready five clean, labeled 13 × 100-mm test tubes. Measure 1 mL of aqueous 1% solutions of D-fructose, D-glucose, D-xylose, and the unknown into four of the test tubes. Measure 1 mL of water into the fifth, for use as a control. To each test tube, add 2 drops of the Molisch reagent and swirl to mix. For each solution, tilt the test tube at an angle and *slowly* pour 1 mL of concentrated sulfuric acid down the side to form a separate layer below the aqueous layer. Observe the interface between the layers and record your observations in your lab notebook.

Have ready five clean, labeled 13 × 100-mm test tubes. Measure 1 mL of aqueous 1% solutions of D-fructose, D-glucose, D-xylose, and the unknown into four of the test tubes, and measure 1 mL of water into the fifth. Add 2 mL of the 1,3-dihydroxybenzene solution to each test tube. Place all five test tubes into a beaker of boiling water [OP-7] and record your observations, including the time required for a color (if any) to develop.

Have ready five clean, labeled 13 × 100-mm test tubes. Measure 1 mL of aqueous 1% solutions of D-fructose, D-glucose, D-xylose, and the unknown into four of the test tubes, and measure 1 mL of water into the fifth. Add 1 mL of the 3,5-dihydroxytoluene solution to each test tube. Place all five test tubes into a beaker of boiling water [OP-7] and record your observations, including the time required for a color (if any) to develop.

Tabulate your results and describe how the tests (except the Molisch test) can be used to distinguish different classes of monosaccharides. Then classify your unknown as an aldopentose, aldohexose, ketohexose, or none of these, and explain how you arrived at your conclusion. Write an equation for the reaction of 2-furaldehyde with 1,3-dihydroxybenzene to form a product analogous to **1**, and propose a mechanism for this reaction.

Take Care! Wear gloves and avoid contact with sulfuric acid and the Molisch reagent.

Take Care! Avoid contact with the solution.

Take Care! The solution contains concentrated HCl; wear gloves and avoid contact.

Waste Disposal: Dispose of all test-tube contents as directed by your instructor.

MINILAB 44

Extraction of Trimyristin from Nutmeg

Before You Begin: Read or review the operations as necessary.

Although most triglycerides found in plants are mixtures containing several different fatty acid residues, the triglyceride derived from nutmeg contains only one, myristic acid (tetradecanoic acid). In this minilab, you will isolate trimyristin from ground nutmeg by hot solvent extraction with diethyl ether.

$$\underset{\text{myristic acid}}{CH_3(CH_2)_{12}\overset{\overset{\displaystyle O}{\|}}{C}OH}$$

Directions

Safety Notes

> **Diethyl ether is extremely flammable and may be harmful if inhaled. Do not breathe its vapors and keep it away from flames and hot surfaces.**

Take Care! Keep flames away from diethyl ether and do not breathe its vapors.

Accurately weigh 0.50 g of finely ground nutmeg and combine it with 3.0 mL of diethyl ether in a 5-mL conical vial. Attach a water-cooled condenser and use a hot water bath to heat the mixture gently under reflux [OP-7] for 45 minutes or more. Filter the mixture by gravity [OP-12], washing the nutmeg residue in the filtering pipet with a small amount of ether and saving the filtrate. Evaporate [OP-16] the ether from the filtrate. Recrystallize [OP-25] the product from 95% ethanol. Dry [OP-23] the trimyristin, weigh it, and measure its melting point [OP-30].

Draw the structure of trimyristin and calculate the percent recovery of trimyristin from nutmeg.

Waste Disposal: Place any recovered diethyl ether in the designated solvent recovery container. Pour the ethanol filtrate down the drain.

Preparation of a Soap Using a Phase-Transfer Catalyst

MINILAB 45

Before You Begin: Read or review the operations as necessary.

Soaps have long been made by heating fats or oils with lye (sodium hydroxide) and other alkaline substances, including potash (potassium carbonate) from wood ashes. This alkaline hydrolysis process is called *saponification*, from the Latin *sapon*, meaning soap.

Alcohol is sometimes used to speed up a soap-making reaction by bringing the base and fat together in solution. However, the soap must then be salted out or the alcohol evaporated. In this minilab, you will use a phase-transfer catalyst to bring the base and fat together, making the alcohol unnecessary. The fat will be a vegetable shortening, such as Crisco, containing mainly palmitic, stearic, oleic, and linoleic acid residues.

Directions

Sodium hydroxide is toxic and corrosive, causing severe damage to skin, eyes, and mucous membranes. Wear gloves and avoid contact with the NaOH solution.

Safety Notes

Measure 2.0 g of vegetable shortening into a 5-mL conical vial and heat [OP-7] it in a boiling-water bath until it melts. Stir in 1.0 mL of 8.0 M (~25%) sodium hydroxide and 1 small drop of tricaprylmethylammonium chloride (Aliquat 336). Attach an air condenser and heat the mixture in the boiling-water bath, while stirring [OP-10], for 20 minutes or more, until all of the oily globules have disappeared. Cool [OP-8] the reaction mixture in an ice/water bath, add 1.0 mL of water, and use a flat-bladed microspatula to break up and wash the soap particles. Collect the soap by vacuum filtration [OP-13], washing it on the filter with several small portions of ice water. Dry [OP-23] the soap in a desiccator and weigh it.

Draw the structures of the major components of the soap you prepared. Write an equation for the reaction of glyceryl tristearate with NaOH.

Take Care! Wear gloves and avoid contact with the NaOH solution.

$$CH_2OCO(CH_2)_{17}CH_3$$
$$CHOCO(CH_2)_{17}CH_3$$
$$CH_2OCO(CH_2)_{17}CH_3$$
glyceryl tristearate

MINILAB 46

Isolation of a Protein from Milk

Before You Begin: Read or review the operations as necessary.

Casein is a milk protein that contains all of the common amino acids and is particularly rich in the essential ones. Cheese-making is based on the precipitation of casein from milk to form solid *curds*. The liquid *whey* is drained from the casein curds, which are concentrated by cooking, pressing, and salting, and then aged to develop the characteristic flavor of the cheese. Casein, which exists in milk as a soluble calcium salt, precipitates at pH values below 4.6, so milk can be curdled by acids such as the lactic acid that forms during the natural souring of milk. In this minilab, you will isolate casein from skim milk by curdling the milk with acetic acid. You will also carry out the *xanthoproteic test*, which is positive for proteins that contain aromatic amino acid residues, such as tyrosine and tryptophan.

Directions

Safety Notes

> Nitric acid is corrosive to eyes, skin, and mucous membranes and can cause severe upper respiratory irritation. Wear gloves, use a hood, avoid contact, and do not breathe its vapors.

Waste Disposal: Pour the whey down the drain.

Mix 0.50 g (accurately weighed) of nonfat dry milk with 5.0 mL of water in a small beaker. Heat [OP-7] the milk to 40°C with magnetic or manual stirring [OP-10]. Then add aqueous 10% (~1.7 *M*) acetic acid drop by drop with stirring until no more precipitate forms. Stir until the casein coagulates into an amorphous mass. Decant (pour out) most of the liquid whey and transfer the wet solid to a vacuum filtration apparatus [OP-13]. With the aspirator running, press the casein with a clean cork to squeeze out as much liquid as possible; then dry it further between filter papers. Transfer the casein to a small beaker, cover it with 95% ethanol, and use a flat-bottomed stirring rod or flat-bladed microspatula to stir and crush it until it is finely divided. Collect the casein by vacuum filtration, wash it with two small portions of acetone, and let it dry [OP-23] at room temperature. Weigh the dry casein. *Under the hood,* add a drop of concentrated nitric acid to a small

Take Care! Wear gloves and do not breathe vapors of HNO_3.

piece of casein on a watch glass and record your observations. (**Stop and Think:** What kind of reaction is occurring?)

Calculate the mass percentage of casein in the dry milk. Interpret the result of the xanthoproteic test and write an equation for the reaction of a tyrosine residue with nitric acid.

Qualitative Organic Analysis

PART IV

P art IV describes how to classify and identify unknown organic compounds by using chemical and spectrometric methods.

Whenever an organic chemist synthesizes a new compound or isolates a previously unknown compound from a natural source, he or she must find out what it is. If the chemist has prepared the compound by a reaction whose outcome is reasonably predictable, its identification may require no more than a melting or boiling point and some spectra. But if the compound has been obtained by a process whose outcome is uncertain, or from a natural product or other source whose composition is unknown, the chemist will ordinarily follow a systematic course of action to identify it.

The systematic identification of organic compounds is called *qualitative organic analysis*, which chemists often shorten to "qual organic." As discussed in Experiment 38, you can identify an organic compound in somewhat the same way that a detective identifies the perpetrator of a crime. You should always be on the lookout for clues to the compound's identity, and for any evidence that will help you narrow down the roster of "suspects"; that is, the list of known compounds that you believe contains your unknown.

There is no single best way to identify an organic compound and no one path to follow. Although a useful approach is outlined in the qualitative analysis scheme in the margin, it is not necessary to follow it inflexibly. You should use your own judgment and initiative in choosing which tests to perform, which physical properties to measure, which derivatives to prepare, and which spectra to record. By keeping your eyes and mind open at all times, you may find clues to the structure of your compound that suggest a more direct route to its identification.

Each qual organic problem should provide an exciting challenge, giving you the opportunity to apply all of your skill and ingenuity to its solution. Besides testing your mastery of operations you have used in the laboratory, qual organic includes practical applications of many concepts you have studied in the organic chemistry lecture course. Throughout the analysis of an unknown compound, you may apply your knowledge of functional group chemistry, nomenclature, acid-base equilibria, structure-property relationships, spectral analysis, organic synthesis, and many other topics in organic chemistry.

An Overview of Qual Organic

If you are doing Experiment 38, your unknown compound will be an alcohol, aldehyde, or ketone. If you are doing Experiment 49, your unknown will be an amine. If you are identifying a general unknown, the unknown will be from one of the following chemical families: *alcohols, aldehydes, ketones, amides, amines, carboxylic acids, esters, halides, aromatic hydrocarbons,* and *phenols.* Your instructor may also assign an unknown from a limited number of these families. An unknown may contain a secondary functional group, such as the carbon-carbon double bond in cinnamic acid or the nitro group in *p*-nitrophenol, but it will be classified as a member of one of the families listed here.

cinnamic acid   *p*-nitrophenol

The qualitative analysis scheme is divided into four parts, as outlined in the margin. The *preliminary work* generally involves purification of the

Qualitative Analysis Scheme

1 Preliminary work
   a Purification
   b Measurement of physical constants
   c Physical examination
   d Ignition test
2 Classification
   a Solubility tests
   b Classification tests
3 Spectral analysis
   a Infrared spectra
   b NMR spectra
4 Identification
   a List of possibilities
   b Additional tests and data
   c Preparation of derivatives

Note: If the unknowns provided are all in a single family or a small group of families, Part 2 can be omitted or modified.

unknown, measurement of its boiling point or melting point, a simple physical examination, and an ignition test. The *classification* phase involves solubility tests, chemical tests to identify functional groups, and additional chemical tests to detect structural features and other characteristics of the unknown. *Spectral analysis*—ordinarily by infrared spectrometry but sometimes by NMR spectrometry as well—can be used to detect or confirm functional groups and to provide information about such structural features as conjugation and the positions of substituents on benzene rings. The *identification* phase involves the preparation of a list of known compounds believed to contain the unknown, accumulation of additional evidence as needed to shorten the list, and synthesis of a derivative to identify the unknown as a compound on the short list. Your list of compounds can be based on the tables in Appendix VI or tables in another source recommended by your instructor.

If you have access to a computer simulation program, such as SQUALOR (a **s**imulation of the **qual**itative **or**ganic analysis laboratory experience), you can learn more about qualitative organic analysis and develop effective problem-solving strategies by identifying some "virtual unknowns." SQUALOR [Bibliography, M6], which is structured like a computer game, offers practice in identifying nearly a hundred unknowns from 11 different families.

Preliminary Work

Purification

Impurities in your unknown may cause inaccurate measurements of its physical constants, misleading results on classification tests, low derivative melting points, and other problems that will make it difficult to correctly identify the unknown. Light or dull colors (light yellow, tan, brown, black, etc.) may suggest that the unknown is impure, whereas intense yellows, oranges, reds, and so on, are usually intrinsic colors. Compounds that are easily oxidized, such as aldehydes, aromatic amines, and phenols, are particularly likely to be impure. If your unknown is a liquid, you should purify it by simple distillation and record its boiling range. A liquid with a distillation boiling-point range greater than 2°C may need to be redistilled and the high-boiling or low-boiling fractions discarded. If your unknown is a solid, you should first obtain its melting point to see whether or not it requires purification. Solids with melting point ranges of 2–3°C or more should ordinarily be purified by recrystallization from a suitable solvent. Purity can also be assessed by gas chromatography for a liquid and by TLC or HPLC for a solid.

Measurement of Physical Constants

The boiling point or melting point of your unknown is a major factor in its identification because it determines what compounds will be on your list of possibilities. If your value is inaccurate, your list may not even contain the name of your unknown. The median distillation boiling point of an unknown liquid may be sufficient for its identification, but obtaining a capillary-tube boiling point for confirmation is always a good idea. The melting point of a pure unknown solid should also be measured as accurately as possible. The dividing line between liquids and solids is generally near a melting point of 25°, but some compounds with melting points over 25° (such as *t*-butyl alcohol) may be liquid at room temperature, particularly when impure. When in doubt, try freezing the compound in an ice-salt bath, and estimate its melting point if it freezes.

Other physical constants, such as the density and refractive index of a liquid, may also be helpful, but you will have to obtain their literature values from an appropriate reference book.

Physical Examination

A simple physical examination of an unknown may provide some useful information about it. For instance, the fact that an unknown is a solid eliminates all organic compounds that are liquids at room temperature, and an intrinsic color suggests that chromophoric groups with conjugated double bonds or rings are present. Odors can also provide clues to the identity of a compound, but since many organic compounds are toxic by inhalation, you should never smell them indiscriminately or in a way that would cause you to inhale appreciable amounts of vapor. Before you smell your unknown, it is a good idea to ask your instructor about its toxicity. The safest procedure is to hold the vial or other container at least 6 inches from your nose with the open end pointed away from you, then use your free hand to gently wave the vapors toward your nose, sniff briefly, and exhale.

Ignition Test

A compound's behavior upon ignition may provide some clues to its identity. As a rule, the higher the oxygen content of a compound, the bluer its flame; as hydrogen content increases, the flame becomes more yellow. For example, methanol burns with a bluer flame than 1-hexanol because it has a higher oxygen/hydrogen ratio. Most aromatic compounds burn with a sooty yellow flame. Organic compounds that do not burn in the ignition test may contain a high ratio of halogen to hydrogen or have a high molecular weight.

Directions

Record all of your observations and data in your laboratory notebook.

Safety Notes

> **Assume that all unknowns are flammable and harmful by inhalation, ingestion, and skin absorption. Do not inhale their vapors and avoid contact with eyes, skin, and clothing.**

A thermometer correction should be applied for temperatures over 200°C (see OP-30).

If your unknown is a liquid, purify it by simple distillation [OP-27]. Collect the main fraction over a range of about 2°C and record its boiling range and median boiling point. If desired, carry out a capillary-tube boiling-point determination [OP-31] on the purified liquid.

If your unknown is a solid, measure its melting point [OP-30] accurately. If the melting-point range is 2–3°C or higher, or if the solid shows other evidence of impurities ("dirty" color, etc.), purify it by recrystallization [OP-25] from a suitable solvent or solvent mixture. Then dry [OP-23] the purified solid and measure its melting point.

Observe and describe the physical state and color of the unknown. With your instructor's permission, observe and describe its odor as well.

Under the hood, carry out an ignition test by placing a drop of an unknown liquid or about 25 mg of an unknown solid in a small evaporating dish or watch glass and igniting it with a burning wood splint. If it burns, observe the color of the flame and whether it is clean or sooty.

Take Care! Make sure there are no flammable liquids nearby.

Classification

The traditional procedure for classifying an unknown involves the use of solubility tests that categorize the unknown into one of several solubility classes, followed by classification tests to detect functional groups and other structural features. Infrared spectrometry can also be used to detect functional groups and structural features, so you may want to read the later section "Spectral Analysis" before you carry out the classification tests. When you think you have determined the main functional group of your unknown, you can use its infrared (IR) spectrum to confirm your initial conclusion. If you have previous experience with infrared spectral analysis (and your instructor approves), you may want to record your unknown's IR spectrum before you carry out any classification tests, use the spectrum to tentatively identify its main functional group, and then use one or more classification tests to confirm its identity.

Some of the classification tests in this section are used mainly to detect secondary functional groups and various structural features. You will find them most useful for shortening the list of possibilities that you construct during the identification phase of your analysis (see the following section "Identification"). Since it is hard to know which of these tests will be useful until you know what compounds are on your list, you should construct the list as soon as you have confirmed your unknown's main functional group (unless it is an amine) and then perform the additional classification tests. If your unknown is an amine, you will have to classify it as primary, secondary, or tertiary before you can prepare your list.

Solubility Tests

The solubility behavior of a compound in appropriate solvents can be used to place it into one of the eight solubility classes listed in Figure Q1. For example, a compound that is insoluble in water and 5% HCl, but soluble in 5% NaOH and 5% $NaHCO_3$, is placed in solubility class A_1, classifying it as a carboxylic acid. Most of the classes are broader than this. For example, class S_n includes a number of families of neutral compounds with relatively low numbers of carbon atoms.

In these tests, compounds that dissolve to the extent of about 30 mg per mL of the solvent are considered soluble. Solubility in water suggests the presence of at least one oxygen or nitrogen atom and a relatively low molecular weight. In the case of monofunctional oxygen and nitrogen compounds, the borderline for water solubility is usually around five carbon atoms. For example, 1-butanol (with four carbon atoms) is soluble in water, whereas 1-pentanol (with five carbon atoms) is not. Cyclic and branched compounds are usually more soluble than straight-chain compounds of the same carbon number. Thus a phenyl group has about the same effect on solubility as a butyl group. Of the ten families of organic compounds considered here, water solubility can be expected for low-molecular-weight alcohols, aldehydes, ketones, amides, amines, carboxylic acids, esters, and phenols. Most water-soluble carboxylic acids and aliphatic amines can be distinguished from the rest by testing their aqueous solution with blue and red litmus paper; the acids are placed in class S_a and the amines in class S_b. Most phenols and aromatic amines are too weakly acidic or basic to give a positive litmus test, although there are some exceptions. Aldehydes that contain carboxylic acid impurities may give a false positive test with blue litmus. All water-soluble compounds that do *not* test positive to red or blue litmus are placed in class S_n.

Water-insoluble compounds that are soluble in 5% hydrochloric acid contain basic functional groups and are placed in class **B**. Of the families considered here, aliphatic and aromatic amines fall into this category. Some amines may form insoluble hydrochloride salts as they dissolve, so solubility behavior should be observed carefully to detect any change in the appearance of the unknown when it is shaken with the solvent.

Compounds that are insoluble in water and 5% hydrochloric acid but soluble in 5% sodium hydroxide contain acidic functional groups that give them pK_a values of ~12 or less. Acids with pK_a values of ~6 or less will also dissolve in 5% sodium bicarbonate, placing them in class A_1, which includes carboxylic acids. Note that carboxylic acids having 12 or more carbon atoms may form relatively insoluble salts that yield a soapy foam when shaken with 5% NaOH.

Compounds that are soluble in 5% NaOH but insoluble in 5% NaHCO$_3$ are placed in class A_2, which includes most phenols. Some other types of compounds, such as reactive esters, may react with 5% NaOH over time to form soluble reaction products, so it is important to avoid delays while carrying out this solubility test.

Compounds that are insoluble in all of the previous solvents but dissolve in or react with cold, concentrated sulfuric acid are placed in solubility class **N**. This class includes high-molecular-weight alcohols, aldehydes, ketones, amides, and esters. Unsaturated compounds and some aromatic hydrocarbons—those containing several alkyl groups on the benzene ring—are also soluble in this reagent. Solution in sulfuric acid is accompanied by protonation of a basic atom (usually nitrogen or oxygen) or by some other reaction, such as sulfonation, dehydration, addition to a multiple bond, or polymerization.

Compounds that are insoluble in all of these solvents (class **X**) include most aromatic hydrocarbons and the halogen derivatives of aliphatic and aromatic hydrocarbons.

It is not necessary to test an unknown with every solvent because if it dissolves in water, for example, it will also dissolve in aqueous solutions of HCl, NaOH, and NaHCO$_3$. In most cases, an unknown is not tested further once it is found to dissolve in a given solvent, since its solubility classification is established at that point. Compounds that are soluble in 5% NaOH, however, are also tested with 5% NaHCO$_3$ to determine whether they are carboxylic acids or phenols. Keep in mind that solubility classifications can be misleading, since there are exceptions and borderline cases. Therefore, the solubility test results should be supported by other chemical or spectral evidence before a definite conclusion is drawn.

Some reactions in cold, concentrated sulfuric acid

$$RCHO \xrightarrow{H_2SO_4} R\overset{\oplus}{C}HOH$$

$$RCOOH \xrightarrow{H_2SO_4} R\overset{\oplus}{C}(OH)_2$$

$$RCH_2OH \xrightarrow{H_2SO_4} RCH_2OSO_3H$$

$$\underset{\underset{R''}{|}}{\overset{\overset{OH}{|}}{RCH_2C}} - R' \xrightarrow{H_2SO_4} \underset{\underset{R''}{|}}{RCH{=}C} - R'$$

$$RCH{=}CHR' \xrightarrow{H_2SO_4} RCH_2\underset{\underset{}{|}}{\overset{\overset{OSO_3H}{|}}{C}}HR'$$

An outline of the solubility scheme is given in Figure Q1.

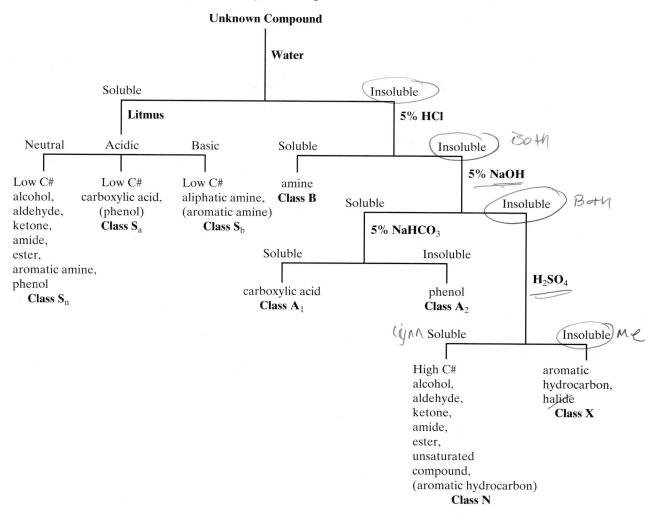

Figure Q1   Flow diagram for solubility tests
Note: Compounds in parentheses are found only rarely in the solubility class indicated.

Directions

Safety Notes

If the unknown is a liquid, use a measuring pipet or syringe to measure each 0.10 mL portion, or estimate the number of drops in 0.10 mL and measure it by drops. Measure 0.10 mL of the liquid into a 13 × 100-mm test tube and add 3.0 mL of water. Stopper and shake vigorously for 10 seconds or so, then observe the mixture carefully. If the mixture is homogeneous, with no separate layer or suspended droplets of liquid, classify the unknown as *soluble*. Do not mistake air bubbles (which may form on shaking, but should soon escape) for suspended liquid droplets. If the mixture looks cloudy or shows suspended droplets when shaken, or if a separate layer is visible above or below the water layer after standing, shake it again for a

If the quantity of unknown is limited, you can scale down these amounts.

Take Care!   Don't use your thumb for a stopper.

short while, then observe it. If the cloudiness, droplets, or separate layer disappear on shaking, classify it as *soluble*; otherwise, classify it as *insoluble*.

If the unknown is a solid, weigh out 90 mg of the solid and grind it to a fine powder on a watch glass using a flat-bottomed stirring rod or a flat-bladed microspatula. Mix the solid with 3.0 mL of water in a 13 × 100-mm test tube and shake the test tube vigorously for about 30 seconds or until the solid dissolves. If some of the solid remains undissolved, grind it against the bottom or sides of the test tube with your stirring rod and continue to shake and stir at intervals until the solid dissolves, or until it is apparent that it will not completely dissolve. (It may take several minutes to dissolve some solids.) If the solid dissolves completely, classify it as *soluble*; if it doesn't dissolve completely even after extensive shaking and grinding, classify it as *insoluble*.

If the unknown is *soluble* in water, test its aqueous solution with red and blue litmus paper. If there is no reaction to litmus but you suspect that the unknown may be an aromatic amine or a phenol, dissolve a small amount of the unknown in 5% HCl (for an amine) or 5% NaOH (for a phenol) and see if its odor disappears. Amines and phenols generally have strong odors, but their salts are odorless. Decide whether the unknown should be in class S_n, S_a, or S_b, and go on to the classification tests.

If the unknown is *insoluble* in water, test its solubility (using 0.10 mL of a liquid or 90 mg of a solid) in 3.0 mL of 5% HCl by the same procedure as described for water. If it does not dissolve completely in 5% HCl, transfer [OP-6] most of the solvent to another test tube, then carefully neutralize the solvent to red litmus with a dilute NaOH solution. If a precipitate, a separate liquid phase, or a cloudy solution occurs upon neutralization, consider the unknown soluble. If it is *soluble* in 5% HCl, place it in class **B** and go on to the classification tests.

If the unknown is *insoluble* in 5% HCl, test its solubility (using 0.10 mL of a liquid or 90 mg of a solid) in 3.0 mL of 5% NaOH and in 3.0 mL of 5% $NaHCO_3$ by the same procedure as before. When testing with NaOH, notice whether the mixture foams on shaking, which could indicate a long-chain carboxylic acid. If the unknown does not dissolve completely, transfer most of the solvent to another test tube and carefully neutralize it to blue litmus with a dilute HCl solution. If a precipitate, a separate liquid phase, or a cloudy solution occurs upon neutralization, consider the unknown soluble. If the unknown is *soluble* in 5% NaOH or in both solvents, place it in class A_1 or A_2 and go on to the classification tests.

If the unknown is *insoluble* in all of the foregoing solutions, carefully test its solubility in 3.0 mL of cold (room temperature or below) concentrated sulfuric acid. Shake this mixture vigorously and look for any evidence of a reaction, such as the generation of heat, a distinct change in color, the formation of a precipitate, or the evolution of a gas. Solubility or a definite reaction is considered a positive test. The nature of any reaction that takes place may provide some clues to the identity of the unknown. Based on the results of this test, place the unknown into class **N** or **X**.

Classification Tests

The results of the solubility tests should reduce the number of families to which your compound may belong, limiting the number of classification tests that are needed to classify it into one of the remaining families. You should choose the tests carefully, so that they will provide the information you need in a minimum number of steps. The tests used to detect functional groups are listed in Table Q1.

Table Q1 Functional-class tests

Family	No.	Test	Comments
alcohols	C-9	chromic acid	negative for 3° alcohols
	C-17	Lucas' test	negative for 1° and high M.W. alcohols
	C-1	acetyl chloride	useful if other tests are inconclusive
aldehydes	C-9	chromic acid	reacts more slowly than alcohols
	C-23	Tollens' test	
aldehydes and ketones	C-11	2,4-Dinitrophenyl-hydrazine	ketones react with C-11, not with C-9 or C-23
amides	C-2	alkaline hydrolysis	best for amides of ammonia or low M.W. amines
amines		solubility tests	soluble in 5% HCl
	C-12	elemental analysis	detects N; use to confirm solubility test results
aromatic hydrocarbons	C-3	aluminum chloride/chloroform	should be confirmed by tests indicating absence of functional groups (such as halogens)
carboxylic acids		solubility tests	soluble in 5% NaOH and 5% NaHCO$_3$
esters	C-14	ferric hydroxamate	
	C-2	alkaline hydrolysis	also used to prepare derivatives
halogenated hydrocarbons	C-5	Beilstein's test	simple, not always reliable
	C-21	silver nitrate/ethanol	negative for aryl, vinyl, and certain other classes of halides
	C-10	density test	negative for monochloroalkanes
	C-12	elemental analysis	can distinguish Cl, Br, and I
phenols	C-13	ferric chloride	positive for most (but not all) phenols
	C-8	bromine water	aromatic amines also react

Within families, functional-class tests are listed in approximate order of simplicity and utility. It may be necessary to perform two (or more) tests for the same functional group in case the first test is inconclusive, or to use an infrared spectrum for confirmation.

You may be able to obtain additional information about your unknown using one or more of the classification tests listed in Table Q2. Some of these tests are used to further classify alcohols, amines, and alkyl halides. Others can distinguish aromatic compounds from aliphatic compounds in the same family, detect unsaturation, and detect certain groups of atoms, such as —COCH$_3$ in methyl ketones.

Tertiary amines are listed separately from primary and secondary amines in Appendix VI, so if you have an amine, you should classify it before you prepare your list of possibilities. If you have any other kind of compound, you can prepare your list first, and then decide which of the tests in Table Q2 will help you shorten it.

Table Q2 Chemical tests that provide structural information

Family or structural feature	No.	Test	Application
alcohols	C-17 C-16	Lucas' test iodoform test	to classify alcohols as 1°, 2°, 3°, etc. to detect —CH(OH)CH$_3$ groupings
aldehydes	C-6	Benedict's test	to distinguish aliphatic from aromatic aldehydes
ketones	C-16	iodoform test	to detect —COCH$_3$ groupings
amines	C-15 C-4 C-20	Hinsberg's test basicity test quinhydrone	to classify amines as 1°, 2°, or 3° to distinguish alkylamines from arylamines complements C-15
carboxylic acids	C-18	neutralization equivalent	to determine the equivalent weight of an acid
halogenated hydrocarbons	C-10 C-22 C-21	density test sodium iodide/acetone silver nitrate/ethanol	to distinguish chlorides, bromides, and iodides; aliphatic and aromatic halides to classify halides as 1°, 2°, 3°, etc. complements C-22
unsaturation	C-7 C-19	bromine test potassium permanganate	to detect C=C and C≡C bonds complements C-7

Directions

Directions for performing all classification tests are given in this section by test numbers that are preceded by **C**. Unless otherwise indicated, solutions used for the classification tests are aqueous.

When each test is performed for the first time, it is advisable to run a *control* and a *blank* at the same time. Only by doing so will you know what to look for in deciding whether the test with the unknown is positive or negative. A control is a known compound that is expected to give a predictable result with the test reagent. A blank is run by combining all reagents as in the actual test, but omitting the unknown. Unless otherwise indicated, all compounds suggested as controls in the following procedures should give a positive test.

Most of the tests are carried out in test tubes, whose size is usually specified. If you don't have a test tube of the specified size (e.g., 10 × 75-mm), use the next size larger (e.g., 13 × 100-mm). In the procedures, the volumes of most liquid reactants are given in drops of the size delivered by a medicine dropper. You can measure drops using Pasteur pipets, reducing the amount of unknown needed by about half. But you must then divide the quantities of all reagents *not* measured with Pasteur pipets in half. Always use different dropping devices for the reagent and the unknown, and clean your own droppers thoroughly after using them. Never insert your own droppers into reagent bottles, as you may contaminate the reagents.

Select the tests you want to perform, and then turn to the appropriate procedures. Always read any Safety Notes before starting a test, and dispose of wastes as directed by your instructor.

C-1 *Acetyl Chloride*

If you think your unknown is a tertiary alcohol, try method **B**; otherwise use method **A**.

Safety Notes

Acetyl chloride is very corrosive and its vapors are irritating and toxic. It reacts violently with water and some alcohols. Use gloves and a hood, avoid contact, do not breathe vapors, and keep it away from water. *N,N*-Dimethylaniline and ammonia (procedure B) are harmful if inhaled or allowed to contact the skin. Use gloves and a hood, avoid contact, and do not breathe vapors.

Reaction: $CH_3COCl + ROH \longrightarrow CH_3COOR + HCl$

Control: 1-butanol

A. *Under the hood*, cautiously add 5 drops of acetyl chloride to 5 drops of the unknown (or 0.20 g, if it is a solid) in a 10 × 75-mm test tube. Observe any evolution of heat; carefully exhale over the mouth of the test tube to see if a cloud of HCl gas is revealed by the moisture in your breath. After a minute or two, add 1.0 mL of water, stopper and shake the test tube, and note any phase separation. Carefully smell the mixture for evidence of an ester aroma, which is usually pleasant and fruity.

Take Care! Possible violent reaction. Wear gloves, and avoid contact with and inhalation of acetyl chloride.

Interpretation: Evidence of reaction (heat, HCl gas), especially if accompanied by phase separation and an ester-like odor, indicates a primary or secondary alcohol. Amines and phenols also react, but amines do not yield pleasant odors. Tertiary alcohols do not form esters by this procedure, but they should by method **B**.

B. *Under the hood*, mix 5 drops of acetyl chloride with 10 drops of *N,N*-dimethylaniline in a conical vial. Cautiously add 5 drops of the unknown (or 0.20 g if it is a solid), attach an air condenser, and warm the mixture in a 50° water bath for 15 minutes; then cool the mixture to room temperature. *Under the hood*, carefully add 1.0 g of ice and 1.0 mL of concentrated ammonia, mix, and let the mixture stand. If an organic layer forms, separate it from the aqueous layer and test it for ester using the ferric hydroxamate test (**C-14**).

Take Care! Wear gloves, and avoid contact with and inhalation of the reactants.

C-2 *Alkaline Hydrolysis*

Use procedure **A** if your unknown may be an amide or procedure **B** if it may be an ester. Esters with boiling points higher than 200°C may be unreactive in aqueous NaOH.

Safety Notes

Sodium hydroxide is toxic and corrosive, causing severe damage to skin, eyes, and mucous membranes. Wear gloves and avoid contact with the NaOH solution.

A. *Amides*

Reactions: $RCONR'_2 + NaOH \longrightarrow RCOONa + R'_2NH$

$RCOONa + H^+ \longrightarrow RCOOH + Na^+$

$R' = $ H, alkyl, or aryl

Control: benzamide

Stir 0.10 g of the unknown (or 3 drops, if it is a liquid) with 4.0 mL of 6 M sodium hydroxide in an 18 × 150-mm test tube. Secure a small piece of filter paper over the top of the tube and moisten it with 2 drops of 10% copper(II) sulfate. Boil the mixture gently for a minute or two and note any color change on the filter paper. Remove the paper and cautiously note the odor of the vapors while the solution is boiling. Acidify the solution with 6 M HCl; if a carboxylic acid precipitate forms, collect it by vacuum filtration and use it as a derivative (see **D-6**).

Interpretation: An intense blue color on the filter paper, accompanied by an ammonia or amine-like odor, indicates an amide. Amides of higher amines that do not turn the paper blue may nevertheless give an amine-like odor. Some amides will yield a precipitate or a separate liquid phase (the carboxylic acid) when the hydrolysis mixture is acidified. The characteristic odor of a carboxylic acid may also be observed. If the test is inconclusive, try increasing the reaction time or repeating the reaction at 200°C using 20% KOH in glycerine.

B. *Esters*

Reactions: $RCOOR' + NaOH \longrightarrow RCOONa + R'OH$

$RCOONa + H^+ \longrightarrow RCOOH + Na^+$

Control: butyl acetate

Mix 0.20 mL of the unknown (or 0.20 g, if it is a solid) with 2.0 mL of 6 M sodium hydroxide in a 5-mL conical vial, add a boiling chip, and heat the mixture under reflux for 30 minutes, or until the solution is homogeneous. Note whether the odor of the unknown is gone and (if the unknown was water insoluble) whether the organic layer has disappeared. If a separate organic layer or residue remains, heat the mixture longer until it disappears or until it is apparent that no reaction is taking place. Most esters with boiling points below 110°C will hydrolyze in 30 minutes; higher-boiling esters may take several hours. Cool the reaction mixture, remove the organic layer (if there is one), and acidify the aqueous solution with 6 M sulfuric acid. If a carboxylic acid precipitates on acidification, collect it by vacuum filtration and use it as a derivative (see **D-16**). If acidification does not yield a solid carboxylic acid, carefully note the odor of the acidified reaction mixture. Then make the solution basic with 6 M NaOH and saturate it with potassium carbonate to see if an organic layer (the alcohol) separates. Note the odor of this layer also.

Interpretation: Evidence for an ester is indicated by the disappearance of the organic layer (if any) and of the odor of the unknown during the reflux period, and the appearance of a precipitate or the odor of a carboxylic acid upon acidification. Formation of a separate liquid layer having an odor different from that of the unknown after treatment of the acidified product mixture with NaOH also suggests an ester.

C-3 *Aluminum Chloride and Chloroform*

Safety Notes

Chloroform (trichloromethane) is harmful if inhaled or absorbed through the skin, and it is a suspected carcinogen in humans. Use gloves and a hood; avoid contact and do not breathe its vapors.

> **Aluminum chloride reacts violently with water; skin and eye contact can cause painful burns, and inhaling its dust or vapors is harmful. Wear gloves, avoid contact, do not breathe its dust or vapors, and keep water away.**

Reaction: $\text{ArH} \xrightarrow{\text{CHCl}_3, \text{AlCl}_3} \text{Ar}_3\text{C}^+$

Controls: toluene, biphenyl

Under the hood, prepare a solution containing 3 drops of a solubility class **X** unknown (or 0.10 g, if it is a solid) in 2.0 mL of dry chloroform (trichloromethane). Place 0.20 g of anhydrous aluminum chloride in a dry 13×100-mm test tube and heat it over a flame, angling the test tube so that the AlCl_3 sublimes onto the inner wall of the tube. Allow the tube to cool until it can be held comfortably in the hand, then pipet a few drops of the solution down the side of the tube so that it contacts the aluminum chloride. Note any color change at the point of contact.

Take Care! Wear gloves, and avoid contact with and inhalation of chloroform and aluminum chloride.

Interpretation: An intense color, such as yellow-orange, red, blue, green, or purple, indicates an aromatic compound. A light yellow color is inconclusive or negative.

C-4 *Basicity Test*

Reaction:

$\text{RNH}_2 + \text{H}^+ \xrightarrow{\text{pH } 5.5} \text{RNH}_3^+$ (and similar reactions for 2° and 3° amines)

Controls: *p*-toluidine, dibutylamine

If the unknown is water soluble, dissolve 4 drops (or 0.10 g, if it is a solid) in 3.0 mL of water in a 13×100-mm test tube and measure the pH of the solution using pH paper or a universal indicator. If the unknown is insoluble in water, add 4 drops (or 0.10 g, if it is a solid) to 3.0 mL of a pH 5.5 acetate–acetic acid buffer in a 13×100-mm test tube, and then stopper and shake the test tube.

Interpretation: Most water-soluble aliphatic amines give pH values above 11, and water-insoluble aliphatic amines should dissolve in the buffer. Most water-soluble aromatic amines give pH values below 10, and water-insoluble aromatic amines do not dissolve in the buffer. Test **C-8** can also be used to test for aromatic amines.

C-5 *Beilstein's Test*

Control: chlorobenzene

Make a small loop in the end of a length of copper wire (10 cm or longer), and heat the loop to redness in a flame. Let the wire cool; then dip the loop into the unknown and heat it in the nonluminous (blue) flame of a burner, near the lower edge.

Interpretation: A distinct green or blue-green flame indicates a halogen compound.

C-6 *Benedict's Test*

Reaction: $RCHO + 2Cu^{2+} + 4OH^- \longrightarrow RCOOH + Cu_2O + 2H_2O$

Control: butanal (butyraldehyde)

Benedict's reagent contains copper(II) sulfate, sodium citrate, and sodium carbonate.

Add 2 drops of the unknown (or 80 mg, if it is a solid) to 2.0 mL of water in a 13 × 100-mm test tube and mix in 2.0 mL of Benedict's reagent. Heat the mixture to boiling. Observe whether a precipitate forms, and if one forms, record its color.

Interpretation: Aliphatic aldehydes generally produce a yellow to orange suspension or precipitate of copper(I) oxide; it may appear greenish in the blue solution. Most ketones and aromatic aldehydes do not react.

C-7 *Bromine Test*

Safety Notes

> **Dichloromethane may be harmful if inhaled or absorbed through the skin, and it is a suspected carcinogen. Avoid contact and do not breathe its vapors. Bromine is toxic and corrosive and its vapors are very harmful. Avoid contact with the bromine solution and do not breathe its vapors.**

Reaction:
$$-\overset{|}{\underset{|}{C}}=\overset{|}{\underset{|}{C}}- \ + \ Br_2 \longrightarrow -\overset{Br}{\underset{|}{\overset{|}{C}}}-\overset{Br}{\underset{|}{\overset{|}{C}}}-$$

Control: cyclohexene

Under the hood, dissolve 1 drop of the unknown (or 40 mg, if it is a solid) in 0.50 mL of dichloromethane in a 10 × 75-mm test tube. Add freshly prepared 1.0 *M* bromine in dichloromethane drop by drop, with shaking. Continue adding until the red-orange color of bromine persists *or* until about 20 drops have been added. Immediately after the addition, carefully exhale over the mouth of the test tube and observe whether a cloud of HBr gas appears.

Take Care!   Avoid contact with and inhalation of dichloromethane and the bromine solution.

Interpretation: Decolorization of more than 1 drop of the bromine solution, without the evolution of HBr, indicates unsaturation (C=C or C≡C bonds). Aldehydes, ketones, amines, and phenols react by substitution to evolve HBr.

C-8 *Bromine Water*

Safety Notes

> **Bromine is toxic and corrosive, and its vapors are very harmful. Avoid contact with the bromine water and do not breathe its vapors.**

Reaction:

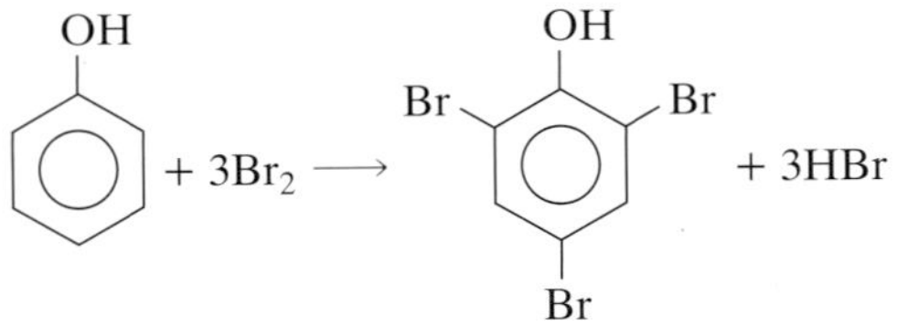

Reaction is for phenol. Aromatic amines and substituted phenols undergo similar reactions

Control: phenol

Dissolve 3 drops of the unknown (or 0.10 g, if it is a solid) in 10 mL of water in an 18 × 150-mm test tube. If it is insoluble in water, add just enough ethanol to bring it into solution. Measure the pH of the solution with pH paper. *Under the hood*, add saturated bromine water drop by drop until the bromine color persists. Watch for evidence of a precipitate.

Take Care! Avoid contact with and inhalation of the bromine solution.

Interpretation: Decolorization of the bromine, accompanied by formation of a white (or nearly white) precipitate, indicates a phenol or aromatic amine. If the unknown is a phenol, the pH of the initial solution should be less than 7.

C-9 *Chromic Acid*

The chromic acid reagent is very corrosive and may be carcinogenic. Avoid contact.

Safety Notes

Reactions:

1°Alcohol:
$$3RCH_2OH + 4CrO_3 + 6H_2SO_4 \longrightarrow 3RCOOH + 2Cr_2(SO_4)_3 + 9H_2O$$

2°Alcohol:
$$3R_2CHOH + 2CrO_3 + 3H_2SO_4 \longrightarrow 3R_2CO + Cr_2(SO_4)_3 + 6H_2O$$

Aldehyde:
$$3RCHO + 2CrO_3 + 3H_2SO_4 \longrightarrow 3RCOOH + Cr_2(SO_4)_3 + 3H_2O$$

Controls: 1-butanol, butanal

Dissolve 1 drop of the unknown (or 40 mg, if it is a solid) in 1.0 mL of reagent-grade acetone in a 10 × 75-mm test tube. (If there is any doubt about the purity of the acetone, test it with a drop of the reagent beforehand.) Add 1 drop of the chromic acid reagent and swirl, noting the time required for a positive test.

The chromic acid reagent contains chromium(VI) oxide and concentrated sulfuric acid.

Take Care! Avoid contact with the chromic acid reagent.

Interpretation: Formation of an opaque blue-green suspension within 2–3 seconds, accompanied by disappearance of the orange color of the reagent, indicates a primary or secondary alcohol. Aldehydes give the same result but react more slowly. With aliphatic aldehydes, the solution turns cloudy in about 5 seconds and the blue-green suspension forms within 30 seconds, while aromatic aldehydes require 30–90 seconds or longer to form the suspension. The generation of some other dark color, particularly with the color of the liquid remaining orange, should be considered a negative test.

C-10 *Density Test*

Controls: 1-chlorobutane, 1-bromobutane

Add 5 drops of a liquid solubility class **X** unknown to 1.0 mL of deionized water in a 10 × 75-mm test tube, stir gently, and note whether the unknown floats or sinks. If it sinks, determine its density by accurately measuring 0.10 mL of the unknown liquid (use an automatic pipet or measuring pipet) into a tared vial and weighing the liquid on an accurate balance.

Interpretation: Of the compounds that are insoluble in cold, concentrated sulfuric acid, most aromatic hydrocarbons and monochloroalkanes will float, whereas aryl chlorides, polychloroalkanes, and all bromides and iodides will sink. Density ranges for some classes of organic halides are listed here.

Approximate density ranges for halogenated hydrocarbons

Halide type	Density range
alkyl chloride (mono)	0.85–1.0
alkyl bromide (mono)	1.1–1.5
alkyl iodide (mono)	>1.4
alkyl chloride (poly)	1.1–1.7
alkyl bromide (poly)	1.5–3.0
aryl chloride	1.1–1.3
aryl bromide	1.3–2.0
aryl iodide	>1.8

Note: Density is in g/mL.

C-11 *2,4-Dinitrophenylhydrazine*

Safety Notes

> **2,4-Dinitrophenylhydrazine (DNPH) is harmful if absorbed through the skin, and it will dye your hands yellow. Wear gloves, avoid contact with the DNPH reagent, and wash your hands after using it.**

Reaction:

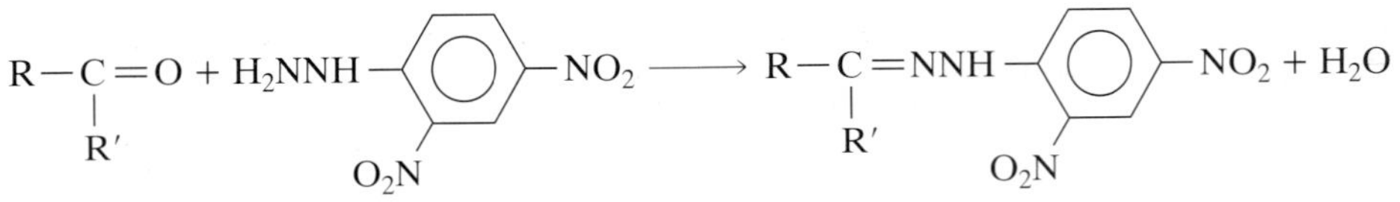

$$R-C(R')=O + H_2NNH-C_6H_3(NO_2)(O_2N)-NO_2 \longrightarrow R-C(R')=NNH-C_6H_3(NO_2)(O_2N)-NO_2 + H_2O$$

R, R′ = alkyl, aryl, or H

Controls: cyclohexanone, benzaldehyde

Dissolve 1 drop of the unknown (or 40 mg, if it is a solid) in 1.0 mL of 95% ethanol; use more ethanol if necessary. Add this solution to 2.0 mL of the DNPH reagent in a 10 × 75-mm test tube. Stopper and shake the test tube, and let the mixture stand for 15 minutes or until a precipitate forms. If no precipitate has formed after 15 minutes, scratch the inside of the test tube.

Take Care!   Wear gloves and avoid contact with the reagent.

Interpretation: Formation of a crystalline yellow or orange-red precipitate indicates an aldehyde or ketone. Some carbonyl compounds initially form oils that may or may not become crystalline; a few may require *gentle* heating, but overheating can cause oxidation of allylic or other reactive alcohols, resulting in a false positive test. Some aromatic compounds (hydrocarbons, halides, phenols, and phenyl esters) may form slightly soluble complexes with the reagent, and some alcohols may be contaminated with

Rinsing the reaction tube with acetone and not drying it thoroughly may also result in a false positive test.

small amounts of the corresponding aldehyde or ketone. In such cases, the amount of precipitate should be quite small, and comparison with a control should show the difference between a positive test and a doubtful one. The color of the precipitate may give a clue to the structure of the carbonyl compound; unconjugated aliphatic aldehydes and ketones usually yield yellow precipitates, while aromatic and α,β-unsaturated aldehydes and ketones yield orange-red precipitates.

C-12 *Elemental Analysis*

> **Sodium can cause serious burns, and the sodium-lead alloy may react violently with some substances. Wear gloves, avoid contact, and keep Na(Pb) away from other chemicals.** *Do not use this procedure with sodium metal*; **it will react violently when water is added.**
> **The PNB reagent is harmful if inhaled or allowed to contact the skin. Wear gloves, avoid contact, and do not breathe its vapors.**

Reactions:

Nitrogen: $[C,N] \xrightarrow{Na(Pb)} NaCN \xrightarrow{PNB}$ purple color

Halogens: $RX, ArX \xrightarrow{Na(Pb)} NaX \xrightarrow{AgNO_3}$ **AgX** $(X = Cl, Br, I)$

Chlorine: $AgCl + 2NH_3 \longrightarrow Ag(NH_3)_2Cl$

Bromine: $2HBr + Cl_2 \longrightarrow 2HCl + Br_2$ (red-orange)

Iodine: $2HI + Cl_2 \longrightarrow 2HCl + I_2$ (purple)

Controls: acetamide (N), bromobenzene (Br)

Under the hood, place 0.25 g of 10% sodium-lead alloy in a *clean, dry* 10×75-mm test tube held vertically by a clamp. Melt the alloy in a burner flame and continue heating until the sodium vapor rises about 1 cm up the inside of the test tube. Add 2 drops of the unknown from a Pasteur pipet (or 10 mg of unknown from a microspatula, if it is a solid) directly onto the molten alloy so that it does not touch the sides of the tube. Heat gently to start the reaction, remove the flame until the reaction subsides, then heat the tube strongly for a minute or two, keeping the bottom a dull red color. Let the tube cool to room temperature. Add 1.5 mL of water and heat gently for a minute or so until the excess sodium has decomposed and gas evolution ceases. Filter the solution through a filtering pipet (see OP-12) containing a cotton plug, wash the cotton with 1 mL of water, and combine the wash water with the filtrate. The filtrate should be colorless or just slightly yellow. If it is darker, repeat the fusion with stronger heating or more of the alloy.

Take Care! Wear gloves, and avoid contact with the sodium-lead alloy.

To test for *nitrogen*, put 5 drops of the sodium fusion solution into a 10×75-mm test tube and add enough solid sodium bicarbonate, with stirring, to saturate it (a small amount of excess solid should be present). Add 1 drop of this solution to another 10×75-mm test tube containing 10 drops of PNB reagent (*p*-nitrobenzaldehyde in dimethyl sulfoxide) and note any color change.

Take Care! Wear gloves, and avoid contact with and inhalation of the PNB reagent.

To test for *halogens*, transfer 10 drops of the sodium fusion solution to a 10 × 75-mm test tube, acidify it with dilute nitric acid, boil it gently *under the hood* for a few minutes, add a drop or two of 0.30 *M* silver nitrate, and note the color and volume of any precipitate that forms. If a voluminous precipitate forms, remove the liquid with a filter-tip pipet (see OP-6). Then add 2.0 mL of 3 *M* ammonia to the solid, shake vigorously, and note your observations. To test further for bromine and iodine, acidify 1 mL of the original sodium fusion solution with 1 *M* sulfuric acid in a 10 × 75-mm test tube, boil gently for a few minutes, and then add 0.50 mL of dichloromethane and a drop of freshly prepared chlorine water. Shake and look for a color in the dichloromethane layer.

Interpretation: In the PNB test, a purple color indicates the presence of nitrogen (green indicates sulfur). In the halogen tests, formation of a voluminous precipitate on addition of silver nitrate indicates that a halogen is present, and the color of the precipitate (a silver halide) may suggest which halogen: white for chloride, pale yellow for bromide, and yellow for iodide. If only a faint turbidity is produced, it may be caused by traces of impurities or by incomplete sodium fusion. If the precipitate is silver chloride, it will dissolve in aqueous ammonia; silver bromide is only slightly soluble and silver iodide is insoluble. In the chlorine water test, a red-orange color is due to elemental bromine and a violet color to elemental iodine.

C-13 *Ferric Chloride*

Control: phenol

In a 10 × 75-mm test tube, dissolve 1 drop of the unknown (or 40 mg, if it is a solid) in 1.0 mL of water, or in a water/alcohol mixture if it does not dissolve in water. Add two drops of 2.5% ferric chloride solution.

Interpretation: Formation of an intense red, green, blue, or purple color suggests a phenol or an easily enolizable compound. A few phenols do not react under these conditions. Many aromatic carboxylic acids form tan precipitates; aliphatic hydroxy acids yield yellow solutions.

C-14 *Ferric Hydroxamate Test*

Safety Notes

Hydroxylamine hydrochloride is toxic and mutagenic, and it can cause a form of anemia. Avoid contact with its solution.
Sodium hydroxide is toxic and corrosive, causing severe damage to skin, eyes, and mucous membranes. Wear gloves and avoid contact with the NaOH solution.

Reactions:

$$\underset{RC-OR'}{\overset{O}{\overset{\|}{}}} + H_2NOH \longrightarrow \underset{RC-NHOH}{\overset{O}{\overset{\|}{}}} + R'OH$$

$$3RCONHOH + FeCl_3 \longrightarrow (RCONHO)_3Fe + 3HCl$$
$$\text{ferric hydroxamate}$$

1/2 · 2/3 = 1/3

Control: butyl acetate

Before you carry out the ferric hydroxamate test, perform the following preliminary test. Dissolve 1 drop of the unknown (or 40 mg, if it is a solid) in 1.0 mL of 95% ethanol in a 10 × 75-mm test tube. Add 1.0 mL of 1 M hydrochloric acid; then add two drops of 2.5% ferric chloride. If a definite color other than yellow results, the ferric hydroxamate test cannot be used. This test eliminates those phenols and enols that give colors with ferric chloride in acidic solution and that would therefore give a false positive result in the ferric hydroxamate test.

Mix 1.0 mL of 0.50 M ethanolic hydroxylamine hydrochloride with 5 drops of 6 M sodium hydroxide in a 13 × 100-mm test tube. Add 1 drop of the unknown (or 40 mg, if it is a solid) and heat the solution to boiling. Allow it to cool slightly and add 2.0 mL of 1 M hydrochloric acid. If the solution is cloudy at this point, add enough 95% ethanol to clarify it. Add 5 drops of 2.5% ferric chloride solution and observe any color produced. If the color does not persist, continue to add the ferric chloride solution until the color becomes permanent.

Take Care!   Avoid contact with the hydroxylamine hydrochloride and NaOH solutions.

Interpretation: A burgundy or magenta color that is distinctly different from the color obtained in the preliminary test indicates an ester.

C-15   *Hinsberg's Test*

Record all observations carefully during this test.

***p*-Toluenesulfonyl chloride is toxic and corrosive. Use gloves and a hood, avoid contact, and do not breathe its vapors.**

Safety Notes

Reactions:

1° Amine:

$$RNH_2 + ArSO_2Cl + 2NaOH \longrightarrow ArSO_2NR^-Na^+ + NaCl + 2H_2O$$
$$ArSO_2NR^-Na^+ + HCl \longrightarrow \textbf{ArSO}_2\textbf{NHR} + NaCl$$

2° Amine:

$$R_2NH + ArSO_2Cl + NaOH \longrightarrow \textbf{ArSO}_2\textbf{NR}_2 + NaCl + H_2O$$

3° Amine:

$$R_3N + ArSO_2Cl \longrightarrow \text{no reaction}$$
$$R_3N + HCl \longrightarrow R_3NH^+Cl^-$$

Controls: butylamine, dibutylamine, tributylamine

Under the hood, mix 3 drops of the unknown (or 0.10 g, if it is a solid) with 5.0 mL of 3 M sodium hydroxide in an 18 × 150-mm test tube, and then add 0.20 g of *p*-toluenesulfonyl chloride. Stopper the test tube and shake it intermittently for 3–5 minutes. Then remove the stopper and heat the solution in a boiling-water bath, with shaking, for 1 minute. The solution should be basic at this point (if not, add more NaOH). If there is a solid or liquid residue in the test tube, separate it from the solution by vacuum filtration, if it is a solid, or with a Pasteur pipet, if it is a liquid. Test the solubility of the

Take Care!   Wear gloves, and avoid contact with and inhalation of *p*-toluenesulfonyl chloride.
Benzenesulfonyl chloride (6 drops) can be used in place of p*-toluenesulfonyl chloride, but it is more hazardous and it tends to form oils.*

residue in water; if it is insoluble in water, separate it as before and test its solubility in 5% hydrochloric acid. Acidify the original solution with 6 M hydrochloric acid and, if no precipitate forms immediately, scratch the sides of the test tube and cool. If you obtain a solid either from the original reaction mixture or after addition of 6 M HCl, save it for possible use as a p-toluenesulfonate derivative (see **D-9**).

Interpretation: Formation of a white precipitate (a p-toluenesulfonamide) when the reaction mixture is acidified indicates a *primary amine*. Most primary amines yield a clear solution after the initial reaction, but some form sodium salts or disulfonyl derivatives that precipitate during the reaction. The sodium salts should be soluble in water, and amines that form disulfonyl derivatives should yield additional precipitate when the reaction mixture is acidified.

Most *secondary amines* yield a white solid that does not dissolve in water or 5% HCl. A liquid residue that is more dense than water and insoluble in 5% HCl may be a secondary amine's arenesulfonamide that has failed to crystallize.

Tertiary amines do not react; any residue will be the original liquid or solid amine, which should dissolve in dilute HCl. Water-soluble tertiary amines yield a clear solution that does not form a separate phase on acidification.

C-16 *Iodoform Test*

Reactions:

1. Methyl carbinols

$$\underset{\text{OH}}{\overset{|}{\text{RCH}}}-\text{CH}_3 + 4\text{I}_2 + 5\text{NaOH} \longrightarrow \text{RC}\overset{\text{O}}{\overset{||}{-}}\text{CI}_3 + 5\text{NaI} + 5\text{H}_2\text{O}$$

$$\text{RC}\overset{\text{O}}{\overset{||}{-}}\text{CI}_3 + \text{NaOH} \longrightarrow \text{RC}\overset{\text{O}}{\overset{||}{-}}\text{ONa} + \underset{\text{(iodoform)}}{\text{CHI}_3}$$

2. Methyl ketones and acetaldehyde

$$\text{RC}\overset{\text{O}}{\overset{||}{-}}\text{CH}_3 + 3\text{I}_2 + 3\text{NaOH} \longrightarrow \text{RC}\overset{\text{O}}{\overset{||}{-}}\text{CI}_3 + 3\text{NaI} + 3\text{H}_2\text{O}$$

$$\text{RC}\overset{\text{O}}{\overset{||}{-}}\text{CI}_3 + \text{NaOH} \longrightarrow \text{RC}\overset{\text{O}}{\overset{||}{-}}\text{ONa} + \text{CHI}_3$$

Control: 2-butanone

Dissolve 3 drops of the unknown (or 0.10 g, if it is a solid) in 2.0 mL of water in a 25 × 150-mm test tube. (If the unknown is insoluble in water, dissolve it in 2.0 mL or more of methanol instead.) Add 1.0 mL of 3 M sodium hydroxide solution, then add 0.50 M iodine–potassium iodide reagent drop by drop until the brown iodine color persists after shaking. If no yellow precipitate appears, place the test tube in a 60°C water bath and add more iodine–potassium iodide solution as necessary until the brown color remains after at least 2 minutes of heating. Then add 3 *M* NaOH drop by drop until the color just disappears (a light yellow color may remain). Remove the test tube from the water bath, add 10 mL of cold water, and let it stand for 15 minutes.

Interpretation: Formation of a yellow precipitate with the characteristic medicinal odor of iodoform is a positive test. If there is any doubt about the identity of the precipitate, its melting point (121°C) can be measured. The test is positive for acetaldehyde, methyl ketones, and methyl carbinols—alcohols that contain a —CH(OH)CH$_3$ grouping. Other compounds that may yield iodoform in this test include certain conjugated aldehydes, such as acrolein and furfural, and some 1,3-dicarbonyl or dihydroxy compounds.

C-17 *Lucas' Test*

This test is not applicable to most alcohols with boiling points higher than 140–150°C, or to solid alcohols.

Safety Notes

The Lucas reagent (ZnCl$_2$ in concentrated HCl) can cause serious burns, and inhaling the vapors is harmful. Use gloves and a hood; avoid contact and do not breathe its vapors.

Reaction: $ROH + HCl \xrightarrow{ZnCl_2} RCl + H_2O$

Controls: 1-butanol (no reaction), 2-butanol, 2-methyl-2-propanol

Under the hood, place 2.0 mL of the Lucas reagent in a 10 × 75-mm test tube. Add 4 drops of the liquid unknown, stopper the test tube immediately, and shake vigorously, taking note of any cloudiness or layer separation. Allow the mixture to stand for 15 minutes or more, observing it periodically for evidence of reaction. If the alcohol appears to be secondary or tertiary, repeat the test using concentrated HCl in place of the Lucas reagent.

Take Care! Wear gloves, and avoid contact with and inhalation of the Lucas reagent.

Take Care! Wear gloves, and avoid contact with and inhalation of HCl.

Interpretation: *Tertiary alcohols* that are soluble in the Lucas reagent should turn the reagent cloudy almost immediately and soon form a separate layer of alkyl chloride. *Secondary alcohols* usually turn the clear solution cloudy in 3–5 minutes and form a distinct layer within 15 minutes. *Primary alcohols* do not react under these conditions. Most allylic and benzylic alcohols give the same result as tertiary alcohols, except that the chloride formed from allyl alcohol is itself soluble in the reagent and separates out only upon addition of ice water. High-boiling alcohols that are insoluble in the Lucas reagent cannot be tested, because they will form a separate layer immediately. When tested with concentrated HCl, a tertiary alcohol should react within minutes, and a secondary alcohol should not react.

C-18 *Neutralization Equivalent*

Reaction: $RCOOH + NaOH \longrightarrow RCOONa + H_2O$

Control: hexanedioic acid (adipic acid)

Accurately weigh (to 3 decimal places) about 0.10 g of an unknown carboxylic acid, transfer it to a 50-mL Erlenmeyer flask, and dissolve it in about 25 mL of water, ethanol, or a mixture of the two, depending on its solubility. (If necessary, you can more solvent to dissolve it.) Add a drop of phenolphthalein indicator (or bromothymol blue indicator, if the solvent is ethanol) and titrate this solution with a standardized solution of ~0.1 M sodium hydroxide. Record the exact concentration of the NaOH solution from its label. Calculate the neutralization equivalent (N.E.) of the acid using this formula:

$$\text{N.E.} = \frac{\text{mass of sample (g)} \times 1000}{\text{volume of NaOH (mL)} \times \text{concentration of NaOH}}$$

Interpretation: The neutralization equivalent (equivalent weight) of a carboxylic acid is equal to its molecular weight divided by the number of carboxyl groups it has. For instance, the N.E. of adipic acid [$HOOC(CH_2)_4COOH$; M.W. = 146] is 73. A solid carboxylic acid that has an unusually low neutralization equivalent for its melting point probably contains more than one carboxyl group.

C-19 *Potassium Permanganate*

Reaction:

$$3 - C = C - + 2KMnO_4 + 4H_2O \longrightarrow$$

$$
\begin{array}{cc}
\text{HO} & \text{OH} \\
| & | \\
3 - C - C - & + 2MnO_2 + 2KOH \\
| & |
\end{array}
$$

Control: cyclohexene

Dissolve 1 drop of the unknown (or 40 mg, if it is a solid) in 2.0 mL of water or 95% ethanol in a 10 × 75-mm test tube. (If ethanol is the solvent, perform the test with a blank as well.) Add 0.10 M potassium permanganate drop by drop until the purple color of the permanganate persists, or until about 20 drops have been added. If a reaction does not take place immediately, shake the mixture and let it stand for up to 5 minutes. Disregard any decolorization that takes place after 5 minutes have elapsed.

Interpretation: Decolorization of more than one drop of the purple permanganate solution, accompanied by the formation of a brown precipitate (or reddish-brown suspension) of manganese dioxide, suggests unsaturation. The test is positive for most compounds containing double and triple bonds, except for conjugated alkadienes. Easily oxidizable compounds, such as aldehydes, aromatic amines, phenols, formic acid, and formate esters, also give positive tests. Most pure alcohols will not react in less than 5 minutes, but alcohols that contain oxidizable impurities may react slightly, so decolorization of only the first drop of potassium permanganate should not be considered a positive test.

C-20 *Quinhydrone*

This test is not applicable to diaminobenzenes or nitro-substituted aromatic amines.

Safety Notes

Controls: butylamine, dibutylamine, tributylamine, aniline, *N*-methylaniline, *N,N*-dimethylaniline

This test should be run in conjunction with Test **C-4** or with an infrared spectrum that shows whether the amine is aliphatic or aromatic. Controls should be run for comparison, since the colors are difficult to describe accurately. Shake 1 drop of an unknown alkylamine (use 30 mg, if it is a solid) or 6 drops of an unknown arylamine (use 0.2 g, if it is a solid) with 6.0 mL of water in an 18 × 150-mm test tube. If the amine dissolves, add 6.0 mL more of water; if not, add 6.0 mL of ethanol. Shake the mixture, add 1 drop of 2.5% quinhydrone in methanol, and let it stand for 2 minutes or more. Compare the color of this solution with those of the controls.

Take Care! Avoid contact with the quinhydrone solution.

Interpretation: Most amines from the following classes give the colors indicated.

    1° aliphatic: violet
    2° aliphatic: rose
    3° aliphatic: yellow
    1° aromatic: rose
    2° aromatic: amber
    3° aromatic: yellow

It is best to use this test in conjunction with the Hinsberg test or an IR spectrum for confirmation.

C-21 *Silver Nitrate in Ethanol*

Safety Notes

Reaction: $RX + AgNO_3 + EtOH \longrightarrow ROEt + AgX + HNO_3$
        (Organic products other than ROEt are also formed.)

Controls: 1-chlorobutane, 2-chloro-2-methylpropane, 1-bromobutane, iodoethane

Add 1 drop of the unknown (or 50 mg, dissolved in a small amount of ethanol, if it is a solid) to 2.0 mL of 0.10 M ethanolic silver nitrate in a 13 × 100-mm test tube. Shake the mixture and let it stand. If no precipitate forms within 5 minutes, heat the solution to boiling and boil it gently for 30 seconds. If a precipitate forms, note its color and see if it dissolves when the mixture is shaken with 2 drops of 1 M nitric acid.

Take Care! Avoid contact with the AgNO_3 solution.

Interpretation: The reaction of a halide with silver nitrate occurs by an S_N1 mechanism, so the reactivity of a halide varies considerably with its

Halides that react with silver nitrate at room temperature:

>*Chlorides: 3°, allyl, benzyl*
>
>*Bromides: 1°, 2°, and 3° alkyl (except gem-di- and tribromides); allyl, benzyl, CBr$_4$*
>
>*Iodides: All aliphatic and alicyclic except vinyl*

Halides that react upon heating:

>*Chlorides: 1° and 2° alkyl*
>
>*Bromides: alkyl gem-di- and tribromides*
>
>*Some activated aryl halides, such as 2,4-dinitrohalobenzenes*

Halides that do not react:

>*Chlorides: alkyl gem-di- and trichlorides, CCl$_4$*
>
>*Most aryl and vinyl halides*

Halides that react with sodium iodide at room temperature:

>*1° alkyl bromides*
>
>*benzyl and allyl halides*
>
>*vic-dihalides*; CBr$_4$**
>
>*α-haloketones, -esters, and -amides*

Halides that react at 50°C:

>*1° and 2° alkyl chlorides*
>
>*2° and 3° alkyl bromides*
>
>*gem-di- and tribromides**

Halides that do not react:

>*3° alkyl chlorides*
>
>*aryl and vinyl halides*
>
>*cyclopropyl, cyclobutyl, and cyclohexyl halides*
>
>*gem-polychloro compounds (except benzyl and allyl)*

*Turns the solution red-brown

structure. Alkyl iodides react faster than the corresponding bromides, which are more reactive than alkyl chlorides. For the same halogen, tertiary, allylic, and benzylic halides react fastest, followed by secondary and primary halides. Most aryl and vinyl halides are unreactive, except for aryl halides activated by two or more nitro groups. Most geminal alkyl halides, which have two or more halogen atoms on the same carbon, are less reactive than the corresponding monohaloalkanes. The types of compounds that react at room temperature, react after boiling the solution, or do not react at all are listed in the margin.

The color of the precipitate formed may indicate the type of halide responsible. Silver chloride is white, silver bromide is pale yellow or cream colored, and silver iodide is yellow. These salts will not dissolve when the mixture is acidified. Some carboxylic acids and alkynes yield silver salts, but these salts should dissolve in the acidic solution.

C-22   *Sodium Iodide in Acetone*

Reactions: $RCl + NaI \xrightarrow{\text{acetone}} RI + \textbf{NaCl}$

$RBr + NaI \xrightarrow{\text{acetone}} RI + \textbf{NaBr}$

Controls: 1-chlorobutane, 2-chloro-2-methylpropane (no reaction), 1-bromobutane

Place 1.0 mL of the sodium iodide/acetone reagent into a 13 × 100-mm test tube. Add 2 drops of a liquid halogen compound (or 0.10 g, dissolved in the minimum volume of acetone, if it is a solid). Shake the mixture, allow it to stand for 3 minutes, and note whether a precipitate or color forms. Disregard any precipitate that forms upon mixing but then dissolves. If there is no precipitate at the end of this time, place the test tube in a 50°C water bath (replenish the acetone if some evaporates), and leave it there for an additional 6 minutes; then cool the mixture to room temperature and record your observations.

Interpretation: Certain alkyl chlorides and bromides react to precipitate sodium chloride or sodium bromide, which are insoluble in acetone. Bromides react faster than comparable chlorides, and the test is not applicable to iodides. Since the reaction involves an S_N2 displacement by iodide ion, halides of the same halogen react in the order methyl > primary > secondary > tertiary > aryl, vinyl. Cycloalkyl halides tend to react more slowly than the corresponding open-chain compounds and may give no precipitate even after heating. Vicinal dihalides (with two halogen atoms on adjacent carbon atoms) and some geminal halides (with two or more halogen atoms on the same carbon atom) undergo oxidation-reduction reactions to liberate iodine—which is red-brown in acetone—while precipitating the sodium halide. A summary of the results with various halides is given in the margin.

C-23   *Tollens' Test*

Safety Notes

> **Silver nitrate is corrosive and toxic and it stains the skin black. Avoid contact with its solutions.**
>
> **The Tollens reagent and the test solution must never be stored—explosive silver salts form on standing.**

Reaction:

$$RCHO + 2Ag(NH_3)_2OH \longrightarrow 2\textbf{Ag} + RCOONH_4 + H_2O + 3NH_3$$

Control: benzaldehyde

Prepare the reagent immediately before use as follows: Measure 2.0 mL of 0.30 *M* silver nitrate into a *thoroughly cleaned* 13 × 100-mm test tube and add 1 drop of 3 *M* sodium hydroxide. Then add 2 *M* aqueous ammonia drop by drop, with shaking, until the precipitate of silver oxide just dissolves (avoid an excess of ammonia). Add 1 drop of the unknown (or 40 mg, if it is a solid) to this solution, shake the mixture, and let it stand for 10 minutes. If no reaction has occurred by this time, heat the mixture in a 35°C water bath for 5 minutes.

Interpretation: Formation of a silver mirror on the inside of the test tube is a positive test for an aldehyde. If the tube is not sufficiently clean, a black precipitate or a suspension of metallic silver may form instead. Certain cyclic ketones (such as cyclopentanone), aromatic amines, phenols, and α-alkoxy- or α-dialkylaminoketones may also give positive tests.

Take Care! Avoid contact with the $AgNO_3$ solution.

Waste Disposal: After the test is completed, dissolve any residue in dilute nitric acid and place the solution in a designated waste container (*important!*).

Spectral Analysis

Infrared Spectra

To anyone proficient in spectral interpretation, an infrared (IR) spectrum provides so much information that it is equivalent to many classification tests. Because infrared absorption bands arise from the vibrational motions of specific chemical bonds, it is usually possible to identify or confirm the functional group(s) present in an organic compound from its IR spectrum. For example, a compound whose IR spectrum has a C=O stretching band at 1715 cm^{-1}, a C—O stretching band at 1240 cm^{-1}, and a very broad O—H stretching band centered at 3000 cm^{-1} is almost certainly a carboxylic acid. Characteristic infrared bands of the most common families of organic compounds are described in the section "Interpretation of Infrared Spectra" in OP-36.

Structural features other than functional groups can be detected from IR spectra. If your unknown compound is an alcohol, amide, or amine, you should be able to classify it as primary, secondary, or tertiary by reading about the IR spectra of these families in OP-36. If your unknown may have a benzene ring or a carbon-carbon double bond, you can confirm this and perhaps determine the kind of substitution on the ring or the C=C bond by reading about the IR spectra of aromatic hydrocarbons and alkenes. You can also find out whether the C=O group of an aldehyde, ketone, ester, or carboxylic acid is conjugated with an aromatic ring or a carbon-carbon double bond.

The best way to become proficient in the interpretation of infrared spectra is to read about and interpret actual spectra. Before attempting to interpret your IR spectrum, you should read "Interpretation of Infrared Spectra" in OP-36 and examine the sample spectra provided there.

Directions

Obtain an infrared spectrum [OP-36] of the unknown as directed by your instructor. Interpret the spectrum as completely as you can (see OP-36).

Use it to confirm any functional group suggested by the classification tests and to detect any additional functional groups and other structural features.

NMR Spectra

Like infrared spectra, nuclear magnetic resonance (NMR) spectra can be used to detect certain functional groups, but they are more often used to provide detailed structural information about organic molecules. With practice, an organic chemist can deduce the structural units present in an organic compound, or even its entire molecular structure, from an NMR spectrum. See the section "Interpretation of ^{1}H NMR Spectra" in OP-37 for help.

Directions

Safety Notes

> **Assume that the NMR solvent is flammable and harmful by inhalation, ingestion, and skin absorption. Some NMR solvents, such as deuterochloroform, are also carcinogenic. Avoid contact and do not breathe their vapors.**

With your instructor's permission and guidance, record an integrated ^{1}H NMR spectrum [OP-37] of your unknown in deuterochloroform or another suitable solvent, using a TMS reference standard. Measure and tabulate the NMR parameters (chemical shift, signal area, signal multiplicity, and coupling constant). Then use them to reconstruct as much of the molecular structure of your unknown as you can.

Identification

After you learn what chemical family your unknown belongs to, you are ready to begin the identification phase of your analysis by preparing a list of possible compounds. This phase will be completed when you have obtained sufficient evidence to eliminate all of the possibilities but one.

List of Possibilities

At your instructor's discretion, you may use either the tables of selected organic compounds provided in Appendix VI of this book or consult a more complete listing such as the *CRC Handbook of Tables for Organic Compound Identification* [Bibliography, G5]. Most tables of this kind separate the compounds into liquids and solids, listing the liquids in order of increasing boiling point and the solids in order of increasing melting point.

Your initial list of possibilities should include all compounds that boil or melt within about ±10° of your measured value (your instructor may suggest a different range). The list should be prepared in the form of a table that shows, for each compound,

- Its name and structural formula
- Its boiling point or melting point
- The listed melting points of its derivatives

If necessary, you can look up some of the structural formulas in *The Merck Index*, *Lange's Handbook of Chemistry*, or another appropriate reference book. If the chemicals for some derivatives are not available in your laboratory, their melting points can be omitted from your list. Some derivative melting points may not be given in published tables, either because the compound does not form that derivative or because the derivative's melting point has not been reported in the literature. Additional properties (such as the densities of organic halides) obtained from reference books can be included in your table as well.

Directions

Referring to the appropriate table in Appendix VI (or another source recommended by your instructor), prepare a table listing each compound whose boiling point or melting point is within the designated range. Give the boiling or melting point of each compound, its molecular structure, any other useful data or information you can find, and the melting points of possible derivatives. If requested, submit your list to your instructor, who may (or may not) tell you if your unknown compound is on the list.

Additional Tests and Data

Once you have prepared your list of possibilities, you should be able to decide which compounds can be eliminated on the basis of spectral data and tests already performed, and what physical properties, chemical tests, or spectral bands will help you shorten the list further. If your list is already a short one, you may decide to prepare a derivative immediately. Otherwise, you may need to shorten it before you can decide on the most suitable derivative. Some ways of shortening your list are described in the following "Directions" section. Don't try to do them all; just do the ones that should be the most useful and that can be performed with the chemicals and equipment available in your laboratory.

Directions

To arrive at your short list, carry out one or more of the following procedures. Use only those that are permitted by your instructor (many instructors will not allow the use of NMR to identify qual organic unknowns).

1 Refer to Table Q2 in the section "Classification Tests" and decide which classification test or tests will help you shorten your list. Then carry out the tests according to the Directions provided.
2 Reexamine your infrared spectrum, looking for bands that may indicate the presence or absence of a subsidiary functional group or a structural feature that is present in some of the compounds on your list but absent in others.
3 Record an ^{1}H NMR spectrum of your unknown, if you haven't already. Try to predict the kinds of NMR spectra that would be produced by the compounds on your list, and compare the predicted spectra with the actual one.
4 Measure the refractive index [OP-32] of your unknown at 20°C, or correct it to that temperature. Compare it with literature values for the refractive indexes of the compounds on your list. If the unknown is very pure, a thermostated refractometer that measures to the fourth decimal place can be used to eliminate most compounds whose refractive index values deviate by more than ±0.001 or so from the observed value.

5 Determine the density of an unknown liquid by accurately weighing a precisely measured volume [OP-5] (0.20 mL or more) of the unknown. Compare it with literature values for the densities of the compounds on your list. Density cannot be measured as accurately as refractive index, but it is useful for distinguishing among compounds that have different structural features. For example, most aromatic compounds are more dense than aliphatic compounds of the same family, and density varies among halogenated hydrocarbons as described in Test **C-10**.

Preparation of Derivatives

In qualitative organic analysis, a *derivative* is a crystalline solid that can be prepared from nearly any compound in a particular family by using a standardized procedure, and that, after purification, should give an accurate melting point that can be used to identify the compound from which it was prepared. For example, an unknown carboxylic acid can be converted to an amide by treating it with thionyl chloride followed by ammonia.

$$RC{-}OH \xrightarrow{\ SOCl_2\ } RC{-}Cl \xrightarrow{\ NH_3\ } RC{-}NH_2$$

If the unknown acid is hexanoic acid (bp = 205°C), its amide (hexanamide) should melt near 101°C, as shown in Table 7 of Appendix VI. The only other carboxylic acid listed that boils near 205°C is 2-bromopropanoic acid, but its amide melts at 123°C, making it easy to distinguish these compounds by means of their derivative melting points.

You will need to prepare at least one derivative to confirm the identity of your unknown. In some cases, it may be necessary to prepare two or more derivatives to be certain of your identification. A suitable derivative should have a melting point between about 60°C and 250°C, because solids melting below 60°C are not easily purified by recrystallization, and melting points higher than 250°C are hard to measure accurately. Its melting point should not be close to the melting point of the unknown itself, since it would then be hard to tell whether the compound obtained from a reaction is the derivative or the unreacted starting material. For example, preparing the amide of a compound believed to be 2-chlorobenzoic acid would not be a good idea because the amide has the same melting point (140°C) as the acid.

If possible, the derivative should be one whose melting-point value will point to just one compound from your list of possibilities. For example, suppose your compound is a ketone whose boiling point is 168°C, and you have prepared the list of possibilities shown in Table Q3.

Table Q3 Derivatives of selected ketones

Compound	bp	Derivative melting points (°C)		
		Oxime	Semicarbazone	2,4-Dinitro-phenylhydrazone
2-methylcyclohexanone	163	43	195	137
2,6-dimethyl-4-heptanone	168	210	121	92
3-methylcyclohexanone	169	...	180	155
4-methylcyclohexanone	169	37	199	130
2-octanone	172	...	122	58

Note: bp is in °C.

The oxime would not be a suitable derivative because melting points have not been reported for two of the possibilities and two of the oximes melt below 60°C. Semicarbazones are reported for all of the ketones, but those for 2,6-dimethyl-4-heptanone and 2-octanone melt within a degree of each other. 2-Octanone might be distinguished by an iodoform test, but that would still leave 2-methyl- and 4-methylcyclohexanone, whose semicarbazones melt within 4°C of each other—a little too close to distinguish them with certainty. (Your derivative might melt at 197°C, for example.) The 2,4-dinitrophenylhydrazone is the preferred derivative because its melting points are well spread out.

The procedures whose page locations are given in Table Q4 are suitable for preparing derivatives of most common organic compounds in the specified class. In some cases, a variation of the procedure (the use of different reaction conditions or recrystallization solvents, etc.) may be required for satisfactory results. If you want to prepare a derivative that is not described in this section, see your instructor for permission and to find out whether the reagents are available. Procedures for additional derivatives, and alternative procedures for some of the derivatives used here, are given in qualitative analysis texts listed under category G of the Bibliography.

Sometimes the same procedure will be used to prepare two related derivatives, such as the phenylurethane and α-naphthylurethane of an alcohol, and will list two different reagents, such as phenyl isocyanate and α-naphthyl isocyanate. In such cases, select the reagent appropriate for the derivative you wish to prepare. Thus you would use phenyl isocyanate to prepare a phenylurethane and α-naphthyl isocyanate to prepare an α-naphthylurethane.

The relative quantities of unknown and reagent are usually not crucial, since any excess reagent is removed during the isolation and purification of the derivative. For cases in which the quantities could affect the results, the quantity of the unknown is given in millimoles so that the required mass can be calculated using an estimate of its molecular weight. Quantities specified in drops are for medicine-dropper-size drops; if you are using a Pasteur pipet, double the number of drops.

Many of the reactions can be run in test tubes, but some of them require heating under reflux [OP-7c]. If you don't have a test tube of the specified size (e.g., 10 × 75 mm), use the next size larger (e.g. 13 × 100 mm). Nearly all of the derivatives must be purified by recrystallization from an appropriate solvent or solvent mixture. If the derivative forms an oil when boiled in the solvent specified, you may be able to substitute a lower boiling solvent of the same type, such as methanol for 95% ethanol. (Unless otherwise indicated, "ethanol" refers to 95% ethanol.) When the recrystallization solvent is described as an ethanol/water mixed solvent or some other mixture, use the procedure for recrystallization from mixed solvents described in OP-25b. The derivative should ordinarily be dissolved in the less polar solvent and the solution saturated by adding the more polar solvent drop by drop. It should be understood that each derivative preparation requires a melting-point determination [OP-30] using the purified product.

Always read the Safety Notes before beginning any derivative preparation, and heed their warnings. Many of the reagents are highly reactive and may react violently with water or other substances. All should be considered toxic, and many of them are corrosive, lachrymatory, or have other

Table Q4 Location of procedures for preparing derivatives

Family	Page number
alcohols	542
aldehydes and ketones	543
amides	544
primary and secondary amines	546
tertiary amines	547
carboxylic acids	548
esters	550
alkyl halides	552
aryl halides	553
aromatic hydrocarbons	554
phenols	555

unpleasant properties. Wear your safety goggles at all times and wear protective gloves when directed by the Safety Notes. Dispose of any wastes or unused reagents as directed by your instructor.

In the following Directions section, basic operations such as heating and weighing are not flagged by "OP" numbers in brackets.

Directions

Derivatives of Alcohols

D-1 *3,5-Dinitrobenzoates, p-Nitrobenzoates*

Safety Notes

> **The acid chlorides are corrosive and lachrymatory. Use a hood and avoid contact with and inhalation of vapors.**

Reaction: $ArCOCl + ROH \longrightarrow ArCOOR + HCl$

Take Care! Avoid contact with the acid chloride and do not breathe its vapors.

Under the hood, mix 0.20 g of pure 3,5-dinitrobenzoyl chloride *or* p-nitrobenzoyl chloride with 0.10 g of the unknown alcohol in a 13 × 100-mm test tube. Heat the mixture carefully in a heating block or over a *small* flame, so that it is just maintained in the liquid state. If you overheat the mixture it may decompose, turning dark brown or black, and you will have to start over. If the alcohol boils below 160°C, heat the liquid mixture for 5 minutes; otherwise, heat it for 10–15 minutes. Allow the melt to cool and solidify. Break it up with a stirring rod and stir in 4 mL of 0.2 *M* sodium carbonate. Heat this mixture to 50–60°C in a hot water bath (measure the temperature in the test tube, not the water bath) and stir it at that temperature for 30 seconds. Then cool it and collect the precipitate by vacuum filtration [OP-13]. Wash the precipitate several times with cold water and recrystallize [OP-25] it from ethanol or an ethanol/water mixed solvent. Derivatives of higher boiling (or melting) alcohols require a higher ethanol/water ratio in the recrystallization solvent.

D-2 *α-Naphthylurethanes, Phenylurethanes*

The alcohol and glassware must be dry because moisture will cause the formation of either diphenylurea or di-α-naphthylurea, depending on the derivative you are preparing (these compounds melt at 241°C and 297°C, respectively). Tertiary alcohols do not form urethanes readily.

Safety Notes

> **The isocyanates are irritants and lachrymators. Use a hood, avoid contact, and do not breathe their vapors.**

Reaction:

$$ArN{=}C{=}O + ROH \longrightarrow ArNH\overset{\displaystyle O}{\overset{\displaystyle \|}{C}}{-}OR$$

Dry a 13 × 100-mm test tube in an oven, then stopper it and let it cool. Unless you are sure that your alcohol is anhydrous, dry it with magnesium sulfate or sodium sulfate. *Under the hood*, mix 0.20 g of the unknown alcohol with 0.20 mL of phenyl isocyanate *or* 0.25 mL of α-naphthyl isocyanate in the reaction tube. If no reaction takes place immediately, heat the mixture in a 60–70°C water bath for 5–15 minutes. Cool the reaction tube in ice and gently scratch its sides, if necessary, to induce crystallization. Collect the precipitate by vacuum filtration [OP-13]. Recrystallize [OP-25] it from about 5 mL of high-boiling petroleum ether or heptane, filtering the hot solution by gravity [OP-12], if necessary, to remove high-melting impurities.

Take Care! Avoid contact with and inhalation of the isocyanate.

Derivatives of Aldehydes and Ketones

D-3 *2,4-Dinitrophenylhydrazones*

Safety Notes

2,4-Dinitrophenylhydrazine is toxic and sulfuric acid is very corrosive. Avoid contact with the DNPH reagent.

Reaction:

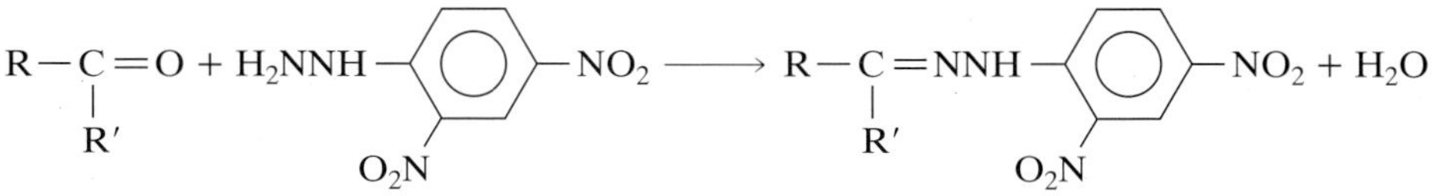

R, R′ = alkyl, aryl, or H

Dissolve 0.10 g of the unknown aldehyde or ketone in 4.0 mL of 95% ethanol in an 18 × 150-mm test tube. Add 3.0 mL of the 2,4-dinitrophenyl-hydrazine–sulfuric acid (DNPH) reagent and allow the solution to stand at room temperature until crystallization is complete. If necessary, warm the solution for a minute in a hot water bath. If no precipitate appears after 15 minutes, add water drop by drop to the warm solution until it just becomes cloudy, heat it until it clears up, and let it cool. Collect the derivative by vacuum filtration [OP-13]. Recrystallize [OP-25] it from 95% ethanol or an ethanol/water mixed solvent. If the derivative does not dissolve in 6 mL of boiling 95% ethanol, add ethyl acetate drop by drop to the boiling mixture until it becomes clear.

Take Care! Avoid contact with the DNPH reagent.

D-4 *Semicarbazones*

Safety Notes

Semicarbazide hydrochloride is a suspected carcinogen. Avoid contact.

Reaction:

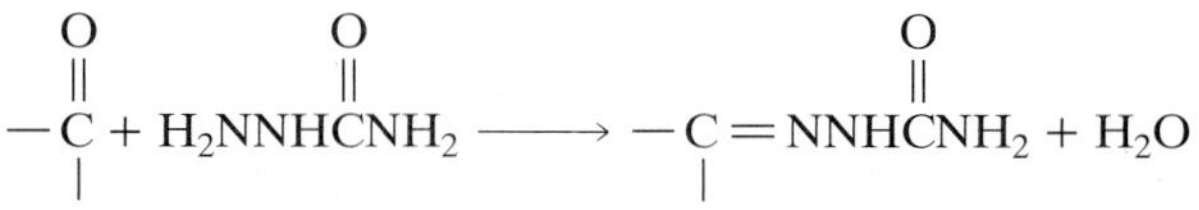

Mix together 0.10 g of semicarbazide hydrochloride, 0.15 g of sodium acetate, 1.0 mL of water, and 1.0 mL of 95% ethanol in a 10 × 75-mm test tube (if the unknown is water soluble, omit the ethanol). Add 0.10 g of the unknown aldehyde or ketone and stir or shake to dissolve. If the mixture is

Take Care! Avoid contact with semicarbazide hydrochloride.

cloudy, add more ethanol until it clears up. Stopper the test tube and shake the mixture for a minute or two. Then let it stand, cooling it in ice if necessary to induce crystallization. If no crystals form, place the test tube in a boiling-water bath for a few minutes, and then let it cool. Collect the product by vacuum filtration [OP-13], wash it with cold water, and recrystallize [OP-25] it from 95% ethanol or an ethanol/water mixed solvent.

D-5 *Oximes*

Oximes are suitable derivatives for most ketones and for some (though not all) aldehydes.

Safety Notes

> **Hydroxylamine hydrochloride is toxic, mutagenic, and can cause a form of anemia. Avoid contact.**

Reaction:

$$
\underset{\underset{|}{\overset{\overset{O}{\|}}{-C}}}{} + H_2N-OH \longrightarrow \underset{\underset{|}{}}{-C}=N-OH + H_2O
$$

Take Care! Avoid contact with hydroxylamine hydrochloride.

Prepare the oxime following the procedure given for semicarbazones (**D-4**), using hydroxylamine hydrochloride in place of semicarbazide hydrochloride. It is usually necessary to heat the reactants in a boiling-water bath for 10 minutes or more. Adding a few milliliters of cold water to the reaction mixture may facilitate precipitation.

Derivatives of Amides

The acid and amine portions of an amide can be obtained by hydrolysis, as described in Procedure **D-6**, and one or both of them characterized by a melting point or derivative preparation. Melting points of the carboxylic acids obtained from amides are given in Table 4 of Appendix VI. Melting points of amine derivatives can be found in Table 5 and Table 6; melting points of carboxylic acid derivatives are in Table 7. You can also prepare a derivative of a primary amide by performing Procedure **D-7**.

D-6 *Hydrolysis Products*

If the amide is known to be primary, omit Procedure **A**.

Safety Notes

> **Arenesulfonyl chlorides are toxic and corrosive. Use a hood, avoid contact, and do not breathe their vapors.**

If the alkaline hydrolysis classification test (C-2) suggests that the amide is difficult to hydrolyze, use 6 M NaOH or a longer reaction time, or both.

Measure 0.30 g of the unknown amide and 5.0 mL of 3 *M* sodium hydroxide into a 10-mL round-bottom flask. Add a boiling chip, attach a water-cooled condenser, and heat the mixture under reflux for 15 minutes or more. If you can't carry out the reaction under a hood, place a gas trap [OP-24] on top of the condenser. Insert a Hickman still between the reaction flask and the condenser (including the gas trap, if used), and distill [OP-27] the reaction mixture, collecting the distillate in an 18×150-mm test tube containing 2.0 mL of 3 M hydrochloric acid. Continue the distillation until about 3 mL of distillate has been collected.

A. Characterization of the Amine. *Under the hood*, add 5.0 mL of 3 M sodium hydroxide to the distillate; then add 0.20 mL of benzenesulfonyl chloride *or* 0.30 g of *p*-toluenesulfonyl chloride. Proceed according to the Directions given in **D-9** (beginning at "Stopper the test tube ...") for preparing arenesulfonamide derivatives of amines. Note that you may not obtain a derivative if the "amine" is ammonia or a gaseous amine that escapes during the reaction or distillation.

Take Care! Wear gloves, and avoid contact with and inhalation of the arenesulfonyl chloride.

B. Characterization of the Carboxylic Acid. Carefully acidify the residue in the boiling flask with 6 M HCl. If a precipitate forms (the carboxylic acid), collect it by vacuum filtration [OP-13], wash it, and recrystallize [OP-25] it from water, ethanol/water, or another suitable solvent. If no precipitate forms, you can prepare a *p*-nitrobenzyl derivative following the Directions in **D-15**.

D-7 N-*Xanthylamides*

This procedure is suitable only for primary amides.

Safety Notes

Reaction:

Under the hood, dissolve 0.20 g of xanthydrol in 2.5 mL of glacial acetic acid in a 5-mL conical vial, cap it securely, and shake until most of the solid has dissolved. Separate the solution from undissolved residue, if there is any. Then add 0.10 g (or as close to 0.75 mmol as you can estimate) of the unknown amide to the solution, attach an air condenser, and heat the reaction mixture in an 85°C water bath for 20 minutes. (If the amide is insoluble in acetic acid, dissolve it in a minimal volume of ethanol before adding it to the solution; then add 0.5 mL of water to the warm reaction mixture after the reaction is complete.) Let the solution cool until precipitation is complete. Collect the solid by vacuum filtration [OP-13] and recrystallize [OP-25] it from an ethanol/water mixed solvent.

Take Care! Wear gloves, and avoid contact with and inhalation of acetic acid.

Derivatives of Primary and Secondary Amines

See **D-12** for directions for preparing picrate derivatives of amines.

D-8 *Benzamides*

> **Benzoyl chloride is corrosive, a lachrymator, and a possible carcinogen. Use gloves and a hood, avoid contact, and do not breathe its vapors.**

Reaction:

$$\underset{\text{O}}{\overset{\text{O}}{\underset{\|}{\text{PhC}}}} - \text{Cl} + \text{R} - \underset{\underset{\text{R}'}{|}}{\text{NH}} \longrightarrow \underset{\overset{\|}{\text{O}}}{\text{PhC}} - \underset{\underset{\text{R}'}{|}}{\text{N}} - \text{R} + \text{HCl}$$

(R′ may be alkyl, aryl, or H.)

Take Care! Wear gloves, and avoid contact with and inhalation of benzoyl chloride.

Under the hood, combine 0.10 g of the unknown primary or secondary amine (or as close to 1.0 mmol as you can estimate) with 1.0 mL of 3 *M* sodium hydroxide in a 13 × 100-mm test tube. Then add 0.40 mL of benzoyl chloride drop by drop, with vigorous shaking. Continue to shake the stoppered test tube for about 5 minutes; then carefully neutralize the solution to a pH of 8 with 3 *M* HCl (use pH paper). Break up the solid mass with a stirring rod, if necessary, and collect the derivative by vacuum filtration [OP-13]. After washing the product thoroughly with cold water, recrystallize [OP-25] it from an ethanol/water mixed solvent.

D-9 *Benzenesulfonamides,* p-*Toluenesulfonamides*

A precipitate that forms in the Hinsberg test (**C-15**) can be purified and used as a *p*-toluenesulfonamide derivative.

> **Arenesulfonyl chlorides are toxic and corrosive. Use a hood, avoid contact, and do not breathe their vapors.**

Reactions:

$$\text{RNH}_2 + \text{ArSO}_2\text{Cl} \xrightarrow{\text{NaOH}} \xrightarrow{\text{HCl}} \text{ArSO}_2\text{NHR}$$
$$\text{R}_2\text{NH} + \text{ArSO}_2\text{Cl} \xrightarrow{\text{NaOH}} \text{ArSO}_2\text{NR}_2$$

Take Care! Wear gloves, and avoid contact with and inhalation of the arenesulfonyl chloride.

Under the hood, combine 0.10 g (or as close to 1.0 mmol as you can estimate) of the unknown primary or secondary amine with 5.0 mL of 3 *M* sodium hydroxide in an 18 × 150-mm test tube. Add 0.20 mL of benzenesulfonyl chloride *or* 0.30 g of *p*-toluenesulfonyl chloride. Stopper the test tube and shake it vigorously over a period of 3–5 minutes. Check the pH of the solution to make sure it is basic; add more 3 *M* NaOH if it is not. Loosen the stopper and heat the test tube in a boiling-water bath for 2 minutes. Let the solution cool to room temperature.

(a) If a precipitate forms, collect it by vacuum filtration [OP-13], wash it with water, and recrystallize [OP-25] it from an ethanol/water mixed solvent.

(b) If no precipitate forms, acidify the solution to pH 6 with 6 *M* HCl and cool it in an ice/water bath. Collect the precipitate by vacuum filtration [OP-13], wash it with water, and recrystallize [OP-25] it from an ethanol/water mixed solvent.

D-10 *Phenylthioureas, α-Naphthylthioureas*

Note that the reagents are iso*thio*cyanates; do not use the corresponding isocyanate by mistake. Protect the reactants and the reaction mixture from water.

The isothiocyanates are corrosive and toxic, and α-naphthylisothiocyanate is a lachrymator. Use gloves and a hood, avoid contact, and do not breathe their vapors.

Reaction:

$$ArN{=}C{=}S \ + \ \underset{\underset{R'}{|}}{R}NH \longrightarrow ArNH\overset{\overset{S}{\|}}{C}\underset{\underset{R'}{|}}{N}R$$

(R′ may be alkyl, aryl, or H)

Under the hood, dissolve 0.10 g (or as close to 1.0 mmol as you can estimate) of the unknown primary or secondary amine in 1.0 mL of *absolute* ethanol in a *dry* 3-mL conical vial. Add 0.10 mL of phenylisothiocyanate *or* 0.15 g of α-naphthylisothiocyanate. Add a boiling chip, attach an air condenser, and use a hot water bath to heat the mixture under gentle reflux for 5 minutes (use a hood or gas trap). Let the reaction mixture cool and scratch the sides of the reaction vial to induce crystallization or cause any oil to crystallize. If no crystals form, continue heating under reflux for another 20 minutes or so and again try to induce crystallization. Adding water drop by drop until the solution becomes cloudy may help.

Take Care! Wear gloves, and avoid contact with and inhalation of the isothiocyanate.

Using a small amount of aqueous 50% ethanol to facilitate transfer, collect the derivative by vacuum filtration [OP-13] and wash it once with 50% ethanol and once with 95% ethanol. If the solid appears quite impure, you can extract impurities by boiling and stirring it with 1.0 mL of petroleum ether. Recrystallize [OP-25] the derivative from ethanol or an ethanol/water mixed solvent.

Derivatives of Tertiary Amines

D-11 *Methiodides*

Iodomethane is toxic and irritant, and it is a suspected human carcinogen and mutagen. Use gloves and a hood, avoid contact, and do not breathe its vapors.

Reaction:

$$RN_3 + CH_3I \longrightarrow R_3NCH_3{}^+I^-$$

(The R groups may be alkyl, aryl, or a combination of the two.)

Take Care! Wear gloves, and avoid contact with and inhalation of iodomethane.

Under the hood, combine 0.10 g of the unknown tertiary amine with 0.10 mL of iodomethane in a 3-mL conical vial, attach an air condenser, and heat the mixture in a 50–60°C water bath for 5 minutes. Cool the mixture in ice, collect the product by vacuum filtration [OP-13], and recrystallize [OP-25] it from ethanol, ethyl acetate, or a mixture of the two.

D-12 *Picrates*

Picrates of primary and secondary amines and aromatic hydrocarbons can also be prepared by this method. Picrates of many aromatic hydrocarbons dissociate when heated and therefore cannot be recrystallized.

Safety Notes

> In solid form, picric acid is unstable and can explode when subjected to heat or shock. It is also toxic and corrosive. Avoid contact with solutions containing picric acid.

Reaction:

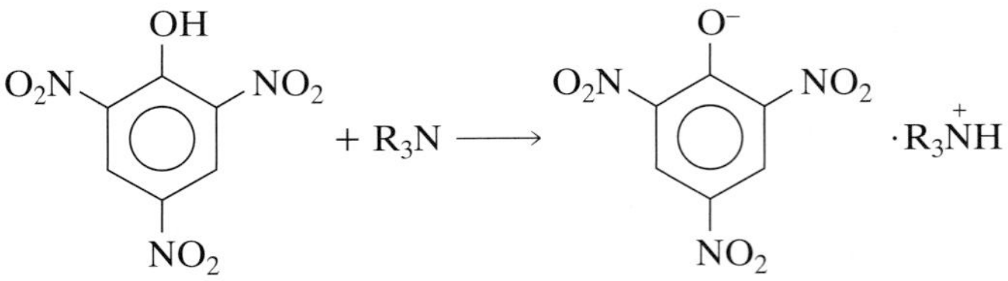

Take Care! Avoid contact with the picric acid solution. Don't let the heated solution boil over.

Dissolve 0.10 g of the unknown amine (or aromatic hydrocarbon) in 2.5 mL of 95% ethanol in a 15-mL conical centrifuge tube and add 2.5 mL of a saturated solution of picric acid in ethanol. Heat the solution just to boiling in a boiling-water bath, then let it cool to room temperature. Collect the crystals by vacuum filtration [OP-13] and wash them with cold ethanol or methanol. Recrystallize [OP-25] the derivative from ethanol, if necessary (some picrates are pure enough to use without recrystallization). Some picrates may explode when heated, so use caution when obtaining the melting point.

Derivatives of Carboxylic Acids

D-13 *Amides*

This derivative is suitable for aromatic acids and for most aliphatic acids that have six or more carbon atoms. Since Procedures **D-13** and **D-14** both involve preparation of the acid chloride, it is convenient to make enough acid chloride to prepare one of the **D-14** derivatives as well. In that case, double the amounts given for the preparation of the acid chloride, add half of it to the aqueous ammonia, and save the other half for the **D-14** derivative.

Safety Notes

> **Thionyl chloride and ammonia can both cause serious burns, and their vapors are highly irritating and toxic. Use gloves and a hood, avoid contact, and do not breathe their vapors.**
> **Acid chlorides are toxic and corrosive and their vapors are harmful. Use gloves and a hood, avoid contact, and do not breathe their vapors.**

Reactions:

$$RCOOH + SOCl_2 \longrightarrow RCOCl + HCl + SO_2$$
$$RCOCl + NH_3 \longrightarrow RCONH_2 + HCl$$

Assemble an apparatus for heating under reflux using a 3-mL conical vial and a water-cooled condenser, and dry it in an oven for 20 minutes or more. Carry out the reaction under the hood or use a gas trap [OP-24c]. Mix 0.20 g of the unknown carboxylic acid with 1.0 mL of thionyl chloride in the reaction vial, add a boiling chip, and heat the mixture with a hot water bath under gentle reflux for 20–30 minutes. Cool the resulting acid chloride in an ice/water bath. Still under the hood, add the acid chloride slowly, with stirring [OP-10], to 2.5 mL of ice-cold concentrated aqueous ammonia in a small Erlenmeyer flask. Let the reaction mixture stand for 5 minutes, collect the derivative by vacuum filtration [OP-13], and recrystallize [OP-25] it from water or an ethanol/water mixed solvent.

Take Care! Wear gloves, and avoid contact with and inhalation of thionyl chloride and the acid chloride.

Take Care! Possible violent reaction! Wear gloves and avoid contact with and inhalation of NH$_3$.

D-14 *Anilides,* p-*Toluidides*

Safety Notes

> **The aromatic amines are toxic if inhaled, ingested, or absorbed through the skin. Use gloves, avoid contact, and do not breathe their vapors.**
> **Acid chlorides are toxic and corrosive and their vapors are harmful. Avoid contact and do not breathe their vapors.**

Reactions:

$$RCOOH + SOCl_2 \longrightarrow RCOCl + HCl + SO_2$$
$$RCOCl + ArNH_2 \longrightarrow RCONHAr + HCl$$

The reaction apparatus used for preparing the acid chloride must be dry. *Under the hood,* prepare the acid chloride of the unknown carboxylic acid as directed in Procedure **D-13** or use the acid chloride saved from the amide preparation in **D-13**. Still under the hood, dilute the acid chloride with 1.0 mL of anhydrous diethyl ether and transfer the solution to a 15-mL conical centrifuge tube. Then add in portions, shaking or stirring [OP-10] after each addition, a solution containing 0.35 mL of aniline *or* 0.40 g of *p*-toluidine dissolved in 5.0 mL of anhydrous diethyl ether. Continue the addition until the odor of the acid chloride has disappeared. If the odor persists after all of the arylamine solution has been added, heat the mixture

Take Care! Wear gloves, and avoid contact with and inhalation of thionyl chloride and the acid chloride.

Take Care! Wear gloves, and avoid contact with and inhalation of the amine.

in a boiling-water bath *under the hood* for a minute or two (replace any ether that evaporates). Wash [OP-21] the ether solution with 2.5 mL of 1.5 *M* HCl, followed by 2.5 mL of water, and remove the aqueous layers. Evaporate [OP-16] the solvent under vacuum. Recrystallize [OP-25] the derivative from water or an ethanol/water mixed solvent.

D-15 p-*Nitrobenzyl Esters*

Preparation of this derivative is recommended only when the previous derivatives are not suitable or when the carboxylic acid is obtained in aqueous solution. It is important that *p*-nitrobenzyl chloride not be present in excess, because it is difficult to remove from the product.

Safety Notes

> *p*-**Nitrobenzyl chloride is corrosive, lachrymatory, and can cause blisters. Use gloves and a hood, avoid contact, and do not breathe its vapors. The derivative may cause blistering. Use gloves and avoid contact.**

Reaction:

$$RCOONa + O_2N-\!\!\bigcirc\!\!-CH_2Cl \longrightarrow RCOOCH_2-\!\!\bigcirc\!\!-NO_2 + NaCl$$

If the carboxylic acid was obtained by hydrolysis of an unknown ester or amide, identify the alcohol or amine portion of the unknown first and estimate the molecular weight of the acid portion from your list of possibilities. Then estimate the amount of acid in the hydrolysis solution, and measure out enough of the solution to provide ~1.0 mmol of the acid.

Take Care! Wear gloves, and avoid contact with and inhalation of *p*-nitrobenzyl chloride.

Mix an estimated 1.0 mmol of the unknown carboxylic acid with 1.2 mL of water (or use the measured hydrolysis solution) in a 10-mL round-bottom flask and add a drop of phenolphthalein indicator solution. Add 3 *M* NaOH drop by drop until the solution turns pink. If necessary, heat to dissolve the acid salt; then add a drop or so of 1 *M* HCl to just discharge the pink color. *Under the hood*, add 0.15 g of *p*-nitrobenzyl chloride dissolved in 4.0 mL of 95% ethanol. Drop in a boiling chip and heat the mixture under gentle reflux for about 90 minutes (di- and triprotic acids should be heated for 2 and 3 hours, respectively). Transfer the solution to a small Erlenmeyer flask and let it cool to room temperature. If crystallization has not occurred by then, add 0.5 mL of water and scratch the sides of the flask. When crystallization is complete, collect the derivative by vacuum filtration [OP-13] and wash it twice with small portions of 5% sodium carbonate, then once with water.

Take Care! Wear gloves, and avoid contact with the derivative.

Recrystallize [OP-25] the derivative from ethanol or an ethanol/water mixed solvent.

Derivatives of Esters

The acid and alcohol portions of an ester can be obtained by hydrolysis as described in **D-16** and one or both of them characterized by a melting point or derivative preparation. Melting points of solid carboxylic acids and alcohols are given in Table 8 of Appendix VI. Melting points of alcohol and carboxylic acid derivatives are given in Table 1 and Table 7. You can also prepare a derivative from the ester itself by following either Procedure **D-17** or Procedure **D-18**.

Procedure **D-17** forms a derivative of its carboxylic acid portion, and Procedure **D-18** forms a derivative of its alcohol portion.

D-16 *Hydrolysis Products*

> **Hydrochloric acid is poisonous and corrosive. Use gloves and a hood, avoid contact, and do not breathe its vapors.**

Safety Notes

Reactions: $RCOOR' + NaOH \longrightarrow RCOONa + R'OH$

$RCOONa + H^+ \longrightarrow RCOOH + Na^+$

Mix 0.50 mL of the unknown ester (or 0.50 g, if it is a solid) with 5.0 mL of 6 *M* sodium hydroxide in a 10-mL round-bottom flask, drop in a boiling chip, and heat the mixture under reflux for 30 minutes or more. If there are two layers, separate them and save the aqueous layer (the larger one). Acidify the aqueous layer *under the hood* by adding concentrated HCl drop by drop. If a solid carboxylic acid precipitates, collect it by vacuum filtration [OP-13], recrystallize [OP-25] it from water or an ethanol/water mixed solvent, and use it as a derivative—*or* convert it to one of the carboxylic acid derivatives. If no precipitate forms, make the solution basic to litmus with dilute sodium hydroxide and distill [OP-27] it until about 1.0 mL of distillate has been collected (save the distillate, which may contain an alcohol). Acidify the solution remaining in the reaction flask and either (a) use the solution to prepare a *p*-nitrobenzyl ester as described in Procedure **D-15** or (b) extract [OP-15] the solution with diethyl ether, and then evaporate [OP-16] the ether to isolate the carboxylic acid for a derivative preparation.

The alcohol portion of the ester may separate if the distillate is saturated with solid potassium carbonate. It can then be isolated by extracting [OP-15] the distillate with diethyl ether and evaporating [OP-16] the ether. The alcohol can be used to prepare an alcohol derivative, if desired.

Take Care! Wear gloves and avoid contact with and inhalation of HCl.

If the unknown may be an ester of a phenol, test a small amount of any precipitate for solubility in 5% NaHCO₃ (see "Solubility Tests").

D-17 *N-Benzylamides*

This procedure works well only with methyl and ethyl esters. If your unknown may be an ester of a higher alcohol, heat 0.50 g of the ester under reflux with 2.5 mL of 5% sodium methoxide in methanol for 30 minutes and then evaporate [OP-16] the methanol. Use all of the resulting methyl ester in the following procedure.

> **Benzylamine is corrosive and lachrymatory. Use gloves and a hood, avoid contact, and do not breathe its vapors.**

Safety Notes

Reactions:

$$RCOOR' + CH_3OH \xrightarrow{CH_3ONa} RCOOCH_3 + R'OH$$

$$RCOOCH_3 + PhCH_2NH_2 \xrightarrow{NH_4Cl} RCONHCH_2Ph + CH_3OH$$

or

$$RCOOCH_2CH_3 + PhCH_2NH_2 \xrightarrow{NH_4Cl} RCONHCH_2Ph + CH_3CH_2OH$$

Under the hood, combine 0.50 g of an unknown methyl or ethyl ester with 1.0 mL of benzylamine and 50 mg of powdered ammonium chloride in a 5-mL conical vial. Add a boiling chip and heat the mixture under reflux for 1 hour. Cool the vial and remove the boiling chip; then wash [OP-21] the reaction mixture with 3 mL of water. If no precipitate forms, add a drop or two of 3 *M* HCl and scratch the sides of the vial. If a precipitate still does not form, boil the mixture for a few minutes to remove any excess ester. Then cool it and try to induce crystallization. Collect the derivative by vacuum filtration [OP-13], wash it with petroleum ether, and recrystallize [OP-25] it from an ethanol/water mixed solvent or from ethyl acetate.

D-18 *3,5-Dinitrobenzoates*

This procedure is not satisfactory for esters of tertiary and some unsaturated alcohols.

Safety Notes

> **Sulfuric acid causes chemical burns that can seriously damage skin and eyes. Use gloves and avoid contact.**
> **3,5-Dinitrobenzoic acid is an irritant. Avoid contact.**

Reaction:

$$O_2N\text{-}C_6H_3(NO_2)\text{-}COOH + RCOOR' \xrightarrow{H_2SO_4} O_2N\text{-}C_6H_3(NO_2)\text{-}COOR' + RCOOH$$

Combine 0.50 g of the unknown ester with 0.40 g of 3,5-dinitrobenzoic acid in a 3-mL conical vial and add a small drop of concentrated sulfuric acid and a boiling chip. If the ester boils below 150°C, heat the mixture under reflux until the 3,5-dinitrobenzoic acid dissolves, then continue heating for an additional 30 minutes. Otherwise heat it, while stirring [OP-10], using a 150°C aluminum block or a sand bath, for the same reaction time. Dissolve the cooled reaction mixture in 8.0 mL of anhydrous diethyl ether in a 15-mL conical centrifuge tube. Wash [OP-21] the ether solution twice with 4-mL portions of 0.5 *M* sodium carbonate, then with 4 mL of water. Evaporate [OP-16] the ether and dissolve the residue (often an oil) in 1–1.5 mL of boiling ethanol. Filter the hot solution by gravity [OP-12], add water until the mixture becomes cloudy, and cool the solution to crystallize the product. Recrystallize [OP-25] the derivative from an ethanol/water mixed solvent.

Derivatives of Alkyl Halides

In addition to the following derivative, density (Test **C-10**) and refractive index values are useful for characterizing alkyl and aryl halides.

D-19 S-*Alkylthiuronium Picrates*

Tertiary alkyl halides will not form this derivative. If the unknown has been shown to be an alkyl chloride, use Procedure **B**.

Safety Notes

In solid form, picric acid is unstable and can explode when subjected to heat or shock. It is also toxic and corrosive. Avoid contact with solutions containing picric acid.
Thiourea is toxic if absorbed through the skin. Avoid contact.

Reaction:

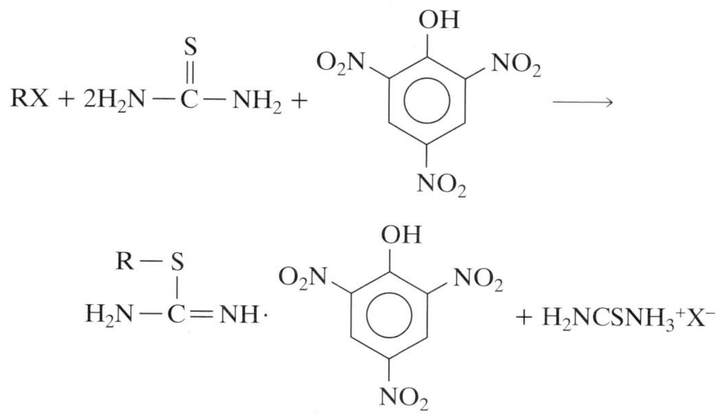

X = Cl, Br, I R = alkyl

A. Mix together 0.25 g of the unknown alkyl bromide or iodide, 0.25 g of powdered thiourea, and 2.5 mL of 95% ethanol in a 10-mL round-bottom flask. Add a boiling chip and heat the mixture under reflux for 20 minutes if the halide is primary or 2–3 hours if it is secondary. Then add 2.5 mL of a saturated solution of picric acid in ethanol and heat the mixture under reflux until a clear (but not colorless) solution is obtained. Allow the solution to cool, collect the derivative by vacuum filtration [OP-13], and recrystallize [OP-25] it from ethanol or an ethanol/water mixed solvent. If a derivative fails to form under these conditions, try Procedure **B**.

Take Care! Avoid contact with thiourea and the picric acid solution. Do not allow any solution containing picric acid to evaporate to dryness, as an explosion may result.

B. Mix together 0.10 g of the unknown alkyl halide, 0.10 g of powdered thiourea, and 2.5 mL of ethylene glycol in a 5-mL conical vial. Add a boiling chip, attach an air condenser, clamp a thermometer with its bulb immersed in the reaction mixture, and heat the solution at 120°C in an aluminum block or sand bath for 30 minutes. Then add 1.0 mL of a saturated solution of picric acid in ethanol and continue heating the solution at the same temperature for 15 minutes. Add 3.0 mL of water and cool the mixture in an ice/water bath to induce crystallization. Collect the derivative by vacuum filtration [OP-13] and recrystallize [OP-25] it from ethanol or an ethanol/water mixed solvent.

Take Care! Avoid contact with thiourea and the picric acid solution.

Derivatives of Aryl Halides

In addition to preparing the following derivative, you can characterize aryl halides that have alkyl side chains by oxidizing the side chain (Procedure **D-21**). Density (Test **C-10**) and refractive index values are also useful.

D-20 *Nitro Compounds*

Procedure **A** yields mononitro derivatives of comparatively reactive aryl halides and aromatic hydrocarbons. Procedure **B** should yield dinitro or

trinito derivatives of reactive compounds and mononitro derivatives of unreactive ones. Most di- and trialkylbenzenes yield trinitro derivatives by
Procedure **B**, while monoalkylbenzenes yield dinitro derivatives. Since di-
and trinitro derivatives are often easier to purify than mononitro derivatives, Procedure **B** is usually preferred.

Reaction: $ArH + HNO_3 \xrightarrow{H_2SO_4} ArNO_2 + H_2O$
(and di- or trinitro derivatives in some cases)

Take Care!   Wear gloves, and avoid
contact with and inhalation of the
acids.

A. *Under the hood,* mix 0.25 g of the unknown aryl halide (or aromatic
hydrocarbon) with 1.0 mL of concentrated sulfuric acid in a 3-mL conical
vial. Cautiously add 1.0 mL of concentrated nitric acid drop by drop, with
magnetic stirring [OP-10], and then attach an air condenser. Heat the mixture, while stirring, in a 60°C water bath for 10 minutes. Cautiously pour the
mixture onto 15 mL of cracked ice in a small beaker, with manual stirring.

Take Care!   Possible violent reaction.

After the ice has melted, collect the precipitate by vacuum filtration [OP-
13], wash it with water, and recrystallize [OP-25] it from an ethanol/water
mixed solvent. Repeated recrystallization may be necessary if the product
melts at a lower temperature than expected, or over a wide range.
B. Follow Procedure **A**, but use 1.0 mL of fuming nitric acid in place of
concentrated nitric acid, and heat the solution in a boiling-water bath for 10
minutes. Add the acid slowly to minimize the generation of brown fumes of

Take Care!   Wear gloves, and avoid
contact with and inhalation of the
acid.

nitrogen dioxide. If the product separates as an oil, it is probably a mixture
of compounds with different numbers of nitro groups.

Derivatives of Aromatic Hydrocarbons

In addition to preparing the following derivative, you can characterize aromatic hydrocarbons by the preparation of picrates (**D-12**) and nitro compounds (**D-20**). The picrates of many aromatic hydrocarbons dissociate
when heated and cannot be recrystallized.

D-21   *Aromatic Carboxylic Acids*

The following procedure is for an aromatic hydrocarbon or aryl halide with
one alkyl side chain. If the unknown may have more than one side chain, increase the quantities of potassium permanganate and 6 *M* NaOH in proportion to the anticipated number of side chains.

Reaction: $\text{ArR} \xrightarrow{\text{KMnO}_4} \text{ArCOOH}$

With magnetic stirring [OP-10], mix 0.30 g of potassium permanganate with 5.0 mL of water in a 10-mL round-bottom flask, and then add 0.10 g of the unknown aromatic hydrocarbon (or aryl halide), followed by 0.10 mL of 6 M sodium hydroxide. Add a stirring device and heat the mixture under reflux, while stirring, for 1 hour or until the purple permanganate color has disappeared. Cool the reaction mixture to room temperature, acidify it with 6 M sulfuric acid, and then heat it under reflux for about 5 minutes. If a brown precipitate of manganese dioxide is present, stir just enough solid sodium bisulfite into the hot solution to remove it, keeping the solution acidic with 6 M H_2SO_4 during this treatment. Cool the reaction mixture, collect the product by vacuum filtration [OP-13], and recrystallize [OP-25] the carboxylic acid from water or an ethanol/water mixed solvent.

Take Care! Wear gloves, avoid contact with NaOH, and keep $KMnO_4$ away from other chemicals.

Derivatives of Phenols

D-22 *Aryloxyacetic Acids*

The equivalent weight of an aryloxyacetic acid (and thus of the unknown phenol) can be determined by measuring its neutralization equivalent (Test **C-18**).

Safety Notes

Chloroacetic acid is corrosive and toxic. Use gloves, avoid contact, and do not breathe its vapors.
Sodium hydroxide is toxic and corrosive, and it may cause severe damage to skin, eyes, and mucous membranes. Use gloves and avoid contact with the NaOH solution.
Some aryloxyacetic acids, such as 2,4-dichlorophenoxyacetic acid (2,4-D), are suspected human carcinogens. Avoid contact with the derivative.

Reaction: $\text{AroH} + \text{ClCH}_2\text{COOH} \longrightarrow \text{ArOCH}_2\text{COOH} + \text{HCl}$

Under the hood, combine 0.10 g of the unknown phenol with 0.14 g of chloroacetic acid and 0.60 mL of 8 M sodium hydroxide in a 3-mL conical vial. Attach an air condenser and heat the mixture in a boiling-water bath for 1 hour. Then cool it to room temperature and add 1.2 mL of water. Acidify the solution to pH 3 with 6 M hydrochloric acid and transfer it to a 15-mL conical centrifuge tube. Extract [OP-15] the solution with 4 mL of diethyl ether, and wash [OP-21] the ether extract with 1 mL of cold water. Extract the derivative from the ether using 2 mL of 0.5 M sodium carbonate, and then acidify the sodium carbonate solution with 6 M hydrochloric acid. Collect the derivative by vacuum filtration [OP-13] and recrystallize [OP-25] it from boiling water.

Take Care! Wear gloves, avoid contact with NaOH and chloroacetic acid, and do not inhale vapors from the acid.

Take Care! Foaming will occur.

Take Care! Avoid contact with the derivative.

D-23 *Bromo Derivatives*

Safety Notes

Bromine is toxic and corrosive, and its vapors are harmful. Use gloves and a hood, avoid contact, and do not breathe its vapors.

Reaction:

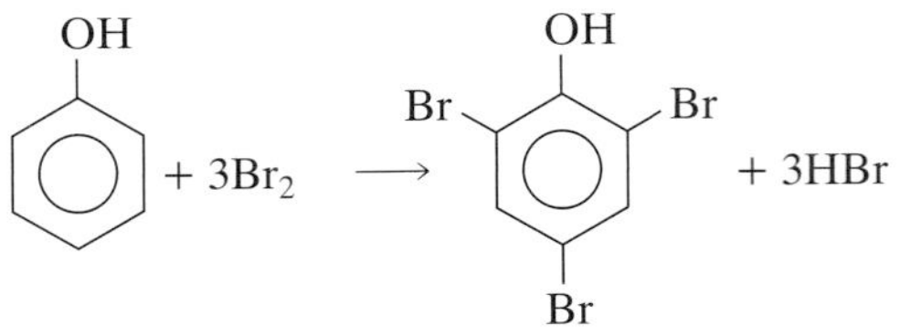

Reaction is for phenol. Substituted phenols undergo similar reactions.

Take Care! Wear gloves, and avoid contact with and inhalation of bromine.

The brominating solution may be prepared beforehand. If not, prepare the brominating solution *under the hood* by dissolving 1.5 g of potassium bromide in 10 mL of water and *carefully* adding 1.0 g (~0.32 mL) of pure bromine. Dissolve 0.10 g of the unknown phenol in 2.0 mL of 50% ethanol (try 95% ethanol or acetone if it doesn't dissolve) in a 25-mL Erlenmeyer flask. Still under the hood, add the brominating solution slowly, with stirring [OP-10], until the bromine color persists for 1 minute (this may require up to 10 mL of brominating solution). Add 5.0 mL of water with stirring. Collect the derivative by vacuum filtration [OP-13], washing it on the filter with 1 *M* sodium bisulfite to remove any unreacted bromine. Recrystallize [OP-25] it from 95% ethanol or an ethanol/water mixed solvent.

D-24 *α-Naphthylurethanes*

Safety Notes

Pyridine is toxic, an irritant, and has an unpleasant odor. Use a hood, avoid contact, and do not breathe its vapors.
α-Naphthyl isocyanate is lachrymatory and a strong irritant. Use gloves and a hood, avoid contact, and do not breathe its vapors.

Reaction:

$$\mathrm{ArN{=}C{=}O + Ar'OH \longrightarrow ArNH\overset{\displaystyle O}{\overset{\|}{C}}{-}OAr'}$$

Take Care! Avoid contact with and inhalation of pyridine.

Under the hood, carry out the procedure for preparing *α*-naphthylurethanes of alcohols (**D-2**), using 0.20 g of your unknown phenol in place of the alcohol, but add a drop of pyridine to catalyze the reaction. For particularly unreactive phenols, it may be helpful to add 1.0 mL of pyridine and a drop of 10% triethylamine in petroleum ether to the reaction mixture, and to heat the mixture at 70°C for about 30 minutes. If the urethane does not precipitate on cooling, add 1.0 mL of 0.5 *M* sulfuric acid.

Report

Your report should include a statement of the problem and an account of how you applied scientific methodology to solve it. It should also include the following information, or other information requested by your instructor:

- Unknown number
- Results of preliminary examination

- Physical constants determined
- Results of solubility tests
- Results of classification tests
- Spectra and spectrometric data
- Interpretation of tests and spectra
- Original list and "short list" of possibilities
- Melting points of derivatives
- Discussion and conclusion
- Answers to assigned Exercises

Your report should reveal the thought processes that led you to your conclusion, as well as the physical and chemical information supporting it. For example, the "Interpretation" section should tell how your observations led you to make tentative assumptions about the nature of the functional group(s) or other structural features. You should also tell how you eliminated compounds from your original list of possibilities to arrive at your short list. When feasible, you should tabulate your data and observations to present them clearly and concisely.

Exercises

1 Write balanced equations for the chemical reactions undergone by your unknown, including those involved in classification tests, derivative preparations, and all solubility tests except the one with water.

2 Indicate the solubility class to which each of the following is most likely to belong: (a) propanoic acid, (b) toluene, (c) *p*-anisidine, (d) *p*-cresol, (e) 3-methylpentanal, (f) *p*-toluic acid, (g) 2-chlorobutane, (h) butylamine, (i) ethylene glycol, (j) acetophenone.

3 Deduce the functional class (chemical family) of each of the following unknowns and give any additional structural information suggested by the results. If sufficient information is provided, give one or more possible structures for the unknown, assuming that it is one of the compounds listed in Appendix VI. Explain the reasoning behind your conclusions. (a) Unknown A is insoluble in water but soluble in 5% NaOH. It decolorizes a dilute, neutral solution of potassium permanganate, yielding a brown precipitate. Addition of bromine water to a solution of the unknown yields a white precipitate. (b) Unknown B is in solubility class **B** and yields an infrared spectrum with no absorption band above 3100 cm^{-1}. The unknown is insoluble in a pH 5.5 buffer, and treatment with *p*-toluenesulfonyl chloride in aqueous sodium hydroxide leaves a liquid residue that is soluble in dilute HCl. (c) Unknown C burns with a sooty, yellow flame and its infrared spectrum has strong bands at 1660, 3180, and 3370 cm^{-1}. When the unknown is heated with 6 *M* sodium hydroxide, a gas is generated that turns red litmus paper blue. Acidification of the alkaline hydrolysis solution yields a white precipitate. (d) Unknown D is insoluble in cold, concentrated sulfuric acid, and a drop of the unknown in water sinks to the bottom. A solution of the unknown in chloroform gives an orange-red color on sublimed aluminum

chloride. The unknown reacts immediately with both ethanolic silver nitrate and sodium iodide in acetone to form precipitates. Adding dichloromethane and chlorine water to its sodium fusion solution yields a red-orange color. (e) Unknown E boils near 100°C, is water soluble, and reacts with acetyl chloride to yield a pleasant-smelling liquid. It dissolves in the Lucas reagent, giving a cloudy solution after 4 minutes. Treating the unknown with iodine in sodium hydroxide forms a yellow precipitate with a medicinal odor. (f) Unknown F is water soluble and its infrared spectrum has a strong band at 1691 cm^{-1}. It gives an orange-red precipitate with 2,4-dinitrophenylhydrazine reagent and a blue-green suspension with chromic acid reagent. (g) Unknown G is insoluble in water and 5% NaHCO$_3$ and gives no precipitate with 2,4-dinitrophenylhydrazine. When the unknown is heated with 6 M NaOH and the reaction mixture acidified, a vinegar-like odor is detected. When the reaction mixture is then made basic and distilled, extraction of the distillate yields a liquid that forms a blue-green precipitate with chromic acid reagent. (h) Unknown H is a water-insoluble solid that dissolves in 5% NaHCO$_3$. It decolorizes a solution of bromine in dichloromethane. An aqueous solution containing 0.195 g of the unknown is titrated to a phenolphthalein endpoint by 32.0 mL of 0.106 M NaOH. (i) Unknown I is in solubility class **X**, burns with a clean yellow flame, and floats on water. It reacts with both silver nitrate/ethanol and sodium iodide/acetone after heating, but not at room temperature. (j) Unknown J is a water-insoluble liquid and has strong infrared bands at 690, 746, and 1688 cm^{-1}. The unknown gives a precipitate with 2,4-dinitrophenylhydrazine reagent, but no precipitate with either chromic acid reagent or iodine in sodium hydroxide.

4 Give reasonable mechanisms for the reactions involved in each of the following classification tests or derivative preparations: (a) the alkaline hydrolysis of N-ethylbenzamide; (b) the reaction of p-cresol with bromine water; (c) the preparation of the semicarbazone of acetophenone; (d) the reaction of ethyl acetate with hydroxylamine in alkaline solution; (e) the reaction of cyclohexylamine with p-toluenesulfonyl chloride in aqueous sodium hydroxide; (f) the reaction of 2-propanol in the Lucas test; (g) the reaction of 2-chloro-2-methylbutane with ethanolic silver nitrate; (h) the reaction of 1-bromobutane with sodium iodide in acetone; (i) the mononitration of toluene by nitric and sulfuric acids; (j) the preparation of the aryloxyacetic acid of p-cresol; (k) the formation of the 3,5-dinitrobenzoate of 2-butanol; (l) the iodoform reaction of acetophenone.

The Operations

Part V contains descriptions of all the laboratory operations used in the experiments in this book. Each operation is designated by a number, which is listed in large type in the top right-hand corner of every odd-numbered page and can be located quickly by thumbing through the pages. The operations are separated into the following categories:

A. Basic Operations
B. Operations for Conducting Organic Reactions
C. Separation Operations
D. Washing and Drying Operations
E. Purification Operations
F. Measuring Physical Constants
G. Instrumental Analysis

There is some overlap between the categories. For example, gas chromatography is essentially a separation method because a gas chromatograph separates the components of a sample, but this operation is used primarily for quantitative analysis of samples so it is placed in section G rather than section C.

A. Basic Operations

Cleaning and Drying Glassware

Cleaning Glassware

Clean glassware is essential for good results in the organic chemistry laboratory. Even small amounts of impurities can sometimes inhibit chemical reactions, catalyze undesirable side reactions, and invalidate the results of chemical tests or rate studies. Always clean your dirty glassware at the end of each laboratory period, or as soon as possible after the glassware is used. This way, your glassware will be clean and dry for the next experiment, and you will be ready to start working when you arrive. If you wait too long to clean glassware, residues may harden and become more resistant to cleaning agents; they may also attack the glass itself, weakening it and making future cleaning more difficult. It is particularly important to wash out strong bases such as sodium hydroxide promptly, because they can etch the glass permanently and cause glass joints to "freeze" tight. When glassware has been thoroughly cleaned, water applied to its inner surface should wet the whole surface and not form droplets or leave dry patches. Used glassware that has been scratched or etched may not wet evenly, however.

You can clean most glassware adequately by vigorous scrubbing with hot water and a laboratory detergent, such as Alconox, using a brush of appropriate size and shape to reach otherwise inaccessible spots. A plastic trough or another suitable container can serve as a dishpan. A tapered centrifuge-tube brush can be used to clean conical vials as well as centrifuge tubes. A nylon mesh scrubber is useful for cleaning spatulas, stirring rods, beakers, and the outer surfaces of other glassware. Pipe cleaners or cotton swabs can be used to clean narrow funnel stems, eyedroppers, Hickman still side ports, and so forth.

Organic residues that cannot be removed by detergent and water will often dissolve in organic solvents such as technical-grade acetone. (Never use reagent-grade solvents for washing.) For example, it is difficult if not impossible to scrub the inside of a porcelain Hirsch funnel, but squirting a little acetone around the inside of the funnel stem and letting it drain through the porous plate should remove chemical residues that have lodged there. Use organic solvents sparingly and recycle them after use, as they are much more costly than water. Be certain that acetone is completely removed from glassware before you return it to a lab kit because it will dissolve a foam liner.

After washing, always rinse glassware thoroughly with water (a final distilled-water rinse is a good idea) and check it to see if the water wets its surface evenly rather than forming separate beads of water. If it doesn't pass this test, scrub it some more or use a cleaning solution such as Nochromix. Note that some well-used glassware may not pass the test because of surface damage, but it may still be clean enough to use after thorough scrubbing.

Drying Glassware

Always dry glassware thoroughly if it will be used with organic reactants and solvents under non-aqueous conditions. Don't waste time drying wet glassware if it will come into contact with water or an aqueous solution during an experiment. Just let it drain for a few minutes before you use it.

The easiest (and cheapest) way to dry glassware is to let it stand overnight in a position that allows easy drainage. You can dry the outer surfaces of glassware with a soft cloth, but don't use a cloth to dry any surfaces that will be in contact with chemicals because of the likelihood of contamination. If a piece of glassware is needed shortly after washing, drain it briefly to remove excess water. Then rinse it with one or two small portions of wash acetone and dry it in a stream of clean, dry air or put it in a drying oven for a few minutes. Compressed air from an air line may contain pump oil, moisture and dirt, so don't use it directly from the line for drying. It can be cleaned and dried as described in OP-24a.

Glassware that is to be used for a moisture-sensitive reaction (such as a Grignard reaction) must be dried very thoroughly before use. If possible, clean the glassware during the previous lab period, let it dry overnight or longer, and then dry it in an oven set at about 110°C for 20–30 minutes. Assemble the apparatus and attach one or more drying tubes (see OP-24b) as soon as possible after oven drying; otherwise, moisture will condense inside it as it cools. If the glassware must be cleaned the same day as it is used, rinse it with acetone after washing and flush it with clean, dry air before you put it in the oven. You can also dry glassware by playing a "cool" Bunsen burner flame over the surface of the assembled apparatus, but this practice should never be used in laboratories where volatile solvents such as diethyl ether are in use. It should be done only with the instructor's permission and according to his or her directions.

Take Care! Use tongs or heat-resistant gloves when handling hot glass.

Using Specialized Glassware

OPERATION **2**

Most specialized glassware components used in organic chemistry have ground-glass joints called *standard-taper* joints. The size of a standard-taper joint is designated by two numbers, such as 14/10, in which the first number is the diameter at the top of the joint and the second is the length of the taper, measured in millimeters (see Figure A1).

Microscale glassware comes in a variety of configurations. Most undergraduate microscale laboratory courses use glassware of the kind developed and tested by students at Bowdoin and Merrimack Colleges under the direction of Professors Dana W. Mayo and Ronald M. Pike. The glassware provided in a typical Mayo-Pike style microscale lab kit is illustrated in Appendix I. There are variations in the construction of this type of glassware, but the components usually have ground-glass joints that are secured with threaded caps.

Figure A1 14/10 standard-taper joint

Most microscale glassware has 14/10 glass joints.

Assembling Glassware

The standard-taper joints of a typical microscale glassware apparatus are held together with plastic compression caps, using an O-ring to hold the cap

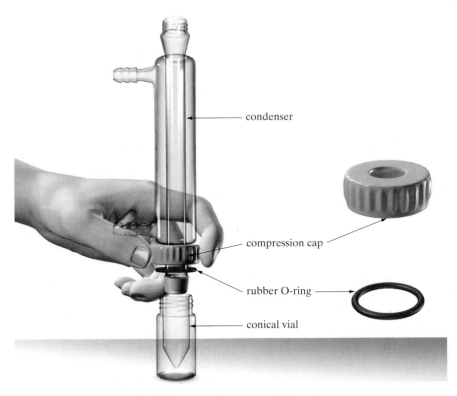

Figure A2 Connecting microscale components

in place and provide a tight, greaseless seal. You can connect such joints as illustrated in Figure A2 for a conical vial and a water-cooled condenser. First, put a compression cap, threaded side down, over the male (outer) joint of the condenser. Hold it in place as you roll an O-ring over the joint onto the clear part of the glass. Make sure that the entire O-ring is above the ground joint; then release the cap, which should be held in place by the O-ring. Now insert the male joint of the condenser into the female (inner) joint of the conical vial and screw the cap over the threads at the top of the vial, tightly enough so that the outer joint cannot be rotated around the inner joint. Be careful, because screwing it down too tightly may break off the threads or cause strains in the glass that will lead to eventual breakage.

Ground-joint glassware can be used to assemble setups for performing a wide variety of laboratory operations. For example, if you want to assemble an apparatus for fractional distillation [OP-29], turn to the operation and locate an illustration of the apparatus, which is in Figure E19.

A microscale setup can be assembled completely on the bench top and then clamped to a ring stand, usually with a single microclamp. Since the reaction vessel (flask or conical vial) is not going to fall off the joint it is connected to, it need not be clamped. Instead, you can clamp the apparatus higher up—on a condenser, for example—to make it more stable. Be sure that the microclamp is held securely to the ring stand and that the clamp jaws are tightened securely around the apparatus so that it doesn't wobble or fall to the bench and break. If you intend to heat a reaction mixture in a flask or conical vial, clamp the apparatus so that, if necessary (as when a reaction mixture is boiling too vigorously), you will be able to loosen the clamp holder at the ring stand and raise the entire apparatus above the heat source quickly.

Figure A3 illustrates the steps in the assembly of a simple distillation setup.

1 Support compression caps on both male joints with O-rings
2 Insert Hickman still joint into conical vial; tighten cap
3 Insert air condenser joint into Hickman still; tighten cap
4 Clamp apparatus securely to ring stand, lower conical vial into heat source
5 Insert thermometer through air condenser, position correctly, clamp in place

The glassware components can be connected in any order, but it is convenient to assemble the apparatus from the bottom up.

Disassembling Glassware

Disassemble (take apart) ground-joint glassware promptly after use, as joints that are left coupled for an extended period of time may freeze together and become difficult or impossible to separate without breakage. Unscrew the

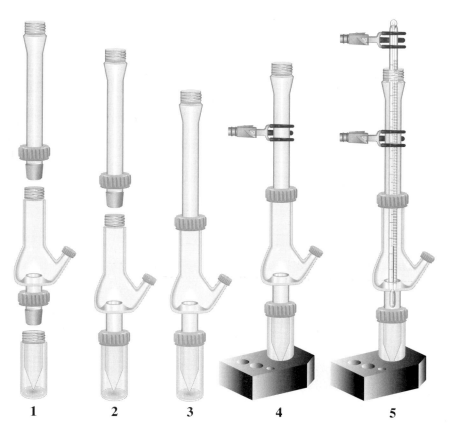

Figure A3 Steps in the assembly of a simple distillation apparatus

Take Care! The glass may break, so protect your hands with heavy gloves.

compression cap completely, then separate the joint by pulling the components apart with a twisting motion. If the joint is frozen, you can sometimes loosen it by tapping it gently with the handle of a wooden-handled spatula or by applying heat to the joint while rotating the apparatus slowly, and then pulling the components apart with a twisting motion. If this doesn't work, see your instructor. Clean [OP-1] the glassware thoroughly and return each component to its proper location in the lab kit or to the stockroom. To reduce assembly time, caps and O-rings are sometimes left on components when they are returned to their lab kit case; this can make the components more difficult to clean thoroughly, so your instructor may prefer that you remove them.

OPERATION **3**   Using Glass Rod and Tubing

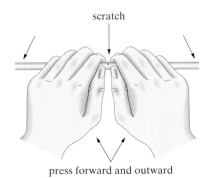

Figure A4 Breaking glass tubing

Glass connecting tubes, stirring rods, and other simple glass items are required for certain operations in organic chemistry. Soft-glass rod and tubing can be worked easily with a Bunsen burner, but borosilicate glass (Pyrex, Kimax, etc.) requires the hotter flame provided by a Meker-type burner or an oxygen torch.

Cutting Glass Rod and Tubing

Glass rods and tubes are *cut* by scoring them at the desired location and snapping them in two. Score the rod or tube by drawing a sharp triangular file (or another glass-scoring tool) across the surface at a right angle to the axis of the tubing. Often, only a single stroke is needed to make a deep scratch in the surface; don't use the file like a saw. To cut a thin, fragile glass tube, such as a melting-point tube or the capillary tip of a Pasteur pipet, it is best to use a special glass scorer, but a sharp triangular file may work if applied carefully so as not to crush the glass. Moisten the scratch with water or saliva. Using a towel or gloves to protect your hands, place your thumbs about 1 cm apart on the side opposite the scratch and, while holding the glass firmly in both hands, press forward against the glass with your thumbs as you rotate your wrists outward (Figure A4).

Working Glass Rod and Tubing

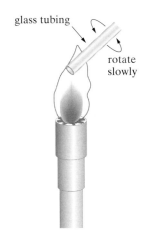

Figure A5 Fire-polishing

The cut ends of a glass rod or tube should always be *fire-polished* to remove sharp edges and prevent accidental cuts. To fire-polish a glass rod or tube, hold it at a 45° angle to a burner flame (see OP-7 for directions on using a burner) and rotate its cut end slowly in the flame until the edge becomes rounded and smooth (Figure A5).

To *round* the cut end of a glass rod, rotate the rod in a burner flame, holding it at a 45° angle with its tip at the inner blue cone of the flame. The end should be hemispherical in shape, not rounded only at the edges and flat on the bottom.

To *flatten* one end of a glass rod, rotate it with its tip at the inner blue cone of the flame until it is incandescent and very soft, but not starting to bend. Then press the softened end straight down onto a hard surface, such as the base of a ring stand. The flattened end should be well centered and about twice the diameter of the rod.

To *seal* one end of a glass tube, hold the tube at a 45° angle to the burner with its end just above the inner blue cone of the flame, and rotate it until the soft edges come together and eventually merge. Remove the tube from the flame as soon as it is closed, and immediately blow into the open end to obtain a sealed end of uniform thickness. Let the tube cool to room temperature. Then check it for leaks by connecting the open end to an aspirator or vacuum line with a length of rubber tubing, placing the closed end in a test tube containing a small amount of dichloromethane, and turning on the vacuum. (**Take Care!** Avoid contact with dichloromethane and do not breathe its vapors.) If the tube is not properly sealed, the liquid will leak into it under vacuum. To seal the end of a thin, fragile tube such as a melting-point capillary, rotate its open end in the *outer* edge of the flame.

To *bend* glass tubing, first place a flame spreader on the barrel of a Bunsen burner (or use a Meker-type burner). Hold the tubing over the burner flame parallel to the long axis of the flame spreader and rotate it constantly at a slow, even rate until it is nearly soft enough to bend under its own weight (Figure A6). (The flame will turn yellow as the glass begins to soften.) Remove the hot tubing from the flame and immediately bend it to the desired shape with a firm, even motion and a minimum of force (if much force is required, the glass is not soft enough). Bend it in a vertical plane, with the ends up and the bend at the bottom; the bend should follow a smooth curve with no constrictions.

Inserting Glass Items into Stoppers

> **Improper insertion of glass tubes and thermometers through rubber stoppers is one of the most frequent causes of laboratory accidents. The resulting cuts and puncture wounds can be very severe, requiring medical treatment and sometimes causing the victim to go into shock. Protect your hands, hold the glass object close to the stopper, do not try to force it through a hole that is too small for it, and avoid applying sideways force or excessive force.**

Chemical supply houses sell rubber stoppers with pre-bored holes, but if a suitable one-hole or two-hole stopper is not available, you may have to bore one or more holes in a solid stopper. To bore a hole in a rubber stopper, first obtain a sharp cork borer that is slightly smaller in diameter than the object to be inserted in the stopper. (If the borer is dull, use a special cork-borer sharpener to sharpen it.) Lubricate its cutting edge with a small amount of glycerine, then twist it through the stopper using a minimum of force (don't try to "punch" out the hole). Rotate the borer and stopper in opposite directions, checking the alignment frequently to make sure that the borer is going in straight. When the borer is about halfway through, twist it out and start boring from the opposite end of the stopper until the holes meet. You can remove the plug left inside the cork borer with a rod that comes with a set of cork borers.

Take Care! Don't burn yourself on the hot end of a glass rod or tube, or lay the glass onto combustible materials.

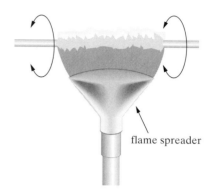

Figure A6 Bending tubing

Safety Notes

Cork borer

To insert a glass tube into a rubber stopper, first lubricate the hole lightly with glycerol or another suitable lubricant; water may work if the hole is not too tight. You can use a cotton swab or an applicator stick to apply the lubricant evenly. Protect your hands with gloves or a towel, then grasp the tube close to the stopper and *twist* it through the hole with firm, steady pressure. (If you grasp the tube too far from the stopper, the glass may break and lacerate your hand.) Apply force directly along the axis of the tube, as any sideways force may cause it to break. Using excessive force or forcing the tube through a hole too small for it can also cause it to break. After the tube is correctly positioned, rinse off any glycerol with water.

Follow the same directions to insert a thermometer through a rubber stopper or a rubber thermometer adapter cap. A typical microscale thermometer adapter uses an O-ring rather than a rubber cap to hold the thermometer in place. Never try to insert a thermometer into an assembled microscale thermometer adapter; instead, roll the O-ring onto the thermometer and then secure the thermometer in the adapter as described in OP-9.

To remove a glass tube from a stopper, lubricate the part of the glass that will pass through the stopper with water or glycerol, protect your hands with gloves or a towel, and twist the tube out with a firm, continuous motion. Hold the tube close to the stopper or cap and avoid applying any sideways force that could cause it to break. If you can't remove the tube by this method, obtain a cork borer of a size that will just fit around it, lubricate its cutting edge, and twist it gently through the stopper until the tube can be pulled out easily.

| OPERATION **4** | # Weighing |

Most modern chemistry laboratories are equipped with electronic balances that display the mass directly, without any preliminary adjustments (see Figure A7). If you will be using a different type of balance, your instructor will demonstrate its operation. For the experiments in this book, chemicals should be weighed on balances that measure to at least the nearest milligram (0.001 g). Most products obtained from a preparation are transferred to vials or other small containers, which should be *tared*—weighed empty—and then reweighed after the product has been added. As a rule, the container should be weighed with its cap and label on and this *tare mass* recorded.

A balance is a precision instrument that can easily be damaged by contaminants, so avoid spilling chemicals on the balance pan or on the balance itself. If spillage does occur, *clean it up immediately*. If you spill a liquid or corrosive solid on any part of the balance, notify your instructor as well. Before you leave the balance area, replace the caps on all reagent bottles, return them to their proper locations if you obtained them elsewhere, and see that the area around the balance is clean and orderly.

Weighing Solids

Solids can be weighed in glass containers (such as vials or beakers), in aluminum or plastic weighing dishes, or on glazed weighing papers. Filter paper

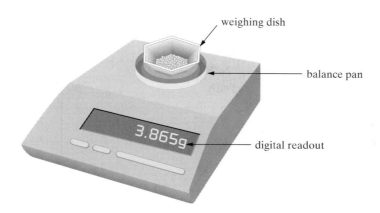

weighing dish

balance pan

3.865g

digital readout

Figure A7 An electronic balance

and other absorbent papers should not be used for weighing, since a few particles will always remain in the fibers of the paper. Hygroscopic solids—those that absorb moisture from the atmosphere—should be weighed in screw-cap vials or other containers that can be capped immediately after the solid is added. For most experiments involving chemical reactions, solid or liquid reactants should be weighed directly into the reaction vessel to avoid losses in transfer.

To weigh a sample of solid that is in a tared container, such as a preweighed screw-cap vial, set the digital readout to zero by pressing the appropriate button, then place the container on the balance pan. Be sure that the draft shield (if there is one) is in place; then read the mass of the container and its contents from the digital display. Wait until the reading remains constant, and then record the mass in your laboratory notebook, including all digits after the decimal point. For example, if the balance reads 3.610 g, do not record the mass as 3.61 g, because zeroes following the decimal point are significant. Then subtract the tare mass to obtain the mass of the solid.

To weigh a sample of solid that is to be transferred to another container, such as a weighing dish or storage vial, place the container on the balance pan, press the tare button to zero the digital display, and transfer the solid to the container. With the draft shield in place, wait until the reading has stabilized and then read the mass of the sample directly from the digital display.

To measure out a specific quantity of a solid, such as a solid reactant, into a reaction flask or other container, first place the container on the balance pan. If the container is a round-bottom flask or a pear-shaped flask, support it in a beaker or on a cork ring. Press the tare button to zero the digital display. Then use a spatula or Scoopula to add the solid in small portions until the desired mass appears on the digital display. With the draft shield in place, wait until the reading has stabilized and then read the mass of the sample directly from the digital display. Ordinarily you need not measure out the exact mass specified in a procedure, but try not to deviate from the specified mass by more than 2% or so, especially for a limiting reactant. Because the theoretical yield of a preparation is based on the actual mass of a starting material, always use your measured mass—not the mass specified in the procedure—for yield calculations.

For example, if a procedure requires 0.250 g of a limiting reactant, you should measure out between 0.245 g and 0.255 g of the reactant.

Weighing Liquids

Organic liquids should be weighed in screw-cap vials or other closed containers to prevent damage to the balance from accidental spillage and losses by evaporation. If liquid must be added to or removed from a weighed container, the container should be removed from the balance pan first. Any excess liquid should be placed in a waste container or otherwise disposed of, *not* returned to a stock bottle. In most experiments involving chemical reactions, liquid reactants should be weighed directly into the reaction vessel to avoid losses in transfer.

To measure the mass of a liquid sample in a tared or untared container, follow the directions for solids, but be sure to keep the container capped while it is on the balance pan. When using a tared container, subtract the tare mass to obtain the mass of the liquid.

To measure out a specific mass of a liquid from a reagent bottle, you should first measure out the approximate quantity of the liquid by volume and then weigh that quantity accurately in an appropriate closed container. If the container is a round-bottom flask or a pear-shaped flask, support it in a beaker or on a cork ring. Use an automatic pipet, a measuring pipet, a graduated syringe, or a bottle-top dispenser to measure the estimated volume of the liquid into the container. Then cap the container, place it in the balance pan, and read off the mass. For example, if you need to obtain 0.37 g of 1-butanol ($d = 0.810$ g/mL) in a conical reaction vial for an experiment, first zero the balance with the empty container and its cap on the balance pan (don't forget to include the cap!). Remove the cap and vial from the balance pan, measure [OP-5] about 0.46 mL (0.37 g ÷ 0.810 g/mL) of the liquid into the vial, cap it, return it to the balance pan, and read the mass of the liquid. (Alternatively, you can tare the cap and vial beforehand and get the mass of the liquid by difference.) If the measured mass is not close enough to 0.37 g, add or remove liquid with a clean Pasteur pipet and reweigh it.

Take Care! Be careful not to spill liquids on the balance pan. If you do, clean up the spill immediately and inform your instructor.

OPERATION 5   Measuring Volume

Several different kinds of volume-measuring devices, described here, are used in the undergraduate organic chemistry laboratory. Microscale experiments seldom require liquid volumes (except for bath liquids) greater than 5–10 mL; volumes of that magnitude can be measured using small graduated cylinders. Smaller volumes of liquids are usually measured with various kinds of pipets and syringes. Reagent bottles containing liquids may be provided with bottle-top dispensers that measure out a preset volume of the liquid. For a few experiments, you may use a buret or a volumetric flask; their use is described in most general chemistry laboratory manuals.

Graduated Cylinders

Graduated cylinders are not highly accurate, so they should never be used to measure liquid reactants. They are most useful for measuring comparatively large volumes of water or other solvents.

Ordinarily, you should not pour liquid from a stock bottle into a 5-mL or 10-mL graduated cylinder, because it's hard to control the amount added and you may spill some liquid in the process. Instead, transfer the liquid to the graduated cylinder from a washing/dispensing bottle or by using a Pasteur pipet or other transfer device. Add just enough liquid to fill the cylinder to the graduation mark corresponding to the desired volume. If you overshoot the mark, remove the excess liquid with a clean Pasteur pipet. Read the liquid volume from the bottom of the meniscus, as shown in Figure A8.

Bottle-Top Dispensers

A typical adjustable bottle-top dispenser (see Figure A9) has a movable plunger that pumps liquid into a glass cylinder, from which it is dispensed through a discharge tube. The cylinder is usually surrounded by a protective sleeve that is raised to fill the cylinder and lowered to dispense the liquid. The dispenser is screwed onto a bottle containing the liquid and can be adjusted to dispense a specified volume of liquid, which is read from a scale on the sleeve or cylinder. Before its initial use, the dispenser must be *primed* by pumping it several times to fill the cylinder and discharge tube and to expel any air bubbles.

To use a bottle-top dispenser, first check to see that there are no air bubbles in the discharge tube (if there are, reprime the dispenser or inform your instructor). Then hold your container underneath the discharge tube outlet and raise the sleeve as high as it will go. Release the sleeve it so that

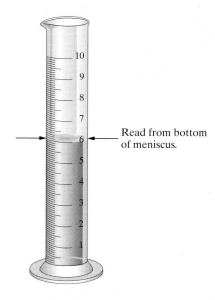

Read from bottom of meniscus.

Figure A8 Reading the volume contained in a graduated cylinder—in this case, 6.0 mL

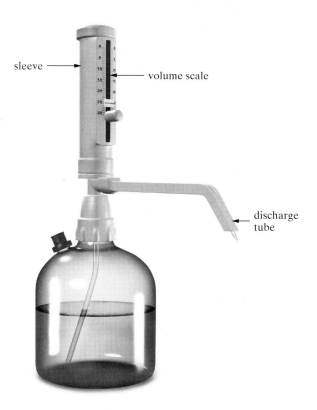

sleeve

volume scale

discharge tube

Figure A9 A bottle-top dispenser

it drops by gravity, and then push it down gently until it moves no further. Touch the tip of the discharge tube to an inside wall of your container to remove the last drop of liquid. If the liquid is the limiting reactant for a preparation, you should then weigh it accurately as described in OP-4.

Measuring Pipets and Volumetric Pipets

A *measuring pipet* has a graduated scale and is used to measure liquid volumes within a range of values; for example, a typical 1-mL measuring pipet can measure volumes up to 1.00 mL to the nearest 0.01 mL. A *volumetric pipet* measures only a single volume, but it is more accurate than a measuring pipet. Suction is required to draw the liquid into a measuring or volumetric pipet, but you should never use mouth suction because of the danger of ingesting toxic or corrosive liquids. A pipet pump is a simple and convenient suction device for filling such pipets. Pipet fillers with rubber bulbs can also be used.

*A convenient "homemade" pipetting bulb is described in J. Chem. Educ. **1974**, 51, 467.*

To use a measuring pipet with a pipet pump of the type shown in Figure A10, first see that the plunger is as far down as it will go. Then insert the wide (untapered) end of the pipet firmly into the opening at the bottom of the pump. Place the tip of the pipet in the liquid and use your thumb to rotate the thumb wheel back toward you until the liquid meniscus rises a few millimeters above the zero graduation mark; be careful not to draw any liquid into the pump itself. Slowly rotate the thumb wheel away from you until the meniscus drops just to the zero mark. Measure the desired volume of liquid into a clean container by placing the pipet tip over the container and rotating the thumb wheel away from you until the meniscus drops to the graduation mark corresponding to the desired volume (see Figure A11). Touch the tip of the pipet to the inside of the container to remove any adherent drop of liquid. If the pipet is one dedicated for use with a particular reagent bottle, the excess liquid can be drained into the bottle by depressing the pump's quick-release lever (if it has one) or by rotating the thumbwheel away from you as far as it will go. Otherwise, the liquid should be drained into another container or disposed of as directed by your instructor.

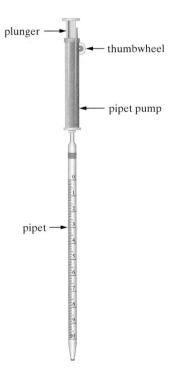

To use a volumetric pipet, obtain a bulb-type pipet filler or a pipet pump with a quick-release lever that allows the liquid to drain by gravity. Use the bulb or pump to fill the volumetric pipet to its calibration mark, hold the pipet tip over a receiving container, and hold down the quick-release lever until only a small amount of liquid is left in the tip. Touch the tip of the pipet to the inside of the container to remove any adherent drop of liquid, but do not expel the liquid in the pipet tip—its volume is accounted for when the pipet is calibrated.

Figure A10 Pipet pump and measuring pipet

Automatic Pipets

Automatic pipets (also called pipetters) provide a quick, convenient way of delivering a specified volume of liquid with a high degree of reproducibility (see Figure A12). A variable-volume automatic pipet can be set to a specified volume within a certain range of volumes, such as 20–200 μL (0.020–0.200 mL). The volume is displayed on a digital display or an analog scale, usually in microliters (μL). To prevent contamination, liquid is drawn into a disposable tip and never inside the pipet itself. Whenever the pipet is used for a different liquid, the volume is reset (if necessary) and a new pipet tip is installed.

The instructor or a lab assistant will ordinarily set the volume of an automatic pipet and designate it for a specific liquid. Do *not* reset the volume or use the pipet for a different liquid without permission from your instructor.

To use an automatic pipet, depress the plunger to the first *detent* (stop) position, when you will feel resistance to further movement. (Using excessive force will move the plunger to its second detent position, causing an inaccurate measurement.) Insert the pipet tip into the liquid to a depth of about 1 cm or less; do not let it touch the bottom of the liquid's container, where impurities may be concentrated. Slowly release the plunger to draw liquid into the pipet tip. Place the tip inside the receiving container and depress the plunger to the first detent position, pause a second or two, then push the plunger down to the second detent position and touch the tip to the inner wall of the receiving container to expel the last droplet of liquid.

Calibrated Pasteur Pipets

A calibrated Pasteur pipet can be used to measure approximate volumes of such liquids as washing and extraction solvents. For measuring volatile liquids, a filter-tip pipet (see OP-6) that has been calibrated should work better than an ordinary calibrated pipet.

To calibrate a Pasteur pipet, first attach a latex rubber bulb to its wide end. Measure 0.50 mL of water into a conical vial using a measuring pipet or another accurate measuring device. Carefully draw all of the liquid into the Pasteur pipet so that there are no air bubbles in its tip (if necessary, squeeze the bulb *gently* to expel any air), and mark the position of the meniscus with an indelible glass-marking pen. Expel all of the water and repeat this operation using 1.00 mL of water, and other volumes as desired.

A quicker, but less accurate, way to calibrate a short ($5\frac{3}{4}$ inch) Pasteur pipet is to use a ruler to mark lines 6 cm, 8 cm, 10 cm, and 12 cm from the narrow (capillary) tip of the pipet (see Figure A13). These lines mark volumes of

Figure A11 Reading the volume delivered from a measuring pipet—in this case, 0.30 mL

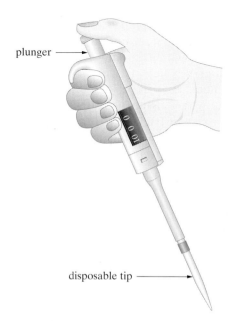

Figure A12 An automatic pipet

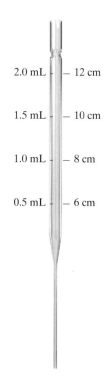

2.0 mL — 12 cm

1.5 mL — 10 cm

1.0 mL — 8 cm

0.5 mL — 6 cm

Figure A13 A calibrated Pasteur pipet

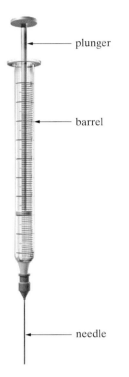

plunger

barrel

needle

Figure A14 A syringe

Take Care! Don't stick yourself with the needle!

Take Care! Don't bend the plunger.

approximately 0.5 mL, 1.0 mL, 1.5 mL, and 2.0 mL. Make sure that the capillary tip is intact; if part of it has broken off, this calibration method will not work.

To use a calibrated pipet, hold the pipet (with its attached bulb) vertically over the liquid to be measured, squeeze the bulb to expel some of the air, and insert the tip in the liquid. (With practice, you should learn how far to squeeze the bulb in order to draw in the desired amount of liquid.) Release the bulb slowly to draw liquid into the pipet until the liquid meniscus is at the level of the appropriate calibration mark. Without delay, raise the pipet tip out of the liquid, move it into position over the receiving container, and squeeze the bulb to expel all of its liquid into the container. Be careful not to lose any liquid during the transfer. (Read OP-6 for information about making transfers with Pasteur pipets.)

Syringes

Syringes (Figure A14) are used to measure and deliver small volumes of liquid, sometimes by inserting the needle through a *septum*—a rubber or plastic disk that can be penetrated by a needle but remain more-or-less airtight after the needle is withdrawn. For example, some air-sensitive liquids are packaged under nitrogen in containers having caps with Teflon-faced plastic liners, so that a syringe can be used to withdraw such liquids without exposing them to air. The cap liners in a microscale lab kit can be used as septa but they may leak air after repeated use, especially when syringes with large-diameter needles are used. Syringes are also used to inject liquid samples into a gas chromatograph and to add liquid reagents to reaction mixtures during a reaction.

To fill a syringe, hold it vertically with the needle pointing down; then place the needle's tip in the liquid and slowly pull out the plunger until the barrel contains a little more than the required volume of liquid. If there are air bubbles in the liquid, try to get rid of them by holding the syringe vertically with the needle pointing up and tapping the barrel with your fingernail, or by expelling the liquid and filling the syringe again, more slowly. Hold the syringe with its needle pointing up and slowly push in the plunger to eject the excess liquid (collect it for disposal if requested) until the bottom of the liquid column is at the appropriate graduation mark. Wipe off the tip of the needle with a tissue, place the needle tip into the receiving vessel (or through a septum), and expel the liquid by gently pushing the plunger in as far as it will go. Clean the syringe immediately after use by flushing it repeatedly with an appropriate solvent, such as acetone, or a soap solution. If you use soap for washing, rinse the syringe thoroughly with water afterward. Dry the syringe by pumping the plunger several times to expel excess solvent. Then remove the plunger to let the barrel dry. If the syringe is to be used again shortly, you can dry it by drawing air through the barrel with an aspirator or a vacuum line.

The plastic 1-mL syringes provided in some microscale lab kits are subject to contamination and are not compatible with certain organic solvents. For some volume-measuring applications, the barrel of such a syringe can be connected to the wide end of a clean Pasteur pipet with a small length of plastic or rubber tubing, so that the measured volume of liquid is drawn into the Pasteur pipet rather than the body of the syringe (see *J. Chem. Educ.* **1993**, *70,* A311).

Making Transfers

In many organic syntheses, losses during transfers constitute a substantial part of the total product loss, so they can have a major impact on your yield. Such losses can occur whenever you transfer a liquid or solid from one container to another, whether the original container is a stock bottle, reaction flask, beaker, or funnel. Making complete transfers is particularly important in microscale work, where losing just a few crystals of a solid product or a drop of a liquid product may reduce your percent yield significantly.

Transferring Solids

Bulk solids (such as those from a lab stock bottle) can be transferred from one container to another with spatulas of various shapes and sizes. Relatively large quantities of solids can be transferred with a Scoopula (see Figure A15), which has a curved surface to help keep the solid from sliding off. A Hayman-type microspatula, which has a U-shaped or V-shaped end, is used to transfer smaller amounts of solids. A flat-bladed microspatula will

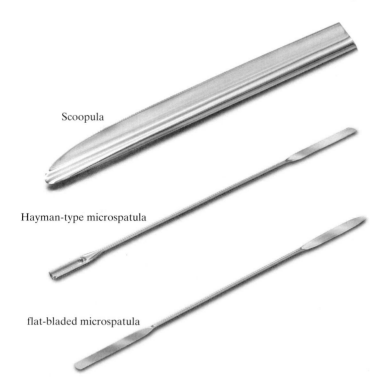

Scoopula

Hayman-type microspatula

flat-bladed microspatula

Figure A15 Spatulas for transferring solids

also work, but unless you are careful, some of the solid may spill over its sides. Small amounts of solids can also be transferred with a plastic microscoop, made by cutting a 1-mL automatic pipet tip in half.

Solids can be transferred to small-mouthed containers such as test tubes and storage vials using a folded square of weighing paper or a square plastic weighing dish as a makeshift funnel. To transfer a solid from a plastic weighing dish, hold opposite corners of the dish between your thumb and middle finger and bend the dish to form a "spout" from which you can pour or scrape out the solid. During such a transfer, set the receiving container on a square of weighing paper or in a weighing dish (or clamp it over the weighing paper or dish) so that any solid that misses the container can be recovered. If the solid you are transferring sticks to the sides of its container (such as a reaction vial or a Buchner funnel), use a flexible flat-bladed spatula or microspatula to scrape as much as you can off the sides. If you need to transfer the last traces of a solid, you can dissolve the residual solid in a volatile solvent, make the transfer, and then evaporate the solvent as described next for liquid transfers.

Transferring Liquids

In standard scale work, liquid transfers are usually accomplished by *decanting* (pouring) the liquid from its original container into another container. In microscale work, most liquids (especially liquid reactants and products) should *not* be decanted, because too much liquid will adhere to the inside of the original container and be lost. Instead, liquids are transferred using pipets or syringes.

A Pasteur pipet fitted with a latex rubber bulb (Figure A16) can be used for most microscale transfers. Volatile liquids such as dichloromethane tend to partially vaporize during a transfer (especially on a warm day or when the pipet is warmed by your hand), causing some of the liquid to spurt out the tip of the pipet. You may be able to avoid this problem by drawing in and expelling the liquid several times to fill the pipet with solvent vapors before you use it for the transfer. Alternatively, you can use a *filter-tip pipet*.

To make a filter-tip pipet, obtain a *very* small wisp of clean cotton, roll it into a loose ball, and use a straight length of thin (20 gauge) copper wire to push it past the narrow neck of a $5\frac{3}{4}$-inch Pasteur pipet into its capillary end. Push the cotton as close to the tip of the pipet as you can (see Figure A16). If the cotton plug is too large, it will get hopelessly stuck in the capillary and you will have to start over with another pipet. You may have to poke the cotton plug repeatedly with the wire to get it in place; the capillary is very fragile, so be careful that you don't break it. Mayo and Pike [Bibliography, D3] recommend that the cotton plug in a filter-tip pipet be washed with 1 mL of methanol and 1 mL of hexanes and then dried before use, but this measure should be necessary only when the liquid transferred must be very pure and dry (when in doubt, follow your instructor's recommendation).

A filter-tip pipet is useful for transferring all types of liquids, not just volatile ones, because the cotton plug helps remove solid impurities from the liquid if any are present. It also gives you better control over the transfer process, reducing the likelihood that some of your product will drip onto the bench top on the way to the collecting container. The main drawbacks of a filter-tip pipet are that (1) it takes some time and practice to prepare

Pipets and syringes are a common cause of contamination, so never allow a liquid to be sucked into a rubber bulb or pipet pump, and clean [OP-1] all pipets and syringes thoroughly after use.

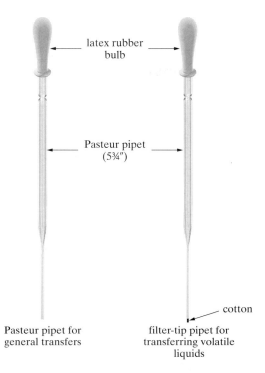

latex rubber
bulb

Pasteur pipet
(5¾")

cotton

Pasteur pipet for
general transfers

filter-tip pipet for
transferring volatile
liquids

Figure A16 Pipets for transferring liquids

one properly, (2) the cotton plug in an improperly prepared pipet may be so tight as to impede liquid flow or so loose that it won't stay in place, and (3) it is difficult to get the cotton plug out without breaking the fragile tip. If you have difficulty preparing and using such a pipet, shorten a $5\frac{3}{4}$-inch Pasteur pipet by cutting off [OP-3] all but about 5 mm of its capillary tip, and then push a wisp of cotton to the end of its shortened tip. A shortened filter-tip pipet is particularly useful for transferring hot recrystallization [OP-25] solutions.

A 1-mL syringe of the kind that comes with some microscale lab kits can also be used for liquid transfers. It is particularly useful for transferring liquids to or from containers that are sealed with septum caps. To inject liquid through a septum with a syringe, fill the syringe with the desired volume of liquid (see OP-5), carefully push its needle through the septum while holding the needle to keep it from bending, and then gently push the plunger in as far as it will go. To withdraw liquid through a septum, push the plunger in as far as it will go, insert the syringe needle through the septum as described previously, and pull the plunger to withdraw the desired amount of liquid. See OP-5 for additional information about the use and care of syringes.

When you are transferring a liquid from one container to another, you can avoid losses by keeping the containers as close together as possible. For example, if you are transferring a liquid from a conical vial to a screw-cap storage vial, hold the vials together in the same hand with their mouths at the same level. Then use your free hand to make the transfer. That way, any dripping liquid should be caught by one or the other container rather than ending up on the bench top. Even after a careful transfer, an appreciable amount of liquid may remain behind in the original container. You can recover that liquid by adding a small amount of a volatile solvent to the original container, tilting and rotating the container to wash all of the liquid off its sides, transferring the resulting solution to the receiving container, and evaporating [OP-16] the solvent with a stream of dry air or nitrogen. The volatile solvent must be one in which the liquid is appreciably soluble. Dichloromethane and diethyl ether are suitable solvents for most organic liquids.

B. Operations for Conducting Chemical Reactions

Heating

a. Heat Sources

Chemical reactions are accelerated by heat, because heat (thermal energy) speeds up the reactant molecules so that they collide more often, and increases the energy of those collisions so that they are more likely to generate product molecules. Therefore, most organic reactions are performed using some kind of heat source so that they can be carried out in a reasonable period of time. Heating devices are used for other purposes as well, such as distilling liquid mixtures and evaporating volatile solvents.

A variety of heat sources are used in the organic chemistry laboratory. The choice of a heat source for a particular application depends on such factors as the temperature required, the flammability of a liquid being heated, the need for simultaneous stirring, and the cost and convenience of the heating device.

Hot Plates and Hot Plate-Stirrers. A *hot plate* (Figure B1) consists of a base with a variable power control and a ceramic or metal platform (sometimes called the top plate) that is heated through a range of temperatures by concealed heating elements. A hot plate can be used to heat most liquids and solutions in flat-bottomed containers, such as Erlenmeyer flasks and beakers. It should *not* be used to heat low-boiling, flammable liquids that could splatter on the hot surface and ignite, or to heat round-bottom flasks directly. Hot plates can also be used to heat water baths, oil baths, sand baths, and heating blocks, which are in turn used to heat reaction mixtures and other liquids or solutions. A *hot plate-stirrer* is simply a hot plate having a built-in magnetic stirrer (see OP-10). It can be used for heating only, for stirring only, and for operations that require simultaneous heating and stirring. Since many microscale experiments require simultaneous heating and stirring, hot plate-stirrers are used to perform both operations in most microscale labs.

If you are using a hot plate or hot plate-stirrer to heat the contents of an Erlenmeyer flask or other container, first plug it into an electrical outlet and adjust the heat control dial to obtain the desired temperature or heating rate. Keep tongs, heat-resistant gloves, or other insulating materials handy so that you can quickly remove the container when necessary. For example, you can fold a rectangle of paper toweling lengthwise several times and loop it around the neck of a hot Erlenmeyer flask to remove the flask from a hot plate.

If you are using the heating device with a sand bath or an aluminum block, it is a good idea to prepare a calibration curve by measuring the equilibrium temperature at each dial setting (wait 10–15 minutes for the temperature to equilibrate at each setting) and plotting the temperature against the dial setting. Then you can adjust the heat control for the desired operating temperature when you use the same sand bath or aluminum

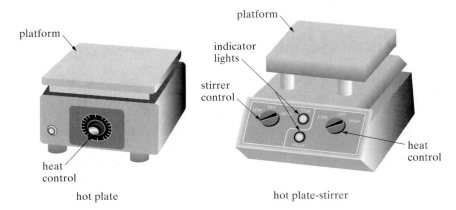

Figure B1 Hot plate and hot plate-stirrer

block again. For a sand bath, the volume of sand must be the same and the thermometer bulb must be in the same location each time the bath is used in order for such a calibration to be useful.

Heating Blocks.   Aluminum heating blocks with holes or wells designed to accommodate small test tubes, round-bottom flasks, reaction vials, and similar containers can be used to conduct microscale reactions and for such operations as recrystallization [OP-25] and distillation [OP-27]. Copper heats and cools much faster than aluminum, so copper heat-transfer plates may also be used. These plates are not commercially available, but they can easily be fabricated [Bibliography, D3].

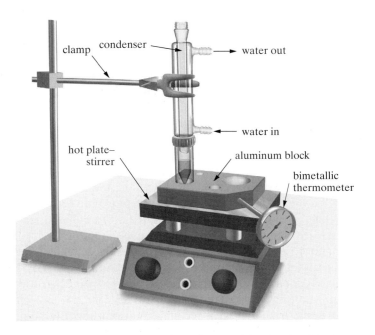

Figure B2 Heating a reaction mixture with a heating block

To use a heating block, set it on a hot plate or hot plate-stirrer and insert a thermometer (such as a nonmercury glass thermometer or a dial thermometer with a metal probe) to monitor its temperature, if desired (see Figure B2). If you will be using a magnetic stirrer [OP-10], position the heat block so that the well holding the container being heated is close to the center of the platform. A glass thermometer should be secured by a three-finger clamp on a ring stand and *carefully* lowered into a small hole (one drilled in the face of the block to accommodate it) until its bulb just rests on the bottom of the hole. Position a glass thermometer in the block *before* you begin heating, or the thermometer bulb may break. The metal probe of a dial thermometer can be inserted into a small hole drilled in one corner of a typical heating block. Note that it is not always necessary to monitor the temperature of the block, which is invariably higher than the temperature inside the container it is heating, but knowing the block temperature can help you control the heating rate more accurately and avoid overheating. Because a heating block takes some time to reach a desired temperature, it's a good idea to start heating it well before it will be needed, using a calibration curve (if you have prepared one) to select an appropriate heat control setting.

Take Care! If the bulb of a mercury thermometer breaks, especially in a heated block, it will release toxic mercury vapors into the atmosphere. Notify your instructor at once if this happens.

Place the container to be heated in a well of the appropriate size. For the most uniform heating, the liquid level in the container should be just below the top of its well. If the liquid level is above the well top, you can insulate the container with glass wool held in place by aluminum foil. (When very high temperatures are required, a conical vial can be provided with a commercially available aluminum collar.) Most glassware setups should be clamped to a ring stand, but a small test tube or a Craig tube can be supported adequately by the walls of its well. Adjust the heat control on the hot plate so that the temperature of the heating block is at least 20°C higher than the temperature you wish to attain inside the container you are heating. Then readjust the heat control, as necessary, until the desired heating rate is attained. For example, if you want to boil a reaction mixture in which water is the solvent, first raise the block temperature to 120°C. If the reaction mixture does not boil when the block is at that temperature, advance the heat control gradually until it does boil. You can control the heating rate to some extent by raising and lowering a container in its well without changing the heat setting. If you need to lower the temperature quickly, as when a recrystallization mixture threatens to boil over, raise the container out of its well first, and then lower the heat setting or clamp the container at a higher level.

Take Care! A hot metal block looks just like a cold one, so never touch a heating block unless you are sure it is cold.

Sand Baths. A typical sand bath consists of a flat-bottomed container, such as a cylindrical crystallization dish, that has been filled with fine sand to a depth of 10–15 mm. (Using a deeper sand layer may overload and damage a hot plate's heating element, especially at high temperatures.) Like a heating block, such a sand bath is usually heated on a hot plate or hot plate-stirrer, using a thermometer to monitor its temperature (see Figure B3). For the most uniform heating, the container being heated should be pushed far enough into the sand so that the liquid level in the container is just below the top of the sand. When necessary, heating can be made more uniform and a higher temperature can be attained at a given heat setting by covering the bath container with aluminum foil. Holes must be cut in the foil to accommodate the thermometer and the container being heated.

Take Care! Don't touch a sand bath or its container unless you are sure it is cold.

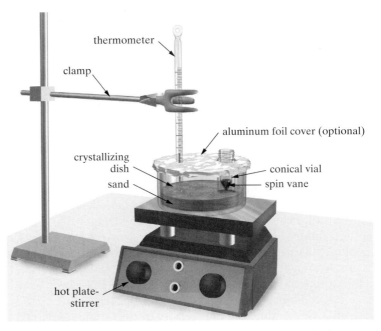

thermometer

clamp

aluminum foil cover (optional)

crystallizing
dish

conical vial

sand

spin vane

hot plate-
stirrer

Figure B3 Heating a reaction mixture with a sand bath

Although sand can be heated to a very high temperature, the glass container of a sand bath may break if it is heated much above 200°C. A convenient sand bath that is usable at high temperatures can be constructed by partly filling the ceramic well of a 100-mL Thermowell heating mantle with sand. (See the next section, "Heating Mantles," for directions.)

Sand baths heat and cool more slowly than aluminum blocks, but it is easier to observe changes in a reaction mixture and to swirl or shake a mixture when using a sand bath, as compared to a heating block. A sand bath also provides a temperature gradient, with lower temperatures near the top of the sand layer and higher temperatures near the bottom. Thus it not necessary to control the measured temperature of a sand bath precisely, since you can vary the amount of heat applied to a container by varying its depth in the sand bath. For example, you can bring a reaction mixture to the boiling point quickly by immersing its container deep in the sand, and then raise the container just enough to keep it boiling gently.

To use a sand bath like the one in Figure B3 (with or without a foil cover), set the sand bath on a hot plate or hot plate-stirrer; then clamp a thermometer to a ring stand and lower it deeply enough into the sand so that its bulb is completely covered, but is not touching the bath container. Try to place the thermometer bulb at the same depth in the sand each time, because the temperature reading will vary with its depth. Begin heating the sand bath well before it will be needed, using a calibration curve (if you have prepared one) to select an appropriate heat setting. Adjust the heat control so that the temperature reading is at least 20°C higher than the temperature you wish to attain inside the container you are heating. Lower the container you are heating into the sand, using clamps for support as necessary. If you will be using a magnetic stirrer, position this container close to the center of the hot plate-stirrer's platform. Raise or lower it in the sand, or readjust the heat control, until the desired heating rate is attained.

Take Care! If the bulb of a mercury thermometer breaks in a heated sand bath, it will release toxic mercury vapors into the atmosphere. Notify your instructor at once if this happens.

Heating Mantles. A heating mantle is generally used to heat a round-bottom flask for a standard scale reaction or distillation. Its primary use in the microscale lab is as a sand bath container and heater, as described previously. A heating mantle is always used in conjunction with a voltage-regulating or time-cycling ("on-off") heat control to vary its heat output (Figure B4). A mantle can be used with a magnetic stirrer (see OP-10), and its heat output can be varied over a wide range.

To use a ceramic-well heating mantle for a sand bath, first fill its well about two-thirds full with high-quality sand. Clamp a thermometer to a ring stand and lower it deeply enough into the sand so that its bulb is completely covered, but is not touching the bath container. See that the heat control unit is set to zero, then plug the mantle into it—*never* directly into an electrical outlet—and adjust the heat control dial until the desired sand temperature is attained. Note that the dial controls only the heating rate and cannot be set to a specific temperature. Because a heating mantle responds slowly to changes in the control setting, it is easy to exceed the desired temperature by setting the dial too high at the start, so it is advisable to prepare a calibration curve as described previously. When you are done heating, adjust the heat control dial to its lowest or "off" setting, and let the mantle and sand cool down before you attempt to move it or remove the sand. Refer to the previous section for additional information about the use of a sand bath.

Never turn on an empty heating mantle because that might burn out its heating element. If you spill any chemicals into the well of a heating mantle, particularly if it is hot, unplug it and notify your instructor.

Figure B4 Heating mantle and heat control

Hot Water Baths. Hot water baths are useful for heating low-boiling reaction mixtures, evaporating [OP-16] volatile solvents, and in other applications that require gentle heating. Although more elaborate bath containers are available, a beaker of an appropriate size is suitable for most purposes. A typical setup for heating with a hot water bath is illustrated in Figure B5. The usual function of the air condenser shown is to return solvent vapors to a boiling reaction mixture (this process is described in OP-7c), but it also makes a convenient "handle" that can be used to lower or raise the container being heated and keeps it from tipping over in the water bath. So when you heat a jointed microscale container in a hot water bath, it's often a good idea to attach an air condenser whether the mixture will be boiling or not.

To prepare a hot water bath, measure an appropriate amount of water into the bath container and set it on a hot plate or hot plate-stirrer. The bath container should be no more than three-fourths full when the container to be heated is lowered into it. For example, a 3-mL conical vial will displace about 10 mL of water when immersed up to its threads, so a 50-mL beaker should contain about 25–30 mL of water when it is used to heat such a container. Secure the container inside the water bath so that the liquid level in the container is below the water level in the bath. Clamp a thermometer with its bulb beneath the water surface and at the same level as the mixture you are heating; its bulb should not touch the walls of either container. If you are using a magnetic stirrer [OP-10], the stirring device inside the container should be close enough to the surface and center of the hot plate-stirrer's platform to allow efficient stirring. If a specific bath temperature is required, adjust the heat setting and observe the thermometer reading until that temperature is reached. If the bath temperature rises above the specified value

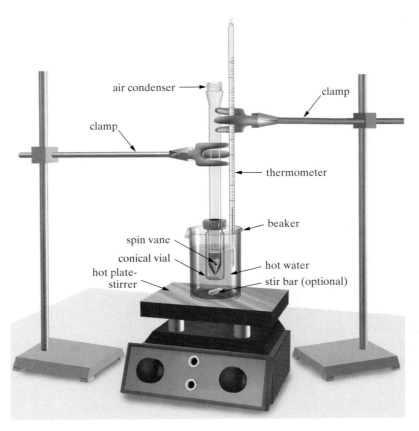

Figure B5 Heating a reaction mixture with a hot water bath

by 5°C or more, withdraw some of the bath water and replace it by an equal volume of cold water. A 10-mL (or larger) pipet equipped with a pipet pump can be used for this purpose. For most applications, the bath temperature can be allowed to vary by ±5°C or so from the specified value. If a boiling-water bath is required, add boiling chips or a stir bar to the water before you heat it to boiling. A stir bar is desirable even when the water is not boiling, as it ensures more uniform heat distribution.

When precise temperature control is not necessary, you can make a simple hot water bath by partly filling a beaker with preheated water from a hot water tap or another source. Adjust the bath temperature by adding cold water, if necessary. As the bath cools, withdraw some of the bath water and replace it with fresh hot water.

Oil Baths.   Like a hot water bath, an oil bath involves the use of a bath liquid in a suitable container, such as a beaker or a porcelain casserole. The bath liquid is a high-boiling liquid such as mineral oil, glycerol, silicone oil, or one of several polyethylene glycols (carbowaxes). The beaker in Figure B5 could be used as an oil bath by simply replacing the water with one of these bath liquids. Although the bath liquid can be heated on a hot plate, internal heating devices such as immersion heaters are also used.

Oil baths are seldom used in undergraduate organic chemistry labs because they are messy to work with, difficult to clean, and somewhat hazardous to operate. Hot oil can cause severe injury if accidentally spilled on the skin—the oil, which is difficult to remove and slow to cool, remains in contact with the skin long enough to produce deep, painful burns. An oil bath liquid can suddenly burst into flames if it is heated above its *flash point* and an ignition source, such as a spark or burner flame, is present. Hot oil can also catch fire if it splatters onto a hot surface, such as the top of a hot plate. Water must be kept away from oil baths, because spilling water into a hot oil bath causes dangerous splattering. Despite all of these drawbacks, there are some things an oil bath does better than any other heat source. It provides very uniform heating and precise temperature control over a wide range of temperatures, making it especially useful for vacuum distillation [OP-28], fractional distillation [OP-29], and sublimation [OP-26], and for conducting reactions when local overheating may cause decomposition and side reactions. Although none of the experiments in this book require the use of an oil bath, many could be performed by using one. If your instructor wants you to use an oil bath for any operation, he or she will tell you how to prepare and operate one safely.

Most oil fires can be extinguished by dry-chemical fire extinguishers or powdered sodium bicarbonate.

Burners.   Bunsen-type burners are simple and convenient to operate, but they present a serious risk of fire in an organic chemistry lab, where highly flammable solvents are often used. For that reason, burners should be used mainly for operations that cannot be conducted with flameless heat sources, such as bending and fire-polishing glass tubing.

Always check to see that there are no flammable liquids in the vicinity before you light a burner. Never use a burner near a heated oil bath or to heat a flammable liquid in an open container. Never leave a burner flame unattended; it may go out and cause an explosion due to escaping gas.

Safety Notes

To operate a typical burner with a needle valve at the base, connect it to a gas outlet with a rubber hose and make sure that the valve at the gas outlet is turned off. Close the needle valve on the burner by rotating the knurled wheel clockwise until you feel resistance (don't close it tightly), and then open it a turn or two. Open the gas valve and—without delay—ignite the burner with a burner lighter. If it doesn't light, rotate the barrel of the burner clockwise (or close the sleeve-type regulator, if it has one) and try again. When the burner is lit, adjust the needle valve and rotate the barrel or sleeve to obtain a flame of the desired size and intensity. Rotating the barrel counterclockwise or opening the sleeve regulator to introduce more air produces a hotter, bluer flame. If you are using a burner to heat a *nonflammable* liquid in a beaker or other container, place the container on a ring support using a ceramic-centered wire gauze to spread out the flame and prevent superheating. The ring support should be positioned so that the bottom of the wire gauze is at the top of the inner blue cone of the flame, where it is hottest. Remember that most organic liquids are flammable; never use a burner to heat a flammable liquid in an open container.

Steam Baths.   A steam bath (Figure B6) is a metal container having metal rings that can be removed or added to accommodate glassware of different sizes. It uses externally generated steam for heating, so it has only one operating temperature, 100°C. This limits its usefulness somewhat—for example,

steam inlet   rings

water outlet

Figure B6 Steam bath

Take Care! Avoid contact with the steam, which can cause serious thermal burns.

Figure B7 Infrared heat lamp

a steam bath cannot be used to boil water or an aqueous solution. Steam baths are often used to heat recrystallization mixtures, evaporate volatile solvents, and heat low-boiling liquids under reflux. Since other heat sources can be used to perform the same operations, steam baths are not a necessary component of a microscale laboratory, but they may come in handy for some purposes. For example, a steam bath can be used to warm up very viscous liquids so that they can be poured easily, or to vaporize iodine crystals to develop the spots on a thin-layer chromatography [OP-19] plate. One can also be used when an aluminum block or other heat source might cause decomposition of a heat-sensitive liquid.

To use a steam bath, obtain two lengths of rubber tubing, attach one to the steam bath's *water outlet* tube and the other to the steam valve outlet over the sink, and place the open ends of both rubber tubes in the sink. (If your steam bath has no water outlet tube, you will have to turn off the steam periodically to empty it of water.) Remove inner rings from the steam bath, leaving enough rings to safely support the container you wish to heat (unless it is supported by a clamp), but providing an opening large enough so that the steam will contact most of the container's bottom. If the container is a round-bottom flask that is clamped to a ring stand, remove enough rings so that the flask can be lowered through the rings to about its midpoint, leaving the smallest possible gap between the innermost ring and the flask. Directing the steam into the sink drain, open the steam valve fully and let it run until little or no water drips from the end of the rubber tube. Close the steam valve, connect it to the steam bath's *steam inlet* tube, and then open it just enough to maintain the desired rate of heating with the container in place. You can adjust the heating rate somewhat by adding or removing rings, raising or lowering a clamped flask, and changing the steam flow rate. If there is excessive condensation of steam in the vicinity of the steam bath, reduce it by turning down the steam or by adding enough rings to bridge any gaps between a flask (or another container) and the steam bath. When you are done heating, turn off the steam valve completely and let the steam bath cool down. Then remove the rubber tubes, drain any water that remains in the steam bath, replace any rings that you removed, and put the steam bath and the rubber tubes back where you found them. (Don't leave rubber tubing in the sink!)

Other Heat Sources. Heating devices such as infrared heat lamps (Figure B7) and electric forced-air heaters (heat guns) can be used in some heating applications. A heat lamp plugged into a variable transformer provides a safe and convenient way to heat comparatively low-boiling liquids. The boiling flask is usually fitted with an aluminum foil heat shield to concentrate the heat on the reaction mixture.

b. Smooth Boiling Devices

When a liquid is heated at its boiling point, it may erupt violently as large bubbles of superheated vapor are discharged from the liquid; this phenomenon is called *bumping*. A porous *boiling chip* prevents bumping by providing nucleating sites on which smaller bubbles can form. Boiling chips (also called boiling stones) are made from pieces of alumina, carbon, glass, marble (calcium carbonate), Teflon, and other materials. Alumina and calcium carbonate boiling chips may break down in strongly acidic or alkaline solutions, so boiling chips made of carbon, Teflon, or other chemically resistant materials should be used with such solutions. Wooden applicator sticks can be broken in two and the broken ends used to promote smooth boiling in nonreactive solvents; they should not be used in reaction mixtures because of the possibility of contamination. Boiling chips are not needed when a liquid being heated is stirred at a moderate rate with a magnetic stir bar or spin vane, because stirring [OP-10] causes turbulence that breaks up the large bubbles responsible for bumping.

Unless you are instructed differently, always add one or more boiling chips to any liquid or liquid mixture that will be boiled without stirring, such as a liquid to be distilled or a reaction mixture to be heated under reflux (see Section **c**). One or two small boiling chips are usually sufficient for microscale work. It is important to add boiling chips *before* heating begins, because the liquid may froth violently and boil over if you add them when it is hot. If you let a boiling liquid cool below its boiling point, add a fresh boiling chip if you reheat it later because liquid fills the pores of a boiling chip and reduces its effectiveness when boiling stops.

c. Heating Under Reflux

Most organic reactions are carried out by heating the reaction mixture to increase the reaction rate. The temperature of a reaction mixture can be controlled in several ways, the simplest and most convenient being to use a reaction solvent that has a boiling point within the desired temperature range for the reaction. Sometimes a liquid reactant itself may be used as the solvent. The reaction is conducted at the boiling point of the solvent, using a *condenser* to return all of the solvent vapors to the reaction vessel so that no solvent is lost. This process of boiling a reaction mixture and condensing the solvent vapors back into the reaction vessel is known as *heating under reflux* (or more informally as "refluxing"), where the word reflux refers to the "flowing back" of the solvent. Usually a reaction time is specified for a reaction conducted under reflux. That interval should be measured from the time the reaction mixture begins to boil, *not* from the time heating is begun.

Several different kinds of condensers are available. A *water-cooled condenser* consists of two concentric tubes, with cold tap water circulating through the outer tube and solvent vapors from a boiling reaction mixture rising up the inner tube. The circulating water cools the walls of the inner tube, cooling the vapors and causing them to condense to liquid droplets that flow back into the reaction vessel. An *air condenser* is ordinarily a single tube whose walls transfer heat to the surrounding air, cooling and condensing the vapors of a boiling liquid. As a rule, air condensers can be used with solvents that boil around 150°C or higher, or with small amounts of lower-boiling solvents that are heated gently. Air condensers can also be used with most aqueous solutions because any loss of water vapor into the laboratory air does not present a safety hazard. When in doubt, use a water-cooled condenser for its higher cooling efficiency.

A typical microscale lab kit contains both a water-cooled condenser and an air condenser. Figure B8a in the next section shows a reflux apparatus with a round-bottom flask and a water-cooled condenser, and Figure B8b illustrates one with a conical vial and an air condenser. Either type of condenser can be used with either kind of reaction vessel, however.

General Directions for Heating Under Reflux

Never heat the reaction flask before the condenser water is turned on; solvent vapors may escape and cause a fire or health hazard.

Equipment and Supplies

heat source
conical vial, round-bottom flask, or pear-shaped flask
air condenser or water-cooled condenser
boiling chip(s) or stirring device
2 lengths of rubber tubing (for water-cooled condenser only)

As the *reaction vessel*, select a conical vial or reaction flask of a size such that the reactants fill it about half full or less. For example, if the total volume of the reaction mixture will be 2.1 mL, use a 5-mL conical vial rather than a 3-mL vial. Transfer [OP-6] the reactants and any specified solvent to the reaction vessel. It is best to weigh limiting reactants directly into the reaction vessel. Add a magnetic stirring device (such as a spin vane; see OP-10) or a boiling chip and mix the reactants by swirling or stirring. Attach an appropriate condenser to the reaction vessel, making sure that the compression cap is screwed on securely (see OP-2). Do *not* stopper the condenser because that will create a closed system, which may shatter violently when heated. Clamp the apparatus to a ring stand and lower it into a heat source, such as a heating block or sand bath on a hot plate-stirrer.

Take Care! Never heat a closed system.

If you are using an air condenser, skip to the next paragraph. If you are using a water-cooled condenser, connect the water inlet (the lower connector) on its jacket to a cold water tap with a length of rubber tubing, and run another length of tubing from the water outlet (the upper connector) to a sink, making sure that it is long enough to prevent splashing when the water is turned on. If the rubber tubing slips off when pulled with moderate force, replace it with tubing of smaller diameter, or secure it with wire or a tubing clamp. Turn on the water carefully so that the condenser jacket slowly fills with water from the bottom up, and adjust the water pressure so that a narrow stream flows from the outlet. The flow rate should be just great enough to (1) maintain a continuous flow of water in spite of pressure changes in the water line and (2) keep the condenser at the temperature of the tap water during the reaction. Excessively high water pressure may force the tubing off the condenser and spray water on you and your neighbors.

If you are using a stirring device, begin stirring at a moderate rate. Adjust the heat setting or the depth of the reaction vessel in the heat source to keep the solvent boiling gently. Measure the reaction time from the time that boiling begins—when a continuous stream of bubbles rises through the liquid. If the liquid bubbles violently or froths up, reduce the heat setting or reposition the reaction vessel in the heat source. If there is sufficient liquid in the reaction vessel, its vapors should form a *reflux ring* of condensing

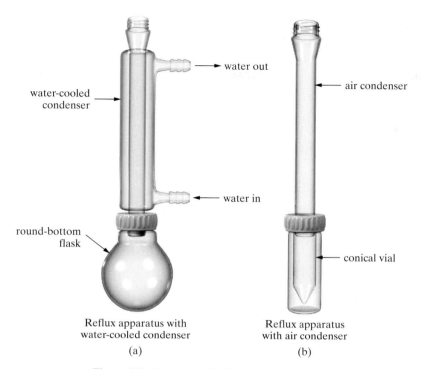

water out

air condenser

water-cooled
condenser

water in

round-bottom
flask

conical vial

Reflux apparatus with
water-cooled condenser
(a)

Reflux apparatus
with air condenser
(b)

Figure B8 Apparatus for heating under reflux

vapors in the condenser. Below this point, solvent will be seen flowing back into the reaction vessel; above it, the condenser should be dry. If the reflux ring rises more than halfway up the condenser, reduce the heating rate or increase the water flow rate to prevent the escape of solvent vapors.

At the end of the reaction period, turn off the hot plate-stirrer (or other heating device) and raise the apparatus on the ring stand, clamping it several inches above the heat source. Do not turn off the condenser water (if used) until the reaction vessel has cooled nearly to room temperature. Unless the next operation will be carried out in the reaction vessel, transfer its contents to a container suitable for that operation. Clean [OP-1] the reaction vessel as soon as possible so that residues do not dry on the glass.

Summary

1 Transfer reactants and solvent to reaction vessel.
2 Add boiling chip or a stirring device.
3 Attach appropriate condenser.
4 Clamp apparatus over heat source.
   IF an air condenser is being used, GO TO 7.
5 Attach tubing to water-cooled condenser.
6 Turn on condenser water and adjust flow rate.
7 Start stirrer (if used), adjust heat so that reaction mixture boils gently.
8 Readjust water flow or heating rate as necessary, boil gently until end of reaction period.
9 Turn off heating device, let reaction vessel cool, transfer reaction mixture.
10 Disassemble and clean apparatus.

OPERATION 8

Cooling

Some reactions proceed too violently to be conducted safely at room temperature, or involve reactants or products that decompose at room temperature. In such cases, the reaction mixture is cooled with some kind of cold bath, which can be anything from a beaker filled with cold water to an electrically refrigerated device. Cold baths are also used to increase the yield of crystals from a reaction mixture or recrystallization mixture.

A setup like the one shown in Figure B5 (page 582) for a hot water bath can also be used for a cold bath. A cold bath can be prepared using any suitable container, such as a beaker of suitable size, a crystallization dish, an evaporating dish, or a pair of nested Styrofoam cups. A beaker can be wrapped with glass wool and placed inside a larger beaker to keep it cold longer, if necessary.

A number of cooling media are used for cold baths. A mixture of ice (or snow) and tap water can be used for cooling in the 0–5°C range. The ice should be finely divided and enough water should be present to just cover the ice, since ice alone is not an efficient heat-transfer medium. An ice-salt bath consisting of three parts of finely crushed ice or snow to one part of sodium chloride can attain temperatures down to −20°C, and mixtures of $CaCl_2 \cdot H_2O$ containing up to 1.4 g of the calcium salt per gram of ice or snow can provide temperatures down to −55°C. In practice, these minimum values may be difficult to attain, as the actual temperature of an ice-salt bath depends on such factors as the fineness of the ice and salt and the insulating ability of the container. Temperatures down to −75°C can be attained by mixing small chunks of dry ice with acetone, ethanol, or another suitable solvent in a vacuum-jacketed container such as a Dewar flask.

Temperatures below −40°C cannot be measured using a mercury thermometer because mercury freezes at that temperature.

Take Care! Never handle dry ice with your bare hands.

General Directions for Cooling

Equipment and Supplies

cold bath container
cooling medium
thermometer
air condenser (optional)

Obtain a suitable cold bath container and fill it with the cooling medium to a level depending on the size of the container to be cooled. When this container is immersed in the cold bath, the cooling medium should fill the cold bath container about three-fourths full. Clamp a thermometer [OP-9] so that its bulb is entirely immersed in the cooling medium but not touching either container. If you are using an ice-salt bath, mix in the appropriate salt in small portions, waiting for the temperature to equilibrate after each addition, until the desired temperature is attained. If the container to be cooled is a conical vial or a ground-joint flask, attach an air condenser (unless experimental conditions preclude this) and clamp the condenser to a ring stand. If not, either clamp the neck of the container to a ring stand or hold the container in your hand so that it does not tip over. Lower the apparatus into the cooling bath so that the liquid level in the container is below the cooling fluid level. Keep the contents of the cold bath mixed by occasional manual stirring. The contents of the container being cooled can also be swirled or stirred for more efficient cooling. Add small portions of ice as needed to keep the temperature in the desired range, removing an equal amount of water (with a 10-mL pipet, for example) to make room for it.

Summary

1 Fill container with cooling medium.
2 Insert thermometer in cooling medium, adjust temperature if necessary.
3 Insert container to be cooled.
4 Keep contents of container and cooling bath mixed, adjust temperature as needed.

Temperature Monitoring

OPERATION **9**

In the organic chemistry lab, thermometers are used to monitor the temperature of heating devices, cooling baths, reaction mixtures, distillations, and for many other purposes. Such thermometers should have a range of at least $-10°C$ to $260°C$, and a wider range is desirable for some purposes. Most broad-range glass thermometers contain mercury, which is toxic and presents a safety hazard if a thermometer is broken, but broad-range non-mercury thermometers are also available. Although standard-length thermometers can be used for most purposes in the microscale lab, short glass thermometers (about 15 cm long) are preferable for measuring the temperature of small volumes of liquids. Bimetallic thermometers with metal probes can be used to measure the temperature of some heat sources, such as aluminum blocks. For the most accurate temperature readings, a thermometer should be *calibrated* and an *emergent stem correction* applied as described in OP-30, but this is generally not necessary for routine temperature monitoring.

You can monitor the temperature of a liquid or solid heating medium (such as sand) using a thermometer clamped with its bulb entirely immersed in the liquid or heating medium. It should be held in place by a three-fingered clamp or a special thermometer clamp. The thermometer should not touch the side or bottom of the container, or anything inside the container.

To monitor the temperature of a reaction mixture that will be stirred [OP-10] in an open container such as an Erlenmeyer flask, clamp a thermometer so that its bulb is completely immersed in the mixture but does not contact the stirring device (a large stir bar could break the thermometer bulb). If continuous mixing is not necessary, you can insert the thermometer each time you stop stirring or shaking the mixture, read it when the temperature has stabilized, and then remove it and resume mixing. Never use the thermometer itself for stirring because the bulb is fragile and breaks easily.

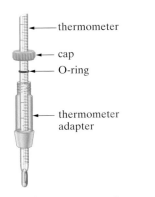

Figure B9 Thermometer adapter assembly

Using a Thermometer Adapter

You will ordinarily need a thermometer adapter to monitor the temperature of an operation (such as distillation) conducted in a ground-joint glassware setup. Be certain that the thermometer adapter, when inserted in the apparatus, does not create a closed system. For example, never put a sealed thermometer adapter assembly on top of a reflux condenser that is attached directly to a reaction flask because heating such a system may cause it to shatter.

To use a typical microscale thermometer adapter (Figure B9), first slide a small rubber O-ring down the thermometer stem to near its center (the O-ring should come with the adapter). Secure the adapter, without its upper compression cap, in the appropriate joint on the apparatus. Insert the thermometer (bulb end down) through the adapter and move the O-ring up or down until its bulb is positioned correctly when the O-ring is resting on the adapter. Then slide the adapter's threaded cap down the thermometer stem and use it to secure the thermometer assembly to the apparatus.

OPERATION **10**

Mixing

Reaction mixtures are often stirred, shaken, or agitated in some other way to promote efficient heat transfer, prevent bumping, increase contact between the components of a heterogeneous mixture, or mix in a reactant as it is being added. Mechanical stirring, which requires a motor that turns a shaft connected to a stirring paddle, is seldom used in undergraduate organic chemistry labs and will not be described here.

Manual Mixing

If you are carrying out a reaction in an Erlenmeyer flask or a test tube, the reactants can be mixed by using a stirring rod or spatula, or by manual shaking or swirling. A motion combining shaking with swirling is more effective than swirling alone. But when more efficient and convenient mixing is required, particularly over a long period of time, it is best to use a magnetic stirrer, as described in the next section.

Manual mixing may be used at times other than during reactions. For example, a liquid can be dried [OP-22] by swirling it with a drying agent in an Erlenmeyer flask. This increases the amount of contact between the liquid and particles of the drying agent, increasing drying efficiency. Small quantities of liquids are often dried by stirring the liquid with the drying agent in a conical vial. You can do this by twirling the pointed end of a microspatula in the bottom of the vial. Similarly, you can twirl the rounded end of a flat-bladed microspatula inside a Craig tube or a small test tube to stir recrystallization solutions and other mixtures.

Magnetic Stirring

A magnetic stirrer (Figure B10) contains a motor that rotates a bar magnet underneath a metal or ceramic platform. As the bar magnet rotates, it in turn spins a stirring device inside a container placed on or above the platform. The most common stirring device, called a *stir bar*, is an oblong (usually cylindrical) Teflon-encased magnet. Because no moving parts extend outside of the container in which stirring occurs, a reaction assembly that is to be stirred magnetically can be completely enclosed if necessary. The stirring rate is controlled by a dial on the magnetic stirrer. For efficient stirring, the vessel (flask, beaker, conical vial, etc.) containing the stirring device should be positioned near the center of the platform and as close to its surface as practicable.

Magnetic stirrers can be used in conjunction with heating mantles, oil baths, and other heat sources that are constructed of nonferrous materials. A hot plate-stirrer resembles a magnetic stirrer except that it has two dials, one to control the heating unit and the other to control the stirrer. Hot plate-stirrers can be used to heat and stir flat-bottomed containers such as Erlenmeyer flasks directly, but they are also used in conjunction with heating devices that require an external heat source, such as aluminum blocks, hot water baths, and sand baths [OP-7]. Figure B5 on page 582 shows a hot plate-stirrer being used to heat a hot water bath and stir the bath liquid and a reaction mixture at the same time.

A hot plate-stirrer is nearly indispensable in the microscale laboratory. The stirring device can be either a *spin vane*, which is used in conical vials, or a small stir bar, which is used in round-bottom flasks, small Erlenmeyer flasks, beakers, and other vessels having flat or gently rounded bottoms. A spin vane consists of a V-shaped piece of Teflon having a small, Teflon-coated bar magnet inserted through its short side (see Appendix I for an illustration). When the spin vane is inserted into a conical vial with the point of the "V" down, the magnetic bar is farther from the platform than a stir bar would be, so the spin vane may not rotate properly unless it is centered correctly and the stirring rate is relatively low. If necessary, you can use a small stir bar to stir the contents of a conical vial. At low speeds, the stir bar wobbles around in the conical well of the vial and doesn't stir very efficiently, but at higher speeds, it should spin horizontally in the wider part of the vial, creating a vortex for more efficient mixing.

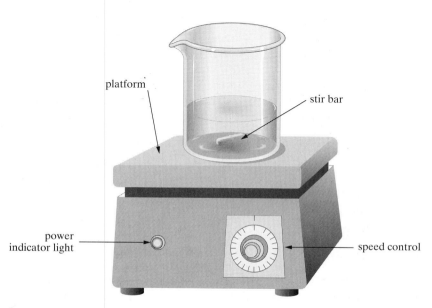

Figure B10 Magnetic stirrer

General Directions for Stirring a Reaction Mixture Magnetically

Equipment and Supplies

magnetic stirrer or hot plate-stirrer

heat source (optional)

reaction vessel (flask, conical vial, etc.)

stir bar or spin vane

Set the heat source (aluminum block, sand bath, etc.), if any, directly on the platform of the magnetic stirrer or hot plate-stirrer. Place a stir bar or spin vane inside the reaction vessel; if you are using a spin vane, position it with its sharply pointed end down. (Don't add boiling chips; the stirring action prevents bumping.) If you need to stir a heating bath as well, use a stir bar that is larger than the one in the reaction mixture. If necessary, attach a condenser or other device to the reaction vessel. Position the reaction vessel in the heat source so that it is close to the center of the platform, and secure the reaction apparatus with a clamp, if necessary. Start water circulating through a water-cooled condenser, if you are using one. Start the magnetic stirrer and adjust the stirring rate dial carefully to obtain an appropriate stirring rate. If the stirring rate is too high or the reaction vessel is not positioned correctly, the stirring device will flop around erratically rather than rotate smoothly. If that happens, reposition the reaction vessel to bring it closer to the center of the stirring unit's platform; then reduce the stirring rate until the device rotates smoothly and increase it gradually until a suitable rate is attained. High stirring rates may be needed for heterogeneous reaction mixtures, such as those involving two immiscible liquids, but in most cases a moderate stirring rate is suitable. If you are heating a reaction mixture also, adjust the heating rate as described in OP-7 for the heating device you are using.

Summary

1 Place heat source (if used) on stirring unit.
2 Put stir bar or spin vane in reaction vessel.
3 Secure reaction apparatus over stirring unit.
4 Adjust stirrer for appropriate stirring rate.

OPERATION **11**   # Addition of Reactants

In many organic preparations, the reactants are not all combined at the start of the reaction. Instead, one or more of them is added during the course of the reaction. This is necessary when a reaction is strongly exothermic or when one of the reactants must be kept in excess to prevent side reactions.

Solid reactants can be added slowly or at regular intervals from a plastic weighing dish that is bent to form a pouring spout. Solids can also be divided into small portions that are added at regular intervals with a spatula.

Liquid reactants can be added, in portions or drop by drop, using a Pasteur pipet. A long (9-inch) Pasteur pipet should be used to add liquids through a condenser. When liquid reactants must be added over an extended period of time, it is more convenient to use a syringe. The syringe is filled with a measured amount of the liquid to be added, and is then inserted through a septum, such as a Teflon-faced cap liner, so that its needle is directly over the reaction mixture. A cap liner may not remain airtight after being punctured, so try to find an intact liner or one that has already been punctured but still fits tightly around the syringe needle. (Your instructor may want you to use a punctured one to keep the others intact.) Remember that a syringe needle is very sharp (unless it has been blunted to prevent injury) and may be contaminated with dangerous biological or chemical substances, so be careful not to stick yourself—or anyone else—with it.

General Directions for Addition Under Reflux

Equipment and Supplies

reaction vessel
Claisen adapter
air condenser or water-cooled condenser
compression cap and cap liner
syringe

Measure the appropriate reactants into the reaction vessel. To assemble the apparatus illustrated in Figure B11, first attach a Claisen adapter to the reaction vessel—a conical vial or a ground-joint flask. Attach an appropriate condenser to the adapter's long arm and screw a compression cap with a

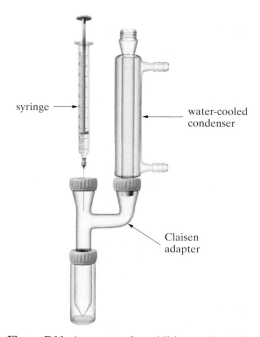

syringe

water-cooled
condenser

Claisen
adapter

Figure B11 Apparatus for addition under reflux

Teflon-faced liner—Teflon side down—onto its short, straight arm. Clamp the apparatus securely to a ring stand. Fill the syringe with the specified volume of the liquid to be added (see OP-5). Hold the syringe vertically with its needle pointing down and carefully insert its needle through the cap liner. During the reaction (see OP-7c), add the liquid drop by drop, or in small portions at designated intervals, by depressing the plunger. If necessary, remove the syringe to refill it, then immediately replace it on the apparatus. When the addition is complete, remove the syringe and clean it (see OP-5) without delay.

Summary

1 Assemble apparatus for addition under reflux.
2 Fill syringe, insert needle through cap liner.
3 Add liquid drop by drop or at designated intervals.
4 Remove and clean syringe.

C. Separation Operations

Gravity Filtration   OPERATION **12**

Filtration is used for two main purposes in organic chemistry:

- To remove solid impurities from a liquid or solution
- To separate an organic solid from a reaction mixture or a crystallization solvent

Gravity filtration is generally used for the first purpose and *vacuum filtration* [OP-13] for the second. *Centrifugation* [OP-14] can be used for either. In a gravity filtration, the liquid component of a liquid-solid mixture drains through a filtering medium (such as cotton or filter paper) by gravity alone, leaving the solid on the filtering medium. The filtered liquid, called the *filtrate*, is collected in a flask or another container. Gravity filtration is often used to remove drying agents from dried organic liquids or solutions and solid impurities from hot recrystallization solutions.

A standard scale gravity filtration of organic liquids is usually carried out with a funnel having a short, wide stem and a relatively coarse filter paper that is fluted (folded) to increase the filtration rate. Such funnels are seldom needed in the microscale lab, but they may come in handy when relatively large quantities of a mixture are to be filtered.

A *filtering pipet*—a Pasteur pipet containing a small plug of cotton or glass wool—is suitable for most microscale gravity filtrations. A filtering pipet (see Figure C1) is usually made with a standard $5\frac{3}{4}$-inch Pasteur pipet. For filtering hot recrystallization [OP-25] solutions, it is best to use a shortened filtering pipet made by cutting off [OP-3] most of the capillary tip to leave a 5-mm stub; otherwise, crystals may form in the narrow capillary and block the liquid flow. A glass-wool plug can be used when large particles are being removed or when the mixture being filtered contains an acid or another substance that may react with cotton. Glass wool does not filter out fine particles, so cotton is preferred for most applications. Very fine particles may pass through cotton also; a filtering pipet containing a layer of chromatography-grade alumina or silica gel on top of a cotton plug can be used to remove such particles from a mixture.

A microscale gravity filter with a larger capacity can be constructed by cutting the rounded top off the bulb of a plastic Beral-type pipet and packing some cotton or glass wool in its neck (see *J. Chem. Educ.* **1993**, *70*, A204). The plastic may not be compatible with some organic solvents, however.

A filter-tip pipet, prepared as described in OP-6, can also be used to filter small amounts of liquids. This technique is not really gravity filtration, but it accomplishes the same purpose—removing a solid from a liquid. It has the disadvantage that some solid may adhere to the cotton and be transferred to the collecting container. When the solid particles are quite coarse, such as the granules of sodium sulfate used for drying liquids [OP-22], you

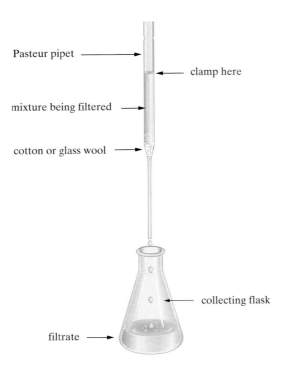

Pasteur pipet

clamp here

mixture being filtered

cotton or glass wool

collecting flask

filtrate

Figure C1 Filtration with a filtering pipet

may be able to use a Pasteur pipet without a filter tip to remove them. Use the pipet tip to push aside any solid that is in the way; then hold it flat against the bottom of the container as you draw liquid into it.

General Directions for Gravity Filtration

Equipment and Supplies

> 2 $5\frac{3}{4}$-inch Pasteur pipets
> rubber bulb (2-mL capacity)
> cotton (or glass wool)
> applicator stick or stirring rod
> collecting containers

Roll a small amount of cotton (or glass wool) between your fingers to form a loose ball, and insert it into the wider end of a $5\frac{3}{4}$-inch Pasteur pipet. (Alternatively, use a Beral-type pipet with the top cut off.) Then use a small stirring rod or a wooden applicator stick to push it down the body of the pipet, forming a cotton plug that ends about where the capillary section of the pipet begins, as shown in Figure C1. Don't pack it too tightly because that will reduce the filtration rate. Clamp the filtering pipet vertically over a small beaker. Rinse the plug with a suitable wash solvent (usually the solvent present in the mixture being filtered) by using a second Pasteur pipet to transfer about 0.5 mL of the solvent to the top of the filtering pipet, letting it drain, and using a rubber bulb to force any remaining solvent through. Replace the beaker with another collecting container, such as a conical vial or a small flask, and use the second Pasteur pipet to transfer the mixture being filtered, in several portions if necessary, to the top of the filtering pipet. Let the liquid drain by gravity into the collecting container. If the filtration rate is very slow, you can use a pipet pump (see OP-5) to apply a gentle, constant pressure to the top of the filtering pipet, depressing the quick-release lever (if it has one) before you remove the pipet. (Don't use excessive pressure because that may force particles into the filtrate.) Wash the solid and the plug with a small amount of wash solvent as described previously, collecting the solvent in the container that holds the filtrate. Use a rubber bulb or a pipet pump to force out the last few drops of liquid. Let the plug dry, then remove it by snagging it with a copper wire bent to form a "J" at one end.

Summary

1 Prepare filtering pipet by inserting cotton or glass-wool plug in Pasteur
   pipet.
2 Clamp filtering pipet over beaker, rinse with solvent.
3 Replace beaker with collecting container.
4 Transfer mixture being filtered to filtering pipet; let drain.
5 Wash filtering pipet with solvent.
6 Remove plug; clean up.

General Directions for Filtration with a Filter-Tip Pipet

Construct a filter-tip pipet as described in OP-6 and rinse it with a suitable solvent (usually the solvent present in the mixture being filtered) by drawing in about 0.5 mL of the solvent and slowly ejecting it into a waste container. Use the pipet and attached bulb to draw up liquid from the mixture being filtered and transfer it to a suitable collecting container, leaving the solid behind in the original container, until all of the liquid has been transferred. To avoid transferring adherent solid along with the liquid, try to keep the pipet's tip away from the solid when you draw liquid into it. To transfer the last few drops of liquid, push the solid aside with the tip of the pipet and hold it flat against the bottom of the original container as you withdraw the liquid. Stir a little wash solvent into the solid, draw it into the filter-tip pipet, and combine it with the rest of the filtrate.

Summary

1 Construct filter-tip pipet; rinse with wash solvent.
2 Draw liquid being filtered into pipet.
3 Transfer liquid to collecting container.
4 Wash solid, transfer wash solvent to collecting container.
5 Clean up.

OPERATION **13**   # Vacuum Filtration

Vacuum filtration (also called suction filtration) provides a fast, convenient method for isolating a solid from a liquid-solid mixture and for removing solid impurities from relatively large quantities of a liquid. In a typical vacuum filtration, a circle of filter paper is laid flat on a perforated plate inside a special funnel, which is attached by an airtight connector to a heavy-walled *filter flask* (see Figure C2) or to a sidearm test tube (see Appendix I). A conical *Hirsch funnel*, which contains a small perforated plate near the bottom of the cone, is generally used for microscale work. A small *Buchner funnel*, which has nearly vertical sides and a larger perforated plate, can come in handy if you need to separate relatively large quantities of solid from a liquid.

A porcelain Hirsch funnel contains an integral perforated plate about 1–2 cm in diameter; plastic Hirsch-type funnels with separate perforated discs are also available. Porcelain Hirsch funnels use very small filter paper circles. These are available commercially, but they can also be cut from ordinary filter paper using a sharp cork borer on a flat cutting surface, such as the bottom of a large cork. The filter paper should be about equal in size to the perforated plate or slightly smaller, but large enough to completely cover all of its holes.

The sidearm of a filter flask or sidearm test tube is connected to the inlet of a *water aspirator* or to a vacuum line, often by way of a trap (described in the next section), with a short length of heavy-walled rubber tubing. In the water aspirator, a rapid stream of water passes by a small hole at the inlet, creating a vacuum there and in the attached filter flask, and exits into a sink. An aspirator should always be run "full blast" because at lower flow rates, its efficiency decreases and the likelihood of water backup increases. When the mixture being filtered is poured into the Hirsch funnel, the liquid is forced through the paper by the unbalanced external pressure and collects in the filter flask or sidearm test tube, while the solid remains on the filter paper as a compact *filter cake*.

Thin-walled rubber tubing would collapse under vacuum.

Experimental Considerations

Filter Traps and Cold Traps.   You should ordinarily interpose a *filter trap* between the filter flask (or sidearm test tube) and a water aspirator to keep water from backing up into the flask when the water pressure changes, as it often does. Using a filter trap is most important when the filtrate is to be saved because any backed-up water will contaminate the filtrate, but it's a good idea to use the trap for all vacuum filtrations with an aspirator. A suitable filter trap can be constructed by wrapping a thick-walled Pyrex jar with transparent plastic tape (to reduce the chance of injury in case of implosion) and inserting a rubber stopper fitted with two L-shaped connecting tubes, as described in Minilab 1. Note that one of the connecting tubes should be longer than the other. A filter trap with a pressure-release valve is illustrated in Figure E13 of OP-28.

Take Care! Never use a thin-walled container, such as an Erlenmeyer flask, as a trap; it may shatter under vacuum.

If you are using a vacuum line connected to a central mechanical vacuum pump, it may be necessary to use a *cold trap* to protect the pump from solvent vapors that might damage it (your instructor will inform you if it is necessary to use a cold trap). A filter trap can function as a cold trap if it is immersed in an appropriate cold bath [OP-8].

Filtering Media. The external pressure on the mixture being filtered could cause fine particles to pass through the filter paper, so a relatively slow (fine grained) grade of filter paper should be used. An all-purpose filter paper, such as Whatman #1, is adequate for filtration of most solids. When filtering finely divided solids from a liquid, it is sometimes necessary to use a *filtering aid* (such as Celite) to keep the solid from plugging the pores in the filter paper. In a stoppered flask, shake the filtering aid vigorously with a suitable solvent to form a slurry (a thick suspension). Without delay, pour the slurry onto the filter paper under vacuum until a bed about 2–3 mm thick has been deposited. Remove the solvent from the filter flask before continuing with the filtration. This method cannot be used when the solid is to be saved, since it would be contaminated with the filtering aid.

Washing. Unless otherwise instructed, you should *wash* the solid on the filter paper with an appropriate solvent, usually the same as the one from which it was filtered. For example, if you filter a solvent from an aqueous solution, use distilled water as the wash solvent. If you filter a solid from a mixture of solvents, as in a mixed solvent recrystallization [OP-25b], you should ordinarily use the solvent in which the solid is least soluble. To reduce losses, the wash solvent should be chilled in an ice/water mixture.

Sometimes a solid is washed with a lower-boiling solvent to help it dry faster. For example, a solid filtered from toluene (bp 111°C) might be washed with low-boiling petroleum ether (bp ~35–60°C) before drying. The wash solvent must be miscible with the solvent from which the solid was filtered, and the solid should not be appreciably soluble in it.

Drying. A solid that has been collected by vacuum filtration is usually *air dried* after the last washing by leaving it on the filter for a few minutes with the vacuum turned on. The vacuum draws air through the solid, which increases the drying rate. If the solid is still quite wet, you can place a *rubber dam* (a thin, flexible rubber sheet) or a sheet of plastic wrap over the mouth of the funnel. The vacuum should cause the sheet to flatten out on top of the filter cake, forcing water out of it. Unless the solid was filtered from a very low-boiling solvent, it ordinarily requires further drying by one of the methods described in OP-23.

General Directions for Vacuum Filtration

Equipment and Supplies

    Hirsch funnel
    filter flask (or sidearm test tube)
    1-hole rubber stopper or neoprene adapter
    trap
    filter paper
    thick-walled rubber tubing
    flat-bladed microspatula
    wash solvent

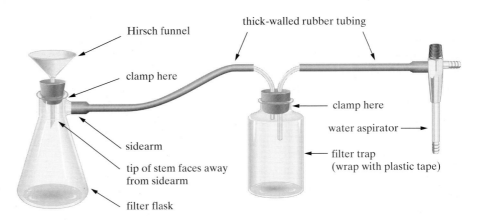

Figure C2 Apparatus for vacuum filtration

Place 2 mL or so of the wash solvent in a small test tube and chill it in a beaker containing ice and water; then set the beaker aside. Clamp the filter flask (or sidearm test tube) and trap (if you are using one) securely to a ring stand and connect them to an aspirator or a vacuum line as shown in Figure C2. Use thick-walled rubber tubing that will not collapse under vacuum for all connections. If you are using a water aspirator, connect the longer glass tube in the filter trap (the tube that extends farther into the trap) to the aspirator and the shorter glass tube to the filter flask. If you are using a vacuum line that must be protected by a cold trap, connect the cold trap to the vacuum line and filter flask and place it inside a cooling bath, as directed by your instructor. Insert a Hirsch funnel into the filter flask using a neoprene filter flask adapter or a snug-fitting rubber stopper to provide a tight seal. Obtain a circle of filter paper of the correct diameter and place it inside the Hirsch funnel so that it covers all of the holes in the perforated plate, but does not extend up the sides of the funnel.

Moisten the filter paper with a few drops of the wash solvent. Open the aspirator tap or vacuum-line valve as far as it will go. With an aspirator, you can direct the water stream into a large beaker or another container to prevent splashing. If the solid is finely divided, let it settle before you pour the liquid into the funnel, and transfer the bulk of the solid near the end of the filtration. Otherwise, stir or swirl the mixture just before pouring it to transfer more of the solid and leave less behind in the decanting vessel. If the volume of the filtration mixture is greater than the capacity of the funnel, add the mixture rapidly enough to keep the funnel about two-thirds full throughout the filtration until it has all been added. Transfer any remaining solid to the filter paper with a flat-bladed microspatula, using a small amount of the filtrate or some cold wash solvent to facilitate the transfer. Leave the vacuum on until only an occasional drop of liquid emerges from the stem of the funnel. If you are using an aspirator with no filter trap, or if there is a possibility that water collecting in a trap will back up into the filter flask, break the vacuum with a pressure-release valve or by disconnecting the rubber tubing at the vacuum source before you turn off the vacuum.

Add enough of the chilled wash solvent to just cover the solid; about 1–2 mL of wash solvent per gram of solid is usually sufficient. For the most efficient washing, the mixture should be stirred gently with a microspatula

Don't attach rubber tubing to the aspirator outlet to reduce splashing because it will also reduce the aspirator's effectiveness.

or flat-bottomed stirring rod until the solid is suspended in the liquid, but this step is sometimes omitted in the microscale lab because—unless performed quickly and very carefully—it can lead to product loss. Without delay, turn on the vacuum to drain the wash liquid. After the last washing, leave the vacuum on for 3–5 minutes to air-dry the solid on the filter and make it easier to handle. Run the tip of a flat-bladed microspatula around the circumference of the filter paper to dislodge the filter cake, then carefully invert the funnel and transfer the filter cake and filter paper to a square of glazed paper, a watch glass, a weighing dish, or another suitable container. Use your spatula to scrape any remaining particles onto the paper or into the container. Dispose of the filtrate as specified by the experimental directions or your instructor, and dry the solid by one of the methods described in OP-23. To reduce losses, dry the filter paper along with the filter cake and scrape off any additional solid after it is dry, being careful not to scrape any filter paper fibers into your product.

Summary

1 Assemble apparatus for vacuum filtration.
2 Position and moisten filter paper, turn on vacuum.
3 Add mixture being filtered to Hirsch funnel.
4 Transfer any remaining solid to funnel.
5 Wash solid on filter with chilled wash solvent.
6 Air-dry solid on filter paper.
7 Transfer solid to container; remove filtrate from filter flask.
8 Disassemble and clean apparatus.

Centrifugation   OPERATION **14**

Centrifugation is used to separate different phases from one another by centrifugal force. When a mixture in a *centrifuge tube* is whirled around a circular path at high speed, the denser phase (often a solid) is forced to the bottom of the tube, leaving the other phase on top. In the microscale laboratory, centrifugation is often used to collect solids that have crystallized from solution in a Craig tube. Centrifugation may also be used to separate a finely divided solid from a liquid or to separate two immiscible liquids sharply after an extraction [OP-15].

General Directions for Centrifugation

Equipment and Supplies

    benchtop centrifuge
    2 centrifuge tubes
    Pasteur pipet (optional)
    flat-bladed microspatula (optional)

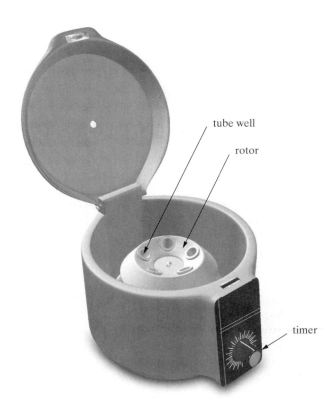

Figure C3 A benchtop centrifuge

Transfer the mixture to be centrifuged to a conical centrifuge tube with a capacity of ~15 mL. (If you are using a Craig tube, see OP-25.) Obtain an empty matched centrifuge tube and add enough water to it so that the two tubes, with their contents, have approximately equal masses. Usually you can estimate the amount of water needed by volume, but you may have to weigh the tubes to ensure proper balance. (Alternatively, find another student who is doing the same operation and use his or her centrifuge tube to balance your own. Label the tubes so you don't mix them up.) Place the two centrifuge tubes directly opposite one another in the rotor of the centrifuge (Figure C3); they fit into tube wells, which may be cushioned to help prevent breakage. Close and lock the centrifuge lid, set the centrifuge's timer (if it has one) to 3–5 minutes, and start the centrifuge. If the centrifuge rattles loudly or stops before the time is up, the tubes are not balanced properly. Balance them and try again. When the time is up (or when you switch off the centrifuge), the rotor will slowly come to a stop. Wait until its whirring sound has stopped; then open the lid and remove the centrifuge tubes. If you are centrifuging a liquid-solid mixture, carefully decant the liquid or remove it with a filter-tip pipet, leaving the solid behind. Use the pointed end of a flat-bladed microspatula to remove the solid. If you are centrifuging a mixture of two immiscible liquids, separate the liquids after centrifugation by removing the lower layer with a Pasteur pipet, as described in OP-15. Use a tapered centrifuge brush, if one is available, to clean the centrifuge tube.

Summary

1 Transfer mixture to centrifuge tube.
2 Place centrifuge tube and a second tube of comparable mass in opposite tube wells.
3 Run centrifuge 3–5 minutes.
4 Let centrifuge stop, remove tubes.
5 Separate liquid layer from solid or second liquid layer.

Extraction

If you shake a bromine/water solution with some dichloromethane, the red-brown color of the bromine fades from the water layer and appears in the organic layer as you shake. These color changes show that the bromine has been transferred from one solvent (water) to another (dichloromethane). The process of transferring a substance from a liquid or solid mixture to a solvent is called *extraction*, and the solvent is called the *extraction solvent*. The extraction solvent is usually a low-boiling organic solvent that can be evaporated [OP-16] to isolate the desired substance after extraction.

Extraction is used for the following purposes in organic chemistry:

- To separate a desired organic substance from a reaction mixture or some other mixture.
- To remove impurities from a desired organic substance, which is usually dissolved in an organic solvent.

The second process is described in OP-21, "Washing Liquids."

a. Liquid-Liquid Extraction

Principles and Applications

Liquid-liquid extraction is based on the principle that, if a substance is soluble to some extent in two immiscible liquids, most of it can be transferred from one liquid to the other by a process that involves thorough mixing of the liquids. For example, acetanilide is partly soluble in both water and dichloromethane. If a solution of acetanilide in water is shaken with a portion of dichloromethane, some of the acetanilide will be transferred to the organic (dichloromethane) layer (see Figure C4). The organic layer, being more dense than water, separates below the water layer and can be removed and replaced with another portion of dichloromethane. When the fresh dichloromethane is shaken with the aqueous solution, more acetanilide passes into the new organic layer. This new layer can then be removed and combined with the first. By repeating this process enough times, virtually all of the acetanilide can be transferred from the water to the dichloromethane.

The ability of an extraction solvent (S_2) to remove a solute (A) from another solvent (S_1) depends on the partition coefficient (K) of solute A in the two solvents, as defined in Equation **1**:

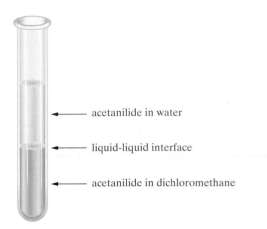

Figure C4 Distribution of a solute between two liquids

$$K = \frac{\text{concentration of A in } S_2}{\text{concentration of A in } S_1} \tag{1}$$

In the example of acetanilide in water and dichloromethane, the partition coefficient is given by

$$K = \frac{[\text{acetanilide}]_{\text{dichl}}}{[\text{acetanilide}]_{\text{water}}}$$

The larger the value of K, the more solute will be transferred to the organic layer with each extraction, and the fewer portions of dichloromethane will be required for essentially complete removal of the solute. A rough estimate of K can be obtained by using the ratio of the solubilities of the solute in the two solvents—that is,

$$K \approx \frac{\text{solubility of A in } S_2}{\text{solubility of A in } S_1}$$

This approximate relationship can be helpful for choosing a suitable extraction solvent.

Extraction Solvents

Most extraction solvents are organic liquids that are used to extract nonpolar and moderately polar solutes from aqueous solutions, but water and aqueous solutions are sometimes used to extract polar solutes from organic solutions. For example, dilute aqueous NaOH can be used to extract carboxylic acids from organic solvents by first converting them to carboxylate salts, which are much more soluble in water and less soluble in organic solvents than the original carboxylic acids.

$$RCOOH + NaOH \rightarrow RCOO^-Na^+ + H_2O$$

If the carboxylic acid is insoluble in water, it can be precipitated by acidifying the aqueous extract and collected by vacuum filtration [OP-13].

Similarly, dilute aqueous HCl is used to extract basic solutes such as amines from organic solvents by first converting the amines to ammonium salts.

$$RNH_2 + HCl \rightarrow RNH_3^+Cl^-$$

A good *organic* extraction solvent should be immiscible with water, dissolve a wide range of organic substances, and have a low boiling point so that it can be removed by evaporation after the extraction. The substance being extracted should be more soluble in the extraction solvent than in water; otherwise, too many steps will be required to extract it. Diethyl ether and dichloromethane (methylene chloride) are the most commonly used organic extraction solvents. Diethyl ether (also called ether or ethyl ether) has a very low boiling point (34.5°C) and can dissolve both polar and non-polar organic compounds, but it is extremely flammable and tends to form explosive peroxides on standing. Dichloromethane is more dense than water, which simplifies the extraction process, and it is not flammable. Dichloromethane has a tendency to form emulsions, which can make it difficult to separate cleanly, and it must be handled with caution because it is a suspected carcinogen.

Some extraction solvents and their properties are listed in Table C1. Organic extraction solvents that have densities less than that of water (1.00 g/mL) will separate as the top layer during the extraction of an aqueous solution; extraction solvents having densities greater than that of water will ordinarily separate as the bottom layer.

Potential hazards should be considered when selecting and using an extraction solvent. For example, solvents such as benzene, trichloromethane (chloroform), and tetrachloromethane (carbon tetrachloride) should not be used as extraction solvents in an undergraduate laboratory because of their toxicity and carcinogenic potential. Precautions must be taken with all organic solvents to minimize skin and eye contact and inhalation of vapors. Flames must not be allowed in the laboratory when highly flammable solvents, such as diethyl ether and petroleum ether, are in use.

Table C1 Properties of commonly used extraction solvents

Solvent	bp (°C)	*d* (g/mL)	Comments
water	100	1.00	for extracting polar compounds, generally using a reactive solute such as NaOH or HCl
diethyl ether	34.5	0.71	good general solvent; absorbs some water; very flammable
dichloromethane	40	1.34	good general solvent; suspected carcinogen
toluene	111	0.87	for extracting aromatic and nonpolar compounds; difficult to remove
petroleum ether	~35–60	~0.64	for extracting nonpolar compounds; very flammable
hexane*	69	0.66	for extracting nonpolar compounds; flammable

*The mixture of C_6H_{14} isomers called *hexanes* is cheaper than pure hexane, and is often used in its place.

Experimental Considerations

Extraction Methods.   In the microscale lab, an extraction is usually performed by shaking the liquids in a conical vial or conical centrifuge tube and separating them with a Pasteur pipet. A 5-mL conical vial can be used with liquid volumes up to ~4 mL, and a 15-mL centrifuge tube with liquid volumes up to ~12 mL. A screw-cap centrifuge tube is preferable to one having a snap-on cap, since it is less likely to leak. A conical vial is usually capped with a compression cap having a Teflon-faced liner; the liner should be inserted in the cap so that its white Teflon side faces down when the vial is capped. A conical vial may leak if the liner is damaged or the rim of the vial is chipped, so shake some water in the capped vial to check for leaks before using the vial for an extraction. Do this for a centrifuge tube also, and replace the cap if it leaks.

Liquid layers are ordinarily separated by removing the *lower* layer with a Pasteur pipet and transferring it to another container. This way, the interface between the layers is at the narrowest part of the container when the last of the lower layer is removed, making a sharp separation possible. It takes some practice and a steady hand to remove all of the bottom layer without including any of the top layer, but it is important that you learn how to do so. Otherwise, you will lose part of your product or the product will be contaminated with material from the layer being extracted. Accurate separation is most important for the last extraction step because any extraction solvent that is not recovered during earlier steps can be recovered in the last one. Most extraction solvents have a high vapor pressure, which can cause them to spurt out of the tip of a Pasteur pipet during transfer. Product loss due to spurting can be reduced or prevented by one or more of the following measures:

- Making sure that the extraction solvent, the liquid being extracted, and the Pasteur pipet are at room temperature or below
- Rinsing the Pasteur pipet with the extraction solvent two or three times to fill it with solvent vapors just before use
- Using a filter-tip pipet

Volume of Extraction Solvent.   The volume of extraction solvent and the number of extraction steps are sometimes specified in an experimental procedure. If they are not, use a volume of extraction solvent about equal to the volume of liquid being extracted, divided into at least two portions. For example, you can extract 3.0 mL of an aqueous solution with two successive 1.5-mL or three 1-mL portions of dichloromethane. Note that it is more efficient to use several small portions of extraction solvent rather than one large portion of the same total volume.

Rule of Thumb: Total volume of extraction solvent ≈ volume of liquid being extracted.

Getting Good Separation.   Under some conditions, the liquid layers do not separate sharply, either because an *emulsion* forms at the interface between the two liquids or because droplets of one liquid remain in the other liquid layer. Emulsions can often be broken up by using a wooden applicator stick to stir the liquids gently at the interface. If that doesn't work, mix in some saturated aqueous sodium chloride solution (or enough solid NaCl to saturate the aqueous layer) and allow the extraction container to stand open and undisturbed for a time. To consolidate the liquid layers, use an applicator stick to rub or stir any liquid droplets that form on the sides or bottom of the

An emulsion usually contains microscopic droplets of one liquid suspended in another.

extraction container. You can also use an applicator stick to remove small amounts of insoluble "gunk" that sometimes form near the interface. Larger amounts of insoluble material can be removed by filtering [OP-12] the mixture through a loose pad of glass wool in a filtering pipet, but this may result in some product loss.

Saving the Right Layer. *Always keep both layers until you are certain which layer contains the desired product!* All too often, a student will unthinkingly discard the extraction layer that contains the product and have to repeat the experiment from the beginning. The safest practice is to keep both layers until you have actually isolated the product from one of them, but you can usually determine which is the right layer before then. In most extractions, the product is extracted from an aqueous solution into an organic solvent, such as dichloromethane or diethyl ether. If you make careful observations when you add the extraction solvent, you can usually tell whether it floats on top of the aqueous layer or sinks below it. Since diethyl ether is less dense than water, it will form the upper layer when it is used to extract an aqueous solution. Dichloromethane is more dense than water, so it should ordinarily form the lower layer with an aqueous solution. However, aqueous solutions are more dense than pure water, and a very concentrated aqueous solution (10 *M* NaOH, for example) may even be more dense than dichloromethane. If you are not sure which layer is the aqueous one, add a drop or two of water to a drop or two of each layer separately. The one in which the water dissolves is the aqueous layer.

General Directions for Extraction

Equipment and Supplies

conical vial or centrifuge tube with cap
support for extraction container
2 Pasteur pipets with bulbs, one calibrated
extraction solvent
wooden applicator stick
1–2 containers (conical vials, screw-cap vials, test tubes, etc.)

Depending on the amount of liquid to be extracted, obtain a 3-mL or 5-mL conical vial with a compression cap and unperforated liner or a 15-mL conical centrifuge tube with a screw cap. Check the extraction container for leaks. To keep a conical vial from tipping over, place it in a small beaker; set a centrifuge tube in a test tube rack or other suitable support. Add the liquid to be extracted, which should be at room temperature or below. If its volume is less than 1 mL, it is advisable to add enough of a suitable solvent (usually water) to give it a total volume of at least 1 mL. Use a calibrated Pasteur pipet (or other measuring device) to add a measured portion of extraction solvent and cap the container tightly. Shake the extraction container gently and unscrew the cap slightly after 5–10 shakes to release any pressure inside the container. Tighten the cap and continue to shake the mixture for a minute or so, but remember that excessively vigorous shaking may cause an emulsion to form. (Alternatively, you can use a spin vane to stir [OP-10] the contents of a conical vial vigorously for at least 1 minute, or use a vortex mixer as directed by your instructor.) Loosen the cap and let

Take Care! Wear gloves during an extraction to protect your hands in case of leakage.

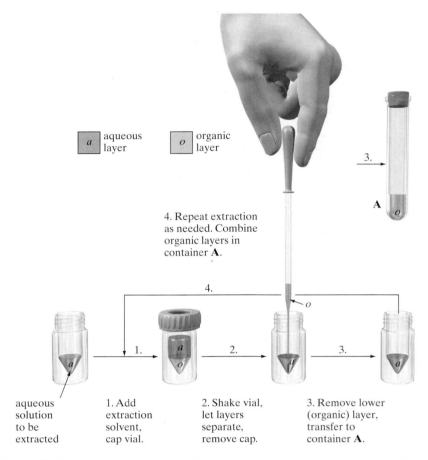

aqueous
layer

organic
layer

3.

A

4. Repeat extraction
as needed. Combine
organic layers in
container **A**.

4.

1.   2.   3.

aqueous
solution
to be
extracted

1. Add
extraction
solvent,
cap vial.

2. Shake vial,
let layers
separate,
remove cap.

3. Remove lower
(organic) layer,
transfer to
container **A**.

Figure C5 Extraction using an extraction solvent that is more dense than the liquid being extracted

the mixture stand until there is a sharp interface between the layers. Use a wooden applicator stick to help consolidate the layers, if necessary. If you are using a centrifuge tube, it can be spun in a centrifuge [OP-14] to facilitate layer separation. If the layers do not separate cleanly, take the appropriate measures described in "Getting Good Separation."

Follow Method **I** (illustrated in Figure C5) if the extraction solvent is *more* dense than the liquid being extracted (forming the lower layer), and Method **II** (illustrated in Figure C6) or **III** if it is *less* dense than the liquid being extracted (forming the upper layer). Method **III** can be used only if all of both liquid layers will fit into the Pasteur pipet. Although this method requires fewer transfers than **II**, it is more difficult to perform proficiently.

I. Squeeze the bulb of a Pasteur pipet to expel air and insert the pipet vertically so that its tip just touches the bottom of the "vee" in the conical extraction container. Slowly withdraw the *bottom* (organic) layer, taking care not to mix the layers, and transfer it to a suitable container (container **A**). Add another measured portion of pure extraction solvent to the liquid remaining in the extraction container and shake to

*The kind of container you use for **A** depends on the next operation (washing, drying, etc.) that you will use.*

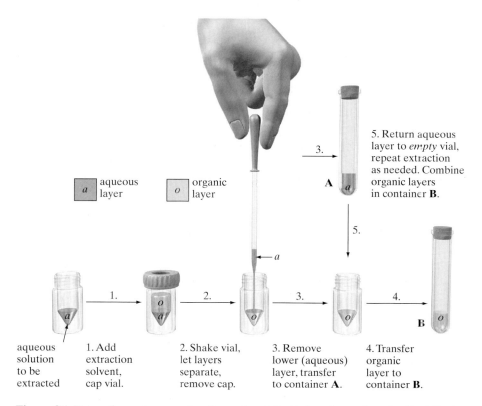

Figure C6 Extraction using an extraction solvent that is less dense than the liquid being extracted

extract as before, transferring the extract (the bottom layer) to **A**. If another extraction is necessary, repeat the process, combining all of the extracts in **A**. Retain the liquid in the extraction container for later disposal.

II. Squeeze the bulb of a Pasteur pipet to expel air and insert the pipet vertically so that its tip just touches the bottom of the "vee" in the conical extraction container. Slowly withdraw the *bottom* (aqueous) layer, taking care not to mix the layers, and transfer it to a test tube or other small container (**A**). Transfer the contents of the extraction container to a different container (**B**) and return the contents of **A** to the extraction container (see Figure C6). Add another measured portion of pure extraction solvent and shake to extract as before, then transfer the lower layer to **A** and combine the upper layer with the extract in **B**. If another extraction is necessary, return the contents of **A** to the extraction container and repeat the process. Retain the liquid in **A** for later disposal.

III. (For small amounts of liquid) Slowly draw all of *both* liquid layers into the Pasteur pipet (avoid drawing in much air), and wait until a sharp interface forms between the layers; then carefully return the bottom layer to the empty extraction container and transfer the top layer to another container (**A**). Add another measured portion of pure extraction solvent to the liquid in the extraction container and shake to extract as before. Again draw both layers into the Pasteur pipet, let them separate, and then return the bottom layer to the empty extraction container and

*The kind of container you use for **B** depends on the next operation (washing, drying, etc.) that you will use.*

transfer the top layer to **A**. Repeat this process for any subsequent extractions, combining all of the extracts in **A**. Retain the liquid in the extraction container for later disposal.

Summary (Methods I and II)

1 Add liquid to be extracted to extraction container.
2 Add extraction solvent to extraction container, cap tightly.
3 Shake gently, vent, and continue shaking again to extract solute into extraction solvent.
4 Loosen cap, let layers separate.
5 Uncap extraction container, transfer lower layer to container **A**.
   IF extraction solvent formed lower layer, GO TO 6.
   IF extraction solvent formed upper layer, GO TO 7.
6 Cap container **A**.
   IF another extraction step is needed, GO TO 2.
   IF extraction is complete, clean up, STOP.
7 Transfer upper layer to container **B** and cap it.
   IF another extraction step is needed, transfer contents of container **A** to extraction container, GO TO 2.
   IF extraction is complete, clean up, STOP.

b. Salting Out

Adding an inorganic salt (such as sodium chloride or potassium carbonate) to an aqueous solution containing an organic solute usually reduces the solubility of the organic compound in the water, and thus promotes its separation. This *salting out* technique is often used to separate a sparingly soluble organic liquid from its aqueous solution. For example, when an ester is hydrolyzed, the product mixture may be distilled to remove the alcohol, which can then be recovered by saturating the aqueous distillate with potassium carbonate and extracting it with diethyl ether. Salting out can also be used during an extraction to increase the amount of an organic solute transferred from the aqueous to the organic layer and to remove excess water from the organic layer.

To salt out an organic liquid from an aqueous solution containing the liquid, add enough of the salt to saturate the aqueous solution, then stir or shake it to dissolve the salt. You can estimate the amount of salt needed by using data from the table "Saturated Solutions" in *The Merck Index* [Bibliography, A10]. If you don't know how much salt to use, add it in small portions until a portion no longer dissolves completely. If any undissolved salt remains, filter the mixture through a plug of glass wool. Transfer the mixture to a conical vial or other suitable extraction vessel for separation or extraction.

To improve separation during an extraction, add enough of the salt to the extraction container to saturate the aqueous layer and shake to dissolve the salt. If necessary, filter the contents of the extraction container through a plug of glass wool to remove any dissolved salt. (You can also add a saturated solution of the salt, rather than the pure salt, to the extraction mixture, but the salting-out process is less complete because the resulting salt solution is more dilute.) Then proceed as for a normal extraction.

c. Liquid-Solid Extraction

Liquid-solid extraction involves the removal of one or more components of a solid by mixing the solid with an extraction solvent and separating the resulting solution from the solid residue. This method is often used to separate substances from natural products and other solid mixtures. For example, the red-orange pigment lycopene can be extracted from tomato products by an acetone-petroleum ether mixed solvent, as described in Experiment 9.

To perform a liquid-solid extraction, shake the solid vigorously with the extraction solvent in a capped centrifuge tube or other suitable container, and then use a flat-bladed microspatula to crush and rub the solid against the sides of the tube. Repeat the shaking and crushing sequence several times. Centrifuge [OP-14] the mixture and transfer the extract to a suitable container, leaving the residue in the centrifuge tube. (Alternatively, use a filter-tip pipet to separate the extract from the solid residue.) Repeat the extraction as many times as necessary, and combine the liquid extracts in a single collecting container.

Evaporation

Evaporation is the conversion of a liquid to vapor at or below the boiling point of the liquid. Evaporation can be used to remove a volatile solvent, such as diethyl ether or dichloromethane, from a comparatively involatile liquid or solid. Volatile solvents are usually removed under vacuum or in a stream of dry air or nitrogen, but they can also be removed by distillation when appropriate. Complete solvent removal is used to isolate an organic solute after such operations as extraction [OP-15] or column chromatography [OP-18]. Partial solvent removal, or *concentration*, can be used to bring a recrystallization solution to its saturation point (see OP-25).

Experimental Considerations

Because of possible health and fire hazards, you should never evaporate an organic solvent by heating an open container outside a fume hood. Even when using the method described here, you should know and allow for the hazards associated with each solvent.

To avoid transfers, you should carry out an evaporation in the container that will be used in the next step, if possible. For example, if a substance in solution is to be purified by distillation, evaporate the solvent in the conical vial from which the residue will be distilled. If that vial is not large enough to hold all of the liquid, you can carry out the evaporation in stages by adding one portion of the liquid and evaporating most of the solvent, adding the next portion and evaporating, and so on. If the residue will be the final product from a preparation, you can carry out the evaporation from a tared storage vial.

Evaporation Under Dry Air or Nitrogen

Relatively small quantities of a volatile solvent can be evaporated by passing a slow stream of dry air or nitrogen over a liquid. Nitrogen is preferred because the oxygen in air may react with easily oxidized solutes, but clean, dry air is suitable for most purposes. The gas stream sweeps solvent molecules away from the surface of the liquid, accelerating the evaporation rate. Evaporation cools the remaining liquid, however, so heating is needed to maintain a rapid evaporation rate and to prevent condensation of water vapor in the product. This operation must be carried out under a fume hood to keep solvent vapors out of the laboratory. Your lab may have a hood with an "evaporation station" where a number of Pasteur pipets are attached to a source of dry air or nitrogen and supported above a large hot plate.

The solution to be evaporated is placed in a conical vial or another suitable container, which is heated gently in a hot water bath or warm sand bath while a stream of the dry gas is directed over the surface of the liquid (see Figure C7). The bath temperature should be about 5–10°C below the boiling point of the solvent so that the solvent evaporates rapidly without boiling. (Boiling can cause product loss by spattering.) You can also use a hot plate as the heat source, but you must be very careful to avoid overheating, which may cause the residue to decompose. The evaporation container should not be placed directly on the hot platform; instead, clamp it above the platform at a level that will allow evaporation without boiling the solvent.

Evaporation Under Vacuum

Evaporation under vacuum is used to remove relatively large amounts of volatile solvent, especially when it is important to recover the solvent.

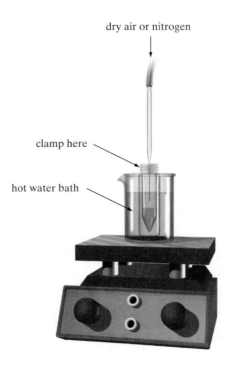

Figure C7 Evaporation under a stream of dry air or nitrogen

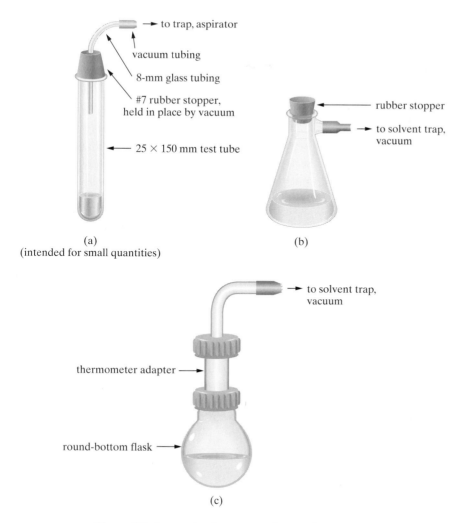

to trap, aspirator

vacuum tubing

8-mm glass tubing

#7 rubber stopper,
held in place by vacuum

25 × 150 mm test tube

rubber stopper

to solvent trap,
vacuum

(a)
(intended for small quantities)

(b)

to solvent trap,
vacuum

thermometer adapter

round-bottom flask

(c)

Figure C8 Apparatus for evaporation under vacuum

Figure C8 illustrates some of the setups that can be used for this purpose.
The test tube or flask containing the liquid to be evaporated is heated [OP-7a]
gently in a hot water bath (or over a steam bath, if one is available), and an
aspirator or vacuum line is used to reduce the pressure inside the apparatus,
increasing the evaporation rate. Swirling or stirring the solution continuously during evaporation accelerates evaporation and reduces foaming and
bumping. Magnetic stir bars, because they tend to retain some of the
residue, should only be used if the residue will later be dissolved in another
solvent or undergo an operation requiring magnetic stirring.

Evaporation under vacuum requires constant attention because excessive heat or a sudden pressure decrease may cause liquid to foam up and
out of the container. One way to control the vacuum and reduce the likelihood of boil-over is to replace the stopper shown in Figure C8b with a
Hirsch funnel assembly. If you hold your thumb over the holes in the porcelain plate of the Hirsch funnel, you can decrease the internal pressure by
pressing down with your thumb or increase the pressure by raising it.

A trap similar to the one pictured in Figure C2 of OP-13 should be interposed between the evaporation container and the aspirator to collect the evaporated solvent, which should then be returned to a solvent recovery container. To recover a low-boiling solvent such as diethyl ether, the solvent trap should be immersed in an appropriate cold bath (see OP-8).

Distillation

High-boiling solvents and relatively large quantities of low-boiling solvents can be removed by simple distillation [OP-27] or vacuum distillation [OP-28]. Distillation is often used to remove a volatile solvent from a higher-boiling liquid that will then be distilled as well. For example, if a liquid product is extracted from a reaction mixture with dichloromethane, the extraction solvent can be removed by distillation into a Hickman still, and the product can then be distilled at a higher temperature. This reduces losses by avoiding unnecessary transfers. Distillation can also be used to concentrate a solution that is then evaporated further by another method. For example, a solution can be concentrated to a volume of 0.5–1.0 mL by distillation into a Hickman still, and the remaining solvent can then be evaporated under a stream of dry air or nitrogen.

Concentration of a Solution

An aluminum block or sand bath can be used to concentrate solutions by boiling, as illustrated in Figure C9 for the concentration of a recrystallization [OP-25] solution in a Craig tube. Twirling the end of a microspatula in the liquid helps prevent bumping and boil-over. Decomposition of the solute is unlikely as long as some solvent remains, but care should be taken to avoid inadvertently evaporating the solution to dryness.

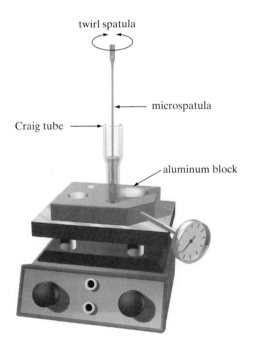

Figure C9 Concentration of a solution

General Directions for Evaporation with Dry Air or Nitrogen

Equipment and Supplies

> evaporation container (conical vial, etc.)
> Pasteur pipet
> rubber tubing
> drying tube with drying agent (optional)
> hot water bath, sand bath, or other heat source
> hot plate

The liquid to be evaporated should be in a tared conical vial or other container suitable for evaporation. Unless an evaporation station is available, prepare a suitable gas-drying tube [OP-24a], and then connect one end to an air line and the other end to a Pasteur pipet (the gas delivery pipet) using short lengths of rubber tubing. If dry nitrogen is used, omit the drying tube. Clamp the gas delivery pipet (your own or one from an evaporation station) vertically above the heat source. Heat the hot water bath or sand bath to a temperature about 5–10°C below the solvent's boiling point and immerse the evaporation container in it. Holding the container in your hand provides more control over the evaporation rate, but clamping or otherwise securing it in the bath can help prevent accidental spillage. Position the gas delivery pipet so that its tip is 1–2 cm above the surface of the liquid, and *slowly* adjust the flow rate so that the gas distorts the liquid surface but causes no spattering. As necessary, adjust the heating rate, the position of the vial, or the position of the gas delivery pipet so that the solvent evaporates quite rapidly but the liquid does not boil appreciably or foam up. If you are holding the vial in your hand, swirl it gently to increase the evaporation rate.

To avoid the likelihood of product loss, you can adjust the gas flow rate using a vial of water or solvent.

Continue evaporating until the residue, if it is a liquid, no longer appears to decrease in volume with time, or if it is a solid, appears dry. Remove the vial from the heat source. If you are not certain that all of the solvent has evaporated, dry the outside of the vial and weigh it when it has cooled. Hold it under the gas stream for another minute or so, and then dry and reweigh it. If the masses are essentially identical, evaporation is complete; if not, continue evaporating until the mass no longer decreases significantly between weighings.

If you only need to concentrate the solution, stop the evaporation when sufficient solvent has been removed.

If you need to transfer [OP-6] the residue to another container, let the evaporation container cool down first. You can transfer the last traces of residue by rinsing the container with a small amount of volatile solvent (such as diethyl ether or dichloromethane) and allowing the solvent to evaporate under a hood or in a stream of dry air or nitrogen.

Summary

1 Transfer liquid being evaporated to evaporation container.
2 Place container in or over heat source.
3 Position gas delivery pipet over liquid and adjust gas flow.
4 Apply gentle heat until evaporation is complete.
5 Transfer residue, if necessary.
6 Clean apparatus.

General Directions for Evaporation Under Vacuum

Equipment and Supplies

evaporation container (Figure C8)
rubber tubing
solvent trap
aspirator
heat source

Assemble one of the setups pictured in Figure C8 using heavy-walled rubber tubing that will not collapse under vacuum. Be sure to check all glassware for cracks, star fractures, and other imperfections that might cause them to implode under vacuum. Add the solution to be evaporated, stopper the evaporation container, and connect the apparatus to a solvent trap and the trap to a vacuum source. Turn on the vacuum and heat the evaporation container gently with a hot water bath or other appropriate heat source, swirling it throughout the evaporation to minimize foaming and bumping. (In some cases, the liquid can be stirred magnetically [OP-10].) Adjust the water bath temperature to attain a satisfactory rate of evaporation; the liquid may boil gently, but it should not foam up. Be ready to remove the evaporation container from the heat source immediately if it starts to foam up; otherwise, your product may be carried over to the trap.

If you only need to concentrate the solution, stop the evaporation when sufficient solvent has been removed.

Continue evaporating until the residue, if it a liquid, no longer appears to decrease in volume with time, or if it is a solid, appears dry. At this time, boiling should have stopped and the odor of the solvent should be gone. When evaporation appears complete, discontinue heating and remove the evaporation container from the heat source. Break the vacuum by detaching the vacuum hose, opening a pressure-release valve on the trap (if it has one), or—for the apparatus in Figure C8a—sliding the stopper off the mouth of the test tube. Then turn off the vacuum source. If you are not certain that all of the solvent has evaporated, dry the outside of the evaporation container to remove moisture from the heating bath, weigh it when it has cooled, and resume evaporation for a few minutes; then dry and weigh it again. Repeat this process as necessary until the mass no longer decreases significantly between weighings.

If you need to transfer [OP-6] the residue to another container, let the evaporation container cool down first. You can transfer the last traces of residue by rinsing the container with a small amount of volatile solvent (such as diethyl ether or dichloromethane) and allowing the solvent to evaporate under a hood or in a stream of dry air or nitrogen. Place the solvent from the trap in a solvent recovery container.

Summary

1 Assemble apparatus for evaporation.
2 Add liquid, stopper evaporation container, connect to trap and vacuum source.
3 Turn on vacuum.
4 Apply heat with swirling or stirring until evaporation is complete.

5 Discontinue heating, turn off vacuum.
6 Transfer residue and recover solvent.
7 Disassemble and clean apparatus.

Steam Distillation   OPERATION **17**

Distillation of a mixture of two (or more) immiscible liquids is called *codist-illation.* When one of the liquids is water, the process is usually called *steam distillation. Internal steam distillation* is carried out by boiling a mixture of water and an organic material in a distillation apparatus, causing vaporized water (steam) and organic liquid to distill into a receiver. *External steam distillation* is carried out by passing externally generated steam (usually from a steam line) into a boiling flask containing the organic material. The vaporized organic liquid is carried over into a receiver along with the condensed steam. External steam distillation is not very practical for the microscale lab, so it will not be discussed further.

Steam distillation is used to separate organic liquids from reaction mixtures and natural products, leaving behind high-boiling residues such as tars, inorganic salts, and other relatively involatile components. Steam distillation is particularly useful for isolating the essential oils of plants from various parts of the plant. For example, clove oil can be steam distilled from clove buds as described in Experiment 10. Steam distillation is not useful for the final purification of a liquid, however, because it cannot effectively separate components having similar boiling points.

Principles and Applications

When a *homogeneous mixture* of two liquids is distilled, the vapor pressure of each liquid is lowered by an amount proportional to the mole fraction of the other liquid present. This usually results in a solution boiling point that is somewhere between the boiling points of the separate components. For example, a solution containing equal masses of cyclohexane (bp = 81°C) and toluene (bp = 111°C) boils at 90°C.

If you are not familiar with the principles of distillation, see OP-27.

When a *heterogeneous mixture* of two immiscible liquids, A and B, is distilled, each liquid exerts its vapor pressure more or less independently of the other. The total vapor pressure over the mixture (P) is thus approximately equal to the sum of the vapor pressures that would be exerted by the separate pure liquids (P_A° and P_B°) at the same temperature.

$$P \approx P_A^\circ + P_B^\circ$$

This has several important consequences. First, the vapor pressure of a mixture of immiscible components will be *higher* than the vapor pressure of its most volatile component. Because raising the vapor pressure of a liquid or liquid mixture lowers its boiling point, the boiling point of the mixture will be *lower* than that of its most volatile (lowest boiling) component. Because the vapor pressure of a pure liquid is constant at a constant temperature, the vapor pressure of the mixture of liquids will be constant as well. Thus the boiling point of the mixture will remain constant throughout its distillation as long as each component is present in significant quantity.

For example, suppose that you are distilling a mixture of the immiscible liquids toluene and water at standard atmospheric pressure (760 torr, 101.3 kPa). The mixture will start to boil when the sum of the vapor pressures of the two liquids is equal to the external pressure, 760 torr. This occurs at 85°C, where the vapor pressure of water is 434 torr and that of toluene is 326 torr. Because the vapor pressures of the two components are additive, the mixture distills well below the normal boiling point of either toluene (bp = 111°C) or water. According to Avogadro's law, the number of moles of a component in a mixture of ideal gases is proportional to its partial pressure in the mixture, so the mole fraction of toluene in the vapor should be about 0.43 (326/760) and that of water should be about 0.57 (434/760). (These calculations are approximate because the vapors are not ideal gases.) In other words, about 43% of the molecules in the vapor are toluene molecules. Because toluene (M.W. = 92) molecules are heavier than water (M.W. = 18) molecules, they make a greater contribution to the total mass of the vapor. In 1.00 mol of vapor, there will be 0.43 mol of toluene and 0.57 mol of water, so the mass of toluene in the vapor will be about 40 g (0.43 mol × 92 g/mol), and the mass of water will be about 10 g (0.57 mol × 18 g/mol). The mass of one mole of the vapor is thus about 50 g, of which toluene makes up 40 g, or 80%. The liquid that collects in the receiver during a distillation—the *distillate*—is merely condensed vapor, so distilling a mixture of toluene and water will yield a distillate that contains about 80% toluene, by mass.

Because of its comparatively low molecular weight and its immiscibility with many organic compounds, water is nearly always one of the liquids used in a codistillation involving an organic liquid. The organic liquid must be insoluble enough in water to form a separate phase, and it cannot react with hot water or steam. As shown in Table C2, the higher the boiling point of the organic liquid, the lower will be its proportion in the distillate, and the closer the mixture boiling point will be to 100°C. Because the distillation boiling point is never higher than 100°C at 1 atm—well below the normal boiling points of most water-immiscible organic liquids—thermal decomposition of the organic component is minimized.

Table C2 Boiling points and compositions of heterogeneous mixtures with water (Component B)

Component A	bp of A (°C)	bp of A–B mixture (°C)	Mass % of A in distillate
toluene	111	85	80
chlorobenzene	132	90	71
bromobenzene	156	95	62
iodobenzene	188	98	43
quinoline	237	99.6	10

Experimental Considerations

An organic liquid can be separated by internal steam distillation using essentially the same procedure as for simple distillation [OP-27], except that additional water may need to be added during the distillation. Because you will probably have to distill a relatively large volume of water to obtain a small volume of the organic liquid, the distillation should be carried out rapidly. Increasing the surface area of a liquid increases its vaporization rate, so you should ordinarily use the largest available round-bottom flask. This also helps prevent a liquid from foaming up into the Hickman still by providing more space for the foam to occupy. If the organic distillate is quite volatile, a thermometer can be used to indicate when the end of the distillation is near. For example, with a toluene-water mixture, the temperature will rise rather rapidly from 85°C to about 100°C when the toluene is nearly gone. With liquids that boil around 180°C or higher, the temperature will be close to 100°C throughout the distillation, so a thermometer will be of little use.

General Directions for Internal Steam Distillation

See OP-27 for more detailed directions for conducting a distillation.

Equipment and Supplies

>heat source (aluminum block or sand bath)
>round-bottom flask
>Hickman still
>water-cooled condenser
>stir bar (optional)
>condenser tubing
>ring stand, clamp
>thermometer (optional)
>$5\frac{3}{4}$-inch Pasteur pipet
>9-inch Pasteur pipet (optional)
>collecting container

If it will be necessary to add more water during the distillation, measure out the approximate amount of water that will be needed. Assemble the apparatus pictured in Figure C10; you can add a thermometer if the organic liquid being distilled has a boiling point below 180°C. Add the mixture to be steam distilled, a stir bar, and enough water to fill the boiling flask about one-half full (unless enough water is already present). In some cases, a stir bar may cause excessive foaming and should be omitted. Turn on the stirrer, if you are using one, and heat the flask with an appropriate heat source to maintain a rapid rate of distillation without allowing any boil-over into the Hickman still. As the well of the Hickman still fills with liquid, use a short Pasteur pipet to transfer the distillate to an appropriate

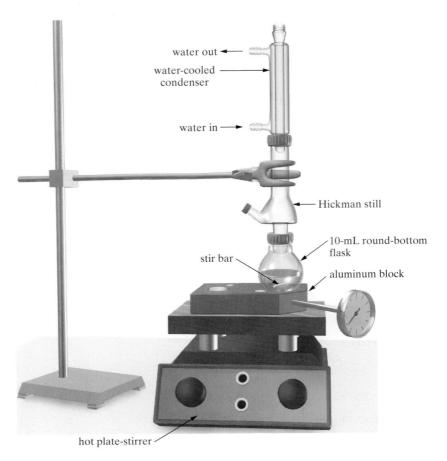

water out

water-cooled
condenser

water in

Hickman still

10-mL round-bottom
flask

stir bar

aluminum block

hot plate-stirrer

Figure C10 Apparatus for internal steam distillation

collecting container. As necessary, use a long Pasteur pipet to add water through the condenser to replace that lost during distillation. Discontinue heating when the fresh distillate appears to contain no more of the organic component *or* after a designated volume of liquid has been collected. When distillation is complete, the distillate should no longer be cloudy or contain droplets of organic liquid. If droplets of the organic component have collected on the inner walls of the Hickman still, use a Pasteur pipet to rinse the walls into the well with some of the distillate (or an appropriate solvent), and transfer the contents of the well to the collecting container.

Separate the organic liquid from the distillate by extraction [OP-15] with a suitable solvent. To improve recovery of the organic liquid, you can saturate the distillate with sodium chloride or another salt to "salt out" [OP-15b] the organic liquid before extracting it.

Summary

1 Assemble apparatus for internal steam distillation.
2 Add organic mixture and water to boiling flask.
3 Turn on stirrer (if used) and condenser water.
4 Distill until distillate is clear, adding water as necessary.
5 Separate organic liquid from distillate.
6 Disassemble and clean apparatus.

Column Chromatography

If you touch the tip of a felt-tip pen to a piece of absorbent paper, such as a coffee filter, and then slowly drip isopropyl rubbing alcohol onto the spot with a medicine dropper, the spot will spread out and separate into rings of different color—the dyes of which the ink is composed. This is a simple example of *chromatography*, the separation of a mixture by distributing its components between two phases. The *stationary phase* (the coffee filter in this example) remains fixed in place, while the *mobile phase* (the rubbing alcohol) flows through it, carrying components of the mixture along with it. The stationary phase acts as a "brake" on most components of a mixture, holding them back so that they move along more slowly than the mobile phase itself. Because of differences in such factors as the solubility of the components in the mobile phase and the strength of their interactions with the stationary phase, some components move faster than others, and the components therefore become separated from one another.

Different types of chromatography can be classified according to the physical states of the mobile and stationary phases. In *liquid-solid* chromatography, which is applied in column chromatography [OP-18] and thin-layer chromatography [OP-19], a liquid mobile phase filters down or creeps up through the solid phase, which may be cellulose, silica gel, alumina, or some other *adsorbent*. The adsorbent is a finely divided solid that attracts solute molecules from the mobile phase onto its surface. In *liquid-liquid chromatography*, which is used in high-performance liquid chromatography [OP-35], the mobile phase is usually an organic solvent and the stationary phase can be a high-boiling liquid that is adsorbed by or chemically bonded to a solid *support*. In *gas-liquid chromatography*—the most common type of gas chromatography [OP-34]—the mobile phase is a gas that passes through a hollow or packed column containing a high-boiling liquid on a solid support. Gas chromatography (GC) and high-performance liquid chromatography (HPLC) are instrumental methods that are used primarily for analyzing mixtures, so they will be discussed in the "Instrumental Analysis" section of the operations.

a. Liquid-Solid Column Chromatography

Principles and Applications

The usual stationary phase for liquid-solid column chromatography is a finely divided solid adsorbent, which is packed into a glass tube called the *column*. The mixture to be separated (the *sample*) is placed on top of the column and *eluted*—washed down the column—by the mobile phase, which is a liquid solvent or solvent mixture. Different components of the sample are attracted to the surface of the adsorbent more or less strongly, depending on their polarity and other structural features. The more strongly a component is attracted to the adsorbent, the more slowly it will move down the column. So as the mobile phase—also called the *eluent*—filters down through the adsorbent, the components of the sample spread out to form separate bands of solute, some passing down the column rapidly and others lagging behind.

Carvone and limonene are major constituents of spearmint oil.

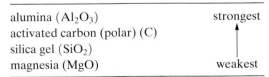

limonene   carvone

Hexanes, a mixture of isomeric six-carbon alkanes, is often used in place of pure hexane.

For example, consider a separation of carvone and limonene on a silica gel adsorbent using hexane as the eluent. At any given time, a molecule of one component will either be adsorbed on the silica gel stationary phase or dissolved in the mobile phase. While it is adsorbed, the molecule will stay put; while dissolved, it will move down the column with the eluent. Molecules with polar functional groups are attracted to polar adsorbents such as silica gel and are relatively insoluble in nonpolar solvents such as hexane. So a molecule of carvone, with its polar carbonyl group, tends to spend more time adsorbed on the silica than dissolved in the hexane. It will therefore pass down the column very slowly with this solvent. On the other hand, a nonpolar molecule of limonene is quite soluble in hexane and only weakly attracted to silica gel, so it will spend less time sitting still and more time moving than will a carvone molecule. As a result, limonene molecules pass down the column rapidly and are soon separated from the slow-moving carvone molecules.

The separation attained by column chromatography depends on a number of factors, including the nature of the components in the mixture, the quantity and kind of adsorbent used, and the polarity of the mobile phase. The lists that follow show how strongly different functional groups are attracted to polar adsorbents and how strongly different adsorbents attract polar molecules. Table C4 shows how different eluents compare in their ability to elute different components.

Approximate strength of adsorption of different functional groups on polar adsorbents

COOH	strongest
OH	
NH_2	
SH	
CHO	
C=O	
COOR	
OR	
C=C	
Cl, Br, I	weakest

Common chromatography adsorbents in approximate order of adsorbent strength

alumina (Al_2O_3)	strongest
activated carbon (polar) (C)	
silica gel (SiO_2)	
magnesia (MgO)	weakest

Note: Adsorbent strength varies with grade, particle size, and other factors

Experimental Considerations

Adsorbents. A number of different adsorbents are used for column chromatography, but alumina and silica gel are the most popular. Adsorbents are available in a wide variety of activity grades and particle size ranges; alumina can be obtained in acidic, basic, and neutral forms as well. The *activity* of an adsorbent is a measure of its attraction for solute molecules, the most active grade of a given adsorbent being one from which all water has been removed. The most active grade may not be the best for a given application, since too active an adsorbent may catalyze a reaction or cause bands to move down the column too slowly. Less active grades of alumina, for example, are prepared by adding different amounts of water to the most active grade (see Table C3). Since all polar adsorbents are deactivated by water, it is important to keep their containers tightly closed and to minimize their exposure to atmospheric moisture. Some samples should not be separated on certain kinds of adsorbents. For example, basic alumina would be a poor choice to separate a mixture containing aldehydes or ketones, which might undergo aldol reactions on the column. Silica gel, which is less active than alumina, is a good all-purpose adsorbent that can be used with most organic compounds.

The amount of adsorbent required for a given application depends on the sample size and the difficulty of the separation. For most separations, the mass of adsorbent should be about 20–50 times the mass of the substances being separated. The more difficult the separation, the more adsorbent is needed.

Table C3 Alumina activity grades (Brockmann scale)

Grade	Mass % water
I	0
II	3
III	6
IV	10
V	15

Eluents. In a column chromatography separation, the eluent acts primarily as a solvent to differentially remove molecules of solute from the surface of the adsorbent. In some cases, polar solvent molecules will also *displace* solute molecules from the adsorbent by becoming adsorbed themselves. If the solvent is too strongly adsorbed, the components of a mixture will spend most of their time in the mobile phase and will not separate efficiently. For this reason, it is usually best to start with a solvent of low polarity, and then (if necessary) increase the polarity gradually to elute the more strongly adsorbed components. Table C4 lists a series of common chromatographic solvents in order of increasing eluting power from alumina and silica gel. Such a listing is called an *eluotropic series.*

Table C4 Eluotropic series for alumina and silica gel

Alumina	Silica gel
pentane	cyclohexane
petroleum ether	petroleum ether
hexane	pentane
cyclohexane	trichloromethane
diethyl ether	diethyl ether
trichloromethane	ethyl acetate
dichloromethane	ethanol
ethyl acetate	water
2-propanol	acetone
ethanol	acetic acid
methanol	methanol
acetic acid	

Elution Techniques.   Many chromatographic separations cannot be performed efficiently with a single solvent, so several solvents or solvent mixtures are used in sequence, starting with the weaker eluents—those near the top of the eluotropic series for the adsorbent being used. Such eluents will wash down only the most weakly adsorbed components, while strongly adsorbed solutes remain near the top of the column. The remaining solute bands can then be washed off the column by more powerful eluents.

In practice, it is best to change eluents gradually by using solvent mixtures of varying composition, rather than to change directly from one solvent to another. In *stepwise elution*, the strength of the eluting solvent is changed in stages by adding varying amounts of a stronger eluent to a weaker one. The proportion of the stronger eluent is increased more or less exponentially. For example, 5% dichloromethane in hexane may be followed by 15% and 50% mixtures of these solvents. According to one rule of thumb, the eluent composition should be changed after approximately three column volumes of the previous eluent have passed through. For example, if the packed volume of the adsorbent is 1.7 mL, the eluent composition should be changed with every 5 mL or so of eluent.

Columns.   There are many different kinds of chromatography columns, ranging from a simple glass tube with a constriction at one end to an elaborate column with a detachable base and a porous plate to support the packing. For simple microscale separations, a $5\frac{3}{4}$-inch Pasteur pipet packed with a suitable adsorbent makes an adequate chromatography column. Such a column provides the best separation when used with no more than 0.1 g of sample, but it can be used to remove small amounts of impurities from larger quantities of sample. Figure C11 shows a packed chromatography column constructed from a Pasteur pipet. A 5-mL or 10-mL microburet may also be used to construct a microscale column. A microburet with an integral funnel top works well because the funnel top functions as a solvent reservoir so that fresh eluent can be added less frequently. A funnel top may also make it easier to pack the column.

Flow Rate.   The rate of eluent flow through the column should be slow enough so that the solute can attain equilibrium, but not so slow that the solute bands broaden significantly by diffusion. Difficult separations require the slowest rates. For a microscale column, a flow rate of about 5–10 drops per minute should be suitable. The flow rate can be varied by adjusting a stopcock or screw clamp at the outlet of the column (if it has either) or by varying the *solvent head*—the depth of the eluent layer above the adsorbent.

For a column constructed using a Pasteur pipet, the flow rate can be increased using a pipet pump with a quick release lever. This method can decrease the separation efficiency, however, and removing the pump during elution without proper pressure release may damage the column, so it should be used only when absolutely necessary. With its plunger fully extended, attach the pipet pump firmly to the top of the column and rotate the thumbwheel until a suitable flow rate is achieved; then periodically rotate it just enough to keep the flow rate relatively constant. When more eluent must be added, or if the plunger is down as far as it will go, hold the quick release lever down as you carefully remove the pipet pump. After adding eluent, pull out the plunger and replace the pipet pump on top of the column.

Packing the Column. To achieve good separation with a chromatographic column, it must be packed properly. The packing must be uniform,
without air bubbles or channels, and its surface must be even and horizontal. Columns using alumina can be packed either with the dry adsorbent or
by pouring the adsorbent through a layer of solvent. Columns using silica
gel are usually packed with a slurry containing the adsorbent suspended in
a solvent. Pasteur-pipet columns cannot easily be slurry packed, so they are
most often used with alumina as the adsorbent. Once the adsorbent is moistened with solvent, it must be kept covered with the solvent at all times; allowing it to dry out creates channels that lead to uneven bands and poor
separation.

General Directions for Column Chromatography

Equipment and Supplies

$5\frac{3}{4}$-inch Pasteur pipet *or* 5–10-mL microburet
column-packing solvent
$5\frac{3}{4}$-inch Pasteur pipet with rubber bulb
plastic weighing dish
glass wool
pencil
clean sand
adsorbent
collectors (vials, small flasks, etc.)
eluent(s)

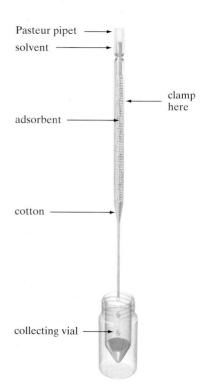

Pasteur pipet
solvent
clamp here
adsorbent
cotton
collecting vial

Figure C11 Apparatus for column chromatography using a Pasteur-pipet
column

Preparing the Column. Use a wooden applicator stick or a glass rod to
push a small plug of glass wool down the body of the pipet or microburet to
the point at which it begins to narrow. Be careful not to tamp it down so
tightly as to restrict the solvent flow. Clamp the column vertically to a ring
stand. Measure the approximate amount of adsorbent you will need (usually
1.5–2.0 g for a Pasteur pipet) and keep it in a tightly closed container. Number and weigh the collectors you will be using to collect the eluent fractions.
Then pack the column by one of the methods that follow. Method **1** should
be used only for packing Pasteur-pipet columns with alumina. Method **2**
gives a more evenly packed column when alumina is used as the adsorbent;
it is not satisfactory for silica gel. If you use Method **2** with a Pasteur-pipet
column, you should be prepared to add your sample immediately after filling the column. Method **3** is suitable for packing a microburet (or a comparable column) with silica gel.

1 *Packing a Pasteur-Pipet Column Using No Solvent.* Measure about 50 mg
   of fine sand into a small plastic weighing dish. Bend the dish to form a
   spout and add the sand through the top of a clamped column, while gently tapping it with the eraser end of a pencil, to form a thin (2–3-mm)
   layer on top of the glass wool. The top of the sand layer should be even
   and horizontal. Then use the weighing dish to add the adsorbent *slowly*
   through the top of the column, while continuously tapping the column
   near its middle with the eraser end of a pencil, until the specified amount

of adsorbent has been added *or* the adsorbent level is about 1 cm below the upper constriction in the Pasteur pipet. The surface of the adsorbent should be as even and horizontal as possible; if necessary, tap gently near the surface to make it so. If desired, you can sprinkle a thin, even layer of clean sand on top to protect the adsorbent layer.

2a *Packing a Pasteur-Pipet Column Using a Solvent.* This method requires constant attention because the adsorbent layer must be completely covered with solvent at all times. Measure about 50 mg of fine sand into a small plastic weighing dish. Bend the dish to form a spout and add the sand through the top of the clamped column, while gently tapping it near its center with the eraser end of a pencil, to form a thin (2–3-mm) layer on top of the glass wool. Put a small flask or beaker under the column and use a Pasteur pipet to fill it about two-thirds full with the least polar eluent to be used in the chromatography (or another suitable solvent). Use the weighing dish to add the adsorbent *slowly* to the top of the column as the solvent drains, while continuously tapping the column near its center with the eraser end of a pencil. If the solvent is draining too fast, stop the flow temporarily by holding your fingertip on top of the column, and allow the adsorbent to settle before adding the next portion of adsorbent. Continue adding adsorbent until the specified amount has been added *or* until the adsorbent level is about 1 cm below the upper constriction in the column. The surface of the adsorbent should be as even and horizontal as possible; if necessary, tap gently near the surface to make it so. Be ready to add your sample *immediately* when the solvent surface drops to the top of the adsorbent layer. (If you are not ready, stopper the column with a size 000 cork to stop the solvent flow.)

2b *Packing a Microburet Column Using a Solvent.* Close the stopcock of the microburet and put a small flask or beaker under it. Fill the column about two-thirds full with the least polar eluent to be used in the chromatography (or another suitable solvent). Use a small plastic weighing dish, bent to form a spout, to add enough fine sand through the top of the column to form a 5-mm layer on top of the glass wool. As the sand filters down through the solvent, tap the column gently and continuously near its center with the eraser end of a pencil so that the sand layer is uniform and level. Use a plastic weighing dish to *slowly* add adsorbent through the top of the column, with tapping, until it forms a layer about 2 cm thick at the bottom of the column. Open the stopcock and add the rest of the alumina as the solvent drains slowly, tapping constantly to help settle and pack the adsorbent. There should be enough solvent in the column so that the solvent level is well above the adsorbent level at all times; if necessary, add more solvent. When all of the adsorbent has been added, close the stopcock. The surface of the adsorbent should be as even and horizontal as possible; if necessary, tap gently near the surface to make it so. Use a Pasteur pipet containing the solvent to rinse down any adsorbent that adheres to the sides of the column. Add enough clean sand to form a protective layer about 0.5 cm thick on top of the adsorbent; the sand surface should also be level. Open the stopcock until the solvent surface drops to within 1–2 cm of the sand surface, and then close the stopcock and stopper the column. Keep the adsorbent covered with solvent at all times.

3 *Slurry Packing a Microburet Column.* Close the stopcock of the microburet and put a small flask or beaker under it. Fill the column about half full with the least polar eluting solvent to be used in the separation, or a solvent recommended in the experimental procedure. Use a small plastic weighing dish, bent to form a spout, to add enough fine sand through the top of the column to form a 5-mm layer on top of the glass wool. As the sand filters down through the solvent, tap the column gently and continuously near its center with the eraser end of a pencil so that the sand layer is uniform and level. In a small beaker, mix the measured amount of silica gel (or another suitable adsorbent) thoroughly with enough of the column-packing solvent to make a fairly thick, but pourable, slurry. Pour a little of this slurry into the column, with tapping, so that the adsorbent gently filters down to form a layer about 2 cm thick at the bottom. (Use a small funnel, if necessary.) Then open the stopcock so that the solvent drains slowly as you add the rest of the slurry, tapping constantly to help settle and pack the adsorbent. If the slurry becomes too thick to pour, add more solvent to it. There should be enough solvent in the column so that the solvent level is always well above the adsorbent level at all times; if necessary, add more solvent. When all of the adsorbent has been added, close the stopcock. The surface of the adsorbent should be as even and horizontal as possible, so continue tapping until it has settled. Use a Pasteur pipet containing the solvent to rinse down any adsorbent that adheres to the sides of the column. Add enough clean sand to form a protective layer about 0.5 cm thick on top of the adsorbent; the sand surface should also be level. Open the stopcock until the solvent surface drops to within 1–2 cm of the sand surface, and then close the stopcock and stopper the column. Keep the adsorbent covered with solvent at all times.

Separating the Sample.   If the sample is a solid, dissolve it in a minimal amount of a nonpolar solvent; use liquid samples without dilution. Be sure that there is a suitable collector under the column. If the column was packed without solvent, use a Pasteur pipet to add enough of the initial eluent to moisten all of the adsorbent and cover the adsorbent layer to a depth of about 1 cm. If you are using a microburet, open the stopcock. When the solvent surface *just* drops to the top of the adsorbent layer (or to the top of a sand layer protecting the adsorbent), add the sample with a Pasteur pipet. Quickly rinse this pipet with a very small amount of the initial eluent and use it to rinse the sides of the column. (If you are using a microburet, you can close the stopcock while you add and rinse down the sample.) When the liquid surface has fallen to the top of the adsorbent (or sand) layer, use a Pasteur pipet to add enough of the initial eluent to nearly fill the column. Be careful not to disturb the surface of the adsorbent as you add it. Continue adding eluent as necessary to keep the solvent level near the top of the column throughout the elution. If you need to change eluents, allow the previous eluent to drain to the level of the adsorbent (or sand) before adding the next one.

If the components are colored or can be observed on the column by some visualization method (such as irradiation with ultraviolet light), change collectors each time a new band of solute begins to come off the column *and* when it has about disappeared from the column. If two or more

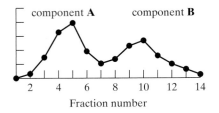

Figure C12 Elution curve

bands overlap, collect the overlapping regions in separate collectors to avoid contaminating the purer fractions. If the components are not visible and the procedure does not specify the fraction volumes, collect equal-volume fractions (~5–10 mL each when using a microburet column) in tared collectors.

Stop the elution when a designated volume of eluent has been added or when the desired sample bands have been collected. Unless directed otherwise, evaporate [OP-16] the solvent from any fractions that contain the desired component(s). If you collected equal-volume fractions, evaporate them all, weigh the tared collectors and their contents, and plot the mass of each residue versus the fraction number to obtain an elution curve such as that illustrated in Figure C12. From the elution curve, you should be able to identify separate components and decide which fractions can be combined.

To remove the contents of the column, let it dry completely, and then tap it as you hold it over a beaker. If that doesn't work, attach a rubber hose from an air line to the column at its narrow end (be careful you don't break the tip of a Pasteur pipet), hold the open end over a beaker, and *slowly* open the air valve until the air blows out the adsorbent. To remove the glass-wool plug, twirl the end of an applicator stick in the glass wool until it snags the fibers, and then pull it out. Dispose of any unused or recovered solvents as directed by your instructor.

Summary

1 Clamp microscale chromatography column vertically.
2 Pack column with adsorbent, put collector in place.
   IF column was packed using a solvent, GO TO 4.
3 Cover adsorbent with eluent.
4 Drain column to top of adsorbent, add sample, rinse, drain again.
5 Add eluent to elute sample, keeping eluent level nearly constant.
6 To change eluents, drain current eluent to top of adsorbent, add next eluent.
   IF components are visible, GO TO 7.
   IF components are not visible, GO TO 8.
7 Change collectors when new band starts or ends and where bands overlap.
   GO TO 9.
8 Change collectors after a predetermined volume has been collected.
9 Stop elution after last fraction has been collected.
10 Evaporate and weigh appropriate fractions.
11 Disassemble and clean apparatus, dispose of solvents as directed.

b. Flash Chromatography

Flash chromatography, a variation of liquid-solid column chromatography, uses a single elution solvent and takes less time than the standard method. Pressurized nitrogen, air, or another gas is applied to the top of the column to force eluent through the adsorbent, which is usually a finely divided silica gel with a particle size of ~40–63 μm. Since only one eluent is used, it must be selected carefully to ensure good separation. Separation is poor if

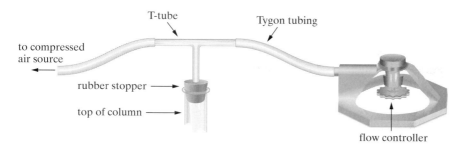

Figure C13 Flow controller for flash chromatography

the sample size is too large; the procedure described next works best with 0.10 g of sample or less, although it may be suitable for larger amounts of an easily separated sample.

Commercial flash chromatography systems may require expensive columns, pumps, and flow controllers, so they are rarely available for use in undergraduate laboratories. An inexpensive flash chromatography system can be constructed using the "homemade" flow controller shown in Figure C13. The flow controller is assembled by inserting the bottom of a small plastic T-tube into a one-hole rubber stopper that fits into the top the column, attaching two lengths of $\frac{3}{8}$-inch i.d. (inner diameter) Tygon tubing to the straight ends of the T, and securing them with copper wires. One tube is attached to the gas inlet on a Bunsen burner base (the barrel can be removed), and the other to a source of clean, dry, compressed air.

To carry out a separation using this flow controller, follow the usual directions for packing a column and separating the sample, with the following modifications. Obtain a 10-mL buret with a funnel top that will accommodate a #3 rubber stopper (other columns may also be suitable) and fill it about one-third full with the appropriate adsorbent, using the eluent as the column-packing solvent. Drain the solvent to the top of the adsorbent layer, introduce the sample with a 9-inch Pasteur pipet, and fill the column nearly to the top with eluent. Insert the stopper of the flow controller into the top of the column with a firm twist to keep it from popping out (it should stay in place at the desired pressure). Open the needle valve on the burner base; then open the column's stopcock with a collector in place. Carefully turn on the air valve to pressurize the system, and adjust the needle valve so that the eluent level decreases at a rate of ~2 cm per minute. Collect and evaporate the fractions by the usual methods.

*This flash chromatography apparatus is described in J. Chem. Educ. **1992**, 69, 939.*

c. Reversed-Phase Column Chromatography

Adsorbents are polar solids that attract polar compounds more strongly than nonpolar ones, so nonpolar solutes are eluted from adsorbent-packed columns more rapidly than polar ones. Another kind of stationary phase can be prepared by coating particles of silica gel with a high-boiling nonpolar liquid. With such a stationary phase, components of the sample are partitioned between the liquid mobile phase and the liquid layer of the stationary phase, where to *partition* a solute means to distribute it between two phases. If the mobile phase is more polar than the stationary phase, the usual order of elution will be reversed; that is, polar compounds will be eluted before nonpolar ones. This general method is called *reversed-phase chromatography*. The mobile phases for reversed-phase column chromatography are usually polar solvents such as water, methanol, and acetonitrile, or mixtures of these solvents. Stationary phases can be prepared by coating a specially treated (silanized) silica gel with a nonpolar liquid phase, such as a hydrocarbon or silicone. Bonded liquid phases, such as the ones described for HPLC [OP-35], can also be used.

To prepare a column using coated silica gel, the liquid mobile and stationary phases are shaken together in a separatory funnel to saturate each phase with the other and the layers are separated. The stationary phase is then stirred with the silica gel and the coated support is made into a slurry with the saturated mobile phase. The column is slurry packed, usually with stirring to make it more uniform and to remove air bubbles. A separation is carried out by eluting with the saturated mobile phase, essentially as previously described for normal-phase column chromatography.

OPERATION 19

Thin-Layer Chromatography

Principles and Applications

Like column chromatography, thin-layer chromatography (TLC) utilizes a solid adsorbent as the stationary phase and a liquid solvent as the mobile phase, but the mobile phase creeps up the adsorbent layer by capillary action rather than filtering down through it by gravity. A *TLC plate* consists of a thin layer of the adsorbent on an appropriate *backing* (solid support) made of plastic, aluminum, or glass. The sample, or several samples, are dissolved in a suitable solvent and applied near the lower edge of the TLC plate as small spots. Ordinarily, the plate is also spotted with a selection of standard solutions for comparison. The plate is then *developed* by immersing its lower edge in a suitable mobile phase, the *developing solvent*. As this solvent moves up the adsorbent layer, it carries with it the components of each spot, which are separated as described in OP-18 for column chromatography.

Although TLC is not useful for separating large amounts of material, it is faster than column chromatography and it can be carried out with very small sample volumes so that little is wasted. TLC provides better separation than the related technique of paper chromatography [OP-20], and it can be applied to a wider range of organic compounds. TLC is categorized under separation operations in this book, but it can also be used for qualitative and quantitative analysis of organic compounds. For example, in Experiment 11 TLC is used to identify an unknown ketone, and in Experiment 15 it is used to separate and identify the components of an unknown analgesic drug.

Other applications of TLC include the following:

- To monitor the course of the reaction and determine the purity of its product by comparing the relative amounts of products, reactants, and by-products on successive chromatograms
- To determine the best solvent for a column chromatography separation (see "Choosing a Developing Solvent")
- To determine the composition of each fraction from a column chromatography separation so that fractions containing the same component can be detected and combined
- To determine whether a substance purified by recrystallization or another method still contains appreciable amounts of impurities

Experimental Considerations

Adsorbents. The most commonly used adsorbents for TLC are silica gel, alumina, and cellulose. The adsorbent is more finely divided than that used in column chromatography, and it is provided with a *binder*, such as poly-acrylic acid, to make it stick to the backing. It may also contain a *fluorescent indicator*, which makes most spots visible under ultraviolet light.

TLC Plates. Small "do-it-yourself" TLC plates can be prepared by dipping a glass microscope slide into a slurry of the adsorbent and binder in a suitable solvent, and allowing the solvent to evaporate. Such plates may give inconsistent results because of variations in the thickness of the adsorbent layer. More uniform TLC plates measuring 20×20 cm or larger are commercially available with a wide variety of adsorbents, backings, and layer thicknesses. For example, a typical TLC plate suitable for use in undergraduate laboratories has a flexible plastic backing coated with a 200-µm-thick layer of silica gel mixed with a binder, and possibly a fluorescent indicator as well. The adsorbent layer of a TLC plate is easily damaged, so it is important to avoid unnecessary contact with its coated surface and to protect the plate from foreign materials. A TLC adsorbent will pick up moisture when exposed to the atmosphere, making it less active; the adsorbent can be *activated* by heating the plate in a 110°C oven for an hour or so.

Spotting. A TLC plate is prepared for development by applying solutions of the sample(s) to be analyzed, and any reference standards, as small spots; this process is called *spotting*. The sample is dissolved in a suitable solvent to make an approximately 1% (mass/volume) solution. The solvent should be relatively nonpolar and have a boiling point in the 50–100°C range. Column chromatography fractions and other solutions can often be used as is, if the solute is present at a concentration in the 0.2–2.0% range.

It is best to wear thin disposable gloves while spotting a TLC plate because, if you touch the surface of the adsorbent with your bare fingers, your fingerprints may hinder development or obscure developed spots. Position the spots accurately because incorrectly placed spots may run into one another or onto the edge of the adsorbent layer. This can be done with a transparent plastic ruler supported just above the surface of the plate so that it does not touch the adsorbent. Mark the starting line with a pencil on both edges, about 1.5 cm from the bottom of the plate (or 1.0 cm for a microscope-slide plate), and position the spots along the starting line at least 1.5 cm from each edge of the plate and 1.0 cm from each other. Thus a 10×10-cm TLC plate can accommodate up to eight spots, as shown in Figure C14.

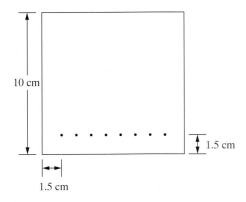

Figure C14 Spotted 10 × 10-cm TLC plate

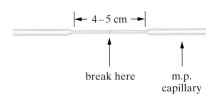

Figure C15 Drawing a capillary micropipet

You can practice your spotting technique on a used or damaged TLC plate.

Large, diffuse spots spread out too much for accurate results, so each spot should be as small and concentrated as possible. The spots are best applied with a microliter syringe or a capillary micropipet. Capillary micropipets such as Drummond Microcaps are commercially available, but suitable micropipets can also be prepared by heating an open-ended melting-point capillary in the middle over a small flame and drawing it out to form a fine capillary about 4–5 cm long. This tube is allowed to cool, and is then scored and snapped apart in the middle to form two micropipets (see Figure C15).

To spot a TLC plate with a micropipet, dip the narrow tip into the solution to draw in a small amount of liquid. Then gently touch the tip to the surface of the TLC plate, at the proper location, for only an instant. Be careful not to dig a hole in the adsorbent surface because this will obstruct solvent flow and distort the chromatogram. Make several successive applications at each location, letting the solvent dry each time, to form a spot 1–3 mm in diameter. Too much solution can result in "tailing" (a zone of diffuse solute following the spot), "bearding" (a zone of diffuse solute preceding the spot), and overlapping of components. Too little solution makes it difficult to detect some of the components. So when there is enough space on the TLC plate, it is a good idea to make one, two, and three applications of a sample at three separate locations and use the best-looking spot (after development) for the R_f calculation, described later. Capillary micropipets can be reused a few times, but a different micropipet should be used for each different solution to avoid cross-contamination.

To spot a TLC plate with a microliter syringe, deliver about 1 μL of solution with each application and make two to three applications for each spot, letting the spot dry between applications. Take care not to touch the adsorbent surface with the syringe needle. Before you fill the syringe with a different solution, rinse it with a suitable solvent and remove excess solvent by pumping the plunger gently a few times. After the last application, clean the syringe with more solvent, remove the plunger, and let it dry.

Choosing a Developing Solvent. Solvents that are suitable eluents for column chromatography are equally suitable as TLC developing solvents; the eluotropic series of Table C4 [OP-18] may help you choose a solvent for a particular application. A quick way to find a suitable solvent is to spot a

TLC plate with the sample, applying as many spots as you have solvents to test (you can use a grid pattern with the spots 1.5–2.0 cm apart), and then apply enough solvent directly to each spot to form a circle of solvent 1–2 cm in diameter. Mark the circumference of each circle before the solvent dries. A solvent whose chromatogram (after visualization) shows well-separated rings, with the outermost ring about 50–75% of the distance from the center to the solvent front, should be satisfactory. It is preferable to use the least polar solvent that gives good separation. Hexanes, toluene, dichloromethane, and methanol or ethanol (alone or in binary combinations) are suitable for most separations. If no single solvent is suitable, choose two mutually miscible solvents whose outermost rings bracket the 50–75% range (one with too little solute migration, the other with too much), and test them in varying proportions.

Development. TLC plates are developed by placing them in a *developing chamber* that contains the developing solvent. A paper wick can be used to help saturate the air in the developing chamber with solvent vapors, which improves reproducibility and increases the rate of development. The developing chamber can be a jar with a screw-cap lid, a beaker covered with plastic film, or a commercial developing tank with a lid. You should use the smallest available container that will accommodate the TLC plate, since a larger container takes longer to fill with solvent vapors. Development should be carried out in a place away from direct sunlight or drafts to prevent temperature gradients. It may take 20 minutes or more to develop a 10 × 10-cm TLC plate; a microscope slide plate can often be developed in 5–10 minutes. Development should be stopped when the *solvent front*—the boundary between the wet and dry parts of the adsorbent—is within 5 mm or so of the top of the plate. If development is allowed to continue until the solvent front reaches the top edge of the plate, the spots will spread out by diffusion. The solvent front should be marked with a pencil before the plate has had time to dry.

Visualization. If the spots are colored, they can be observed immediately; otherwise, they must be *visualized* (made visible) by some method. The simplest way to visualize many spots is to observe the TLC plate under ultraviolet light. Fluorescent compounds produce bright spots, and if the adsorbent contains a fluorescent indicator, compounds that quench fluorescence will show up as dark spots on a light background. You should mark the center of each spot immediately with a pencil because the spots will disappear when the ultraviolet light is removed. If a spot is irregular, mark its center of concentration—the midpoint of its most densely shaded region—instead. It is also a good idea to outline each spot with your pencil.

Another general visualization procedure is to place the dry plate in a closed chamber, such as a wide-mouthed jar with a screw-cap lid, add a few crystals of iodine, and heat the chamber gently (use a steam bath or a hot plate on a very low heat setting) so that the iodine vapors sublime onto the adsorbent. Most organic compounds, except saturated hydrocarbons and halides, form brown spots with iodine vapor. Unsaturated compounds may show up as light spots against the dark background. The iodine color fades in time, so mark the spots shortly after visualization.

Spots can also be made visible by applying a visualizing reagent to the TLC plate. The visualizing reagent can be applied by spraying it onto the

*See J. Chem. Educ. **1985**, 62, 156 for an alternative method of visualizing TLC plates with iodine.*

See J. Chem. Educ. **1996**, *73, 358 for a description of the cotton-ball procedure.*

plate, dipping the plate into the reagent, or wiping the plate with a cotton ball that has been saturated with the reagent. A solution of phosphomolybdic acid in ethanol can be used to visualize most organic compounds; the spots appear after the plate is heated with a heat gun or in an oven. Other visualizing reagents are used for specific classes of compounds, such as ninhydrin reagent for amino acids and 2,4-dinitrophenylhydrazine reagent for aldehydes and ketones. All spraying should be done under a hood in a "spray box," which can be made from a large cardboard box with the top and one side removed. A thin spray is applied to the TLC plate from about 2 feet away, using an aerosol can or spraying bottle containing the visualizing reagent. Large plates are sprayed by crisscrossing them with horizontal and vertical passes.

Analysis.   The ratio of the distance a component travels up a TLC plate to the distance the solvent travels is called its R_f *value.*

$$R_f = \frac{\text{distance traveled by spot}}{\text{distance traveled by solvent}}$$

The R_f (ratio to front) value of a spot is determined by measuring the distance from the starting line to the center of the spot and dividing that value by the distance from the starting line to the solvent front, both distances being measured along a line extending from the starting point of the spot to its final location. If the spot is irregular, its center of concentration is used instead. With a given mobile and stationary phase, the R_f value of a substance depends on the polarity of its functional groups and other structural features, so it is a physical property of the substance that can be used in its identification. R_f values for some compounds have been reported in the literature, using specified mobile and stationary phases. Reported R_f values can seldom be used to establish the identity of a substance, however, since they depend on a number of factors that are difficult to standardize, such as the sample size, the thickness and activity of the adsorbent, the purity of the solvent, and the temperature of the developing chamber. The only way to be reasonably sure that a TLC unknown is identical to a known compound is to spot a solution of the known compound on the same TLC plate as the unknown. Even then, the identity of the unknown may have to be confirmed by an independent method. An unknown substance or mixture is often analyzed on the same plate with a series of standard solutions, each containing a substance that may be identical to the unknown—or to one of its components, if it is a mixture.

Directions for Preparing Microscope-Slide TLC Plates

Several students should work together so they can use the same slurry. Using detergent and water, clean as many microscope slides as you need TLC plates, and then rinse them with distilled water and 50% aqueous methanol. After a slide has been cleaned, do not touch the surface that will be coated; hold the slide by the edges or at the top. *Under the hood,* measure 100 mL of dichloromethane into a 4-oz. (125-mL) screw-cap jar, add 35 g of Silica Gel G with vigorous stirring or swirling, and shake the capped jar vigorously for about a minute to form a smooth slurry. Stack two clean microscope slides

Take Care! Avoid contact with dichloromethane and do not breathe its vapors.

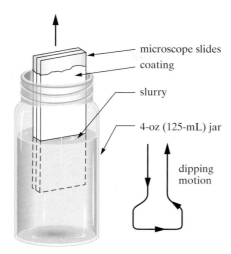

Figure C16 Dipping a pair of stacked microscope slides

back to back, holding them together at the top. Without allowing the slurry to settle (shake it again, if necessary), dip the stacked slides into the slurry for about 2 seconds, using a smooth, unhurried, paddlelike motion (see Figure C16) to coat them uniformly with the adsorbent. Immerse the slides deeply enough so that only the top 1 cm or so remains uncoated. Touch the bottom of the stacked slides to the jar to drain off the excess slurry, and let them air dry for a minute or so to evaporate the solvent. Then separate them and wipe off the excess adsorbent from the edges with a tissue paper. Repeat with more slides, as needed; if the slurry becomes too thick, dilute it with dichloromethane. Activate the coated slides by heating them in a 110°C oven for 15 minutes. Slides having streaks, lumps, or thin spots in the coating should be wiped clean and redipped. If the coating is too fragile, try adding some methanol (up to one-third by volume) to the slurry. Coated slides can be stored in a microscope-slide box inside a dessicator.

General Directions for Thin-Layer Chromatography

Equipment and Supplies

TLC plate(s)
developing chamber
paper wick
developing solvent
pencil and ruler
capillary micropipets or microliter syringe
solutions to be spotted
visualizer (sprayer, iodine chamber, UV lamp, etc.)

Obtain a beaker, screw-cap jar, or another container large enough to hold the TLC plate. A 4-oz screw-cap jar is suitable for a microscope-slide TLC plate; a 400-mL beaker will hold a 6.7 × 10-cm plate; and a 1-L beaker will

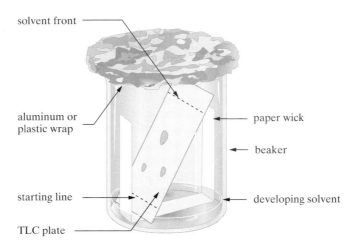

Figure C17 Development of a TLC plate

A wick should not be necessary if the developing chamber is allowed to stand for an hour or more after the solid has been added.

hold a 10 × 10-cm plate. If necessary, prepare and insert a paper wick made from filter paper or chromatography paper. The wick can be prepared by cutting a rectangular strip of paper 3–5 cm wide and long enough to extend in a "U" down one side of the developing chamber, along the bottom, and up the opposite side. Pour in enough developing solvent to form a liquid layer about 1 cm deep on the bottom of the developing chamber. Cover the chamber with a screw cap, a square of plastic food wrap, or another suitable closure. Tip the chamber and slosh the solvent around to soak the wick with developing solvent; then put it in a place where it can sit undisturbed away from drafts and direct sunlight. Let the developing chamber stand for 30 minutes or more to saturate the atmosphere inside the chamber with solvent vapors.

Mark the starting line on a TLC plate and spot it with the solution(s) to be analyzed and any standard solutions required, making spots 1–3 mm in diameter. Except on a microscope-slide TLC plate, which can accommodate up to three evenly spaced spots, the spots should be at least 1.0 cm apart and the outermost spots should be 1.5 cm from the edges of the plate. When the spots are dry, place the TLC plate in the developing chamber, with its spotted end down so that it leans *across* the wick (perpendicular to it) with its top against the glass wall of the chamber. No part of a plastic- or aluminum-backed TLC plate should touch the exposed part of the wick, because solvent can diffuse onto the adsorbent at that point. Cover the developing chamber without delay, and do not move it during development.

Observe the development frequently, and when the solvent front is within about 5 mm of the top of the plate, remove the plate from the developing chamber. Before the solvent evaporates, use a pencil to trace a line along the solvent front and mark the centers (or centers of concentration) of any visible spots. Let the plate dry thoroughly, preferably under a hood. Then visualize the spots, if necessary, by one of the methods described previously. Outline each spot and mark its center (or center of concentration) with a pencil. For each spot, place a ruler along an imaginary line—perpendicular to the starting line—that goes through the center of that spot. Then measure the distance in millimeters (1) from the starting line to the point marking the center (or center of concentration) of the spot, and (2) from the starting line to the solvent front. Calculate the R_f value for each spot. If requested, make a permanent record of the chromatogram by photocopying or photographing it.

The distance from the starting line to the solvent front may be different for different spots.

Summary

1 Put wick and developing solvent in developing chamber, cover, and let stand.
2 Obtain or prepare TLC plate.
3 Spot TLC plate with solutions of the unknown(s) and any standards.
4 Place TLC plate in developing chamber, cover, and observe.
5 Remove TLC plate when solvent front nears top of plate.
6 Mark solvent front and any visible spots, let plate dry.
7 Visualize and mark spots, as needed.
8 Measure R_f values.
9 Clean up, dispose of solvent as directed.

Paper Chromatography

OPERATION **20**

Principles and Applications

Paper chromatography is similar to thin-layer chromatography [OP-19] in practice, but quite different in principle. As for TLC, the paper is spotted with solutions of the sample and standards, and the chromatogram is developed with a suitable mobile phase. Although paper consists mainly of cellulose, the stationary phase is not cellulose itself but the water that is adsorbed by it. Chromatography paper can adsorb up to 22% water, and the developing solvents used contain enough water to keep it saturated. During development, the comparatively nonpolar mobile phase creeps up through the cellulose fibers, partitioning the components of the spots between the bound water and the mobile phase. Paper chromatography thus operates by a liquid-liquid partitioning process rather than by adsorption on the surface of a solid.

Since only polar compounds are appreciably soluble in water, paper chromatography is most frequently used to separate polar substances such as amino acids and carbohydrates. Manufactured chromatography paper is quite uniform, and the activity of cellulose doesn't vary as much as the activity of most TLC adsorbents, so R_f values obtained by paper chromatography may be more reproducible than those from thin-layer chromatography. However, the resolution of spots is often poorer and the development times are usually much longer. Nevertheless, paper chromatography is a useful analytical technique that can accomplish a variety of separations.

Experimental Considerations

Since many of the experimental aspects of paper chromatography are similar or identical to those for thin-layer chromatography, you should read the appropriate parts of OP-19 for additional information about experimental techniques.

Paper.   Various grades of chromatography paper, such as Whatman #1 Chr, are manufactured in rectangular sheets, strips, and other convenient shapes. For good results, the chromatography paper must be kept clean. The benchtop or other surface on which the paper is handled should be covered with a sheet of butcher paper or some other liner. The chromatography paper should be held only by the edges or along the top; alternatively, an extra strip of paper can be left on one or both ends of the chromatography paper, used for handling it, and then cut off just before development. The paper should be cut so that its grain is parallel to the direction of development. For most purposes, the paper should be 10–15 cm high (in the direction of development) and wide enough to accommodate the desired number of spots.

Spotting.   As for thin-layer chromatography, the substance(s) to be analyzed should be dissolved in a suitable solvent, usually at a concentration of about 1% (mass/volume). The starting line should be marked in pencil about 2 cm above the bottom edge; the positions of the spots can also be marked lightly with pencil. The spots tend to spread out more with paper chromatography than with TLC, so they should be applied farther apart— about 1.5–2.0 cm from each other and 2.0 cm from the edges of the paper, as shown in Figure C18. Spotting can be performed by the same techniques as for TLC using a capillary micropipet or syringe (see OP-19), but a round wooden toothpick is adequate for most purposes. During the spotting, nothing should touch the underside of the paper at the point of application. Each spot should be 2–5 mm in diameter. It should be made using several consecutive applications, allowing the spot to dry after each application. Use no more than 10 μL of solution, in all, per spot; too little is better than too much. You can practice your spotting technique on a piece of filter paper before attempting to spot the chromatogram.

Developing Solvents.   As a rule, paper chromatography developing solvents contain water, which is needed to maintain the composition of the

Paper Chromatography

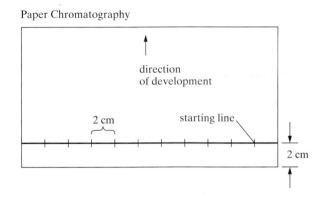

Figure C18 11 × 22-cm chromatography paper marked for spotting

Table C5 Some monophase solvent mixtures for paper chromatography

Solvents	Composition
2-propanol/ammonia/water	$9 : 1 : 2$
1-butanol/acetic acid/water	$12 : 3 : 5$
phenol/water	500 g phenol, 125 mL water
ethyl acetate/1-propanol/water	$14 : 2 : 4$

Note: All solvent ratios are by volume.

aqueous stationary phase on the cellulose. They also contain an organic solvent and—if necessary—one or more additional components to increase the solubility of the water in the organic solvent or provide an acidic or basic medium. A developing solvent for paper chromatography may be prepared by saturating the organic solvent(s) with water in a separatory funnel, separating the two phases, and using the organic phase for development. Alternatively, a monophase (single-phase) mixture with essentially the same composition as the organic phase of the saturated mixture can be used. Some typical monophase solvent mixtures are listed in Table C5. Most solvent mixtures should be made up fresh each time they are used and not kept for more than a day or two.

Development. Narrow paper strips accommodating two or more spots can be developed in test tubes, bottles, cylinders, or Erlenmeyer flasks. A wider sheet can be rolled into a cylinder and developed in a beaker. A typical paper chromatogram is developed in much the same way as for TLC, but more time is required to saturate the developing chamber with solvent vapors.

Visualization. Visualization of spots by ultraviolet light is quite useful in paper chromatography, since paper fluoresces dimly in a dark room and many organic compounds will quench its fluorescence, producing dark spots on a light background. Paper chromatograms can also be visualized by spraying them or by dipping them into a solution of a suitable visualizing reagent.

Analysis. A substance responsible for a spot on a developed chromatogram is characterized by its R_f value, which is determined as illustrated in Figure C19. Spot migration distances are customarily measured from the starting line to the *front* of each spot, rather than to its center as for TLC. It is usually necessary to run one or more standards along with an unknown to identify the unknown. Whenever possible, the solvent and concentration should be the same for the standard as for the unknown.

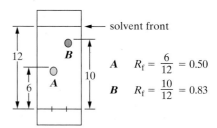

A $R_f = \dfrac{6}{12} = 0.50$

B $R_f = \dfrac{10}{12} = 0.83$

$$R_f = \frac{\text{Distance to leading edge of spot}}{\text{Distance to solvent front}}$$

Figure C19 Measuring R_f values on a paper chromatogram

General Directions for Paper Chromatography

Equipment and Supplies

    chromatography paper
    developing chamber
    developing solvent
    pencil and ruler
    capillary micropipets (or toothpicks, etc.)
    solutions to be spotted
    visualizer (sprayer, UV lamp, etc.)

The following procedure can be used when up to 13 spots are to be applied. For just a few spots, a strip of chromatography paper can be spotted, folded in the middle (or hung from a wire embedded in a cork), and developed in a test tube, a jar, or another appropriate container.

Add enough developing solvent to a 600-mL beaker (or another suitable developing chamber) to provide a liquid layer about 1 cm deep. Cover the chamber tightly with plastic food wrap (or cap it), slosh the solvent around in it for about 30 seconds, and then put it in a place where it can sit undisturbed away from drafts and direct sunlight. Allow sufficient time (usually an hour or more) for the solvent to saturate the developing chamber. While the developing chamber is equilibrating, obtain a sheet of chromatography paper and cut it to form an 11 × 22-cm rectangle (or another appropriate size). Without touching the surface of the paper with your fingers, use a pencil to draw a starting line 2 cm from one long edge (the bottom edge) and lightly mark the positions for spots with a pencil, spacing them about 2 cm from each side and 1.5–2.0 cm apart. Spot the paper with all of the solutions to be chromatographed; then bend it into a cylinder and staple the ends together, leaving a small gap between them (see Figure C20). Uncover the developing chamber, place the paper cylinder inside (spotted end down), and cover it without delay. The paper must not touch the sides of the chamber, and the chamber should not be moved during development. Development may take an hour or more, depending on the developing solvent and the distance traveled.

When the solvent front is a centimeter or less from the top of the paper, remove the cylinder and separate its edges; then accurately draw a line along the entire solvent front with a pencil. If any spots are visible at this time, outline them with a pencil (carefully, to avoid tearing the wet paper), as they may fade in time. Again bend the paper into a cylinder and stand it on edge to air dry, preferably under a hood. When the paper is completely

(cover omitted for clarity)

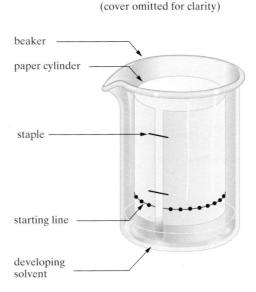

Figure C20 Developing a paper chromatogram

dry, visualize the spots by an appropriate method, if necessary, and outline them with a pencil. For each spot, place a ruler along an imaginary line—perpendicular to the starting line—that goes through the center of that spot. Then measure the distance in millimeters (1) from the starting line to the leading edge of the spot, and (2) from the starting line to the solvent front. Calculate the R_f value of each spot.

Summary

1 Add developing solvent to developing chamber, cover, and let stand.
2 Obtain chromatography paper and cut it to size.
3 Spot chromatography paper with solutions.
4 Place paper in developing chamber, cover, and observe.
5 Remove paper chromatogram when solvent front nears top of paper.
6 Mark solvent front and any visible spots, let chromatogram dry.
7 Visualize and mark spots, as needed.
8 Measure R_f values.
9 Clean up, dispose of solvent as directed.

D. Washing and Drying Operations

OPERATION **21**   ## Washing Liquids

In practice, the process of washing liquids is identical to liquid-liquid extraction [OP-15], but its purpose is different. *Extraction* separates a desired substance (such as the product of a reaction) from an impure mixture; *washing* removes impurities from a desired substance. The liquid being washed may be a neat liquid (one without solvent) or a solvent containing the desired substance. In either case, that substance must not dissolve appreciably in the wash liquid or it will be extracted along with the impurities.

The *wash liquid* is usually water or an aqueous solution, although organic solvents such as diethyl ether can be used to remove low-polarity impurities from aqueous solutions containing polar solutes. Both water and saturated aqueous sodium chloride remove water-soluble impurities, such as salts and polar organic compounds, from organic liquids. Saturated sodium chloride is frequently used for the last washing before a liquid is dried [OP-22] because it removes excess water from the organic liquid by the salting-out effect described in OP-15b. It is preferred to water in some other cases because it helps prevent the formation of emulsions at the interface between the liquids.

Some aqueous wash liquids contain chemically reactive solutes that convert water-insoluble impurities to water-soluble salts, which then dissolve in the wash solvent. Aqueous solutions of bases remove acidic impurities, as illustrated for sodium bicarbonate.

$$NaHCO_3 + HA \text{ (acidic impurity)} \longrightarrow Na^+A^- \text{ (soluble salt)} + H_2O + CO_2$$

Aqueous solutions of acids remove alkaline impurities, as illustrated for hydrochloric acid.

$$HCl + B \text{ (basic impurity)} \longrightarrow BH^+ Cl^- \text{ (soluble salt)}$$

When a chemically reactive wash liquid is used, it is usually advisable to perform a preliminary washing with water or aqueous NaCl to remove most of the water-soluble impurities. This may prevent a potentially violent reaction between the reactive wash liquid and the impurities. Sometimes a follow-up washing with water or aqueous NaCl is used to remove traces of the chemically reactive solute from the product.

The effectiveness of washing with a given total volume of wash liquid increases if it is carried out in several steps. Unless otherwise indicated in the experimental procedure, a liquid should be washed in two or three stages using equal volumes of wash liquid for each stage, and the total volume of wash liquid should be roughly equal to the volume of the liquid being washed. For example, if you are washing 3 mL of a liquid, you can use two 1.5-mL or three 1-mL portions of the wash liquid. When several different wash liquids are used in succession, one washing with each wash liquid

Rule of Thumb: Total volume of wash liquid ≈ volume of liquid being washed.

may be sufficient, but the total volume of all of the wash liquids should usually equal or exceed the volume of the liquid being washed.

The procedure for washing liquids is essentially the same as that for liquid-liquid extraction [OP-15], except that the wash liquid is discarded and the liquid being washed is saved. As for extraction, it is important to save *both* layers until you are absolutely certain you are working with the right layer. Refer to the General Directions in OP-15 for illustrations and experimental details.

General Directions for Washing Liquids

Equipment and Supplies

washing container (conical vial or centrifuge tube) with cap
support for washing container
stirring rod (optional)
2 Pasteur pipets with bulbs, one calibrated
wash solvent
1 or 2 containers (conical vials, screw-cap vials, test tubes, etc.)

Use a calibrated Pasteur pipet to transfer the desired volume of wash liquid to a conical vial or conical centrifuge tube containing the liquid being washed. If the wash liquid contains sodium carbonate, sodium bicarbonate, or another reactive solute that generates a gas, use a stirring rod to stir it vigorously with the liquid being washed until gas evolution subsides. Then cap the washing container securely, shake it gently 5–10 times, and unscrew the cap slightly to allow gases to escape. Continue shaking for a minute or so, venting occasionally if you are using a gas-generating wash liquid. Then loosen the cap and let it stand until the layers separate sharply.

If the liquid being washed is *less* dense than the wash liquid (for example, if it is a diethyl ether solution), use a Pasteur pipet to transfer the lower layer (the wash liquid) to another container (**A**) after each washing. Then add the next portion of wash liquid to the liquid being washed, which remains in the washing container, and repeat the washing as needed. Set the wash liquid aside for later disposal. After the last washing, try to remove all of the wash solvent from the liquid being washed.

If the liquid being washed is *more* dense than the wash liquid (for example, if it is a dichloromethane solution), use a Pasteur pipet to transfer the lower layer to an appropriate container (**A**), and then transfer the wash liquid to a different container (**B**) after each washing. Return the lower layer from **A** to the washing container, add the next portion of wash liquid, and repeat the washing as needed. Set the wash liquid aside for later disposal. After the last washing, try to not to include any of the wash solvent when you remove the lower layer.

Summary

1 Add liquid to be washed to washing container.
2 Add wash solvent, cap tightly, shake gently, vent.
3 Shake to extract impurities into wash solvent.
4 Loosen cap, let layers separate.
5 Uncap washing container, transfer lower layer to container **A**.

IF liquid being washed formed upper layer, GO TO 6.
IF liquid being washed formed lower layer, GO TO 7.
6 Cap container **A**.
IF another washing step is needed, GO TO 2.
IF washing is complete, clean up, STOP.
7 Transfer washing container contents to container **B** and cap it.
IF another washing step is needed, transfer contents of **A** to washing container, GO TO 2.
IF washing is complete, clean up, STOP.

OPERATION 22   # Drying Liquids

When an organic substance (or a solution containing it) is extracted from an aqueous reaction mixture, washed with an aqueous wash liquid, steam distilled from a natural product, or comes into contact with water in some other way, the substance or its solution will retain traces of water that must be removed before such operations as evaporation and distillation are carried out. Organic liquids and solutions can be dried in bulk by allowing them to stand in contact with a *drying agent*, which is then removed by decanting or filtration. They may also be dried by passing them through a *drying column*, such as a Pasteur pipet packed with a suitable drying agent.

Drying Agents

Most drying agents are anhydrous (water-free) inorganic salts that form hydrates by combining chemically with water. For example, a mole of anhydrous magnesium sulfate can combine with up to seven moles of water to form hydrates of varying composition.

$$MgSO_4 + nH_2O \rightleftharpoons MgSO4 \cdot nH_2O \qquad (n = 1\text{--}7)$$

The effectiveness and general applicability of a drying agent depends on the following characteristics:

• Speed—how fast drying takes place
• Capacity—the amount of water absorbed per unit of mass
• Intensity—the degree of dryness attained
• Chemical inertness—unreactivity with substances being dried
• Ease of removal

Ideally, a drying agent should be very fast and have both a high capacity and a high intensity. It should not react with or dissolve in a substance being dried, and it should be easy to remove when drying is complete. Table D1 summarizes the properties of some common drying agents. As you can see from the table, there is no ideal drying agent, but anhydrous magnesium sulfate is perhaps the best all-around drying agent.

The choice of a drying agent depends, in part, on the properties of the liquid being dried and the degree of drying required. Solvents such as diethyl ether and ethyl acetate retain appreciable quantities of water, so their

Table D1 Properties of commonly used drying agents

Drying agent	Speed	Capacity	Intensity	Comments
magnesium sulfate	fast	medium	medium	good general drying agent, suitable for nearly all organic liquids
calcium sulfate (Drierite)	very fast	low	high	fast and efficient, but low capacity
sodium sulfate	slow	high	low	good for predrying; loses water above 32.4°C
calcium chloride	slow to fast	low to medium	medium	removes traces of water quickly, larger amounts slowly; reacts with many organic compounds
silica gel	medium	medium	high	good general drying agent, more expensive than most
potassium carbonate	fast	low	medium	cannot be used to dry acidic compounds
potassium hydroxide	fast	very high	high	used to dry amines, reacts with many other compounds; caustic

solutions are often washed [OP-21] with saturated aqueous sodium chloride to salt out some of the water before further drying. They should then be dried by a drying agent with a relatively high capacity, such as magnesium sulfate or sodium sulfate. Powdered anhydrous magnesium sulfate tends to adsorb organic solutes onto its surface, so it should always be washed with a suitable pure solvent to recover adsorbed material. Granular anhydrous sodium sulfate is often used when thorough drying is not necessary because it doesn't adsorb much of the substance being dried and is easy to remove. Because of its high capacity, it is also used to predry very wet solutions that are later dried with a high-intensity drying agent such as calcium sulfate.

Relatively nonpolar solvents, such as petroleum ether and dichloromethane, retain little water and are often dried with lower capacity drying agents such as calcium sulfate (Drierite) or calcium chloride. These and other drying agents come in different particle sizes, as indicated by their mesh number. Fine particles with mesh numbers of 20 and higher dry liquids most effectively, but coarse particles make it easier to separate the liquid after drying. Calcium chloride reacts with many organic compounds that contain oxygen or nitrogen, including alcohols, aldehydes, ketones, carboxylic acids, phenols, amines, amides, and some esters, so it should not be used to dry such compounds or their solutions. To dry a dichloromethane solution that will later be evaporated, it may be sufficient to filter the solution through a cotton plug or a layer of sodium sulfate to remove water droplets (if there are any). The remaining water forms an azeotrope with dichloromethane and evaporates along it.

The mesh number of a granular substance is the number of meshes per inch in the finest screen that will allow the substance's particles to pass through. Thus the finer the particles, the larger the mesh number.

Carrier Solvents

Small quantities of neat liquids should be dissolved in a *carrier solvent*, such as diethyl ether or dichloromethane, before drying. Otherwise, you may lose much of your product by adsorption on the drying agent. After drying, the solvent is removed by evaporation [OP-16].

Drying Liquids in Bulk

The *quantity* of drying agent needed for bulk drying depends on the capacity and particle size of the drying agent and on the amount of water present.

Rule of Thumb: Use about 50 mg of drying agent per milliliter of liquid.

Usually about 50 mg (0.05 g) of drying agent is needed per milliliter of liquid. More may be needed if the drying agent has a low capacity or large particle size or if the liquid has a high water content. It is best to start with a small amount of drying agent and then add more if necessary, since using too much results in excessive losses by adsorption of liquid on the drying agent. For microscale work, you can usually start with the amount of drying agent that will fit in the V-shaped groove of a Hayman-style microspatula, and then add more if most of the drying agent clumps up or shows other evidence of hydration, as described next.

The appearance of the drying agent when the drying time is up often suggests whether more drying agent is needed. As they become hydrated, powdered magnesium sulfate clumps together in visible crystals, calcium chloride displays a glassy surface appearance, and blue indicating Drierite changes color to pink. If most of a drying agent has changed as described after the initial drying period, more drying agent should be added *or* the spent drying agent should be removed and replaced with fresh drying agent.

Spent drying agent contains hydrates that reduce drying efficiency.

The time required for bulk drying depends on the speed of the drying agent and the amount of water present. Most drying agents attain at least 80% of their ultimate drying capacity within 15 minutes, so longer drying times are seldom necessary—5 minutes is usually sufficient for magnesium sulfate or Drierite, while 15 minutes is recommended for calcium chloride and sodium sulfate. When more complete drying is required, it is better to replace the spent drying agent or to use a more efficient drying agent than to extend the drying time.

The drying agent should be removed as completely as possible when the drying period is over. For most microscale work, the liquid being dried is separated from the drying agent with a filter-tip pipet [OP-6] or by passing the liquid through a glass-wool plug in a filtering pipet [OP-12]. With small amounts of a coarse-grained drying agent, such as 8-mesh Drierite, the large grains can sometimes be removed with a clean tweezers or microspatula, but granular calcium chloride usually contains a fine powder that remains behind.

When a solution (an ether extract, for example) is being dried, the spent drying agent should be washed with a small amount of the pure solvent to recover adsorbed solute that would otherwise be discarded with the drying agent. When a neat liquid is being dried, the drying agent should ordinarily be washed with dichloromethane or diethyl ether and the solvent evaporated to recover the adsorbed liquid.

Drying Columns

A column suitable for drying relatively small quantities of liquid can be constructed by supporting a drying agent on a glass-wool plug inside a Pasteur pipet. When thorough drying is not necessary, granular anhydrous sodium sulfate is the preferred drying agent. This form of sodium sulfate comes in relatively large grains, allowing liquids to flow through it easily, and a sodium sulfate column can absorb relatively large amounts of water without plugging up. The powdered form of anhydrous magnesium sulfate is not suitable for a drying column because liquids won't flow through it, but a granular form of the drying agent can be used for that purpose. Although magnesium sulfate is a more efficient drying agent than sodium sulfate, a magnesium sulfate column is more likely to plug up if the liquid being dried is quite wet. Thus it is important to make sure that the liquid does not contain any visible water droplets, such as those resulting from imperfect separation during an extraction. If it does, a cotton plug can be inserted into the drying column above the layer of drying agent to remove most of the water.

General Directions for Bulk Drying

Equipment and Supplies

> drying container (conical vial, small Erlenmeyer flask, etc.)
> drying agent
> filtering pipet with glass-wool plug, *or* filter-tip pipet

If you are drying a small amount of a neat liquid, it is advisable to dissolve the liquid in a carrier solvent, such as diethyl ether or dichloromethane, and evaporate the solvent after drying. If the liquid to be dried contains water droplets or a separate aqueous layer, remove the water using a Pasteur pipet. Alternatively, pass the wet liquid through a filtering pipet [OP-12] containing a cotton plug, using a rubber bulb to force the last bit of liquid through.

Select a conical vial, a small Erlenmeyer flask, or another suitable drying container that will hold the liquid with plenty of room to spare, and add the liquid. Obtain the estimated quantity needed of a suitable drying agent (protect it from atmospheric moisture until you are ready to use it) and cap its original container tightly. Add the drying agent to the liquid, stopper or cap the container, and stir the contents or swirl and shake the container for a few seconds. If you are using a conical vial, you can stir the mixture by twirling the pointed end of a flat-bladed spatula in it. Let it dry for 5–15 minutes, using the longer drying time with calcium chloride, sodium sulfate, and other slow drying agents. Stir or shake and swirl the mixture occasionally during the drying period. If a second (aqueous) phase forms during drying, remove it with a Pasteur pipet and add more drying agent.

When the drying period is over, examine the drying agent carefully. If most of it is spent, add more drying agent (or remove it and replace it with fresh drying agent), and continue drying for 5 minutes or more. When drying appears to be complete, withdraw the liquid with a dry filter-tip pipet and transfer it to another container, or filter it by gravity through a filtering pipet containing glass wool. If you are drying a solution, rinse the drying agent with a small volume of the pure solvent and combine the rinse liquid with the dried liquid. If you are drying a neat liquid, you can rinse the drying agent with a suitable solvent, combine it with the dried liquid, and evaporate the solvent.

Summary

1 Select drying agent and obtain estimated quantity needed.
2 Add drying agent to liquid in drying container.
3 Stopper container and mix contents, set aside.
4 Stir or shake and swirl occasionally until drying time is up.
   IF drying agent is spent or aqueous phase separates, GO TO 5.
   IF not, GO TO 6.

5 Remove aqueous phase or spent drying agent if necessary; add fresh drying agent. GO TO 3.

6 Separate liquid from drying agent.
   IF liquid being dried is a neat liquid, GO TO 7.
   IF liquid being dried is a solution, GO TO 8.

7 Wash drying agent with appropriate solvent, combine solvent with dried liquid, evaporate solvent. GO TO 9.

8 Wash drying agent with fresh solvent, combine with dried solution.

9 Clean up, dispose of spent drying agent as directed.

General Directions for Using a Drying Column

Equipment and Supplies

2 $5\frac{3}{4}$-inch Pasteur pipets
latex rubber bulb
cotton
applicator stick or stirring rod
drying agent
collecting container

If you are drying a neat liquid, dissolve it in approximately 2 mL of a carrier solvent, such as diethyl ether or dichloromethane (evaporate the solvent after drying). To construct a drying column, obtain a $5\frac{3}{4}$-inch Pasteur pipet and use an applicator stick or a thin glass stirring rod to gently tamp a small ball of clean glass wool into the column until it lodges where the pipet narrows. Add enough anhydrous sodium sulfate or *granular* anhydrous magnesium sulfate (usually about 0.7–1.0 g) through the top of the Pasteur pipet to provide a 2–3-cm layer of the drying agent (see Figure D1). If the liquid to be dried contains visible water droplets or a separate aqueous layer, remove most of the water (if you can) with a Pasteur pipet, and then tamp a loose plug of cotton into the drying column so that it rests on top of the layer of drying agent. Clamp the drying column vertically to a ring stand, place a suitable collecting container under the outlet, and use a Pasteur pipet (or a filter-tip pipet) to transfer the liquid being dried to the column. If the liquid flow stops or becomes very slow, attach a latex rubber bulb to the top of the column and squeeze it just enough to force the liquid *slowly* through the column. If that doesn't work, obtain a pipet pump, pull out its plunger, attach it to the top of the column, and rotate the thumbwheel enough to force the liquid *slowly* through the column. When all of the liquid has passed through the drying agent, add about 0.5 mL of fresh solvent (the solvent present in the liquid being dried) to the top of the drying column and let it drain into the container holding the dried liquid. Use a latex rubber bulb to expel any liquid remaining on the column into the container.

After allowing time for the drying agent to dry thoroughly, remove it by inverting the column in a beaker and tapping gently, and remove the cotton plug by snagging it with a length of copper wire bent to form a "J" at one end. If that doesn't work, attach a rubber hose from an air line to the column at its narrow end (be careful you don't break the tip), hold the open end over a beaker, and *slowly* open the air valve until the air blows out the drying agent.

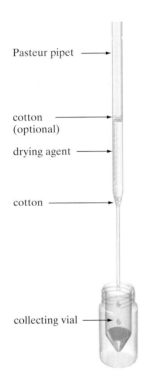

Pasteur pipet

cotton (optional)

drying agent

cotton

collecting vial

Figure D1 A drying column

Summary

1 Select drying agent and measure out quantity needed.
2 Prepare drying column containing drying agent supported on glass-wool plug.
3 Clamp drying column to ring stand over collecting container.
4 Transfer liquid to column with Pasteur pipet, drain into collecting container.
5 Rinse drying agent with fresh solvent, combine with dried liquid.
6 Clean up, dispose of spent drying agent as directed.

Drying Solids

OPERATION **23**

Solids that have been separated from a reaction mixture or isolated from other sources usually retain traces of water or other solvents used in the separation. Solvents can be removed by a number of drying methods, depending on the nature of the solvent, the amount of material to be dried, and the melting point and thermal stability of the solid compound.

Experimental Considerations

Solids that have been collected by vacuum filtration [OP-13] are usually air dried on the filter by leaving the vacuum on for a few minutes after filtration is complete. Unless the solvent is very volatile, further drying is then required.

Comparatively volatile solvents can be removed by simply spreading the solid on a watch glass or evaporating dish (covered to keep out airborne particles) and placing the container in a location with good air circulation, such as a hood, for a sufficient period of time (see Figure D2). Clamping an inverted funnel above the container and passing a gentle stream of dry air or nitrogen over the wet solid will accelerate the drying rate.

A very wet solid can be partially dried by transferring it to a filter paper on a clean surface and blotting it with another filter paper to remove excess solvent. It should then be dried completely by one of the methods described next. Some of the solid will adhere to the filter paper, however, so this method should not be used with comparatively small amounts of solid products.

Many wet solids can be spread out in a shallow ovenproof container and dried in a laboratory oven set at 110°C or another suitable temperature. The expected melting point of the solid should be at least 20°C above the oven temperature, and the solid should not be heat sensitive or sublime readily at the oven temperature. It is not unusual for a student to open an oven door and discover that the product he or she worked many hours to prepare has just turned into a charred or molten mass, or disappeared entirely. Aluminum weighing dishes and other commercially available containers made of heavy aluminum foil are usually suitable for oven drying because the aluminum conducts heat well, cools quickly, and is not likely to

folded filter paper

watch glass

Figure D2 Covered watch glass for drying solids

Take Care! Don't blow the crystals away.

If you are not sure whether your product can be oven dried safely, consult your instructor.

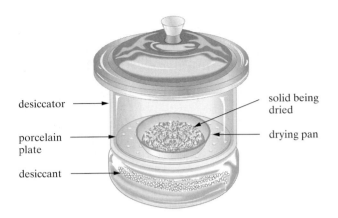

Figure D3 Commercial desiccator

burn your fingers. Aluminum reacts with acids and bases, so acidic and basic solids, or solids that may be wet with acidic or basic liquids, should be oven dried in Pyrex or porcelain containers, such as watch glasses and evaporating dishes.

A *vacuum oven* combines the use of heat with low air pressure for very fast, efficient drying. A simple vacuum oven can be constructed by clamping a sidearm test tube horizontally and connecting it to a vacuum source (see *J. Chem. Educ.* **1988**, *65,* 460). The sample, in an open vial, is inserted in the sidearm test tube, which is stoppered and heated with a heat lamp while the vacuum is turned on.

When time permits, the safest way to dry a solid is to leave it in a *desiccator* overnight or longer. A desiccator consists of a tightly sealed container partly filled with a *desiccant* (drying agent) that absorbs water vapor, creating a moisture-free environment in which the solid should dry thoroughly. Desiccators such as the one shown in Figure D3 are available commercially.

A simple "homemade" desiccator for drying small amounts of solid, made from an 8-oz. (~250-mL) wide-mouth jar with a screw cap, is illustrated in Figure D4. Enough of a solid desiccant (about 50 mL, measured in a graduated beaker) is added to form a layer of desiccant on the bottom. This desiccator is well suited to microscale work, since several samples in small open vials can be dried at the same time. The wet solid, in an appropriate container, is set inside the desiccator, which is capped and allowed to

A similar desiccator with a polyethylene storage rack is available commercially.

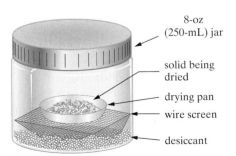

Figure D4 "Homemade" desiccator

stand undisturbed until drying is complete. If there is any danger of the container tipping over, a wire screen can be cut to fit on top of the desiccant layer and provide a more stable surface. Except when a product is being added or removed, the desiccator must be kept tightly closed at all times to keep the dessicant active.

Anhydrous calcium chloride is a good (if rather slow working) desiccant because it is inexpensive and has a high water capacity. Drierite (anhydrous calcium sulfate) is faster and more efficient than calcium chloride but has a much lower capacity. A combination of calcium chloride with a small amount of blue indicating Drierite works better than either desiccant separately. The blue Drierite removes traces of moisture that calcium chloride cannot, and it turns pink when the desiccant is spent and needs to be replaced. Unless your instructor indicates otherwise, spent desiccant should be placed in a designated container for reactivation. Drierite can be reactivated by heating it in a 225°C oven overnight; calcium chloride should be heated overnight at 250–350°C.

General Directions for Drying Solids in an Oven

Obtain a wide, shallow, ovenproof container, such as a small evaporating dish or an aluminum weighing dish. Weigh it, and label it to prevent mixups. Spread the solid on the bottom of the container in a thin, uniform layer and place it in the oven, preferably where it is well separated from other containers. After 30 minutes or so, remove the container (wear gloves when handling a glass container), let it cool to room temperature, and weigh it. Then put it back in the oven for 5–10 minutes, let it cool, and weigh it again. If the mass has decreased by 1% or more, repeat this process until the mass doesn't change significantly between weighings.

Summary

1 Put wet sample in tared, ovenproof drying container, label container.
2 Leave container in oven about 30 minutes or until dry.
3 Weigh container and sample; dry further, and reweigh if necessary.
    IF weight does not change significantly between weighings, STOP.
    IF weight changes by more than 1% between weighings, GO TO 2.

General Directions for Drying Solids in a Desiccator

If the sample will remain in the desiccator until the next lab period, you can dry it in a tared, labeled storage vial with the cap removed. (At the beginning of the next lab period, you should cap the vial and weigh it.) If the sample will only be in the desiccator overnight or for a day or two, spread it out in a shallow tared container, such as a polystyrene or aluminum weighing dish or a drying tray made by folding a square of heavy aluminum foil. Label the container if the desiccator will be used by other students or if you are drying several samples at the same time. Set the container inside a desiccator containing fresh desiccant, taking care to place it securely so that it won't tip over. With a homemade desiccator, vials can be pushed down into the desiccant layer and shallow containers set on a rack made of wire screen or another material (see Figure D4). A commercial desiccator should have a porcelain plate or another kind of rack to hold the samples. Cap or cover the desiccator securely and let it stand overnight or longer (preferably longer). Remove and weigh the container with the sample in it, then return it to the desiccator and reweigh it after an hour or so. If the mass has decreased by 1% or more, return the container to the desiccator and leave it there until the mass doesn't change significantly between weighings. Replace the used desiccant with fresh desiccant if necessary.

Summary

1 Put wet sample in tared drying container.
2 Set open container inside desiccator, cover desiccator, let stand overnight or until dry.
3 Weigh container and sample; dry further and reweigh if necessary.
  IF weight does not change significantly between weighings, STOP.
  IF weight changes by more than 1% between weighings, GO TO 2.

OPERATION 24   # Drying and Trapping Gases

Gases such as air and nitrogen should be dry for such applications as evaporating solvents and providing an inert atmosphere for a nonaqueous reaction mixture. Certain reactions, including Grignard reactions, must be carried out in a dry atmosphere to prevent decomposition or side reactions brought about by contact with atmospheric moisture. Other reactions generate toxic gases, which may have to be *trapped* to keep them out of the laboratory. Gases can be dried and moisture can be excluded from reaction mixtures using desiccants like the ones described in OP-23 for drying solids. Toxic gases can often be trapped using water, activated charcoal, or reactive solutions that convert reactive gases to soluble salts.

a. Drying Gases

Many gases, such as nitrogen, are available in cylinders and can be purchased in a form that is dry enough for most applications. Other gases, especially compressed air obtained from a laboratory air line, may have to be dried before use. Air from an air line (and other wet gases) can be dried by passing the gas slowly through a standard scale drying tube or a U-tube filled with a suitable desiccant (see Figure D5). Indicating silica gel and granular alumina are very efficient desiccants; indicating Drierite and calcium chloride are satisfactory for many purposes. Plugs of cotton or glass wool should be placed at both ends of the drying tube to support the desiccant and remove particles and other impurities from the air line. One end of the drying tube should have a connector that is inserted in a rubber or plastic tube going to the air line; the other end should have a connecter that is inserted in a similar flexible tube going to a Pasteur pipet or gas delivery tube. If necessary, you can

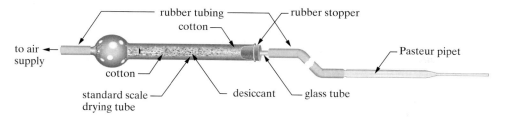

Figure D5 Apparatus for drying gases

make a connector by inserting a short length of fire-polished glass tubing (see OP-3) into a one-hole rubber stopper of a size that will fit in the drying tube.

b. Excluding Moisture from Reaction Mixtures

A *drying tube* containing a suitable desiccant (drying agent) is used when it is necessary to exclude moisture from an experimental setup during an operation such as refluxing or distillation. Calcium chloride, calcium sulfate (Drierite), granular alumina, and silica gel can be used for this purpose. Calcium chloride is the least efficient of these, but it is adequate in many cases. When a very dry atmosphere is required, the apparatus should be swept out by passing dry nitrogen or another dry gas through it using a Pasteur pipet or gas delivery tube.

To fill a drying tube like the one shown in Figure D6, use an applicator stick or glass rod to push a plug of cotton or glass wool through its long end nearly to the bend in the tube. Holding the drying tube with its long arm upright (open end up), use a plastic weighing dish to add enough calcium chloride ($\sim$1.5 g) or other drying agent to form a layer 3–4 cm deep; then insert another plug of cotton or glass wool to keep the desiccant in place. Attach the drying tube to the top of a reflux condenser or any other part of an apparatus that is open to the atmosphere.

Unless your instructor directs otherwise, put used desiccant in a designated container (*not* the container you got it from) when you are done. If you leave calcium chloride in a drying tube exposed to the atmosphere, its granules will eventually clump together in a solid mass that can only be removed by immersing the drying tube in water overnight or longer until the solid dissolves.

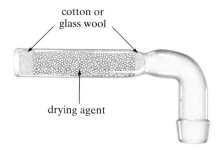

Figure D6 Drying tube

Take Care! Never stopper the drying tube, as this will result in a closed system that might explode or fly apart when heated.

c. Trapping Gases

The best way to keep toxic gases out of the laboratory air is to conduct all reactions under an efficient fume hood. If that is not possible, or if the hood is not adequate, a gas trap containing a suitable gas-absorbing liquid or solid should be used. Water alone will dissolve some gases effectively, but dilute aqueous sodium hydroxide is generally used for acidic gases, such as HBr or SO_2. It converts them to salts that dissolve in the water.

$$HBr + NaOH \longrightarrow NaBr + H_2O$$
$$SO_2 + NaOH \longrightarrow NaHSO_3$$

Similarly, dilute aqueous HCl can be used to trap ammonia and other alkaline gases. Activated charcoal, in the form known as pelletized Norit, is quite effective for removing small quantities of gases during reactions. It removes neutral gases as well as acidic and basic ones, so it can also be used to keep strong or offensive odors out of the lab. Glass wool moistened with water can remove some odors, but it is only effective for the odors of water-soluble organic compounds.

For most microscale work you can use a Norit gas trap, which consists of a layer of pelletized Norit in a drying tube (see Figure D6). To construct the gas trap, push a plug of cotton or glass wool nearly to the bend in the drying tube; then hold it with the long arm upright (open end up) and use a plastic weighing dish to add enough pelletized Norit (~0.5 g) to form a layer 3–4 cm deep. Insert another plug of cotton or glass wool to keep the Norit in place. Insert the drying tube in any part of the reaction apparatus that is exposed to the atmosphere, usually the top of a reflux condenser. If the reaction is to be conducted in a dry atmosphere, you can add a layer of calcium chloride on top of the Norit layer before inserting the second cotton plug.

A gas trap consisting of a plug of moistened glass wool in a drying tube can effectively trap some water-soluble gases, but care must be taken to keep water from dripping into the reaction apparatus. Such a trap is prepared by filling most of the long arm of the drying tube with glass wool and using a Pasteur pipet to drop water onto it until it is moist but not dripping wet.

Another kind of gas trap that can be used to trap water-soluble gases and reactive gases such as hydrogen chloride or sulfur dioxide is shown in Figure D7. It is constructed by (1) inserting two short glass tubes into both ends of a 30-cm (or longer) length of rubber tubing, (2) clamping one end over a suitable gas-absorbing liquid (such as dilute aqueous sodium hydroxide) in a test tube, and (3) connecting the other end to a thermometer adapter (see OP-9). The thermometer adapter is then attached to the top of a reflux condenser or any other part of an apparatus that is open to the air. The free end of the trap's glass tube should be clamped about a millimeter (no more) *above* the surface of the liquid to keep liquid from backing up into the reaction apparatus when gas evolution ceases or heating is discontinued.

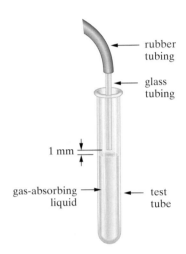

rubber tubing

glass tubing

1 mm

gas-absorbing liquid

test tube

Figure D7 Small-scale gas trap

E. Purification Operations

Recrystallization

The simplest and most widely used operation for purifying organic solids is *recrystallization*. Recrystallization is so named because it involves dissolving a solid that (in most cases) had originally crystallized from a reaction mixture or other solution and then causing it to *again* crystallize from solution. In a typical recrystallization procedure, the crude solid is dissolved by heating it in a suitable *recrystallization solvent*. The hot solution is then filtered by gravity and the filtrate is allowed to cool to room temperature or below, whereupon crystals appear in the saturated solution and are collected by vacuum filtration. The crystals are ordinarily much purer than the crude solid because most of the impurities either fail to dissolve in the hot solution, from which they are separated by gravity filtration or transfer, or remain dissolved in the cold solution, from which they are separated by vacuum filtration or centrifugation.

Recrystallization is based on the fact that the solubility of a solid in a given solvent increases with the temperature of the solvent. Consider the recrystallization from boiling water of a 0.500-g sample of salicylic acid contaminated by 0.025 g of acetanilide. The solubility of salicylic acid in water at 100°C is 7.5 g per 100 mL, so the amount of water required to just dissolve 0.500 g of salicylic acid at the boiling point of water is 6.7 mL.

$$0.500 \text{ g} \times \frac{100 \text{ mL}}{7.5 \text{ g}} = 6.7 \text{ mL of water}$$

All of the acetanilide impurity will also dissolve in the boiling water. If the solution is cooled to 20°C, at which temperature the solubility of salicylic acid is only 0.20 g per 100 mL, about 0.013 g of salicylic acid will remain dissolved.

$$6.7 \text{ mL} \times \frac{0.20 \text{ g}}{100 \text{ mL}} = 0.013 \text{ g of salicylic acid}$$

The dissolved salicylic acid will end up in the filtrate during the vacuum filtration. The remaining 0.487 g (0.500 g − 0.013 g) will crystallize from solution, if sufficient time is allowed, and be collected on the filter. The solubility of acetanilide in water is 0.50 g per 100 mL at 20°C, so up to 0.035 g of acetanilide can dissolve in 6.7 mL of water at 20°C.

$$6.7 \text{ mL} \times \frac{0.50 \text{ g}}{100 \text{ mL}} = 0.035 \text{ g of acetanilide}$$

This means that all 0.025 g of acetanilide in the crude product should remain in solution and end up in the filtrate. Therefore, under ideal conditions, the recrystallization should yield a 97% recovery of salicylic acid uncontaminated by acetanilide.

$$\frac{0.487 \text{ g}}{0.500 \text{ g}} \times 100 = 97\% \text{ recovery}$$

The recovery could be increased by cooling the mixture in an ice/water bath to further lower the solubility of salicylic acid, but even then, some of the salicylic acid would remain in solution. You can never recover all of your product after a recrystallization, but by allowing plenty of time for the product to crystallize and making careful transfers, you should be able to minimize your losses.

This is a simplified description of a rather complex process; a number of factors may bring about results different from those calculated.

- Crystals of the desired solid may adsorb impurities on their surfaces or trap them within their crystal lattice.
- The solubility of a solute in a saturated solution of a different solute may not be the same as its solubility in the pure solvent.
- Using only enough recrystallization solvent to dissolve a solid can result in premature crystallization, so additional solvent may be added to prevent this.

a. Recrystallization from a Single Solvent

Experimental Considerations

Most experimental procedures that involve recrystallization specify a suitable recrystallization solvent. If the solvent is not specified, see Section c, "Choosing a Recrystallization Solvent."

In its simplest form, the recrystallization of a solid is carried out by dissolving the impure solid in the hot (usually boiling) recrystallization solvent and letting the resulting solution cool to room temperature or below to allow crystallization to occur. Additional steps, such as filtering or decolorizing the hot solution, may also be necessary. Sometimes it is desirable to collect a second or third crop of crystals by concentrating (see OP-16) the *mother liquor*—the liquid from which the crystals are filtered—from the previous crop. These crystals will contain more impurities than the first crop and may require recrystallization from fresh solvent. A melting-point determination [OP-30] or TLC analysis [OP-19] can be used to assess the purity of a recrystallized solid.

Recrystallization of about 0.1–1 g of solid can be carried out using test tubes or small (10–25-mL) Erlenmeyer flasks. A sand bath or an aluminum block [OP-7a] is generally used for heating. If necessary, the hot solution is filtered using a preheated filter-tip pipet or filtering pipet, and the crystals are collected by vacuum filtration on a Hirsch funnel.

For quantities smaller than about 0.1 g, a specialized crystallization tube called a *Craig tube* can be used. A Craig tube looks much like a small test tube except that it widens at the top, where it has a ground-glass inner surface that accommodates a glass or Teflon plug (see Figure E2 in the "General Directions" section). The impure solid is dissolved in the recrystallization solvent and allowed to crystallize in the Craig tube. Then the Craig tube and its plug are inverted in a centrifuge tube and spun in a centrifuge, forcing liquid past the plug into the centrifuge tube and leaving the solid behind on the plug. This procedure reduces the number of transfers needed for recrystallization and therefore reduces product losses. Craig tubes are fragile and expensive, and they can easily roll off a benchtop and break, so they should be handled with care.

Dissolving the Impure Solid.   The impure solid is dissolved by heating it with a sufficient amount of recrystallization solvent at the boiling point of the solvent. Except for recrystallization from a Craig tube, the pure recrystallization solvent is heated to boiling before being added to the solid. If you know the solubility of the solid substance at the boiling point of the solvent you are using, you can calculate the approximate volume of boiling solvent that will be needed to dissolve it. Otherwise, you will have to determine the necessary amount of recrystallization solvent by trial and error; adding a measured amount of the hot solvent, boiling the mixture for a minute or so to see if it dissolves, adding more solvent if it doesn't, and continuing to add and boil fresh portions of hot solvent until it eventually goes into solution. Since the solid is most soluble at the boiling point of the solvent, it is important to bring the mixture back to the boiling point after each solvent addition and to boil it for a minute or more before making the next addition. Most solids dissolve fairly rapidly in a boiling solvent, but some do not. Slow-dissolving solids will usually dissolve when heated under reflux (to prevent solvent loss) for 5–10 minutes after each addition of fresh solvent.

If the solid's melting point is below the boiling point of the solvent, the solvent should not be heated to boiling.

Filtering the Hot Solution.   Some impurities in a substance being crystallized may be insoluble in the boiling solvent and should be removed after the desired substance has dissolved. Don't mistake such impurities for the substance being purified and add too much solvent in an attempt to dissolve them, because excess solvent will reduce the yield of crystals and may even prevent the substance from crystallizing at all. If, after most of the solid has dissolved, addition of another portion of hot solvent does not appreciably reduce the amount of solid in the flask, that solid is probably an impurity—particularly if it is different in appearance from the solid that dissolved previously.

Excess solvent can be removed by evaporation [OP-16] if necessary.

In a standard scale recrystallization, the hot solution is usually filtered through coarse filter paper, supported in a funnel, into an Erlenmeyer flask. As the liquid passes through the funnel, it cools down somewhat, and this may cause the solid to crystallize prematurely on the filter paper. Small quantities of liquid and small-scale filtering devices cool more rapidly than larger ones, making premature crystallization even more likely during a microscale crystallization. To prevent it, you should dilute the hot solution with about 30–50% more recrystallization solvent than was needed to dissolve the solid, and preheat everything that will contact the solution when it is being filtered. With relatively small amounts of liquid, you can filter the diluted hot solution with a shortened filter-tip pipet—one having all but ~5 mm of the capillary tip cut off (see OP-6). Preheat the filter-tip pipet by drawing in and expelling several portions of the boiling recrystallization solvent (*not* the solution you are filtering). Without delay, use it to transfer the hot solution—while it is just below its boiling point—to another preheated container such as a test tube or Craig tube. To remove the excess solvent, continuously twirl the rounded end of a microspatula in the test tube or Craig tube as you boil the resulting solution (see Figure C9 in OP-16), in a heating block or sand bath. Stop boiling when solid begins to form on the spatula just above the liquid level, indicating that the solution is near the saturation point. Then set the solution aside to cool and crystallize as described later.

Filtration of the hot solution can also be accomplished with a similarly shortened filtering pipet (see OP-12). Clamp the preheated filtering pipet over the crystallization container (test tube, Craig tube, etc.). Add 25–50% more recrystallization solvent than was needed to dissolve the solid and heat the solution to boiling. Then use a preheated Pasteur pipet to transfer the recrystallization solution to the filtering pipet. Work fast, as the pipets and their contents will cool down rapidly. If necessary, you can increase the filtration rate through a filtering pipet by applying gentle pressure with a rubber bulb or pipet pump. Boil off the excess solvent, as described previously, before cooling.

If premature crystallization occurs in either kind of filtration apparatus, you may be able to redissolve the crystals in the solvent by blowing hot air from a heat gun over the pipet while its outlet is over the crystallization container, adding a little more solvent if necessary. If this doesn't work, return the contents of the pipet to the hot recrystallization mixture, add more recrystallization solvent, and try again. If necessary, use a freshly prepared and preheated filtering device to complete the filtration.

Removing Colored Impurities.   If a crude sample of a compound known to be white or colorless yields a recrystallization solution with a pronounced color, activated carbon (Norit) can often be used to remove the colored impurity. Pelletized Norit, which consists of small cylindrical pieces of activated carbon, is preferable to finely powdered Norit. Powdered Norit, although it is somewhat more efficient than the pelletized form, obscures the color and is difficult to filter out completely.

Take Care! Never add Norit to a solution at or near the boiling point—it may boil up violently.

To use pelletized Norit, let the boiling recrystallization solution cool down for a minute or so, then stir in a *small* amount of the Norit. Unless otherwise directed, start with about 20 mg of Norit. Stir or swirl the mixture for a few minutes, keeping it hot enough to prevent your product from crystallizing, but not boiling it. Then let it settle and observe the color of the solution. If much color remains, you can add more pelletized Norit and repeat the process. Avoid adding too much, because the excess Norit may adsorb your product as well as the impurities, and some color may remain no matter how much you use. Then heat the solution just to boiling and separate it from the Norit by one of the methods described in the previous section, "Filtering the Hot Solution." Note that even pelletized Norit contains some fine particles that can plug up a filter-tip pipet, so a filtering pipet is preferable.

Inducing Crystallization.   If no crystals form after a hot recrystallization solution is cooled to room temperature, the solution may be supersaturated. If so, crystallization can often be induced by one or more of the following methods:

- Dip the end of a glass stirring rod into the liquid; then remove it and let the solvent evaporate to leave a thin coating of the solid. Reinsert the glass rod in the liquid and stir gently.
- Rub the tip of a glass stirring rod against the inside of the recrystallization container, just above the liquid surface, for a minute or so. Use an up and down motion, with the rod tip just touching the liquid on the

Rubbing with a glass rod that has not been fire-polished works best, but it will also scratch the glass, so don't use one without your instructor's permission.

downstroke. If you are using a Craig tube, rub gently so as not to scratch or break the tube.

• Cool the solution in an ice/water bath, then continue rubbing as described previously for several minutes.
• If any are available, drop a few *seed crystals* (crystals of the pure compound) into the solution with cooling and stirring.

If crystals still do not form, you may have used too much recrystallization solvent. In that case try one or both of the following procedures, in order:

• Concentrate the solution by evaporation [OP-16] until it becomes cloudy or crystals appear, heat it until the cloudiness or crystals disappear (add a little more recrystallization solvent if necessary), and then let it cool. If necessary, use one or more of the previous methods to induce crystallization.
• Heat the solution back to boiling and add another solvent that is miscible with the first, and in which the compound should be less soluble (for example, try adding water to an ethanol solution). Add just enough of the second solvent to induce cloudiness or crystal formation at the boiling point; then add enough of the original solvent to cause the cloudiness or crystals to disappear from the boiling solvent, and let it cool.

As a last resort, remove all of the solvent by evaporation and try a different recrystallization solvent—but see your instructor for advice first.

Cooling the Hot Solution. The size and purity of the crystals formed depends on the rate of cooling; rapid cooling yields small crystals and slow cooling yields large ones. The medium-sized crystals obtained from moderately slow cooling are usually the best, because larger crystals tend to *occlude* (trap) impurities, while smaller ones adsorb more impurities on their surfaces and take longer to filter and dry. You can reduce the rate of cooling by setting the recrystallization container on a surface that is a poor heat conductor and inserting it inside or covering it with another container. Set an Erlenmeyer flask on the benchtop and cover it with a beaker. Support a test tube in a small Erlenmeyer flask and invert a beaker over the apparatus. Support a Craig tube, with its plug, in a 25-mL Erlenmeyer flask and set the flask inside a 100-mL beaker; if desired, invert a larger beaker over this assembly.

You can increase the yield of crystals somewhat by cooling the mixture in an ice/water bath once a good crop of crystals is present, but their purity may decrease slightly as a result. Cooling the mixture before well-formed crystals are present may result in small, impure crystals that take longer to filter and dry.

Dealing with Oils and Colloidal Suspensions. When the solid being recrystallized is quite impure or has a low melting point, it may separate as an *oil* (a second liquid phase) upon cooling. Oils are undesirable because, even if they solidify on cooling, the solid retains most of the original impurities.

If the solid to be purified has a melting point below the boiling point of a particular recrystallization solvent, it may be possible to prevent oiling by substituting a lower-boiling solvent with similar properties. For example, methanol (bp 65°C) might be substituted for ethanol (bp 78°C), or acetone

Using a lower-boiling solvent usually results in a lower percent recovery.

(bp 57°C) for 2-butanone (bp 80°C). Oiling may also be prevented by using more recrystallization solvent, adding seed crystals, or both. Seed crystals can sometimes be obtained by dissolving a small amount of the oil in an equal volume of a volatile solvent in a small, open test tube and letting the solvent evaporate slowly.

If oiling occurs, try the following remedies, in order:

1 Heat the solution until the oil dissolves completely, adding more solvent if necessary; then cool it slowly while rubbing the inside of the container (see "Inducing Crystallization").
2 Add an amount of pure recrystallization solvent equal to about 25% of the total solvent volume and repeat the process described in Step **1**.
3 Follow the procedure in Step **1**, but add a seed crystal or two at the approximate temperature where oiling occurred previously.
4 Try to crystallize the oil by either (a) cooling the solution in an ice-salt bath, rubbing the oil with a stirring rod, and adding seed crystals if necessary, or (b) removing all of the oil with a Pasteur pipet, dissolving it in an equal volume of a volatile solvent, and letting the solvent evaporate slowly in an open test tube. Then collect the solid by vacuum filtration and recrystallize it from the same solvent or a more suitable one. If necessary, use one or more of the previous methods to prevent further oiling.

A *colloid* is a suspension of very small particles dispersed in a liquid or other phase. Colloids generally have a cloudy appearance and cannot be filtered with ordinary filtering media because the particles pass right through. If a solid separates from a cooled solution as a colloidal suspension, the colloid can often be coagulated to form normal crystals by extended heating in a hot water bath, or (if the solvent is polar) by adding an electrolyte such as sodium sulfate. Colloid formation can sometimes be prevented by treating a recrystallization solution with Norit, as described previously, or by cooling the solution very slowly.

General Directions for Single-Solvent Recrystallization

Safety Notes

> **Unless you are informed otherwise, consider all recrystallization solvents (except water) to be flammable and harmful by ingestion, inhalation, and contact. Avoid contact with and inhalation of such solvents and keep them away from flames and hot surfaces.**

Use Method **A** when you have approximately 0.1–1 g of crude solid. Use Method **B** when you expect the recrystallization to require a comparatively large volume of solvent. Use Method **C** when you have approximately 0.1 g or less of crude solid or know that the recrystallization will require no more than ~2 mL of recrystallization solvent.

A. Recrystallization Using a Test Tube

Equipment and Supplies

2 test tubes, 13 × 100 mm or larger
recrystallization solvent
calibrated Pasteur pipet (preferably filter tipped)
flat-bladed microspatula
Hirsch funnel with filter paper
small filter flask
cold washing solvent
small Erlenmeyer flask (to hold test tube)
beaker (to cover flask and test tube)
shortened filter-tip pipet or filtering pipet (optional)
pelletized Norit (optional)

If you know the solubility of the solid in the boiling recrystallization solvent, calculate the approximate volume of solvent you will need to recrystallize it and measure out that amount plus 20–50% extra (use the larger amount if you think you will need to filter the hot solution). Otherwise, start with about 1 mL per 0.1 g of solid and use more if needed. Use a test tube large enough so that the solvent will fill it no more than half full. Measure the recrystallization solvent into the test tube (the *solvent tube*) with a calibrated Pasteur pipet, add a boiling chip or applicator stick, and heat it to boiling (see Figure E1) using a heating block or sand bath. Place the solid to be purified in an identical test tube (the *boiling tube*) and clamp this test tube *above* the heat source (not in it, as the solid may melt) so that you can control the heating rate quickly by raising or lowering it. Use the calibrated Pasteur pipet to transfer about one-quarter of the liquid in the solvent tube to the boiling tube (use about two-thirds of the liquid if you have calculated the approximate volume needed). Lower the boiling tube and heat the mixture *at the boiling point* while continuously stirring it by twirling the rounded end of a flat-bladed microspatula in the test tube, until it appears that no more solid will go into solution. If undissolved solid remains, add more portions of hot solvent, about 10% of the total each time, heating the solution *at the boiling point* with swirling or stirring after each addition. Continue this process until (1) the solid has completely dissolved *or* (2) no more solid dissolves when a fresh portion of solvent is added and it appears that only solid impurities remain.

If you will need less than ~2 mL of recrystallization solvent, it is not necessary to preheat the solvent.

If the solution has an intense color but the pure product should not be colored, see the section "Removing Colored Impurities." If the boiling solution does not contain solid impurities, use the boiling tube as a crystallization tube and skip to the next paragraph. If it does contain solid impurities (including Norit used for decolorizing), add 30–50% as much recrystallization solvent as you used to dissolve the solid and heat the mixture back to boiling. Use a preheated filter-tip pipet to transfer the hot solution to a *crystallization tube* (use the emptied solvent tube), leaving the solid impurities behind. (Alternatively, filter the solution into the crystallization tube through a preheated filtering pipet.) If crystals begin to form in the pipet, redissolve them as described in the section "Filtering the Hot Solution."

See the section "Filtering the Hot Solution" for additional information.

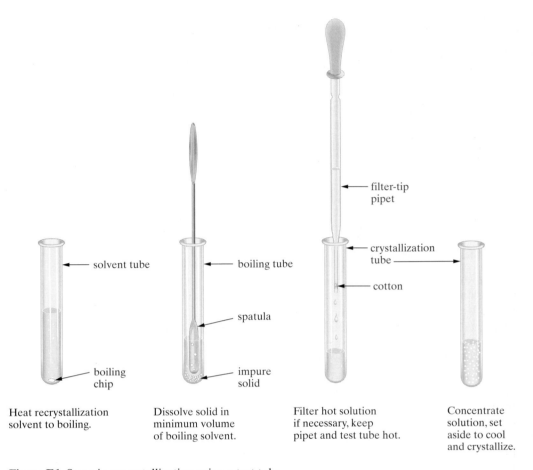

Heat recrystallization
solvent to boiling.

Dissolve solid in
minimum volume
of boiling solvent.

Filter hot solution
if necessary, keep
pipet and test tube hot.

Concentrate
solution, set
aside to cool
and crystallize.

Figure E1 Steps in recrystallization using a test tube

Concentrate the filtered solution by boiling it while stirring with a flat-bladed microspatula until traces of solid begin to form on the microspatula just above the solvent level.

Set the crystallization tube in a small Erlenmeyer flask, cover it with an inverted beaker, and let the solution cool slowly to room temperature. If no crystals form by the time the solution reaches room temperature, see "Inducing Crystallization." If an oil separates or the solution becomes cloudy but no solid precipitates, see "Dealing with Oils and Colloidal Suspensions." Once crystals have begun to form, allow at least 10 minutes (sometimes much longer) for complete crystallization. If desired, cool the test tube further in an ice bath for 5 minutes or more to increase the yield.

Collect the product by vacuum filtration [OP-13] on a Hirsch funnel. Transfer any crystals remaining in the crystallization flask to the funnel with a small amount of ice-cold recrystallization solvent (or another appropriate solvent) and use more cold solvent to wash the solid on the filter (see OP-13). Air-dry the crystals by leaving the vacuum on for a few minutes after the last washing, then dry [OP-23] them further as necessary.

Summary

1 Measure recrystallization solvent into solvent tube, heat to boiling.
2 Add some hot solvent to solid in boiling tube, boil with stirring.

3 Add more hot solvent in portions (as necessary) until solid dissolves.
IF solution contains colored impurities, GO TO 4.
IF solution contains undissolved impurities, add more hot solvent, GO TO 5.
IF not, GO TO 6.

4 Cool below boiling point, heat with pelletized Norit, bring to boiling point.

5 Filter and concentrate hot solution.

6 Cover hot solution and set aside to cool until crystallization is complete.

7 Collect crystals by vacuum filtration, wash and air-dry on filter.

8 Clean up, dispose of solvent as directed.

B. Recrystallization Using an Erlenmeyer Flask

Follow the directions for Method **A** with the following modifications. Use two small (10–25-mL) Erlenmeyer flasks—a *boiling flask* and a *solvent flask*—in place of the two test tubes. Heat the flasks using a hot plate or hot plate-stirrer (or a steam bath, if available) as the heat source. Use a stirring rod rather than a flat-bladed microspatula to stir the recrystallization mixture in the boiling flask, or swirl the flask continuously. You can use the flattened end of a flat-bottomed stirring rod to break up the solid, if necessary. If you need to filter the hot solution and the amount of solution is too large to be easily filtered with a filtering pipet or filter-tip pipet, support a small, preheated powder funnel on the mouth of the emptied solvent flask and filter the solution through a coarse fluted filter paper into the *crystallization flask* (the emptied solvent flask). When using a powder funnel, you should not need to add more than about 20–30% excess recrystallization solvent to prevent premature crystallization. To cool the crystallization flask, set it on the bench top and cover it with a beaker of appropriate size.

C. Recrystallization Using a Craig Tube

Equipment and Supplies

Craig tube with plug
recrystallization solvent
calibrated Pasteur pipet (preferably filter tipped)
flat-bladed microspatula
25-mL Erlenmeyer flask (to hold Craig tube)
100-mL beaker
copper wire
centrifuge tube (preferably plastic)
centrifuge
13 × 100-mm test tube (optional)
shortened filter-tip pipet (optional)
pelletized Norit (optional)

It may be necessary to filter the hot solution because of the likelihood of insoluble or colored impurities; if so, put the impure solid in a 13 × 100-mm test tube and dissolve it in the boiling recrystallization solvent (adding 50% more solvent than is needed to dissolve it), as described in the directions for

The experimental procedure or your instructor may indicate whether the hot solution will have to be filtered.

Method **A**. If necessary, decolorize the solution with pelletized Norit as described in "Removing Colored Impurities." Then transfer the hot solution to a Craig tube using a preheated filter-tip pipet. Concentrate the solution by boiling it while stirring with a flat-bladed microspatula until traces of solid begin to form on the microspatula just above the solvent level. Skip the next paragraph and go on to the following one.

If it will not be necessary to filter the hot solution, place the impure solid in a Craig tube and add just enough room-temperature recrystallization solvent to cover it. Heat the mixture *at the boiling point* with a heating block or sand bath while continuously stirring it by twirling the rounded end of a flat-bladed microspatula in the Craig tube, until it appears that no more solid will go into solution. If undissolved solid remains, add more solvent drop by drop with stirring, keeping the solution at the boiling point, until the solid is completely dissolved.

*Don't add solvent to the level where the Craig tube widens; if it takes that much solvent to dissolve the solid, transfer the hot solution to a test tube for recrystallization as described in Method **A**.*

Insert the plug into the Craig tube, support it in a 25-mL Erlenmeyer flask, and set the flask inside a 100-mL beaker. If desired, cover the apparatus with a large inverted beaker. Let the solution cool slowly to room temperature. If no crystals form by the time the solution reaches room temperature, see "Inducing Crystallization." If an oil separates or the solution becomes cloudy but no solid precipitates, see "Dealing with Oils and Colloidal Suspensions." Once crystals have begun to form, allow at least 10 minutes (sometimes much longer) for complete crystallization. If desired, cool the Craig tube further in an ice bath for 5 minutes or more to increase the yield.

Obtain a length of thin copper wire that is about as long as the combined length of the assembled Craig tube and its plug. Make a loop in the copper wire that is large enough to fit over the narrow stem of the plug and bend it so that it is at right angles to the straight part of the wire (see Figure E2). Holding the Craig tube vertically, slip the loop over the plug stem and slide it down as far as it will go, with the straight part of the wire parallel to the Craig tube. Hold the wire against the bottom of the Craig tube between your thumb and forefinger, then invert a conical centrifuge tube over the Craig tube assembly until the narrow end of the plug is nested in (or nearly touching) the V-shaped end of the tube. Pressing the wire firmly against the bottom of the Craig tube to keep the plug in place, take the centrifuge tube in your other hand and carefully invert the entire apparatus so that the centrifuge tube is upright. Bend the wire over the lip of the centrifuge tube and down the outside so that it does not stick out. Place the centrifuge tube in a centrifuge [OP-14] opposite another tube containing enough water to balance it *or* a centrifuge tube containing another student's Craig tube assembly. If the wire scrapes against any part of the centrifuge, cut it shorter, but not so short that you will not be able to grasp it to lift out the Craig tube. Run the centrifuge for 3 minutes or more. When the centrifuge has stopped turning, remove the centrifuge tubes and use the copper wire to lift the Craig tube assembly out of its centrifuge tube. All of the solvent should be in the bottom of the centrifuge tube, and most of the solid should be on the plug or near the ground joint where the plug fits into the Craig tube. If there is still solvent in the Craig tube, replace the assembly in the centrifuge tube and centrifuge it again. Use a microspatula to transfer the solid to another container for further drying, if necessary.

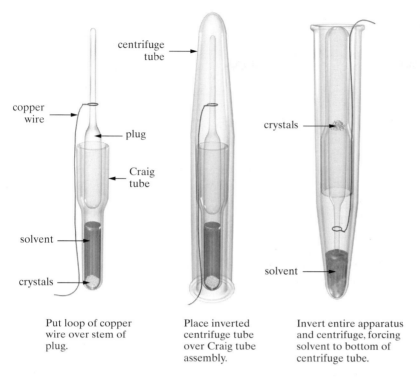

Figure E2 Steps in recrystallization using a Craig tube

Put loop of copper wire over stem of plug.

Place inverted centrifuge tube over Craig tube assembly.

Invert entire apparatus and centrifuge, forcing solvent to bottom of centrifuge tube.

copper wire

plug

Craig tube

solvent

crystals

centrifuge tube

crystals

solvent

Summary

1 IF filtering the hot solution will be necessary, dissolve solid in excess hot recrystallization solvent in test tube.
   IF filtering the hot solution should not be necessary, dissolve solid in boiling recrystallization solvent in Craig tube. GO TO 3.
2 Transfer hot solution to Craig tube with filter-tip pipet and boil to concentrate.
3 Insert plug, cover hot solution, cool until crystallization is complete.
4 Collect crystals by centrifugation.
5 Clean up, dispose of solvent as directed.

b. Recrystallization from Mixed Solvents

Solids that cannot be recrystallized readily from any single recrystallization solvent can usually be purified by recrystallization from a mixture of two compatible solvents such as those listed here.

Some compatible solvent pairs

ethanol-water

methanol-water

acetic acid-water

acetone-water

diethyl ether-methanol

ethanol-acetone

absolute ethanol-petroleum ether

ethyl acetate-cyclohexane

chloroform-petroleum ether

The solvents must be miscible in one another, and the solid compound should be quite soluble in one solvent and relatively insoluble in the other. If the composition of a suitable solvent mixture is known beforehand (such as 40% ethanol in water, for example), the recrystallization can be carried out with the premixed solvent in the same manner as for a single-solvent recrystallization. But a mixed-solvent recrystallization is usually performed by heating the compound in the solvent in which it is most soluble (which we will call solvent A) until it dissolves, and then adding enough of the second solvent (solvent B) to bring the solution to the saturation point.

If the compound is *very* soluble in solvent A, the total volume of solvent may be quite small compared to that of the crystals, which may then separate as a dense slurry. In such a case, you should use more of solvent A than is needed to just dissolve the compound, and add correspondingly more of solvent B to bring about saturation. Be careful to avoid adding so much of solvent A that *no* amount of solvent B will result in saturation. If that occurs, you will have to concentrate the solution (see OP-16) or remove all of the solvent and start over.

General Directions for Mixed-Solvent Recrystallization

Most of the steps in a mixed-solvent recrystallization are the same as for single-solvent recrystallization; refer to the previous "General Directions" for experimental details.

Place the crude solid in a boiling tube or flask and add solvent A (previously heated to boiling) in portions, while boiling and stirring, until the solid dissolves or only undissolved impurities remain. If necessary, the solution can be decolorized with pelletized Norit at this point. If the hot solution contains undissolved impurities (including Norit), add enough solvent A to prevent premature crystallization, heat it to boiling, and filter it. After filtration, concentrate the solution to remove most of the excess solvent.

Add hot solvent B in small portions, with stirring, keeping the mixture boiling after each addition, until a persistent cloudiness appears or a precipitate start to form. Then add just enough hot solvent A drop by drop to the boiling mixture, with stirring, to clear it up or dissolve the precipitate. Set the mixture aside to cool to room temperature, allow sufficient time for crystallization, and cool it further in an ice/water bath, if desired. When crystallization is complete, collect the product by vacuum filtration, or by centrifugation if you are using a Craig tube. After vacuum filtration, wash the product on the filter with cold solvent B or another suitable solvent.

c. Choosing a Recrystallization Solvent

A solvent suitable for recrystallization of a given solid should meet the following criteria, or as many of them as is practical:

- Its boiling point should be in the 60–100°C range.
- Its freezing point should be well below room temperature.

- It must not react with the solid.
- It should not be excessively hazardous to work with.
- It should dissolve between 50 mg and 250 mg of the solid per milliliter at the boiling point, and less than 20 mg per milliliter at room temperature, with at least a 5 : 1 ratio between the two values.

As a rule, the recrystallization solvent should be either somewhat more or somewhat less polar than the solid, since a solvent of very similar polarity will dissolve too much of it. Solubility information from reference books listed in Category A of the Bibliography may help you choose a suitable solvent. For example, a solvent in which the compound is designated as *sparingly soluble* or *insoluble* when cold and *very soluble* or *soluble* when hot may be suitable for recrystallization. If solubility data for the compound are not available, a solvent may have to be chosen by trial and error from a selection such as that in Table E1. Once you have identified some possible solvents, test them as described next. You can test several solvents at the same time and choose the best one.

J. Chem. Educ. **1989**, *66, 88 describes another microscale method for determining recrystallization solvents.*

General Directions for Testing Recrystallization Solvents

Measure the number of drops in a measured volume (0.5–1 mL) or so of the solvent, using a short Pasteur pipet. Weigh about 10 mg of the finely divided solid into a 10×75-mm test tube and add 6 drops of the solvent. Stir the mixture by twirling the round end of a flat-bladed microspatula in it and

Table E1 Properties of common recrystallization solvents

Solvent	bp	fp	Comments
water	100	0	solvent of choice for polar compounds; crystals dry slowly
methanol	64	−94	good solvent for relatively polar compounds; easy to remove
95% ethanol	78	−116	excellent general solvent; usually preferred over methanol because of higher boiling point
2-butanone	80	−86	good general solvent; acetone is similar but its boiling point is lower
ethyl acetate	77	−84	good general solvent
toluene	111	−95	good solvent for aromatic compounds; high boiling point makes it difficult to remove
petroleum ether (high-boiling)	~60–90	low	a mixture of hydrocarbons; good solvent for less polar compounds
hexane	69	−94	good solvent for less polar compounds; easy to remove
cyclohexane	81	6.5	good solvent for less polar compounds; freezes in some cold baths

Note: bp and fp (freezing point) are in °C. Solvents are listed in approximate order of decreasing polarity.

carefully observe what happens. If the solid dissolves in the cold solvent, the solvent is unsuitable. If the solid does not dissolve, heat the mixture to boiling, with stirring. If it dissolves after heating, try to induce crystallization as described next. If it does not dissolve completely, add more solvent in 2-drop portions, gently boiling and stirring after each addition, until it dissolves *or* until the total volume of added solvent is about 20 drops. If the solid *does not* dissolve in that amount of hot solvent, the solvent is probably not suitable (unless it is water). If it *does* dissolve at any point, record the volume of boiling solvent required to dissolve it. Let the solution cool while rubbing the inside of the tube with a glass rod to see whether crystallization occurs. If crystals separate, examine them for apparent yield and evidence of purity (absence of extraneous color, good crystal structure).

If no single solvent is satisfactory, choose one solvent in which the compound is quite soluble and another in which it is comparatively insoluble (the two solvents must be miscible). Dissolve the specified amount of solid in the first solvent with stirring and boiling, recording the amount of solvent required. Add the second hot solvent drop by drop with boiling and stirring until saturation occurs. Then cool the mixture and try to induce crystallization.

Once a suitable solvent or solvent pair has been identified, use the volume of solvent (converted to mL) you used to estimate the volume of solvent needed to dissolve the entire sample to be purified.

OPERATION 26   Sublimation

Principles and Applications

Sublimation is a phase change in which a solid passes directly into the vapor phase without going through an intermediate liquid phase. Many solids that have appreciable vapor pressures below their melting points can be purified by (1) heating the solid to sublime it (convert it to a vapor), (2) condensing the vapor on a cold surface, and (3) scraping off the condensed solid. This method works best if impurities in the crude solid do not sublime appreciably. Sublimation is not as selective as recrystallization or chromatography, but it has some advantages in that no solvent is required and losses in transfer can be kept low.

Experimental Considerations

Sublimation is usually carried out by heating the *sublimand* (the solid before it has sublimed) with a suitable heat source and collecting the *sublimate* (the solid after it has sublimed and condensed) on a cool surface. For best results, the sublimand should be dry and finely divided, and the distance between the sublimand and the condensing surface should be minimized. A simple but effective sublimator consists of two nested beakers of appropriate sizes. For example, a 100-mL beaker can be nested inside a 150-mL beaker, with the inner beaker rotated so as to leave a gap of about 1 cm at the bottom. In some cases, it may be necessary to place separators made of folded-over strips of filter paper or paper toweling between the beakers to get the right spacing. The sublimand is spread out on the bottom of the outer beaker, and the condensing (inner) beaker is partially filled with crushed ice or ice water. As the outer beaker is heated, crystals of sublimate collect on the bottom of the condensing beaker. Figure E3 illustrates a sublimator that operates on the same principle, except that an Erlenmeyer flask is used as a condenser and the temperature is controlled by flowing water.

Solids that do not sublime rapidly at atmospheric pressure may do so under vacuum. Figure E4a illustrates a vacuum sublimator that can be assembled by fitting a 15 × 125-mm test tube snugly inside an 18 × 150-mm

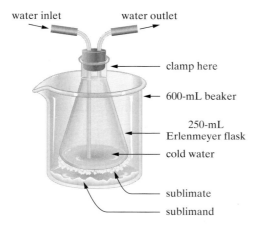

Figure E3 Sublimation apparatus

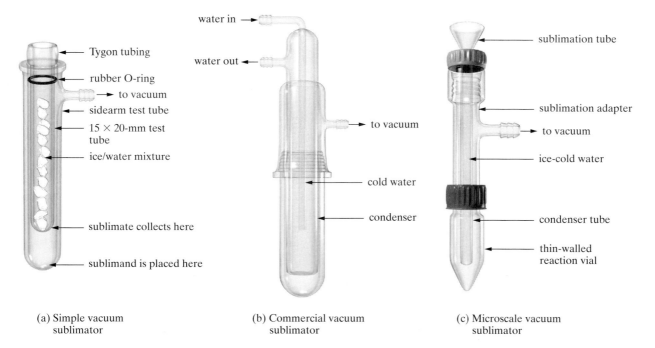

(a) Simple vacuum sublimator

(b) Commercial vacuum sublimator

(c) Microscale vacuum sublimator

Figure E4 Apparatus for vacuum sublimation

sidearm test tube, using a rubber O-ring from a microscale lab kit to act as a vacuum seal. A short section of 15-mm i.d. Tygon tubing (or several layers of masking tape) is placed around the lip of the inner tube to keep it from slipping inside the sidearm test tube (see *J. Chem. Educ.* **1991**, *68*, A63). A commercial vacuum sublimator, such as the one in Figure E4b, is more efficient because the condenser has a flat, wide bottom that is close to the sublimate. The microscale vacuum sublimation apparatus pictured in Figure E4c can be constructed using a special sublimation tube and adapter.

Depending on the temperature required and the nature of the sublimation apparatus, heat sources such as a hot water baths, oil baths, steam baths, hot plates (for nested beakers), heating blocks, and sand baths can be used for sublimation. An oil bath provides the most uniform heating, but oil baths are messy and somewhat hazardous to work with. With some of these heat sources, crystals tend to collect on the sides of the sublimation container as well as the condenser. Wrapping the base of the sublimation container with aluminum foil or other insulation will help prevent this.

General Directions for Sublimation

Safety Notes

> **Because of the possibility of an implosion, safety glasses must be worn during a vacuum sublimation. It is best to work behind a hood sash or safety shield while the apparatus is under vacuum.**

Equipment and Supplies

  sublimation apparatus
  heat source
  cooling fluid
  flat-bladed microspatula

Assemble one of the sublimation setups described or one suggested by your instructor. Powder the dry sublimand finely and spread it in a thin, uniform layer over the bottom of the sublimation container. If you are using a vacuum sublimator, attach the apparatus to a trap and vacuum source, and then turn on the vacuum. Turn on the cooling water *or* partly fill the condensing tube or beaker with crushed ice or ice water. Heat the sublimation container with an appropriate heat source until sublimate begins to collect on the condenser; then adjust the temperature to attain a suitable rate of sublimation without melting or charring the sublimand. As needed, add small pieces of ice to the condenser to replace melted ice. When the condenser is well covered with sublimate (but before sublimate crystals begin to drop into the sublimand), stop the sublimation and remove the sublimate as described in the next paragraph, and then resume sublimation. If the sublimand hardens or becomes encrusted with impurities, stop the sublimation, grind it to a fine powder, and then resume sublimation.

When all of the compound has sublimed or only a nonvolatile residue remains, remove the apparatus from the heat source, break the vacuum if you are using a vacuum sublimator, and let the apparatus cool. Carefully remove the condenser (avoid dislodging any sublimate) and scrape the crystals into a suitable tared container using a flat-bladed spatula.

Summary

1 Assemble sublimation apparatus, add sublimand.
2 Turn on vacuum if necessary, add cooling mixture or turn on cooling water.
3 Heat until sublimation begins, adjust heating to maintain good sublimation rate.
4 When sublimation is complete, stop heating, break vacuum if necessary, let cool.
5 Scrape sublimate into tared container.
6 Clean sublimation apparatus.

Simple Distillation

OPERATION **27**

a. Distillation of Liquids

A pure liquid in a container open to the atmosphere boils when its vapor pressure equals the external pressure, which is usually about 1 atmosphere (760 torr, 101.3 kPa). The vapor contains the same molecules as the liquid, so its composition is identical to that of the pure liquid. A mixture of two (or more) liquids with different vapor pressures will boil when the total vapor pressure over the mixture equals the external pressure, but the composition of the vapor will be different than that of the liquid itself, being richer in the more volatile component (the one with the higher vapor pressure). If this vapor is condensed into a separate receiving vessel, the condensed liquid will have the same composition as the vapor; that is, it will also be richer in the more volatile component.

The process of vaporizing a liquid mixture in one vessel and condensing the vapors into another is called *distillation*. The liquid mixture being distilled (the *distilland*) is heated in a boiling vessel (the *pot*), the vapors are condensed on a cool surface, and the resulting liquid (the *distillate*) is collected in a suitable *receiver*.

In the conventional distillation apparatus illustrated in Figure E5, the distilland is boiled in a round-bottom flask, the resulting vapors pass through a water cooled-condenser where they are condensed to a liquid, and the liquid distillate drips into the receiver. If the components of the distilland have sufficiently different vapor pressures, most of the more volatile component will end up in the receiver and most of the less volatile component(s) will remain behind in the pot.

The purity of the distillate increases with the number of vaporization-condensation cycles it experiences—the number of times it is vaporized and condensed—on its way to the receiver. *Simple distillation* involves only a single vaporization-condensation cycle. It is most useful for purifying a liquid that contains either involatile impurities or small amounts of higher- or

The distilland is usually a solution of two or more miscible liquids, but it may also be a liquid-solid solution. The distillation of immiscible liquids is discussed in OP-17.

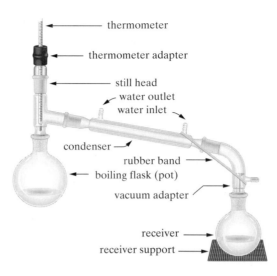

Figure E5 Conventional distillation apparatus

lower-boiling impurities. *Fractional distillation* [OP-29] allows for several vaporization-condensation cycles in a single operation. It can be used to separate liquids with comparable volatilities and to purify liquids that contain relatively large amounts of volatile impurities. *Vacuum distillation* [OP-28] is carried out under reduced pressure, which reduces the temperature of the distillation. It is used to purify high-boiling liquids and liquids that decompose when distilled at atmospheric pressure.

Principles and Applications

To understand how distillation works, consider a mixture of two ideal liquids with ideal vapors, which obey both Raoult's law (Equation **1**) and Dalton's law (Equation **2**).

$$\text{Raoult's law: } P_A = X_A \cdot P_A^o \tag{1}$$

$$\text{Dalton's law: } P_A = Y_A \cdot P \tag{2}$$

See your general chemistry textbook or another source for a discussion of Raoult's law and Dalton's law.

Presumably, entane and orctane exist only in J. R. R. Tolkien's Middle-Earth, where they are used for fuel by Ents and Orcs, respectively.

In these expressions, P_A is the partial pressure of component A over the mixture, P_A^o is the equilibrium vapor pressure of pure A at the same temperature, X_A is the mole fraction of A in the liquid, and Y_A is the mole fraction of A in the vapor. Unfortunately, there are no real liquids that obey these laws perfectly, so we shall consider the behavior of two imaginary hydrocarbons, *entane* (bp = 50°C) and *orctane* (bp = 100°C), that obey them both. If a mixture of entane and orctane is heated at normal atmospheric pressure, it will begin to boil at a temperature that is determined by the composition of the liquid mixture, producing vapor of a different composition. For example, an equimolar mixture of entane and orctane will start to boil at a temperature just above 66°C, and the vapor will contain more than four moles of entane for every mole of orctane. The liquid and vapor composition of an entane-orctane mixture at any temperature can be calculated using Equations **3** and **4**, which are derived from Dalton's law and Raoult's law.

Table E2 Equilibrium vapor pressures and mole fractions of entane and orctane at different temperatures

T, °C	Entane			Orctane		
	$P°$, torr	X	Y	$P°$, torr	X	Y
50°	760	1.00	1.00	160	0.00	0.00
60°	1030	0.67	0.90	227	0.33	0.10
70°	1370	0.42	0.76	315	0.58	0.24
80°	1790	0.24	0.57	430	0.76	0.43
90°	2300	0.11	0.32	576	0.89	0.68
100°	2930	0.00	0.00	760	1.00	1.00

Note: $P°$ = equilibrium vapor pressure of the pure liquid; X = mole fraction in liquid mixture; Y = mole fraction in vapor.

$$X_A = \frac{P - P_B^o}{P_A^o - P_B^o} \tag{3}$$

$$Y_A = \frac{P_A^o}{P} X_A \tag{4}$$

P is the total pressure over the mixture, assumed here to be 1 atm (760 torr).

For example, at 70°C the vapor pressure of orctane is 315 torr and that of entane is 1370 torr (see Table E2), so the mole fraction of entane in a distilland that boils at 70°C will be (from Equation **3**):

$$X_{entane} = \frac{760 - 315}{1370 - 315} = 0.422$$

Its mole fraction in the vapor will be (from Equation **4**):

$$Y_{entane} = \frac{1370}{760} \times 0.422 = 0.761$$

showing that the vapor (and thus the distillate) is considerably richer in entane than is the liquid. The vapor pressures and approximate liquid-vapor compositions for entane and orctane at this and other temperatures are given in Table E2.

The key to an understanding of distillation is this: *The vapor over any mixture of volatile liquids contains more of the lower-boiling component than does the liquid mixture itself.* So at any time during a distillation, the liquid condensing into the receiver contains more of the lower-boiling component than does the liquid in the pot. As more of the lower-boiling component distills away, the pot liquid becomes richer in the higher-boiling liquid, so by the end of the distillation, most of the lower-boiling liquid is in the receiver and most of the higher-boiling liquid is in the pot.

The purification process is diagrammed in Figure E6, in which the liquid and vapor compositions are plotted against the boiling temperatures of entane-orctane mixtures. Suppose we distill a mixture having a composition of 67 mole percent (mol%) entane—about 2 moles of entane for every mole of orctane. From the graph (and Table E2), you can see that such a mixture will boil at 60°C (point A) and that its vapor will contain 90 mol% entane (point A′). Thus the distillate that is condensed from this vapor (point A″)

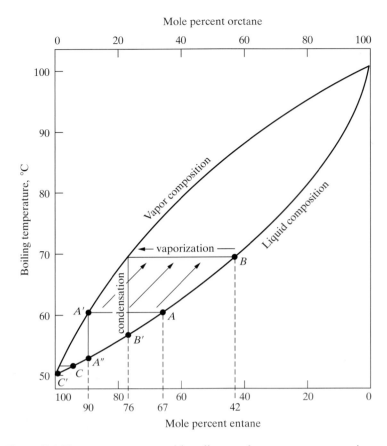

Figure E6 Temperature-composition diagram for entane-orctane mixtures

will be much richer in entane than was the original mixture in the pot. As the distillation continues, however, the more volatile component will boil away faster and the pot will contain progressively less entane. Therefore, the vapor will also contain less entane and the boiling temperature will rise. When the percentage of entane in the pot has fallen to 42 mol% (point B), the boiling temperature will have risen to 70°C and the distillate will contain only 76 mol% entane (point B′). Only if the distillation were to be continued after nearly all of the entane had distilled would the distillate contain more of the less volatile component; at 90°C, for example, more than two-thirds of the distillate would be orctane.

This example shows that the purification effected by simple distillation of a mixture of volatile liquids may be very imperfect. In the example, the distillate never contains more than 90 mol% entane, and it may be considerably less pure than that, depending on the temperature range over which it is collected. If we start with a mixture containing only 5 mol% orctane (point C), however, considerably better purification can be accomplished. The initial distillate will be 99 mol% entane (point C′) at 51°C, and if the distillation is continued until the temperature rises to 55°C, the final distillate will be 95 mol% entane. The average composition of the distillate will lie somewhere between these values, so most of the orctane will remain in the boiling flask, along with some undistilled entane, and the distillate will be relatively pure entane.

Thus simple distillation can be used to purify a liquid containing *small* amounts of volatile impurities if (1) the impurities have boiling points appreciably higher or lower than that of the liquid, and (2) the distillate is collected over a narrow temperature range (usually 4–6°C), starting at a temperature that is within a few degrees of the liquid's normal boiling point.

Experimental Considerations

Apparatus.   The size of the glassware used for distillation should be consistent with the volume of the distilland; otherwise, excessive losses will occur. For example, suppose you recovered 1.5 mL of crude isopentyl acetate from Experiment 5 and decided to distill it from a 25-mL round-bottom flask over a 137–143°C boiling range. At the end of the distillation, the flask would be filled with 25 mL of undistilled vapor at a temperature of 143°C (416 K). An ideal-gas law calculation shows that this is about 7.3×10^{-4} mol of vapor.

$$n = \frac{PV}{RT} = \frac{(1.00 \text{ atm})(0.025 \text{ L})}{(0.0821 \text{ L atm mol}^{-1} \text{ K}^{-1})(416 \text{ K})} = 7.3 \times 10^{-4} \text{ mol}$$

Assuming that the flask contains only isopentyl acetate (M.W. = 130, $d = 0.876$ g/mol), this corresponds to about 0.11 mL of the liquid.

$$7.3 \times 10^{-4} \text{ mol} \times \frac{130 \text{ g}}{1 \text{ mol}} \times \frac{1 \text{ mL}}{0.876 \text{ g}} = 0.11 \text{ mL}$$

Thus the vapor in the boiling flask would condense to about 0.11 mL of liquid isopentyl acetate, which would stay behind in the flask, reducing your recovery by more than 7%. By comparison, a 3-mL conical vial will retain only 0.013 mL of liquid from condensed vapor, or less than 1% of the total. Liquid can also be lost by surface adsorption on glass, so the lower surface area of the smaller boiling container will also reduce losses.

Additional liquid losses occur in the still head, condenser, and any other parts of the apparatus that can trap vapors or adsorb liquid.

In a conventional distillation apparatus such as the one in Figure E5, the vapor has a long way to travel between pot and receiver. This increases the likelihood that distillate will be lost by sticking to the glass surfaces it encounters along the way or being diverted into dead ends such as the top of the still head. For a microscale distillation, the path between the pot and the receiver should be kept as short as is practicable. The *Hickman still* (also called a *Hickman head*) illustrated in Figure E7 is the most commonly used short-path receiver for microscale distillation. Vapors from the pot (usually a conical vial or small round-bottom flask) rise through the narrow neck of the still and collect in a *well* (also called a *reservoir*), which functions as the receiver. The well resembles a hollow doughnut sliced in half and has a capacity of 1–2 mL. Vapors of high-boiling liquids condense on the upper walls of the Hickman still, from which the condensed liquid drains into the well. For lower-boiling liquids, a condenser must be attached to the top of the still (see Figure E8) to keep the vapors from escaping. The following rough guidelines tell you what kind of condenser you should use for a given distillate boiling point *if* the distillation is conducted slowly:

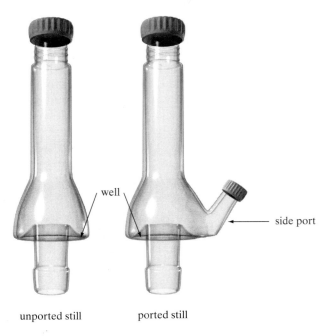

well

side port

unported still   ported still

Figure E7 Hickman stills

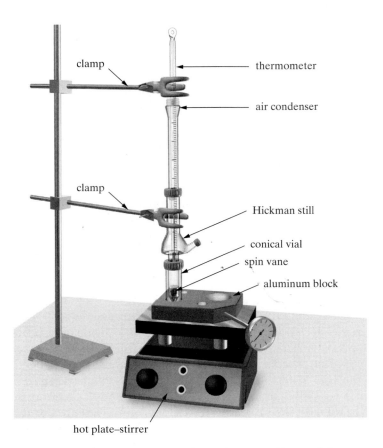

clamp

thermometer

air condenser

clamp

Hickman still

conical vial

spin vane

aluminum block

hot plate–stirrer

Figure E8 Distillation apparatus with air condenser and internal thermometer

<100°C	water-cooled condenser
100–150°C	air condenser
>150°C	no external condenser

Of course, you can *always* use a water-cooled condenser, just to be on the safe side. If, during a distillation, you smell any escaping distillate or see distillate droplets collecting near the top of the condenser (or the top of the still, with no external condenser), either slow the distillation rate, attach a more efficient condenser, or both. If you are distilling a high-boiling liquid, it is advisable to insulate the narrow neck of the still by wrapping it with glass wool, which can be held in place with aluminum foil.

Two types of Hickman still are illustrated in Figure E7. The ported still has a capped *side port*, allowing distillate to be withdrawn from the well with a Pasteur pipet and transferred to a tared screw-cap vial or another collecting vessel. To remove distillate from a ported still during the course of a distillation, carefully remove the port cap, remove the liquid without delay, and quickly recap the port. It is important to avoid leaving the port uncapped for long because distillate vapors will escape through it.

With an unported still, the distillate is collected by inserting a 9-inch Pasteur pipet through the top of the still. To remove distillate from an unported still during a distillation, you may need to remove the condenser and thermometer, if you are using them, and replace them immediately after you withdraw the liquid. Before you remove a condenser, raise the apparatus above the heat source so that the liquid in the pot stops boiling. If the Pasteur pipet doesn't reach all of the distillate, try attaching a short length of fine plastic tubing (of the type used for gas collection) to its capillary tip or bending the tip in a microburner flame. With a longer plastic tube attached to the pipet, you may be able to withdraw liquid without removing the condenser.

A thermometer is inserted through the condenser (if there is one) and the Hickman still and held in place with a small clamp, as shown in Figure E8. It should *never* be secured by a thermometer adapter (unless the adapter's cap is removed) because that makes the distillation apparatus a closed system, which could shatter when heated and cause severe injury due to flying glass. The thermometer should be placed so that the *top* of its bulb is even with the *bottom* of the well, or slightly below it, as illustrated in Figure E9. The bulb should not touch the sides of the well.

Take Care! Never heat a closed system!

Heating. Almost any of the heating devices described in OP-7a can be used for simple distillation. An aluminum block or a sand bath is generally used in the microscale lab.

The heating rate should be adjusted to maintain gentle boiling in the pot and a suitable distillation rate. For most microscale distillations, a distillation rate sufficient to fill the well of a Hickman still in 5–15 minutes or more should be suitable. Distilling at faster rates will decrease the purity of the product and may make it difficult to record the boiling temperature accurately.

If you are distilling a liquid over a wide boiling range, it may be necessary to increase the heating rate gradually to maintain the same distillation rate. Otherwise, the temperature at the heat source should be kept relatively constant throughout a simple distillation. If you are using an aluminum block or a sand bath, you can monitor its temperature with an external thermometer. Keep in mind that the temperature an external thermometer

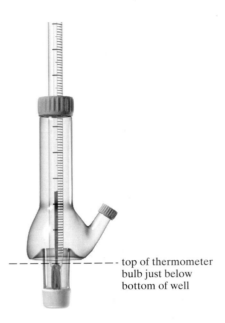

- - - - - - - - - - top of thermometer
bulb just below
bottom of well

Figure E9 Thermometer placement in Hickman still

records will always be higher than the temperature inside the Hickman still, so it is no substitute for an internal thermometer, which records the actual vapor temperature of the liquid being distilled. But an external thermometer can help you maintain a suitable heating rate. For example, if the temperature at the heat source is 50°C higher than the expected boiling point of the distilland, the heating rate is probably too high and you should reduce it. An excessive heating rate during distillation, in addition to reducing distillation efficiency, may cause mechanical carryover or decomposition of the distilland. An insufficient heating rate during distillation can cause the still-head temperature to fluctuate and distillation to slow down or stop altogether.

Boiling Range.   An approximate boiling range for a distillation may be specified in an experimental procedure. For example, when isopentyl acetate is prepared by the reaction of isopentyl alcohol and acetic acid (see Experiment 5), the product is collected over a boiling range of 136°C to 143°C.

$$
\underset{\substack{\text{acetic acid}}}{\overset{O}{\underset{\|}{CH_3COH}}} + \underset{\substack{\text{isopentyl alcohol}\\ \text{bp } 130°C}}{\overset{CH_3}{\underset{|}{HOCH_2CH_2CHCH_3}}} \longrightarrow \underset{\substack{\text{isopentyl acetate}\\ \text{bp } 142°C}}{\overset{O \qquad\quad CH_3}{\underset{\| \qquad\quad |}{CH_3COCH_2CH_2CHCH_3}}} + H_2O
$$

Because the major impurity in this preparation is the more volatile isopentyl alcohol, most of the isopentyl acetate should distill below its normal boiling point of 142°C. (Using 143° as the end of the range allows for experimental error.) If the boiling range for a distillation is not specified, you should collect the *main fraction* (the distilled liquid containing the desired component) over a relatively narrow boiling range, usually 4–6°C, that brackets the boiling point of the desired component.

A liquid fraction that distills below the expected boiling range for the main fraction is called a *forerun*. It should be collected in a different container and saved until distillation is complete, and then disposed of as directed. Sometimes, because of improper thermometer placement or an excessive heating rate, the internal thermometer will record a temperature lower than the actual vapor temperature. In that case, some or all of the liquid collected as forerun will actually be part of the expected main fraction, and should *not* be discarded. So if you collect more "forerun" than expected, save it and redistill it later after readjusting the heating rate or thermometer position. It may be advisable to redistill *all* of the liquid that distilled previously to get an accurate boiling-point range, since the rest of the distillate may also have been collected over the wrong temperature range.

General Directions for Simple Distillation with a Hickman Still

Equipment and Supplies

> heat source
> microclamps, support
> conical vial *or* round-bottom flask (pot)
> boiling chip *or* stirring device and magnetic stirrer
> Hickman still
> thermometer(s)
> condenser (omit for high-boiling liquids)
> collecting container(s)

Transfer the distilland to a clean, dry conical vial or round-bottom flask of appropriate size (the pot) and add a boiling chip, stir bar, or spin vane. Use the smallest container for which the liquid volume will be roughly half or less of the container's capacity; for example, use a 3-mL conical vial if the liquid volume is 1.5 mL or less. Assemble [OP-2] a distillation setup as illustrated in Figure E8. Depending on the boiling point of the liquid to be distilled, you can use a water-cooled condenser, an air condenser, or no condenser (see the "Apparatus" section). Clamp the apparatus securely to a ring stand with the pot positioned correctly in the heat source (usually a sand bath or heating block). Insert the thermometer through the condenser (if you are using one) and into the Hickman still so that its bulb is positioned correctly in the neck of the still (see Figure E9), and clamp it in place. If desired, place a thermometer in the heat source to monitor its temperature as well. If you are using a water condenser, turn on the condenser water to provide a slow but steady stream of coolant. Note that the condenser water should flow *in* the lower end of the condenser and *out* the upper end.

Start the stirrer (if you are using one) and adjust the heat control so that the liquid boils *gently* and condensing vapors rise *slowly* into the neck of the Hickman still. If vapors begin rising rapidly into the still shortly after boiling begins, raise the apparatus above the heat source to reduce the heating rate; then carefully reposition the apparatus and readjust the heating rate for slow distillation (be careful not to bump the thermometer when you move the apparatus). Shortly after the vapors reach the thermometer bulb, the temperature reading should rise rapidly and liquid droplets should begin to appear on the sides of the still and drain into the well. During this time, the thermometer reading should rise to an equilibrium value and stabilize at that value, and the entire thermometer bulb should be bathed in a thin film of condensing liquid that drips off the end of the bulb into the pot. Record the temperature when the reading stabilizes and continue to monitor it during the distillation.

If the initial thermometer reading is below the expected boiling range, carry out the distillation until the lower end of the range is reached, and then transfer any liquid in the well to a small beaker or another container (save this forerun, if any, for later disposal or possible redistillation). The distillate should collect in the well slowly, taking 5–15 minutes (or even more) to fill the well completely, if there is enough liquid to fill it. If it distills more rapidly than this, especially if the temperature reading never stabilizes or stabilizes at the wrong temperature, return the distillate to the pot and redistill it more slowly. If the well becomes filled (or nearly so) with distillate, use a Pasteur pipet to transfer it to a collecting container, such as a tared screw-cap vial. (If you don't collect the distillate when the well is full, the excess will simply overflow and drip down into the pot.) Continue distilling and transferring the distillate as needed until the upper end of the expected boiling range is reached *or* until only a drop or less of liquid remains in the pot. Turn off the heat and raise the apparatus away from the heat source before the pot is completely dry. Let the apparatus cool down, then transfer any remaining distillate to the collecting container. You should be able to recover more distillate by tilting the Hickman still so that the liquid flows to one side of the well as you collect it. Gently tapping the sides of the still or cooling it with air or cold water may cause more distillate to collect in the well. You can also rinse the still with a volatile solvent in which the distillate is soluble, and then evaporate [OP-16] the solvent. Disassemble and clean the apparatus as soon as possible after the distillation is completed.

Summary

1 Add distilland and boiling chips or stirring device to pot.
2 Assemble distillation apparatus.
3 Turn on condenser water and stirrer (if used).
4 Start heating; adjust heating rate so that vapors rise slowly into Hickman still.
5 Record temperature after distillation begins and thermometer reading stabilizes.
   IF initial temperature is below expected boiling range, GO TO 6.
   IF initial temperature is within expected boiling range, GO TO 7.
6 Remove forerun when temperature reaches low end of boiling range.
7 Distill until temperature reaches high end of boiling range or until pot is nearly dry, transferring distillate to collector as well fills up.
8 Stop heating and let apparatus cool.

9 Transfer any remaining distillate to collector.

10 Disassemble and clean apparatus, dispose of forerun (if any) and residue in pot.

b. Distillation of Solids

A Hickman still distillation setup can be used for the distillation of small amounts of low-melting solids. The apparatus is constructed as illustrated in Figure E8, except that no condenser is attached, and the solid—which liquefies when the pot is heated—is distilled as described in the section "General Directions for Simple Distillation with a Hickman Still." Most low-melting solids will not crystallize during or immediately after distillation and can be transferred to a suitable container with a Pasteur pipet while the distillation apparatus is still warm. If the solid crystallizes in the Pasteur pipet, it can be melted with a heat gun or dissolved in a volatile solvent. If the solid crystallizes in the well of a Hickman still, it can be dissolved by pipetting a small amount of a volatile solvent, such as diethyl ether or dichloromethane, into the well. The sides of the well should be rinsed down with the liquid to dissolve any adherent solid. The solution is then removed with a Pasteur pipet and the solvent is evaporated to recover the solid. Alternatively, the solid in the well can be melted with a heat gun or heat lamp and then transferred.

c. Water Separation

During some reactions that yield water as a by-product, it may be necessary to remove the water to prevent decomposition of a water-sensitive product or to increase the yield. For standard scale work a *water separator*, such as the Dean-Stark trap shown in Figure E10, is often used for this purpose. The Dean-Stark trap is inserted in a reaction flask, filled to the level of its sidearm with the reaction solvent, and fitted with a reflux condenser. The reaction solvent must be less dense than water and immiscible with it; toluene, which forms an azeotrope with water, is often used for this purpose. As water forms during the reaction, its vapors and those of the reaction solvent condense inside the reflux condenser and drip down into the water separator. The organic solvent overflows through the sidearm and returns to the reaction flask, while the water collects in the bottom of the separator. The theoretical yield of water from the reaction can be calculated, so the volume of water in the separator is monitored to determine when the reaction is nearing completion.

For microscale work, a Hickman still can function as a water separator in an apparatus such as one pictured in Figure E8. The well of the Hickman still, which corresponds to the solvent reservoir of a Dean-Stark trap, can be previously filled with toluene or another suitable organic solvent. Alternatively, an amount of excess organic solvent equal to the well's capacity (1–2 mL) can be added to the pot, from which it distills into the well during the reaction. Any water that forms during the reaction distills into the Hickman still and collects in the bottom of its well, while the organic solvent overflows into the pot. Although the volume of water in the well cannot be estimated with much accuracy, observing no change in its volume over a period of 10 minutes or more may suggest that the reaction is over, or nearly so.

Figure E10 Dean-Stark trap

OPERATION 28   # Vacuum Distillation

Principles and Applications

The boiling point of a liquid decreases when the external pressure is reduced, so under reduced pressure a liquid distills at a temperature that is lower than its normal boiling point. For example, a liquid that boils at 200°C at a pressure of 1 atmosphere (760 torr) will boil near 100°C at 25 torr, as shown in Table E3. Reducing the boiling point of a liquid reduces the likelihood that thermal decomposition and other high-temperature reactions will occur during distillation. As a rule, most liquids that are heat sensitive or have boiling points of 200°C or higher at atmospheric pressure should be purified by distillation under reduced pressure, which is also called *vacuum distillation*.

Vacuum distillation has certain inherent features that can cause potential hazards and experimental difficulties. According to Boyle's law ($V \propto 1/P$), the volume of vapor generated by boiling a given amount of liquid increases when the pressure is reduced. For example, a vapor bubble that occupies a volume of 1 μL at 760 torr will expand to 30 μL at 25 torr. This can cause excessive bumping in the boiling flask and mechanical carryover of liquid to the receiver. Microporous boiling chips and efficient magnetic stirring help to maintain smooth boiling, and using the largest available round-bottom flask as the pot reduces carryover. Reducing the pressure also increases the vapor velocity because there are fewer molecules around to bump into; this can cause superheating of the vapor in the neck of a Hickman still and a pressure differential throughout the system, making the observed distillation temperature too high and the pressure reading too low. These problems can be circumvented by using the right kind of apparatus and by carrying out the distillation slowly. Even then, the separation attainable under vacuum distillation does not equal that possible at atmospheric pressure, and there is always the danger that the apparatus may implode because of the unbalanced external pressure on the system. Accordingly, a vacuum distillation must be performed with great care and attention to detail to obtain satisfactory results and prevent accidents.

Experimental Considerations

Apparatus. Although vacuum distillation can be used successfully in some microscale experiments, it is not very practical for liquid volumes much less than 0.5 mL. The apparatus for microscale vacuum distillation is similar to that for simple distillation except that a Claisen adapter or multipurpose adapter is needed to connect it to the vacuum source, as shown in Figure E13 in the "General Directions" section. The glassware used to assemble the apparatus must be free of cracks, star fractures, and other imperfections that might cause it to shatter under reduced pressure. All rubber connecting tubing should be thick-walled to prevent collapse under vacuum and as short as possible to reduce the pressure differential between the vacuum source and the system. The rubber tubes should be stretched or bent to

Table E3 Approximate boiling points of liquids at 25 torr

Normal bp	bp at 25 torr
150°C	60°C
200°C	100°C
250°C	140°C
300°C	180°C

see that they are pliable and free of cracks. Connections between rubber and glass, as well as ground-joint connections, must be secure and airtight.

Vacuum grease, which is used to make joints air-tight in standard scale apparatus, is generally *not* used on ground joints held together by compression caps, because an O-ring should provide a tight seal. Ungreased joints that are under vacuum for some time may freeze, however, so your instructor may suggest that you grease them lightly to prevent this. If so, he or she will show you how to grease them and clean them after use. All O-rings should be inspected to see that they are flexible and free of cracks or nicks that might allow air to enter. Microporous boiling chips are generally used to reduce bumping, but a stir bar may work with relatively large quantities of distilland. To help prevent mechanical carryover of liquid, it is advisable to use the largest distilling vessel available; thus a 10-mL round-bottom flask is preferable to a conical vial. If you are distilling a high-boiling liquid, insulate the narrow neck of the Hickman still by wrapping it with glass wool, which can be held in place with aluminum foil.

Vacuum Sources.   Most organic chemistry laboratories are provided with either vacuum lines that are connected to a central vacuum pump or water aspirators. In principle, an aspirator should be able to attain a reduced pressure equal to the vapor pressure of the water flowing through it, which is a function of the water temperature, as shown in Table E4. In practice, the pressure is often 5–10 torr higher because of insufficient water pressure, leaks in the system, or deficiencies of the aspirator itself. Thus with cold tap water at a temperature of 14°C running through it, a typical aspirator may provide a vacuum of around 20 torr.

An aspirator must be provided with a trap to prevent backup of water into the receiving flask due to changes of water pressure and to reduce pressure fluctuations throughout the system. Either a thick-walled glass bottle wrapped with plastic tape (see OP-13) or a thick-walled filter flask of the largest convenient size can be used for the trap. The trap should be provided with a pressure-release valve and hooked up to the aspirator and distillation apparatus as illustrated in Figure E13. If a vacuum line is used, a trap of the kind described here can be immersed in an ice-salt bath, dry ice in acetone, or some other efficient coolant to keep organic vapors out of the vacuum pump.

Pressure Measurement.   To know with any certainty when the desired substance is distilling, you need to know (or be able to estimate) the pressure inside the system so that you can estimate the boiling point of the desired component at that pressure. If a device such as the closed-end manometer shown in Figure E11 is attached to an evacuated system, the

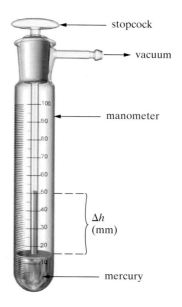

Figure E11 A closed-end manometer

Take Care! Never use a thin-walled container, such as an Erlenmeyer flask, as a trap; it may shatter under vacuum.

Table E4 Vapor pressure of water below 30°C

T, °C	P, torr	T, °C	P, torr
30	31.8	18	15.5
28	28.3	16	13.6
26	25.2	14	12.0
24	22.4	12	10.5
22	19.8	10	9.2
20	17.5	8	8.0

pressure (in torr) inside the system will be equal to the vertical distance (Δh, in millimeters) between the mercury levels in the inner tube and the cylinder. The mercury in a manometer can present a hazard if the vacuum is broken suddenly—air rushing in can push the mercury column forcefully to the closed end of the tube, breaking it and releasing toxic mercury into the laboratory. For this and other reasons, the vacuum must always be released *slowly*. It is advisable to open the valve connecting a manometer to an evacuated system only when a pressure reading is being made.

When the boiling points of any impurities are quite different from that of the main fraction, you should get by without a manometer if you can make a rough estimate of the pressure. For example, if you're using a water aspirator you can measure the temperature of the water, estimate its vapor pressure from Table E4, and estimate the minimum boiling temperature of the liquid as described in the next paragraph. The product should distill somewhat *above* that temperature; its actual boiling range will depend on the efficiency of the aspirator and the air tightness of your apparatus. You can also use a vacuum gauge to measure the pressure at the outlet of an aspirator or vacuum line, keeping in mind that the pressure in your distillation apparatus will be somewhat higher than the measured pressure.

Boiling Points Under Reduced Pressure. If the boiling point of a substance at a given pressure is not known, it can be estimated using the vapor pressure-temperature nomograph shown in Figure E12. More precise estimates can be made using tables such as those in R. R. Dreisbach, *Pressure–Volume–Temperature Relationships of Organic Compounds* (New York: McGraw-Hill, 1952), or by using various empirical relationships. The main fraction should be collected over a range that brackets the expected boiling point, keeping in mind that the distillation temperature of a liquid may vary by 10°C or more under vacuum because of pressure fluctuations and other factors.

Heat Sources. The heat source should be capable of providing constant, uniform heating to prevent bumping and superheating and to maintain a constant distillation rate. Although an oil bath works best, a heating block or sand bath is generally used for a microscale vacuum distillation. See "Heating" in OP-27 for additional information about the use of heat sources in a distillation.

General Directions for Vacuum Distillation

Safety Notes

Because of the possibility of an implosion, safety glasses *must* be worn during a vacuum distillation. Work behind a hood sash or safety shield while the apparatus is under vacuum. A rapid pressure increase, accompanied by a thick fog in the distilling flask, indicates decomposition of the distilland. If this occurs, remove the heat source *immediately* and get away (warning others to do so also) until the flask cools. Report the incident to your instructor.

Before beginning a vacuum distillation, you should make a rough estimate of the boiling temperature of your sample under vacuum, or be ready to estimate

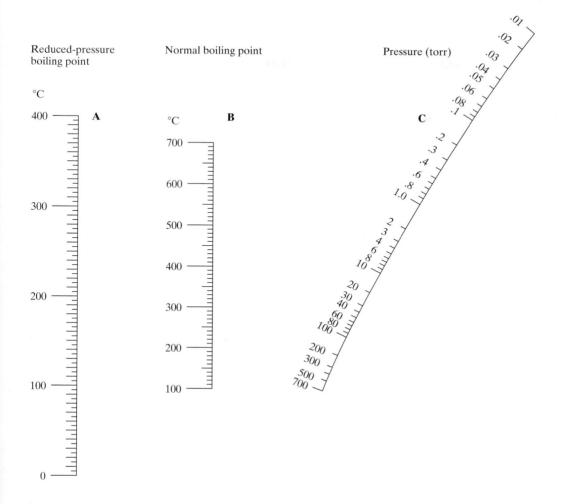

To estimate the boiling point at pressure P given the boiling point at another pressure P': (a) connect pressure P' in **C** with the boiling point at that pressure in **A** using a ruler, and place a sharp pencil point where the ruler intersects line **B**; (b) pivot the ruler around the pencil point until it reaches the desired pressure (P) in **C**, and then read the boiling point at that pressure from **A**.
Example: To estimate the boiling point of dibutyl phthalate at 10 torr from its reported boiling point of 236° at 40 torr, place a ruler at 40 torr in **C** and 236° in **A**, causing it to intersect line **B** at about 345°. Then hold a pencil point at 345° on **B**, pivot the ruler about that point to 10 torr in **C**, and read the boiling point from **A**. This yields an estimated boiling point of 197° at 10 torr.

Figure E12 Reduced-pressure boiling point nomograph

it using the nomograph in Figure E12 once you obtain a manometer reading.
Italicized instructions apply only if you are using a manometer; otherwise, disregard them.

Equipment and Supplies

heat source
microclamps, support
boiling flask (pot)

smooth-boiling device

Hickman still

*Claisen adapter

thermometer

*2 thermometer adapters (or 1 adapter and a 1-hole rubber stopper)

*bent glass tube (8 mm o.d.)

thick-walled rubber tubing

screw clamps

trap with pressure-release valve

manometer (optional)

3-way tubing connector (optional)

*These parts can be replaced by a multipurpose adapter, which is provided in some microscale lab kits.

Inspect all glassware, O-rings, and rubber tubing and replace any damaged items; if you have any doubt about their condition, see your instructor. Add the liquid to be distilled and a few microporous boiling chips (or a stir bar) to the boiling flask; then assemble the apparatus illustrated in Figure E13. If you are using a multipurpose adapter, connect it directly to the Hickman still. If you are using a Claisen adapter and have only one thermometer adapter, use it for the thermometer and insert the bent glass tube into a one-hole #0 rubber stopper (it must fit snugly). Make sure that the top of the thermometer bulb is just below the well of the Hickman still, as shown in Figure E9 [OP-27]. If a manometer is not used, omit the 3-way tubing connector shown. Connect the apparatus to a trap having a pressure-release valve, and connect the trap to the vacuum source. If you are using a vacuum line rather than an aspirator, the trap can be cooled in a suitable cold bath, as directed by your instructor.

Make sure that all joints and connections are tight. (If the liquid contains a volatile solvent, remove it by distilling at atmospheric pressure and let the apparatus cool before proceeding.) Open screw clamp R *and screw clamp M* and turn on the vacuum fully. Slowly close clamp R. (If bumping and foaming occur, there may be some residual solvent in the distilland; open clamp R and then close it down to a point at which the solvent will evaporate without excessive bumping. If a rubber tube collapses, replace it with heavy-walled tubing after releasing the vacuum.) *Wait a minute or two until the pressure equilibrates, then read the manometer and close clamp M. If the observed pressure is more than ~10 torr above the estimated pressure, check the system for leaks caused by loose joints, cracked tubing, and so forth. If you find any, release the vacuum by opening clamp R and fix them. If the pressure is satisfactory, use the nomograph shown in Figure E12 to estimate the boiling range at that pressure.*

If you are using a magnetic stirrer, turn it on. Begin heating to bring the mixture to the boiling point. If excessive foaming occurs upon boiling, reduce the heating rate. If bubbles form around any joint during the distillation, that joint is leaking air into the system; remove the heat source, release the vacuum by opening clamp R, and replace the O-ring or (if necessary) the glassware part containing the joint. When liquid begins to appear in the

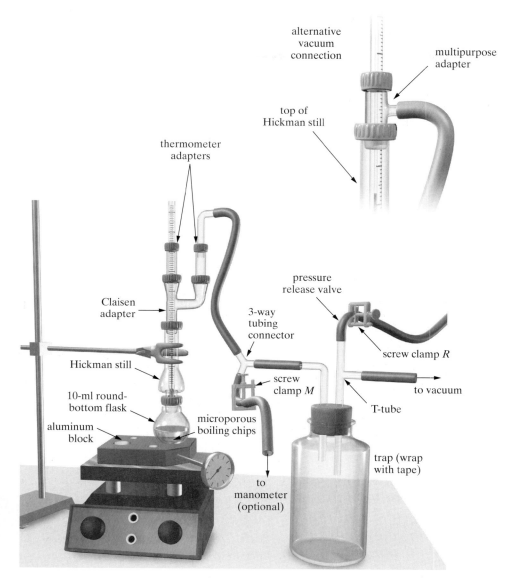

Figure E13 Apparatus for vacuum distillation

well of the Hickman still, record the temperature reading. *Open clamp M, let the pressure equilibrate, and record the pressure reading; then close clamp M and leave it closed except when you need to make another pressure reading.* Adjust the heat control to maintain a slow distillation rate without excessive bumping. If the temperature jumps up or down while the liquid is distilling, the pressure may be fluctuating due to changes in the aspirator flow rate; adjust the heating rate as necessary to maintain a suitable distillation rate. (To minimize such fluctuations, only a few students should use aspirators at the same time.)

If the initial distillation temperature is markedly lower than the estimated boiling range for the product, you are probably distilling a volatile forerun. Continue distilling until the temperature reaches the low end of the expected boiling range. Then remove the forerun by the following procedure:

1 Raise the apparatus away from the heat source and let the system cool down.
2 Open clamp R slowly until the system is at atmospheric pressure, then turn off the vacuum.
3 Remove the liquid in the well of the Hickman still with a Pasteur pipet. If you are using microporous boiling chips, add another chip.
4 Turn on the vacuum and slowly close clamp R.
5 Heat until distillation begins.
6 Record the boiling temperature, *open clamp M, record the pressure, close clamp M.*

Continue distilling until the upper end of the estimated boiling range is reached or until a significant drop in temperature indicates that the product is completely distilled. If the well of the Hickman still becomes full (or if you will be collecting a higher-boiling fraction), remove the liquid in the well and resume distillation by following Steps 1–6 above. Stop the distillation before the boiling flask is completely dry.

When the distillation is completed, follow Steps 1 and 2 for bringing the system back to atmospheric pressure; then remove any distillate. Disassemble the apparatus, clean the glassware promptly, and dispose of any residue and forerun as directed by your instructor.

Summary

1 Inspect glassware and tubing, add distilland and smooth-boiling device.
2 Assemble apparatus, check connections and joints.
3 Open clamp R *and clamp M*, turn on vacuum.
4 Close R, let pressure equilibrate.
5 *Read pressure, close M, estimate boiling range.*
6 Heat until distillation begins.
7 Record temperature *and pressure.*
   IF temperature is below estimated boiling range, GO TO 8.
   IF temperature is within estimated boiling range, GO TO 12.
8 Adjust heat and distill until temperature reaches lower end of expected boiling range.
9 Raise apparatus above heat source, let cool, open R.
10 Remove distillate, close R.
11 Heat until distillation begins, record temperature *and pressure.*
12 Distill until upper end of temperature range is attained or only a little distilland remains.
13 Raise apparatus above heat source, let cool, open R.
14 Turn off vacuum and remove distillate.
15 Disassemble and clean apparatus, dispose of residue and forerun.

Fractional Distillation

Principles and Applications

Although simple distillation can purify organic liquids that contain small amounts of volatile impurities, it is not a very effective means of separating the components of a mixture when each component makes up a substantial fraction of the mixture, unless the boiling points of the components are far apart. With mixtures of closer-boiling liquids, the distillate composition and boiling point will change continually throughout the distillation, as illustrated in Figure E14, and most of the distillate will be a relatively impure mixture of the components.

The separation could be improved by redistilling portions of the initial distillate and subsequent distillates. Such a process is diagrammed in Figure E15, in which the initial distillate is delivered directly to a second distilling flask, which redistills the condensed vapors and delivers its distillate to a third distilling flask, which distills that liquid into a receiver. (Note that the apparatus is purely hypothetical; there are more practical ways of accomplishing the same result.) This process can be understood by referring to the temperature-composition diagram in Figure E16 for the imaginary entane-orctane mixture discussed in OP-27.

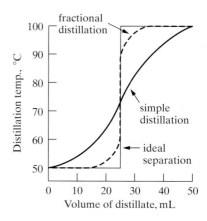

Figure E14 Separation efficiency of simple and fractional distillations

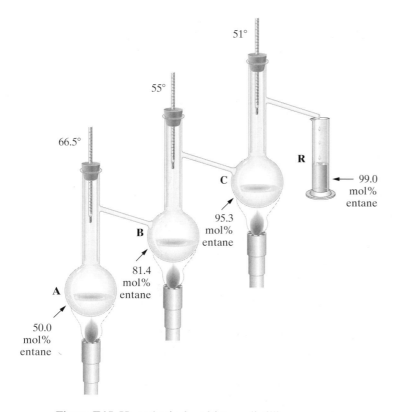

Figure E15 Hypothetical multistage distilling apparatus

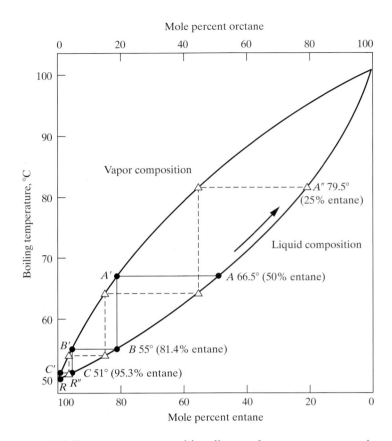

Figure E16 Temperature-composition diagram for entane-orctane mixture

When the mixture containing 50.0 mol% of both entane (bp = 50°C) and orctane (bp = 100°C) is boiled in flask **A** (point *A* on the diagram in Figure E16), the vapor over the mixture has a composition of 81.4 mol% entane and only 18.6 mol% orctane (point *A'*). This is because entane has a considerably higher vapor pressure at the boiling point of the mixture (66.5°C) than does orctane. The vapor is condensed into flask **B** (line *A'* – *B*), where it boils at 55°C to yield a vapor containing 95.3 mol% entane (line *B* – *B'*), which is condensed (line *B'* – *C*) into flask **C**. The condensed liquid in **C** boils at 51°C to yield a vapor that is 99.0 mol% entane (line *C* – *C'*), which is delivered to receiver **R** as a liquid of the same composition (line *C'* – *R*).

Only the first few drops of distillate will attain this degree of purity because as the more volatile entane is removed, the less volatile orctane will accumulate in the distilling flasks, reducing the proportion of entane in the vapor. For example, when the amount of entane in flask **A** has decreased to 25.0 mol% (point *A"*), the vapor condensing into the receiver (point *R"*) will be only 96.6 mol% entane, as shown by the broken lines in Figure E16. In other words, during a distillation, the distilland climbs inexorably up the temperature-composition graph and the distillate composition changes accordingly. Nevertheless, the separation effected by several distillation stages is considerably better than by only one, as can be seen by comparing the curves for simple distillation and fractional distillation shown in Figure E14.

The boiling point decreases with each subsequent distillation because the distilland becomes richer in the more volatile component.

Fractional distillation refers to a distillation process involving several concurrent vaporization-condensation cycles. During a fractional distillation, the distillate is collected in several separate *fraction collectors* (receivers), the contents of each being a different *fraction*. Each fraction is collected over a different temperature range, with the lower boiling fractions containing a greater percentage of the more volatile component and the higher boiling fractions containing more of the less volatile component. For an entane-orctane fractional distillation having the distillation curve illustrated by the broken line in Figure E14, the first 30 percent (15 mL) or so of distillate would be nearly pure entane and the last 30 percent would be nearly pure orctane. The middle fraction would be a mixture of the two, which could be redistilled if desired.

Like the hypothetical distillation diagrammed in Figure E15, fractional distillation is a multistage distillation process performed in a single operation. A vertical *distilling column* performs the function of boiling flasks **B** and **C** in the hypothetical apparatus, in effect redistilling the original distillate from flask **A**. The column is filled with some kind of *column packing*, which provides a large surface area from which repeated vaporization and condensation cycles can take place.

Suppose our 50.0 mol% mixture of entane and orctane is distilled through such a column, as diagrammed in Figure E17. The vapor leaving the pot will have the same composition (81.4 mol% entane) as it did in the hypothetical apparatus, but as it passes onto the column it will cool, condense onto the packing surface, and begin to trickle down the column on its way back to the pot. Since the temperature is higher near the pot, part of the condensate will vaporize on the way down, yielding a vapor richer in entane. This vapor will rise up the column until, at a higher level than before (since its boiling point is lower), it cools enough to recondense. This process of vaporization and condensation may be repeated a number of times on the way to the top of the column, so that, when the vapor finally arrives at that point, it is nearly pure entane. Although this is a continuous process involving the simultaneous upward flow of vapor and downward flow of liquid, the net result is the same as that produced by successive discrete distillations.

Each section of the distillation apparatus that provides a separation equivalent to one cycle of vaporization and condensation (one "step" on the temperature-composition graph) is called a *theoretical plate*. The first cycle occurs in the pot, which provides one theoretical plate; the column illustrated in Figure E17 contributes two more. Note that there is a continuous variation in both temperature and vapor composition as one proceeds up the column, and that the temperature is fixed by the liquid-vapor composition. For instance, the entane-rich liquid near the top of the column boils at a lower temperature than the original mixture, so the average temperature of plate 3 is lower than that of the plates below it.

The efficiency of a fractional distillation apparatus can be determined using the *Fenske equation*,

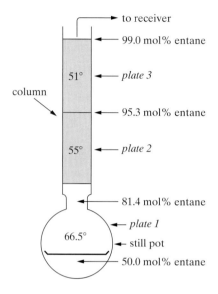

Figure E17 Operation of a fractionating column

In certain multistage columns, the "theoretical" plates are real. A Bruun column consists of a series of horizontal plates stacked at intervals inside a vertical tube; one vaporization-condensation cycle occurs on each plate.

$$n = \frac{\log \dfrac{Z_A}{X_A} - \log \dfrac{Z_B}{X_B}}{\log \alpha}$$

where n is the total number of theoretical plates, X_A and X_B are the mole fractions of liquids A and B in the distilland, and Z_A and Z_B are the mole fractions of the same components in the vapor that emerges from the top of the column. The term α is the *relative volatility* of the two liquids, where the volatility of a liquid in a mixture is the ratio of its mole fraction in the vapor to its mole fraction in the liquid. The pot provides one theoretical plate, so the number of theoretical plates provided by the column is $n - 1$. The efficiency of a particular kind of column or column packing is given by its *HETP* (height equivalent to a theoretical plate), which is equal to the height of the column divided by the number of theoretical plates it provides. For example, a 24-cm column that provides four theoretical plates has an HETP of 6 cm. The lower its HETP, the more efficient is the column.

In practice, a number of factors limit the efficiency of a given column. Efficiency is highest under the equilibrium condition of *total reflux*, in which all of the vapors are returned to the pot. In practice, some of the vapors are continually distilling into the receiver, which disturbs the equilibrium and reduces the column's efficiency. The efficiency of a distillation depends on the *reflux ratio (R)*, the ratio of liquid returned to liquid distilled measured over the same time interval.

$$R = \frac{\text{liquid volume returning to pot}}{\text{liquid volume distilled}}$$

According to one rule of thumb, the reflux ratio should at least equal the number of theoretical plates for efficient operation. Reflux ratios of 5–10 are common for routine separations. To a lesser extent, the distillation rate and the *holdup* of a column (the amount of liquid that adheres to the column's surface and packing) can also affect the efficiency of a distillation.

To this point, we have considered only ideal liquids. No real liquid systems display ideal behavior, although some may come very close. The greatest deviations from ideal behavior occur with liquid mixtures that form azeotropes, where an *azeotrope* is a solution of two or more liquids whose composition does not change during distillation. Azeotrope formation can make it difficult or impossible to purify certain liquids by distillation. For example, ethanol and water form a *minimum-boiling azeotrope* called 95% ethanol, which contains 95.6% ethanol and 4.4% water by mass and boils at 78.15°C—lower than the boiling point of either ethanol (78.5°C) or water. Distilling ethanol–water mixtures that contain a higher percentage of water eventually yields 95% ethanol, but it is impossible to obtain pure ethyl alcohol by distilling 95% ethanol, because its vapor has the same composition as the liquid in the pot. In this case, the problem caused by an azeotrope can be solved by a different azeotrope. Benzene and water form an azeotrope with ethanol that boils at a lower temperature (64.9°C) than the ethanol-water azeotrope, so by distilling 95% ethanol with some benzene, the benzene/water/ethanol azeotrope can be distilled off until all of the water is removed and absolute (100%) ethanol is obtained.

Experimental Considerations

Apparatus.   The column is the most important part of any fractional distillation apparatus. One of the simplest is the *Vigreux column*, which has a series of indentations in a spiral arrangement down the length of the column

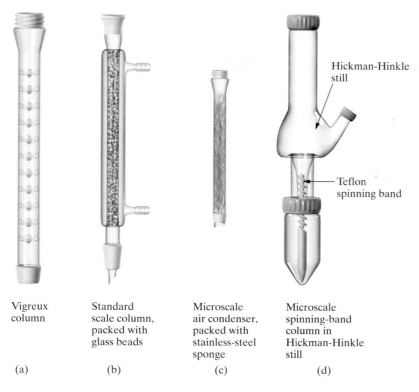

Vigreux
column

(a)

Standard
scale column,
packed with
glass beads

(b)

Microscale
air condenser,
packed with
stainless-steel
sponge

(c)

Microscale
spinning-band
column in
Hickman-Hinkle
still

(d)

Hickman-Hinkle
still

Teflon
spinning band

Figure E18 Some columns for fractional distillation

(see Figure E18a). The surface area for condensation inside the column is comparatively small, so Vigreux columns are not very efficient, having HETP values in the 8–12-cm range. At the other extreme, a *spinning-band column*, which contains a spiral of Teflon or some other material that is rotated inside the column, can have HETP values around 0.5 cm or lower. The Hickman-Hinkle still shown in Figure E18d is a modified Hickman still with a long neck that functions as a column and a Teflon spinning band that is rotated by a magnetic stirrer.

The simplest type of column for fractional distillation is simply a straight tube filled with a suitable packing material. Packed columns are more efficient than Vigreux columns, but their holdup is higher and distillation rates are lower. Highly efficient packing materials, such as glass helices, may provide HETP ratings down to about 1 cm, but they are quite expensive. Glass beads and stainless-steel sponge are more practical for use in most undergraduate laboratories. The jacketed standard scale column shown in Figure E18b is packed with glass beads, while the microscale column in Figure E18c is constructed of an air condenser packed with stainless-steel sponge. A stainless-steel sponge pad can be stretched and cut into 6–8 inch lengths, each weighing about 1.5 g, which are then pulled into the air condenser with a copper wire. With stainless-steel packing, the column is sometimes deliberately flooded by strong heating to wet the packing (see "Flooding"). Then the heat is reduced to drain the column before distillation is begun. Stainless-steel packing should not be used to distill halogen compounds, which corrode it.

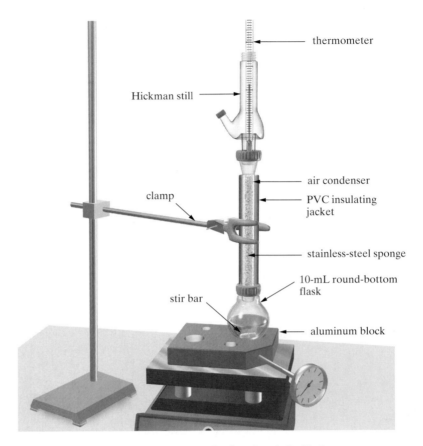

Figure E19 Apparatus for fractional distillation

The parts from a typical microscale lab kit can be used to construct fractional distillation setups like the one illustrated in Figure E19. Unless you are using a spinning-band column, it is seldom practical to purify less than 3–5 mL of liquid by fractional distillation, so the pot should ordinarily be a round-bottomed flask with a capacity of 10 mL or more. A microscale column made from an air condenser should be insulated to reduce heat loss. The column can be insulated with two concentric layers of clear PVC tubing, as shown in Figure E19, or with glass wool and aluminum foil. To insulate a column with PVC tubing, obtain two split 20-cm lengths of tubing of different diameters; then wrap the smaller (in diameter) tube around the column and the larger tube around the smaller one. For an air condenser having an outer diameter (o.d.) at its narrowest of $\frac{1}{2}$ inch, use $\frac{1}{2}$-inch inner diameter (i.d.) by $\frac{5}{8}$-inch o.d. and $\frac{5}{8}$-inch i.d. by $\frac{7}{8}$-inch o.d. tubing.

A Hickman still is set atop the column to collect the condensing vapors. The narrow neck of the still should be insulated by wrapping it with glass wool, which can be held in place with aluminum foil. If the distilland contains components that boil much below 100°C, it is a good idea to attach a condenser to the Hickman still. For most fractional distillations, however, no condenser is needed if the heating rate is kept low enough to prevent the escape of distillate vapors from the top of the still. If desired, an *uncapped* (important!) thermometer adapter or a multipurpose adapter with a sidearm can be inserted on top of the Hickman still to support the thermometer and reduce the likelihood of vapor escape.

Heat Sources. For good results, the heat source should provide constant, uniform heating. An oil bath works best, but a heating block or sand bath is suitable for microscale work if the heating rate is carefully controlled. (See "Heating" in OP-27 for additional information about the use of heat sources.)

Flooding. One problem often encountered during a fractional distillation is *flooding*, in which the column becomes partly or entirely filled with liquid. Flooding is usually caused by an excessive heating rate, but it may also be caused by poor insulation, an unsuitable packing support, or improper packing. For example, sponge packing that is too tightly compressed may hold up enough liquid to cause flooding. If flooding occurs, the apparatus should be removed from the heat source until all of the excess liquid has returned to the pot. Heating can then be resumed at a lower rate. Flooding will greatly decrease the efficiency of a separation, because it reduces the surface area of packing available for the separation.

Chasers. A column with a high holdup will retain a relatively large amount of distillate. To improve the distillate recovery, the held-up liquid can be driven off the column by a suitable high-boiling substance, called a *chaser*. For example, *p*-xylene (bp 138°C) is a good chaser for liquids boiling around 100° or below. After the last fraction has distilled, the chaser is added to the pot and heated to boiling, and the temperature is monitored as it climbs up the column. When the temperature begins to rise sharply, indicating that the chaser has reached the still head, distillation is immediately stopped and the recovered distillate is collected.

Take Care! Capping the thermometer adapter will create a closed system, which could shatter and cause serious injury.

General Directions for Fractional Distillation

Equipment and Supplies

heat source
microclamps, support
boiling flask (pot)
boiling chips *or* stir bar and stirrer
air condenser
column packing
insulating material
Hickman still
thermometer
condenser (optional)
condenser tubing (optional)
fraction collectors

If you are using column packing, pack the column (an air condenser) to within a centimeter or less of the upper ground joint. To pack a column with stainless-steel sponge, *pull* it into the column using a copper wire bent into a hook at one end, making sure that it is as uniform as possible. Wear gloves, or you might cut yourself on the sharp edges of the packing. Number and tare as many small screw-cap vials as there are fractions to collect. Add the distilland and one or two boiling chips or a stir bar to the boiling flask (usually a 10-mL or larger round-bottom flask), and assemble the apparatus illustrated in Figure E19. Make sure that all joints are tight, the column is perpendicular to the benchtop, and the apparatus and thermometer are supported securely. Position the thermometer correctly, as illustrated in Figure E9 [OP-27]. Insulate the column and the narrow neck of the Hickman still.

Position the boiling flask in the heat source, start the stirrer if you are using one, and begin heating to bring the mixture to the boiling point. When it boils, adjust the heating rate so that the reflux ring of condensing vapor passes up the column at a slow, even rate—it should take 5 minutes or more to reach the top of the column. Watch the packing at the bottom of the column closely for evidence of flooding. If flooding occurs, raise the apparatus away from the heat source immediately and let the liquid drain into the boiling flask; then resume heating at a lower rate. If flooding is still a problem, you may need to reinsulate or repack the column. When distillate begins to collect in the well of the Hickman still, read the thermometer after the temperature reading stabilizes. Distill the liquid so that the liquid level in the well rises very slowly; it should take 10 minutes or more to fill the well.

If the column is not very efficient or the components' boiling points are close together, the temperature may rise only gradually throughout the distillation. In that case, it is best to collect fractions continuously at regular temperature intervals and redistill them. Otherwise, continue distilling until the still-head temperature begins to rise sharply or until a predetermined target temperature is reached, then transfer the liquid from the well to the first fraction collector. Close the port (of a ported still) immediately after the liquid has been withdrawn. If you are collecting more than one fraction, transfer each fraction to a different collector and record the temperature range over which it was collected. If necessary, increase the heating rate to maintain a suitable distillation rate. Continue to collect fractions over appropriate temperature ranges until the boiling flask is nearly dry, the temperature drops sharply, or the final target temperature is reached. (Note that the temperature may also drop if the heating rate is too low, so that the hot vapors no longer reach the thermometer bulb.) Then remove the apparatus from the heat source and let the column drain. Any fractions collected while the temperature is rising rapidly are impure; unless you wish to redistill them, they should be placed in a solvent recovery container. Disassemble the apparatus and clean it promptly.

Summary

1 Pack column, add distilland and boiling chips or stir bar to pot.
2 Assemble and insulate apparatus.
3 Turn on stirrer (if used) and heat source.
4 Adjust heat so that reflux ring passes slowly up the column.
5 Record temperature when distillation begins and thermometer reading stabilizes.
6 Distill until temperature rises sharply or target temperature is reached, transfer distillate to collector, record temperature range.

IF more fractions are to be collected, change collector, REPEAT 6.
IF you are distilling the last (or only) fraction, CONTINUE.
7 Distill until temperature drops sharply or final target temperature is reached.
8 Remove apparatus from heat source, drain column.
9 Disassemble and clean apparatus, dispose of residue and any impure fractions.

F. Measuring Physical Constants

Melting Point

Principles and Applications

The *melting point* of a pure substance is defined as the temperature at which the solid and liquid phases of the substance are in equilibrium at a pressure of 1 atmosphere. At a temperature slightly lower than the melting point, a mixture of the two phases solidifies; at a temperature slightly above the melting point, the mixture liquefies. Melting points can be used to characterize organic compounds and to assess their purity. The melting point of a pure compound is a unique property of that compound, which is essentially independent of its source and method of purification. This is not to say that no two compounds will have the same melting point; many compounds have melting points that differ by no more than a fraction of a degree. If two pure samples have *different* melting points, however, they are almost certainly different compounds.

The melting point of an organic solid is usually measured by grinding the solid to a powder and packing the powder inside a *melting-point tube*, a capillary tube that is closed at one end. The melting-point tube is then placed in an appropriate heating device and the *melting-point range* of the sample—the range of temperatures over which the solid is converted to a liquid—is observed and recorded. A pure substance usually melts within a range of no more than 1–2°C; that is, the transition from a crystalline solid to a clear, mobile liquid occurs within a degree or two, if the rate of heating is sufficiently slow and the sample is properly prepared.

The presence of impurities in a substance *lowers* its melting point and *broadens* its melting-point range. To better understand the effects of impurities on melting points, consider the phase diagram for phenol (P) and diphenylamine (D) in Figure F1. Pure phenol melts at 41°C and pure diphenylamine at 53°C. If a sample of phenol contains diphenylamine as an impurity, its melting point will be lower than 41°C by an amount that depends on the mole percent of diphenylamine present. Similarly, the melting point of diphenylamine will be lower than 53°C if it contains phenol as an impurity. For example, the melting point of phenol containing 10 mol% diphenylamine is given by point M on the phase diagram, and that of diphenylamine containing 20 mol% phenol is given by point N.

Pure phenol and pure diphenylamine both have sharp melting points, meaning that the transition from solid to liquid occurs over a narrow temperature range. Mixtures of the two (except the *eutectic mixture* at the minimum in the diagram) exhibit broader melting-point ranges that depend on their composition. The approximate melting-point range for a mixture is given by the distance between the broken line connecting points P and E or points E and D and the solid lines connecting the same points on the phase diagram. For example, the melting-point range of phenol containing 10 mol% diphenylamine

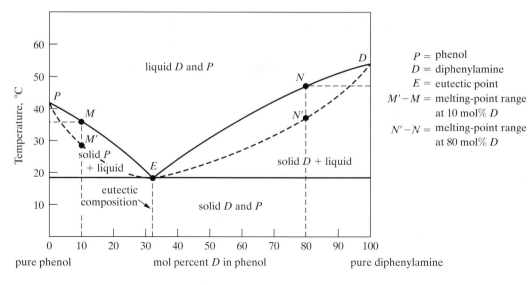

Figure F1 Phase diagram for the phenol-diphenylamine system

is given by the distance between M' and M, and the melting-point range of diphenylamine containing 20 mol% phenol is given by the distance between N' and N.

Since the melting point decreases in a nearly linear fashion as the amount of impurity increases (up to a point, at least), the difference between the observed and expected values can make it possible to estimate a compound's purity, as is done for camphor in Experiment 7. The melting-point lowering effect can also be used to confirm the identity of a substance, such as the product of a reaction, when it is thought to be a certain known compound. The compound in question is mixed with a sample of the known compound and the melting point of the mixture is measured. If the two compounds are identical, the mixture melting point will be essentially the same as that of the known compound. If they are not identical, the known compound will act as an impurity in the unknown, so the melting point of the mixture will be lower and its melting-point range broader than for the known compound.

Experimental Considerations

Melting Behavior.   The melting-point range of a sample is reported as the range between (1) the temperature at which the sample first begins to liquefy and (2) the temperature at which it is completely liquid, called the *liquefaction point.* When a single melting point is to be reported, the liquefaction point is generally used. A compound may also be characterized by its *meniscus point,* the temperature at which the liquid meniscus is barely clear of the solid below it. Some automatic melting-point devices report melting points that are closer to the meniscus point than to the liquefaction point.

If traces of solvent remain in a sample due to insufficient drying or other causes, you may observe "sweating" of solvent from the sample or bubbles in the molten sample, which may resolidify when all of the solvent is driven off. If a sample does show this behavior, it should be dried and the

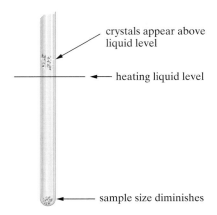

Figure F2 Sublimation of a sample in a melting-point tube

crystals appear above liquid level

heating liquid level

sample size diminishes

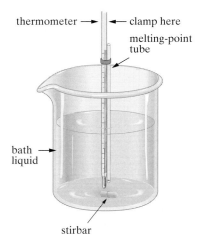

thermometer

clamp here

melting-point tube

bath liquid

stirbar

Figure F3 Stirred heating bath

melting point remeasured. Even a dry sample will tend to soften and shrink before it begins to liquefy; this process begins at the *eutectic temperature* (point *E* in Figure F1). In any case, softening, shrinking, and sweating should not be mistaken for melting behavior. The melting-point range does not begin until the first free liquid is clearly visible, at which time you should see movement of both solid and liquid in the melting-point tube.

Some compounds sublime—change directly from the solid to the vapor state—when they are heated in an open container. Sublimation can be detected during a melting-point determination by a pronounced shrinking of the sample, accompanied by the appearance of crystals higher up inside the melting-point tube (see Figure F2). The melting point of a sample that sublimes at or below its melting temperature can be measured using a sealed melting-point tube. An ordinary melting-point tube should be cut (see OP-3) short enough so that it does not project above the block of a melting-point apparatus such as the one in Figure F4 (see the "General Directions" section), or so that it can be entirely immersed in a heating-bath liquid. The sample is introduced and the open end is sealed in a burner flame (see OP-3). Then the melting point is measured by one of the methods described in the next section.

If a sample becomes discolored and liquefies over an unusually broad range during a melting-point determination, it is probably undergoing thermal decomposition. Compounds that decompose on heating usually melt at temperatures that vary with the rate of heating. The approximate melting point of such a compound should be measured by heating the melting-point apparatus to within a few degrees of the expected melting temperature before inserting the sample, and then raising the temperature at a rate of about 3–6°C per minute.

Apparatus for Measuring Melting Points. Melting points can be determined with good accuracy using a Thiele tube or Thiele-Dennis tube filled with a heating-bath liquid such as mineral oil, as illustrated in Figure F5 in the "General Directions" section. (Refer to OP-7a for precautions to be followed when using oil baths.) The design of Thiele-type tubes promotes good circulation of the heating liquid without stirring. A melting-point tube containing the solid is secured to a thermometer, which is then immersed in the bath liquid. The solid is observed carefully for evidence of melting as the apparatus is heated. Melting points can also be measured in a small beaker if the bath liquid is stirred constantly (Figure F3), preferably with a magnetic stirrer [OP-10].

Melting-point instruments that use capillary melting-point tubes, such as the Mel-Temp illustrated in Figure F4, are available commercially. Such melting-point devices are quite accurate if operated properly, and they can be used to make several measurements at once. This feature is useful when a mixture melting point is being determined, since the melting points of the unknown compound, the known, and the mixture can be measured and compared at the same time. The heating rate is adjusted by a dial controlling the voltage input to a heating coil. The dial reading required to attain the desired heating rate at the sample's melting point can be estimated from a heating-rate chart furnished with the instrument. The dial is initially set higher than this to bring the temperature to within 20°C or so of the expected melting point, and then reduced to the value estimated from the chart.

Either kind of apparatus can be used in the microscale lab, but commercial melting-point instruments are easier to use and provide more uniform heating than melting point baths.

Thermometer Corrections.   The observed melting point of a compound may be inaccurate because of defects in the thermometer used or because of the "emergent stem error" that results when a thermometer is not immersed in a heating bath to its intended depth. Many thermometers are designed to be immersed to the depth indicated by an engraved line on the stem, usually 76 mm from the bottom of the bulb; slight deviations from this depth will not result in serious error. Other thermometers are designed for total immersion; if they are used under other circumstances, the temperature readings will be in error. An emergent stem error can be compensated for by placing the bulb of a second thermometer opposite the middle of the exposed part of the thermometer's mercury column when the sample is melting, recording the temperature readings of both thermometers, and calculating an *emergent stem correction* with the following equation:

$$\text{emergent stem correction} = 0.00017 \cdot N(t_1 - t_2)$$

N = length in degrees of exposed mercury column
t_1 = observed melting temperature
t_2 = temperature at middle of exposed column

The emergent stem correction is added to the observed melting temperature, t_1. Melting points that have been corrected in this way should be reported as, for example, "mp 123–124° (corr.)."

Errors arising from thermometer defects as well as an emergent stem can be corrected by *calibrating* the thermometer under the conditions in which it is to be used. A thermometer used for melting-point determinations, for example, is calibrated by measuring the melting points of a series of known compounds, subtracting the observed melting point from the true melting point of each compound, and plotting this difference—the *melting-point correction*—as a function of temperature. The correction at the melting point of any other compound can then be read from the graph and added to its observed melting point, taking into account the sign (+ or −) of the correction. A list of pure compounds that can be used for melting-point calibrations is given in Table F1. Sets of pure calibration substances are available from chemical supply houses.

Mixture Melting Points.   A mixture melting point is obtained by grinding together approximately equal quantities of two solids (a few milligrams of each) until they are thoroughly intermixed, and then measuring the melting point of the mixture by the usual method. Usually one of the compounds (X) is an "unknown" that is believed to be identical to a known compound, Y. A sample of pure Y is mixed with X and the melting points of this mixture and of pure Y (and sometimes of X as well) are measured. If the melting points of pure Y and of the mixture are the same, to within a degree or so, then X is probably identical to Y. If the mixture melts at a lower temperature and over a broader range than pure Y, then X and Y are different compounds.

Table F1 Substances used for melting-point calibrations

Substance	mp (°C)
water ice	0
diphenylamine	54
m-dinitrobenzene	90
benzoic acid	122.5
salicylic acid	159
3,5-dinitrobenzoic acid	205

General Directions for Melting-Point Measurement

> **Mineral oil begins to smoke and discolor below 200°C and can burst into flames at higher temperatures. Oil fires can be extinguished with solid-chemical fire extinguishers or powdered sodium bicarbonate.**

Equipment and Supplies   (Starred items are for Method **B** only):

Mel-Temp (Method **A** only)
watch glass
flat-bladed spatula
1-m length of glass tubing
thermometer
capillary melting-point tube
*Thiele tube or Thiele-Dennis tube
*clamp, ring stand
*heating oil
*burner
*cut-away cork
*3-mm rubber ring

A melting-point tube can be constructed by sealing one end of a 10-cm length of 1-mm (i.d.) capillary tubing.

Put a few milligrams of the dry solid on a small watch glass and grind it to a fine powder with a flat-bladed spatula. Use the spatula to make a small pile of powder near the middle of the watch glass. Press the open end of a capillary melting-point tube into the pile until enough has entered the tube to form a column 1–2 mm high. Using too much sample can result in a melting-point range that is too broad and a melting-point value that is too high. Tap the closed end of the tube gently on the benchtop (or rub its sides with a small file); then drop it (open end up) through a 1-meter length of small-diameter glass tubing onto a hard surface, such as a bench top. Repeat this process several times to pack the sample firmly into the bottom of the tube. Use one of the methods that follow to measure the melting-point range of the sample. If you don't know the sample's expected melting point, determine its approximate value with rapid heating (6°C per minute or more); then carry out a more accurate melting-point measurement with a second sample as described here.

A. Mel-Temp Method.   Place the melting-point tube (sealed side down) in one of the channels on the Mel-Temp's heating block. Use the heating-rate chart to estimate the heating control dial setting that will cause the temperature to rise at a rate of 1–2°C per minute at the expected melting point of the sample. If the thermometer reading is well below the expected melting point, adjust the dial setting to raise the temperature quite rapidly until it is about 20°C below the expected melting point; then reduce the dial setting to the value estimated from the chart. As the temperature nears the expected melting point, adjust the dial as necessary so that the temperature rises at a rate of 1–2°C per minute while the sample is melting. Observe the sample through the eyepiece and record (as the limits of the melting-point range) the

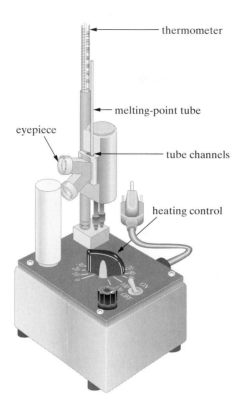

Figure F4 Apparatus for measuring melting points (Mel-Temp method)

temperatures (1) when the first free liquid appears in the melting-point tube and (2) when the sample is completely liquid. For best results, do at least two measurements on each compound. Let the block cool to 15–20°C below the melting point before you do another measurement. Cooling can be accelerated by passing an air stream over the block.

B. Thiele-Tube Method.   (You can use a similar method with other melting-point baths, such as the one shown in Figure F3, but stirring and a hot plate or other flameless heat source will be required in that case.) Clamp the Thiele tube or Thiele-Dennis tube securely to a ring stand and add enough mineral oil (or other appropriate bath liquid) to just cover the top of the sidearm outlet, as shown in Figure F5a. Secure the melting-point tube to a broad-range thermometer as follows:

1 Cut a 3-mm-thick rubber ring from $\frac{1}{4}$-inch i.d. thin-walled rubber tubing (rubber rings may be provided).
2 Place the rubber ring around the thermometer about 9 cm from its bulb end.
3 Pinch the rubber ring between your fingers to create a gap and insert the open end of the melting-point tube into the gap.
4 Move the melting-point tube until the sample is adjacent to the middle of the thermometer bulb (see Figure F5c).

Snap the thermometer into the center of a cutaway cork (Figure F5b) at a point above the rubber ring, with the melting-point tube and the degree

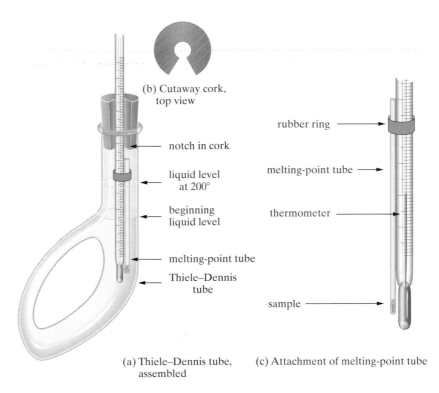

(b) Cutaway cork, top view

notch in cork

liquid level at 200°

beginning liquid level

melting-point tube

Thiele–Dennis tube

rubber ring

melting-point tube

thermometer

sample

(a) Thiele–Dennis tube, assembled

(c) Attachment of melting-point tube

Figure F5 Apparatus for measuring melting points (Thiele-tube method)

The cork is bored out to accommodate the thermometer, and then cut with a single-edged razor blade or a sharp knife so that the thermometer can be snapped into place from the side rather than inserted from the top.

markings on the same side as the opening in the cork. Insert this assembly into the bath liquid and move the thermometer, if necessary, so that its bulb is centered in the tube about 3 cm below the sidearm junction and the temperature can be read through the opening in the cork (Figure F5a). The rubber ring should be 2–3 cm above the liquid level so that the bath liquid, as it expands on heating, will not contact it. If the hot oil contacts the rubber, it may soften and allow the melting-point tube to drop out.

If the expected melting point of the compound is known, heat the bottom of the Thiele tube with a burner flame (or use a microburner, for more precise heat control) until the temperature is about 15°C below the expected value. Reduce the heating rate by turning down the flame and using it to heat the sidearm of the Thiele tube, so that the temperature rises at a rate of 1–2°C per minute near the melting point. Continue heating at that rate as you observe the sample closely, and record (as the limits of the melting-point range) the temperatures (1) when the first free liquid appears in the melting-point tube and (2) when the sample is completely liquid. For best results, do at least two measurements on each compound. Let the heating bath cool to 15–20°C below the melting point before you do another measurement. Cooling can be accelerated by passing an air stream over the tube.

Clean up the apparatus and either place the oil in a container for recycling or, if the oil is not too dark, store it in the Thiele tube. Mineral oil can be removed from glassware by rinsing the glassware with petroleum ether, followed by acetone, and then washing it with a detergent and water.

Summary

1 Assemble apparatus for melting-point determination, if necessary.
2 Grind solid to powder, fill melting-point tube(s) to depth of 1–2 mm.
   IF you are using Method **A**, GO TO 4.
3 Secure melting-point tube to thermometer; insert assembly in heating bath. GO TO 5.
4 Insert sample in Mel-Temp heating block.
5 Heat rapidly to ~15–20°C below mp, then reduce heating rate to 1–2°C/min.
6 Observe and record melting-point range.
7 Disassemble apparatus as needed, clean up.

Boiling Point

The *boiling point* of a liquid is defined as the temperature at which the vapor pressure of the liquid is equal to the external pressure at the surface of the liquid, and also as the temperature at which the liquid is in equilibrium with its vapor phase at that pressure. These definitions are the basis for various methods for measuring boiling points. For example, the boiling point of a liquid can be determined by distilling a small quantity of the liquid and observing the temperature at the still head, where the liquid and its vapors are assumed to be in equilibrium. It can also be determined by measuring the temperature at which the liquid's vapor, trapped inside a capillary tube immersed in the liquid, exerts a pressure equal to the external pressure. Like the melting point of a solid, the boiling point of a liquid can be used to help identify it and assess its purity.

Boiling-Point Corrections

The *normal boiling point* of a liquid is its boiling point at an external pressure of 1 atmosphere (760 torr, 101.3 kPa). Since the atmospheric pressure at the time of a boiling-point determination is seldom exactly 760 torr, observed boiling points may differ somewhat from values reported in the literature and should be corrected. If a laboratory boiling-point determination is carried out at a location reasonably close to sea level, atmospheric pressure will rarely vary by more than 30 torr from 760. For deviations of this magnitude, a *boiling-point correction*, Δt, can be estimated using Equation **1**.

$$\Delta t \approx y(760 - P)(273.1 + t) \qquad \textbf{(1)}$$

Δt = temperature correction, to be added to observed boiling point
P = barometric pressure, in torr
t = observed boiling point, in °C

The value 1.0×10^{-4} is used for the constant y if the liquid is water, an alcohol, a carboxylic acid, or another associated liquid; otherwise y is assigned the value 1.2×10^{-4}. For example, the boiling point of water at 730 torr is 98.9°C. Use of Equation **1** leads to a correction factor of $[(1.0 \times 10^{-4})(760 - 730)(273.1 + 98.9)]$ °C = 1.1°C, which yields the correct normal boiling point of 100.0°C.

At high altitudes, the atmospheric pressure may be considerably lower than 1 atmosphere, resulting in observed boiling points substantially lower than the normal values. For example, water boils at 93°C on the campus of the University of Wyoming at Laramie (elevation 7520 feet). For major deviations from atmospheric pressure, Equation 2 can be used in conjunction with approximate entropy of vaporization values obtained from the section "Correction of Boiling Points to Standard Pressure" in older editions of the *CRC Handbook of Chemistry and Physics*.

Equation 2 is a simplified form of the Hass-Newton equation found in the CRC Handbook, 64th Edition, p. D-189.

$$\Delta t \approx \frac{(273 + t)}{\phi} \log \frac{760}{P} \qquad (2)$$

$\phi = $ (entropy of vaporization at normal boiling point)/$2.303R$

For example, hexane boils at 49.6°C at 400 torr. The value of ϕ for alkanes is found (from the *CRC Handbook*) to be about 4.65 at that temperature. Substituting these values into Equation 2 yields a correction factor of

$$\Delta t \approx \frac{(273°C + 49.6°C)}{4.65} \log \frac{760 \text{ torr}}{400 \text{ torr}} = 19.3°C$$

which gives a normal boiling point of 68.9°C. A second approximation using the value of ϕ at the corrected boiling point gives 68.8°C, which compares very favorably to the reported normal boiling point of 68.7°C. As for melting-point determinations, it may be necessary to correct a boiling point for thermometer error, especially when working with high-boiling liquids. See "Thermometer Corrections" in OP-30 for instructions.

a. Distillation Boiling Point

During a carefully performed distillation of a pure liquid, the vapors surrounding the thermometer bulb are in equilibrium (or nearly so) with the liquid condensing on the bulb. Thus the vapor temperature recorded during the distillation of a pure liquid should equal its boiling point. If the liquid is contaminated by impurities, the distillation boiling point may be either too high or too low, depending on the nature of the impurity. Volatile impurities in a liquid lower its boiling point, whereas involatile impurities raise it; in either case, its boiling point will vary throughout its distillation. Therefore, if there is any doubt about the purity of a liquid, it should be distilled or otherwise purified prior to a boiling-point determination.

General Directions for Distillation Boiling-Point Measurement

Assemble an apparatus like the one in Figure E8 [OP-27] using a conical vial or 10-mL round-bottom flask and placing the thermometer as shown in Figure E9. If the liquid boils below 100°C, use a water-cooled condenser; otherwise you can use an air condenser. For high-boiling liquids, be sure that the still head or Hickman still is well insulated from the heat source to prevent superheating of the vapors. Add 2–5 mL of the pure liquid to the flask or conical vial and drop in a boiling chip or a stirring device. Distill the liquid slowly, recording the temperature readings when (1) the liquid has begun to collect in the well and the temperature has equilibrated; (2) about half of the liquid has distilled; (3) the pot is nearly empty but the temperature has not begun to drop. The lowest and highest readings (1 and 3) represent the boiling-point range, and the middle reading is the *median boiling point*. The median value should provide the best estimate of the actual boiling point. A boiling-point range greater than 2°C suggests that the liquid may need to be purified before further work is done. Record the barometric pressure so that you can make a pressure correction, if necessary.

b. Capillary-Tube Boiling Point

When a liquid is heated to its boiling point, the pressure exerted by its vapor becomes just equal to the external pressure at the liquid's surface. If a tube that is closed at one end is filled with a liquid and immersed (open end down) in a reservoir containing the same liquid, the tube will begin to fill with vapor as the liquid is heated to its boiling point. At the boiling point, the vapor pressure inside the tube will balance the pressure exerted on the liquid surface by the surrounding atmosphere, so that the liquid levels inside and outside the tube will be equal (Figure F6). If the temperature is raised above the boiling point, vapor will escape in the form of bubbles; if the temperature is lowered below the boiling point, the tube will begin to fill with liquid.

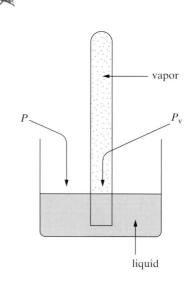

Figure F6 Boiling-point principle. At the boiling point, $P_v = P$, where P_v = pressure exerted by vapor on liquid surface, and P = pressure exerted by atmosphere on liquid surface.

This behavior is the basis of semimicroscale and microscale methods for measuring boiling points. The semimicroscale method, which uses 2–4 drops of liquid, can be used in the microscale lab when the amount of available liquid is reasonably large. It requires a *bell*—a capillary tube sealed at one end—that is inverted inside a *boiling-point tube* containing the liquid. A bell can be made by cutting [OP-3] an ordinary melting-point tube in half, saving the sealed half. The boiling tube is constructed by sealing one end of a piece of 5-mm o.d. soft glass tubing, cutting it to a length of 8–10 cm, and fire-polishing the open end (see OP-3). This assembly is then secured to a thermometer and heated in a Thiele or Thiele-Dennis tube. Alternatively, you can use a stirred oil bath as illustrated in Figure F3. (Refer to OP-7a for precautions to be followed when using oil baths.)

The microscale method, which requires only about 5 µL of liquid, is similar in practice except that a melting-point tube is used as the boiling tube and the bell is constructed from a capillary micropipet. Two bells can be made by cutting a 10-µL capillary micropipet (such as a Drummond Microcap) in half and sealing one end of each half (see OP-3).

It is important to avoid overheating during a boiling-point measurement; otherwise the liquid will boil away. It may be necessary to add more liquid if the sample is very volatile. If the bell moves up the capillary tube when the liquid boils, try using a longer, heavier bell. For best results, you should repeat the boiling-point measurement at least once. If repeated determinations on the same sample give appreciably different (usually higher) values for the boiling point, the sample is probably impure and should be distilled.

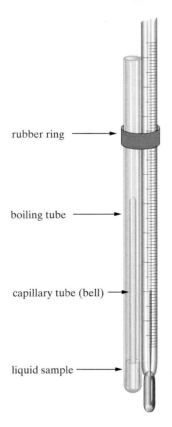

rubber ring →

boiling tube →

capillary tube (bell) →

liquid sample →

Figure F7 Semimicroscale boiling-point assembly

General Directions for Capillary-Tube Boiling-Point Measurement

Use Method **A** if the necessary supplies are available and you can spare a few drops of liquid. Otherwise, use Method **B**.

A. Semimicroscale Method

Equipment and Supplies

> boiling tube
> bell (made from mp tube)
> thermometer
> rubber ring
> Thiele-type tube
> mineral oil
> burner

Add 2–4 drops of the liquid to a boiling tube constructed of 5 mm o.d. glass tubing. Insert a bell (cut from a melting-point tube) into the boiling tube with its open end down, and secure the assembly to a thermometer by means of a rubber ring cut from thin-walled rubber tubing, as illustrated in Figure F7. Insert this assembly into a Thiele tube or Thiele–Dennis tube as for a melting-point determination (see Figure F5 [OP-30]), with the rubber ring 2 cm or more above the liquid level. Heat the tube with a burner flame (or use a microburner, for more precise heat control) until the temperature is within a few degrees of the expected boiling point of the liquid (if it is known). Then continue heating more slowly until you see a *rapid, continuous* stream of bubbles emerging from the bell. (The expansion of heated air in the bell will cause a *slow* evolution of bubbles; if you stop heating before rapid bubbling occurs, your reported boiling point will be too low.) Immediately remove the heat source and let the bath liquid cool slowly until the bubbling stops; then record the temperature when liquid *just* begins to enter the bell. Let the temperature drop a few degrees so that the liquid partly fills the bell. Then heat very slowly until the first bubble of vapor emerges from the mouth of the bell. Record the temperature at this point also. The two temperatures represent the boiling-point range; they should be within a degree or two of each other. Record the barometric pressure so that you can make a pressure correction, if necessary. Clean and dry the boiling tube, and save it for future boiling-point measurements. (See the Summary on the following page.)

B. Microscale Method

Equipment and Supplies

> boiling tube (capillary melting-point tube)
> bell (made from 10-μL micropipet)
> syringe
> Mel-Temp (or other capillary melting-point apparatus)

Construct a pair of bells by cutting a 10-μL micropipet in half using a sharp triangular file or glass-scoring tool, trying to cut it as squarely as possible.

Using a burner flame, carefully seal [OP-3] the cut end of each half. Use a syringe with a thin needle to introduce enough liquid into the boiling tube (a melting-point capillary tube) to fill it to a depth of approximately 5 mm. Holding the open end of the boiling tube securely, carefully "shake down" the liquid, as you would shake down the mercury column in a clinical thermometer, until it is all at the bottom of the tube and there are no air bubbles. (Alternatively, you can secure the boiling tube in a centrifuge tube using glass-wool packing and centrifuge it.) Insert a bell, open end down, into the boiling tube, and use a thin copper wire to push it all the way to the bottom, as shown in Figure F8. Insert the boiling tube into a Mel-Temp or other capillary melting-point apparatus. Heat until the temperature is within a few degrees of the expected boiling point of the liquid (if it is known). Then continue heating more slowly until you see a *rapid, continuous* stream of bubbles emerging from the bell. (The expansion of heated air in the bell will cause a *slow* evolution of bubbles; if you stop heating before rapid bubbling occurs, your reported boiling point will be too low.) Immediately lower the heat setting so that the temperature drops slowly. When the bubbling stops, record the temperature when liquid *just* begins to enter the bell. Then heat very slowly until the first bubble of vapor emerges from the mouth of the bell. Record the temperature at that point also. The two temperatures represent the boiling-point range; they should be within a degree or two of each other. Record the barometric pressure so that you can make a pressure correction, if necessary.

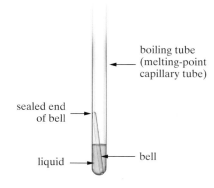

Figure F8 Microscale boiling-point assembly

Summary

1 Construct boiling-point bell.
2 Add liquid to boiling tube, insert inverted bell.
   IF you are using Method **B**, GO TO 4.
3 Secure boiling tube assembly to thermometer.
4 Insert assembly in Mel-Temp or heating bath.
5 Heat until continuous stream of bubbles emerges from bell, lower heat setting.
6 Record temperature when liquid just enters bell.
7 Heat slowly until first vapor bubble emerges from bell, record temperature.
8 Let cool below boiling point.
   IF another measurement is needed, add more liquid if necessary, GO TO 5.
9 Record barometric pressure, correct observed boiling point.
10 Disassemble apparatus, clean up.

Refractive Index   OPERATION **32**

The *refractive index* (index of refraction) of a substance is defined as the ratio of the speed of light in a vacuum to its speed in the substance in question. When a beam of light passes into a liquid, its velocity is reduced, causing it to bend downward. The refractive index is related to the angles that

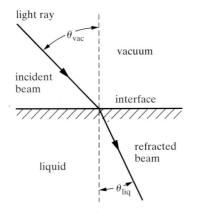

Figure F9 Refraction of light in a liquid

the incident and refracted (bent) beams make with a line perpendicular to the liquid surface, as shown in Figure F9 and Equation **1**.

$$n_\lambda{}^t = \frac{c_{vac}}{c_{liq}} = \frac{\sin \theta_{vac}}{\sin \theta_{liq}} \qquad (1)$$

$n_\lambda{}^t$ = refractive index at temperature t using light of wavelength λ

c = velocity of light

The refractive index is a unique physical property that can be measured with great accuracy (up to eight decimal places), and it is thus very useful for characterizing pure organic compounds. Refractive-index measurements are also used to assess the purity of known liquids and to determine the composition of solutions.

Experimental Considerations

It is much easier to make measurements in air than in a vacuum, so most refractive-index measurements are made in air, which has a refractive index of 1.0003, and the small difference is corrected by the instrument. The refractive-index value for a liquid depends on both the wavelength of the light used for its measurement and the density of the liquid, which varies with its temperature. Most values are reported with reference to light from the yellow D line of the sodium emission spectrum, which has a wavelength of 589.3 nm. Thus a refractive index value may be reported as $n_D{}^{20}$ = 1.3330, for example, where the superscript is the temperature in °C and "D" refers to the sodium D line. Since most refractive index readings are made at or corrected to this wavelength, the "D" is sometimes omitted. If another light source is used, its wavelength is specified in nanometers.

The Abbe refractometer (Figure F10) is a widely used instrument for measuring refractive indexes of liquids. It measures the *critical angle* (the

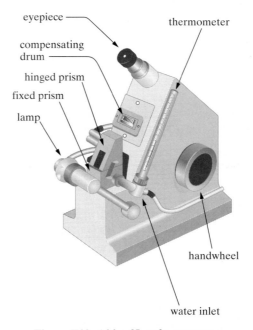

Figure F10 Abbe-3L refractometer

smallest angle at which a light beam is completely reflected from a surface) at the boundary between the liquid and a glass prism and converts it to a refractive-index value. The instrument employs a set of compensating prisms so that white light can be used to give refractive-index values corresponding to the sodium D line. A properly calibrated Abbe-3L refractometer should be accurate to within ±0.0002. The calibration of the instrument can be checked by measuring the refractive index of distilled water, which should be 1.3330 at 20°C and 1.3325 at 25°C.

The sample block of an Abbe refractometer can be kept at a constant temperature of 20.0°C by water pumped through it from a thermostatted water bath. If a refractive index is measured at a temperature other than 20.0°C, the temperature should be read from the thermometer on the instrument and the refractive index corrected by using the following equation:

$$\Delta n = 0.00045 \times (t - 20.0)$$

where t is the temperature of the measurement in °C. The correction factor (including its sign) is added to the observed refractive index. For example, if the refractive index of an unknown liquid is found to be 1.3874 at 25.1°C, its refractive index at 20.0°C should be about $1.3874 + [0.00045 \times (25.1 - 20.0)] = 1.3897$.

Alternatively, the refractive index of a reference liquid similar in structure and properties to the unknown liquid can be measured at the same temperature (t) as the unknown and a correction factor calculated from the equation

$$\Delta n = n^{20} - n^{t}$$

where n^{t} is the measured refractive index of the reference liquid at temperature t and n^{20} is the reported value of its refractive index at 20°C. The correction factor is then added to the measured refractive index of the unknown at temperature t. This method can correct for experimental errors (improper calibration of the instrument, etc.) as well as temperature differences.

Small amounts of impurities can cause substantial errors in the refractive index. For example, the presence of just 1% (by mass) of acetone in chloroform reduces the refractive index of the latter by 0.0015. Such errors can be critical when refractive-index values are being used for qualitative analysis, so it is essential that an unknown liquid be pure when its refractive index is measured.

General Directions for Refractive-Index Measurements

These directions apply to an Abbe-3L refractometer; if you are using a different instrument, your instructor will demonstrate its operation.

Equipment and Supplies

Abbe-3L refractometer
dropper
washing liquid
soft tissues

Raise the *hinged prism* of the refractometer and, using an eyedropper, place 2–3 drops of the sample in the middle of the *fixed prism* below it. Never allow the tip of a dropper or another hard object to touch the prisms—they are easily damaged. (If the liquid is volatile and free flowing, it may be introduced into the channel alongside the closed prisms.) With the prism assembly closed, switch on the *lamp* and move it toward the prisms to illuminate the visual field as viewed through the eyepiece. Rotate the *handwheel* until two distinct fields (light and dark) are visible in the eyepiece, and reposition the lamp for the best contrast and definition at the borderline between the fields. Rotate the *compensating drum* on the front of the instrument until the borderline is sharp and achromatic (black and white) where it intersects an inscribed vertical line. If the borderline cannot be made sharp and achromatic, the sample may have evaporated. Rotate the handwheel (or the fine adjustment knob, if there is one) to center the borderline exactly on the crosshairs (Figure F11).

Depress and hold down the *display switch* on the left side of the instrument to display an optical scale in the eyepiece. Read the refractive index from this scale (estimate the fourth decimal place). Record the temperature if it is different from 20.0°C. Open the prism assembly and remove the sample by gently *blotting* it with a soft tissue (do not rub!). Wash the prisms by moistening a tissue or cotton ball with a suitable solvent (acetone, methanol, etc.) and blotting them gently. When any residual solvent has evaporated, close the prism assembly and turn off the instrument.

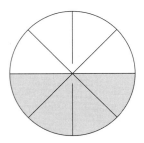

Figure F11 Visual field for properly adjusted refractometer

Summary

1 Insert sample between prisms.
2 Switch on and position lamp.
3 Rotate handwheel until two fields are visible, reposition lamp for best contrast.
4 Rotate compensating drum until borderline is sharp and achromatic.
5 Rotate handwheel or fine adjustment knob until line is centered on crosshairs.
6 Depress display switch, read refractive index, record temperature.
7 Clean and dry prisms, close prism assembly.
8 Make temperature correction if necessary.

OPERATION 33   # Optical Rotation

Principles and Applications

Light can be considered a wave phenomenon with vibrations occurring in an infinite number of planes perpendicular to the direction of propagation. When a beam of ordinary light passes through a *polarizer*, such as a Nicol prism, only the light-wave components that are parallel to the plane of the polarizer can pass through. The resulting beam of *plane-polarized light* has light waves whose vibrations are restricted to a single plane (see Figure F12).

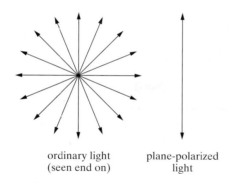

ordinary light   plane-polarized
(seen end on)   light

Figure F12 Schematic representations of ordinary and plane-polarized light

When a beam of plane-polarized light passes through an optically active substance, molecules of the substance interact with the light so as to rotate its plane of polarization. The angle by which a sample of an optically active substance rotates the plane of that beam is called its *observed rotation*, α. The observed rotation of a sample depends on the length of the light path through the sample and the concentration of the sample, as well as its identity. The first two variables are not intrinsic properties of the sample itself, so the observed rotation is divided by these variables to yield its *specific rotation*, $[\alpha]$, as shown in this equation.

$$[\alpha] = \frac{\alpha}{lc} \tag{1}$$

$[\alpha]$ = specific rotation
α = observed rotation
l = length of sample, in decimeters (dm)
c = concentration of solution, in g solute per mL solution
    (for a neat liquid, substitute the density in g/mL)

The specific rotation of a pure substance is an intrinsic property of the substance and can be used to characterize it.

The composition of a mixture of two optically active substances can be calculated from its specific rotation using the equation:

$$[\alpha] = [\alpha]_A X_A + [\alpha]_B (1 - X_A) \tag{2}$$

In this equation, $[\alpha]$ is the specific rotation of the mixture, $[\alpha]_A$ the specific rotation of component A, X_A the mole fraction of component A, and $[\alpha]_B$ the specific rotation of component B. For example, an equilibrium mixture of α-D-glucose ($[\alpha]$ = 112°) and β-D-glucose ($[\alpha]$ = 18.7°) has a specific rotation of 52.7°. The mole fraction of α-D-glucose in the mixture can be calculated by substituting these values into Equation **2** and solving for X_A.

$$52.7° = (112°)X_A + (18.7°)(1 - X_A)$$
$$X_A = 0.364$$

Since both forms of glucose have the same molecular weight, the equilibrium mixture contains 36.4% α-D-glucose and 63.6% β-D-glucose by mass.

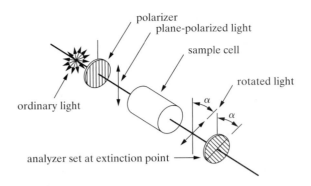

polarizer
plane-polarized light
sample cell
rotated light
ordinary light
α
α
analyzer set at extinction point

Figure F13 Schematic diagram of a polarimeter

Experimental Considerations

The specific rotation of an optically active sample can be determined using an instrument called a *polarimeter*. A polarimeter consists of a polarizer, a *sample cell* to hold the sample, and an *analyzer*—a second polarizing prism that can be rotated at an angle to the polarizer (see Figure F13). When the axis of the analyzer is perpendicular to that of the plane-polarized light, it blocks out the light completely; this condition is called the *extinction point*. In a precision polarimeter, the analyzer contains two or more prisms set at a small angle to each other, and it is rotated until the prisms bracket the extinction point, transmitting dim light of equal intensity. This feature is necessary because it is easier for the human eye to match two intensities than to estimate a point of minimum intensity. The angle by which the analyzer must be rotated to reach this point equals the angle at which the sample rotated the beam of plane-polarized light—its observed rotation, α. Using Equation **1**, we can convert the observed rotation of the sample to its specific rotation given the sample's concentration and the length of the sample cell. The most commonly used sample cells have lengths of 1 dm (10 cm) and 2 dm (20 cm).

The optical rotation of a substance is usually measured in solution. Water and ethanol are common solvents for polar compounds, and dichloromethane can be used for less polar ones. Often a suitable solvent will be listed in the literature, along with the light source and temperature used for the measurement. The volume of solution required depends on the size of the polarimeter cell, but 10–25 mL is usually sufficient. A solution for polarimetry ordinarily contains about 1–10 g of solute per 100 mL of solution and should be prepared using an accurate balance and a volumetric flask. If the solution contains particles of dust or other solid impurities, it should be filtered [OP-12]. When possible, the concentration should be comparable to that reported in the literature. For example, *The Merck Index* reports the specific rotation of (+)-menthol as "+49.2 (al, $c = 5$)," where the concentration (c) is given in grams per 100 mL of an alcoholic (al) solution (c is defined differently here than in Equation **1**). Thus, a menthol solution suitable for polarimetry could be prepared by accurately weighing about 0.50 g of (+)-menthol, dissolving it in ethyl alcohol in a 10-mL volumetric flask, and adding more alcohol up to the calibration mark. Less menthol than this can be used, but the results will be somewhat less accurate.

General Directions for Polarimetry

Equipment and Supplies

> polarimeter
> light source
> 1-dm sample cell
> 10-mL volumetric flask
> sample and solvent

This procedure applies to a Zeiss-type polarimeter with a split-field image. Experimental details may vary somewhat for different instruments. If the light source is a sodium lamp, make sure that it has ample time to warm up (some require 30 minutes or more).

Weigh the sample accurately and use a volumetric flask (or another appropriate volumetric container) to prepare about 10 mL of solution in a suitable solvent. Remove the screw cap and glass end plate from one end of a clean 1-dm sample cell (do not get fingerprints on the end plate). Rinse the cell with a small amount of the solvent you used to prepare the solution, and let it dry. (Alternatively, use a small amount of the solution—if you have enough—to rinse the sample cell, and fill the cell without delay.) Stand the polarimeter cell vertically on the bench top and fill it with the solution, rocking the cell if necessary to shake loose any air bubbles. Add the last milliliter or so with a Pasteur pipet so that the liquid surface is convex. Carefully slide on the glass end plate, trying not to leave any air bubbles trapped beneath it. If the cell has a bulge at one end, a small bubble can be tolerated; in that case, tilt the cell so that the bubble migrates to the bulge and stays out of the light path. Screw on the cap just tightly enough to provide a leakproof seal—overtightening it may strain the glass and cause erroneous readings.

Place the sample cell in the polarimeter trough, close the cover, and see that the light source is oriented to provide maximum illumination in the eyepiece. Set the analyzer scale to zero and (if necessary) rotate it a few degrees in either direction until a dark and a light field are clearly visible. You may see a vertical bar down the middle and a background field on both sides, as shown in Figure F14, or two fields divided down the middle. Focus the eyepiece so that the line(s) separating the fields are as sharp as possible. Rotate the analyzer scale away from zero in a clockwise direction, then (if necessary) in a counterclockwise direction, until you reach a point at which both fields are of nearly equal intensity. (If you have rotated it about 90° and see a very bright visual field, you have gone too far; return to zero and

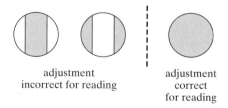

adjustment
incorrect for reading

adjustment
correct
for reading

Figure F14 Split-field image of polarimeter

try again.) Then back off a degree or so toward zero and use the fine adjustment knob (if there is one) to rotate the scale *away* from zero until the entire visual field is as uniform as possible and the dividing line(s) between fields have all but disappeared. If you overshoot the final reading, move the scale back a few degrees so that you again approach it going away from zero. Read the rotation angle from the analyzer scale, using the Vernier scale (if there is one) to read fractions of a degree, and record the direction of rotation, + for clockwise or − for counterclockwise. Obtain another reading of the rotation angle, this time approaching the final reading going *toward* zero. For accurate work, you should take a half-dozen readings or more, reversing direction each time to compensate for mechanical play in the instrument, and average them.

Rinse the cell with the solvent used in preparing the solution. Fill the cell with that solvent (or use a solvent blank provided) and determine the solvent's rotation angle by the same procedure as before. Remove the solvent and let the cell drain dry, or clean it as directed by your instructor. Subtract the average observed rotation of the solvent blank (note + and − signs) from that of the sample to obtain the observed rotation of the sample, and calculate its specific rotation using Equation **1**. When reporting the specific rotation, specify the temperature, light source, solvent, and concentration.

Summary

1 Prepare solution of compound to be analyzed.
2 Rinse sample cell and fill with solution, place in polarimeter.
3 Adjust light source, focus eyepiece.
4 Rotate analyzer scale until optical field is uniform, read rotation angle.
5 Repeat Step 4 several times, reversing direction each time, and average the readings.
  IF solvent blank has been run, GO TO 7.
  IF not, GO TO 6.
6 Place solvent blank in polarimeter, GO TO 4.
7 Clean cell, drain dry, dispose of solution and solvent as directed.
8 Calculate observed rotation and specific rotation of sample.

G. Instrumental Analysis

Gas Chromatography

Gas chromatography (GC) affords a powerful method for the separation and analysis of volatile components of mixtures. Like all chromatographic methods, its operation is based on the distribution of the sample components between a mobile phase and a stationary phase. *Gas-liquid chromatography* is the most useful form of gas chromatography. The mobile phase for gas-liquid chromatography is an unreactive *carrier gas* such as helium, and the stationary phase consists of a high-boiling liquid on a solid support, contained within a heated column. The components of the sample must have reasonably high vapor pressures so that their molecules will spend enough time in the vapor phase to travel through the column with the carrier gas. Thus gas chromatography can be used to separate gases, liquids that vaporize without decomposing when heated, and some volatile solids; the latter must first be dissolved in a suitable solvent. Less volatile liquids and solids can be separated by high-performance liquid chromatography [OP-35].

  Analytical gas chromatography is used for the qualitative and quantitative analysis of mixtures; that is, to identify the components of a mixture and find out how much of each component is present. An analytical gas chromatograph requires only a tiny amount of sample, usually a microliter or so. *Preparative gas chromatography* is used to separate the components of a mixture and recover the pure components, or to purify a major component of a mixture. Because even a large GC column will not accommodate sample sizes greater than about 0.5 mL, preparative GC cannot be used to separate large volumes of liquids. In the undergraduate organic chemistry lab, it is most useful for purifying the products of some microscale reactions.

Principles and Applications

In gas-liquid chromatography, the components of a mixture are partitioned between the liquid stationary phase and the gaseous mobile phase, the carrier gas. The time it takes for a component to pass through the column, called its *retention time*, depends on its relative concentrations in the liquid and gas phases. While the molecules of a component are in the gas phase, they pass through the column at the speed of the carrier gas. While they are in the liquid phase, they remain stationary. The more time a component spends in the gas phase, the sooner it will get through the column, and the lower its retention time will be. Conversely, the more time a component spends in the liquid phase, the longer its retention time will be. The time a component spends in the gas phase depends on its volatility (and thus on its boiling point) and the temperature of the column, while the time it spends in the liquid phase depends on the strength of the attractive forces between its molecules and the molecules of the liquid phase.

In order for any two components of a mixture to be separated sharply, they must have significantly different retention times, and their bands (the regions of the column they occupy) must be narrow enough so that they do not overlap appreciably as they exit the column. The degree of separation depends on a number of factors, including the length and efficiency of the column, the carrier gas flow rate, and the temperature at which the separation is carried out. Modern gas chromatographs provide the means to vary these factors precisely, and they offer an almost endless variety of applications, from analyzing automobile emissions for noxious gases to detecting PCBs in lake trout.

There are several important differences between gas-liquid chromatography and liquid-solid chromatographic methods such as column chromatography [OP-18] and thin-layer chromatography [OP-19]. Unlike the mobile phase in a liquid-solid separation, the mobile phase in gas chromatography does not interact with the molecules of the sample; its only purpose is to carry them through the column. Therefore, the separation of the components of a mixture depends mainly on (1) how strongly they are attracted to the stationary phase and (2) how volatile they are. If the components have similar polarities, they will be attracted to the stationary phase to about the same extent, so they will tend to be separated in order of relative volatility, with the lower-boiling components having the shorter retention times. If the components have different polarities, their retention times will depend in part on their relative polarities and the polarity of the stationary phase. A polar stationary phase will attract polar components more strongly than nonpolar ones, giving polar components longer retention times than nonpolar components, while a nonpolar stationary phase will attract nonpolar components more strongly, giving them the longer retention times.

A column chromatography or TLC separation is usually conducted at room temperature, and the temperature stays about the same throughout the separation. In contrast, a GC separation is usually carried out at an elevated temperature, and the temperature can be changed throughout the separation according to a preset program. Increasing the temperature of the column increases the fraction of each component in the gas phase, thus decreasing their retention times and reducing the time required for analysis. The column temperature also affects the separation efficiency; at excessively high temperatures, components will tend to spend most of their time in the vapor phase, causing their bands to overlap.

Instrumentation

In addition to the column and carrier gas, a gas chromatograph must have some means of vaporizing the sample, controlling the temperature of the separation, detecting each constituent of the sample as it leaves the column, and recording data from which the composition of the sample can be determined. A typical gas chromatograph includes the components diagrammed in Figure G1 and described here.

Injection Port.   The injection port is the starting point for a sample's passage through the column. The sample, in a microliter syringe, is injected into the column by inserting the needle through a rubber or silicone septum and depressing the plunger. The sample size for a packed column is typically

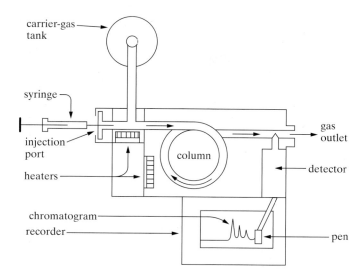

Figure G1 Schematic diagram of a gas chromatograph

about 1–2 μL, but it may vary from a few tenths of a microliter to 20 μL. With an open tubular column (described later), a sample splitter delivers about 1 nL (10^{-3} μL) of the injected sample to the column; the rest is discarded. The injector port is heated to a temperature sufficient to vaporize the sample, usually about 50°C above the boiling point of its least volatile component.

Packed Columns. A packed column is a long tube packed with a finely divided solid support whose particles are coated with the high-boiling liquid phase. The tube is usually made of stainless steel or glass. A typical column may be 1.5–3.0 meters long with an inner diameter of 2–4 mm (or $\frac{1}{8}$–$\frac{1}{4}$ inch o.d.), and it is usually bent into a coil to fit inside a column oven, where its temperature is controlled. Although such columns can be packed by the user, most are purchased ready made. The composition of the packing is described on a tag attached to each column. For example, a column packing described as "10% DEGS/Chromosorb W 80/100" contains 10% by mass of the liquid phase, diethylene glycol succinate (DEGS), coated onto a Chromosorb W (crushed diatomaceous earth) support having a particle-size range from 80 to 100 mesh (0.17–0.15 mm). As for a distillation column [OP-29], the efficiency of a chromatography column can be measured in terms of the number of theoretical plates it provides. Most packed columns have efficiencies ranging from 500 to 1000 theoretical plates per meter.

Open Tubular Columns. An *open tubular column* or *capillary column* consists of an open tube coated on the inside with the liquid phase. A very thin film of the liquid phase (ranging from 0.05 μm to 1.0 μm or more in thickness) is adsorbed on or chemically bonded to the inner wall of the column, which is a long capillary tube made of glass or fused silica. Fused silica open tubular (FSOT) columns are flexible enough to be bent into coils 15 cm or so in diameter. The columns range in size from "microbore" columns that may have an inner diameter of 0.05–0.10 mm and a length of 20 meters to "megabore" columns having an i.d. of about 0.50 mm and a

The mesh number is the number of meshes per inch in the finest sieve that will let the particles pass through.

length of 100 meters or more. Packed columns are generally cheaper, less fragile, and easier to use than open tubular columns, and they accommodate much larger sample sizes. But open tubular columns are faster and less likely to react with the sample, and they provide unparalleled resolution of sample components—a microbore column may have an efficiency of more than 10,000 theoretical plates per meter. Since an open tubular column may be 20–50 times longer than a typical packed column, it is usually hundreds of times more efficient than a packed column containing the same liquid phase.

Column Oven.   The column is mounted inside a heated chamber, the *column oven*, which controls its temperature. Under *isothermal operation*, the oven temperature is kept constant throughout a separation; under *program operation*, it is varied according to a programmed sequence. As a general rule, the column temperature for isothermal operation should be approximately equal to or slightly above the average boiling point of the sample. If the sample has a broad boiling range, it is best separated by program operation. The more volatile components are eluted during the low-temperature end of the program, and less volatile components are eluted at its high-temperature end.

Carrier Gas.   A chemically unreactive gas is used to sweep the sample through the column. Helium is the most widely used carrier gas, but nitrogen and argon are used with some detectors. Carrier gases usually come in pressurized gas cylinders, and various devices are used to measure and control their flow rates.

Detector.   The detector is a device that detects the presence of each component as it leaves the column and sends an electronic signal to a recorder or other output device. A separate oven is used to heat the detector enclosure so that the sample molecules remain in the vapor phase as they pass the detector. The most widely used detectors for routine gas chromatography are *thermal conductivity* (*TC*) detectors and *flame ionization* (*FI*) detectors. A TC detector responds to changes in the thermal conductivity of the carrier gas stream as molecules of the sample pass through. With an FI detector, the component molecules are pyrolyzed in a hydrogen-oxygen flame, producing short-lived ions that are captured and used to generate an electrical current. Thermal conductivity detectors are comparatively simple and inexpensive, and they respond to a wide variety of organic and inorganic species. Flame ionization detectors are much more sensitive than TC detectors, so they are used with open tubular columns (for which TC detectors are unsuited) as well as packed columns. An FI detector is somewhat inconvenient to use because it requires hydrogen and air to produce the flame, and it cannot be used for preparative gas chromatography because it destroys the sample.

Data Display.   A typical gas chromatograph intended for student use may have a mechanical *recorder* that records a peak on moving chart paper as each component band passes the detector. The resulting *gas chromatogram* consists of a series of peaks of different sizes, each produced by a different component of the mixture. The recorder may be provided with an *integrator* that measures the area under each component peak. Modern research-grade

Table G1 Selected liquid phases for gas chromatography

Stationary phase	Maximum T, °C	McReynolds numbers		
		X'	Y'	Z'
squalane	150	0	0	0
dimethylpolysiloxane (OV-1)	350	16	55	44
diphenyl/dimethylpolysiloxane (OV-17)	350	119	158	162
polyethylene glycol (Carbowax 20M)	250	322	536	368
diethylene glycol succinate (DEGS)	225	496	746	590
dicyanoallylpolysiloxane (OV-275)	275	629	872	763

Note: McReynolds numbers apply to the commercial stationary phase in parentheses. Commercial designations for the same kind of stationary phase vary widely; for example. AT-1, DB-1, Rtx-1, SE-30, and DC-200 are all similar to OV-1.

gas chromatographs are interfaced with computers and other hardware devices that display the peaks on a monitor screen, print or plot copies of the gas chromatogram, and provide digital readouts of peak areas, retention times, and various instrumental parameters.

Liquid Phases

The factor that most often determines the success or failure of a gas chromatographic separation is the choice of a liquid phase. In general, polar liquid phases are best for separating polar compounds and nonpolar liquid phases for separating nonpolar compounds. However, there are hundreds of liquid phases available, and choosing the right one for a particular separation may not be an easy task. Some typical liquid phases and their maximum operating temperatures are listed in Table G1 in order of polarity (lower to higher). General-purpose liquid phases such as dimethylpolysiloxane (dimethylsilicone) can separate a variety of compounds successfully, usually in approximate order of their boiling points. Other liquid phases have more specialized applications; for example, DEGS is a high-boiling ester that is used primarily to separate the esters of long-chain fatty acids.

The selection of a liquid phase can be facilitated by the use of *McReynolds numbers*, which indicate the affinity of a liquid phase for different types of compounds; the higher the number, the greater is the affinity. X' is a McReynolds number measuring the relative affinity of a liquid phase for aromatic compounds and alkenes; Y' measures its affinity for alcohols, phenols, and carboxylic acids; and Z' measures its affinity for aldehydes, ketones, ethers, esters, and related compounds. For example, the polar liquid phase DEGS has a much higher Z' value (590) than OV-1 (44), so it will retain an ester much longer. McReynolds numbers are particularly useful for selecting a liquid phase to separate compounds of different chemical classes, such as alcohols from esters. For example, a Carbowax 20M column should separate isopentyl acetate from isopentyl alcohol satisfactorily because its Y' and Z' values are very different, but an OV-17 column would not be a good choice for such a separation. Chromatography supply companies often provide detailed information about the kinds of separations their columns can accomplish.

Heating a column above its maximum operating temperature will cause the liquid phase to vaporize.

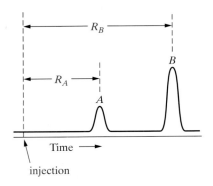

Figure G2 Retention times of two components, A and B

Qualitative Analysis

The retention time of a component is the time it spends on the column, from the time of injection to the time its concentration at the detector reaches a maximum. The retention time corresponds to the distance on the chromatogram, along a line parallel to the baseline, from the injection point to the top of the component's peak, as shown in Figure G2. This distance can be converted to units of time if the chart speed is known. If all instrumental parameters (temperature, column, flow rate, etc.) are kept constant, the retention time of a component will be characteristic of that compound and may be used to identify it. You can seldom (if ever) identify a compound with certainty from its retention time alone, but you can sometimes use retention times to confirm the identity of a compound whose identity you suspect. For example, finding that an unknown and a likely known compound have the same retention times on two or more different stationary phases, under the same operating conditions, suggests that the compounds are identical.

When you carry out a reaction and obtain a gas chromatogram of the reaction mixture, you may be able to guess which GC peak corresponds to which reactant or product, especially if the components have significantly different boiling points and the stationary phase is known to separate compounds in order of boiling point. It is always a good idea to back up your guess with proof, however. One way to identify the components of a reaction mixture is to "spike" the mixture with an authentic sample of a possible component and see which GC peak increases in relative area after spiking. For example, suppose that you carry out a Fischer esterification of acetic acid by ethanol and record a gas chromatogram of the reaction mixture.

$$CH_3COOH + CH_3CH_2OH \rightleftharpoons CH_3COOCH_2CH_3 + H_2O$$
$$\text{acetic acid} \qquad \text{ethanol} \qquad\qquad \text{ethyl acetate}$$

Since this is an equilibrium reaction, the reaction mixture would be expected to contain some unreacted starting materials as well as the product. To detect the presence of unreacted ethanol, you can add a small amount of pure ethanol to the sample and record a second gas chromatogram. The peak that increases in relative area is the ethanol peak, as shown in Figure G3. The process can be repeated using different pure compounds to identify the other components, if necessary.

Another way to identify the components of a mixture using GC is to obtain spectra of the components after they pass through the column. As described in OP-39, a mass spectrometer coupled to a gas chromatograph can act as a "superdetector," identifying the components as they come off the column. If the GC column has a large enough capacity that individual components can be collected at the column outlet, they can often be identified from their IR or NMR spectra.

Quantitative Analysis

The use of gas chromatography for quantitative analysis is based on the fact that, over a wide range of concentrations, a detector's response to a given component is proportional to the amount of that component in the sample. Therefore, the area under a component's peak can be used to determine its mass percentage in the sample. If a peak on a gas chromatogram is symmetrical, its

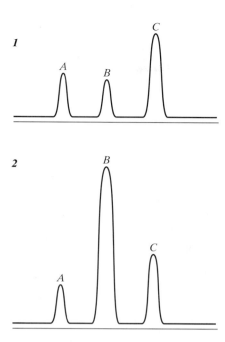

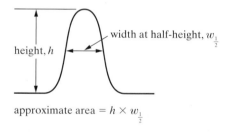

Figure G3 Use of gas chromatography for qualitative analysis: (1) chromatogram before adding ethanol; (2) chromatogram after adding ethanol

Figure G4 Measurement of approximate peak area

area can be calculated with fair accuracy by multiplying its height (h) in millimeters (measured from the baseline) by its width at a point exactly halfway between the top of the peak and the baseline $\left(w_{\frac{1}{2}}\right)$, as shown in Figure G4.

$$\text{approximate peak area} = h \times w_{\frac{1}{2}}$$

Many gas chromatographs are equipped with an integrator that automatically calculates the peak areas and displays them digitally. Peak areas can also be measured by making a photocopy of the chromatogram, accurately cutting out each peak with a sharp knife or razor blade (using a ruler to cut a straight line along the baseline), and weighing the peaks to the nearest milligram (or tenth of a milligram) on an accurate balance.

From the peak areas for a sample, you can sometimes estimate the percentage of each component in the mixture by dividing its area by the sum of the areas and multiplying by 100%. For example, suppose a gas chromatogram has three peaks, corresponding to three components, A, B, and C.

If the area of peak A is 1227 mm^2, the area of peak B is 214 mm^2, and the area of peak C is 635 mm^2, the sum of the areas is 2076 mm^2 so the mass percentages of the components are approximately

$$\%A = \frac{1227}{2076} \times 100\% = 59.1\%$$

$$\%B = \frac{214}{2076} \times 100\% = 10.3\%$$

$$\%C = \frac{635}{2076} \times 100\% = \underline{30.6\%}$$

$$100.0\%$$

This kind of calculation is valid only if the detector responds similarly to each component, which may be the case if you are analyzing very similar compounds, such as a mixture of long-chain fatty acid esters. More often, however, a detector will respond differently to different compounds. Flame ionization detectors respond mainly to ions produced by certain reduced carbon atoms, such as those in methyl and methylene groups. For example, ethanol is only about 52% carbon by mass while heptane is 84% carbon, so the response of an FI detector to a gram of ethanol is only about three-fifths as great as its response to a gram of heptane. Thermal conductivity detectors also respond differently to different substances, but the variations are usually not as large as with FI detectors.

For accurate quantitative analysis of most mixtures, you should multiply each component's peak area by a *detector response factor* to obtain a corrected area that is proportional to its mass. Detector response factors can be obtained from the literature in some cases, or by direct measurement. For example, detector response factors for cyclohexane and toluene using a TC detector are reported to be 0.942 and 1.02, respectively (relative to benzene). Suppose that the gas chromatogram of a cyclohexane-toluene mixture is recorded using such a detector and has peak areas of 78.0 mm^2 for cyclohexane and 14.6 mm^2 for toluene. The corrected peak areas are then $(0.942 \times 78.0 \text{ mm}^2) = 73.5$ mm^2 for cyclohexane and $(1.02 \times 14.6 \text{ mm}^2) = 14.9$ mm^2 for toluene. Dividing each corrected area by the sum of the corrected areas and multiplying by 100 gives mass percentages of 83.1% for cyclohexane and 16.9% for toluene.

If detector response factors for the components of a mixture are not known, you can determine them by the following method, which is described for two components, A and R. Component R is a reference compound, arbitrarily assigned a detector response factor of 1.00.

- Obtain a gas chromatogram of a mixture containing carefully weighed amounts of A and R.
- Measure the peak area for each component.
- Calculate the detector response factor for A using the following relationship:

$$\text{detector response factor for A} = \frac{\text{mass}_A}{\text{area}_A} \times \frac{\text{area}_R}{\text{mass}_R}$$

This method can be used to calculate detector response factors for any number of components simultaneously if their peaks are well resolved. Each component (including the reference compound) is carefully weighed, a gas chromatogram of the mixture is recorded, all of the peak areas are measured, and the detector response factors are calculated from the resulting masses and areas using the preceding equation.

Preparative Gas Chromatography

Most preparative gas chromatographs are too expensive to be used routinely in undergraduate laboratories, but some inexpensive analytical gas chromatographs, such as the Gow-Mac 69-350, can be operated with preparative columns using a metal adapter that is connected to the exit port on the gas chromatograph. A special GC collection tube, provided in some microscale lab kits, is inserted into the adapter when the desired component's peak begins to appear on the chromatogram, and removed just after all of the peak has been recorded. The condensed liquid is then transferred to a small (~100 μL) conical vial by centrifugation. Cotton or glass-wool packing is used to support the conical vial and to center the top of the collection tube in the centrifuge tube. The liquid can then be used to obtain an infrared spectrum [OP-36] or nuclear magnetic resonance spectrum [OP-37] of the compound.

If an adapter and specialized collection glassware are not available, a 2-mm o.d. glass tube about 8 cm in length can be packed with glass wool and inserted into the GC exit port while the desired component's peak is being recorded. The component should condense on the surface of the glass wool, from which it can be washed off with a small amount of solvent (if the component is quite volatile, the tube should be chilled first). An IR or NMR spectrum of the compound can then be obtained in a solution of the solvent used, or the solvent can be evaporated.

General Directions for Recording a Gas Chromatogram

Equipment and Supplies

> gas chromatograph and recorder
> 10-μL microsyringe
> screw-cap vial containing sample
> lab tissues
> syringe-washing solvent

Do not attempt to operate the instrument without prior instruction and proper supervision. The directions that follow are for a typical student-grade gas chromatograph connected to a mechanical recorder. For other kinds of instruments, follow your instructor's directions. Consult the instructor if the instrument does not seem to be working properly or if you have questions about its operation. It will be assumed that all instrumental parameters have been preset, that an appropriate column has been installed, and that the column oven will be operated isothermally. If not, the instructor will show you what to do. Before you begin, be sure you have read the section in OP-5 about the use of syringes.

If the sample is a volatile solid, dissolve it in the minimum volume of a suitable low-boiling solvent; otherwise, use the neat liquid. Rinse a microsyringe with the sample a few times, and then partially fill it with the sample. A microsyringe is a very delicate instrument, so handle it carefully and avoid using excessive force that might bend the needle or plunger. If there are air bubbles inside the syringe, tap the barrel with the needle pointing up, or eject the sample and refill the syringe more slowly. Hold the syringe with the needle pointing up and expel excess liquid until the desired volume of sample (usually 1–2 μL) is left inside. Wipe the needle dry with a tissue, and pull the plunger back a centimeter or so to prevent prevaporization of the sample.

Take Care! The injection port is hot! Don't touch it.

Set the chart speed, if necessary, and switch on the recorder-chart drive (and the integrator, if there is one). Carefully insert the syringe needle into the injection port by holding the needle with its tip at the center of the septum and pushing the barrel slowly, but firmly, with the other hand until the needle is as far inside the port as it will go. Inject the sample by *gently* pushing the plunger all the way in just as the recorder pen crosses a chart line (the starting line); use as little force as possible to avoid bending the plunger. Withdraw the needle and promptly mark the starting line, from which all retention times will be measured. Let the recorder run until all of the anticipated component peaks have been recorded. Then turn off the chart drive and tear off the chart paper using a straightedge. Inspect the chromatogram carefully; if it is unsuitable because the significant peaks are too small, poorly resolved, or off-scale, repeat the analysis after taking measures to remedy the problem. If there is evidence of prevaporization (as indicated by an extraneous small peak preceding each major peak at a fixed interval), be sure that the plunger is pulled back before you inject the sample for the next run. Injecting the sample immediately after you insert the needle in the injector port will also prevent prevaporization, but this technique requires good timing.

Before you analyze a different sample, clean the syringe and rinse it thoroughly with that sample. When you are finished, rinse the syringe with an appropriate low-boiling solvent such as methanol or dichloromethane, remove the plunger, and set it on a clean surface to dry.

Summary

1 Rinse syringe, fill with designated volume of sample.
2 Start chart drive and integrator (if applicable).
3 Insert syringe needle into injector port, inject sample when pen crosses chart line.
4 Withdraw needle, mark starting line.
5 Stop chart drive after last peak is recorded.
6 Remove chromatogram, clean and dry syringe.

OPERATION 35

High-Performance Liquid Chromatography

Principles and Applications

High-performance liquid chromatography (HPLC) can be regarded as a hybrid of column chromatography and gas chromatography, sharing some features of both methods. As in column chromatography, the mobile phase is a solvent or solvent mixture (the eluent) that carries the sample through a column packed with fine particles that interact with the components of the sample to different extents, causing them to separate. As in gas chromatography, the sample is usually injected onto the column and detected as it leaves the column, and its passage through the column is recorded as a series of peaks on a chromatogram. The stationary phase may be a solid adsorbent, as in column chromatography, but it is more often an organic phase that is bonded to tiny beads of silica gel. The silica beads are, in effect, coated with a very thin layer of a liquid organic phase. Each component of the sample is partitioned between a liquid mobile phase and the liquid stationary phase according to a ratio—the *partition coefficient*—that depends on its solubility in each liquid. The components of a mixture generally have different partition coefficients, so they pass down the column at different rates.

Instrumentation

HPLC was developed as a means of improving the efficiency of a column chromatographic separation by reducing the particle size. Most column chromatography packings contain particles with diameters in the 75–175 μm range, while most modern HPLC packings have particle sizes in the 3–10 μm range, increasing separation efficiency dramatically. But solvents will not easily flow through such small particles by gravity alone, so a powerful pump is needed to force them through the column at pressures up to 6000 psi (~400 atm).

The basic components of an HPLC system are diagrammed in Figure G5. The instrument ordinarily has several large solvent reservoirs, each of which can be filled with a different eluting solvent. *Isocratic elution* is elution using a single solvent. *Gradient elution* utilizes two or more solvents of different polarity and varies the solvent ratio throughout a separation according to a programmed sequence. Gradient elution can reduce the time needed for a separation and increase the separation efficiency.

Preparative HPLC systems, which require special wide-bore columns, are equipped with fraction collectors to collect the eluent as it comes off the column. The fractions are then evaporated to yield the pure components. Analytical HPLC systems, which are used to determine the compositions of mixtures, require much smaller samples and the components are not recovered. A typical analytical HPLC column is a straight stainless steel tube with a length of 10–25 cm and an inner diameter of 2.1–4.6 mm. A microbore analytical column may have an inner diameter of 1 mm, while a large preparative column may have an inner diameter of 50 mm or so. Samples are introduced onto the column by means of a syringe or sampling valve and are carried through it by the pressurized eluent mixture. Because the particles of the stationary phase are so small, a column can easily be plugged by particulate matter or adherent solutes introduced by either the eluent or the sample. To remove anything that might harm the column, the eluent is forced through one or more filters and the sample may be introduced through a short *guard column*.

As each component of a sample leaves the column, the detector responds to some property of the component, such as its ability to absorb ultraviolet radiation. The detector then sends an electronic signal to a recorder, which

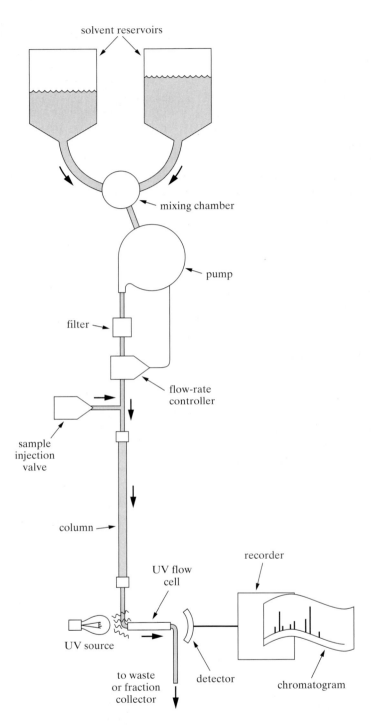

Figure G5 Schematic diagram of HPLC system

traces the component's peak on a chart. The resulting chromatogram is a graph of some property of the components, such as UV absorbance, plotted against the volume of the mobile phase. Because different solutes may have greatly different UV absorptivities at the wavelength used by an *ultraviolet absorbance detector*, for example, a detector response factor (see OP-34) must be determined for each component before its percentage in the mixture can be calculated. Components that do not absorb ultraviolet radiation will not be detected by a UV detector; in that case, the components can be converted to derivatives that do absorb, or a different type of detector can be used. Any component whose refractive index is different from that of the eluent can be detected by a *refractive index (RI) detector*, but RI detectors are less sensitive than UV detectors.

High-performance liquid chromatography can be used to separate mixtures containing proteins, nucleic acids, steroids, antibiotics, pesticides, inorganic compounds, and many other substances whose volatilities are too low for gas chromatography. Because HPLC separations can be run at room temperature, there is little danger of decomposition reactions or other chemical changes that sometimes occur in the heated column of a gas chromatograph. These advantages have made HPLC the fastest growing separation technique in chemistry, but its use in undergraduate laboratories is limited because of the high cost of the instruments, columns, and high-purity solvents required (HPLC-grade water costs about $50 a gallon!).

Stationary Phases

Some HPLC columns contain a solid adsorbent such as silica gel or alumina. Such a stationary phase is much more polar than the eluent, so nonpolar components are eluted faster than polar ones. Most modern HPLC columns contain a *bonded liquid phase*—an organic phase that is chemically bonded to particles of silica gel. Silica gel contains silanol ($-Si-OH$) groups to which long hydrocarbon chains, such as octadecyl groups, can be attached by reactions such as the following:

$$-Si-OH \xrightarrow{R_2SiCl_2} -Si-O-\overset{\displaystyle R}{\underset{\displaystyle R}{\overset{|}{\underset{|}{Si}}}}-Cl \xrightarrow{H_2O} \xrightarrow{(CH_3)_3SiCl}$$

$$-Si-O-\overset{\displaystyle R}{\underset{\displaystyle R}{\overset{|}{\underset{|}{Si}}}}-O-Si(CH_3)_3 \qquad R = \text{octadecyl, } CH_3(CH_2)_{17}-$$

The bonded stationary phase in this example is *less* polar than the eluent, which may be a mixture of water with another solvent such as methanol, acetonitrile, or tetrahydrofuran. Therefore, polar components will spend more time in the eluent than in the stationary phase and will be eluted faster than nonpolar ones, reversing the usual order of elution. This mode of separation is called *reversed-phase chromatography*.

Other stationary phases operate by still different mechanisms. The stationary phase for *size-exclusion chromatography* is a porous solid that separates molecules based on their effective size and shape in solution. Small molecules can enter even the narrowest openings in the porous structure, larger molecules find fewer openings they can get into, and still larger molecules may be completely excluded from the solid phase. As a result, large, bulky molecules pass down the column faster than smaller molecules. A stationary phase for *ion-exchange chromatography* has ionizable functional groups that carry a negative or positive charge and therefore attract certain ionic solutes. In organic chemistry, such stationary phases are used mainly to separate ionizable organic compounds such as carboxylic acids, amines, and amino acids. Chiral stationary phases that can separate enantiomers and determine their optical purity are also available.

Hundreds of different stationary phases are used for HPLC separations. Some of the most popular reverse-phase packings contain silica gel bonded to methyl ($-CH_3$), phenyl ($-C_6H_5$), octyl [$-(CH_2)_7CH_3$], octadecyl [$-(CH_2)_{17}CH_3$], cyanopropyl [$-(CH_2)_3CN$], and aminopropyl [$-(CH_2)_3NH_2$] groups. Size-exclusion stationary phases contain silica, glass, and polymeric gels of varying porosity. Ion-exchange stationary phases include (a) styrene-divinylbenzene copolymers to which ionizable functional groups (such as $-SO_3H$ and $-NR_4^+OH^-$) are attached; (b) beads with a thin surface layer of ion-exchange material; and (c) bonded phases on silica particles.

General Directions for High-Performance Liquid Chromatography

Do not attempt to operate the instrument without prior instruction and proper supervision. Because HPLC systems vary widely in construction and operation, only a very general outline of the procedure is provided here. It will be assumed that all instrumental parameters have been preset, that an appropriate reverse-phase column has been installed, and that the elution will be isocratic. If not, the instructor will provide additional directions.

Prepare an approximately 0.1% stock solution of the sample in the same solvent or solvent mixture as the one being used for the elution, and dilute an aliquot of this solution further, if necessary. Be certain that you are using HPLC-grade solvents for preparing the solution and eluting it through the column. The solvents should be purged with helium prior to elution to remove dissolved gases. Filter the solution through a 1.0-μm membrane filter to remove any particulate matter. Degas it, if necessary, as directed by your instructor. Inject the sample through the injection port, or use a sampling valve to introduce it into the system. (Using a guard column to protect the analytical column from particles is advisable.) Wait until no more peaks appear on the chromatogram, and then inspect the chromatogram to see if the components are well resolved. If they are not, repeat the determination using a higher percentage of the more polar solvent (usually water) in the solvent mixture.

Infrared Spectrometry

Principles and Applications

The atoms of a molecule behave as if they were connected by flexible springs rather than by rigid bonds resembling the connectors of a ball-and-stick molecular model (see Figure G6). A molecule's component parts can oscillate in different *vibrational modes*, which are vividly described by such terms as stretching, rocking, scissoring, twisting, and wagging. When infrared (IR) radiation is passed through a sample of a pure compound, its molecules can absorb radiation of the energy and frequency needed to bring about transitions between vibrational ground states and vibrational excited states. For example, if a molecule contains a C—H bond that vibrates 90 trillion times a second in its vibrational ground state, the molecule must absorb infrared radiation having exactly that frequency (9.0×10^{13} Hz) to jump to an excited state in which the C—H bond vibrates twice as fast. The frequency of infrared radiation is usually given in *wave numbers* ($\bar{v}$). The wave number in cm^{-1} of a vibration is the number of peak-to-peak waves per centimeter. The relationship between wave number and frequency (v) in hertz (s^{-1}) is given by

$$v = c \cdot \bar{v}$$

where c is the speed of light ($\sim 3.0 \times 10^{10}$ cm/s). For example, the wave number of 9.0×10^{13} Hz radiation is

$$\bar{v} = \frac{v}{c} = \frac{9.0 \times 10^{13} \ s^{-1}}{3.0 \times 10^{10} \ cm \ s^{-1}} = 3000 \ cm^{-1}$$

The wave number of an infrared band is the inverse of its wavelength, which is generally measured in micrometers (μm); the two can be interconverted using the following relationship:

$$\text{wave number (in } cm^{-1}) = \frac{10^4 \ \mu m/cm}{\text{wavelength (in } \mu m)}$$

For example, the wave number of a 5.85 μm C=O bond vibration is $(10,000/5.85) \ cm^{-1} = 1710 \ cm^{-1}$.

The *infrared spectrum* of a compound plots the amount of infrared radiation the compound absorbs over a broad range of wavelengths, usually about 2.5–17 μm (4000–600 cm^{-1}). An IR spectrum is obtained by placing a sample of the compound in an *infrared spectrometer*, which measures the sample's response to infrared radiation of different wavelengths. In a typical IR spectrum, wavelengths increase and wave numbers decrease going from left to right, so wave number ranges are usually reported with the higher value first. When a sample absorbs IR radiation of a given wavelength, the recorder pen on the spectrometer moves downward a distance that depends on the amount of IR radiation absorbed. Therefore, an IR spectrum consists of a series of inverted peaks, each corresponding to a different kind of bond

Figure G6 "Ball-and-spring" model of a chemical bond

A spectrometer is an instrument that measures and records the components of a spectrum in order of wavelength, mass, or some other property.

vibration. Because vibrational transitions are usually accompanied by rotational transitions in the frequency region scanned, the inverted peaks appear as comparatively broad "valleys" called *IR bands*, rather than sharp peaks like those seen in NMR spectra [OP-37].

Infrared spectrometry is most often used to detect the presence of specific functional groups and other structural features from band positions and intensities, and to show whether an unknown compound is identical to a known compound whose IR spectrum is reproduced in literature sources. The *fingerprint region* of an IR spectrum ($1250-670$ cm^{-1}) is best for establishing that two substances are identical, since the bands found in this region are often characteristic of the molecule as a whole and not of isolated bonds. Infrared spectrometry may also be used to assess the purity of a compound, monitor the rate of a reaction, measure the concentration of a solution, and study hydrogen bonding and other phenomena.

Instrumentation

In a conventional *dispersive infrared spectrometer*, infrared radiation is broken down into its component wavelengths by a diffraction grating or prism and beamed through the sample, one wavelength interval at a time. As the spectrum is scanned from lower to higher wavelength (higher to lower wave number), the radiation that passes through the sample at each wavelength is detected and its intensity is recorded on a chart. The resulting infrared spectrum is a graph of *percent transmittance* (the percentage of radiation that passes through the sample) on the *y*-axis versus wavelength and wave number on the *x*-axis (see Figure G15 for an example).

In a *Fourier-transform infrared (FTIR) spectrometer*, the sample is irradiated with a brief, intense beam that contains the entire spectrum of infrared radiation in the instrument's range. The multiple-wavelength radiation that exits the sample is then converted by a microprocessor to an infrared spectrum that closely resembles a scanned spectrum. To understand how an FTIR spectrometer works, refer to the diagram in Figure G7. The heart of the instrument is a *Michelson interferometer*, which causes two beams of infrared radiation to interfere with one another. The radiation

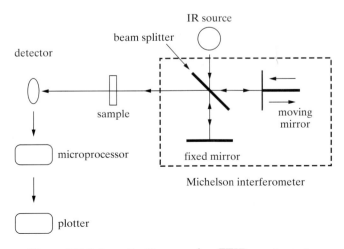

Figure G7 Schematic diagram of an FTIR spectrometer

generated by a heated filament or other IR source passes through a *beam splitter* that sends half of the radiation to a fixed mirror and half to a moving mirror. The beam that reflects off the fixed mirror always travels the same distance from the source to the sample; the distance the other beam travels is varied continuously by the moving mirror. The beams from both mirrors are recombined before the infrared radiation passes through the sample. Suppose for a moment that the IR beam coming from the source is of a single wavelength, λ. When the moving and fixed mirrors are the same distance from the beam splitter, both beams will travel the same distance before they recombine. As a result, the peaks and troughs of their waveforms will be aligned and will interfere constructively with one another, sending a beam of high intensity to the sample. When the moving and fixed mirrors are not the same distance from the beam splitter, the beams will travel different distances and their waveforms will usually not be aligned peak to peak when they recombine. For example, if the peaks of one beam exactly align with the troughs of the other, the beams will interfere destructively with one another, sending no radiation to the sample. In other alignments, the intensity of the combined beam will be somewhere between these two extremes.

As a result of constructive and destructive interference, the intensity of a light beam of wavelength λ will vary, with a sinusoidal wave pattern, at the frequency of the moving mirror. This pattern is called an *interferogram*. Because light beams of a different wavelength will align differently when they recombine, each different wavelength of IR radiation will generate a different interferogram. Interferograms are additive, so the interferogram that reaches the sample will be the sum of all the interferograms of all of the wavelengths. If IR radiation of a particular wavelength is absorbed as it passes through the sample, the intensity of that wavelength's interferogram—and thus its contribution to the overall interferogram—will decrease, changing the waveform of the overall interferogram that exits the sample. As a result, the interferogram that exits the sample will differ from the one that entered it in a way that depends on the amount by which the intensity of each of its component interferograms was reduced by the sample. In other words, this interferogram contains all of the intensity information for all of the different wavelengths that passed through the sample—there is no need to break down the infrared radiation into its component wavelengths and measure the intensity of the transmitted radiation at each individual wavelength, as for a dispersive IR spectrometer. When the interferogram impinges on a detector, it generates an electronic signal that is sent to a microprocessor. The microprocessor uses a mathematical technique called Fourier-transform analysis to "decode" the interferogram and recover the intensity data for each wavelength. These data are then plotted as an FTIR spectrum.

Unlike a continuous-wave IR spectrometer, which takes 5 minutes or more to record a spectrum, an FTIR spectrometer can acquire a complete spectrum in a fraction of a second, making it possible to accumulate many spectra of the same sample in a minute or less. The data from these spectra are averaged to yield a composite spectrum that is much cleaner and better resolved than a spectrum obtained by a dispersive instrument, because it lacks the electronic background noise that can distort IR bands. This averaging capability makes it possible to obtain good spectra using very small samples (1 mg or less). In addition, a helium–neon laser is used as a standard against which the frequencies of the radiation are measured, so the wave numbers displayed by an FTIR spectrometer are much more accurate than those on a scanned spectrum.

Experimental Considerations

An infrared spectrometer is a precision instrument that may cost more than a Jeep Grand Cherokee, but does not respond as well to rough handling. Treat it with respect and follow instructions carefully to prevent damage and avoid the need for costly repairs. Never unplug or switch off the instrument unless directed otherwise—some infrared spectrometers must be kept on continually to prevent damage to their optical parts. Keep water and aqueous solutions away from an IR spectrometer, since some of its components may be water sensitive. Never move the chart holder, drum, or pen carriage on a dispersive instrument while the instrument is in operation or before it has been properly reset at the end of a run—this can damage mechanical components. Do not leave objects lying on the bed of a flat-bed recorder, as they might jam the chart drive.

FTIR Spectrometers.   A typical Fourier-transform infrared spectrometer is illustrated in Figure G8. To obtain an infrared spectrum using such an instrument, the user places a cell containing the sample into the sample compartment and selects the desired number of scans (usually 4–16). Unless a larger number of scans is selected, the spectrum should appear on the monitor in less than a minute. A copy of the spectrum can then be obtained by

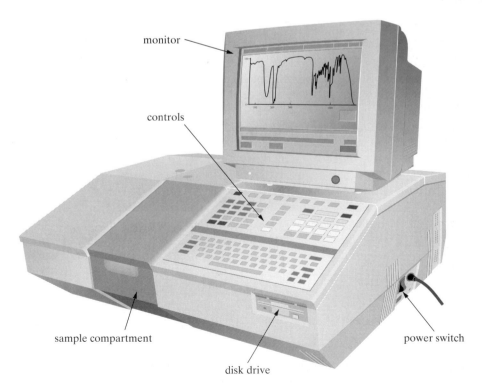

Figure G8 Perkin-Elmer Spectrum RXI FTIR spectrometer

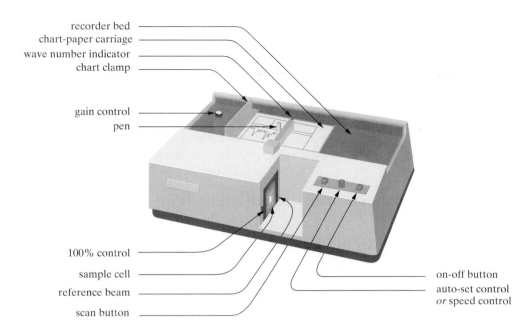

recorder bed
chart-paper carriage
wave number indicator
chart clamp

gain control
pen

100% control
sample cell
reference beam
scan button

on-off button
auto-set control
or speed control

Figure G9   Perkin-Elmer Model 710b infrared spectrometer

using a plotter or a computer printer. Some instruments print the wave number of each band directly on the spectrum; with other instruments, you may need to determine the wave numbers by using a cursor and then write them on the spectrum.

Dispersive IR Spectrometers.   A typical dispersive infrared spectrometer with a flat-bed recorder is shown in Figure G9. The chart paper on which spectra are recorded, which has wave number and wavelength values printed on it, is clipped to a movable chart-paper carriage. With other instruments, it may be wrapped around a movable drum. The user may need to align the chart paper by matching a wave number on the chart (for instance, 4000 cm^{-1}) with an alignment mark on the recorder bed or drum. If it is not aligned properly, the IR bands will appear at the wrong locations on the chart paper, and their wave numbers will be incorrect. The user then moves the carriage or drum to its initial position (with the pen at the far left of the chart paper), moves the pen to a point near the top of the chart paper with an attenuator control, and scans the spectrum.

Sample Preparation

Most infrared spectra are obtained by using sample cells into which the sample is introduced with a syringe or by some other means. Certain sampling accessories, such as attenuated total reflectance devices (ATRs), make it possible to record the IR spectrum of a liquid or solid without using a sample cell. Since sample cells and other sampling accessories vary widely in construction and application, most sampling techniques should be demonstrated by the instructor.

Care of Infrared Windows.   In most sample cells, the sample is placed between two *windows* made by compressing a metal halide (sodium chloride, silver chloride, etc.) to form a transparent round or rectangular crystal.

These windows are very fragile and most are water soluble. A window should be touched only on the edges with clean, dry hands or gloves, and handled with great care to avoid damage. Sodium chloride and potassium chloride windows must not be exposed to moisture; even breathing on such a window can cause some etching because of moisture in your breath. Therefore, all samples and solvents that will come into contact with such windows must be dry, and the windows must be kept dry when not in use. After use, sodium chloride and potassium chloride windows are washed with a dry, volatile solvent such as dichloromethane, sodium chloride-saturated absolute ethanol, or a 1:1 mixture of absolute ethanol and low-boiling petroleum ether. They are then allowed to air dry or carefully blotted dry with nonabrasive tissues, and stored in a desiccator. Silver chloride windows are unaffected by water, but they must be stored in the dark, because AgCl eventually turns black when exposed to light. Silver chloride windows can be washed with ethanol or acetone.

Thin Films.   Thin films cannot be used for very volatile liquids, because they may evaporate before the spectrum is complete; with such liquids, use a spacer as described in the next paragraph. To prepare a thin film of a neat (undiluted) liquid using a *demountable cell* such as the one illustrated in Figure G10, layer a gasket and a window on the back plate and place 1–2 drops of the liquid on the window. If the cell has round windows, use an O-ring rather than a gasket. Then position the upper window by touching an edge to the corresponding edge of the lower window and carefully lowering it into place. Press the windows together so that the liquid fills the space between them, taking care to exclude air bubbles. Place another gasket (or O-ring) on the top window, add the front plate, and tighten the cinch nuts just enough to hold the cell components together securely. Be careful not to overtighten

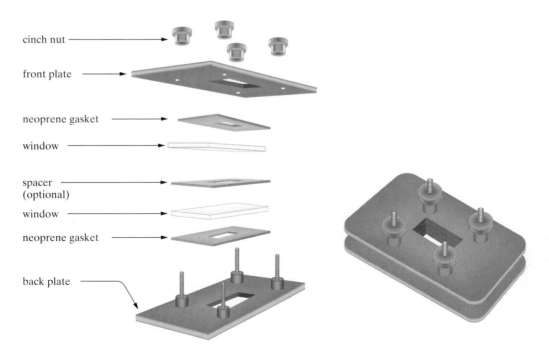

cinch nut

front plate

neoprene gasket

window

spacer
(optional)

window

neoprene gasket

back plate

Figure G10 Demountable cell for rectangular windows

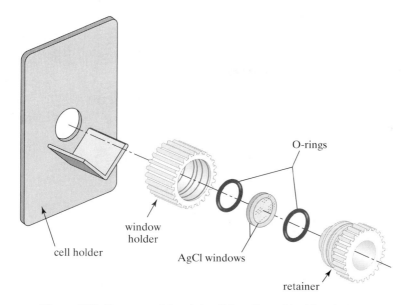

Figure G11 Demountable mini-cell for silver chloride windows

the nuts, as this may make the liquid film too thin or even break a window. After you have recorded a spectrum, disassemble the cell and wash, dry, and store the windows as described previously.

The demountable silver chloride mini-cell illustrated in Figure G11 is especially convenient for microscale work because less sample is required; a small drop from a Pasteur pipet is usually sufficient. After the sample is spread between the flat sides of the recessed AgCl windows, the cell is assembled by placing the windows, protected by two O-rings, inside the plastic window holder and carefully screwing on the plastic retainer. When an infrared spectrum is recorded, the cell is supported in the special cell holder shown.

Volatile Neat Liquids. If a liquid is comparatively volatile, its spectrum can be run in a demountable, sealed, or sealed-demountable cell, using a spacer approximately 0.015–0.030 mm thick. Fill a demountable cell by placing the spacer on the lower window, adding sufficient liquid to fill the cavity in the spacer, positioning the upper window, and reassembling the cell as described in the previous section. After you have recorded a spectrum, disassemble the cell and wash, dry, and store the windows as described previously.

A *sealed cell* or *sealed-demountable cell* (Figure G12) is filled by injecting the sample into one of its filling ports with a Luer-Lok syringe body. If you are using a sealed-demountable cell that has not been assembled, assemble it using Figure G12 as a guide, starting with the back plate. Remove both plugs from the filling ports, draw about 0.5 mL of liquid into the syringe, and carefully insert the syringe tip into one of the ports with a slight twist. Holding the cell upright with the syringe port at the bottom, depress the syringe's plunger until the space between the windows is filled and a small amount of liquid appears at the upper port. If there is much resistance to filling, try the push-pull method described next. If there are air bubbles between the windows, they can sometimes be removed by tapping gently on

Take Care! The syringes are fragile and break easily.

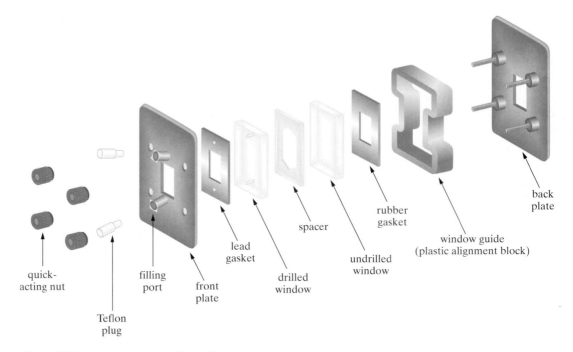

quick-
acting nut

Teflon
plug

filling
port

front
plate

lead
gasket

drilled
window

spacer

undrilled
window

rubber
gasket

window guide
(plastic alignment block)

back
plate

Figure G12 Sealed-demountable cell

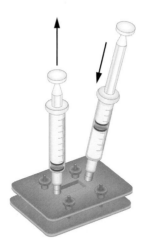

Figure G13 Push-pull method for
flushing a sealed or sealed-demountable cell

the metal frame of the cell. Put the cell on a flat surface, remove the syringe,
and insert a plug in the upper port with a slight twist. Remove any excess
solvent in the lower port with a piece of tissue paper or cotton, and close
that port with another plug.

After the spectrum has been run, clean the cell by removing most of
the liquid with a syringe and flushing the cell several times with a dry,
volatile solvent. To flush a cell using the push-pull method illustrated in
Figure G13, lay it on a flat surface and then insert a syringe filled with solvent in one port and an empty syringe in the other port. Twist the syringes
slightly as you insert them so that they will not pull out easily. Slowly push
on one plunger while pulling on the other, as shown in the figure, to draw
washing liquid through the cell from one syringe to the other. Do this several times; then remove the excess liquid with an empty syringe and dry the
cell by passing clean, dry air or nitrogen between the windows. An ear syringe or a special cell-drying syringe can be used to force air through the
cell, or it can be dried by attaching a trap and aspirator to one port and a
drying tube filled with desiccant [OP-24a] to the other. Store the cell in a
desiccator.

Solutions.   Solutions of solids or liquids in a suitable solvent can be analyzed in a sealed or sealed-demountable cell that has a spacer 0.1 mm or
more in thickness. An inexpensive solution cell, constructed as described in
J. Chem. Educ. **1991**, *68*, A124, is particularly useful for microscale work, as
it requires only a small volume of solution.

The solvent should be dry, relatively nonpolar, unreactive with the
solute, and transparent to IR radiation in the regions of interest. It should
ordinarily dissolve enough of the solute to yield a 5–10% solution; more dilute solutions can be used with FTIR spectrometers. Only a few solvents,

such as carbon tetrachloride, chloroform, and carbon disulfide, meet these criteria with a wide range of solutes. These solvents (particularly carbon tetrachloride) are all hazardous to human health, and carbon disulfide is extremely flammable, so they are to be used only with appropriate safety equipment (hood, gloves, safety goggles, etc.) and under close supervision by an instructor. It is best to use spectral-grade solvents so that impurities will not give rise to extraneous peaks. When the sample is run, the infrared spectrum of the solvent is, in effect, subtracted from the solution spectrum to give the spectrum of the solute. Even then, strong solvent peaks will obscure certain portions of the spectrum. For example, the spectrum of a solute run in chloroform will yield no useful information in the 1250–1200 cm^{-1} and 800–650 cm^{-1} regions because the chloroform absorbs nearly all of the infrared radiation there. Sometimes spectra are run separately in two solvents that absorb at different wave numbers, such as chloroform and carbon disulfide, to obtain the equivalent of a complete spectrum.

With an FTIR spectrometer, an IR cell is filled with the pure solvent and the solvent's spectrum is run as a *background*; then the spectrum of the sample is run in the same cell. The background spectrum is stored in the instrument until another background is recorded, so a number of samples can be run in the same cell using the same background. With a dispersive IR spectrometer, one cell is filled with the solution and another identical cell is filled with the solvent, using spacers of the same thickness. The spectrum is then run with the solvent cell in the reference beam.

With most FTIR spectrometers, you can also record both spectra under the same conditions and subtract the solvent spectrum from the solution spectrum.

To prepare an IR cell for a solution spectrum, first make up a 5–10% solution of the substance to be analyzed. Usually, 0.1–0.5 mL of solution will be required. Fill a ported IR cell using a Luer-Lok syringe body as described under "Volatile Neat Liquids." A 0.1-mm spacer can be used unless the solution is quite dilute. (A silver chloride mini-cell can be used for solutions by filling the recessed side of one AgCl window with the solution and covering it with the flat side of the other window, so as to exclude air bubbles.) When the spectrum has been recorded, clean the cell by removing the excess solution and flushing it (as described under "Volatile Neat Liquids") with the pure solvent used in preparing the solution. If necessary, rinse the cell with a more volatile solvent before drying it, and store it in a desiccator.

Mulls.   A solid sample can be prepared as a *mull* in Nujol (a kind of mineral oil) or another mulling oil. The sample spectrum should be compared with a spectrum of the mulling oil so that peaks due to the oil can be identified and disregarded during interpretation. When Nujol is used, the aliphatic C—H stretching and bending regions (3000–2850, 1470, 1380 cm^{-1}) cannot be interpreted, but most functional groups and other structural features can be identified. If it is necessary to examine the entire spectrum, another sample can be prepared using a complementary mulling oil such as Fluorlube. Nujol is essentially transparent at wave numbers lower than 1300 cm^{-1}, while Fluorlube is transparent above 1300 cm^{-1}. Some scattering of infrared radiation by particles of the solid will reduce transmittance at the high wave-number end of the spectrum, so the baseline of a dispersive IR spectrometer should not be set at that end, but wherever the transmittance is highest.

To prepare a mull, grind about 10–20 mg of the solid in an agate or mullite mortar until it coats the entire inner surface of the mortar and has a glassy appearance (about 5–10 minutes of grinding will be required). The particles must be ground to an average diameter of about 1 μm or less to avoid excessive radiation loss by scattering. Add a drop or two of mulling oil and grind the mixture until it has about the consistency of petroleum jelly. Transfer most of the mull to the lower window of a demountable cell using a rubber policeman. Spread the mull evenly with the top window, taking care to exclude air bubbles. Assemble the cell, run the spectrum, and clean the windows as for a neat liquid, using low-boiling petroleum ether or another suitable solvent. Then dry and store the windows as described previously.

Potassium Bromide Discs.   A potassium bromide (KBr) disc is prepared by mixing a solid with *dry* spectral-grade potassium bromide and using a die to press the mixture into a more or less transparent wafer. Potassium bromide absorbs moisture from the atmosphere, so it must be kept in a tightly capped container and stored in an oven or desiccator. It is advisable to dry it before use by heating it in a 110°C oven for several hours. Grind 0.5–2.0 mg of the solid very finely in a dry agate or mullite mortar as if you were preparing a mull; then add about 100 mg of dry potassium bromide and mix it thoroughly with the sample.

If you are using a Mini-Press (Figure G14), screw in the bottom bolt of the Mini-Press five full turns and introduce about half of the sample mixture into the barrel. Keeping the open end of the barrel pointed up, tap it gently against the bench top to level the mixture, brush down any material on the threads with a soft brush, and screw in the top bolt. Alternately tap the bottom bolt on the bench top and screw in the top bolt with your fingers several times to level the sample further. When the bolt is finger tight, clamp the bottom bolt in a vise and gradually tighten the top bolt as far as you can with a heavy wrench (or to 20 ft-lbs with a torque wrench). Leave the die under pressure for a minute or two. Remove both bolts, leaving the KBr disc in the center of the barrel, and check to see that it is reasonably transparent and homogeneous; if not, prepare a new disc. A disc may be cloudy or inhomogeneous because the components are wet or not completely mixed, the pellet is too thick, the sample size is too large, or the die was not tightened enough.

Place the block containing the disc on a holder (provided with the Mini-Press) in the spectrometer sample beam. As for a mull, scattering will reduce transmittance at the left end of the spectrum, so for a dispersive instrument, set the baseline wherever the transmittance is highest. After you have run the spectrum, punch out the KBr disc using the eraser end of a pencil. Wash the barrel and bolts with water, rinse them with acetone or methanol, and store the clean, dry press in a desiccator.

Melts.   Infrared spectra of low-melting solids can sometimes be obtained by one of the following procedures.

A.   Spread a thin, uniform layer of the finely powdered solid on a silver chloride window, cover it with another silver chloride window, and heat the assembly *slowly* on a hot plate. As soon as the solid melts to produce a uniform film between the windows, remove the windows, press them together with forceps until the sample solidifies, install them in a demountable cell, and run the spectrum as for a thin film. If the spectrum is distorted because of excessive light scattering, try Method **B**.

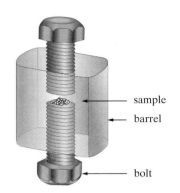

Figure G14 Potassium bromide Mini-Press

If another kind of press is to be used, follow the manufacturer's or your instructor's directions for preparing the disc.

See J. Chem. Educ. **1977**, *54,* 287 *for additional suggestions regarding the preparation of KBr discs.*

B. Heat a pair of sodium chloride windows in an oven to a temperature at least 20°C above the melting point of the solid, then set one window on the back plate of a demountable cell (don't forget to use a gasket or O-ring). Without delay, place about 0.1 g of the solid on that window; position the other window so that the solid, as it melts, fills the space between the windows; and assemble the cell. If possible, run the spectrum while the sample is still liquid. You may still be able to get a good spectrum if the liquid crystallizes as a glassy film or very small crystals, but large crystals produce excessive light scattering.

General Directions for Recording an Infrared Spectrum

Equipment and Supplies

IR spectrometer

sample

Pasteur pipet

IR sample cell(s) and windows (or Mini-Press)

*Luer-Lok syringe, spacer (volatile neat liquids)

*spectral-grade solvent, Luer-Lok syringe, spacer (solutions)

*mortar and pestle, mulling oil, glass rod with rubber policeman (mulls)

*dry potassium bromide, mortar and pestle, Mini-Press, vise, wrench (KBr disks)

*hot plate and forceps, or oven (melts)

solvent for cleaning windows

tissues

desiccator

*Needed for only the sample type indicated in parentheses.

Do not attempt to operate the instrument without prior instruction and proper supervision. The construction and operation of commercial infrared spectrometers vary widely, so the following is meant only as a general guide to assist you in recording an infrared spectrum. Specific operating techniques must be learned from the instructor, the operating manual, or both. It will be assumed that the necessary operational parameters have been set beforehand; if not, your instructor will show you what to do. If you are using an FTIR spectrometer, follow Procedure **A**; for a dispersive IR spectrometer, follow Procedure **B**.

A. FTIR Spectrometer. Your instructor will tell you which keys to use for such processes as scanning a spectrum, using a cursor, and printing or plotting a spectrum. If a background spectrum has not been recorded recently, run the background with an empty sample compartment, unless your compound is in solution. In that case, run the background with a cell containing the pure solvent in the sample compartment and use the same cell for the sample. Prepare a sample cell or KBr disc containing the sample by one of the methods described previously, and place the cell or KBr disc holder in the sample cell holder. Select the desired number of scans (four is usually sufficient for a routine spectrum) and start scanning the spectrum. Wait until a spectrum appears on the monitor. If the spectrum doesn't look right—for example, if the low or high wave-number end is missing or you are seeing only a small part of the total spectrum, display the "normal" spectrum using the appropriate key(s) (some instruments use *Rerange* and *Rescale* keys for this purpose). The strongest bands should extend nearly to the bottom of your spectrum; if they do not, use a vertical-scale expansion key to improve the appearance of the spectrum. See that the printer or plotter is turned on and properly adjusted, and then press the appropriate key to print or plot the spectrum. If the instrument you are using does not record wave numbers directly on the spectrum, use the cursor arrow keys to move the cursor to the significant bands, and write the displayed wave numbers below the corresponding IR bands on your spectrum. Reset the instrument to display a normal spectrum, if necessary. Then remove the sample cell and close the sample compartment. Clean, dry, and store the cell windows as described previously.

B. Dispersive IR Spectrometer.   Prepare a sample cell or KBr disc containing the sample by one of the methods described previously, and place the cell or KBr disc assembly in the sample cell holder. If you are running a solution spectrum, place an identical cell containing the solvent in the reference compartment; otherwise, leave it empty. If necessary, place chart paper on the chart-paper carriage (or wrap it around the drum); then align the paper properly and move the carriage or drum to the starting position. Set the 100% transmittance control so that the pen is at 85–90%T, lower the pen onto the chart paper, and scan the spectrum at "normal" or "fast" speed. Examine the spectrum to see that the absorption bands show satisfactory intensity and resolution. Ideally, the strongest absorption band should have a maximum transmittance of 5–10%. If your first spectrum is not acceptable, try varying the following parameters, depending on the sample preparation method:

Take Care! Make sure the instrument has been reset correctly before attempting to move the drum or carriage.

- *Neat liquid*—Vary cell path length or film thickness.
- *Solution*—Vary concentration or cell path length.
- *KBr disc*—Vary amount of sample or thickness of disc.
- *Mull*—Vary amount of sample or film thickness.

Then run another spectrum, using a slower speed if desired. Remove the sample and spectrum and reset the instrument, if necessary. Disassemble the cell if needed; then clean, dry, and store the cell windows as described previously.

Summary
1 Put sample in sample cell or prepare KBr disc.
2 Put sample cell or KBr disc holder in instrument's cell holder.
    IF you are using an FTIR spectrometer, GO TO 5.
3 Align chart paper, move drum or carriage to starting position.
4 Adjust 100% transmittance control, lower pen to paper.
5 Scan spectrum.
    IF you are using a dispersive IR spectrometer, GO TO 7.
6 Print or plot spectrum, record wave numbers if necessary.
7 Disassemble cell, clean, dry, and store cell windows.
Note: Additional steps are required for running solution spectra.

Interpretation of Infrared Spectra

Because most infrared bands are associated with specific chemical bonds, it is usually possible to deduce the functional class of an organic compound from its infrared spectrum. The *stretching* vibrations of chemical bonds resemble the vibrations of springs in that stronger bonds have higher vibrational energies and frequencies than weaker ones. Thus, triple bonds generally absorb at higher wave numbers than double bonds, and double bonds absorb at higher wave numbers than single bonds. Because of the comparatively low mass of a hydrogen atom, however, single bonds to hydrogen (C—H, O—H, N—H, etc.) have even higher vibrational frequencies than double and triple bonds. Stretching bands involving single bonds to hydrogen occur at the high frequency (left) end of an infrared spectrum, in the region between 3700 and 2700 cm^{-1} (2.7–3.7 μm). Triple bonds usually absorb between 2700 and 1850 cm^{-1} (3.7–5.4 μm), and double bonds and aromatic bonds absorb between 1950 and 1450 cm^{-1} (5.1–6.9 μm). Most IR bands between 1500 and 600 cm^{-1} (6.7–16.7 μm) are produced by *bending* vibrations or single-bond stretching vibrations. It takes less energy to bend a bond than to stretch it, so bending vibrations tend to have comparatively low frequencies. An absorption band in the 1500–600 cm^{-1} region may be associated with more than one bond; for example, the so-called acyloxygen stretching band of an ester arises from the vibration of C—C—O units rather than isolated C—O bonds.

What to Look for in an IR Spectrum

Consider the infrared spectrum of 2-methyl-1-propanol (isobutyl alcohol) in Figure G15. At first glance, it may seem indecipherable—just a series of dips and rises in a graph. But each "dip" (IR band) arises from a stretching or bending vibration of one or more bonds in the 2-methyl-1-propanol molecule, and some of the bands can tell you a great deal about the molecules that gave rise to them. First look at the WAVENUMBERS scale at

The infrared spectra in this book are reproduced from the Aldrich Library of FT-IR Spectra, *2nd ed., with permission.*

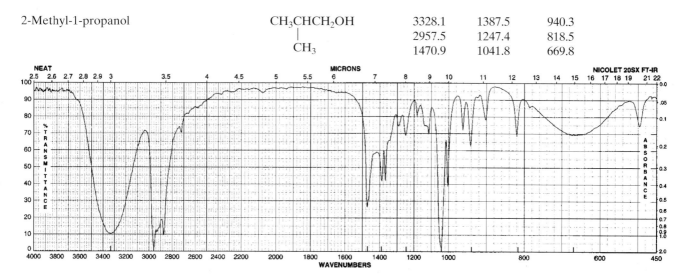

Figure G15 Infrared spectrum of 2-methyl-1-propanol

the bottom of the spectrum; from the scale you can read off the wave number, in cm^{-1}, of each band. For example, the first large band in the spectrum is between 3400 and 3200 cm^{-1}, and its minimum is at approximately 3330 cm^{-1}. Bands designated by tick marks (short lines) at the bottom of a spectrum have their exact wave numbers listed, so you can find a more accurate wave number for this band, 3328.1 cm^{-1}, in the list of numbers above the right-hand end of the spectrum.

Another numerical scale at the top of the spectrum indicates the wavelength in micrometers (microns). The numbers on the left side of the spectrum are transmittance values. For example, the minimum in the 3328 cm^{-1} band has a transmittance of ~10%, meaning that only 10% of the 3328 cm^{-1} IR radiation passed through the sample. The other 90% was absorbed during a change in the bond vibrational frequency of some bond in the compound—but which bond? Since the band is at the left end of the spectrum, the bond must have a high vibrational energy, and we have already noted that bands in the 3700–2700 cm^{-1} region of the spectrum arise from vibrations of bonds to hydrogen atoms. There are only two bonds of this type in the molecule; C—H bonds and O—H bonds. And there are two bands in the 3700–2700 cm^{-1} region; the strong, broad, symmetrical one on the left, and a rather ragged band with several minima (centered around 2900 cm^{-1}) on the right. The ragged band is actually composed of several overlapping bands, arising from vibrations of several different bonds of the same general type. Since there is only one O—H bond in the molecule while there are nine C—H bonds, it is reasonable to assume that the overlapping bands on the right arise from C—H stretching vibrations and that the band on the left is the O—H band. Note that the size of a band has little to do with the number of bonds that give rise to it; the single O—H bond has a much broader band than the nine C—H bonds.

So the bond responsible for a given IR band can often be identified from its *location* on the spectrum (as indicated by its wave number), its *intensity* (relative strength), and its *shape*. Most O—H bands, like the one in the previous spectrum, are very broad and strong (a *strong* band is one whose minimum is near the bottom of the spectrum). Most C—H bands are relatively strong and give rise to a ragged array of overlapping bands. Another very strong band appears at 1042 cm^{-1} in the Figure G15 spectrum. This band, which is in the wave-number region for single-bond stretching vibrations (except those involving hydrogen), arises from stretching vibrations of the C—O single bond. The presence of both a C—O and an O—H band in the IR spectrum of a compound is good evidence that the compound is an alcohol (or possibly a phenol), since all alcohols contain a C—O—H grouping in their molecules.

Although we have now located bands corresponding to every kind of bond in the 2-methyl-1-propanol molecule (except C—C single bonds, which don't give prominent bands), there are still a number of bands left. This is because the same kind of bond can undergo different kinds of vibrations. For example, most of the bands just to the left of the C—O band arise from *scissoring*, *wagging*, and *twisting* vibrations of CH_2 and CH_3 groups (see Figure G16), and the very broad, weak band centered at 670 cm^{-1} arises from an O—H bending vibration.

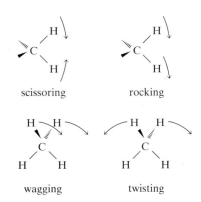

Figure G16 Some carbon-hydrogen vibrations

Spectral Regions

You can see that some infrared bands, particularly the stretching bands we have discussed, are more easily recognized than others, and are more useful in revealing the presence of functional groups. Many of the other bands can be ignored for the time being, although they may provide useful information to a chemist skilled in spectral interpretation. The key to efficient IR spectral interpretation is *knowing where to look* for the more useful bands. Examining the following regions of the infrared spectrum will help you locate the most useful infrared bands quickly:

Region 1: $3600-3200$ cm^{-1} ($2.8-3.1$ μm). Bands in this region can arise from O—H and N—H stretching vibrations of alcohols, phenols, amines, and amides. O—H bands are generally very strong and broad; N—H bands are somewhat weaker, and in the case of primary amines and amides, they have two peaks.

Region 2: $3100-2500$ cm^{-1} ($3.2-4.0$ μm). This region contains most of the C—H stretching vibrations. A strong band in the $3000-2850$-cm^{-1} region, arising from C—H bonds to sp^3 carbon atoms, is present for most organic compounds. The sp^2 C—H bonds associated with aromatic hydrocarbons and alkenes absorb at higher frequencies ($3100-3000$ cm^{-1}), and the C—H bonds of aldehyde (CHO) groups absorb at lower frequencies. The O—H bond of a carboxylic acid gives rise to a very broad absorption band in this region.

Region 3: $1750-1630$ cm^{-1} ($5.7-6.1$ μm). This region contains most of the carbonyl (C=O) stretching bands of aldehydes, ketones, carboxylic acids, amides, and esters. The carbonyl band is usually strong and quite unmistakable. Unsaturated compounds may have a C=C stretching band in the $1670-1640$-cm^{-1} region, but this band is nearly always weaker and narrower than a carbonyl band.

Region 4: $1350-1000$ cm^{-1} ($7.4-10.0$ μm). This region is usually cluttered with many C—H bending bands and other bands, but it is usually possible to identify the C—O stretching bands of alcohols, phenols, carboxylic acids, and esters, and some C—N stretching bands of amines and amides.

2981.9	1367.2	1108.5
1718.5	1275.8	1028.5
1451.4	1175.2	710.3

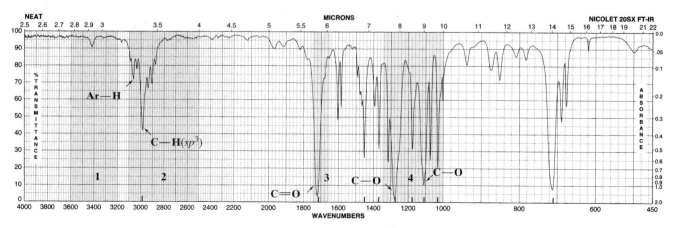

Figure G17 Classification of a compound from its IR spectrum

These four spectral regions are shaded on the infrared spectrum in Figure G17. In this spectrum, the absence of any band in Region 1 (or a broad band in Region 2) eliminates from consideration all compounds containing O—H and N—H bonds, including alcohols, phenols, and primary or secondary amines and amides. In Region 2, the appearance of a weak "shoulder" on the C—H band at 3050 cm^{-1} indicates an sp^2 C—H bond associated with either an aromatic ring or a carbon–carbon double bond. Region 3 has a strong C=O band at 1719 cm^{-1}, and Region 4 shows a strong C—O band at 1276 cm^{-1} as well as a weaker one at 1109 cm^{-1}. The absence of an O—H or N—H band and the presence of the C=O and two C—O bands suggest that the compound responsible for this spectrum is an ester. The compound is, in fact, the aromatic ester ethyl benzoate, whose structure and bond vibrational frequencies are shown in the margin.

Table G2 summarizes the locations and characteristics of important absorption bands from the four spectral regions, and tells you where to look for other bands that may help you confirm the presence of a particular functional group.

ethyl benzoate

Table G2 Important bands in Regions 1–4 of infrared spectra

Region	Frequency range (cm^{-1})	Bond type	Family	Comments
1	3500–3200	N—H	amine, amide	Weak–medium. 1°: 2 bands; 2°:1 band; 3°: no bands. See also Region 3.
	3600–3200	O—H	alcohol, phenol	Broad, strong. See Region 4.
2	3300–2500	O—H	carboxylic acid	Very broad, strong, centered around 3000. See Regions 3 and 4.
	3100–3000	C—H	aromatic hydrocarbon, alkene	May be shoulder on stronger sp^3 C—H band.
	2850–2700	C—H	aldehyde	Weak to medium, usually two sharp bands. See Region 3.
3	1740–1685	C=O	aldehyde	Strong. See Region 2.
	1750–1660	C=O	ketone	Strong.
	1725–1665	C=O	carboxylic acid	Strong. See Regions 2 and 4.
	1775–1715	C=O	ester	Strong. See Region 4.
	1695–1615	C=O	amide	Strong. See Region 1.
4	1350–1210	C—O	carboxylic acid	Medium–strong. See Regions 1 and 3.
	1300–1180	C—O	phenol	Strong. See Region 1.
	1200–1000	C—O	alcohol	Strong. See Region 1. Wave number decreases in the order 3° > 2° > 1°.
	1310–1160	C—O	ester	Strong. See Region 3. Accompanied by weaker C—O band as for alcohol.

Note: Tentative classifications must be confirmed by referring to the following descriptions of specific families.

The best way to become proficient at identifying IR bands is to study spectra that contain those bands, such as the spectra in your lecture textbook, in the "Characteristic Infrared Bands" section that follows, and in collections of spectra described in Section F of the Bibliography. When you have learned to recognize the most important IR bands, you can take some shortcuts that will help you identify functional groups quickly—or at least eliminate the functional groups that aren't there. The following flow chart should help you do that. Just start at the top and work your way down, following the yes–no arrow that answers each question.

Possible family   C=O present?   *Possible family*

carboxylic ⟵ᵧₑₛ O—H present?   O—H present? —ʸᵉˢ→ alcohol
  acid   or phenol

amide (1°, 2°) ⟵ᵧₑₛ N—H present?   N—H present? —ʸᵉˢ→ amide (1°, 2°)

ester   ⟵ᵧₑₛ C—O present?   C=C present? —ʸᵉˢ→ alkene

aldehyde ⟵ᵧₑₛ C—H at ~ 2700 cm⁻¹?   Ar—H present? —ʸᵉˢ→ aromatic
                                                              hydrocarbon

ketone   ⟵ᵧₑₛ None of the above?   None of the above? —ʸᵉˢ→ alkane
or 3° amide   or 3° amine

Flow chart for determining functional classes from IR bands

This flow chart is intended as a rapid screening device and is not infallible; some bands (such as the C=C stretching band) are hard to identify with certainty, and the locations of other bands may vary widely. And some compounds may contain more than one functional group; thus hydroxyacetone (CH_3COCH_2OH) has both an O—H and a C=O band in its spectrum, but it is not a carboxylic acid, as you could tell from the location of its O—H band. When you arrive at a tentative conclusion about the nature of the compound responsible for an IR spectrum, you should refer to Table G2 to see whether other bands in the spectrum are consistent with your initial choice. Then study the spectral characteristics of the appropriate class of compounds to confirm (or disprove) your tentative classification. The IR correlation chart on the back endpaper of this book may also help you identify some infrared spectral bands.

For example, suppose that the flow chart suggests that your compound may be an alcohol or phenol. You can first check Table G2 to see if any other bands characteristic of alcohols and phenols appear in its spectrum, such as a C—O band. If so, you should read the "Characteristic Infrared Bands" sections about alcohols and phenols to find out whether your compound is an alcohol or a phenol. If you find that your compound is an alcohol, you should then study its spectrum for clues to its structure. The frequency of its C—O band may tell you whether it is primary, secondary, or tertiary. By consulting the sections on aromatic hydrocarbons and alkenes (which also apply to other compounds that contain aromatic rings and C=C bonds), you can find out whether your alcohol is aromatic or contains a carbon–carbon double bond. Of course, if you find that your compound is *not* an alcohol or phenol, you should continue down the chart or start back at the beginning.

Characteristic Infrared Bands

This section contains information about the most useful infrared bands of alkanes, alkenes, and the families of compounds that can be identified as described in Part IV, "Qualitative Organic Analysis,"—aromatic hydrocarbons, alcohols, phenols, aldehydes, ketones, carboxylic acids, esters, amines, amides, and organic halides. Compounds other than hydrocarbons may contain bands characteristic of alkanes, alkenes, or aromatic hydrocarbons, so you can check the sections for these hydrocarbons when you are interpreting the spectra of other kinds of compounds. For each family of organic compounds, a summary of the main spectral features that characterize the family is followed by a description of individual bond vibrations and a representative infrared spectrum. The wave-number ranges given are for solids (in Nujol mulls or KBr discs) or neat liquids; values for solutions may differ somewhat. Although the wave-number ranges apply to most of the organic compounds in each class, compounds with certain structural features (such as highly strained rings) may have bands outside of the ranges indicated. On the spectra, absorption bands are designated either as stretching (v) or bending (δ) bands. Only those bands that are most useful for identifying functional groups or structural features are labeled. Note that the exact wave numbers of significant bands (designated by tick marks along the lower edge of the spectra) are listed with each spectrum.

Alkanes.   Alkanes are identified primarily by the absence of any IR bands characteristic of functional groups. Their spectra are quite simple, containing only the C—H stretching and bending vibrations characteristic of sp^3 hybridized carbon atoms. Since nearly all other organic compounds contain such C—H bonds, their spectra will also contain some or all of the bands described for alkanes. (See Figure G18.)

C—H *stretch:* 3000–2800 cm^{-1} (multiple overlapping bands, strong to weak). CH$_3$ bands are near 2960 cm^{-1} and 2870 cm^{-1}. CH$_2$ bands are near 2925 cm^{-1} and 2850 cm^{-1}. Nearly always to the *right* of 3000 cm^{-1}.

2,2,4-Trimethylpentane

$$
\begin{array}{cc}
CH_3 & CH_3 \\
| & | \\
\end{array}
$$
CH$_3$CHCH$_2$CCH$_3$
$$
\begin{array}{c}
| \\
CH_3
\end{array}
$$

2956.0	1247.6
1468.6	1168.4
1365.7	979.6

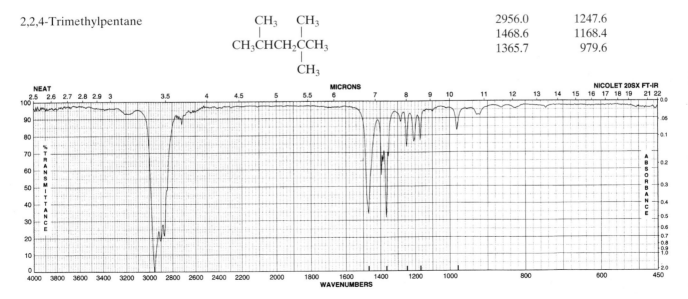

Figure G18 IR spectrum of an alkane, 2,2,4-trimethylpentane

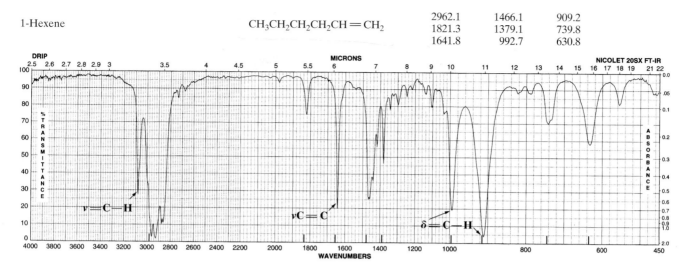

Figure G19 IR spectrum of an alkene, 1-hexene

C—H *bend:* 1465–720 cm^{-1} (moderate to weak). CH$_3$ bands are near 1450 cm^{-1} and 1375 cm^{-1}. CH$_2$ bands occur near 1465 cm^{-1}, between 1350 cm^{-1} and 1150 cm^{-1} (several weak bands), and sometimes around 720 cm^{-1}. The 720 cm^{-1} band is characteristic of unbranched alkanes having seven or more carbon atoms.

Alkenes: Most alkenes contain the same kinds of bands as alkanes, plus additional bands associated with carbon-carbon double bonds and vinylic (=C—H) carbon-hydrogen bonds. The presence of one or two strong bands in the 1000–650 cm^{-1} region and a sharp band near 1650 cm^{-1} suggests an alkene functional group, especially if the compound is not aromatic. (See Figure G19.)

=C—H *stretch:* 3125–3030 cm^{-1} (moderate to weak). May appear as a shoulder on a stronger sp^3 C—H band, but nearly always to the *left* of 3000 cm^{-1}.

C=C *stretch:* 1675–1600 cm^{-1} (moderate to weak; narrow). May be absent for symmetrical alkenes. Conjugation moves band to lower wavelengths.

=C—H *out-of-plane bend:* 1000–650 cm^{-1} (usually strong). Position depends on type of substitution: RCH=CH$_2$ has bands at 995–985 and 915–905 cm^{-1}; *cis*-RCH=CHR a band at 730–665 cm^{-1}; *trans*-RCH=CHR a band at 980–960 cm^{-1}; and R$_2$C=CH$_2$ a band at 895–885 cm^{-1}. (R = alkyl or aryl substituent)

Aromatic Hydrocarbons. Most compounds containing benzene rings are characterized by (1) aromatic C—H (Ar—H) stretching bands near 3070 cm^{-1}, (2) a distinctive pattern of weak bands in the 2000–1650-cm^{-1} region, (3) two sets of bands near 1600 cm^{-1} and 1515–1400 cm^{-1}, and (4) one or more strong absorption bands in the 900–675-cm^{-1} region. The presence of such bands and the absence of absorption bands characteristic of functional groups suggest an aromatic hydrocarbon. (See Figure G21.)

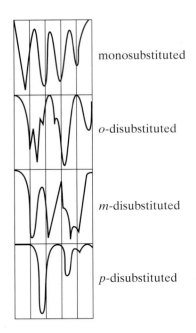

Figure G20 Typical absorption patterns of substituted aromatic compounds in the 2000–1650-cm⁻¹ region

Ar—H *stretch*: 3100–3000 cm⁻¹ (moderate to weak). May appear as a shoulder on a stronger sp³ C—H band, but nearly always to the *left* of 3000 cm⁻¹.

Overtone-combination vibrations: 2000–1650 cm⁻¹ (multiple bands, weak). The band pattern is related to the kind of ring substitution, as shown in Figure G20.

C═C *stretch*: 1615–1585 cm⁻¹ and 1515–1400 cm⁻¹ (variable).

Ar—H *out-of-plane bend*: 910–730 cm⁻¹ (strong). The band frequency varies with the number of adjacent ring hydrogens:

two adjacent hydrogens: 855–800 cm⁻¹

three adjacent hydrogens: 800–765 cm⁻¹

four or five adjacent hydrogens: 770–730 cm⁻¹

Monosubstituted, *meta*-disubstituted, and some trisubstituted benzenes show an additional ring-bending band around 715–680 cm⁻¹. For example, a *meta*-disubstituted benzene has three adjacent ring hydrogens, so it should have bands in the 800–765-cm⁻¹ and 715–680-cm⁻¹ regions.

Alcohols. The presence of a strong, broad band centered around 3300 cm⁻¹ and a strong C—O band in the 1200–1000-cm⁻¹ region is good evidence for an alcohol. (See Figure G22.) A C—O band above 1200 cm⁻¹ may suggest a phenol (see Figure G23), particularly when it is accompanied by bands indicating an aromatic ring.

O—H *stretch*: 3600–3200 cm⁻¹ (strong, broad). Usually centered near 3300 cm⁻¹.

C—O *stretch*: 1200–1000 cm⁻¹ (strong to moderate). Most saturated aliphatic alcohols absorb near 1050 cm⁻¹ if they are primary, near 1110 cm⁻¹ if they are secondary, and near 1175 cm⁻¹ if they are tertiary. Alicyclic alcohols and alcohols with aromatic rings or vinyl groups on

Isopropylbenzene

2961.3	1383.7	698.4
1604.0	1027.9	534.7
1493.7	760.6	404.5

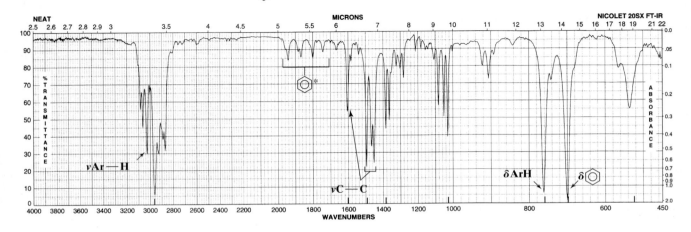

Figure G21 IR spectrum of an aromatic hydrocarbon, isopropylbenzene. The bands marked ⬡* are aromatic overtone-combination bands.

2-Methyl-1-propanol

$$CH_3CHCH_2OH$$
$$|$$
$$CH_3$$

3328.1	1387.5	940.3
2957.5	1247.4	818.5
1470.9	1041.8	669.8

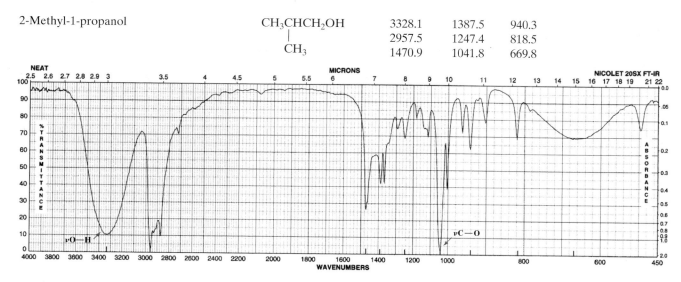

Figure G22 IR spectrum of a primary alcohol, 2-methyl-1-propanol

Phenol

OH

3372.6	1224.4	751.8
1595.3	1168.0	689.8
1498.9	809.8	506.0

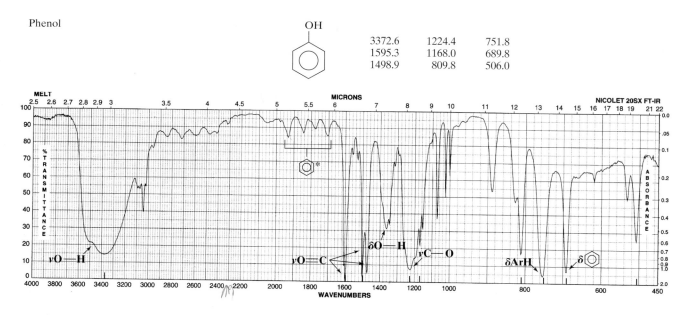

Figure G23 IR spectrum of phenol

Note: The bands marked ⬡* are aromatic overtone-combination bands.

the carbon that is bonded to OH absorb at wave numbers about $25–50 \ cm^{-1}$ lower than these.

Phenols. Phenols are characterized by a strong, broad band centered around $3300 \ cm^{-1}$ and a strong band near $1230 \ cm^{-1}$, accompanied by bands indicating an aromatic ring. (See "Aromatic Hydrocarbons" and Figure G23.)

O—H *stretch:* $3600–3200 \ cm^{-1}$ (strong, broad).
O—H *bend:* $1390–1315 \ cm^{-1}$ (moderate).

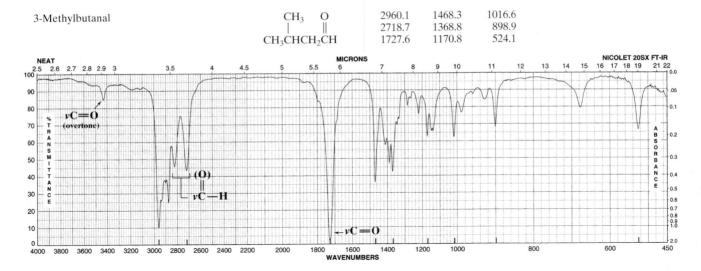

Figure G24 IR spectrum of an aldehyde, 3-methylbutanal

C—O *stretch:* 1300–1180 cm^{-1} (strong). Usually close to 1230 cm^{-1}. This band may be split, with several distinct peaks.

Aldehydes. The presence of a sharp, medium-intensity band near 2720 cm^{-1} and a strong carbonyl band near 1700 cm^{-1} is good evidence for an aldehyde. (See Figure G24.)

C—H *stretch:* 2850–2700 cm^{-1} (moderate to weak). From the carbonyl C—H bond. Most aldehydes have two bands near 2850 and 2720 cm^{-1}, with the low-frequency band well separated from other aliphatic C—H bands.

C=O *stretch:* 1740–1685 cm^{-1} (strong). Most unconjugated aldehydes absorb near 1725 cm^{-1}; conjugation of the carbonyl group with an aromatic ring or other unsaturated system shifts the band to the 1700–1685 cm^{-1} region. A weak overtone of this band may appear near 3400 cm^{-1}.

Ketones. The presence of a strong carbonyl band around 1700 cm^{-1} is good evidence for a ketone if other bands described in Table G2 (O—H, N—H, C—O, and aldehyde C—H) are absent. One or more bands in the 1300–1100-cm^{-1} region arise from C—C—C vibrations involving the carbonyl carbon. Such bands are generally weaker and narrower than C—O bands, for which they might be mistaken. (See Figure G25.)

C=O *stretch:* 1750–1660 cm^{-1} (strong). Most unconjugated aliphatic ketones absorb around 1715 cm^{-1}, and conjugated ketones absorb near 1670 cm^{-1}. A weak C=O overtone band is usually evident near 3400 cm^{-1}.

C—C—C *stretch–bend:* 1300–1100 cm^{-1} (moderate). Often multiple bands. Unconjugated ketones absorb around 1230–1100 cm^{-1}; conjugated ketones absorb around 1300–1230 cm^{-1}.

2-Pentanone

$$\underset{\displaystyle \text{CH}_3\text{CCH}_2\text{CH}_2\text{CH}_3}{\overset{\displaystyle \overset{\text{O}}{\|}}{}}$$

2963.9	1366.0	1170.7
1717.4	1295.5	727.0
1422.9	1235.5	591.9

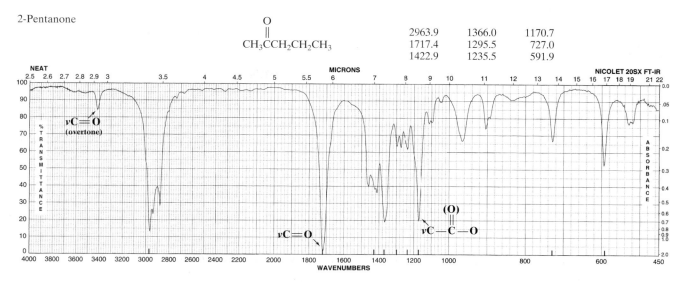

Figure G25 IR spectrum of a ketone, 2-pentanone

Hexanoic acid

$$\underset{\displaystyle \text{CH}_3(\text{CH}_2)_3\text{CH}_2\text{COH}}{\overset{\displaystyle \overset{\text{O}}{\|}}{}}$$

3191.2	1710.7	1293.4
2959.4	1467.5	1213.2
2669.9	1413.8	939.2

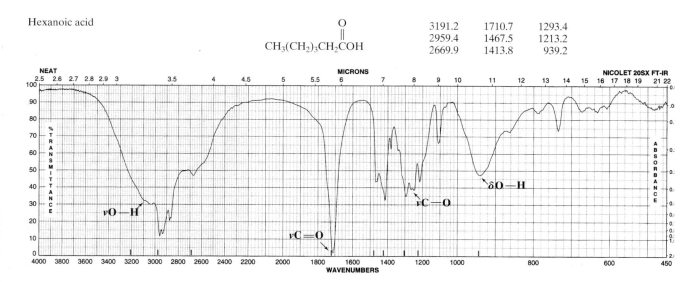

Figure G26 IR spectrum of a carboxylic acid, hexanoic acid

Carboxylic Acids. The presence of a very broad band centered near 3000 cm^{-1} and a carbonyl band around 1700 cm^{-1} is good evidence for a carboxylic acid. (See Figure G26.)

O—H *stretch:* $3300–2500 \text{ cm}^{-1}$ (strong, very broad). C—H stretching bands are usually superimposed on this band.

C＝O *stretch:* $1725–1665 \text{ cm}^{-1}$ (strong). Unconjugated acids absorb around $1725–1700 \text{ cm}^{-1}$; conjugated acids absorb around $1700–1665 \text{ cm}^{-1}$.

C—O *stretch:* $1350–1210 \text{ cm}^{-1}$ (strong). Long-chain acids may have a number of sharp peaks in this region.

O—H *bend:* $950–870 \text{ cm}^{-1}$ (moderate, broad).

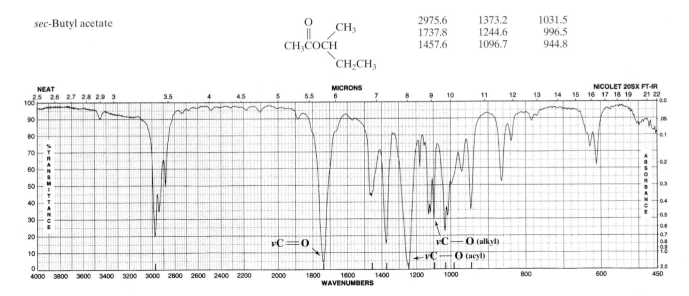

Figure G27 IR spectrum of an ester, *sec*-butyl acetate

Esters.   The presence of a strong carbonyl band around 1740 cm^{-1} and an unusually strong C—O band in the 1310–1160-cm^{-1} region is good evidence for an ester, especially if there is no O—H band. (See Figure G27.)

C=O *stretch:* 1775–1715 cm^{-1} (strong). Near 1770 cm^{-1} for phenyl esters (RCOOAr) and vinyl esters, 1740 cm^{-1} for most unconjugated esters, and 1730–1695 cm^{-1} for formates and conjugated esters.

C—O *stretch (acyl-oxygen):* 1310–1160 cm^{-1} (strong, broad). Occurs near 1310–1250 cm^{-1} for conjugated esters, 1240 cm^{-1} for unconjugated acetates, and 1210–1165 cm^{-1} for most other unconjugated esters. Both the "acyl-oxygen" and "alkyl-oxygen" bands arise from coupled vibrations involving C—C—O groupings.

C—O *stretch (alkyl-oxygen):* 1200–1000 cm^{-1} (moderate). Occurs in the same region as alcohol C—O bands and varies in the same way with changes in the alkyl group's structure. Esters of phenols absorb at higher wave numbers.

Amines.   Primary amines are characterized by a medium-intensity, two-pronged band near 3350 cm^{-1} and two medium-strong bands near 1615 and 800 cm^{-1}, the latter one being very broad. Secondary amines have a single weak band near 3300 cm^{-1} and a broad band near 715 cm^{-1}. Tertiary amines can sometimes be distinguished by the presence of a C—N band. (See Figure G28.)

N—H *stretch:* 3500–3200 cm^{-1} (moderate to weak, broad). Primary aliphatic amines give rise to a two-pronged band centered near 3350 cm^{-1}, secondary aliphatic amines have one weak band near 3300 cm^{-1}, and tertiary amines have none. Primary and secondary aromatic amines absorb near 3400 and 3450 cm^{-1}, respectively.

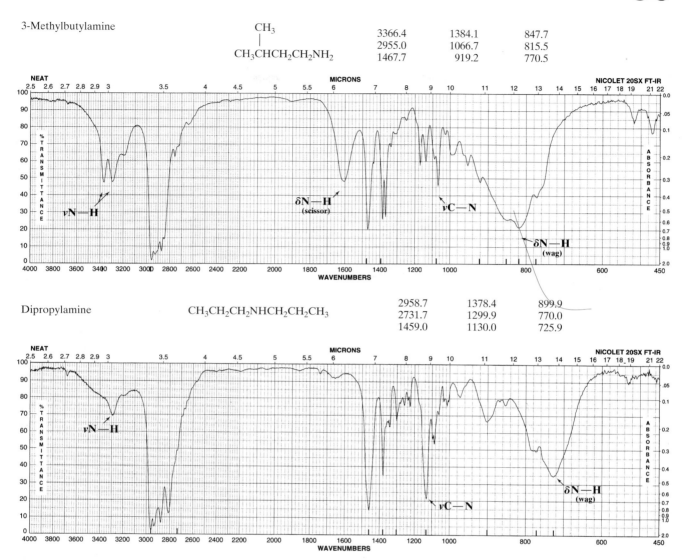

3-Methylbutylamine

$$CH_3$$
$$|$$
$$CH_3CHCH_2CH_2NH_2$$

3366.4	1384.1	847.7
2955.0	1066.7	815.5
1467.7	919.2	770.5

Dipropylamine

$$CH_3CH_2CH_2NHCH_2CH_2CH_3$$

2958.7	1378.4	899.9
2731.7	1299.9	770.0
1459.0	1130.0	725.9

Figure G28 IR spectra of a primary amine, 3-methylbutylamine, and a secondary amine, dipropylamine

N—H *bend (scissoring):* 1650–1500 cm^{-1} (strong to moderate). Usually near 1615 cm^{-1} for primary amines. Seldom observed for secondary aliphatic amines; secondary aromatic amines absorb near 1515 cm^{-1}.

N—H *bend (wagging):* 910–660 cm^{-1} (strong to moderate, broad). Often strong and very broad; around 910–770 cm^{-1} for primary amines, and near 715 cm^{-1} for secondary amines.

C—N *stretch:* 1340–1020 cm^{-1} (strong to moderate). Around 1340–1250 cm^{-1} for aromatic amines, and 1250–1020 cm^{-1} for aliphatic amines. As with an alcohol C—O band, the frequency of an aliphatic C—N band varies with changes in the structure of the attached alkyl group.

2-Methylpropanamide

$$CH_3 \quad O$$
$$| \qquad ||$$
$$CH_3CH — CNH_2$$

3352.3	1296.5	655.8
1640.2	1147.1	625.8
1401.0	1090.6	511.7

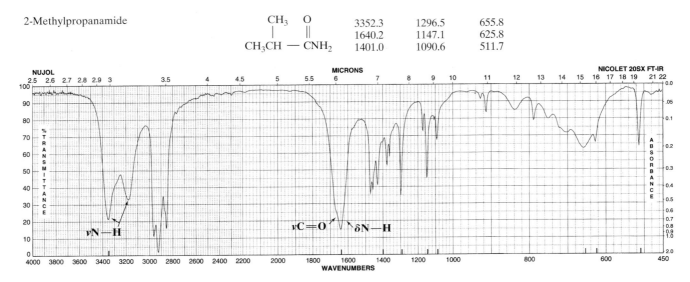

Figure G29 IR spectrum of an amide, 2-methylpropanamide

Amides. The presence of a carbonyl band near 1640 cm^{-1} and two bands (or peaks) in the 3500–3000 cm^{-1} region is good evidence for an amide. (See Figure G29.)

> N—H *stretch:* 3450–3300 cm^{-1} and 3225–3180 cm^{-1} (one or two bands, strong to moderate). Primary amides have two bands (or two prongs on a broad band) near 3400 and 3200 cm^{-1}. Secondary amides have a single N—H stretching band near 3340 cm^{-1}, with an N—H bending overtone near 3080 cm^{-1}. Tertiary amides have no N—H stretching bands.

> C=O *stretch:* 1695–1615 cm^{-1} (strong). Usually centered near 1640 cm^{-1}.

> N—H *bend:* 1655–1615 cm^{-1} (primary) or 1570–1515 cm^{-1} (secondary) (strong to moderate). This band usually overlaps the carbonyl band on spectra of primary amides obtained using KBr discs or mulls; it appears at lower frequencies on spectra obtained in solution. The band is near 1540 cm^{-1} for most secondary amides, and an overtone can sometimes be seen at about 3080 cm^{-1}.

Alkyl Halides. Alkyl chlorides and bromides show fairly strong absorption between 800 and 500 cm^{-1}. Additional chemical evidence is usually needed to characterize organic halides. (See Figure G30.)

> C(X)—H *bend:* 1300–1150 cm^{-1} (moderate). Observed only for halides with terminal halogen atoms (–CH$_2$X).

> C—Cl *stretch:* 850–550 cm^{-1} (strong to moderate). Two bands near 725 and 645 cm^{-1} when the chlorine is terminal, and below 625 cm^{-1} otherwise—unless several chlorine atoms are on the same or adjacent carbons. Ar—Cl bonds absorb around 1175–1000 cm^{-1}.

> C—Br *stretch:* 760–500 cm^{-1} (strong to moderate). Near 645 cm^{-1} when the bromine is terminal. Ar—Br bonds absorb around 1175–1000 cm^{-1}.

1-Chloropentane

$CH_3(CH_2)_4Cl$

2959.3	1282.1	789.3
1467.0	1037.2	730.8
1380.3	925.7	653.7

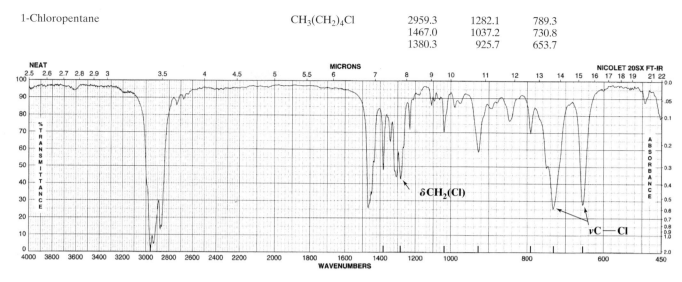

Figure G30 IR spectrum of an alkyl chloride, 1-chloropentane

Nuclear Magnetic Resonance Spectrometry

The theoretical principles underlying nuclear magnetic resonance (NMR) spectrometry can be found in most textbooks of organic chemistry and in appropriate sources listed in Category F of the Bibliography in this book. Here, we will review only those principles that are needed to gain a working knowledge of NMR spectrometry.

Nuclear magnetic resonance spectrometry is based on the magnetic properties of certain nuclei that possess a quality known as *spin*. The nucleus of a ^1H atom, which is a single proton, has spin. The nuclei of ^{13}C atoms also have spin, but the nuclei of ^{12}C atoms, which are nearly 100 times more abundant than ^{13}C atoms, do not. An atom with spin behaves like a tiny bar magnet. When placed in a strong magnetic field, it tends to become aligned with the field. For convenience, we will refer to a nucleus that is aligned *with* the external magnetic field as being in an **up** spin state and a nucleus that is aligned *against* the field as being in a **down** spin state. If a sample containing magnetic nuclei is exposed to radio frequency (RF) radiation of just the right frequency, some of its **up** nuclei will flip over, into the **down** spin state. This transition is illustrated in Figure G31. Since a nucleus in the **down** state is less stable (contains more energy) than a nucleus in the **up** state, the spin transition results in an absorption of energy by the nucleus. Such a transition is possible only if the energy of a radio frequency photon, $h\nu$, is exactly equal to the energy of the transition, ΔE, so that $\nu = \Delta E/h$. When this is the case, the *resonance condition*—the condition under which nuclei of a given kind can undergo spin transitions—is fulfilled. The transition energy, ΔE, is directly proportional to the strength of the external magnetic field, H_0, so ν is also

If you are not familiar with the principles and terminology of NMR spectrometry, read the section "Interpretation of ^1H NMR Spectra" or consult your lecture text.

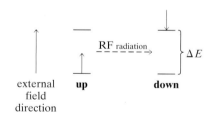

Figure G31 Spin transition of a magnetic nucleus

proportional to H_o. This means that the resonance condition for a nucleus can be attained either by adjusting the frequency of the RF radiation or by adjusting the strength of the external field.

a. ^{1}H NMR Spectrometry

Instrumentation

There are two fundamentally different ways of obtaining an NMR spectrum. With a *continuous-wave (CW) NMR spectrometer*, the sample is irradiated continuously with radio frequency waves as the magnetic field or RF frequency is varied, and the electromagnetic signals generated by nuclei as they change spin states are converted to peaks on a moving chart. With a *Fourier-transform NMR (FT-NMR) spectrometer*, the sample is irradiated with intense pulses of full-spectrum RF radiation that displace the nuclei from their equilibrium distribution. Their response to the displacement is monitored, generating data that is converted by a microprocessor to an NMR spectrum.

Continuous-Wave NMR.   In a continuous-wave NMR spectrometer, a glass tube containing the sample is placed between the poles of a magnet and irradiated with RF radiation from a transmitter coil as the magnetic field is "swept" (varied continuously) over a preset range. In an instrument of the type diagrammed in Figure G32, the magnetic field is swept from low to high field by varying the strength of an electric current passing through the sweep coils. When the resonance condition for a particular kind of nucleus in the sample is met, nuclei of that kind flip from the **up** state to the **down** state. As they do so, they generate a small fluctuating magnetic field that can be detected by a receiver coil encircling the sample tube. The receiver coil sends an electronic signal to an RF receiver, which amplifies and modifies the signal so that it can be displayed on a recorder as an NMR spectrum.

An NMR spectrum is a record of all the signals generated by all of the different kinds of nuclei in the sample that absorb RF radiation over the range swept by the instrument. If the sweep range is one in which the resonance conditions for ^{1}H nuclei (protons) are met, the spectrum should display different

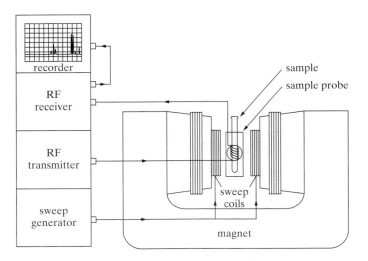

Figure G32 Schematic diagram of a continuous-wave NMR spectrometer

signals for protons that are in different molecular environments. For example, protons on the benzene ring in *para*-xylene are in a different molecular environment than protons on the methyl groups, so the NMR spectrum of *p*-xylene will display two signals, one for each kind of proton, at different positions on the spectrum. The position of a signal relative to the position of a reference signal, usually that of tetramethylsilane (TMS), is called its *chemical shift* (see Figure G38 in the section "Interpretation of ^{1}H NMR Spectra"). A chemical shift, represented by the Greek letter δ, is ordinarily measured in parts per million (ppm). Since the TMS signal ($\delta = 0$) is on the right side of a spectrum, chemical shifts increase from right to left.

A typical continuous-wave NMR spectrometer suitable for use by undergraduate students may operate at an RF frequency of 60 MHz and a magnetic field strength of approximately 1.4 tesla (14,000 gauss). When a ^{1}H NMR spectrum is recorded using a 60-MHz spectrometer, the magnetic field is swept over a range of about 1.4×10^{-5} tesla (0.14 gauss), which is only 10 millionths of the external field strength, or 10 parts per million (ppm). This sweep range can be extended to 15 ppm or so to detect protons whose resonance conditions occur outside this range.

Fourier-Transform NMR.   A Fourier-transform NMR spectrometer is capable of producing spectra with better resolution and a much higher signal-to-noise ratio than any CW instrument. In an FT-NMR instrument, the nuclei are irradiated with a short ($\sim$10 μs) pulse of radio frequency radiation that covers the entire frequency range of interest. The pulse is so intense that it raises all of the absorbing nuclei into the high-energy **down** state. As the nuclei return to their equilibrium state, they generate a *free induction decay (FID)* signal that contains information about the nuclei whose resonance conditions were met by any of the RF frequencies in the pulse. The FID signal, which is equivalent in information content to a complete NMR spectrum, is detected and sent to a microprocessor that accumulates and averages the FID signals from a series of pulses. The microprocessor then "decodes" the averaged FID signal by Fourier-transform analysis and converts it to a conventional NMR spectrum. The FID signal generated by one pulse takes less than a second to acquire, so an FT-NMR spectrometer can accumulate and average the equivalent of several hundred NMR spectra in the 2–5 minutes it takes a CW instrument to record a single spectrum. The resulting averaged spectrum has very little electronic noise and is thus much "cleaner" than a conventional CW spectrum.

A typical research-grade FT-NMR spectrometer uses a very powerful electromagnet that is cooled with liquid helium. At the temperature of liquid helium (4 K, $-269°$C), the coils of wire that generate the magnetic field are electrical superconductors, making it possible to attain very high field strengths of 14 tesla or more. Increasing the field strength of an NMR spectrometer causes the NMR signals to spread out, reducing overlap between adjacent signals. This and the high signal-to-noise ratio make complex ^{1}H NMR spectra generated on an FT instrument easier to interpret than those obtained with a CW instrument.

Chemical-Shift Reagents

The amount of structural information that can be obtained from a CW-NMR spectrum is often limited by the presence of overlapping signals. Increasing the magnetic-field strength reduces overlapping by increasing the

Adding the data generated by successive pulses improves the quality of the NMR spectrum because a signal increases in intensity with each addition, while electronic noise, being random, tends to cancel out.

p-xylene

chemical shifts (in Hz) of all of the signals by the same amount. Using a *chemical-shift reagent* also changes the chemical shifts of NMR signals, but it affects different signals differently; some are shifted more than others, and some may not be shifted at all. Nevertheless, an appropriate chemical-shift reagent can often be used to separate the signals of interest and facilitate spectral interpretation.

Chemical-shift reagents are organometallic complexes of certain paramagnetic rare earth metals. These complexes can coordinate with the oxygen and nitrogen atoms of alcohols, amines, carbonyl compounds, and other Lewis bases. The local magnetic field produced by the paramagnetic metal atom shifts the signals of nearby protons to an extent that varies with distance; the closer a nucleus is to the metal atom, the more its chemical shift will change. Different chemical-shift reagents have different effects on a spectrum; thus tris(dipivaloylmethanato)europium(III) [Eu(dpm)$_3$] causes signals to shift to the left (downfield) on an NMR spectrum, while the corresponding complex of praseodymium [Pr(dpm)$_3$] causes signals to shift to the right (upfield).

Sample Preparation

Most substances analyzed by NMR are first dissolved in a suitable solvent. Liquids that are no more viscous than water can sometimes be analyzed neat, but neat liquids may give broadened peaks and other spectral distortions due to intermolecular interactions. FT-NMR spectrometers yield good proton NMR spectra with solution concentrations as low as 0.1% (w/v). CW–NMR instruments usually require concentrations on the order of 5–20% (w/v), although satisfactory spectra may be obtained with lower concentrations when the amount of sample is limited. The liquid or solution is placed in a special thin-walled *NMR tube*, which is closed with a tight-fitting cap to prevent evaporation. A typical NMR tube has an o.d. of 5 mm, a length of 17.5 cm, and is both fragile and expensive. The NMR tube should be straight and uniform; a tube that wobbles when it is rolled down a slightly inclined glass plate will give large spinning sidebands, as discussed under the heading "Sample Spinning." For a routine ^{1}H NMR analysis on a CW instrument, you can prepare the sample as described here.

- Dissolve about 50 mg of the compound in 0.5–0.8 mL of a suitable solvent.
- Filter the solution directly into the NMR tube through a Pasteur pipet containing a small plug of tightly packed glass wool (see Figure G33).
- Add 5–10 μL of a reference standard, ordinarily TMS (tetramethylsilane).
- Cap the NMR tube carefully.
- Invert the tube several times to mix the components thoroughly.

If the amount of sample is limited, you may be able to use as little as 10 mg per 0.5 mL of solvent. The NMR tube should be filled to a depth of at least 2.5 cm, but should be no more than three-fourths full. If the NMR solvent already contains added TMS, omit the third step. TMS boils near room temperature, so it should be kept in a refrigerator and added with a *cold* syringe or fine-tipped dropper. For very high-resolution spectra, the sample should be *degassed* by bubbling a fine stream of pure nitrogen through it for 1 minute; degassing is not necessary for routine spectra.

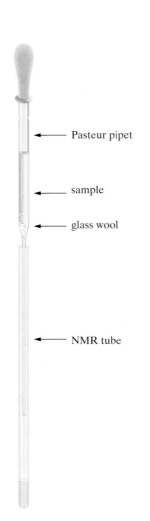

Pasteur pipet

sample

glass wool

NMR tube

Figure G33 Filtering an NMR solution

NMR Solvents

A solvent suitable for ^{1}H NMR analysis should have no protons that produce intense signals of their own, since they might obscure signals from the sample. Therefore, hydrogen-containing solvents such as chloroform and acetone are used in their completely deuterated forms. Deuterium (^{2}H) undergoes resonance at about $6\frac{1}{2}$ times the field strength required for ^{1}H, so an isotopically pure deuterated solvent does not interfere with a proton NMR spectrum. Most deuterated solvents, however, contain a significant amount of the protic form, giving rise to one or more small signals. For example, the NMR spectrum of a deuterochloroform (chloroform-d, $CDCl_3$) solution has a small $CHCl_3$ signal at 7.27 ppm, but this signal usually doesn't interfere with the solute's signals. A solvent for FT-NMR analysis *must* contain deuterium because the instrument locks onto the resonance signal of deuterium to help the user adjust the controls for maximum spectral resolution.

A good NMR solvent should have a low viscosity, a high solvent strength, and no appreciable interactions with the solute. Deuterochloroform is the most widely used NMR solvent because its polarity is low enough to prevent significant solute-solvent interactions and most organic compounds are sufficiently soluble in it for NMR analysis. When a more polar solvent is required, dimethyl-d_6 sulfoxide can be used, often in mixtures with deuterochloroform. It is convenient to add 1–3% TMS to the bulk solvent so that it does not have to be added during sample preparation (some commercial solvents already contain TMS).

Table G3 compares the properties of some deuterated NMR solvents and gives the approximate chemical shifts (δ) of their ^{1}H NMR signals.

Instrumental Parameters

Spinning Rate. To average out the effect of magnetic-field variations in the plane perpendicular to the axis of the NMR tube, an NMR sample is rotated at a rate of 30–60 revolutions per second while its NMR spectrum is being recorded. It is important to use an appropriate spinning rate—excessively high rates create a vortex that may extend into the region of the receiver coil, and low rates can cause large *spinning sidebands* or signal distortion (see Figure G34).

Table G3 Properties of some NMR solvents

Solvent	^1H δ, ppm	Solvent strength	Freedom from interactions	Viscosity
carbon disulfide	none	good	good	low
cyclohexane-d_{12}	1.4	poor	good	medium
acetonitrile-d_3	2.0	good	fair	low
acetone-d_6	2.1	good	poor	low
dimethyl-d_6 sulfoxide	2.5	very good	poor	high
1,4-dioxane-d_8	3.5	good	fair	medium
deuterium oxide	~5.2 (v)	good	poor	medium
chloroform-d	7.3	very good	fair	low
pyridine-d_5	7.0–8.7	good	poor	medium
trifluoroacetic acid-d	~12.5 (v)	good	poor	medium

Note: δ is the chemical shift of the protic form of the solvent; v = variable; solvent strength refers to the ability to dissolve a broad spectrum of organic compounds.

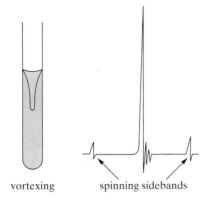

vortexing   spinning sidebands

Figure G34 Effects of spinning rate

Spinning sidebands are small peaks that are symmetrically spaced on either side of a main peak at a distance equal to the spinning rate; thus an NMR tube spun at 30 cycles per second can give rise to sidebands 30 Hz from each main peak. Spinning sidebands can also be caused by field inhomogeneity and wobbling NMR tubes or sample spinners. To find out whether small signals are spinning sidebands or impurity peaks, change the spinning rate and scan over the peaks again to see if their positions change.

Field Homogeneity. Recording a good NMR spectrum requires that the magnetic field be homogeneous (uniform) at the sample. The most important homogeneity control, usually called the Y control, is adjusted to produce a uniform field along the axis of the sample tube. For routine work on a previously tuned CW-NMR spectrometer, the Y control can be set by placing a blank sheet of paper over the chart paper and repeatedly scanning a strong peak in the spectrum of the sample (or of a standard acetaldehyde solution), each time making small adjustments of the Y control until the peak is as tall and narrow as possible and shows a good "ringing" (beat) pattern. Figure G35 shows an excellent ringing pattern for the quartet of acetaldehyde, characterized by the high amplitude, long duration, and exponential decay of the "wiggles" following the main peaks. An FT-NMR spectrum does not show a ringing pattern; magnetic-field homogeneity is adjusted by maximizing the intensity of a deuterium lock signal. Ordinarily the instructor or a lab technician performs such adjustments.

Signal Amplitude. The amplitude (height) of the signals on an NMR spectrum is adjusted with two controls. The *spectrum amplitude* control changes the amplitude of both the signals and the baseline noise. The *RF power* control increases signal height without increasing baseline noise up to the point at which *saturation* begins. At that point, the number of nuclei in both spin states is so nearly equal that increasing the intensity of the RF radiation no longer increases the number of transitions; instead it can cause distortion and reduce the signal size. Unless high sensitivity is required, the RF power level is usually set to a value at which there is little likelihood of

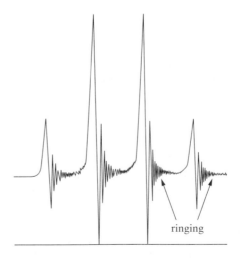

ringing

Figure G35 Ringing pattern for the quartet of acetaldehyde

saturation (usually about midrange), and the spectrum amplitude control is then adjusted so that the strongest peak in the spectrum extends nearly to the top of the chart paper.

Sweep. There are four sweep controls on a typical continuous-wave NMR spectrometer; these control the reference point of the spectrum, the sweep rate (sweep width divided by sweep time), and the portion of the spectrum to be scanned. The *sweep zero* control is used to set the signal for the reference compound to the proper value (zero for TMS). The *sweep width* control sets the total chemical-shift range to be scanned, usually 600–1000 Hz when the entire spectrum is being scanned on a 60-MHz instrument. A 600-Hz (10-ppm) range can be used if the sample is known to contain no protons that absorb downfield of 10 δ. A CW-NMR spectrum is often scanned at a rate of 1 Hz per second, so the *sweep time* can be set numerically equal to the sweep width (for example, 600 seconds for a 600-Hz sweep width). The *sweep offset* control is used when only a specific portion of the spectrum is to be scanned; it sets the upfield limit of the scan. For example, if a scan between 350 and 500 Hz is desired, the sweep offset should be 350 Hz and the sweep range 150 Hz.

Phasing. The *phasing* control should be adjusted to obtain a straight baseline before and after a signal (see Figure G36). Correct phasing is much more important when an NMR spectrum is being integrated than when it is being recorded.

Integrating a Spectrum

While an NMR spectrum is being *integrated*, the recorder pen traces a horizontal line until it reaches a signal; then it rises a distance that is proportional to the signal's area as it crosses the signal. Since the area of a signal on a proton NMR spectrum is proportional to the number of protons responsible for the signal, integrating the spectrum makes it possible to determine how many protons gave rise to each signal.

Here is a summary of the steps in the integration of a typical ^{1}H NMR spectrum. If you will be expected to integrate your NMR spectrum, your instructor will provide more detailed directions.

1 The RF power is optimized to provide an acceptable signal-to-noise ratio.
2 The instrument is switched to the integral mode.
3 While the spectrum is scanned rapidly, the integral amplitude control is adjusted until the integrator trace spans the vertical axis of the chart.
4 With the sweep offset and sweep width controls set to scan a region free from NMR signals, the balance control is adjusted during a slow scan of that region to give a horizontal line.
5 While scanning a signal, the phasing control is adjusted to make the integrator traces before and after the signal as nearly horizontal as possible.
6 The integral over the entire spectrum is recorded (preferably once in each direction) using a sweep time about one-fifth to one-tenth that for the normal spectrum. The pen should be returned to the baseline after each scan.
7 The relative peak areas are determined by measuring the vertical distances between the integrator traces before and after each signal, and the results for successive scans are averaged (see Figure G37).

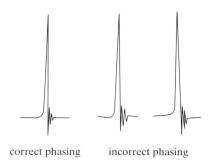

correct phasing   incorrect phasing

Figure G36 Effects of phasing on baseline

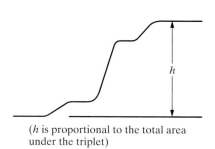

(*h* is proportional to the total area under the triplet)

Figure G37 Measuring signal areas

General Directions for Operating a CW-NMR Spectrometer

Equipment and Supplies

NMR spectrometer
sample
deuterated NMR solvent
tetramethylsilane (TMS)
NMR tube
Pasteur pipet
glass wool
tissues
washing solvent

Do not attempt to operate the instrument without prior instruction and proper supervision. Do not make any adjustments other than the ones specified, except at the instructor's request and under his or her supervision. Some of the adjustments described here may be made in advance by the instructor or a lab technician. The following procedure applies to a typical 60-MHz continuous-wave NMR spectrometer; operating procedures for other instruments may vary considerably. A computer program that simulates the operation of an NMR spectrometer may be available for your use.

Take Care! Handle NMR tubes with great care; they are fragile and may break.

Prepare a solution of the sample in a suitable NMR solvent as described under "Sample Preparation" and add 1–3% TMS, if necessary. Fill the NMR tube to a depth of about 3 cm with this solution, and cap the tube. Wipe the outside of the NMR tube carefully with a tissue paper or lint-free cloth, insert it in the sample spinner using a depth gauge to adjust its position, wipe it again, and carefully place the assembly into the sample probe between the magnet pole faces. Adjust the air flow to spin the sample at 40–50 Hz; it may need to be readjusted later to minimize spinning sidebands. Align the chart paper on the recorder and cover it with a sheet of scrap paper. Set the sweep controls to scan the desired range at a suitable rate. Typical settings for a 60-MHz instrument are: sweep offset, 0; sweep width, 600 Hz; and sweep time, 600 s. Set the RF power to about midrange and the filter response time to 1 s or less. Set the spectrum amplitude control to about midrange, and then scan the spectrum to find the tallest peak. Readjust the spectrum amplitude to keep that peak on scale near the top of the chart during a scan. If necessary, optimize peak shape and ringing and adjust the phasing as directed by your instructor. Set the TMS peak to 0.0 δ with the sweep zero control; you may have to sweep through the TMS signal several times, adjusting the control each time, until it is lined up with the zero on the chart paper. Set the recorder baseline, if necessary, to a convenient location near the bottom of the spectrum. Remove the blank paper and record the spectrum. If you are to integrate your spectrum, cover it with scrap paper while you make the adjustments described under "Integrating a Spectrum"; then remove the paper and record the integral on the original spectrum.

When you have finished scanning and integrating your spectrum, remove the spectrum and record the control settings and other relevant information on it. Then remove the sample tube as demonstrated by your instructor; follow directions carefully or the tube may break. Dispose of the solution as directed by your instructor. Clean the sample tube immediately and thoroughly using a Pasteur pipet and an appropriate solvent. The protic form of the solvent in which the sample was dissolved is generally used for cleaning. For example, if the solvent was $CDCl_3$, rinse the tube with $CHCl_3$—*not* the much more expensive deuterated solvent. Invert the tube in a suitable rack and let it drain dry. Before being reused, an NMR tube should be dried in an oven to remove all traces of the wash solvent.

Summary

1 Prepare solution, add TMS if necessary.
2 Transfer solution to NMR tube, cap tube.
3 Wipe tube, insert in sample spinner.
4 Insert spinner in probe, adjust spinning rate.
5 Align chart paper, cover with scrap paper.
6 Set sweep, RF power, and filter response time controls.
7 Scan spectrum and adjust spectrum amplitude control.
8 Zero TMS signal.
9 Set baseline, remove scrap paper.
10 Scan spectrum.
11 Integrate spectrum.
12 Remove and clean sample tube, dispose of solution.

Interpretation of ^{1}H NMR Spectra

A proton NMR spectrum provides numerical data in the form of chemical shifts, signal areas, signal multiplicities, and coupling constants. Working out the structure of a molecule from these numbers is a fascinating mental exercise comparable to the work of a cryptographer who reconstructs meaningful messages from coded symbols.

The *chemical shift* (δ) is the distance, measured in hertz or parts per million, from the center of a signal to some reference signal, usually that of tetramethylsilane (TMS). The TMS signal occurs farther upfield (to the right) than nearly all other proton signals, so the chemical shift of a signal is usually measured as its distance downfield (to the left) from that of TMS, as shown in Figure G38.

The *signal area*, which is the sum of the areas under all of the peaks in a signal, is proportional to the number of protons giving rise to the signal. Signal areas are determined by using an electronic integrator that traces a line across each proton signal after it is recorded. The area of the signal is proportional to the vertical rise of the integrator pen as it crosses the signal; that is, to the height of the "steps" drawn by the integrator pen, as shown in Figure G37. Integrated signal areas can be converted to proton numbers using the following relationship:

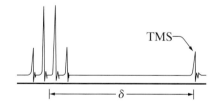

Figure G38 Chemical shift of a proton NMR signal

$$\text{number of protons responsible for signal} =$$
$$\text{total number of protons} \times \frac{\text{area under signal}}{\text{area under all signals}}$$

For example, suppose that a compound with the molecular formula $C_{10}H_{14}$ has four signals with relative areas of 42, 7, 14, and 35. The sum of the areas is 98, so the number of protons responsible for the first signal is

$$14 \times \frac{42}{98} = 6$$

Stop and Think: What is the most likely structure for this compound if it contains a benzene ring and one alkyl side chain?

By similar calculations, it can be shown that 1, 2, and 5 protons, respectively, are responsible for the other three signals. If the molecular formula of a compound is not known, relative proton numbers can be obtained by reducing the signal areas to the lowest ratio of integers.

The signal generated by a given set of protons may be split into several peaks as a result of *coupling* interactions with nearby proton sets (refer to your lecture textbook or see your instructor for an explanation of coupling). The *multiplicity* of a signal is simply the number of separate peaks it contains; its *coupling constant* is the distance between two adjacent peaks in the signal, measured in hertz (Hz). Figure G39 shows the signals of two sets of protons that are interacting with one another; the protons of set *a* have split the signal of the protons of set *b* into four peaks (a quartet), and the *b* protons have split the signal of the *a* protons into three peaks (a triplet). The coupling constant, which is equal for the two signals, is represented by J_{ab}. In the simplest case, the number of protons responsible for splitting the signal of a neighboring set of protons can be determined by subtracting 1 from the number of peaks in that signal. Thus the three peaks in the *a* signal are produced by two neighboring *b* protons, and the four peaks in the *b* signal by three neighboring *a* protons. An interacting triplet-quartet grouping of this kind is good evidence for an ethyl (CH_3CH_2—) group.

Ideal triplets and quartets should be symmetrical, having relative peak area ratios of 1 : 2 : 1 and 1 : 3 : 3 : 1, respectively. As shown in Figure G39, however, the signals in an actual spectrum are often somewhat distorted, giving paired peaks of unequal height. Note that the two signals in the figure are not perfectly symmetrical but appear to "lean" toward one another, with the peaks on the side facing the other signal being higher than predicted. This and the fact that their coupling constants are equal provide additional evidence that the protons responsible for the two signals are, in fact, coupling with one another and not with some other proton sets in the molecule.

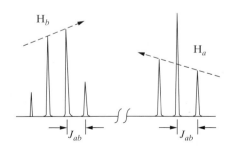

Figure G39 Signals of nearest neighbor protons

The following general procedure should help you derive structural information from a proton NMR spectrum.

1 Measure the integrated area of each signal and use it to determine the number of protons responsible for the signal. Each set of equivalent protons (protons in the same molecular environment) gives rise to a signal, and the relative signal areas can tell you how many protons are in each set. For example, 3,3-dimethyl-2-butanone has nine hydrogen atoms on the three equivalent methyl groups to the left of the carbonyl group and three on the other methyl group, so its ^{1}H NMR spectrum has two signals with an area ratio of 3:1.

2 Determine the chemical shift of each signal by measuring the distance, in parts per million, from the center of the signal to the TMS reference peak. The chemical shift of a signal may indicate what kind of protons are responsible for the signal or suggest their relative locations in the molecule. For example, alkyl hydrogen atoms that are remote from electron-withdrawing substituents should have a chemical shift of approximately 0.9 ppm if they are primary, 1.3 ppm if they are secondary, and 1.5 ppm if they are tertiary. Electron-withdrawing groups containing oxygen, nitrogen, or halogens tend to move ^{1}H NMR signals downfield, thereby increasing their chemical shifts. Benzene rings give rise to large downfield shifts, making it quite easy to recognize aromatic compounds from their ^{1}H NMR spectra. The correlation chart in Figure G40 summarizes chemical-shift data for a number of proton types. Chemical-shift values for compounds from the common families of organic compounds are given in Table G4.

3 Determine the multiplicity of each signal by counting the number of distinguishable peaks in the signal. (Very small peaks may be obscured by baseline noise.) If the signal of a proton set is reasonably symmetrical and contains evenly spaced peaks, it may be possible to determine how many nearby protons are coupled with the protons in that set by subtracting 1 from the multiplicity of the signal. Irregular signals and signals that have been split by several dissimilar proton sets must be analyzed by more advanced methods.

4 Measure the coupling constant of each signal that contains more than one peak and try to determine from the resulting values and the way each signal "leans" what other signals might be coupled with it.

All of this information can be used to build up the structure of a molecule piece by piece. For example, consider the ^{1}H NMR spectrum in Figure G41 of a ketone having the molecular formula $C_7H_{14}O$. The spectrum has only two signals, *a* and *b*, which have a relative area ratio of 6:1. Since the compound contains 14 protons, $\frac{6}{7}$ of them, or 12, must be responsible for signal *a*, and $\frac{1}{7}$ of them, or 2, for signal *b*. Signal *a* has only two peaks, indicating that the *a* protons have only one neighboring proton. The seven peaks in signal *b* (visible under a magnifying glass) indicate that the *b* protons have six neighbors, and the higher chemical shift of their signal suggests that they are close to the electron-withdrawing carbonyl group. The only alkyl group in which a single proton has six equivalent protons for neighbors is the isopropyl group, $(CH_3)_2CH-$. Two such groups provide the required total of 12 *a* and 2 *b* protons, and attaching them both to a carbonyl group gives the

$$CH_3-\overset{\overset{\textstyle CH_3}{|}}{\underset{\underset{\textstyle CH_3}{|}}{C}}-\overset{\overset{\textstyle O}{||}}{C}-CH_3$$

3,3-dimethyl-2-butanone

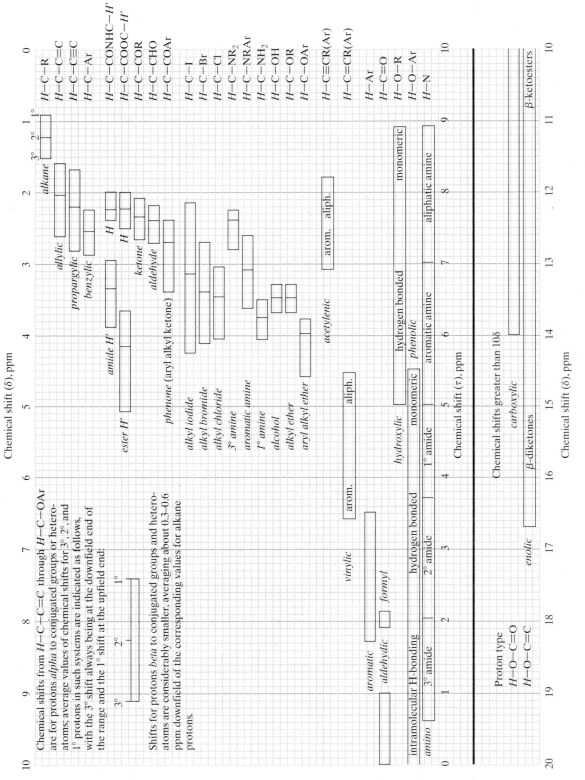

Figure G40 Correlation chart relating ^1H NMR chemical shifts to proton environments

Table G4 Approximate ^{1}H NMR chemical-shift values for different types of protons

Family	Proton type	Chemical shift (δ, ppm)
alcohol	H—O—R	1–5.5
	H—C—OH	3.4–4
phenol	H—O—Ar	4–12
	H—Ar	6–8.5
aldehyde	H—C=O	9–10
	H—C—CHO	2.2–2.7
ketone	H—C—COR	2–2.5
carboxylic acid	H—O—C=O	10.5–12
	H—C—COOH	2–2.6
ester	H—C—COOR	2–2.5
	H—C—OC=O	3.5–5
amine	H—N—R (aliphatic)	1–3
	*H—N—R (aromatic)	3–5
	H—C—N	2.2–4
amide	*H—N—C=O	5–9.5
	H—C—CO—N	2–2.4
	H—C—NC=O	3–4
halide	H—C—Br	2.5–4
	H—C—Cl	3–4
aromatic hydrocarbon	H—Ar	6–8.5
	H—C—Ar	2.2–3

$$CH-\overset{O}{\overset{\|}{C}}\text{-}OH \qquad CH\text{-}O\text{-}\overset{O}{\overset{\|}{C}}$$

Note: Signals of proton types marked with an asterisk are often very broad.

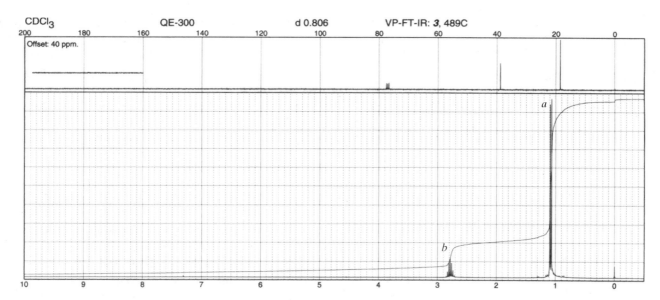

Figure G41 ^{13}C and ^{1}H NMR spectra of compound with molecular formula $C_7H_{14}O$ (The ^{13}C spectrum is in the narrow strip at the top.)

complete structure of the ketone, which is 2,4-dimethyl-3-pentanone. Note that the structure in the margin accounts for all of the features of the ^{1}H NMR spectrum: the 6 : 1 area ratio for the *a* and *b* protons; the higher chemical shift of the signal for the *b* protons resulting from their proximity to the carbonyl group; the splitting of the *a* signal into two peaks by each

2,4-dimethyl-3-pentanone

neighboring *b* proton; the splitting of the *b* signal into seven peaks by each group of six neighboring *a* protons; and the identical coupling constants of the two signals.

For further information about the interpretation of ^{1}H NMR spectra, refer to your lecture textbook or to appropriate sources in Category F of the Bibliography.

b. ^{13}C NMR Spectrometry

Except where noted here, most of the principles and experimental methods described previously for ^{1}H NMR spectrometry also apply to ^{13}C NMR spectrometry. Because carbon-13 nuclei are much less abundant than hydrogen nuclei, signals from the ^{13}C nuclei in a typical sample are about 6000 times weaker than those from its ^{1}H nuclei. A continuous-wave NMR spectrometer can hardly distinguish such weak signals from electronic baseline noise, so carbon-13 NMR spectra are ordinarily recorded using Fourier-transform NMR instruments. Since such instruments are highly complex and are seldom available for use by undergraduate students, no attempt will be made here to describe their operation.

A typical ^{13}C NMR spectrum is usually simpler and easier to interpret than the ^{1}H NMR spectrum of the same compound. Carbon-carbon splitting is unimportant because carbon-13 nuclei cannot couple with nonmagnetic carbon-12 nuclei and because there is only a slight chance that two carbon-13 atoms will be next to one another in the same molecule. However, hydrogen nuclei couple strongly with carbon-13 nuclei—not only with the nuclei of the carbon atoms that they are bonded to, but with those of more distant carbons as well. Since such coupling results in very complex spectra, it is usually prevented by various *decoupling* techniques. For *broad-band proton decoupling*, the sample is subjected to continuous broad-spectrum RF radiation covering the resonance frequencies of its protons, causing all of the protons to flip over (change spin states) so rapidly that their coupling effects on the adjacent carbon atoms average out to zero. In a broad-band–decoupled ^{13}C NMR spectrum, each ^{13}C signal is a single peak rather than a multiplet. For an example, see the broad-band–decoupled ^{13}C NMR spectrum in Figure G41, which is displayed in a narrow strip above the ^{1}H NMR spectrum. (The three closely spaced peaks near 78 ppm arise from the solvent, chloroform-*d*.)

Another decoupling technique, *off-resonance decoupling*, yields *proton-coupled* spectra, in which only those hydrogens that are attached directly to a carbon atom split the signal of that carbon atom. Thus the three protons of a methyl group will split the signal of the methyl carbon into a quartet, but will not split the signal of any other carbon in the molecule. Proton-coupled spectra are less common and harder to interpret than broad-band–decoupled spectra, so we will not consider them further.

The area of a ^{13}C signal, unlike that of a ^{1}H signal, is not directly proportional to the number of carbon atoms responsible for the signal, so ^{13}C NMR spectra are not integrated. Consequently, three of the four parameters that can be obtained from ^{1}H NMR spectra are not present in a broad-band–decoupled ^{13}C NMR spectra. Since there is no integration, there are no signal areas, and since all of the signals are singlets, there are no multiplicities or coupling constants to interpret. That leaves only the chemical shifts, which, as it happens, are extraordinarily useful.

The ^{13}C NMR spectrum of a compound gives us direct information about the compound's carbon "backbone," information that is often not available from its ^{1}H NMR spectrum. A broad-band–decoupled ^{13}C NMR spectrum of a compound contains one sharp peak for each kind of carbon atom in its molecules. Since many compounds have few, if any, magnetically equivalent carbon atoms, most of the peaks in a typical ^{13}C NMR spectrum are one-carbon peaks, each arising from a different carbon atom. Thus the absence of a linear relationship between signal area and the number of carbon atoms is not a serious handicap.

Carbon-13 chemical shifts cover a much broader range than proton chemical shifts, about 250 ppm compared to 15 ppm or so for protons. As a result, the peaks on a ^{13}C NMR spectrum are usually well separated. The chemical shift of a carbon-13 atom is very sensitive to changes in its hybridization and molecular environment. Carbon atoms that are sp^2 hybridized have much higher chemical shifts ($100–160\ \delta$) than sp^3 carbons ($0–60\ \delta$), and the chemical shifts of sp carbons are somewhere in between ($65–105\ \delta$). Electron-withdrawing groups cause downfield chemical shifts at nearby carbon atoms, but they can cause upfield shifts at more distant carbon atoms. For example, a chlorine atom increases the chemical shift of an α-carbon atom by about 30 ppm and of a β-carbon atom by about 10 ppm, but it *decreases* the chemical shift of a γ–carbon atom by about 5 ppm.

$$\begin{array}{ccc} \gamma & \beta & \alpha \\ \text{C} - \text{C} - \text{C} - \text{Cl} \\ -5 & +10 & +30 \end{array}$$

Other electron-withdrawing groups have similar effects. Electron-donating groups have the opposite effect, decreasing the chemical shifts of α- and β-carbon atoms and increasing the chemical shifts of γ-carbon atoms.

Table G5 shows chemical-shift ranges for some kinds of ^{13}C atoms. Such tables can be used to assign the peaks in a ^{13}C NMR spectrum to specific carbon atoms. For example, the ^{13}C NMR spectrum of methyl methacrylate has five peaks with the chemical shifts shown here.

J. Chem. Educ. **1987**, *64*, 915 *describes a method for estimating* ^{13}C *chemical shifts.*

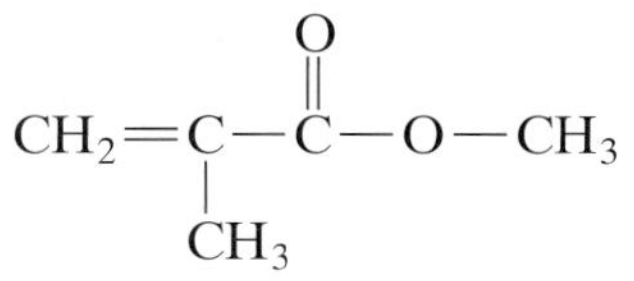

$$CH_2{=}C{-}\overset{\displaystyle O}{\overset{\|}{C}}{-}O{-}CH_3$$
$$\underset{CH_3}{|}$$

1	18 ppm	4	137 ppm
2	52 ppm	5	167 ppm
3	125 ppm		

From Table G5, we find the following information for carbon atoms like those in methyl methacrylate:

ester carbonyl carbon: 160–185 ppm

alkene carbon: 100–160 ppm

carbon bonded to oxygen (**C**—O): 40–90 ppm

primary alkyl carbon: 0–40 ppm

Note that the **C**—O chemical-shift range from the table is for alcohols and ethers, but a carbon atom on the alcohol portion of an ester is in a similar environment. From this information, it is easy to match the peak at 167 ppm

Table G5 ^{13}C NMR chemical-shift ranges for different types of carbon atoms

Type of carbon atom	Chemical-shift range (δ, ppm)
1° alkyl, $\mathbf{R}CH_3$	0–40
2° alkyl, $R_2\mathbf{C}H_2$	10–50
3° alkyl, $R_3\mathbf{C}H$	15–50
alkene, $\mathbf{C}=C$	100–160
alkyne, $\mathbf{C}\equiv C$	60–90
aryl, ⬡$\mathbf{C}$—	100–170
alkyl halide, $\mathbf{C}$—X (X = Cl, Br)	5–75
alcohol or ether, $\mathbf{C}$—O	40–90
amine, $\mathbf{C}$—N	10–70
aldehyde or ketone, $\mathbf{C}=O$	180–220
carboxylic acid or ester, O=$\mathbf{C}$—O	160–185
amide, O=$\mathbf{C}$—N	150–180

$$\underset{\substack{|\\ \underset{18}{CH_3}}}{\overset{\overset{O}{\|}}{\underset{125}{CH_2}=\underset{137}{C}-\underset{167}{C}-O-\underset{52}{CH_3}}}$$

with the carbonyl carbon, the peaks at 125 ppm and 137 ppm with the alkene carbons, the peak at 52 ppm with the OCH$_3$ methyl group, and the peak at 18 ppm with the remaining methyl group. Since the alkene carbon *beta* to the two oxygen atoms would be expected to have a higher chemical shift than the alkene carbon *gamma* to them, we can assign chemical-shift values to the carbon atoms as shown in the margin. Such assignments can often be confirmed by a technique called *distortionless enhanced polarization transfer (DEPT)*, which can pinpoint the carbon atom that is responsible for a particular signal.

OPERATION 38

Ultraviolet-Visible Spectrometry

Principles and Applications

Ultraviolet-visible (UV-VIS) spectrometers, which detect the absorption of radiation in the visible (~400–800 nm) and near ultraviolet (~200–400 nm) regions of the electromagnetic spectrum, are useful for both qualitative and quantitative analysis of organic compounds. Radiation in these regions may induce molecules to undergo transitions from an electronic ground state to one or more excited states. The energy required for an electronic transition is much greater than that needed to induce a vibrational or nuclear magnetic transition, so the wavelength of the radiation used is much shorter— 200–800 nm compared to about 2.5–50 μm for infrared spectrometry and several meters for NMR. Most transitions involving electrons in single

1 nanometer (nm) = 10^{-9} m. The older unit "millimicrons" (mμ) is sometimes used for nm.

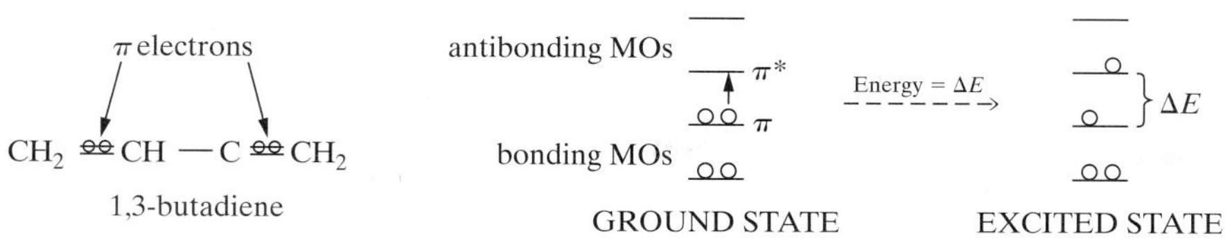

π electrons

$$CH_2 \overset{\ominus\ominus}{=} CH - C \overset{\ominus\ominus}{=} CH_2$$

1,3-butadiene

antibonding MOs

π^*

π

bonding MOs

GROUND STATE

Energy = ΔE

$\Big\} \Delta E$

EXCITED STATE

Figure G42 Electronic energy transition in 1,3-butadiene

bonds and isolated double bonds require wavelengths shorter than 200 nm, but conventional UV-VIS spectrometers are not designed to scan this region because oxygen from the air also absorbs UV radiation below 200 nm. Therefore, the electronic transitions of organic compounds that give rise to UV-VIS spectral bands usually involve pi electrons in aromatic and conjugated aliphatic systems and certain nonbonded electrons. An example of such a transition is illustrated in Figure G42, in which a pi electron in the ground-state electron configuration of 1,3-butadiene jumps from its bonding molecular orbital (MO) to an unoccupied antibonding molecular orbital.

An ultraviolet-visible spectrum is often quite featureless compared to an infrared or NMR spectrum, and may consist of only one or two broad *absorption bands*. The broad band structure is caused by rotational and vibrational transitions that accompany each electronic transition; each different combination of rotational and vibrational transitions has a different energy, so collectively they span a broad range of wavelengths. The height of an absorption band above the baseline of a UV-VIS spectrum is measured in units of *absorbance, A*. The position of an absorption band is given by its wavelength of maximum absorbance, λ_{max}, which is measured from the top of the band. For example, the absorption band illustrated in Figure G43 has an absorbance of 0.80 and a λ_{max} of 350 nm.

Absorbance is related to *transmittance*, the fraction of incident radiation transmitted through a sample, by this equation:

$$A = \log(1/T) = -\log T \qquad \textbf{(1)}$$

Thus the transmittance (T) of a UV-VIS band is equal to 10^{-A}. For a band with $A = 0.80$, $T = 10^{-0.80} = 0.16$, meaning that about 16% of the light entering the sample passes through unchanged and the remaining 84% is absorbed by the sample.

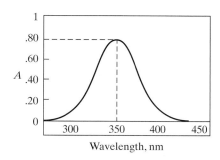

Figure G43 An ultraviolet absorption band

a. UV-VIS Spectra

Sample Preparation

Routine UV-VIS spectra are nearly always obtained in solution. The solvent must be transparent (or nearly so) in the regions to be scanned. Water, 95% ethanol, methanol, dioxane, acetonitrile, and cyclohexane are suitable down to about 210–220 nm; many other solvents can be used at higher wavelengths. The preferred solvent is 95% ethanol, in part because it does not require additional purification; most other solvents must be purified or purchased as spectral-grade solvents. The solvent must not, of course, react with the solute. For example, alcohols should not be used as solvents for aldehydes.

If possible, the solution to be analyzed should produce a maximum absorbance of about 1 when the solute's strongest absorption band is scanned. Using Beer's law (see Equation **2,** p. 753) we can show that the molar concentration of such a solution, if analyzed in a 1-cm sample cell, should be less than or equal to $1/\varepsilon_{max}$, where we define ε_{max} as the maximum molar absorptivity of the solute over the wavelength range to be scanned. For example, if the solute's strongest band has a molar absorptivity of 10,000 at its λ_{max} value, the solution concentration should be about $1 \times 10^{-4}\ M$. To make up such dilute solutions accurately, it may be necessary to prepare a stock solution that is too concentrated by several powers of ten, and then measure an aliquot of this solution and dilute it. For example, to prepare a $\sim 1.0 \times 10^{-4}\ M$ solution of cinnamic acid (M.W. = 148), you could measure 15 mg (~ 0.10 mmol) of the solid into a 10-mL volumetric flask and fill it to the mark with solvent; then transfer a 0.10-mL aliquot of this 0.010 M solution into another 10-mL volumetric flask and fill it to the mark with solvent. For qualitative work, when knowing the exact concentration of the solution is not necessary, you can omit the dilution step and use a microliter syringe to measure a specified or calculated amount of solute into 10–25 mL of the solvent. If the molar absorptivity of the solute is not known, it may be necessary to find the optimum concentration by trial and error, starting with a more concentrated solution and diluting it as needed to bring all of the absorption bands on scale.

Sample Cells

The most commonly used *spectrophotometer cell* is a transparent rectangular container with a square cross section, having a path length of 1.00 cm and a capacity of about 3 mL. Cells with capacities of about 1 mL and 0.5 mL are also available for microscale work. Two sides of a typical cell are nontransparent (usually frosted) and the other two are transparent. Silica or quartz cells are used for the ultraviolet region; optical glass or plastic cells are suitable in the visible region. Cells must be scrupulously cleaned; they should never be touched on their transparent sides, since even a fingerprint can yield a spectrum.

Recording a Spectrum

The construction of ultraviolet-visible spectrometers varies widely. Both single- and double-beam instruments are available, with and without recording capability. For a double-beam recording instrument, two identical spectrophotometer cells are filled about two-thirds full with (1) a solution of the compound being analyzed (the sample) and (2) the solvent used to prepare the solution. For a single-beam instrument, the same cell is used for both the sample and the pure solvent. A spectrophotometer cell filled with the sample or solvent is held by its nontransparent sides and inserted into the appropriate cell holder inside the instrument's sample compartment, oriented so that the light beam will pass through its transparent faces.

Before a spectrum is recorded, the user selects the wavelength region to be scanned and may also select a radiation source appropriate for that region. A tungsten lamp can be used between 300 and 800 nm and a hydrogen lamp between 190 and 350 nm. The absorbance range of some instruments can be preset; a typical range is from zero to one or two absorbance units. Modern computerized instruments allow the operator to apply a baseline correction, so that the instrument automatically subtracts any absorption due to the solvent as the spectrum is scanned. Such instruments have a monitor to display the spectrum and a printer or plotter to record it. Most older instruments record UV-VIS spectra on a roll of chart paper that feeds onto a flat recorder bed as the spectrum is scanned.

Once a spectrum has been recorded, the data it contains can be presented as a tabulation of λ_{max} values giving either the absorbance, molar absorptivity (ε), or log ε at each wavelength specified. For example, the ultraviolet spectrum of cinnamic acid is reported in one reference book as "λ^{al} 210 (4.24), 215 (4.28), 221 (4.18), 268 (4.31)." The numbers in parentheses are log ε values for peaks having the λ_{max} values (in nanometers) given, and "al" indicates that the spectrum was run in ethyl alcohol.

General Directions for Operating a Recording UV-VIS Spectrometer

Equipment and Supplies

UV-VIS spectrometer
sample
solvent(s)
sample cell(s)
lens paper

Do not attempt to operate the instrument without prior instruction and proper supervision. Ultraviolet-visible spectrometers vary widely in construction and operation, so the following is intended only as a general guide and may not be applicable to the instrument you will be using. Specific operating instructions should be learned from in-class demonstrations or the operator's manual.

Prepare a solution of the sample in an appropriate solvent as described under "Sample Preparation." Unless otherwise instructed, clean a sample cell (or two for a double-beam instrument) by wiping its surfaces with a lens paper moistened with spectral-grade methanol or another appropriate solvent, and then let the solvent evaporate. This should leave the cell surfaces free of contaminants that may have accumulated since the cell was last used. Be sure that the instrument, recorder or printer, and source lamps are on and have had sufficient time to warm up. If necessary, select the appropriate radiation source for the desired wavelength range and set the absorbance range to 0–1 or another appropriate value. Set the starting and ending wavelengths. If the instrument scans from high to low wavelength, the starting wavelength will be the highest wavelength of the range to be scanned. If there is no provision for setting the ending wavelength, you will have to end the scan manually. Some instruments require manual adjustment of zero and 100% transmittance values (or infinite and zero absorbance values) before a spectrum or baseline is run; if so, make the adjustments as directed by your instructor. If the instrument provides for a baseline correction, fill the cell with the solvent, cap it, and place it in the sample compartment with a transparent side facing the light source. Double-beam instruments require two such cells, one in the sample beam and the other in the reference beam. Then close the compartment door and record the baseline as directed by your instructor.

For a single-beam instrument, remove the solvent from the sample cell, rinse it with a small amount of the solution to be analyzed, fill it with that solution, cap it, and place it in the cell holder in the sample compartment. For a double-beam instrument, place one capped cell, filled with the sample, in the sample-cell holder and the other, filled with the solvent, in the reference-cell holder. Close the sample compartment door. If the instrument uses chart paper, position it so that the scan starts on an ordinate (vertical) line, label this line with the starting wavelength, lower the pen to the paper, and begin to scan the spectrum. For a computerized instrument, start the scan as directed by your instructor. If any absorption band goes off scale so that its top is "chopped off," change the absorbance range or dilute the sample. If both ultraviolet and visible regions are to be scanned, change the radiation source if necessary (many instruments do this automatically) and scan the spectrum in the other region. If the scan does not stop automatically when the end of the wavelength range has been reached, stop it manually. If the instrument has a chart recorder, raise the pen from the chart and tear off the chart paper; then write down the wavelength and absorbance ranges along its x- and y-axes and record the wavelength interval between chart units. If the instrument has a printer or plotter, initiate printing or plotting of the spectrum. Measure and write down the λ_{max} and absorbance values of all significant bands.

Dispose of the used solvent and solution as directed by your instructor. Rinse the sample cell several times with an appropriate solvent. If necessary, clean it further using a liquid detergent or a special cleaning solution. Never use a *dry* lens paper, an abrasive cleanser, or any scrubbing implement (such as a pipe cleaner with a wire core) that might scratch the cell. Drain both cells of excess solvent and dry them as directed by your instructor.

Summary

1 Clean sample cell(s).
2 Select source, if necessary.
3 Set absorbance and wavelength ranges.
4 Record baseline, if necessary.
5 Scan spectrum of sample.
6 Record spectral parameters.
7 Dispose of used solvent and solution, clean sample cell(s).

b. Colorimetry

A cuvette looks like a small test tube— but it should never be used as one!

Inexpensive nonrecording single-beam UV-VIS spectrometers, often called *colorimeters*, are frequently used for routine quantitative analysis of compounds in solution. The Spectronic 20 illustrated in Figure G44 is a widely used colorimeter, and most other colorimeters are operated similarly. The solution to be analyzed is prepared as described for a recording spectrometer and placed in a *cuvette* that is inserted into the sample compartment. The wavelength control is set to a wavelength at which the sample absorbs strongly and, after some preliminary adjustments, the absorbance (or percent transmittance) of the solution is read from the scale.

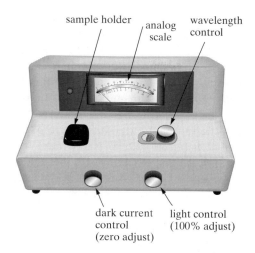

Figure G44 Spectronic 20 spectrophotometer

The concentration of a solution can be determined from its absorbance value using either Beer's law (Equation **2**) or a calibration curve of absorbance versus molar concentration.

Beer's law:

$$c = \frac{A}{\varepsilon \cdot b} \tag{2}$$

c = molar concentration
A = absorbance
b = cell path length, in cm
ε = molar absorptivity, in L mol^{-1} cm^{-1}

Equation **2** can be used to determine concentrations only when solutions of the solute obey Beer's law over the appropriate concentration range and when the absorptivity of the solute is known.

General Directions for Operating a Colorimeter

Equipment and Supplies

colorimeter
sample
solvent
cuvette(s)

Do not attempt to operate the instrument without prior instruction and proper supervision. These directions are for operation of the B & L Spectronic 20 or a similar instrument and may not be applicable to all such instruments.

Obtain or prepare the solution to be analyzed. Make sure that the instrument has been switched on and that adequate time has been allowed for warm-up. Set the wavelength to the desired value. Rotate the zero-adjust control until the digital readout or the pointer of an analog scale indicates

0% transmittance. To read an analog scale, position your eyes so that the pointer is directly over its reflection in the mirror; this prevents parallax errors. Insert a clean cuvette containing the pure solvent (the one used to prepare the solution being analyzed) into the sample holder, making sure that the alignment mark on the cuvette is opposite the mark on the cell holder. Adjust the 100% control until the transmittance reading is 100%. Rinse the cuvette with a little of the solution to be analyzed, and then fill it with that solution. Replace it in the sample holder, being careful to position it in exactly the same alignment as before. (Different cuvettes are sometimes used for the solvent and sample, but errors will result if the cuvettes are not well matched; for precise work, it is best to use the same cuvette for all measurements.) Read the percent transmittance or absorbance as accurately as possible. With an analog scale, it is more accurate to read $\%T$ and convert it to A, if necessary, than to read the absorbance directly.

If another solution containing the same solvent and solute is to be analyzed, empty the cuvette and rinse it with that solution before you make the next measurement. After the last measurement, rinse the cuvette with pure solvent, clean it, and let it air dry. Dispose of the used solvent and solution(s) as directed by your instructor. If necessary, convert the percent transmittance values to absorbance values using the equation $A = \log(100/\%T)$.

Summary

1 Prepare or obtain solution.
2 Clean cuvette.
3 Set wavelength and zero-adjust control.
4 Fill cuvette with solvent, set 100% control.
5 Fill cuvette with solution, record $\%T$.
6 Clean cuvette, dispose of used solvent and solution(s).

OPERATION 39

Mass Spectrometry

The other kinds of spectrometry described in this "Instrumental Analysis" section use some kind of electromagnetic radiation—radio frequency, infrared, ultraviolet, or visible—to gently probe the molecules of a sample and induce them to reveal their secrets. No molecules are damaged; once the spectrum is recorded, they return to their former states. By contrast, mass spectrometry uses a brute-force approach to determine molecular structures. Molecules that enter a mass spectrometer are pummeled by high-energy electrons and shattered into fragments, which are pushed and pulled along a curved path until they smash into an ion collector at journey's end. The fragments cannot be put back together to form the original molecule, so mass spectrometry is a destructive method of analysis.

Mass spectrometry is not really a spectroscopic method in the usual sense, in that no electromagnetic radiation is absorbed. But a mass spectrum does resemble a conventional spectrum, in that it consists of a series of peaks of different amplitude plotted along a numerical scale. Although mass spectra are not easily interpreted and mass spectrometers are costly and complex instruments, mass spectrometry has become an increasingly valuable analytical tool for scientists and technologists in a variety of fields.

Principles and Applications

When a compound is bombarded with a beam of high-energy electrons in a mass spectrometer, each of its molecules (M) can lose an electron to form a *molecular ion*, $M \cdot^+$.

$$M \rightarrow M \cdot^+ + e^-$$

Because a molecular ion contains both an unpaired electron and a positive charge, it is called a *radical cation*. If the energy of the electron beam is high enough, many of the molecular ions will have enough excess vibrational and electronic energy to break apart into fragments. Each pair of fragments consists of another positive ion (A^+), called a *daughter ion*, and a neutral molecule (X).

$$M \cdot^+ \rightarrow A^+ + X$$

The unpaired electron from the molecular ion may end up on either A^+ or X, depending on the kind of fragmentation. Each daughter ion may in turn break down, losing a neutral fragment to form yet another daughter ion, and so on.

$$A^+ \rightarrow B^+ + Y$$

For example, the fragmentation of an ammonia molecule takes place as shown, forming ions (cations and radical cations) having approximate masses of 17, 16, 15, and 14 atomic mass units.

ammonia   molecular ion   daughter ions

The positive ions formed during these transformations are accelerated into an evacuated chamber in which they are separated according to their mass to charge ratios (m/e), usually by means of strong electric and magnetic fields (see Figure G45). As a beam of ions with a given m/e value impinges on an *ion collector*, it gives rise to an electrical current that is amplified and displayed on a monitor as a peak whose amplitude is proportional to the number of ions striking the detector. A *mass spectrum* is a record, usually printed out as a table of data or a computer-generated bar graph, of the relative abundances of all the ions arranged in order of their m/e values. Because most daughter ions have a charge of +1, the m/e value associated with a peak is nearly always equal to the mass of the ion that gave rise to that peak—or in rare cases, to one-half of its mass. A large molecule may be fragmented into several hundred different ions with different m/e values and relative abundances, so mass spectrometers are provided with microprocessors that record, store, and process the data.

Mass spectrometry is an extremely valuable tool for structural analysis; it can be used to identify or characterize a host of organic (and inorganic) compounds, including biologically active substances with very complex molecular structures. The mass spectrum of a compound usually gives the mass of its molecular ion, which is essentially equal to its molecular weight. It also provides the masses of smaller pieces of its molecules, which can often be identified with the help of published tables of molecular fragments. A high-resolution mass spectrum provides data that can be used to determine a compound's molecular formula. Such information often makes it possible to piece together the compound's molecular structure, or at least to learn more about its structural features.

Different compounds yield distinctively different patterns of ion fragments, so an unknown compound can sometimes be identified by a comparison of its mass spectrum with mass spectral data from the scientific literature. Unfortunately, peak intensities on mass spectra are very sensitive to instrumental parameters, such as the energy of the electron beam. Therefore, there may be significant differences in the mass spectra recorded on different instruments, or even by different operators using the same instrument. Nevertheless, it is often possible to make a tentative identification from a literature comparison and then to confirm it by recording mass spectra of the unknown and the most likely known compounds under identical operating conditions. This process can be facilitated by using a computer to compare the spectrum of the unknown with the mass spectra in a memory bank, which may contain tens of thousands of such spectra.

Instrumentation

Mass spectrometers come in a wide variety of sizes and configurations, ranging from high-resolution mass spectrometers that may take up most of an instrument room to compact "tabletop" mass spectrometers that can be used as detectors for gas and liquid chromatographs. Although research-grade mass spectrometers are very expensive, tabletop mass spectrometers are within the equipment budgets of some undergraduate chemistry programs.

In a conventional *single-focusing* mass spectrometer (diagrammed in Figure G45) the sample is introduced into a sample inlet system that is heated to keep some or all of its molecules in the vapor state. These molecules find their way into the *ionizing chamber* through an aperture called a *molecular leak*, which may be a tiny hole in a piece of gold foil. The ionizing chamber is kept at a pressure of about 10^{-6} torr to minimize collisions between particles and interference from ionized air. Molecules that wander into the path of the *electron beam* are ionized to molecular ions, some of which undergo fragmentation to yield daughter ions. These ions are pushed toward a slit by a positively charged *repeller plate*, and then accelerated to a high velocity by electrically charged *accelerator plates* and directed into a *magnetic separator*. In the magnetic separator, a powerful magnetic field deflects each kind of ion into a curved path whose radius depends on the ion's mass-to-charge

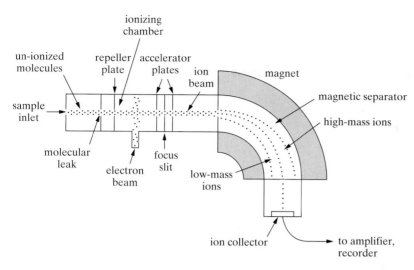

Figure G45 Schematic diagram of a single-focusing mass spectrometer

ratio. Lighter (lower m/e) ions are deflected more than heavier (higher m/e) ions. At a given magnetic-field strength, only ions of a given mass (or m/e value) can pass through the slit that leads to the *ion collector*. As the field strength is increased, ions of progressively higher mass reach the ion collector and are detected. When a beam of ions strikes the ion collector, an electrical signal is generated whose intensity is proportional to the number of ions in the beam—that is, to their abundance. The signals are amplified and displayed by a monitor screen or recorder to produce a mass spectrum.

In a single-focusing mass spectrometer, variations in the kinetic energies of ions having the same mass cause their ion beam to broaden as it passes through the magnetic separator, reducing the instrument's resolving power. In a high-resolution *double-focusing* mass spectrometer, the ions are initially directed along a curved path by an electrostatic field, which allows only particles of the same kinetic energy to pass through a slit leading to the magnetic separator. Double-focusing instruments can often resolve such ions as CH_2N^+ and N_2^+, whose mass numbers (28.0187 and 28.0061) differ by less than 0.05%.

Other kinds of mass spectrometers separate ions by very different methods. In a *quadrupole* mass spectrometer, the ions are introduced between four parallel metal rods that create a rapidly oscillating magnetic field between them. Ions whose mass-to-charge ratio is compatible with the frequency of the field will oscillate along a straight path toward the ion collector, while other ions will follow a different path and be removed. The frequency or intensity of the oscillating field is varied so that ions of all mass-to-charge ratios eventually reach the ion collector.

In a *time-of-flight* mass spectrometer, ion beams are produced by brief pulses of electrons and accelerated by an electrical field pulse that gives all ions, regardless of mass, the same kinetic energy. Ions of a given mass (or m/e value) then pass through a *drift tube* at a speed that is inversely proportional to their mass, so lighter ions reach the ion collector sooner than heavier ions.

In a *Fourier-transform* mass spectrometer (sometimes called an *ion trap* mass spectrometer), ions are forced into a circular path by a strong magnetic

In mass spectrometry, two adjacent peaks of equal amplitude are said to be resolved when the height of the valley between them is no more than 10% of their height.

field and subjected to a radio frequency (RF) pulse. If the frequency of the radiation is equal to the frequency at which ions of a given m/e value move around their circular path (their *cyclotron frequency*), the ions will be accelerated and spiral outward. At the end of the RF pulse, they will be moving around a larger circle in a coherent packet. This packet of revolving ions generates an *image current* that decays with time after the pulse ends, producing a signal that is similar to the free induction decay signal generated by a Fourier-transform NMR spectrometer. The frequency of the RF pulse is varied to match the cyclotron frequencies of all ions in a sample, causing each kind of ion to produce its own image current. The resulting image currents are then detected and converted to a conventional mass spectrum by Fourier-transform analysis. Some Fourier-transform mass spectrometers are capable of resolving ions whose masses differ by only 0.001% or so.

Gas Chromatography/Mass Spectrometry

Rapid-scan mass spectrometers—including some quadrupole, Fourier-transform, and time-of-flight instruments—can be interfaced with gas chromatographs and used to analyze the components of a mixture as they emerge from the GC column. This combination of instruments is, not surprisingly, called a *gas chromatograph/mass spectrometer (GC/MS)*. The output of a GC capillary column (see OP-34) can often be introduced directly into the ionization chamber of a mass spectrometer. With a packed column, however, most of the carrier gas must be removed to maintain a sufficiently low pressure in the evacuated ionizing chamber. The mass spectra of the components can be displayed by a screen or recorder in "real time" (as each component exits the gas chromatogram) or stored to be printed out later.

Instruments in which mass spectrometers are interfaced with high-performance liquid chromatographs (LC/MS) and even with other mass spectrometers (MS/MS) are also available.

Gas chromatograph/mass spectrometers are particularly useful for identifying the components of natural products, biological systems, and ecosystems. For example, flavor and odor components of essential oils, physiologically active components of plants, and chemical pollutants in the environment can be characterized by GC/MS. Forensic chemists can use GC/MS to identify drug metabolites in the body fluids of a suspect, and physicians can diagnose certain illnesses on the basis of a GC/MS analysis of a patient's breath.

Experimental Considerations

Samples prepared for mass spectrometry should be very pure, because traces of impurities can make interpretation of a mass spectrum difficult. Sample sizes range from less than a microgram to several milligrams, and no special sample preparation is required. Liquids are inserted directly into the sample inlet with a syringe, micropipet, or break-off device, and solids can be introduced by means of a melting-point capillary. The samples vaporize in the sample inlet system, after which their molecules flow through a molecular leak into the evacuated ionization chamber.

A typical single-focus mass spectrometer is prepared for operation by turning on the magnet current and adjusting controls that set the potential of the repeller plates, the accelerating voltage, the ionizing current (the number of electrons in the electron beam), and the energy of the electron beam. Increasing the repeller potential reduces the time that ions spend in the ionization chamber; increasing the accelerating voltage increases the speed the ions attain by the time they enter the mass separator; increasing the ionizing current increases the number of ions that are formed; and increasing the energy of the electron beam increases the amount of fragmentation. Since the settings of these controls determine the appearance of a mass spectrum, it is important to set them within ranges that are appropriate for a given analysis. The user also selects the range of masses to be scanned and the scan time. The mass spectrum is then scanned and recorded as a chart or computer printout.

Other types of instruments may operate quite differently, so no operating procedures will be given here. These must be learned by special instruction and by studying the manufacturer's operating manual. A computer program that simulates the operation of a mass spectrometer may also be available for your use.

Interpretation of Mass Spectra

Each ion recorded on a mass spectrum is characterized by its mass-to-charge ratio and relative abundance (intensity), the latter being proportional to the height of its peak. This information can be displayed by a variety of output devices, including oscilloscopes, strip-chart recorders, and computer printers. The most intense peak in a mass spectrum, called the *base peak*, is assigned an intensity of 100, and the intensities of all other peaks are reported as percentages of the base-peak intensity. The molecular-ion peak may be the base peak, but often it is not.

Molecular Weight and Molecular Formula

For most organic compounds, the molecular weight of the compound is virtually equal to the mass of its strongest molecular-ion peak. This molecular-ion peak is usually the last strong peak on the spectrum, since no daughter ion should have a higher mass than the original molecular ion. However, the molecular-ion peak is usually followed by at least two low-intensity peaks corresponding to isotopic variations of the molecular ion, because most of the elements in organic compounds have at least one isotope of higher mass number than the common form.

Just over one carbon atom in a hundred (1.08%) is a carbon-13 atom; the rest are carbon-12 atoms. Since benzene, for example, contains six carbon atoms, the chance that any one of the six will be carbon-13 is $6 \times 1.08\%$, or 6.48%. Thus the molecular-ion peak of benzene (C_6H_6, M.W. = 78) should be followed by a peak for a "heavy" form of benzene ($C_5{}^{13}CH_6$, M.W. = 79) with an intensity that is 6.48% of the parent peak intensity. Actually, the m/e = 79 peak, called the *M + 1 peak*, has a slightly higher intensity than this because benzene also contains minute quantities of benzene-d_1 (C_6H_5D), which also has an approximate molecular weight of 79. Likewise, oxygen-18 occurs naturally to the extent of about 0.20 atom for every 100 atoms of oxygen-16, so formaldehyde shows an *M + 2 peak* (corresponding to $CH_2{}^{18}O$) with an intensity that is 0.20% of the molecular-ion peak's intensity.

Intensities of the M + 1 and M + 2 peaks (relative to the molecular-ion peak's intensity) for a compound $C_wH_xN_yO_z$ can be calculated using Equations **1** and **2**:

$$\%(M + 1) = 1.08w + 0.015x + 0.37y + 0.037z \qquad \textbf{(1)}$$
$$\%(M + 2) = 0.006w(w-1) + 0.0002wx + 0.004wy + 0.20z \qquad \textbf{(2)}$$

For example, quinine (as the hydrate) has the molecular formula $C_{20}H_{30}N_2O_3$. Using Equations **1** and **2** gives the intensity of its M + 1 and M + 2 peaks as 22.9% and 3.16%, respectively, of the M peak intensity.

$$\%(M + 1) = 1.08(20) + 0.015(30) + 0.37(2) + 0.037(3) = 22.9$$
$$\%(M + 2) = 0.006(20)(19) + 0.0002(20)(30)$$
$$+ 0.004(20)(2) + 0.20(3) = 3.16$$

No non-identical sets of atoms are likely to yield M + 1 and M + 2 peaks of exactly the same relative intensity. Therefore, if the intensities of these peaks in the mass spectrum of an unknown compound can be measured to two or more decimal places, its molecular formula (or several possible formulas) can be determined using published formula mass tables, such as those in *Spectrometric Identification of Organic Compounds* [Bibliography, F19]. Here is a general procedure for determining molecular formulas.

1 Locate the molecular-ion peak and determine its mass.
2 Measure the intensities of the M, M + 1, and M + 2 peaks, and express the latter two as a percentage of the intensity of the M peak.
3 Find the formula (or formulas) listed under its M value that give M + 1 and M + 2 intensities close to the experimental values and that make sense from a chemical standpoint.

Some formulas can be eliminated immediately because they do not correspond to stable molecules or because compounds with those formulas would be impossible to obtain from a given reaction or source. Others can be eliminated because they do not have the expected *index of hydrogen deficiency* (*IHD*), where the IHD of a compound is equal to the number of rings plus the number of pi bonds (or their aromatic equivalent) in a molecule of the compound. For compounds with the general formula $C_wH_xN_yO_z$, the IHD can be calculated using Equation **3**.

$$IHD = \tfrac{1}{2}(2w - x + y + 2) \qquad \textbf{(3)}$$

For example, the calculated IHD of Compound **1** is 7, which is consistent with its molecular structure—one ring and six pi bonds.

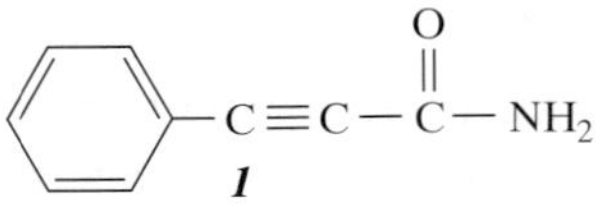

$$IHD = \tfrac{1}{2}(18 - 7 + 1 + 2) = 7$$

A useful generalization that applies to most organic compounds is the *nitrogen rule*, which states that a stable compound whose molecular weight is an even number can have only zero or an even number of nitrogen atoms,

Table G6 Formulas of some neutral species lost from molecular ions

Mass of neutral species	Mass of resulting ion	Possible formulas
1	$M - 1$	H·
15	$M - 15$	·CH$_3$
16	$M - 16$	·NH$_2$
17	$M - 17$	·OH, NH$_3$
18	$M - 18$	H$_2$O
26	$M - 26$	C$_2$H$_2$, ·CN
27	$M - 27$	·C$_2$H$_3$, HCN
28	$M - 28$	CO, C$_2$H$_4$
29	$M - 29$	·CHO, ·C$_2$H$_5$
30	$M - 30$	H$_2$CO, NO
31	$M - 31$	CH$_3$O·, ·CH$_2$OH
35	$M - 35$	Cl·
36	$M - 36$	HCl
42	$M - 42$	CH$_2$CO
43	$M - 43$	CH$_3$CO·, ·C$_3$H$_7$
44	$M - 44$	CO$_2$, ·CONH$_2$
45	$M - 45$	·CO$_2$H, C$_2$H$_5$O·
46	$M - 46$	·NO$_2$
49	$M - 49$	·CH$_2$Cl
57	$M - 57$	CH$_3$COCH$_2$·
59	$M - 59$	·CO$_2$CH$_3$
77	$M - 77$	C$_6$H$_5$·
79	$M - 79$	Br·

whereas one whose molecular weight is an odd number can have only an odd number of nitrogen atoms.

Fragmentation Patterns

The use of fragmentation patterns to determine molecular structures is a broad subject covered in detail elsewhere (refer to Category F of the Bibliography), so only a few generalizations will be given here. A molecular ion or daughter ion often breaks down by eliminating small neutral molecules or free radicals such as CO, H$_2$O, HCN, C$_2$H$_2$, H·, or CH$_3$·, yielding ions with masses equal to $M - X$, where M represents the mass of the molecular ion or a daughter ion undergoing fragmentation, and X is the mass of the neutral species. Table G6 gives the masses and postulated structural formulas of some neutral species that are often lost by fragmentation.

To see how such data can be used to interpret mass spectra, consider the mass spectrum of methyl benzoate in Figure G46. The m/e value of the molecular-ion peak in this spectrum is 136, that of the base peak is 105, and there are other strong peaks at $m/e = 77$ and 51. The molecular ion is believed to be a radical cation with the structure shown in the margin. Since the mass numbers of the base peak and molecular-ion peak differ by 31 mass units, a neutral species with a mass number of 31 must have been lost from the molecular ion to produce the base-peak ion. Table G6 shows two neutral fragments having that mass number, CH$_3$O· and ·CH$_2$OH. Since the molecular ion has a methoxyl group, it must have lost CH$_3$O· to form the base-peak ion, which must therefore be a benzoyl cation. The difference

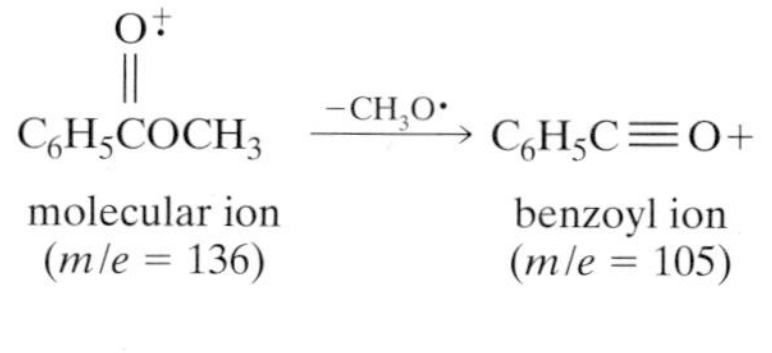

$$C_6H_5COCH_3 \xrightarrow{-CH_3O\cdot} C_6H_5C\equiv O+$$

molecular ion
($m/e = 136$)

benzoyl ion
($m/e = 105$)

$$C_6H_5C\equiv O+ \xrightarrow{-CO} C_6H_5^+$$

benzoyl ion
($m/e = 105$)

phenyl ion
($m/e = 77$)

$$C_6H_5^+ \xrightarrow{-C_2H_2} C_4H_3^+$$

phenyl ion
($m/e = 77$)

($m/e = 51$)

Table G7 Some common fragment ions

m/e	Possible formulas
15	CH_3^+
17	OH^+
18	H_2O^+, NH_4^+
26	$C_2H_2^+$
27	$C_2H_3^+$
28	$CO^+, C_2H_4^+$
29	$CHO^+, C_2H_5^+$
30	$CH_2NH_2^+, NO^+$
31	CH_2OH^+, CH_3O^+
35	Cl^+ (also 37)
39	$C_3H_3^+$
41	$C_3H_5^+$
43	$CH_3CO^+, C_3H_7^+$
44	$CO_2^+, C_3H_8^+$
45	$CH_3OCH_2^+, CO_2H^+$
46	NO_2^+
49	CH_2Cl^+
51	$C_4H_3^+$
57	$C_4H_9^+, C_2H_5CO^+$
59	$COOCH_3^+$
65	$C_5H_5^+$
66	$C_5H_6^+$
71	$C_5H_{11}^+, C_3H_7CO^+$
76	$C_6H_4^+$
77	$C_6H_5^+$
78	$C_6H_6^+$
79	Br^+ (also 81)
91	$C_7H_7^+, C_6H_5N^+$
93	$C_6H_5O^+$
94	$C_6H_6O^+$
105	$C_6H_5CO^+$

between the mass number of the benzoyl ion and that of the next major peak is $105 - 77 = 28$. The two neutral species having mass numbers of 28 are CO and C_2H_4; loss of CO from the benzoyl cation should yield a phenyl cation with the expected mass number of 77. Formation of the next major species ($m/e = 51$) requires a loss of 26 mass units from the phenyl ion. From Table G6, two species having that mass are $\cdot CN$ and C_2H_2 (acetylene), of which only the latter is a possibility. Loss of acetylene from the phenyl ion yields an ion with the formula $C_4H_3^+$; this species is often encountered in the mass spectra of aromatic compounds. There are a number of smaller peaks in the methyl benzoate spectrum that may provide additional structural information, but usually it is not possible (or necessary) to characterize all of the peaks in a mass spectrum.

In the previous example, the interpretation was simplified because each major ion was produced from the preceding one along a single reaction path. Often there are fragmentation paths leading directly from the molecular ion to a number of different daughter ions. It is therefore a common practice to compare the mass number of the molecular ion with those of the significant daughter ions before attempting to compare the mass numbers of individual daughter ions.

The fragment ions described previously and other common fragment ions are listed in Table G7. Structures have been determined for some (but not all) of these cations. For example, the $C_7H_7^+$ ion with mass number 91, which results from the cleavage of alkylbenzenes, has been formulated as either a benzyl cation or a tropylium ion; in most cases, it appears to have the latter structure. Before you can interpret a mass spectrum proficiently using data like that in Tables G6 and G7, you need to learn about the characteristic fragmentation patterns and mechanisms for different classes of organic compounds. This kind of information and some general rules for interpretation of mass spectra are given in references on mass spectrometry listed in Category F of the Bibliography.

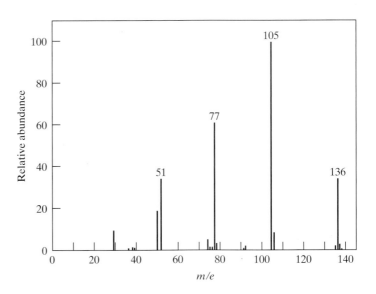

Figure G46 Mass spectrum of methyl benzoate

Appendixes and Bibliography

APPENDIX I

Laboratory Equipment

Chemical Glassware

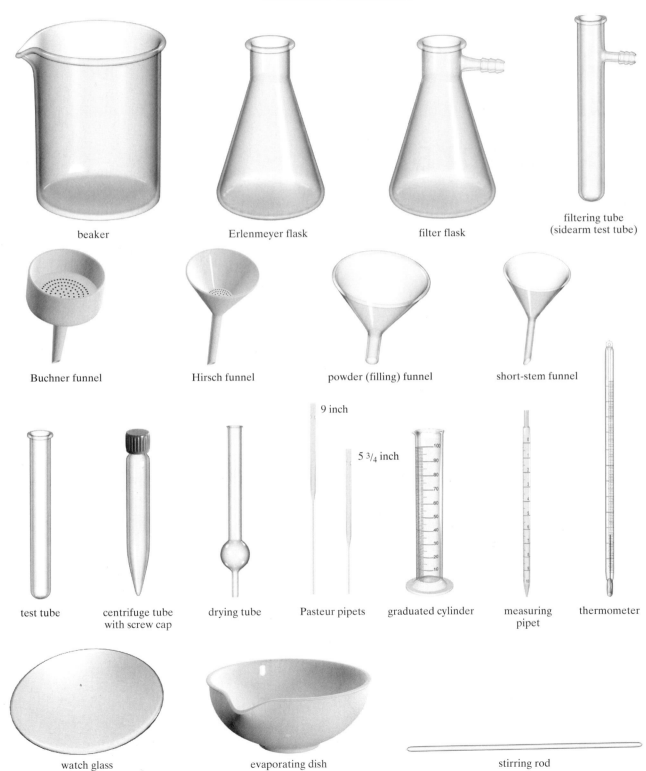

beaker

Erlenmeyer flask

filter flask

filtering tube
(sidearm test tube)

Buchner funnel

Hirsch funnel

powder (filling) funnel

short-stem funnel

test tube

centrifuge tube
with screw cap

drying tube

9 inch

5 3/4 inch

Pasteur pipets

graduated cylinder

measuring
pipet

thermometer

watch glass

evaporating dish

stirring rod

Lab Kit Components

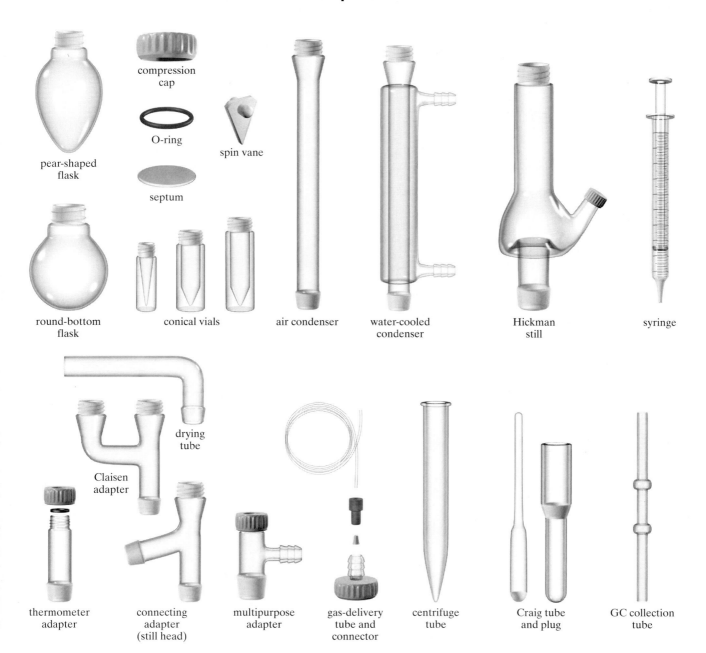

pear-shaped
flask

compression
cap

O-ring

spin vane

septum

round-bottom
flask

conical vials

air condenser

water-cooled
condenser

Hickman
still

syringe

drying
tube

Claisen
adapter

thermometer
adapter

connecting
adapter
(still head)

multipurpose
adapter

gas-delivery
tube and
connector

centrifuge
tube

Craig tube
and plug

GC collection
tube

Hardware

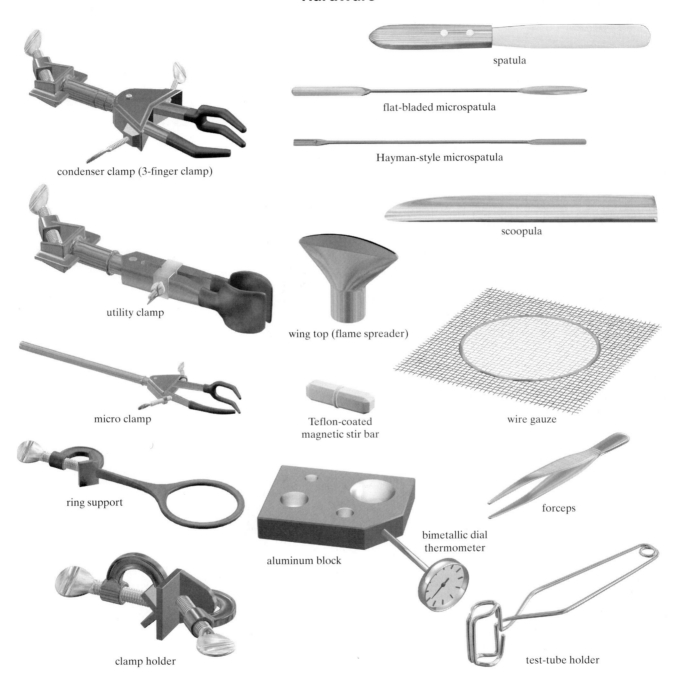

condenser clamp (3-finger clamp)

spatula

flat-bladed microspatula

Hayman-style microspatula

scoopula

utility clamp

wing top (flame spreader)

wire gauze

micro clamp

Teflon-coated
magnetic stir bar

ring support

aluminum block

bimetallic dial
thermometer

forceps

clamp holder

test-tube holder

Keeping a Laboratory Notebook

Your instructor may require that you maintain a laboratory notebook, write formal laboratory reports, or both. In either case, you should write an experimental plan before you begin most of the experiments in this book. This appendix tells you how to keep a laboratory notebook. Information about writing lab reports and experimental plans will be found in Appendix III and Appendix V, respectively.

A laboratory notebook is essentially a factual account of work performed in the laboratory. It may also include the writer's interpretation of the results. Your laboratory work may not require the documentation expected of a research chemist, but you should at least be aware of the characteristics of a good laboratory notebook. Although any kind of notebook can be used to record data and information, a notebook suitable for organic chemistry labs should be bound and have quadrilled (square-ruled) pages. If your instructor requires you to turn in a copy of your notes for each experiment, your notebook should also have duplicate pages. The first page or two should be reserved for a table of contents, and the pages should be numbered sequentially so that you can find information quickly. Whenever possible, each entry should be written immediately after the work is performed, and it should be dated and signed by the experimenter (notebooks of research chemists are usually signed by a witness as well). Each section of the notebook should have a clear, descriptive heading, and the writing should be grammatically correct and sufficiently legible to be read and understood by any knowledgeable individual.

Before each experiment (except Experiments 1–3), you should write an experimental plan (see Appendix V) in your lab notebook, summarizing what you expect to do in the laboratory and how you intend to go about it, and providing other relevant information. Your prelab write-up should include items from the following list (your instructor will tell you which are necessary):

- The experiment number and title
- A clear, concise statement of the scientific problem the experiment is designed to solve
- A brief statement telling how you will attempt to solve the problem
- When appropriate, a working hypothesis predicting the outcome of the experiment
- For a preparation, balanced equations for all significant reactions, including possible side reactions that might reduce the yield
- A table listing relevant properties (M.W., bp, mp, solubility, hazardous characteristics, etc.) of the reactants, products, solvents, and any other chemicals involved in the experiment
- A calculation of the theoretical yield of product and any other necessary stoichiometric calculations, when applicable
- A list of the materials (chemicals, supplies, equipment) needed for the experiment
- A checklist, flow diagram (see Appendix V), or other kind of outline summarizing the experimental procedure

Some of this information, such as reaction equations and physical properties, will be given in the experiment itself, and certain items may not apply to some kinds of experiments, such as kinetic studies or qualitative analyses.

During the experiment, you should keep a detailed account of your work, reporting everything of importance that you actually did and saw. Your account should not simply restate the textbook procedure, but should describe in your own words how you carried out the experiment. You should include all relevant data, such as the quantities of materials that you actually used (not the quantities calculated or given in the procedure, unless they are exactly the same) and the results of any analyses you performed. Raw data should be recorded with particular care; if you forget to record data at the time you measure it, or if you record it incorrectly or illegibly, the results of an entire experiment may be invalidated. As you gather evidence pertaining to the problem, you can write down one or more tentative hypotheses and describe how you tested them. You should also report any observations that might have a bearing on the problem, such as those providing clues to the nature of the reaction or the identity of the product.

You may find it helpful to think of your lab notebook as telling a story about your accomplishments (or misadventures) in the laboratory. Although most professional journal articles are written in a dry, impersonal style, this need not be true of a lab notebook. The American Chemical Society's publication *Writing the Laboratory Notebook* [Bibliography, L42] suggests a more personal approach, stating that "the use of the active voice in the first person tells the story and clearly indicates who did the work." (For example, "I recorded the infrared spectrum of benzaldehyde....") Other scientists prefer to write in the passive voice, avoiding the use of personal pronouns. (For example, "The infrared spectrum of benzaldehyde was recorded....") If your instructor has a strong preference either way, you should follow his or her recommendation.

After you have finished an experiment, summarize your major results, and then write down your conclusion(s) and explain how you arrived at it. If your notebook pages are to be turned in as your laboratory report, include answers to assigned exercises and any other information requested by your instructor. See Appendix III for additional suggestions about reporting and interpreting your observations and results.

APPENDIX **III**   Writing a Laboratory Report

Depending on your instructor's preference, you will be expected to submit a report for each experiment, either (1) from your laboratory notebook, (2) on a report form provided, or (3) on blank sheets of paper, bound or stapled together. Handwritten reports are usually acceptable if they are written legibly in ink. Your report should include information under some or all of the following headings:

1 *Prelab Assignments:* Your experimental plan (see Appendix V), calculations, and any other information requested in the section "Before You Begin." This may also include a prelab write-up of the kind described in Appendix II.

2 *Observations:* Any significant observations made during the course of the experiment. You should record all observations that might be of help to you (or another experimenter) if you were to repeat the experiment at a later time. These include quantities of solvents or drying agents used, reaction times, distillation ranges, and a description of any experimental difficulties you encounter. You should also make note of phenomena that might provide clues about the nature of chemical or physical transformations taking place during the experiment, such as color changes, phase separations, tar formation, and gas evolution. If you keep a detailed record of your observations in a laboratory notebook, it may not be necessary to include them in a separate report.

3 *Raw Data:* All numerical data obtained directly from an experiment, before it is graphed, used in calculations, or otherwise processed. This can include quantities of reactants and products, titration volumes, kinetic data, gas chromatographic (GC) retention times, integrated GC peak areas, and spectrometric parameters.

4 *Calculations:* Yield and stoichiometry calculations and any other calculations based on the raw data. If a number of repetitive calculations are required, one or two sample calculations of each type may suffice.

5 *Results:* A list, graph, tabulation, or verbal description of the significant results of the experiment. For a preparation, this should include a physical description of the product (color, physical state, evidence of purity, etc.), the percent yield, and all significant physical constants, spectral data, and analytical data obtained for the product. For the qualitative analysis of an unknown compound, the results of all tests and derivative preparations should be included, along with spectral data and physical constants. The results of calculations based on the raw data should be reported in this section as well.

6 *Discussion:* A description of the scientific problem the experiment was designed to solve and your interpretation of the results as they pertain to the problem. In this section you should tell how you applied scientific methodology to solve the problem. You should include statements of any working hypotheses you formulated and describe how they were tested, reporting and interpreting the relevant experimental evidence. Your discussion should also describe possible sources of error and, where relevant, discuss the significance of the results. When appropriate, this section may include your interpretation of spectra or chromatograms, which should be attached to the report.

7 *Conclusions:* A statement of your final conclusion (or conclusions) relating to the problem and an explanation of how you arrived at your conclusion, showing clearly how it is supported by the experimental evidence.

8 *Exercises:* Your answers to all exercises assigned by your instructor, showing your calculations and describing your reasoning when applicable.

Each report should also include (on the first page or cover) the name and number of the experiment, your name, and the date when the report was submitted to the instructor.

Your instructor should tell you what kind of report he or she prefers. The following example illustrates the kind of report that might be prepared by a conscientious student. Note that the experimental plan is not included here; see Appendix V for directions on writing experimental plans.

Experiment 58: Preparation of Tetrahedranol

Name: Cynthia Sizer
Date: Jan. 23, 2004

(*Attached*: Experimental plan, IR spectrum, gas chromatograph, and worked exercises)

Prelab Calculations

Tetrahedryl acetate required:

$$\text{mass} = 0.0150 \text{ mol} \times \frac{110 \text{ g}}{1 \text{ mol}} = 1.65 \text{ g}$$

$$\text{volume} = 1.65 \text{ g} \times \frac{0.951 \text{ g}}{1 \text{ mL}} = 1.57 \text{ mL}$$

6.0 *M* sodium hydroxide required:

$$\text{volume} = 0.060 \text{ mol} \times \frac{1 \text{ L}}{6.0 \text{ mol}} \times \frac{1000 \text{ mL}}{1 \text{ L}} = 10 \text{ mL}$$

Observations

The reaction of 1.644 g of the ester (tetrahedryl acetate) with 10 mL of 6.0 *M* NaOH was carried out under reflux for 60 minutes, during which time the organic (top) layer dissolved slowly to give a homogeneous solution and the "fruity" odor of the ester disappeared. The organic layer was no longer visible after 45 minutes of heating. The acidified reaction mixture was extracted with two 10-mL portions of diethyl ether, the ether extracts were washed with two 10-mL portions of saturated aqueous sodium chloride, and the ether solution was dried over 0.70 g of anhydrous magnesium sulfate. The residue was distilled over a 4°C boiling range and the distillate solidified on cooling. After drying the product for 1 week in a desiccator, its mass was 0.854 g. Approximately 0.10 g of the product was dissolved in 0.50 mL of dichloromethane, and gas chromatograms were recorded for (1) the original solution and (2) the solution spiked with an authentic sample of tetrahedryl acetate. The relative area of the second (72 s) product peak was considerably larger in the second chromatogram than in the first. The gas chromatograms were obtained using a 2-meter, $\frac{1}{8}$ inch i.d., OV-101/Chromosorb W packed column. The infrared spectrum of the product was recorded using a thin film between heated silver chloride plates.

Data

Mass of tetrahedryl acetate: 1.644 g
Mass of dry product: 0.854 g
Distillation boiling range of product: 78–82°C
Micro boiling point of product: 81°C
Gas chromatography data:

Column temperature: 110°C
Injector temperature: 150°C
Detector temperature: 150°C
Helium flow rate: 25 cm^3/minute

Component	Retention time	Peak area
tetrahedranol	47 s	320 mm^2
tetrahedryl acetate	72 s	12 mm^2

Calculations

Percentage of tetrahedranol in product:

$$\frac{320 \text{ mm}^2}{332 \text{ mm}^2} \times 100\% = 96.4\%$$

Mass of tetrahedranol in product:

$$0.854 \text{ g} \times \frac{96.4\%}{100\%} = 0.823 \text{ g}$$

Theoretical yield of tetrahedranol (TetOAc = tetrahedryl acetate; TetOH = tetrahedranol):

$$1.644 \text{ g TetOAc} \times \frac{1 \text{ mol TetOAc}}{110.1 \text{ g TetOAc}} \times \frac{68.1 \text{ g TetOH}}{1 \text{ mol TetOH}} = 1.017 \text{ g TetOH}$$

Percent yield of tetrahedranol:

$$\frac{0.823 \text{ g}}{1.017 \text{ g}} \times 100\% = 80.9\%$$

Results

Tetrahedranol was obtained in 80.9% yield from the alkaline hydrolysis of tetrahedryl acetate. At room temperature, tetrahedranol was a colorless, almost transparent solid with a mild "spirituous" odor. It had a distillation boiling range of 78–82°C and a micro boiling point of 81°C. Its infrared spectrum contained significant absorption bands at 3370, 2975, 2885, and 1194 cm^{-1}.

Discussion

Statement of the problem: Can tetrahedranol with a purity of 95% or more be prepared by the alkaline hydrolysis of tetrahedryl acetate?

Working Hypothesis 1: The alkaline hydrolysis of tetrahedryl acetate *will* yield tetrahedranol as one of the products.

I based this hypothesis on (1) our textbook's statement that the alkaline hydrolysis of an ester yields the corresponding hydroxy compound (alcohol or phenol) and the salt of the corresponding carboxylic acid, and (2) the results of Experiment 4, in which I obtained salicylic acid (a phenol) from the alkaline hydrolysis of methyl salicylate. Observations in support of Hypothesis 1 include the following:

1 The slow disappearance of the organic layer during the reaction suggests that the water-insoluble ester was being converted to a water-soluble product. According to Table 58.1 in the lab textbook, tetrahedranol is soluble in water.
2 The disappearance of the "fruity" odor of tetrahedryl acetate also suggests that the ester was reacting.
3 The solidification of the distillate and easy melting of the solid between heated AgCl plates are consistent with the hypothesis, since Table 58.1 indicates that tetrahedranol has a melting point of 27°C.

Hypothesis 1 was tested by obtaining a micro boiling point and infrared spectrum of the product, with the following results:

1 The observed micro boiling point of 81°C is consistent with the hypothesis, since tetrahedranol has a reported boiling point of 82°C.

2 The infrared spectrum, suggesting a tertiary alcohol, is consistent with the hypothesis.

Wave number/cm^{-1}	Assignment	Interpretation
3370 (3368.4)	O—H stretch	Alcohol or phenol
2975, 2885	C—H stretch	Absence of C—H absorption above 3000 cm^{-1}
(2975.1, 2884.2)		eliminates a phenol as a possibility
1194 (1193.6)	C—O stretch	Consistent with tertiary alcohol

The wave numbers of these bands are nearly identical to those on a published FTIR spectrum of tetrahedranol (shown in parentheses), and the band shapes and positions in the fingerprint region matched those of the published spectrum. My spectrum did have a weak band at 1739 cm^{-1} that was not present on the published spectrum, but this is probably the carbonyl (C═O) band of the tetrahedryl acetate impurity that was detected by gas chromatography.

Working Hypothesis 2: The purity of the product *will not* be 95% or better. I based this hypothesis on (1) the statement in the lecture textbook that some ester hydrolysis reactions, especially those involving esters of bulky alcohols, take more than an hour to reach completion, (2) the fact that the ester and alcohol boiling points are only 15°C apart, suggesting that (as we learned in Experiment 5) simple distillation will not remove all of the impurity.

Hypothesis 2 was tested by obtaining a gas chromatogram of the product, identifying the peaks, and calculating the percentage of tetrahedranol in the product from the peak areas. Since the small 72 s peak became larger when the sample was spiked with tetrahedryl acetate, it must be the tetrahedryl acetate peak. Since the IR spectrum showed that the major product was tetrahedranol, the much larger peak at 47 s must be the tetrahedranol peak. The calculated mass percentage of tetrahedranol, 96.4%, is <u>not</u> consistent with my hypothesis. I am not disappointed with this result, since it was nice to know that the experiment came out better than expected! My assumptions that the reaction would not go to completion in an hour and that the impurity would not be completely removed from the product by simple distillation were both correct, but the reaction was more nearly complete than I had guessed.

The yield of tetrahedranol (0.823 g) was 0.194 g less than the theoretical value. Of this at least 0.019 g resulted from incomplete reaction of tetrahedryl acetate, based on the 0.031 g (0.854 g–0.823 g) of ester in the distillate. The remaining losses could have arisen from (1) losses during transfers, (2) losses during the extraction and washing operations, and (3) losses during the distillation. During each transfer I used additional solvent (ether or water) to rinse out the vessel from which the product was transferred, so losses during transfers should not be a major factor. About 0.1 mL of residue remained in the conical vial used for distillation; this can account for no more than 0.1 g of the loss, since some of the residue must have been tetrahedryl acetate. Most of the remaining losses must have occurred during the extraction and washing operations, due to the partial solubility of tetrahedranol in water. Such losses might have been reduced by saturating the aqueous layer with potassium carbonate to salt out the alcohol or by carrying out several more ether extractions.

Conclusion

I conclude that tetrahedranol with a purity of 95% or more can be prepared by the hydrolysis of tetrahedryl acetate. I base this conclusion primarily on the infrared spectrum of the product, whose resemblance to the published spectrum leaves little doubt that the product is tetrahedranol, and on the gas chromatographic analysis, which indicates a purity of 96.4 %. My conclusion that the product is tetrahedranol is supported by the other evidence cited in the discussion. My conclusion that the purity of the product is greater than 95% might possibly be in error, since the GC peak areas were not corrected by applying detector response factors.

Calculations for Organic Synthesis   APPENDIX **IV**

In this book, the quantities of many reactants are given in units of "chemical amount," moles or millimoles. You will have to convert such quantities to units of mass or volume before you can begin a synthetic experiment. In any experiment, it is the relationship between these *chemical* quantities that is significant, not the relationship between such *physical* quantities as mass and volume. Because there are no "mole meters" that measure molar amounts directly, we are forced to use balances and volumetric glassware for that purpose. That should not obscure the fact that the chemical units are fundamental; only by knowing the chemical amounts of reactants involved in a preparation, for example, can you recognize the stoichiometric relationships between them or predict the yield of the expected product.

It is helpful to regard a chemical calculation as a process by which a given quantity is "converted" to the required quantity. This can be accomplished by multiplying the given quantity by a series of ratios used as unit or dimensional *conversion factors*. Unit conversions are carried out by using conversion factors, such as 454 g/lb, that are written as ratios between two quantities whose quotient is unity (for example, 454 g = 1 lb, so 454 g/1 lb = 1). Dimensional conversions are carried out by using conversion factors that are ratios of quantities in different *dimensions*, such as mass, volume, and chemical amount (amount of substance). For example, the density of a substance can be regarded as a conversion factor linking the two dimensions of mass and volume, so it is used to convert the mass of a given quantity of the substance to units of volume, and vice versa. A conversion factor can be inverted when necessary. For example, the molar mass of butyl acetate, written as 116 g/1 mol, will convert moles of butyl acetate to grams; the inverse ratio, 1 mol/116 g, will convert grams of butyl acetate to moles. All calculations should be checked by making sure that the units involved cancel to yield the correct units in the answer. This does not ensure that your answer is correct, but if the units do *not* cancel, the answer is almost certainly wrong.

The following examples illustrate some fundamental types of calculations that you can expect to encounter in an organic chemistry lab course.

Chemical Amount and Mass.   The chemical amount (in moles or millimoles) of a substance is converted to its mass by multiplying by its molar mass. Remember that the molar mass of a substance is obtained by simply appending the units g/mol to its molecular weight, which is a dimensionless quantity. For example, the mass of 15.0 mmol of butyl acetate (M.W. = 116) is 1.74 g.

$$15.0 \text{ mmol} \times \frac{1 \text{ mol}}{1000 \text{ mmol}} \times \frac{116 \text{ g}}{1 \text{ mol}} = 1.74 \text{ g}$$

Note that the chemical amount in millimoles must be converted to moles before the conversion factor is applied; otherwise, the units will not cancel. Mass can be converted to chemical amount by inverting the conversion factor before multiplying.

Chemical Amount and Volume.   The chemical amount (in moles or millimoles) of a pure liquid is converted to volume by multiplying by the liquid substance's molar mass and by the inverse of its density. For example, the volume of 2.50 mmol of acetic acid (M.W. = 60.1; d = 1.049 g/mL) is 0.143 mL.

$$2.50 \text{ mmol} \times \frac{1 \text{ mol}}{1000 \text{ mmol}} \times \frac{60.1 \text{ g}}{1 \text{ mol}} \times \frac{1 \text{ mL}}{1.049 \text{ g}} = 0.143 \text{ mL}$$

The volume of a solution needed to provide a specified chemical amount of solute is calculated by multiplying the number of moles required by the inverse of the solution's molar concentration. For example, the volume of 6.0 M HCl (which contains 6.0 mol of HCl per liter of solution) needed to provide 3.6 mmol of HCl is 0.60 mL.

$$3.6 \text{ mmol} \times \frac{1 \text{ mol}}{1000 \text{ mmol}} \times \frac{1 \text{ L}}{6.0 \text{ mol}} \times \frac{1000 \text{ mL}}{1 \text{ L}} = 0.60 \text{ mL}$$

Note that concentrations expressed in mol/L and mmol/mL have the same numerical value. Thus a 6.0 M solution also has a concentration of 6.0 mmol/mL; using these units simplifies the previous calculation considerably:

$$3.6 \text{ mmol} \times \frac{1 \text{ mL}}{6.0 \text{ mmol}} = 0.60 \text{ mL}$$

Theoretical Yield.   The maximum quantity of a product (usually expressed in mass units) that could be attained from a reaction is called the *theoretical yield* of the product. Theoretical yields can be calculated using *stoichiometric factors*—ratios derived from the coefficients (expressed in moles) of the products and reactants in a balanced equation for the reaction. For example, the stoichiometric factors relating the chemical amount of the organic product to the chemical amounts of the two reactants in the following reaction are (1 mol dibenzalacetone)/(1 mol acetone) and (1 mol dibenzalacetone)/(2 mol benzaldehyde).

$$\underset{\substack{\text{benzaldehyde (B)} \quad \text{acetone (A)}}}{2\text{PhCHO} + \text{CH}_3\overset{\displaystyle \overset{\text{O}}{\|}}{\text{C}}\text{CH}_3} \xrightarrow{\text{NaOH}} \underset{\text{dibenzalacetone (DBA)}}{\text{PhCH}=\text{CH}\overset{\displaystyle \overset{\text{O}}{\|}}{\text{C}}\text{CH}=\text{CHPh} + 2\text{H}_2\text{O}}$$

Suppose you were trying to prepare dibenzalacetone (M.W. = 234.3) starting with 0.500 g of benzaldehyde (M.W. = 106.1) and 0.150 g of acetone (M.W. = 58.1). (For convenience, we will abbreviate the names as dibenzalacetone = DBA, benzaldehyde = B, and acetone = A.) You can calculate the maximum chemical amount of product that could be formed from each reactant by converting the given quantity to moles, and then applying the appropriate stoichiometric factor:

$$0.500 \text{ g B} \times \frac{1 \text{ mol}}{106.1 \text{ g B}} \times \frac{1 \text{ mol DBA}}{2 \text{ mol B}} = 2.36 \times 10^{-3} \text{ mol DBA}$$

$$0.150 \text{ g A} \times \frac{1 \text{ mol A}}{58.1 \text{ g A}} \times \frac{1 \text{ mol DBA}}{1 \text{ mol A}} = 2.58 \times 10^{-3} \text{ mol DBA}$$

Since there is only enough benzaldehyde to produce 2.36×10^{-3} mol of dibenzalacetone, it is impossible to obtain more than that from the specified quantities of reactants. Once that much product has been formed, the reaction mixture will have run out of benzaldehyde, and the *excess* (leftover) acetone will have nothing to react with. Therefore, benzaldehyde is the *limiting reactant* upon which the yield calculations must be based. The theoretical yield of dibenzalacetone, in grams, is then

$$2.36 \times 10^{-3} \text{ mol DBA} \times \frac{234.3 \text{ g DBA}}{1 \text{ mol DBA}} = 0.553 \text{ g DBA}$$

Remember that the limiting reactant is always the one that would produce the least amount of product, which is not necessarily the one present in the lowest amount. In this example, benzaldehyde is the limiting reactant even though the mass and chemical amount of benzaldehyde are much greater than the mass and chemical amount of acetone.

Percent Yield. It is seldom, if ever, possible to attain the theoretical yield of product from an organic preparation. The reaction may not go to completion during the designated reaction period, leaving unreacted starting materials. There may be side reactions that reduce the yield of product, as in the reaction of benzaldehyde with acetone, where some benzalacetone ($PhCH{=}CHCOCH_3$) is formed as a by-product. And there are invariably material losses when the product is separated from the reaction mixture and purified. The *percent yield* of a preparation compares the actual yield to the theoretical yield as defined here:

$$\text{Percent yield} = \frac{\text{actual yield}}{\text{theoretical yield}} \times 100\%$$

For example, if you prepared 0.409 g of dibenzalacetone from 0.500 g of benzaldehyde and 0.150 g of acetone (theoretical yield = 0.553 g), the percent yield of your synthesis would be

$$\text{Percent yield} = \frac{0.409 \text{ g DBA}}{0.553 \text{ g DBA}} \times 100\% = 74.0\%$$

In many experiments, you will estimate the theoretical yield of a preparation based on the amounts of reactants given in the "Before You Begin" section. This will help you assess your performance by comparing your actual yield with an estimate of the "ideal" yield. But the percent yield that you *report* should be based on the amounts of reactants that you actually used in the synthesis, not on the amounts given (unless they are exactly the same).

Planning an Experiment   APPENDIX **V**

Before starting any project, whether you are making a bookshelf, duck à l'orange, or isopentyl acetate, you must have a plan. An *experimental plan* should summarize what you expect to do in the laboratory and how you intend to go about it. You should state how the work is to be done in short phrases, without excessive detail. You can always refer to the "Directions" section of the experiment and the "Operation" descriptions for the details, but as you become more proficient in the laboratory, you should find yourself relying less on the textbook and more on your experimental plan. A good plan should give you quick access to the essential information you will need while performing the experiment. Quantities of chemicals (including wash solvents, drying agents, etc.), reaction times, physical properties, hazard warnings, and other useful data should be included. You can list the supplies and equipment you will need for each operation (these are specified in most operation descriptions) so that you can have them cleaned and ready when you need them. You may also wish to sketch the apparatus you will be using so that you can assemble it quickly in the laboratory.

An experimental plan should help you organize your time efficiently by listing tasks in the approximate order in which you expect to accomplish them. For example, whenever a reflux period is specified in a procedure, you will have some free time to set up the apparatus for the next step, reorganize your work area, review an operation, start a minilab, take a melting point, record the spectrum of a previous product, or tie up other loose ends. Your plan should be flexible enough that you can alter it or deviate from it during the experiment, if there is good reason to do so. One simple and effective way of organizing your time is to use a laboratory checklist, such as the one shown in Experiment 4. You should refer to relevant sections of the experiment (especially "Understanding the Experiment" and the "Directions") as you prepare your checklist, as well as the appropriate "Operation" descriptions. Leave enough space between the items on your checklist so that you can add new ones, as necessary, during the experiment. As you complete each task in the laboratory, simply check it off the list and go on to the next one.

A flow diagram, such as the one in Figure 5.1 of Experiment 5, can help you organize your time by giving you a quick overview of the procedure, showing the purpose of each step. To create such a flow diagram, first list all the substances that you know to be present in the reaction mixture before the reaction starts (reactants, solvents, catalysts), as shown in the following general flow diagram:

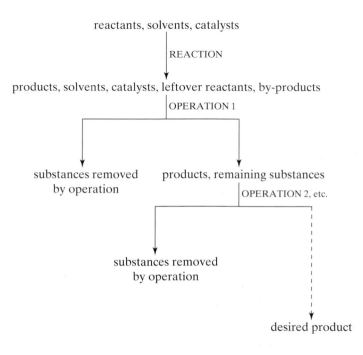

The reaction equation for the experiment tells you what substances will form as a result of the reaction. At the end of the reaction period, the reaction mixture will contain these products as well as the reaction solvent (if any), catalyst (if any), leftover reactants, and usually some by-products formed by various side reactions. The flow diagram should show how all of the unwanted substances in the reaction mixture are separated from the desired product. Each separation or purification operation is represented by a branch in the flow diagram, with the substance(s) being removed on one side and the desired product, along with any remaining substances, on the other side. After the last operation, the product stands alone, with all of the impurities eliminated—on paper, at least!

Properties of Organic Compounds APPENDIX **VI**

The tables in this appendix are to be used in conjunction with the procedures described in Part IV, "Qualitative Organic Analysis." Compounds that melt below ordinary ambient temperature (about 25°C) are listed in order of increasing boiling point (bp); those that (when pure) are generally solid at room temperature are listed in order of increasing melting point (mp). Both melting and boiling points are specified for some borderline cases; if either value is out of sequence, it is shown in italics.

Derivative preparations are described in Part IV, "Preparation of Derivatives," and are referred to by number at the heads of the appropriate columns. Melting points in parentheses are for derivatives that exist in more than one crystalline form or for which significantly different melting points have been reported in the literature. Sometimes the recrystallization solvent will determine the form in which a derivative crystallizes, so a significant deviation from a listed melting point should not be considered conclusive proof that a product is not the expected derivative. If a given compound can form more than one product (for instance, mononitro and dinitro derivatives of an aromatic hydrocarbon), the reaction conditions for the preparation may determine the derivative isolated (a mixture of derivatives may also result). A dash (—) in a derivative column indicates either that the derivative has not been reported in the literature or that it is not suitable for identification (it may be a liquid, for example). See references from Section G in the Bibliography for physical constants and derivative melting points not listed in these tables.

Most compounds are listed by their systematic (IUPAC) names, except when those names would be too lengthy.

Abbreviations Used in Tables

d = decomposes on melting
s = sublimes at or below melting point
m = monosubstituted derivative (such as a mononitrated aromatic hydrocarbon)
di = disubstituted derivative
t = trisubstituted derivative
tet = tetrasubstituted derivative

List of tables

Table 1 Alcohols

Compound	bp	mp	3,5-Dinitro-benzoate	4-Nitro-benzoate	1-Naphthyl-urethane	Phenyl-urethane
methanol	65		108	96	124	47
ethanol	78		93	57	79	52
2-propanol	82		123	110	106	88
2-methyl-2-propanol*	83	26	142	—	—	136
2-propen-1-ol	97		49	28	108	70
1-propanol	97		74	35	105	57
2-butanol	99		76	26	97	65
2-methyl-2-butanol	102		116	85	72	42
2-methyl-1-propanol	108		87	69	104	86
3-pentanol	116		101	17	95	48
1-butanol	118		64	36	71	61
2-pentanol	120		62	24	74	—
3-methyl-3-pentanol	123		94 (62)	69	104	43
3-methyl-1-butanol	132		61	21	68	57
4-methyl-2-pentanol	132		65	26	88	143
1-pentanol	138		46	11	68	46
cyclopentanol	141		115	62	118	132
2-ethyl-l-butanol	148		51	—	60	—
1-hexanol	157		58	5	59	42
cyclohexanol*	161	25	113	50	129	82
furfuryl alcohol	172		80	76	130	45
1-heptanol	177		47	10	62	60
2-octanol	179		32	28	63	114
1-octanol	195		61	12	67	74
1-phenylethanol	202		95	43	106	92
benzyl alcohol	205		113	85	134	77
2-phenylethanol	219		108	62	119	78
1-decanol	231		57	30	73	60
3-phenylpropanol	236		45	47	—	92
1-dodecanol*	259	24	60	45 (42)	80	74
1-tetradecanol		39	67	51	82	74
(−)-menthol		44	153	62	119	111
1-hexadecanol		49	66	58	82	73
1-octadecanol		59	77	64	89	79

continued

Table 1 Alcohols *continued*

Compound	bp	mp	3,5-Dinitro-benzoate	4-Nitro-benzoate	1-Naphthyl-urethane	Phenyl-urethane
diphenylmethanol		68	141	132	136	139
cholesterol		148	—	185	176	168
(+)-borneol		208	154	137 (153)	132 (127)	138

*May be solid at or just below room temperature.
Note: All temperatures are in °C.

Table 2 Aldehydes

Compound	bp	mp	2,4-Dinitrophenyl-hydrazone	Semicarbazone	Oxime
ethanal	21		168 (157)	162	47
propanal	48		148 (155)	154	40
propenal	52		165	171	—
2-methylpropanal	64		187 (183)	125 (119)	—
butanal	75		123	106	—
3-methylbutanal	92		123	107	48
pentanal	103		106 (98)	—	52
2-butenal	104		190	199	119
2-ethylbutanal	117		95 (30)	99	—
hexanal	130		104	106	51
heptanal	153		106	109	57
2-furaldehyde	162		212 (230)	202	91
2-ethylhexanal	163		114 (120)	254d	—
octanal	171		106	101	60
benzaldehyde	179		239	222	35
4-methylbenzaldehyde	204		234	234 (215)	80
3,7-dimethyl-6-octenal	207		77	84 (91)	—
2-chlorobenzaldehyde	213		213 (209)	229 (146)	76 (101)
4-methoxybenzaldehyde	248		253d	210	133
phenylethanal	*195*	33	121 (110)	153 (156)	99
2-methoxybenzaldehyde		38	254	215	92
4-chlorobenzaldehyde		48	265	230	110 (146)
3-nitrobenzaldehyde		58	290	246	120
4-nitrobenzaldehyde		106	320	221 (211)	133 (182)

Note: All temperatures are in °C.

Table 3 Ketones

Compound	bp	mp	2,4-Dinitrophenyl-hydrazone	Semicarbazone	Oxime
acetone	56		126	187	59
2-butanone	80		118	146	—
3-methyl-2-butanone	94		124	113	—
2-pentanone	102		143	112 (106)	58
3-pentanone	102		156	138	69
3,3-dimethyl-2-butanone	106		125	157	75 (79)
4-methyl-2-pentanone	117		95 (81)	132	58
2,4-dimethyl-3-pentanone	124		95 (88)	160	34
2-hexanone	128		110	125	49
4-methyl-3-penten-2-one	130		205	164 (133)	48

continued

Table 3 Ketones *continued*

Compound	bp	mp	2,4-Dinitrophenyl-hydrazone	Semicarbazone	Oxime
cyclopentanone	131		146	210 (203)	56
4-heptanone	144		75	132	—
2-heptanone	151		89	123	—
cyclohexanone	156		162	166	91
2,6-dimethyl-4-heptanone	168		92	122	210
2-octanone	173		58	124	—
cycloheptanone	181		148	163	23
2,5-hexanedione	194		257 (di)	185 (m); 224 (di)	137 (di)
acetophenone	202	*20*	238	198 (203)	60
2-methylacetophenone	214		159	205	61
propiophenone	218	*21*	191	182 (174)	54
3-methylacetophenone	220		207	203	57
2-undecanone	228		63	122	44
4-phenyl-2-butanone	235		127	142	87
3-methoxyacetophenone	240		—	196	—
2-methoxyacetophenone	245		—	183	83 (96)
4-methylacetophenone	*226*	28	258	205	88
4-methoxyacetophenone		38	228	198	87
4-phenyl-3-buten-2-one		42	227 (223)	187	117
benzophenone		48	238	167	144
2-acetonaphthone		54	262d	235	145
3-nitroacetophenone		80	228	257	132
9-fluorenone		83	283	234	195
(−)-camphor		179	177	237	118

Note: All temperatures are in °C.

Table 4 Amides

Compound	bp	mp	Carboxylic acid	*N*-Xanthylamide
formamide	195d		—	184
propanamide		81	—	211
ethanamide		82	17	240
heptanamide		96	—	155
nonanamide		99	12	148
hexanamide		100	—	160
hexadecanamide		106	63	142
pentanamide		106	—	167
octadecanamide		109	69	141
butanamide		115	—	187
chloroacetamide		120	61 (53)	209
4-methylpentanamide		121	—	160
succinimide		126	185	247
2-methylpropanamide		129	—	211
benzamide		130	122	223
3-methylbutanamide		136	—	183
o-toluamide		143	104 (108)	200
furamide		143	133	210
phenylacetamide		156	77	195
p-toluamide		159	180	225
4-nitrobenzamide		201	240	233
phthalimide		238	210d	177

Note: All temperatures are in °C.

Table 5 Primary and secondary amines

Compound	bp	mp	Benzamide	p-Toluene-sulfonamide	Phenylthio-urea	Picrate
t-butylamine	44		134	—	120	198
propylamine	48		84	52	63	135
diethylamine	56		42	60	34	155
sec-butylamine	63		76	55	101	140
2-methylpropylamine	69		57	78	82	150
butylamine	77		42	—	65	151
diisopropylamine	84		—	—	—	140
pyrrolidine	89		—	123	—	112 (164)
3-methylbutylamine	95		—	65	102	138
pentylamine	104		—	—	69	139
piperidine	106		48	96	101	152
dipropylamine	109		—	—	69	75
morpholine	128		75	147	136	146
pyrrole	131		—	—	143	69d
hexylamine	132		40	—	77	126
cyclohexylamine	134		149	—	148	—
diisobutylamine	139		—	—	113	121
n-methylcyclohexylamine	147		86	—	—	170
dibutylamine	159		—	—	86	59
n-ethylbenzylamine	181		—	95	—	118
aniline	184		163	103	154	198 (180)
benzylamine	185		105	116 (185)	156	199 (194)
n-methylaniline	196		63	94	87	145
o-toluidine	200		144	108	136	213
m-toluidine	203		125	114	104 (92)	200
n-ethylaniline	205		60	87	89	138
2-chloroaniline	209		99	105 (193)	156	134
2,6-dimethylaniline	215		168	212	204	180
2-methoxyaniline	225		60	127	136	200
2-ethoxyaniline	229		104	164	137	—
3-chloroaniline	230		120	138 (210)	124 (116)	177
4-ethoxyaniline	250		173	106	136	69
dicyclohexylamine	255d		153 (57)	—	—	173
n-benzylaniline		37	107	149	103	48
p-toluidine		44	158	118	141	182d
diphenylamine		54	180 (109)	141	152	182
4-methoxyaniline		58	154	114	157 (171)	170
4-bromoaniline		66	204	101	148	180
2-nitroaniline		71	110 (98)	142	—	73
4-chloroaniline		72	192	95 (119)	152	178
1,2-diaminobenzene		102	301 (di)	260 (di)	—	208
3-nitroaniline		114	155 (di)	138	160	143
1,4-diaminobenzene		142	300 (di)	266 (di)	—	—
4-nitroaniline		147	199 (di)	191	—	100

Note: All temperatures are in °C. See other references in the Bibliography (Category G) for melting points of benzenesulfonamides and 1-naphthylthioureas.

Table 6 Tertiary amines

Compound	bp	mp	Methiodide	Picrate
triethylamine	89		280	173
pyridine	115		117	167
2-methylpyridine	129		230	169
3-methylpyridine	143		92	150
4-methylpyridine	143		152	167
tripropylamine	157		207	116
2,4-dimethylpyridine	159		113	183 (169)
n,n-dimethylbenzylamine	183		179	93
n,n-dimethylaniline	193		228d	163
tributylamine	216 (211)		186	105
n,n-diethylaniline	217		102	142
quinoline	237		133 (72)	203
isoquinoline	243	*26*	159	222
tribenzylamine	—	91	184	190
acridine	—	111	224	208

Note: All temperatures are in °C.

Table 7 Carboxylic acids

Compound	bp	mp	Amide	*p*-Toluidide	Anilide	*p*-Nitrobenzyl ester
formic acid	101		43	53	50	31
acetic acid	118		82	148	114	78
propenoic acid	139		85	141	104	—
propanoic acid	141		81	124	103	31
2-methylpropanoic acid	154		128	107	105	—
butanoic acid	164		115	75	95	35
3-methylbutanoic acid	176		135	107	109	—
pentanoic acid	186		106	74	63	—
2-chloropropanoic acid	186		80	124	92	—
dichloroacetic acid	194		98	153	125	—
2-methylpentanoic acid	196		79	80	95	—
hexanoic acid	205		101	75	95	—
2-bromopropanoic acid	205d	*24*	123	125	99	—
octanoic acid	239		107	70	57	—
nonanoic acid	254		99	84	57	—
decanoic acid		32	108	78	70	—
2,2-dimethylpropanoic acid	*164*	35	178 (154)	—	129 (133)	—
dodecanoic acid		44	110 (99)	87	78	—
3-phenylpropanoic acid		48	105	135	98	36
tetradecanoic acid		54	103	93	84	—
hexadecanoic acid		62	106	98	90	42
chloroacetic acid		63	120	162	137	—
octadecanoic acid		70	109	102	95	—
trans-2-butenoic acid		72	160	132	118	67
phenylacetic acid		77	156	136	118	65
2-methoxybenzoic acid		101	129	—	131	113
oxalic acid (dihydrate)		101	419d (di)	268 (di)	254 (di)	204(di)
2-methylbenzoic acid		104	142	144	125	91
nonanedioic acid		106	175 (di)	201 (di)	186 (di)	44

continued

Table 7 Carboxylic acids *continued*

Compound	bp	mp	Amide	*p*-Toluid-ide	Anilide	*p*-Nitrobenzyl ester
3-methylbenzoic acid		112	94	118	126	87
benzoic acid		122	130	158	163	89
maleic acid		130	181 (m) 266 (di)	142 (di)	187 (di)	91
decanedioic acid		133	170 (m) 210 (di)	201 (di)	122 (m) 200 (di)	73 (di)
cinnamic acid		133	147	168	153	117
propanedioic acid		135	50 (m) 170 (di)	86 (m) 253 (di)	132 (m) 230 (di)	86
2-chlorobenzoic acid		140	140	131	118	106
3-nitrobenzoic acid		140	143	162	155	141
diphenylacetic acid		148	167	172	180	—
2-bromobenzoic acid		150	155	—	141	110
hexanedioic acid		152	125 (m) 224 (di)	238	151 (m) 241 (di)	106
4-methylbenzoic acid		180s	160	160 (165)	145	104
4-methoxybenzoic acid		184	167 (163)	186	169	132
butanedioic acid		188	157 (m) 260 (di)	180 (m) 255 (di)	143 (m) 230 (di)	—
3,5-dinitrobenzoic acid		205	183	—	234	157
phthalic acid		210d	220 (di)	201 (di)	253 (di)	155
4-nitrobenzoic acid		240	201	204	211	168
4-chlorobenzoic acid		242	179	—	194	129
terephthalic acid		>300s	—	—	337	263 (di)

Note: All temperatures are in °C.

Table 8 Esters

Compound	bp	mp	Carboxylic acid	Alcohol or phenol	*N*-Benzyl-amide	3,5-Dinitro-benzoate
ethyl formate	54		8	—	60	93
methyl acetate	57		17	—	61	108
ethyl acetate	77		17	—	61	93
methyl propanoate	80		—	—	43	108
methyl acrylate	80		13	—	237	108
isopropyl acetate	91		17	—	61	123
tert-butyl acetate	98		17	26	61	142
ethyl propanoate	99		—	—	43	93
methyl 2,2-dimethylpropanoate	101		35	—	—	108
propyl acetate	102		17	—	61	74
methyl butanoate	102		—	—	38	108
ethyl 2-methylpropanoate	111		—	—	87	93
sec-butyl acetate	112		17	—	61	76
methyl 3-methylbutanoate	117		—	—	54	108
isobutyl acetate	117		17	—	61	87
ethyl butanoate	122		—	—	38	93
butyl acetate	126		17	—	61	64
methyl pentanoate	128		—	—	43	108
ethyl 3-methylbutanoate	135		—	—	54	93
3-methylbutyl acetate	142		17	—	61	61
ethyl chloroacetate	145		63	—	—	93
pentyl acetate	149		17	—	61	46
ethyl hexanoate	168		—	—	53	93
hexyl acetate	172		17	—	61	58

continued

Table 8 Esters *continued*

Compound	bp	mp	Carboxylic acid	Alcohol or phenol	N-Benzyl-amide	3,5-Dinitro-benzoate
cyclohexyl acetate	175		17	25	61	113
dimethyl malonate	182		135	—	142	108
diethyl oxalate	185		101*	—	223	93
heptyl acetate	192		17	—	61	47
phenyl acetate	197		17	42	61	146
methyl benzoate	199		122	—	105	108
diethyl malonate	199		135	—	142	93
o-tolyl acetate	208		17	31	61	135
m-tolyl acetate	212		17	12	61	165
ethyl benzoate	213		122	—	105	93
p-tolyl acetate	213		17	36	61	189
methyl *o*-toluate	215		104	—	—	108
benzyl acetate	217		17	—	61	113
diethyl succinate	218		188	—	206	93
isopropyl benzoate	218		122	—	105	123
methyl phenylacetate	220		77s	—	122	108
diethyl maleate	223		137	—	150	93
ethyl phenylacetate	228		77s	—	122	93
propyl benzoate	230		122	—	105	74
diethyl adipate	245		152	—	189	93
butyl benzoate	250		122	—	105	64
ethyl cinnamate	271		133	—	226	93
dimethyl phthalate	284		210d	—	179	108
(+)-bornyl acetate	*226*	27	17	208	61	154
methyl *p*-toluate		33	180s	—	133	108
methyl cinnamate		36	133	—	226	108
benzyl cinnamate		39	133	—	226	113
1-naphthyl acetate		49	17	94	61	217
ethyl *p*-nitrobenzoate		56	240	—	—	93
phenyl benzoate		69	122	42	105	146
2-naphthyl acetate		71	17	123	61	210
p-tolyl benzoate		71	122	36	105	189
methyl *m*-nitrobenzoate		78	140	—	101	108
methyl *p*-nitrobenzoate		96	240	—	142	108

Note: All temperatures are in °C. Additional derivatives of the acid and alcohol portions of most esters can be found in Tables 1 and 7.
*Dihydrate; the anhydrous acid melts at 190°C.

Table 9 Alkyl halides

Compound	bp	Density, d^{20}	S-Alkylthiuronium picrate
bromoethane	38	1.461	188
2-bromopropane	60	1.314	196
1-chloro-2-methylpropane	69	0.879	167(174)
3-bromopropene	71	1.398	155
1-bromopropane	71	1.354	177
iodoethane	72	1.936	188
1-chlorobutane	78	0.884	177
2-iodopropane	89	1.703	196

continued

Table 9 Alkyl halides *continued*

Compound	bp	Density, d^{20}	S-Alkylthiuronium picrate
1-bromo-2-methylpropane	93	1.264	167 (174)
1-chloro-3-methylbutane	100	0.875	173
1-bromobutane	101	1.274	177
1-iodopropane	102	1.749	177
3-iodopropene	102	1.848	155
1-chloropentane	108	0.882	154
1-bromo-3-methylbutane	119	1.207	173 (179)
2-iodobutane	119	1.595	166
1-iodo-2-methylpropane	120	1.606	167 (174)
1-bromopentane	129	1.218	154
1-iodobutane	131	1.617	177
1-chlorohexane	134	0.876	157
1-iodo-3-methylbutane	148	1.503	173
1-iodopentane	155	1.516	154
1-bromohexane	155	1.173	157
1-iodohexane	181	1.439	157
1-bromooctane	201	1.112	134
1-iodooctane	225	1.330	134

Note: All temperatures are in °C; density is in g/mL, at 20°C.

Table 10 Aryl halides

Compound	bp	mp	Density, d^{20}	Nitro derivative	Carboxylic acid
chlorobenzene	132		1.106	52	—
bromobenzene	156		1.495	75 (70)	—
2-chlorotoluene	159		1.083	63	140
3-chlorotoluene	162		1.072	91	158
4-chlorotoluene	162		1.071	38 (m)	242
1,3-dichlorobenzene	173		1.288	103	—
1,2-dichlorobenzene	181		1.306	110	—
2-bromotoluene	182		1.423	82	150
3-bromotoluene	184		1.410	103	155
iodobenzene	188		1.831	171 (m)	—
2,6-dichlorotoluene	199		1.269	50 (m)	139
2,4-dichlorotoluene	200		1.249	104	164
3-iodotoluene	204		1.698	108	187
2-iodotoluene	211		1.698	103 (m)	162
1-chloronaphthalene	259		1.191	180	—
4-bromotoluene	*184*	28	—	—	251
4-iodotoluene		35	—	—	270
1,4-dichlorobenzene		53	—	106, 54 (m)	—
2-chloronaphthalene		56	—	175	—
1,4-dibromobenzene		89	—	84	—

Note: All temperatures are in °C; density is in g/mL, at 20°C. Nitro derivatives signified by (m) are mononitro compounds; all others are dinitro derivatives.

Table 11 Aromatic hydrocarbons

Compound	bp	mp	Nitro derivative	Carboxylic acid	Picrate
benzene	80		89 (di)	—	84u
toluene	111		70 (di)	122	88u
ethylbenzene	136		37 (t)	122	96u
1,4-xylene	138		139 (t)	300s	90u
1,3-xylene	139		183 (t)	330s	9lu
1,2-xylene	144		118 (di)	210d	88u
isopropylbenzene	152		109 (t)	122	—
propylbenzene	159		—	122	103u
1,3,5-trimethylbenzene	165		86 (di) 235 (t)	350 (t)	97u
t-butylbenzene	169		62 (di)	122	—
4-isopropyltoluene	177		54 (di)	300s	—
1,3-diethylbenzene	181		62 (t)	330s	—
1,2,3,4-tetrahydronaphthalene	206		96 (di)	210d	—
diphenylmethane	262	26	172 (tet)	—	—
1,2-diphenylethane		53	180 (di) 169 (tet)	—	—
naphthalene		80	61 (m)	—	149
triphenylmethane		92	206 (t)	—	—
acenaphthene		96	101 (m)	—	161
fluorene		114	199 (di) 156 (m)	—	87 (77)
anthracene		216	—	—	138u

Note: All temperatures are in °C. Picrates designated u are unstable and cannot easily be purified by recrystallization.

Table 12 Phenols

Compound	bp	mp	Aryloxyacetic acid	Bromo derivative	1-Naphthyl-urethane
2-chlorophenol	176		145	49 (m) 76 (di)	120
3-methylphenol	202		—	84 (t)	128
2-methylphenol	192	31	152	56 (di)	142
4-methylphenol	232	36	135	49 (di) 108 (tet)	146
phenol	182	42	99	95 (t)	133
4-chlorophenol		43	156	33 (m) 90 (di)	166
2-nitrophenol		45	158	117 (di)	113
4-ethylphenol		47	97	—	128
5-methyl-2-isopropylphenol		50	149	55 (m)	160
3,4-dimethylphenol		63	163	171 (t)	142
4-bromophenol		64	157	95 (t)	169
2,5-dimethylphenol		75	118	178 (t)	173
1-naphthol		94	194	105 (di)	152
3-nitrophenol		97	156	91 (di)	167
4-t-butylphenol		100	86	50 (m)	110
1,2-dihydroxybenzene		105	—	193 (tet)	175
1,3-dihydroxybenzene		110	195	112 (di)	206
4-nitrophenol		114	187	142 (di)	150
2-naphthol		123	154	84 (m)	157
1,2,3-trihydroxybenzene		133	198	158 (di)	—
1,4-dihydroxybenzene		172	250	186 (di)	—

Note: All temperatures are in °C.

The Chemical Literature

APPENDIX **VII**

The literature of chemistry consists of *primary*, *secondary*, and *tertiary* sources. Most primary sources in chemistry contain descriptions of original research carried out by professional chemists. They include scientific periodicals (such as the *Journal of Organic Chemistry*), patents, dissertations, technical reports, and government bulletins. Secondary sources contain material from the primary literature that has been systematically organized, condensed, or restated to make it more accessible and understandable to users. Secondary sources include most monographs, textbooks, dictionaries, encyclopedias, reference works, review publications, and abstracting journals dealing with chemistry. Tertiary sources are intended to aid users of the primary and secondary sources or to provide facts about chemists and their work. Tertiary sources include guides to the chemical literature, directories of scientists and scientific organizations, bibliographies, trade catalogs, and publications devoted to the financial and professional aspects of chemistry. Some sources may combine several different functions; for example, *Chemical & Engineering News* prints articles about chemical research as well as financial and professional information, and thus serves as both a secondary and a tertiary source.

Although a few secondary sources critically evaluate primary material and correct errors before printing it, primary sources are generally used when it is important to obtain the most accurate and detailed information available on a topic. Errors can always occur when the material reappears in a secondary source, and important information may be left out. Because primary sources are not organized in any systematic way, it is generally necessary to refer to other sources to determine where the desired information can be found. The Bibliography following this Appendix lists a number of secondary and tertiary sources that can be used to obtain information directly, to gain access to the primary literature, or both. For example, *Beilstein's Handbook of Organic Chemistry* (hereafter referred to as *Beilstein*) gives detailed, reliable information about organic compounds and also provides citations to the literature in which the information was first reported. References to the Bibliography will be given in the form (C7), where the letter indicates a category (such as "Laboratory Safety") and the number indicates a specific work in that category.

Using *Chemical Abstracts* and *Beilstein*

Most of the reference works cited in the Bibliography are limited in scope; they make no attempt to cover the entire field of chemistry or to list all of the known organic compounds. The two major works that do attempt that kind of coverage are *Beilstein* (A3) and *Chemical Abstracts* (J3). *Beilstein* summarizes all of he important information published about specific compounds, but it is many years behind the current literature in most areas. The earlier volumes of *Beilstein* are available only in German, so at least a rudimentary knowledge of that language is necessary to make good use of this resource. *Chemical Abstracts (CA)* prints *abstracts* (brief summaries) of scientific papers, patents, and other printed material related to chemistry shortly after publication.

Before using *Chemical Abstracts* for the first time, read the introduction that appears in Issue 1 of each volume (two volumes are published each year), which describes the layout of the abstracts. Each abstract of a scientific paper (or other article) contains an abstract number, the title and author of the paper, a citation that tells where the original paper can be located, and a concise summary of the important information in the paper. The contents of *CA* can be searched by computer (as described in "On-Line Searches") or by consulting the indexes. Although each weekly issue of *CA* has its own indexes, the semiannual and collective indexes are far more useful for literature searches. A Collective Index is published every 5 years; prior to 1957, these indexes came out every 10 years. In 1972, the Subject Index was divided into two parts: the Chemical Substances Index and the General Subject Index. Author, formula, and patent indexes are also published, along with ancillary materials such as the Ring Systems Handbook, Registry Handbook, Index Guide, and Service Source Index, which are updated periodically.

The first thing you must do before searching *Chemical Abstracts* for information about a particular compound or subject is to find the *CA index name* of the compound or the *CA index heading* for the subject. In some cases, it may be possible to derive (or guess) the index name, but that is not always possible because *CA* follows its own rules of nomenclature, which often differ from IUPAC rules. The Index Guide, which now appears with each Collective Index and at intervals in between, gives cross-references from alternative names of substances to the *CA* index name. Thus the entry under "aniline" lists the index name benzeneamine, followed by the *CA registry number* [62-53-3]. The Index Guide does not list every compound indexed or give every synonym for the compounds it does list, so you may have to try different approaches to find what you are looking for. If you can't locate a specific compound, try looking up a possible parent (unsubstituted) compound under its trivial name; the index name of this parent compound should begin with a root name under which you will find your compound listed in the Chemical Substances Index (or the Subject Index, prior to 1972). For example, suppose that you are searching for the following compound:

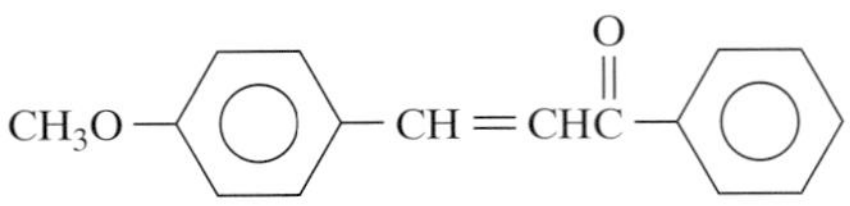

The unsubstituted compound (PhCH=CHCOPh) is known by such names as chalcone and benzalacetophenone. Looking up "chalcone" in a recent Index Guide provides the *CA* index name "2-propen-1-one, 1,3-diphenyl," so you will find the substituted compound listed—under the same root name—as "2-propen-1-one, 3-(4-methoxy)phenyl-1-phenyl." Keep in mind that the index name of a compound may change from time to time. This compound was listed under "chalcone, 4-methoxy" during the 8th Collective Index period (1967–1971) and before. Between the 8th and 9th Collective Index periods some major changes were made in the *CA* nomenclature rules, which now require rigorously systematic names for most chemical substances.

If you have trouble finding the *CA* index name for a compound using the Index Guide, you might look for it in another source such as *The Merck Index* (A10) or the *Dictionary of Organic Compounds* (A11). If you can locate the registry number for a compound, you can easily find its index name in the Registry Handbook. If you have a fairly good idea what the index name for a compound might be, you may be able to locate it in the formula indexes.

Once you locate the index name in use during a particular index period, you can locate abstracts listed under that name in the Chemical Substances Index or Subject Index for that period. Abstract citations in indexes from 1967 on are given in the form **80**:12175e, where the first number is the *CA* volume number and the second is the abstract number. In an index from prior to 1967, an abstract citation such as **51**:4321^b refers to the volume and column number (there are two columns on each page) in which the abstract appears; the superscript (either a letter from "a" through "i" or a number from 1 through 9) indicates the location of the abstract in that column. The information you are looking for may appear in the abstract itself, or you may have to read the original article cited in the abstract. Citations for such articles now appear in the form *Tetrahedron Lett.* **1996**, 37(37), 6767-6770 (Eng.), where the abbreviated name of the publication appears first, followed by the date, volume and issue number, page numbers, and language in which the paper is written. (Earlier citations were given with the volume number first, followed by the pages and year.) The full name of the publication will be found in the *Chemical Abstracts Service Source Index* (CASSI), which also provides a brief publication history of each source and a list of the libraries that carry it.

To carry out a thorough index search of the print version of *Chemical Abstracts*, it is best to start with the most recent Collective Index and all semiannual indexes published since then and work your way back through the previous Collective Indexes. If you are looking for information about a specific compound, you may find it easier to search *Beilstein* through the most recent supplemental series that lists the compound, and *Chemical Abstracts* from that time to the present. For further information about the use of *Chemical Abstracts*, refer to one or more of the literature guides in Category K of the Bibliography.

Beilstein's Handbook of Organic Chemistry (Handbuch der organischen Chemie) is by far the most comprehensive source of organized information about organic compounds. *Beilstein* provides information on the structure, characterization, natural occurrence, preparation, purification, energy parameters, physical properties, and chemical properties of organic compounds. It also cites the primary sources from which the information was obtained. *Beilstein*, unlike *Chemical Abstracts*, evaluates its sources critically and corrects errors that appeared in previous series. Beginning with the fifth supplemental series, covering the period from 1960 to 1979, Beilstein is available in an English-language edition. The basic series (*Hauptwerke*) and four previous supplemental series (*Erganzungswerke* I–IV, abbreviated E I–IV) are available only in German and cover the literature through 1959.

To obtain all of the information about a particular compound in *Beilstein*, you must search the basic series and all of the available supplemental series. The enormous size of this so-called "handbook"—along with the language barrier—could make that seem a monumental task, but there is really no reason to be intimidated by *Beilstein*. The amount of German you need to

The letter "e" in this citation is a check letter. If you looked up abstract number 12715 by mistake, you would find that its check letter is "b."

know is quite limited and can be learned with the help of the *Beilstein Dictionary* (A2), a slim dictionary written specifically for *Beilstein* users. And *Beilstein* is so well organized that it is not difficult to find the information you seek. If a compound has been around for some time, you can locate its *Beilstein* entries by the following procedure:

1 Write the molecular formula of the compound with C and H first, followed by other elements in alphabetical order.
2 Locate the formula in volume (*Band*) 29 of the second supplemental series (*General–Formelregister, Zweites Erganzungswerke*) and look for the name of the compound among those listed under that formula. Although the names are in German, many are similar or identical to the English names. (If you need help, use a German-English dictionary.)
3 Write down the volume number and pages on which information about the compound appears in the basic series (H) and the first and second supplemental series (E I and E II), and look up the appropriate entries in those volumes. (Each volume may include several individually bound subvolumes.)
4 Once you know the index name of the compound and its page number in the basic series, you can locate its entry in the corresponding volume of any later series. Alternatively, you can look it up in the cumulative subject index for that volume. Each compound is also assigned a system number, which can be used to locate its entries in the same way.

For example, the notation for indigo ($C_{16}H_{10}N_2O_2$) in volume 29 of E II reads "Indigo **24**, 417, I 370, II 233." So you will find entries for indigo on page 417 of volume 24 in the basic series and on pages 370 and 233 of volume 24 in the first and second supplemental series, respectively. The page number in the basic series, written as "**H**, 417," is called its *coordinating reference*; to find indigo's entry in volume 24 of a later series, you can locate the pages with **H**, 417 printed at the top and leaf through them until you find the entry for indigo. Knowing the E II index name of a compound may also help you locate it in the current cumulative subject index (*Sachsregister*) for the appropriate volume. The cumulative indexes for some volumes are combined; thus the listing for indigo is found in the volume 23–25 subject index, and it reads "**24** 417 d, I 370 d, II 233 e, IV 469." The letters refer to the location of an entry on the page; for example, "II 233 e" means that information about indigo will be found under the fifth entry on page 233 of E II. There is no listing for E III because supplementary series III and IV were issued jointly for volumes 17–27. Entries in the joint series are designated by E IV rather than by E III/IV.

Locating entries for a compound such as adamantane, which does not appear in the E II formula indexes, may take a little more time. All *Beilstein* entries are organized according to a detailed system, and learning that system is the best way to get complete access to the information contained in this work. However, you can usually locate such entries by either (1) finding the volume number in which a structurally similar compound appears and searching the cumulative indexes of that volume or (2) locating a *Beilstein* reference from another source. The E II indexes indicate that cyclohexane appears in volume 5, which contains all cyclic compounds lacking functional groups, so you will find adamantane listed in the cumulative indexes for that volume.

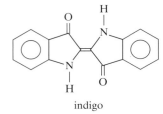

indigo

adamantane

The Beilstein *system is described in the "Notes for Users" (in English) at the beginning of each volume in the recent supplemental series.*

You can also find *Beilstein* references for many compounds in certain reference books (A1, A5, A10, and A14, for example). The entry for adamantane in the *CRC Handbook* gives the notation "B5^4, 469," referring to a *Beilstein* entry on page 469 in volume 5 of E IV. If you look up that entry, you will find a back reference (E III 393) and a coordinating reference (**H**, 165) that will help you find information about adamantane in the other series. For more detailed information about searching *Beilstein*, see reference K2 or other sources in Category K of the Bibliography.

On-Line Searches

Both *Chemical Abstracts* and *Beilstein* can be searched electronically using a variety of on-line search services. Limited *Chemical Abstracts* searching capabilities are available through FirstSearch CA Student Edition, with coverage of the most commonly held journals at academic libraries. More sophisticated search options are available using the on-line service STN (Scientific & Technical Information Network) International, which is operated by the Chemical Abstracts Service (CAS) and provides a variety of scientific and technical databases. The fundamental CAS database, called *CAplus*, includes all entries from the printed *Chemical Abstracts* since 1970 (and some back to 1967), plus additional bibliographic information. Other CAS databases available through STN International include *CAOLD*, which includes abstracts from *CA* prior to 1967; the *REGISTRY* File, a list of virtually all currently known chemicals with their *CA* registry numbers; *CJACS*, which gives the complete texts of articles published in selected American Chemical Society Journals since 1982; *CASREACT*, a database of recent organic reactions; *CHEMSOURCES*, a database of information on chemical products and their suppliers; and *CHEMLIST*, a listing of regulated and other hazardous substances. STN databases can be accessed with any computer that is connected to a telecommunications network, and are often accessible through university libraries. A simplified search option called STN*Easy* provides access to a variety of STN databases through the World Wide Web and features a graphical interface that doesn't require special training to use. Basic and advanced searches are available on STN*Easy*, and on-line help is provided when needed. For a basic search, the user simply enters a category that determines the databases to be searched, types in the words to be searched, and selects a search strategy. Search strategies include "any of these terms," which retrieves references that contain any or all of the words listed, and "all of these words," which only retrieves references that contain all of them.

Access to *Beilstein* is available through the BEILSTEIN on-line database (provided by STN International), DIALOG, and other vendors. The database is intended to cover not only the contents of the printed work, but also information from Beilstein file cards and primary literature up to the current date. For detailed information about on-line searching of *Beilstein* or *CA*, see reference K3 or K6 in the Bibliography.

Using the Bibliography

A number of books, articles, and other literature sources in organic chemistry are listed in the following Bibliography under 12 general categories:

A. Reference Works
B. Organic Reactions and Syntheses
C. Laboratory Safety
D. General Laboratory Techniques
E. Chromatography
F. Spectrometry and Structure Analysis
G. Qualitative Organic Analysis
H. Reaction Mechanisms and Advanced Topics
J. Reports of Chemical Research
K. Guides to the Chemical Literature
L. Sources on Selected Topics
M. Software for Organic Chemistry

Each source is referred to here by the category letter and its number within the category.

Category A: Reference Works

While *Beilstein* attempts to provide all of the important information about the millions of organic compounds mentioned in the chemical literature, the reference books that follow provide selected information about a much smaller number of compounds, usually numbering in the tens of thousands. The *CRC Handbook of Chemistry and Physics* (A14) and *Lange's Handbook of Chemistry* (A5) tabulate physical properties and other data for many common organic compounds and contain a large amount of useful information about chemistry. *The Merck Index* (A10) is an excellent source of information on approximately 10,000 organic and inorganic compounds. It describes their uses and hazardous properties, provides detailed physical and structural data, and gives literature references for the isolation and synthesis of many compounds. The *CRC Handbook of Data on Organic Compounds (HODOC)* (A8) contains data and references to published spectra for more than 27,000 organic compounds. The *Aldrich Catalog* (A1) lists the chemicals manufactured by the Aldrich Chemical Company and gives their physical properties, hazard warnings, procedures for safe disposal, references to published Aldrich spectra, and references to listings in *The Merck Index*, *Beilstein*, and *Fieser* (B9). The *Dictionary of Organic Compounds (DOC)* (A11) is an important multivolume set, updated by annual supplements, that gives structures, physical constants, hazard descriptions, sources, uses, derivatives, and bibliographic references for more than 145,000 organic compounds. It is also available on CD-ROM. Figure 1 shows the level of information provided by three of these reference works.

Before you use a reference work to find information about organic compounds, always read the introduction or explanatory material at the beginning of the work or preceding the table you intend to use. The introductory section of a reference work will usually (1) describe the content and organization of the material, (2) list symbols and abbreviations, and (3) describe the system of nomenclature used. Different sources often use very different naming systems. For example, *The Merck Index* emphasizes therapeutic uses of compounds, so it lists aspirin under that name; but in *Lange's Handbook* you will find aspirin listed as acetylsalicylic acid, and in the *CRC Handbook of Chemistry and Physics* it appears as salicylic acid acetate. To locate most compounds in the *CRC Handbook*, you must also know that they are entered under the name of the parent compound; thus 2,4-dinitrobenzene is listed as "benzene, 2,4-dinitro." Often, the index of a reference work provides the quickest and most reliable access to a given entry. When using *The Merck Index* or the *Dictionary of Organic Compounds*, you should first consult the name index to locate the entry for a given compound. If you can't find the compound in the name index, you may be able to locate it in a formula

No.	Name	Formula	Formula weight	Beilstein reference	Density, g/mL	Refractive index	Melting point, °C	Boiling point, °C	Flash point, °C	Solubility in 100 parts solvent
b44	Benzoic acid	C_6H_5COOH	122.12	9, 92	1.321		122.4	249	121 (CC)	0.29 aq^{25}; 43 alc; 10 bz; 22 chl; 33 eth; 33 acet; 30 CS_2

A. *Lange's Handbook of Chemistry.* (Reprinted with permission from *Lange's Handbook of Chemistry,* 15th ed., by N. A. Lange, edited by J. A. Dean. Copyright McGraw-Hill, Inc., New York, 1999.)

1092. Benzoic Acid. [65-85-0] Benzenecarboxylic acid; phenylformic acid; dracylic acid. $C_7H_6O_2$; mol wt 122.12. C 68.85%, H 4.95%, O 26.20%. Occurs in nature in free and combined forms. Gum benzoin may contain as much as 20%. Most berries contain appreciable amounts (around 0.05%). Excreted mainly as hippuric acid by almost all vertebrates, except fowl. Mfg processes include the air oxidation of toluene, the hydrolysis of benzotrichloride, and the decarboxylation of phthalic anhydride: *Faith, Keyes & Clark's Industrial Chemicals,* F. A. Lowenheim, M. K. Moran, Eds. (Wiley-Interscience, New York, 4th ed., 1975) pp 138-144. Lab prepn from benzyl chloride: A. I. Vogel, *Practical Organic Chemistry* (Longmans, London, 3rd ed, 1959) p 755; from benzaldehyde: Gattermann-Wieland, *Praxis des organischen Chemikers* (de Gruyter, Berlin, 40th ed, 1961) p 193. Prepn of ultra-pure benzoic acid for use as titrimetric and calorimetric standard: Schwab, Wicher, *J. Res. Nat. Bur. Standards* **25,** 747 (1940). *Review:* A. E. Williams in *Kirk-Othmer Encyclopedia of Chemical Technology* vol. 3 (Wiley-Interscience, New York, 3rd ed., 1978) pp 778-792.

COOH

Monoclinic tablets, plates, leaflets. d 1.321 (also reported as 1.266). mp 122.4°. Begins to sublime at ~100°. bp_{760} 249.2°; bp_{400} 227°; bp_{200} 205.8°; bp_{100} 186.2°; bp_{60} 172.8°; bp_{40} 162.6°; bp_{20} 146.7°; bp_{10} 132.1°. Volatile with steam. Flash pt 121°C. pK (25°) 4.19. pH of satd soln at 25°: 2.8. Soly in water (g/l) at 0° = 1.7; at 10° = 2.1; at 20° = 2.9; at 25° = 3.4; at 30° = 4.2; at 40° = 6.0; at 50° = 9.5; at 60° = 12.0; at 70° = 17.7; at 80° = 27.5; at 90° = 45.5; at 95° = 68.0. Mixtures of excess benzoic acid and water form two liquid phases beginning at 89.7°. The two liquid phases unite at the critical soln temp of 117.2°. Composition of critical mixture: 32.34% benzoic acid, 67.66% water: *see* Ward, Cooper, *J. Phys. Chem.* **34,** 1484 (1930). One gram dissolves in 2.3 ml cold alc, 1.5 ml boiling alc, 4.5 ml chloroform, 3 ml ether, 3 ml acetone, 30 ml carbon tetrachloride, 10 ml benzene, 30 ml carbon disulfide, 23 ml oil of turpentine; also sol in volatile and fixed oils, slightly in petr ether. The soly in water is increased by alkaline substances, such as borax or trisodium phosphate, *see also* Sodium Benzoate.

Barium salt dihydrate. Barium benzoate. $C_{14}H_{10}$-$BaO_4.2H_2O$. Nacreous leaflets. *Poisonous!* Soluble in about 20 parts water; slightly sol in alc.

Calcium salt trihydrate. Calcium benzoate. $C_{14}H_{10}$-$CaO_4.3H_2O$. Orthorhombic crystals or powder. d 1.44. Soluble in 25 parts water; very sol in boiling water.

Cerium salt trihydrate. Cerous benzoate. $C_{21}H_{15}$-$CeO_6.3H_2O$. White to reddish-white powder. Sol in hot water or hot alc.

Copper salt dihydrate. Cupric benzoate. $C_{14}H_{10}$-$CuO_4.2H_2O$. Light blue, cryst powder. Slightly soluble in cold water, more in hot water; sol in alc or in dil acids with separation of benzoic acid.

Lead salt dihydrate. Lead benzoate. $C_{14}H_{10}O_4Pb.2H_2O$. Cryst powder. *Poisonous!* Slightly sol in water.

Manganese salt tetrahydrate. Manganese benzoate. C_{14}-$H_{10}MnO_4.4H_2O$. Pale-red powder. Sol in water, alc. Also occurs with $3H_2O$.

Nickel salt trihydrate. Nickel benzoate. $C_{14}H_{10}$-$NiO_4.3H_2O$. Light-green odorless powder. Slightly sol in water; sol in ammonia; dec by acids.

Potassium salt trihydrate. Potassium benzoate. C_7H_5-$KO_2.3H_2O$. Crystalline powder. Sol in water, alc.

Silver salt. Silver benzoate. $C_7H_5AgO_2$. Light-sensitive powder. Sol in 385 parts cold water, more sol in hot water; very slightly sol in alc.

Uranium salt. Uranium benzoate; uranyl benzoate. $C_{14}H_{10}$-O_6U. Yellow powder. Slightly sol in water, alc.

Caution: Mild irritant to skin, eyes, mucous membranes.

USE: Preserving foods, fats, fruit juices, alkaloidal solns, etc; manuf benzoates and benzoyl compds, dyes; as a mordant in calico printing; for curing tobacco. As standard in volumetric and calorimetric analysis. Pharmaceutic aid (antifungal).

THERAP CAT (VET): Has been used with salicylic acid as a topical antifungal.

B. *The Merck Index: An Encyclopedia of Chemicals, Drugs, and Biologicals,* Thirteenth Edition, Maryadele J. O'Neil, Ann Smith, Patricia E. Heckelman, John R. Obenchain Jr., Eds. (Reproduced with permission from *The Merck Index,* Thirteenth Edition. Copyright ©2001 by Merck & Co., Ind., Whitehouse Station, NJ, USA. All rights reserved.)

Figure 1 Entries for benzoic acid from some reference works *(continued)*

Benzoic acid, 9CI   B-0-00650

Benzenecarboxylic acid
[65-85-0]

PhCOOH

$C_7H_6O_2$ M 122.1
Widespread in plants esp. in essential oils, mostly in esterified form. Obt. in 17th Century by sublimation of *Styrax* spp. resin. Produced industrially mainly by oxidation of toluene. Preservative in the food industry. Used in manuf. of preservatives, plasticisers, alkyd resin coatings and caprolactam. Antiseptic and expectorant. Used as alkalimetric standard; in photometric detn. of U and Zr (anionic complexes associated with basic dyes). Reference material used in elemental microanalysis. Leaflets or needles (H_2O). V. spar. sol. H_2O. Mp 122°. Bp 249°, Bp_{10} 133°, Subl. *ca.* 100°. Steam-volatile.

▶ Fl. p. 121°, autoignition temp. 570°. Eye, skin and mucous membrane irritant. Hypersensitivity reactions reported. Low systemic toxicity. DG0875000.

Na salt: [532-32-1].
Used as food preservative, anticorrosion agent. Cryst.
▶ DH6650000.

K salt: [582-25-2].
Cryst.

Me ester: [93-58-3]. *Methyl benzoate*
$C_8H_8O_2$ M 136.1   Used in perfumery and flavourings. Liq. d_{25}^{25} 1.09. Fp −12.3°. Bp 199.6°, Bp_{24} 96-98°.
▶ Fl. p. 83°. Skin and eye irritant. LD_{50} (rat, orl) 1350 mg/kg. DH3850000.

Et ester: [93-89-0]. *Ethyl benzoate*
$C_9H_{10}O_2$ M 150.1   Polymerisation catalyst. Used in perfumery and flavourings. Liq. d_4^{25} 1.04. Fp −34°. Bp 212.9°, Bp_{10} 87.2°.
▶ Fl. p. 88°, autoignition temp. 490°. Skin and eye irritant. LD_{50} (rat, orl) 2100 mg/kg. DH0200000.

Vinyl ester: see *Ethenol*, E-0-00320

Propyl ester: [2315-68-6]. *Propyl benzoate*
$C_{10}H_{12}O_2$ M 164.2   Flavour ingredient. d_{15}^{15} 1.03. Bp 230°.

Isopropyl ester: [939-48-0]. *Isopropyl benzoate*
Polymerisation catalyst, flavour ingredient. d_{15}^{15} 1.02. Bp 218-219°.
▶ Fl. p. 89/99°. Skin and eye irritant. LD_{50} (rat, orl) 3730 mg/kg. DH3150000.

Butyl ester: [136-60-7]. *Butyl benzoate*
$C_{11}H_{14}O_2$ M 178.2   Dye carrier; used in perfumery. d_{15}^{15} 1.01. Bp 248-249°.
▶ Fl. p. 107° (oc). Eye and skin irritant. DG4925000.

tert-Butyl ester: [774-65-2]. *tert-Butyl benzoate*
$C_{11}H_{14}O_2$ M 178.2 Bp_2 96°.

Benzyl ester: [120-51-4]. *Benzyl benzoate*, USAN. *Ascabin. Benylate. Vanzoate.* Many other names

$C_{14}H_{12}O_2$ M 212.2   Contained in Peru balsam. Isol. from other plants e.g. *Jasminum* spp., ylang-ylang oil. Insect repellant component. Acaricide and pediculicide. Used in perfumery as fixative and in food flavouring. Leaflets. d^{18} 1.11. Mp 21° (19.5°). Bp 323-324° (316-317°), $Bp_{0.1}$ 80-82°. Spar. steam-volatile.

▶ Fl. p. 148°, autoignition temp. 480°. Eye, mucous membrane, and possible skin irritant. Hypersensitivity reactions reported. LD_{50} (rat, orl) 500 mg/kg. DG4200000.

Ph ester: see *Phenyl benzoate*, P-0-01360

Fluoride: [455-32-3]. *Benzoyl fluoride*
C_7H_5FO M 124.1   Fuming liq. Bp 159-161°. Hydrolysed by hot H_2O.
▶ Highly irritant, causes burns, violent reaction with DMSO. Fl. p. 72/102°.

Chloride: [98-88-4]. *Benzoyl chloride*
C_7H_5ClO M 140.5   Polymerisation catalyst, benzoylating agent. Can be used for synth. of aliphatic acid chlorides. Used to derivatise steroids and carbohydrates for chromatog. Fuming liq. d_{15}^{15} 1.22. Fp −1°. Bp 197°.
▶ Fl. p. 72/102°. Violent reaction with DMSO. Corrosive and irritating to all tissues. Potent lachrymator. DM6600000.

Bromide: [618-32-6]. *Benzoyl bromide*
C_7H_5BrO M 185.0   Fuming liq. d^{15} 1.57. Fp −24°. Bp 218-219°, $Bp_{0.05}$ 48-50°.

Iodide: [618-38-2]. *Benzoyl iodide*
C_7H_5IO M 232.0   Needles. Mp 3°. Bp_{20} 128°.

Amide: see *Benzamide*, B-0-00069
Anilide: see *Benzanilide*, B-0-00074
Azide: *Benzazide. Benzoylazimide*
$C_7H_5N_3O$ M 147.1   Plates. Mp 32°.
▶ Explodes on heating.

Hydrazide: [613-94-5]. *Benzoylhydrazine*
$C_7H_8N_2O$ M 136.1   Used as $0.2M$ aq. soln. for photometric detn. of V (λ_{max} 400 nm, ε 9000); as $0.1M$ aq. soln. for photometric detn. of $IO_4^{\ominus}$. Cryst. (H_2O). Sol. H_2O, acids, EtOH, C_6H_6, Me_2CO. Mp 112.5°.
▶ DH1575000.

Hydroxamate: see *N-Hydroxybenzamide*, H-0-01671

Nitrile: see *Benzonitrile*, B-0-00747

Anhydride: [93-97-0]. *Benzoic anhydride*
$C_{14}H_{10}O_3$ M 226.2   Cross-linking agent for polymers. Acylation and decarboxylating agent, can be used in polymer-linked form. Can be used to prep. derivs. of e.g. glycosphingolipids for hplc. Rhombic prisms. d^{15} 1.99. Mp 42°. Bp 360°.
▶ Mild irritant and allergen.

Aldrich Library of ^{13}C and ^1H FT NMR Spectra, 2, 1063B, 1199A, 1240A, 1240B, 1241A, 1241B, 1244A, 1337C, 1411A (nmr)

Aldrich Library of FT-IR Spectra, 1st edn., 2, 186A, 271B, 291B, 291C, 292D, 340A, 340B, 340C, 380D (ir)

Aldrich Library of FT-IR Spectra: Vapor Phase, 3, 181D, 1322C, 1357D, 1358A, 1358B, 1358C, 1360A, 1389D, 1390A, 1390B (ir)

Org. Synth., Coll. Vol., 1, 1932, 75, 361 (synth, deriv)

Jesson, J.P. *et al*, *Proc. R. Soc. London, A*, 1962, 268, 68 (Raman)

Beynon, J.H. *et al*, *Z. Naturforsch., A*, 1965, 20, 883 (ms)

Moeken, H.H. *et al*, *Anal. Chim. Acta*, 1967, 37, 480 (detn, U)

Fieser and Fieser's Reagents for Organic Synthesis, Wiley, 1967, 1, 49, 1004; 1975, 5, 23, 24, 249; 1979, 7, 405 (use)

Evans, H.B. *et al*, *J. Phys. Chem.*, 1968, 72, 2552 (pmr)

Escarrilla, A.M. *et al*, *Anal. Chim. Acta*, 1969, 45, 199 (use)

Analyst (London), 1972, 97, 740 (microanal)

Bel'tyukova, S.V. *et al*, *Zh. Anal. Khim.*, 1972, 27, 191 (detn, Zr)

Fitzpatrick, F.A. *et al*, *Anal. Chem.*, 1973, 45, 2310 (chloride, use)

Morris, W.W., *J. Assoc. Off. Anal. Chem.*, 1973, 56, 1037 (ir)

Dubey, S.C. *et al*, *Talanta*, 1977, 24, 266 (detn, U)

Pilipenko, A.T. *et al*, *Zh. Anal. Khim.*, 1977, 32, 1369 (hydrazide, detn, V)

Fauvet, G. *et al*, *Acta Cryst. B*, 1978, 34, 1376 (cryst struct, nitrile)

White, C.A. *et al*, *Carbohydr. Res.*, 1979, 76, 1 (chloride, use)

Opdyke, D.L.J., *Food Cosmet. Toxicol.*, 1979, 17, 715 (rev, tox)

Hassan, M.M.A. *et al*, *Anal. Profiles Drug Subst.*, 1981, 10, 55 (rev, benzyl ester)

Rama Rao, A.V. *et al*, *Chem. Ind. (London)*, 1984, 270 (synth)

Ullmann's Encycl. Ind. Chem., 5th Ed, VCH, Weinheim, 1985, A3, 555 (rev)

Lewandowski, W., *Can. J. Spectrosc.*, 1987, 32, 41 (salts, ir)

Ullman, M.D. *et al*, *Methods Enzymol.*, 1987, 138, 117 (use, anhydride)

Negwer, M., *Organic-Chemical Drugs and their Synonyms*, 6th edn., Akademie-Verlag, Berlin, 1987, 3041.

Cook, L.B., *Aust. J. Chem.*, 1989, 42, 1493 (cmr)

Lewis, R.J., *Food Additives Handbook*, Van Nostrand Reinhold International, New York, 1989, BCL750, BCM000, EGR000, MHA500.

Merck Index, 11th edn., 1989, No. 1107, No. 1141 (nitrile, benzyl ester)

Kirk-Othmer Encycl. Chem. Technol., 4th edn., Wiley, New York, 1991, 4, 103 (rev)

Martindale, The Extra Pharmacopoeia, 30th edn., Pharmaceutical Press, London, 1993, 1124, 1132.

Lewis, R.J., *Sax's Dangerous Properties of Industrial Materials*, 8th edn., Van Nostrand Reinhold, 1992, BBV250, BCL750, BCM000, BCQ250, BDM500, BQK250, EGR000, IOD000, MHA750, PKW760, SFB000.

Bretherick, L., *Handbook of Reactive Chemical Hazards*, 4th edn., Butterworth, London and Boston, 1990, 2511.

Luxon, S.G., *Hazards in the Chemical Laboratory*, 5th edn., Royal Society of Chemistry, Cambridge, 1992, 117.

Chemical Hazards of the Workplace, (eds. Proctor, N.H. *et al*), 3rd edn., VNR, 1991, 107.

C. *Dictionary of Organic Compounds.* (Reprinted with permission from *Dictionary of Organic Compounds*, 6th ed., edited by P. H. Rhodes. Copyright Chapman & Hall, London, 1995.)

Figure 1 *(continued)*

index. In most formula indexes, carbon and hydrogen are listed first, followed by the other elements in alphabetical order.

The *Beilstein Dictionary* (A2) is an invaluable aid to understanding the parts of *Beilstein* (A3) that are available only in German. *The Organic Chemist's Desk Reference* (A12) includes a user's guide to the *Dictionary of Organic Chemistry* as well as a discussion of nomenclature in *Chemical Abstracts*, a list of reference works in organic chemistry, and other useful information. *The Chemist's Companion* (A6) and *The Chemist's Ready Reference Handbook* (A13) provide practical information about a variety of theoretical and experimental topics. *Organic Chemistry: An Alphabetical Guide* (A9) discusses and defines many terms used in organic chemistry. *Kirk-Othmer* (A7) is an excellent source of comprehensive, up-to-date articles on a variety of chemical topics. It offers particularly good coverage of industrial chemistry and commercial products, but also includes entries on natural products, such as coffee, terpenoids, and vitamins.

Category B: Organic Reactions and Syntheses

The works in this category are intended primarily to help chemists and chemistry students design and carry out organic syntheses. Although you may not be required to work out synthetic procedures on your own, an understanding of the strategy and techniques of organic synthesis will help you perform better in your lab course and get more out of it. *Organic Synthesis: The Disconnection Approach* (B28) describes strategies for planning an organic synthesis. At a more advanced level, *The Logic of Chemical Synthesis* (B8) deals with the analysis of complex synthetic problems, while *Principles of Organic Synthesis* (B18) discusses such topics as thermodynamics, kinetics, and stereochemistry as they apply to organic synthesis.

A number of works can help the experimenter select the type of reaction that will best accomplish a given synthetic transformation. *Synthetic Organic Chemistry* (B32) and *Modern Synthetic Reactions* (B12) survey many important synthetic reactions, with references to the earlier literature. More comprehensive coverage of organic transformations, including the more recent synthetic reactions, can be found in *Modern Methods of Organic Synthesis* (B6), *Compendium of Organic Synthetic Methods* (B11), *Comprehensive Organic Transformations* (B13), *Comprehensive Organic Synthesis* (B31), and *Theilheimer's Synthetic Methods of Organic Chemistry* (B30). Reaction files from Theilheimer and other works can be searched on-line using the *REACCS* database from Molecular Design, Ltd.

General works on organic synthetic reactions range from the *Reaction Guide for Organic Chemistry* (B15), a basic compilation of most reactions covered in the sophomore-level organic chemistry course, to *Organic Reactions* (B19), a multivolume set containing very comprehensive monographs on a large variety of reactions. *Organic Reactions* describes each reaction's mechanism, scope, limitations, and experimental conditions. It also provides several detailed experimental procedures and a table that lists examples of each reaction with references to the original sources. *Named Organic Reactions* (B14) and *Name Reactions and Reagents in Organic Synthesis* (B16) describe those synthetic reactions that—like the Hell-Volhard-Zelinskii reaction—are identified by the names of one or more discoverers. *Comprehensive Organic Chemistry* (B4) and *Rodd's Chemistry of Carbon*

Compounds (B7) describe the reactions of different classes of organic compounds in considerable depth. Each volume of *Chemistry of Functional Groups* (B22) covers the chemical reactions of a different functional group. *Asymmetric Synthesis* (B3) deals with the synthesis of chiral compounds and *Stereoselective Synthesis* (B1) with the use of stereoselective reactions in organic synthesis. Many books cover only one type of synthetic reaction, such as cycloaddition (B5); a search of your library's catalog should reveal similar works on specific reactions. *Protective Groups in Organic Synthesis* (B10) describes the use of protective groups in syntheses involving multifunctional reactants. In addition to the works dedicated to organic reactions, several advanced textbooks, such as March (H5) and Carey-Sundberg (H1), describe the most important synthetic reactions and give literature references to specific synthetic procedures.

Two multivolume sources of information about chemical reagents are Fieser's *Reagents for Organic Syntheses* (B9) and the *Encyclopedia of Reagents for Organic Synthesis* (B21). Fieser provides information about the preparation, purification, handling, and hazards of many chemical reagents, as well as examples of their use, with literature citations. Only the individual volumes are indexed, but if you locate the entry for a reagent in a recent volume, it will provide back references to the previous volumes. The *Encyclopedia of Reagents* reviews nearly 3500 reagents, listed alphabetically, and gives a critical assessment of each reagent. *Borane Reagents* (B23) covers the applications of boranes in organic synthesis. *Organic Solvents* (B24) describes physical properties and purification methods for many solvents used in organic synthesis.

Although the primary chemical literature is the most important source of experimental procedures, a number of secondary sources (in addition to *Organic Reactions*) provide relatively detailed, reliable procedures. *Organic Syntheses* (B20) is a continuing series that contains an excellent selection of carefully tested synthetic procedures. Other works that provide experimental procedures include *Organicum* (B2), several works by Sandler and Karo (B25, B26, B27), and *Vogel's Textbook of Practical Organic Chemistry* (B29). Vogel is also a good source of information about laboratory techniques. *Houben-Weyl* (D4), which is listed under "General Laboratory Techniques," includes many synthetic procedures (in German) as well.

Category C: Laboratory Safety

Accidents can happen in the organic chemistry lab, so it is important to know how to prevent accidents and what to do in case of an accident. *Working Safely with Chemicals in the Laboratory* (C2) is a short booklet about laboratory safety written for students. *Prudent Practices in the Laboratory* (C7) is an authoritative guide to safe laboratory practices; it also includes procedures for the safe handling and disposal of chemicals. *Hazards in the Chemical Laboratory* (C6) describes the toxic effects of hazardous substances and reviews recent developments in the safe design and operation of chemical laboratories. The *CRC Handbook of Laboratory Safety* (C1) deals with the recognition and control of hazards and compliance with safety regulations, and includes a chapter on responding to laboratory emergencies. The *First Aid Manual for Chemical Accidents* (C3) gives first aid procedures for accidents caused by specific chemicals and classes of chemicals. The *Sigma-Aldrich Library of Chemical Safety Data* (C4) and *Sax's Dangerous Properties of Industrial Materials* (C5) provide detailed health and safety data for many common chemicals.

Bretherick's Handbook of Reactive Chemical Hazards (C8) describes the properties of chemicals that are hazardous by virtue of their instability or their tendency to react with other chemicals. The ninth edition of *The Merck Index* (see A10) contains a section on first aid for poisoning and chemical burns; however, this section is not included in more recent editions.

Category D: General Laboratory Techniques

Although this textbook covers the techniques you are most likely to use in your organic chemistry lab course, sources from this category and the following two categories may provide more detailed practical and theoretical information about specific lab techniques, information about more advanced techniques, or a different approach to the methods described here. *The Organic Chem Lab Survival Manual* (D11) describes many of the lab techniques used by organic chemistry students and tells you what things *not* to do, such as plugging a heating mantle directly into a wall socket. *Guide for the Perplexed Organic Experimentalist* (D2) deals with the practical aspects of laboratory work for anyone intending to do research in organic chemistry. Weissberger's *Technique of Organic Chemistry* (D7) and *Techniques of Chemistry* (D8) are multivolume sets that cover a wide variety of experimental methods. Information on classical laboratory techniques, such as distillation and recrystallization, can be found in Volume I of reference D7, which is subtitled *Physical Methods of Organic Chemistry*. The first four volumes of *Houben-Weyl* (D4) describe many laboratory methods for organic chemistry, in German. Microscale techniques using Mayo-Pike and Williamson type glassware, respectively, are described in *Microscale Techniques for the Organic Laboratory* (D3) and *Macroscale and Microscale Organic Experiments* (D9). The *Encyclopedia of Separation Technology* (D6) and *Encyclopedia of Separation Science* (D10) provide comprehensive up-to-date descriptions of separation techniques, including modern microscale techniques. *Purification of Laboratory Chemicals* (D5) includes methods for the purification of more than 4000 common chemicals. *Natural Products* (D1) describes laboratory techniques and gives specific procedures for the isolation and structure determination of natural products.

Category E: Chromatography

During your organic chemistry lab course you will probably use a variety of chromatographic methods for the separation and analysis of organic compounds. These methods include thin-layer chromatography (TLC), paper chromatography (PC), gas chromatography (GC) and high-performance liquid chromatography (HPLC). *Chromatography Today* (E6) and *Principles and Practice of Chromatography* (E7) are good general sources of information on the theory and practice of all types of chromatography. *Gas Chromatography* (E1) and *High Performance Liquid Chromatography* (E3) are "open learning" texts designed for self-study. The remaining works provide up-to-date coverage of GC, HPLC, and TLC techniques and applications. *Principles of Instrumental Analysis* (F20) in the next section has chapters on instrumental chromatographic methods.

Category F: Spectrometry and Structure Determination

During your organic chemistry lab course you will probably be required to record and interpret various kinds of spectra of organic compounds, such as infrared (IR) spectra, nuclear magnetic resonance (NMR) spectra, ultraviolet-visible (UV-VIS) spectra, and mass spectra (MS). *Principles of Instrumental Analysis* (F20) is an excellent source of information about the principles and applications of all important kinds of spectrometric methods. An article in the *Journal of Chemical Education* (F8) covers the basics of IR and NMR spectral interpretation. The works by Silverstein (F19), Feinstein (F5), Kemp (F10), Pavia (F13), and Whittaker (F22) are good one-volume introductions to the interpretation of the spectra of organic compounds. More comprehensive coverage of specific spectrometric methods is provided for infrared spectrometry by references F4 and F21; for nuclear magnetic resonance spectrometry by F1, F2, and F6; for mass spectrometry by F3, F7, F9, F11, and F12; and for ultraviolet-visible spectrometry by F14. The *Sadtler Standard Spectra* (F18) series consists of a large number of IR, NMR, and UV-VIS spectra in ring binders; although they are not arranged systematically, individual spectra can be located by using the index volumes. Spectra in the Aldrich collections (F15–F17) are arranged by functional class and in order of increasing molecular complexity within a functional class, making it possible to observe the effect of various structural features on the spectra.

Category G: Qualitative Organic Analysis

During your organic chemistry lab course you may be required to identify one or more unknown organic compounds using either "wet-chemistry" methods (involving chemical tests and derivative preparations) or spectrometric methods, or both. *Organic Structure Determination* (G4), *The Systematic Identification of Organic Compounds* (G6), and *Spectral and Chemical Characterization of Organic Compounds* (G2) cover the traditional wet-chemistry methods, but include chapters on spectrometric methods as well. *Qualitative Organic Analysis* (G3) emphasizes spectral methods of identification, and *Organic Structure Analysis* (G1) focuses on the use of multiple spectrometric methods to identify a molecule's major structural elements. The *CRC Handbook of Tables for Organic Compound Identification* (G5) lists the properties and derivative melting points for many organic compounds in the most important functional classes.

Category H: Reaction Mechanisms and Advanced Topics

Although reaction mechanisms are more often explored in an organic chemistry lecture course than in the laboratory course, you may be expected to write and understand mechanisms for some of the reactions you perform in the lab. *Electron Flow in Organic Chemistry* (H9) teaches an intuitive approach to organic chemistry by breaking down reaction mechanisms into elementary electron-flow pathways. *A Guidebook to Mechanism in Organic Chemistry* (H10) by Sykes is an excellent survey of reaction mechanisms suitable for advanced students; his *Primer* (H11) is a more basic introduction to mechanisms based on a simplified classification scheme. Other how-to books include *Writing Reaction Mechanisms in Organic Chemistry* (H6) and *Reaction Mechanisms at a Glance* (H7). *Mechanism and Theory in Organic Chemistry* (H4) and *Perspectives on Structure and Mechanism in Organic Chemistry* (H3) are advanced textbooks that present the theoretical aspects of organic chemistry and provide up-to-date information about important reaction mechanisms. The *Advanced Organic Chemistry* textbooks by Carey-Sundberg (H1) and March (H5) provide good coverage of the mechanisms and synthetic applications of a large number of organic reactions, giving numerous references to the primary literature. *Determination of Organic Reaction Mechanisms* (H2) describes experimental techniques for studying reaction mechanisms, and *Organic Reaction Mechanisms* (H8) is an annual survey of recent developments in the field.

Category J: Reports of Chemical Research

Virtually all professional chemists do *chemical research*—experimental or theoretical work designed to discover new facts about the various forms of matter, develop new techniques that can be used to study matter, or provide new insights about the fundamental nature of matter. Papers describing the results of their research are reported in such a large number of professional journals and other publications that it impossible for anyone to investigate them all. For that reason, scientists consult various reports of chemical research to locate the papers that deal with their own research interests. *Chemical Abstracts* (J3), previously described in detail, is the most comprehensive single source of information about research in chemistry. The *Science Citation Index* (J7) is an index of literature citations to papers, patents, and books published in the past. For example, if you find an interesting paper by Linus Pauling in a chemistry journal, you can look up the paper in the *Science Citation Index* to find later articles that were based, in part, on Pauling's original paper. In this way, you can sometimes trace the development of an idea or a method from its origin to the present day. A similar index for chemistry, the *Chemistry Citation Index*, is available on CD-ROM. *Chemical Titles* (J4) and *Current Contents* (J5) reproduce the current tables of contents of the most important chemistry journals to inform chemists quickly of recent research in their fields. *Index Chemicus* (J6), a weekly guide to new organic compounds and their chemistry, is available in print and on a searchable database. The other works in this category (J1, J2) provide annual summaries and reviews of research in organic chemistry.

Category K: Guides to the Chemical Literature

If you need specific information to complete a lab report or write a research paper, you have to know where to look for it. *Information Sources in Chemistry* (K1) and *How to Find Chemical Information* (K4) are general guides to the chemical literature that list and describe a large number of information sources. *Library Handbook for Organic Chemists* (K5) is an up-to-date guide to the use of library information resources. *The Beilstein System* (K2) and *The Beilstein Online Database* (K3) tell how to search and use *Beilstein's* print and on-line versions, respectively. *From CA to CAS Online* (K6) serves the same function for *Chemical Abstracts*. The three articles by Somerville (K7) describe the contents and uses of some major works on organic reactions and syntheses. A brief but useful guide to information sources for organic chemistry can be found in Appendix A of March (H5).

Category L: Sources on Selected Topics

This category includes a number of books and articles that can be used as resources for library research papers. Others can simply be read for enjoyment and enlightenment, on everything from beer and perfumes to the O. J. Simpson trial. Most are about topics related to the experiments or minilabs, such as reference L18 on sweetness (see Experiments 20 and 50), L20 on indigo (see Minilab 42), and L55 on perfumes (see Experiment 13). Some works are listed here because they don't fit into any of the other categories. For example, L14 and L42 are guides for writing scientific papers and laboratory notebooks, respectively, and L22 describes organic nomenclature in depth.

Category M: Software for Organic Chemistry

A number of software titles are designed to be used in preparation for or during an organic chemistry laboratory. *Organic Chemistry Laboratory* (M8) is a set of interactive tutorials intended to help students learn about selected lab techniques, including distillation, extraction, melting-point determination, and qualitative organic analysis. *SQUALOR* and *MacSQUALOR* (M6) allow the user to identify simulated unknowns for qualitative organic analysis. *Introduction to Spectroscopy* (M3) and *SpectraBook/SpectraDeck* (M7) help the user analyze and interpret spectral data. *Proton NMR Spectrum Simulator* (M1) generates a simulated NMR spectrum from a molecular structure entered by the user. *MassSpec* (M4) helps the user identify the structural fragments that correspond to peaks on a mass spectrum. *SynTree* (M5) helps the user work out retrosynthetic pathways leading from a selected target compound back to a readily available starting material. *Beaker* (M2) allows the user to draw structures and predict properties and spectra of organic molecules. *PCSpartan* (M9) and its Macintosh version provide sophisticated molecular modeling and computational chemistry features.

Bibliography

A. Reference Works

1. *Aldrich Catalog Handbook of Fine Chemicals.* Milwaukee, WI: Aldrich Chemical Co., 2003–04 (and other years).
2. *Beilstein Dictionary: German–English: For the Users of the Beilstein Handbook of Organic Chemistry.* Ft. Worth, TX: Saunders, 1992.
3. *Beilstein's Handbook of Organic Chemistry.* New York: Springer-Verlag, 1918 to date.
4. *Chemical Abstracts Ring Systems Handbook.* Washington, DC: American Chemical Society, 1993.
5. Dean, J. A., ed., *Lange's Handbook of Chemistry*, 15th ed. New York: McGraw-Hill, 1999.
6. Gordon, A. J., and Ford, R. A., *The Chemist's Companion.* New York: Wiley-Interscience, 1972.
7. *Kirk-Othmer Encyclopedia of Chemical Technology*, 4th ed., New York: Wiley, 1993–1998.
8. Lide, D. R., and Milne, G. W. A., eds., *CRC Handbook of Data on Organic Compounds,* 3rd ed. Boca Raton, FL: CRC Press, 1994.
9. Mundy, B. P., and Ellerd, M. G., *Organic Chemistry: An Alphabetical Guide.* New York: Wiley, 1996.
10. O'Neil, M. J. et al., eds., *The Merck Index: An Encyclopedia of Chemicals, Drugs, and Biologicals*, 13th ed. Whitehouse Station, NJ.: Merck & Co., 2001.
11. Rhodes, P. H., ed., *Dictionary of Organic Compounds,* 6th ed. London: Chapman and Hall, 1995.
12. Rhodes, P. H., *The Organic Chemist's Desk Reference: A Companion Volume to the Dictionary of Organic Compounds,* 6th ed. London: Chapman and Hall, 1995.
13. Shugar, G. J., and Dean, J. A., *The Chemist's Ready Reference Handbook.* New York: McGraw-Hill, 1990.
14. Weast, R. C., ed., *CRC Handbook of Chemistry and Physics*, 81st ed. (and other editions). Boca Raton, FL: CRC Press, 2000.

B. Organic Reactions and Syntheses

1. Atkinson, R. S., *Stereoselective Synthesis.* New York: Wiley, 1995.
2. Becker, H. et al., *Organicum: Practical Handbook of Organic Chemistry*, trans. by B. J. Hazzard. Reading, MA: Addison-Wesley, 1973.
3. Proctor, R.G., *Asymmetric Synthesis*, New York: Oxford, 1996.
4. Barton, D. H., and Ollis, W. D., eds., *Comprehensive Organic Chemistry: The Synthesis and Reactions of Organic Compounds.* New York: Pergamon Press, 1979.
5. Carruthers, W., *Cycloaddition Reactions in Organic Synthesis.* New York: Pergamon Press, 1990.
6. Carruthers, W., *Modern Methods of Organic Synthesis,* 3rd ed. New York: Cambridge, 1987.

7. Coffey, S. (1964–1989), Ansell, M. F. (1973–), Sainsbury (1991–), eds., *Rodd's Chemistry of Carbon Compounds,* 2nd ed. and supplements. New York: Elsevier, 1964–.
8. Corey, E. J. and Cheng, X.-M., *The Logic of Chemical Synthesis.* New York, Wiley, 1995.
9. Fieser, L. F., and Fieser, M. (1967–1986) Fieser, M., and Smith, J. G. (1988–), *Reagents for Organic Synthesis.* New York: Wiley, 1967–.
10. Greene, T. W. and Wuts, P. G. M., *Protective Groups in Organic Synthesis*, 3rd ed. New York: Wiley, 1999.
11. Harrison, I. T., and Harrison, S., *Compendium of Organic Synthetic Methods.* New York: Wiley, 1971–.
12. House, H. O., *Modern Synthetic Reactions,* 2nd ed. Menlo Park, CA: Benjamin, 1972.
13. Larock, R. C., *Comprehensive Organic Transformations: A Guide to Functional Group Preparations*, 2nd ed. New York: Wiley, 1999.
14. Laue, T. and Plagens, A., *Named Organic Reactions.* New York: Wiley, 2000.
15. Millam, M. J., *Reaction Guide for Organic Chemistry.* Lexington, MA: D.C. Heath, 1989.
16. Mundy, B. P., and Ellerd, M. G., *Name Reactions and Reagents in Organic Synthesis.* New York: Wiley, 1988.
17. Nicolaou, K. C., et al., "The Art and Science of Organic and Natural Product Synthesis." *J. Chem. Educ.* **1998**, 75, 1225.
18. Norman, R. O. C., and Coxon, J. M., *Principles of Organic Synthesis*, 3rd ed., Cheltenham, England: Stanley Thornes, 1993.
19. *Organic Reactions.* New York: Wiley, 1942–.
20. *Organic Syntheses, Collective Volumes.* New York: Wiley, 1941–.
21. Paquette, L. A., ed., *Encyclopedia of Reagents for Organic Synthesis*, New York: Wiley, 1994.
22. Patai, S., ed., *Chemistry of Functional Groups.* New York: Wiley, 1964–.
23. Pelter, A., Smith, K., and Brown, H. C., *Borane Reagents.* London: Academic Press, 1988.
24. Riddick, J. A., and Bunger, W. M., *Organic Solvents: Physical Properties and Methods of Purification*, 4th ed. New York: Wiley, 1986.
25. Sandler, S. R., and Karo, W., *Organic Functional Group Preparations*, 2nd ed. Orlando, FL: Academic Press, 1983, 1986, 1989.
26. Sandler, S. R., and Karo, W., *Polymer Syntheses*, 2nd ed. Orlando, FL: Academic Press, 1997.
27. Sandler, S. R., and Karo, W., *Sourcebook of Advanced Organic Laboratory Preparations.* San Diego, CA: Academic Press, 1992.
28. Stuart, W., *Organic Synthesis: The Disconnection Approach.* New York: Wiley, 1982.

29. Tatchell, A. R., et al., *Vogel's Textbook of Practical Organic Chemistry*, 5th ed. New York: Wiley, 1989.

30. Theilheimer, W. (1948–81), Finch, A. F. (1982–), eds., *Theilheimer's Synthetic Methods of Organic Chemistry*. Basel: Karger, 1946–.

31. Trost, B. M. et al., ed., *Comprehensive Organic Synthesis: Selectivity, Strategy & Efficiency in Modern Organic Chemistry*. Elmsford, NY: Pergamon Press, 1991.

32. Wagner, R. B., and Zook, H. D., *Synthetic Organic Chemistry*. New York: Wiley, 1953.

C. Laboratory Safety

1. Furr, A. K., ed., *CRC Handbook of Laboratory Safety*, 4th ed. Boca Raton, FL: CRC Press, 1995.

2. Gorman, C. E., ed. *Working Safely with Chemicals in the Laboratory*, 2nd ed. Schenectady, NY: Genium, 1995.

3. Lefèvre, M. J., *First Aid Manual for Chemical Accidents*, 2nd ed. New York: Van Nostrand Reinhold, 1989.

4. Lenga, R. E., ed., *The Sigma-Aldrich Library of Chemical Safety Data*, 2nd ed. Milwaukee, WI: Sigma-Aldrich, 1988.

5. Lewis, R. J., Sr., *Sax's Dangerous Properties of Industrial Materials*, 8th ed. New York: Van Nostrand Reinhold, 1992.

6. Luxon, S. G., ed., *Hazards in the Chemical Laboratory*, 5th ed. Cambridge: Royal Society of Chemistry, 1992.

7. National Research Council, *Prudent Practices in the Laboratory: Handling and Disposal of Chemicals*. Washington, DC: National Academy Press, 1995.

8. Urben, P. G., ed., *Bretherick's Handbook of Reactive Chemical Hazards*. Stoneham, MA: Butterworths, 1995.

D. General Laboratory Techniques

1. Ikan, R., *Natural Products: A Laboratory Guide*, 2nd ed. San Diego, CA: Academic Press, 1991.

2. Loewenthal, H. J. E., *Guide for the Perplexed Organic Experimentalist*, 2nd ed. New York: Wiley, 1992.

3. Mayo, D. W., et al. *Microscale Techniques for the Organic Laboratory*, 2nd ed. New York: Wiley, 2000.

4. *Methoden der Organischen Chemie, Houben-Weyl*, 4th ed. Stuttgart: Georg Thieme, 1952–.

5. Perrin, D. D., and Armarego, W. L. F, *Purification of Laboratory Chemicals*, 3rd ed. New York: Pergamon Press, 1988.

6. Ruthven, D., *Encyclopedia of Separation Technology*. New York: Wiley, 1997.

7. Weissberger, A., ed., *Technique of Organic Chemistry*, 3rd ed. New York: Wiley, 1959–.

8. Weissberger, A., ed., *Techniques of Chemistry*. New York: Wiley, 1971–.

9. Williamson, K. L., *Macroscale and Microscale Organic Experiments*, 3rd ed. Boston, MA: Houghton-Mifflin, 1999.

10. Wilson, Ian D. et al., eds., *Encyclopedia of Separation Science*. Orlando, FL: Academic Press, 2000.

11. Zubrick, J. W., *The Organic Chem Lab Survival Manual: A Student's Guide to Techniques*, 4th ed. New York: Wiley, 1997.

E. Chromatography

1. Fowlis, I. A., *Gas Chromatography*, 2nd ed. New York: Wiley, 1995.

2. Grob, R. L., ed., *Modern Practice of Gas Chromatography*, 3rd ed. New York: Wiley, 1995.

3. Lindsay, S., *High Performance Liquid Chromatography*, 2nd ed. New York: Wiley, 1992.

4. McNair, H. M. and J. M. Miller, *Basic Gas Chromatography*. New York: Wiley, 1997.

5. Meyer, V. R., *Practical High-Performance Liquid Chromatography*, 3rd ed. Chichester, England: Wiley, 1999.

6. Poole, C. F., and Poole, S. K., *Chromatography Today*. New York: Elsevier, 1991.

7. Ravindranath, B., *Principles and Practice of Chromatography*. New York: Halsted, 1989.

8. Schomburg, G., *Gas Chromatography: A Practical Course*. New York: VCH, 1990.

9. Sherma, J. and B. Fried, *Thin-Layer Chromatography: Techniques and Applications*, 3rd ed. New York: Marcel Dekker, 1996.

10. Touchstone, J. C., *Practice of Thin Layer Chromatography*, 3rd ed. New York: Wiley, 1992.

F. Spectrometry and Structure Analysis

1. Bovey, F. A., *Nuclear Magnetic Resonance Spectroscopy*, 2nd ed. San Diego, CA: Academic Press, 1988.

2. Breitmaier, E. *Structure Elucidation by NMR in Organic Chemistry: A Practical Guide*. New York: Wiley, 1993.

3. Chapman, J. R., *Practical Organic Mass Spectrometry: A Guide for Chemical and Biochemical Analysis*, 2nd ed. New York: Wiley, 1995.

4. Colthup, N. B., Daly, L. H., and Wiberley, S. E., *Introduction to Infrared and Raman Spectroscopy*, 3rd ed. Orlando, FL: Academic Press, 1990.

5. Feinstein, K., *Guide to Spectroscopic Identification of Organic Compounds*. Boca Raton, FL: CRC Press, 1995.

6. Günther, H., *NMR Spectroscopy: Basic Principles, Concepts, and Applications in Chemistry*, 2nd ed. New York: Wiley, 1995.

7. Hoffmann, Edmond de, *Mass Spectrometry: Principles and Applications*. New York: Wiley, 1996.

8. Ingham, A. M., and Henson, R. C., "Interpreting Infrared and Nuclear Magnetic Resonance Spectra of Simple Organic Compounds for the Beginner." *J. Chem. Educ.* **1984**, *61*, 704.

9. Johnstone, R. A. W., and Rose, M. E., *Mass Spectrometry for Chemists and Biochemists*, 2nd ed. New York: Cambridge, 1996.

10. Kemp, W., *Organic Spectroscopy*, 3rd ed. New York: W. H. Freeman, 1991.

11. Lee, T. A., *A Beginners Guide to Mass Spectral Interpretation.* New York: Wiley, 1998.

12. McLafferty, F. W., and Turecek, F., *Interpretation of Mass Spectra*, 4th ed. Mill Valley, CA: University Science Books, 1993.

13. Pavia, D. L., et al., *Introduction to Spectroscopy: A Guide for Students of Organic Chemistry*, 3rd ed. Ft. Worth, TX: Saunders, 2000.

14. Perkampus, H.-H., *UV-VIS Spectroscopy and its Applications.* New York: Springer-Verlag, 1992.

15. Pouchert, C. J., and Behnke, J., *The Aldrich Library of ^{13}C and ^{1}H FT-NMR Spectra.* Milwaukee, WI: Aldrich Chemical Co., 1992.

16. Pouchert, C. J., and Campbell, J. R., *The Aldrich Library of NMR Spectra*, 2nd ed. Milwaukee, WI: Aldrich Chemical Co., 1983.

17. Pouchert, C. J., *The Aldrich Library of FT-IR Spectra*, 2nd ed. Milwaukee, WI: Aldrich Chemical Co., 1997.

18. *Sadtler Standard Spectra.* (Collections of infrared, ultraviolet, and NMR spectra.) Philadelphia, PA: Sadtler Research Laboratories.

19. Silverstein, R. M., and Webster, F. X., *Spectrometric Identification of Organic Compounds*, 6th ed. New York: Wiley, 1998.

20. Skoog, D. A., et al., *Principles of Instrumental Analysis*, 5th ed. Philadelphia, PA: Saunders, 1998.

21. Smith, B. C., *Fundamentals of Fourier Transform Infrared Spectroscopy.* Boca Raton, FL: CRC Press, 1996.

22. Whittaker, D. *Interpreting Organic Spectra.* New York: Springer-Verlag, 2000.

G. Qualitative Organic Analysis

1. Crews, P., et al., *Organic Structure Analysis.* New York: Oxford, 1998.

2. Criddle, W. J., *Spectral and Chemical Characterization of Organic Compounds: A Laboratory Handbook*, 3rd ed. New York: Wiley, 1990.

3. Kemp, W., *Qualitative Organic Analysis: Spectrochemical Techniques*, 2nd ed. New York: McGraw-Hill, 1986.

4. Pasto, D. J., and Johnson, C. R., *Organic Structure Determination.* Englewood Cliffs, NJ: Prentice Hall, 1969.

5. Rappoport, Z., ed., *CRC Handbook of Tables for Organic Compound Identification*, 3rd ed. Cleveland, OH: Chemical Rubber Co., 1967.

6. Shriner, R. L., et al., *The Systematic Identification of Organic Compounds*, 7th ed. New York: Wiley, 1998.

H. Reaction Mechanisms and Advanced Topics

1. Carey, F. A., and Sundberg, R. J, *Advanced Organic Chemistry*, 4th ed. New York: Plenum, 2000.

2. Carpenter, B. K., *Determination of Organic Reaction Mechanisms.* New York: Wiley, 1984.

3. Carroll, F. A., *Perspectives on Structure and Mechanism in Organic Chemistry.* Belmont, CA: Brooks/Cole, 1998.

4. Lowry, T. H., and Richardson, K. S., *Mechanism and Theory in Organic Chemistry*, 3rd ed. New York: Harper & Row, 1987.

5. Smith, M. and J. March, *March's Advanced Organic Chemistry: Reactions, Mechanisms and Structure*, 5th ed. New York: Wiley, 2001.

6. Miller, A. and P. H. Solomon, *Writing Reaction Mechanisms in Organic Chemistry*, 2nd ed. San Diego, CA: Harcourt/Academic Press, 2000.

7. Moloney, M. G., *Reaction Mechanisms at a Glance: A Stepwise Approach to Problem-Solving in Organic Chemistry.* Oxford; Malden, MA: Blackwell Science, 2000.

8. *Organic Reaction Mechanisms.* New York: Wiley, 1965–.

9. Scudder, P. H., *Electron Flow in Organic Chemistry.* New York: Wiley, 1992.

10. Sykes, P., *A Guidebook to Mechanism in Organic Chemistry*, 6th ed. Essex, UK: Longman, 1986.

11. Sykes, P., *A Primer to Mechanism in Organic Chemistry.* Essex, UK: Longman, 1995.

J. Reports of Chemical Research

1. *Annual Reports in Organic Synthesis.* New York: Academic Press, 1970–.

2. *Annual Reports on the Progress of Chemistry, Section B: Organic Chemistry.* London: Royal Society of Chemistry, 1904–.

3. *Chemical Abstracts.* Columbus, OH: CA Service, American Chemical Society, 1907–.

4. *Chemical Titles.* Columbus, OH: CA Service, American Chemical Society, 1961–.

5. *Current Contents: Physical, Chemical & Earth Sciences.* Philadelphia, PA: ISI Press, 1967–.

6. *Index Chemicus.* Philadelphia, PA: ISI Press, 1960–.

7. *Science Citation Index.* Philadelphia, PA: ISI Press, 1961–.

K. Guides to the Chemical Literature

1. Bottle, R. T., and Rowland, J. F. B., eds., *Information Sources in Chemistry*, 4th ed. New Providence, NJ: Bowker-Saur, 1993.

2. Heller, S. R., ed., *The Beilstein System: Strategies for Effective Searching.* New York: Oxford, 1997.

3. Heller, S. R., ed., *The Beilstein Online Database: Implementation, Content, and Retrieval.* New York: Oxford, 1990.

4. Maizell, R. E., *How to Find Chemical Information: A Guide for Practicing Chemists, Educators, and Students*, 3rd ed. New York: Wiley, 1998.

5. Poss, A. J., *Library Handbook for Organic Chemists.* New York: Chemical Pub., 2000.

6. Schulz, H., and Georgy, U., *From CA to CAS Online: Databases in Chemistry*, 2nd ed. New York: Springer-Verlag, 1994.

7. Somerville, A. N., "Information Sources for Organic Chemistry." *J. Chem. Educ.* **1991**, *68*, 553, 843; **1992**, *69*, 379.

L. Sources on Selected Topics

1. Agosta, W. C., *Bombardier Beetles and Fever Trees: A Close-up Look at Chemical Warfare and Signals in Animals and Plants.* Reading, MA: Addison-Wesley 1996.

2. Agosta, W. C., "Medicines and Drugs from Plants." *J. Chem. Educ.* **1997**, *74*, 857.

3. Agosta, W. C., *Chemical Communication: The Language of Pheromones.* New York: Scientific American Library, 1992.

4. Anastas, P. T., and Williamson, T. C., eds., *Green Chemistry: Designing Chemistry for the Environment.* Washington, DC: American Chemical Society, 1996.

5. Atkins, P. W., *Molecules.* New York: Scientific American Library, 1987.

6. Bauer, K, et al., *Common Fragrance and Flavor Materials: Preparation, Properties, and Uses*, 2nd ed. Deerfield Beach, FL: VCH, 1990.

7. Belitz, H. D. and W. Grosch, *Food Chemistry*, 2nd ed. New York: Springer-Verlag, 1999.

8. Benfey, O. T., *From Vital Force to Structural Formulas.* Philadelphia, PA: Beckman Center for the History of Chemistry, 1992.

9. Buxton, S. R., and Roberts, S. M., *Guide to Organic Stereochemistry: From Methane to Macromolecules.* Menlo Park, CA: Benjamin/Cummings, 1996.

10. Carraher, C. E., Jr., *Polymer Chemistry*, 5th ed. New York: Marcel Dekker, 2000.

11. Cole, L. A., *The Eleventh Plague: The Politics of Biological and Chemical Warfare.* New York: W. H. Freeman, 1997.

12. Dehmlow, E. V., and Dehmlow, S. S., *Phase Transfer Catalysis*, 3rd ed. New York: VCH, 1993.

13. Djerassi, C., *From the Lab into the World: A Pill for People, Pets, and Bugs.* Washington, DC: American Chemical Society, 1994

14. Dodd, J. S., ed., *The ACS Style Guide: A Manual for Authors and Editors*, 2nd ed. Washington, DC: American Chemical Society, 1997.

15. Donnelly, T. H., "The Origins of the Use of Antioxidants in Foods." *J. Chem. Educ.* **1996**, *73*, 159.

16. DuPré, D. B., "Blood or Taco Sauce? The Chemistry behind Criminalists' Testimony in the O. J. Simpson Murder Case." *J. Chem. Educ.* **1996**, *73*, 60.

17. Eliel, E. L., and Wilen, S. H., *Stereochemistry of Organic Compounds.* New York: Wiley, 1994.

18. Ellis, J. W., et al., "Symposium: Sweeteners and Sweetness Theory." *J. Chem. Educ.* **1995**, *72*, 671, 676, 680.

19. Emsley, J. *Molecules at an Exhibition: Portraits of Intriguing Materials in Everyday Life.* New York: Oxford, 1998.

20. Fernelius, W. C., and Renfrew, E. E., "Indigo." *J. Chem. Educ.* **1983**, *60*, 633.

21. Fossey, J., et al., *Free Radicals in Organic Chemistry.* New York: Wiley, 1995.

22. Fox, R. B. and W. H. Powell, *Nomenclature of Organic Compounds: Principles and Practice*, 2nd ed. Washington, DC: American Chemical Society, 2000.

23. French, L. G., "The Sassafras Tree and Designer Drugs: from Herbal Tea to Ecstasy." *J. Chem. Educ.* **1995**, *72*, 484.

24. Foye, W. O., et al., *Principles of Medicinal Chemistry*, 4th ed. Baltimore, MD: Lippincott Williams & Wilkins, 1995.

25. Garratt, P. J., *Aromaticity.* New York: McGraw-Hill, 1986.

26. Gerber, S. M., ed., *Chemistry and Crime: from Sherlock Holmes to Today's Courtroom.* New York: Oxford, 1983.

27. Gerber, S. M., and Saferstein, R., eds. *More Chemistry and Crime: from Marsh Arsenic Test to DNA Profile.* New York: Oxford, 1997.

28. Gilchrist, T. L., *Heterocyclic Chemistry*, 3rd ed. New York: Wiley, 1997.

29. Glidewell, C., and Lloyd, D., "The Arithmetic of Aromaticity." *J. Chem. Educ.* **1986**, *63*, 306.

30. Goldsmith, R. H., "A Tale of Two Sweeteners." *J. Chem. Educ.* **1987**, *64*, 954.

31. Goodman, J. M., *Chemical Applications of Molecular Modeling.* New York: Springer-Verlag, 1998.

32. Gribble, G. W., "Natural Organohalogens: Many More than You Think!" *J. Chem. Educ.* **1994**, *71*, 907.

33. Gunstone, F. D., *Fatty Acid and Lipid Chemistry.* New York: Blackie, 1996.

34. Hammond, G. S., and Kuck, V. J., *Fullerenes: Synthesis, Properties, and Chemistry of Large Carbon Clusters.* Washington, DC: American Chemical Society, 1992.

35. Hill, J. W., and Jones, S. W., "Consumer Applications of Chemical Principles: Drugs." *J. Chem. Educ.* **1985**, *62*, 328.

36. Hirsch, A., *The Chemistry of Fullerenes.* New York: Georg Thieme, 1994.

37. Hocking, M. B., "Vanillin: Synthetic Flavoring from Spent Sulfite Liquor." *J. Chem. Educ.* **1997**, *74*, 1055.

38. Hosler, D. M., and Mikita, M. A., "Ethnobotany: The Chemist's Source for the Identification of Useful Natural Products." *J. Chem. Educ.* **1987**, *64*, 328.

39. Jacques, J., *The Molecule and Its Double.* New York: McGraw-Hill, 1993.

40. James, L. K., ed., *Nobel Laureates in Chemistry, 1901–1992.* Washington, DC: American Chemical Society, 1993.

41. Jandacek, R. J., "The Development of Olestra, a Noncaloric Substitute for Dietary Fat." *J. Chem. Educ.* **1991**, *68*, 476.

42. Kanare, H. M., *Writing the Laboratory Notebook.* Washington, DC: American Chemical Society, 1985.

43. Kauffman, G. B., "Wallace Hume Carothers and Nylon, the First Completely Synthetic Fiber." *J. Chem. Educ.* **1988**, *65*, 803.

44. Kauffman, G. B., and Seymour, R. B., "Elastomers: I. Natural Rubber." *J. Chem. Educ.* **1990**, *67*, 422.

45. Kent, J. A., ed., *Riegel's Handbook of Industrial Chemistry*, 9th ed. New York: Van Nostrand Reinhold, 1992.

46. Kikuchi, S., "A History of the Structural Theory of Benzene—the Aromatic Sextet Rule." *J. Chem. Educ.* **1997**, *74*, 194.

47. Kimbrough, D. R., "Hot and Spicy vs. Cool and Minty as an Example of Organic Structure-Activity Relationships." *J. Chem. Educ.* **1997**, *74*, 861.

48. Kimbrough, D. R., "The Photochemistry of Sunscreens." *J. Chem. Educ.* **1997**, *74*, 51.

49. King, F. D., ed., *Medicinal Chemistry: Principles and Practice.* Cambridge: Royal Society of Chemistry, 1994.

50. Kopecky, Jan, *Organic Photochemistry: A Visual Approach.* New York: Wiley, 1992.

51. Laing, M., "Beware—Fertilizer can EXPLODE!" *J. Chem. Educ.* **1993**, *70*, 393.

52. Leung, A.Y. and Foster, S., *Encyclopedia of Common Natural Ingredients Used in Foods, Drugs, and Cosmetics.* New York: Wiley, 1995.

53. Mann, J., et al., *Natural Products: Their Chemistry and Biological Significance.* New York: Wiley, 1994.

54. Milgrom, L. R., *The Colours of Life: An Introduction to the Chemistry of Porphyrins and Related Compounds.* New York: Oxford, 1997.

55. Morris, E. T., *Fragrance: The Story of Perfume from Cleopatra to Chanel.* New York: Scribners, 1990.

56. Nicholson, J. W. and Anstice, H. M., "The Chemistry of Modern Dental Filling Materials." *J. Chem. Educ.* **1999**, *76*, 1497.

57. Ohloff, G., *Scent and Fragrances: The Fascination of Odors and Their Chemical Perspectives.* New York: Springer-Verlag, 1994.

58. Pauli, G. H., "Chemistry of Food Additives." *J. Chem. Educ.* **1984**, *61*, 332.

59. Perrine, D. M., *The Chemistry of Mind-Altering Drugs: History, Pharmacology, and Cultural Context.* New York: Oxford, 1996.

60. Robinson, M. J. T., *Organic Stereochemistry.* New York: Oxford, 2000.

61. Sbrollini, M. C., "Olfactory Delights." *J. Chem. Educ.* **1987**, *64*, 799.

62. Seymour, R. B., and Kauffman, G. B., "The Ubiquity and Longevity of Fibers." *J. Chem. Educ.* **1993**, *70*, 449.

63. Selinger, B., *Chemistry in the Marketplace*, 5th ed. Orlando, FL: Academic Press, 1998.

64. Silverman, R. B., *The Organic Chemistry of Drug Design and Drug Action.* San Diego, CA: Academic Press, 1992.

65. Sotheeswaran, S., "Herbal Medicine: The Scientific Evidence." *J. Chem. Educ.* **1992**, *69*, 444.

66. Spessard, G. O., and Miessler, G. L., *Organometallic Chemistry.* Upper Saddle River, NJ: Prentice Hall, 1997.

67. Starks, C. M., et al., *Phase Transfer Catalysis: Fundamentals, Applications, and Industrial Perspectives.* New York: Chapman & Hall, 1994.

68. Stocker, J. H., *Chemistry and Science Fiction.* New York: Oxford, 1998.

69. Sundberg, R. J., *Indoles.* San Diego, CA: Academic Press, 1996.

70. Teranishi, R., et al., *Flavor Chemistry: Trends and Development.* Washington, DC: American Chemical Society, 1989.

71. Vartanian, P. F., "The Chemistry of Modern Petroleum Product Additives." *J. Chem. Educ.* **1991**, *68*, 1015.

72. Vogler, A., and Kunkley, H., "Photochemistry and Beer." *J. Chem. Educ.* **1982**, *59*, 25.

73. Waddell, T. G., et al., "Legendary Chemical Aphrodisiacs." *J. Chem. Educ.* **1980**, *57*, 341.

74. Walling, C., "The Development of Free Radical Chemistry." *J. Chem. Educ.* **1986**, *63*, 99.

75. Walters, E. E., et al., eds., *Sweeteners: Discovery, Molecular Design, and Chemoreception.* Washington, DC: American Chemical Society, 1991.

76. Waring, D. R., and Hallas, G., eds., *The Chemistry and Application of Dyes.* New York: Plenum Press, 1990.

77. Watson, James D., *The Double Helix.* New York: Atheneum Publishers, 1968.

78. Weissermel, K., and Arpe, H.-J., *Industrial Organic Chemistry*, 3rd ed. New York: Wiley-VCH, 1997.

79. White, M. A., "The Chemistry behind Carbonless Copy Paper." *J. Chem. Educ.* **1999**, *75*, 1061.

80. Wigfield, D.C., *Environmental Aspects of Organic Chemistry.* Winnipeg, Canada: Wuerz, 1997.

81. Zanger M., et al., "The Aromatic Substitution Game." *J. Chem. Educ.* **1993**, *70*, 985.

82. Zielinski, T. J. and M. L. Swift, eds., *Using Computers in Chemistry and Chemical Education.* New York: Oxford, 1997.

M. Software for Organic Chemistry

1. Black, K., *Proton NMR Spectrum Simulator* (Macintosh). Madison, WI: JCE Software.

2. Brockwell, J. C., et al., *Beaker: An Expert System for the Organic Chemistry Student* (Windows or Macintosh). Pacific Grove, CA: Brooks-Cole.

3. Clough, F. W., *Introduction to Spectroscopy: IR, NMR, CMR, and Mass Spec* (DOS or Macintosh). Campton, NH: Trinity Software.

4. Figueras, J., *MassSpec v3.0: A Graphics-Based Mass Spectrum Analyzer* (Windows or Macintosh). Campton, NH: Trinity Software.

5. Figueras, J., *SynTree 2.0: A Program for Exploring Organic Synthesis* (DOS/Windows or Macintosh). Campton, NH: Trinity Software.

6. Pavia, D. L. and F. W. Clough, *SQUALOR: A Simulation of the Qualitative Organic Analysis Laboratory Experience* (DOS) and *MacSQUALOR* (Macintosh). Campton, NH: Trinity Software.

7. Schatz, P. F., *SpectraBook 1.0* (Windows) and *SpectraDeck 2.2* (Macintosh). Wellesley, MA: Falcon Software.

8. Smith, S. G., *Organic Chemistry Laboratory* (DOS). Wellesley, MA: Falcon Software.

9. *PCSpartan* (PCs) and *MacSpartan* (Macintosh). Orlando, FL: Academic Press.

Index